AF802935

Handbook of

Nonmedical Applications of Liposomes

From Design to Microreactors

VOLUME III

Edited by

Yechezkel Barenholz

Department of Biochemistry
Hebrew University–Hadassah Medical School
Jerusalem, Israel

Danilo D. Lasic

MegaBios Corporation
Burlingame, California

CRC Press
Taylor & Francis Group
Boca Raton London New York

CRC Press is an imprint of the
Taylor & Francis Group, an informa business

First published 1996 by CRC Press
Taylor & Francis Group
6000 Broken Sound Parkway NW, Suite 300
Boca Raton, FL 33487-2742

Reissued 2018 by CRC Press

© 1996 by CRC Press, Inc.
CRC Press is an imprint of Taylor & Francis Group, an Informa business

No claim to original U.S. Government works

This book contains information obtained from authentic and highly regarded sources. Reasonable efforts have been made to publish reliable data and information, but the author and publisher cannot assume responsibility for the validity of all materials or the consequences of their use. The authors and publishers have attempted to trace the copyright holders of all material reproduced in this publication and apologize to copyright holders if permission to publish in this form has not been obtained. If any copyright material has not been acknowledged please write and let us know so we may rectify in any future reprint.

Except as permitted under U.S. Copyright Law, no part of this book may be reprinted, reproduced, transmitted, or utilized in any form by any electronic, mechanical, or other means, now known or hereafter invented, including photocopying, microfilming, and recording, or in any information storage or retrieval system, without written permission from the publishers.

For permission to photocopy or use material electronically from this work, please access www.copyright.com (http://www.copyright.com/) or contact the Copyright Clearance Center, Inc. (CCC), 222 Rosewood Drive, Danvers, MA 01923, 978-750-8400. CCC is a not-for-profit organization that provides licenses and registration for a variety of users. For organizations that have been granted a photocopy license by the CCC, a separate system of payment has been arranged.

Trademark Notice: Product or corporate names may be trademarks or registered trademarks, and are used only for identification and explanation without intent to infringe.

Library of Congress Cataloging-in-Publication Data

Handbook of nonmedical applications of liposomes / edited by Yechezkel Barenholz
and Danilo D. Lasic.
p. cm.
Includes bibliographical references and index.
ISBN 0-8493-4012-8 (v. 3)
ISBN 0-8493-4731-9 (v. 1)
ISBN 0-8493-4011-X (v. 2)
ISBN 0-8493-4013-6 (v. 4)
1. Liposomes--Biotechnology. I. Barenholz, Yechezkel. II. Lasic, D. D.
TP248.65.L57H36 1995
660'.63—dc20 95-426

A Library of Congress record exists under LC control number: 95000426

Publisher's Note
The publisher has gone to great lengths to ensure the quality of this reprint but points out that some imperfections in the original copies may be apparent.

Disclaimer
The publisher has made every effort to trace copyright holders and welcomes correspondence from those they have been unable to contact.

ISBN 13: 978-1-315-89362-4 (hbk)
ISBN 13: 978-1-351-07272-4 (ebk)

Visit the Taylor & Francis Web site at http://www.taylorandfrancis.com and the
CRC Press Web site at http://www.crcpress.com

GENERAL PREFACE

To introduce liposomes, we shall look backwards to the origin of the name liposome. As shown in the last chapter of this series, in the fourth volume, liposomes were produced in scientific laboratories at least from 1847 on. The colloidal properties of lipid dispersions (emulsions, sols, micelles, etc.) were well known. However, it was only Bangham in the early 1960s who realized that they encapsulate part of the solvent and that their membrane presents a permeability barrier. So, they were at that time called "banghasomes", while the name "liposome" can occasionally be found in the literature between the World Wars as a description of colloidal fat (mostly triglyceride) particles. In the late 1960s, Weismann proposed the name liposomes for these structures. Due to the very general meaning of this word (i.e., fat body), Papahadjopoulos and others introduced the classification according to size and number of lamellae. In addition, and in contrast to this topological and structural classification, and following the development of liposomes as drug delivery vehicles, a new — functional — liposome classification could be proposed. With respect to functional characterization and the interaction characteristics, one can divide liposomes into those with nonspecific reactivity (conventional liposomes which are exposed to interactions of components from the suspending medium), inert liposomes (sterically stabilized or Stealth™ liposomes which have minimal reactivity with the medium), liposomes with specific reactivity (Stealth™ liposomes with attached specific ligands in which nonspecific repulsion is upgraded with specific antibodies or lectins), and liposomes that are very reactive under particular conditions. The latter are polymorphic liposomes, or "activosomes", which are topologically stable, but change their structure when triggered by a particular stimulus, such as polyions (with examples of cationic liposomes for DNA and antisense molecule delivery) or pH for pH-sensitive liposomes. Fusogenic and target-specific liposomes also fall into this category.

Liposomes are lipid bilayer-based colloidal particles which can encapsulate part of the aqueous medium into their interior, forming a "milieu interne" (internal microenvironment) which can be different from the extraliposomal medium. Nowadays, these assemblies are becoming a very important system in basic research, from theoretical concepts to serving as a tool, model, or reagent in basic studies of many processes. Parallel to numerous medical uses, a subject to which many books and reviews are dedicated, many other applications are being investigated. The aim of these four books is to present these lesser-known applications in depth. This series, *Handbook of Nonmedical Applications of Liposomes,* is divided into four rather different collections of related papers, so that many researchers will be able to find the most pertinent articles concentrated in a single volume or two.

The first volume presents theoretical studies on thermodynamics and molecular dynamics simulations, as well as some engineering applications of liposomes. The second volume presents liposomes as model systems for many biophysical, biochemical, and cellular processes, as well as protein studies using reconstitution into bilayer membranes. It concludes with a group of chapters related to the origin of life. The third volume commences with the "nuts and bolts" of "liposomology", from rational design, raw materials, large-scale preparation and active agent loading, stability, and stabilization, to sterility. It concludes with the use of liposomes as bioreactors and in catalysis. The fourth volume presents additional liposome applications in life sciences, with a group of chapters on gene therapy, cosmetics, diagnostics, and ecology.

Although these four volumes deal with nonmedical applications of liposomes, their content is also very relevant to their medical application, through many specific subjects such as all the design, preparation, loading, preservation, and characterization of liposomal formulations, gene delivery and diagnostics, as well as comparison with other recently developed colloidal carriers. Therefore, we believe that this series should appeal to everyone who has an interest in liposomes, whether from academia or from industry.

It was a pleasure for both of us to undertake the daunting project of putting this series together because obtaining such a broad spectrum of expertise required working with top academic and industrial researchers. We hope that this series will be useful to scientists from many disciplines and to students who are looking for new frontiers. From our personal experience as teachers, we know that such a series is missing, and we hope that this will fill the gap.

We would like to thank the many authors of this series who did such an excellent job in writing and submitting their papers to us. It was enjoyable to interact with them, and there is no question that the pressure we put on many of them was worthwhile.

Finally, we would like to thank Mrs. Beryl Levene for her excellent secretarial work and for sending so many letters to so many scientists and Mr. Sigmund Geller for his editorial assistance. CRC personnel should be acknowledged for their encouragement and patience, as well as their high level of proficiency.

Yechezkel Barenholz
Jerusalem, Israel

Danilo D. Lasic
Burlingame, California

June 1995

INTRODUCTION TO VOLUME III

Glancing through the papers of Volumes III and IV makes it clear that liposomes, as well as other amphiphile assemblies, may have a broad spectrum of nonmedical applications with enormous ecological and economical implications. The first part of Volume III deals with the "nuts and bolts" of applied liposomology. It includes the subject of how to select the optimal amphiphile-based assembly for a specific use: Is it liposomes, micelles, emulsions, or others (Barenholz, Chapter 1) What are the raw materials available for such an assembly and how to characterize it (Burgess et al., Chapter 2) What are the approaches and the issues involved in large-scale production of liposomes (Barenholz and Lasic, Chapter 3) Long-term liposome stability and sterility are two of the main issues that slowed down applications of liposomes. Leung and Lasic (Chapter 4) go into the details of stability, while Zuidam et al. describe the problems and the prospects of sterilization in Chapter 6.

In many cases a major limitation to liposome application is their efficient and stable loading with an active agent. This can be overcome by "active" remote loading, in which the intraliposomal concentration of the agent exceeds manyfold its concentration in extraliposomal medium. The theory behind such loading, as well as its quantitative effects, are described by Čeh (Chapter 5). If and when liposomes will be used in various industrial and ecological applications, it will be very important to be aware of the environmental implications — a subject discussed by Parnham et al. (Chapter 7). Nonmedical applications of liposomes may need the development of lipids which do not exist in nature. Fluorinated amphiphiles are a good example of such a group; their properties are described by Reiss et al. (Chapter 8). As discussed by Barenholz in the first chapter in this volume, there are situations in which nonliposomal amphiphile-based assemblies may be superior. Three such systems are described in this volume.

Emulsions and microemulsions differ from liposomes, having a rather large oily phase which is dispersed in an amphiphile monolayer coating. The oily phase can host agents having a high oil/water partition coefficient, for which liposomes serve as a poor carrier. The microemulsions and emulsions differ very much in their thermodynamic characterization. While the microemulsions are thermodynamically stable, the emulsions are not. The properties of these two assemblies and their applications (especially in the food industry) are reviewed by Garti (Chapter 9).

In addition to liposomes and emulsions, one can have an oil/water emulsion droplet surrounded by several bilayers, an analog to multilamellar liposomes. Two such systems are described in the papers by Florence and by Amselem (Chapters 10 and 11). Puvvada (Chapter 12) discusses bicontinuous phases — an important lyotropic structure with important consequences for vesicles because it is believed that in some systems vesicles can be spontaneously formed by entropic dissolution of ordered or disordered sponge phases.

Chemical modification of liposomal lipids *in situ* is important in order to understand composition–structure–function relationships. Joó and Vígh (Chapter 14) use homo- and heterogeneous hydrogenation to induce such changes. Liposomes are characterized by large surface areas per mass of material. Furthermore, they can compartmentalize the relatively small aqueous (fluid or semi-solid jelly) environment, the "milieu interne". These characteristics, and the ease with which one can tailor surface and interface properties of the liposomes including their hydrophilic–hydrophobic nature, offer possible uses in catalysis. Although, mainly due to stability problems, these effects have not been fully exploited yet, much interesting work has been done, as described by Cuccovia and Chaimovich (Chapter 13). Theoretical considerations of enzyme-containing liposomes are discussed by Cioci and Lavecchia (Chapter 16), while Mossa et al. (Chapter 15) describe the use of liposomes as biomimetic microreactors.

Components of the liposomal membranes, such as cholesterol, can be enzymatically transformed into other products of interest very efficiently, as demonstrated by Bar (Chapter 17). A novel area is the magnetoliposomes described by De Cuyper (Chapter 18) which can be used in various separation processes.

Yechezkel Barenholz
Jerusalem, Israel

Danilo D. Lasic
Burlingame, California

June 1995

THE EDITORS

Yechezkel Barenholz, Ph.D., is Professor of Biochemistry and head of the Department of Biochemistry at the Hebrew University–Hadassah Medical School, Jerusalem, Israel. Professor Barenholz received his undergraduate, M.Sc., and Ph.D. degrees in Biochemistry from the Hebrew University, Jerusalem in 1965, 1968, and 1971, respectively. While working on his Ph.D. thesis he studied at the Animal Research Council Institute, Babraham, Cambridge, England with Drs. R.M.C. Dawson and A.D. Bangham. Professor Barenholz has been on the permanent academic staff of the Hebrew University since 1974 and a professor there since 1982. He has also been on the staff of the University of Virginia, Charlottesville since 1983. Professor Barenholz was a Donders Chair professor at the University of Utrecht in the Netherlands in 1992. His interests are in basic and applied science. In basic research he is involved in many fields related to the biochemistry and biophysics of lipids, especially sphingolipids, phospholipids, and sterols, and including synthesis, chemical and physical characterization, and the relationship between membrane lipid composition and function. In applied research, Professor Barenholz's main interests are in amphiphile-based drug carriers, especially liposomes—from basic aspects of the design of the drug carrier, through animal studies, to clinical trials. He was involved in clinical trials of Amphocil®, an amphotericin B micelle-like assembly, and in the development of DOX-SL®, a doxorubicin remote-loaded, sterically-stabilized liposome. At present, he is also studying the application of liposomes for vaccination and in gene therapy. Professor Barenholz is author of 200 publications and was editor of a special issue of *Chemistry and Physics of Lipids on Quality Control of Liposomes.* He is on the editorial boards of *Chemistry and Physics of Lipids* and *Journal of Liposome Research.* Professor Barenholz is a recipient of the 1995 Kay Award for innovation.

Danilo D. Lasic, Ph.D., is Principal Scientist at MegaBios, San Carlos, California, where he is currently working with cationic liposomes in gene delivery. Previously, he was a Senior Scientist at Liposome Technology, Inc., (now Sequus) where he led studies for theoretical understanding of long-circulating liposomes as well as developed the first formulations for preclinical studies. In addition, he also actively participated in the scale-up of the preparation of stealth liposomes laden with the anticancer agent doxorubicin.

Dr. Lasic graduated from the University of Ljubljana, Slovenia, in 1975 with a degree in physical chemistry. He received his M.S. in 1977 from the University of Ljubljana. He obtained his Ph.D. at the Institute J. Stefan, Solid State Physics Department, Ljubljana, in 1979. After postdoctoral work with Dr. Charles Tanford at Duke University, North Carolina, and Dr. Helmut Hauser at ETH Zurich, he was a research fellow at the Institute J. Stefan, a visiting lecturer in the Department of Chemistry, and a visiting scientist in the Department of Physics at the University of Waterloo in Canada. Dr. Lasic then joined Liposome Technology, Inc., Menlo Park, California. Currently he is studying cationic liposomes, DNA–liposome interactions, and DNA–lipid complexes for gene delivery and gene therapy.

Dr. Lasic has published more than 120 research papers as well as a monograph on liposomes, *Liposomes: From Physics to Applications* (Elsevier, 1993) and co-edited a book, *Stealth Liposomes* (CRC Press, 1995). His best-known papers are the ones dealing with thermodynamics and the mechanism of vesicle formation, stability of liposomes, the origin of liposome stability in biological environments, as well as the applications of drug-laden liposomes.

CONTRIBUTORS

Shimon Amselem
Pharmos Ltd.
Rehovot, Israel

M. Cristina Annesini
Department of Chemical Engineering
Faculty of Engineering
University of Rome "La Sapienza"
Rome, Italy

Raphael Bar
Department of Biological Chemistry
The Hebrew University
Jerusalem, Israel

Yechezkel Barenholz
Department of Membrane Biochemistry
The Hebrew University Medical School
Jerusalem, Israel

Stephen W. Burgess
Avanti Polar Lipids
Alabaster, Alabama

Boris Čeh
Faculty for Chemistry and Chemical Technology
University of Ljubljana
Ljubljana, Slovenia

Hernan Chaimovich
Departamento de Bioquimica
Instituto de Quimica
Universidade de São Paulo
São Paulo, Brazil

Federico Cioci
Department of Chemical Engineering
Faculty of Engineering
University of Rome "La Sapienza"
Rome, Italy

Daan J. A. Crommelin
Department of Pharmaceutics
Utrecht Institute for Pharmaceutical Sciences
Utrecht University
Utrecht, The Netherlands

Iolanda M. Cuccovia
Departamento de Bioquimica
Instituto de Quimica
Universidade de São Paulo
São Paulo, Brazil

Marcel De Cuyper
Interdisciplinary Research Centre
Katholieke Universiteit Leuven–Campus Kortrijk
Kortrijk, Belgium

Luisa Di Marzio
Institute of Experimental Medicine
National Research Council
Rome, Italy

Alexander T. Florence
Centre for Drug Delivery Research
University of London
School of Pharmacy
London, England

Doron Freidman
Vitamed
Tel-Aviv, Israel

Frédéric Frézard
Unité de Chemie Moléculaire
CNRS
Université de Nice–Sophia Antipolis
Nice, France

Nissim Garti
Casali Institute of Applied Chemistry
The Hebrew University
Jerusalem, Israel

Jacques Greiner
Unité de Chemie Moléculaire
CNRS
Université de Nice–Sophia Antipolis
Nice, France

Miklós Ghyczy
Application Technology
Nattermann Phospholipid GmbH
Cologne, Germany

Ferenc Joó
Institute of Physical Chemistry
Lajos Kossuth University
Debrecen, Hungary

Marie Pierre Krafft
Unité de Chemie Moléculaire
CNRS
Université de Nice–Sophia Antipolis
Nice, France

Danilo D. Lasic
MegaBios Corporation
Burlingame, California

Roberto Lavecchia
Department of Chemical Engineering
Faculty of Engineering
University of Rome "La Sapienza"
Rome, Italy

Daisy Leung
MegaBios Corporation
Burlingame, California

Jeff D. Moore
Avanti Polar Lipids
Alabaster, Alabama

Giuseppe Mossa
Institute of Experimental Medicine
National Research Council
Rome, Italy

Michael J. Parnham
Parnham Advisory Services
Bonn, Germany

Sudhakar Puvvada
Lever Brothers Company
Edgewater, New Jersey

Banahalli R. Ratna
Center for Bio/Molecular Science
and Engineering
Naval Research Laboratory
Washington, D.C.

Jean Riess
Unité de Chimie Moléculaire
CNRS
Université de Nice–Sophia Antipolis
Nice, France

Catherine Santaella
Unité de Chemie Moléculaire
CNRS
Université de Nice–Sophia Antipolis
Nice, France

Walter A. Shaw
Avanti Polar Lipids
Alabaster, Alabama

Herre Talsma
Department of Pharmaceutics
Utrecht Institute for Pharmaceutical Sciences
Utrecht University
Utrecht, The Netherlands

Pierre Vierling
Unité de Chemie Moléculaire
CNRS
Université de Nice–Sophia Antipolis
Nice, France

László Vígh
Institute of Biochemistry
Biological Research Center
Hungarian Academy of Sciences
Szeged, Hungary

Armin Wendel
Application Technology
Nattermann Phospholipid GmbH
Cologne, Germany

Ayala Yogev
Pharmos Ltd.
Rehovot, Israel

Toshimitsu Yoshioka
Technical Research Institute
Snow Brand Milk Products Company
Saitama, Japan

Leila Zarif
Unité de Chemie Moléculaire
CNRS
Université de Nice–Sophia Antipolis
Nice, France

Eduardo Zawoznik
Pharmos Ltd.
Rehovot, Israel

Nicolaas J. Zuidam
Department of Pharmaceutics
Utrecht Institute for Pharmaceutical Sciences
University of Utrecht
Utrecht, The Netherlands

CONTENTS

Liposomes in Catalysis
Danilo D. Lasic and Yechezkel Barenholz

Chapter 1

Optimization of Amphiphile-Based Colloidal Carriers

Yechezkel Barenholz

Some amphiphiles tend to self-aggregate; in others, aggregation to defined structures can be induced.[1] This, together with our ability to control the composition, type, shape, and size of the amphiphile-based assembly,[1-3] makes these molecules excellent candidates to form suitable carriers for a broad spectrum of agents including water-soluble, amphipathic, and hydrophobic molecules of low, medium, and high molecular mass.[3-5] These amphiphile-based assemblies include micelles and mixed micelles of various aggregation numbers and shapes,[1-3,6] micelle-like assemblies,[7] liposomes of various types,[3-5,8] emulsions, and microemulsions.[9,10] The enormous built-in versatility due to large differences in the properties of each of the different assemblies — and the ability to control the physicochemical properties of each assembly by its composition and method of preparation — make them suitable to fit controlled-release and passive and active targeting of molecules associated with the assembly in a broad spectrum of applications, such as medical,[3,4,8] cosmetic,[11] agricultural, ecological, food, and other industrial applications.[3,6,8] When using the same agent with various types of amphiphile-based carrier it becomes clear that the exact type of the colloidal assembly has a major impact on the performance of the agent associated with the assembly.[12,13] To a large extent, the ability to form a suitable assembly regarding efficiency and stability of agent-loading, and the ability of the assembly to affect the performance of the agent, is dependent on the physicochemical properties of the agent and its mode of interaction with the molecules forming the carrier.

Using a simplified approach, all relevant agent properties can be expressed through assortment into three classes. This classification is determined by two partition coefficients (K_p's): (a) oil/aqueous phase and (b) octanol/aqueous phase. For example, molecules that have low values for both K_p's are defined as water soluble (Class III) and are suitable only for assemblies that have an internal aqueous phase, such as liposomes and double emulsions. Agents having a high oil/aqueous phase K_p, i.e., hydrophobic agents (Class I), are most suitable for emulsions (oil phase). Agents having low oil/aqueous phase K_p and medium-to-high K_p (Class II, amphipathic agents) are most suitable for assemblies which have interfaces between polar lipid (or solvent) and aqueous phases, i.e., emulsions, microemulsions, and liposomes (Table 1).

The strategy to design a suitable colloidal carrier is also dependent on other requirements, such as size, surface area, charge, media pH, media ionic strength, etc. The performance of the agent relates to factors which determine the release profile of the agent from the colloidal assembly irrespective of whether the activity requires its release from, or its association with, the carrier. The agent release can be determined by the stability of the assembly itself in the medium in which it should perform. For example, micelles and microemulsions — being equilibrium structures — may dissociate[1,6] (micelles) or change location in the phase diagram[9,10] (microemulsions) upon dilution or changing medium composition.

For the metastable assemblies, such as emulsions and some of the liposome types, the assembly itself in most cases is unaffected (or less affected) by dilution, although the release of the agent may be strongly affected. For the latter systems, agent release is determined to a large extent by the agent's (i) assembly/medium partition coefficient ($K_{p,c}$) and (ii) the rate of desorption from the assembly into the medium (k_{off}). In many cases the medium also contains a "sink" to which the released agent can be partitioned. The latter is especially relevant for agents which have very low solubility in the aqueous phase (as monomer or aggregate). For drug delivery (medical or veterinary use) an additional parameter, (iii), the clearance rate (k_c) of the intact assembly (which is not discussed here), also plays an important role. These three parameters are defined in Table 2.

All the above theoretical considerations are exemplified in these four volumes of the *Handbook of Nonmedical Applications of Liposomes* series,[5] although in many of the chapters there is no direct reference to these principles.

For assemblies which do not have an internal aqueous phase, the main factor which determines the agent release level is the $K_{p,c}$: the lower the $K_{p,c}$ the higher is the level of agent release. Similar considerations should hold for liposome membrane-associated agents (Class I and II molecules, Table 1).

0-8493-4012-8/96/$0.00+$.50
© 1996 by CRC Press, Inc.

Table 1 Suitability of Colloidal Carriers to Fit Agents' Properties

Agent class	K_p (oil/water)	K_p (octanol/water)	Suitability of colloidal carrier(s)
I	High	—	Assemblies having oil phase (i.e., emulsions)
II	Low	Medium–High	Assemblies having polar surface (i.e., emulsions, microemulsions, liposomes, micelles)
III	Low	Low*	Assemblies having internal aqueous phase (i.e., liposomes, double emulsions)

* Under specific conditions in which the internal aqueous compartment is different from the external medium (such as pH gradient) an agent belonging to Class II which is translocated into the internal aqueous phase, becomes charged, and behaves like a Class III agent. This is one of the principles of liposome remote-loading.

Table 2 Parameters which Play a Major Role in the Performance of a Colloidal Carrier

Parameter	Meaning
$K_{p,c}$	Carrier-to-medium partition coefficient (thermodynamic parameter)
K_{off}	Desorption rate of the agent of the assembly — an umbrella of several constants (kinetic parameter)
K_c*	Clearance rate of the intact carrier assembly from the compartment of interest (kinetic parameter)

* K_c is relevant only to systems in which the carrier is cleared from the site or the pool of administration.

For such assemblies, one should also check for the rate of desorption from the assembly (k_{off}).[14] For assemblies which have an aqueous compartment, the release rate of a water-soluble agent which has a very low $K_{p,c}$ value is determined mainly by the desorption rate (k_{off}). In the case of water-soluble agents this is a complex rate constant which relates mainly to the permeability coefficient through the lipid phase. More special systems are those in which the agent is aggregated or precipitated in the internal aqueous phase. The driving force for this precipitation/aggregation can be a pH gradient[15] or ammonium sulfate gradient.[16] Although both gradients induce remote loading of liposomes by amphiphathic weak bases, the ammonium sulfate gradient also improves stability of the loading due to the association of the weak base with the sulfate anion to form a gel-like precipitate[16–18] or ion-pairing. The ability of such precipitation to affect release kinetics is dependent to a large extent on the solubility product and its relationship to the permeability coefficient of the agent forming the intraassembly precipitate. In general, it is important to clarify which of the factors is rate-limiting in the release of an agent from a precipitate in the internal aqueous phase of the carrier — the precipitate dissolution rate or the permeability coefficient. Also, the lower the solubility product, the more important is the permeability coefficient.[19] The role of kinetics of precipitate dissolution has not yet been studied systematically. A good example of such a system in which the intraliposomal precipitation plays an important role in the stability of encapsulation is loading of the weak base doxorubicin (an anticancer drug) into liposomes by an ammonium sulfate gradient in which the intraliposomal ammonium sulfate concentration is higher than its extraliposomal concentration. The conditions used in the loading cause the drug to behave as if water soluble (a class III molecule, Table 1[18]). This approach enables one to encapsulate sufficient levels of drug into small liposomes in a stable way, which permits clinical use of sterically-stabilized liposomes loaded with doxorubicin as a pharmaceutically acceptable dosage form.[20] Also, this procedure enables the loading of sterically-stabilized ligandoliposomes which permit targeting of the liposomes.[18]

To sum up, our current level of knowledge is sufficient to enable rational design of a colloidal-agent carrier to fit many needs and applications. It is recommended that prior to investing resources in labor-intense and expensive studies, a simple test should be performed which can predict the performance of the carrier system under real-life conditions. A good example of such a test is a dilution release assay in which both the level of agent release (an equilibrium value) and the release rates are determined.[21–23] The information obtained from such an experiment includes the $K_{p,c}$ and k_{off} (Table 2). When such an experiment is done in the medium in which the carrier should perform and this is compared with a standard salt/buffer medium, the impact of the medium on the carrier/agent association as expressed by $K_{p,c}$ and k_{off} will be determined.

REFERENCES

1. **Israelachvili, J.,** Fluid-like structures and self-assembling systems: micelles, bilayers and biological membranes, in *Intermolecular and Surface Forces*, 2nd ed., Part 3, Academic Press, London, 1991, 341.
2. **Lichtenberg, D. and Barenholz, Y.,** Liposomes: preparation, characterization and preservation, in *Methods in Biochemical Analysis*, Glick, D., Ed., John Wiley & Sons, New York, 1988, 337.
3. **Lasic, D. D.,** *Liposomes: From Physics to Applications*, Elsevier, Amsterdam, 1993.
4. **Gregoriadis, G., Ed.,** *Liposome Technology*, Vols. I–III, 2nd ed., CRC Press, Boca Raton, FL, 1993.
5. **Barenholz, Y. and Lasic, D. D., Eds.,** *Handbook of Nonmedical Application of Liposomes*, Vols. I–IV, CRC Press, Boca Raton, FL, 1996.
6. **Elworthy, P. H., Florence, A. T., and Macfarlane, C. B.,** *Solubilization by Surface Active Agents*, Chapman and Hall, London, 1968.
7. **Lasic D. D.,** Mixed micelles in drug delivery, *Nature*, 355, 279, 1992.
8. **Barenholz, Y. and Crommelin, D. J. A.,** Liposomes as pharmaceutical dosage forms, in *Encyclopedia of Pharmaceutical Technology*, Vol. 9, Swarbrick, J. and Boylan, J. C., Eds., Marcel Dekker, New York, 1994, 1.
9. **Sharma, M. K. and Shah, D. O.,** Introduction to macro- and microemulsions, in *Macro- and Microemulsions*, Shah, D. O., Ed., American Chemical Society, Washington, D.C., 1985, 1.
10. **Qutubuddin, S., Miller, C. A., Benton, W. J., and Fort, T., Jr.,** Effects of polymers, electrolytes and pH on microemulsion phase behavior, in *Macro- and Microemulsions*, Shah, D. O., Ed., American Chemical Society, Washington, D.C., 1985, 235.
11. **Braun-Falco, O., Korting, H. C., and Maibach, H. I., Eds.,** *Liposome Dermatics*, Springer-Verlag, Berlin, 1992.
12. **Takino, T., Nagahama, E., Sakaaeda, T., Yamashita, F., Takakura, Y., and Hashida, M.,** Pharmacokinetic disposition analysis of lipophilic drugs injected with various lipid carriers in a single-pass rat liver perfusion systems, *Int. J. Pharm.*, 114, 43, 1995.
13. **Barenholz, Y. and Cohen, R.,** Rational design of amphiphile-based drug carriers and sterically stabilized carriers, *J. Liposome Res.*, October, 1995.
14. **Bar, L. K., Barenholz, Y., and Thompson, T. E.,** Dependence of the fraction of cholesterol undergoing spontaneous exchange between small unilamellar vesicles and their phospholipid composition, *Biochemistry*, 26, 5460, 1987.
15. **Mayer, L. D., Cullis, P. R., and Bally, M. B.,** The use of transmembrane pH gradient-driven drug encapsulation in the pharmacodynamic evaluation of liposomal doxorubicin, *J. Liposome Res.*, 4, 529, 1994.
16. **Haran, G., Cohen, R., Bar, L. K., and Barenholz, Y.,** Transmembrane ammonium sulfate gradients in liposomes produce efficient and stable entrapment of amphipathic weak bases, *Biochim. Biophys. Acta*, 1151, 201, 1993.
17. **Lasic, D. D., Frederik, P. M., Stuart, M. C. A., Barenholz, Y., and McIntosh, T. J.,** Gelation of liposome interior. A novel method for drug encapsulation, *FEBS Lett.*, 312, 255, 1992.
18. **Bolotin, E. M., Cohen, R., Bar, L. K., Emanuel, N., Ninio, S., Lasic, D. D., and Barenholz, Y.,** Ammonium sulfate gradients for efficient and stable remote loading of amphipathic weak bases into liposomes and ligandoliposomes, *J. Liposome Res.*, 4, 455, 1994.
19. **Lasic, D. D., Čeh, B., Stuart, M. C., Guo, L., Frederik, P. M., and Barenholz, Y.,** Transmembrane driven phase transition within vesicles; lessons for drug delivery, *Biochim. Biophys. Acta*, 1995, in press.
20. **Gabizon, A., Catane, R., Uziely, B., Kaufman, B., Safra, T., Cohen, R., Martin, F., Huang, A., and Barenholz, Y.,** Prolonged circulation time and enhanced accumulation in malignant exudates of doxorubicin encapsulated in polyethylene-glycol coated liposomes, *Cancer Res.*, 54, 987, 1994.
21. **Barenholz, Y. and Amselem, S.,** Quality control assays in the development and clinical use of liposome-based formulations, in *Liposome Technology: Liposome Preparation and Related Techniques*, Vol. 1, 2nd ed., Gregoriadis, G., Ed., CRC Press, Boca Raton, FL, 1993, 527.
22. **Amselem, S., Cohen, R., and Barenholz, Y.,** *In vitro* tests to predict *in vivo* performance of liposomal dosage forms, *Chem. Phys. Lipids*, 64, 219, 1993.
23. **Levy, M. Y., Polacheck, I., Barenholz, Y., and Benita, S.,** Efficacy evaluation of a novel submicron amphotericin B emulsion in murine candidiasis, *J. Med. Vet. Mycol.*, 31, 207, 1993.

Chapter 2

Lipids as Raw Materials for Pharmaceutical Products: Properties and Characteristics

Stephen W. Burgess, Jeff D. Moore, and Walter A. Shaw

CONTENTS

I. INTRODUCTION

In the past 25 years, lipid membrane research has progressed from identifying lipid species in crude tissue extracts and preparing simple lipid aggregates in aqueous buffers, to exploring the role of lipid membranes in basic biological processes. The areas of lipid membrane research are diverse and include membrane fusion, lipid–protein interactions, and lipid second messenger systems. As our understanding of lipids

0-8493-4012-8/96/$0.00+$.50
© 1996 by CRC Press, Inc.

increased, this knowledge was transferred to clinical and diagnostic applications. Lipids are widely used now as diagnostic reagents, drug carrier systems, and most recently as DNA transfection reagents.

With the expansion of lipid research came a need for new lipid species and membrane probes. No longer would the crude tissue extract suffice; this new generation of researchers demanded highly specialized, well-defined, and above all, high purity lipids. Companies formed to satisfy this niche market, specializing in phospholipids, glycolipids, sphingolipids, or neutral lipids. Some companies targeted the clinical applications market, providing bulk quantities of a select group of lipids suitable for human pharmaceuticals, while others chose to provide materials for research use only. A few companies support both research and clinical applications.

During the developmental years, analytical departments in these companies consisted of a thin layer chromatography (TLC) tank, a gas chromatograph (GC), and perhaps a high performance liquid chromatography (HPLC) system. However, the evolution of the field required additional analytical methods be developed to completely characterize these compounds. Partially driven by the pharmaceutical industry and their requirement for materials produced according to FDA guidelines, lipid analytical departments now employ a battery of equipment and methods designed to provide the customer with a comprehensive analytical package on their products. When trying to identify a potential lipid supplier or when seeking analysis of lipid products, one should be aware of the types of analytical information available to the consumer and what this information means. We will describe each of the major analytical methods suitable for characterization of lipids, identify the potential problems associated with these methods, and provide a detailed protocol for a few methods which can be easily incorporated into most lab procedures for qualifying lipid suppliers or analyzing in-house lipid preparations. In addition, we will discuss issues important to consider when selecting lipids for investigational purposes, as well as storage, handling, and sample preparation after receiving lipids.

In the remainder of the chapter and for the purpose of simplifying our discussion, we will limit our comments to phospholipids. The information provided may extend to other classes of lipids, although that is for the reader to determine. If information is given which is related to another class of lipids, the lipid class will be identified.

II. ANALYTICAL METHODS

The diversity and power of analytical instrumentation available for identification and quantitation of lipid products enables most laboratories to adequately characterize any commercially available lipid. Instruments range from commonly available chromatography systems and spectrometers capable of serving multiple functions, to highly specialized equipment providing a limited type of data. When selecting procedures and equipment, one should acquire only the instrumentation which will have the greatest utility in the laboratory environment. Many costly and specialized instruments can be found in contract analytical services which provide selected analysis for a reasonable fee. The types of analysis discussed below describe procedures which are generally applicable to lipid products and many of which are suitable for the general laboratory. We will address these analysis in discreet sections, namely, structural analyses, chromatographic analysis, biological analysis, and miscellaneous analysis.

A. STRUCTURAL ANALYSIS

1. Nuclear Magnetic Resonance

Various structural information can be obtained from nuclear magnetic resonance (NMR) analysis of lipids.[1–5] The most useful information is derived from proton and phosphorus NMR. Figure 1 (Panel A and C) shows a typical proton and phosphorus NMR for dipalmitoyl phosphatidyl choline (DPPC). Phosphatidylcholine (PC) is easily identified by its characteristic 9 protons on the trimethylammonium group of choline (3.35 ppm). The hydrocarbon region can be characterized as well by monitoring the region between 0.75 and 2.50 ppm. Also, the presence of double bonds in the acyl chains can be detected by the presence of protons at 5.30 and 2.00 ppm (Panel B) corresponding to protons on the carbons α and β to the double bond, respectively. Phosphorus NMR provides almost no information about the hydrocarbon region of phospholipids; however, it can be quite helpful in identifying lipid headgroup species. The phosphate is quite sensitive to its environment and the presence of a polar headgroup causes the signal to shift dramatically. Loss of an acyl group is also sufficient to shift the signal. Panel D shows a PC sample containing 1% lyso PC, demonstrating the usefulness of phosphorus NMR not only as a structural identification tool, but also a purity indicating method as well.

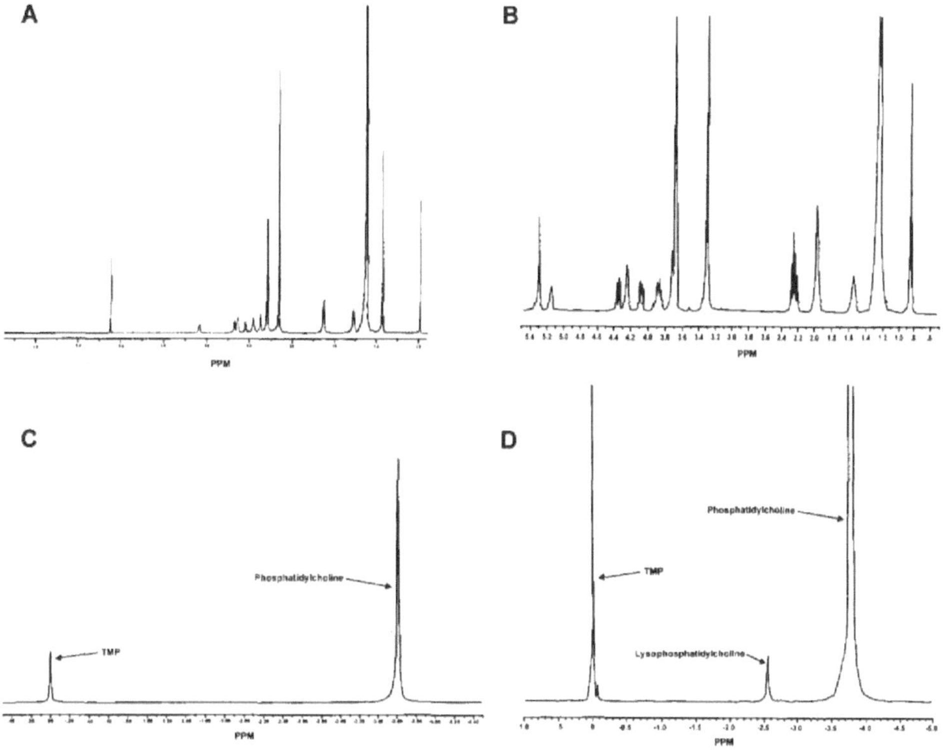

Figure 1 Proton and phosphorus NMR spectra for (A) 1,2-dipalmitoyl-*sn*-glycero-3-phosphocholine (DPPC), proton NMR, (B) 1-palmitoyl(d_{31})-2-oleoyl-*sn*-glycero-3-phosphocholine, proton NMR, (C) DPPC, phosphorus NMR, and (D) 1,2-distearoyl-*sn*-glycero-3-phosphocholine (DSPC) spiked with 1% 1-stearoyl-2-hydroxy-*sn*-glycero-3-phosphocholine (18:0 lyso PC), phosphorus NMR.

2. Mass Spectrometry

Mass spectrometry of lipids has become a prevalent analytical application due to technological advances in interfacing mass spectrometers with one of the primary tools of lipid analysis, the high performance liquid chromatograph. Nebulization and ion formation at room temperature and atmospheric pressure, acceptance of moderate HPLC mobile phase flows (1 µl/min to 1 ml/min), high ionization efficiency and multiple charge formation of large biomolecules make the electrospray interface ideal for the introduction of polar lipid compounds in a quadrapole mass spectrometer. Electrospray ionization is a "soft" ionization process where ions generated and subsequently identified by their mass/charge ratio (m/z) in the mass spectrometer are generally [M+H] or [M–H] respectively for positive and negative ionization operation. Millimolar concentrations of volatile buffer salts such as ammonium acetate, formic acid, acetic acid, and sodium hydroxide form corresponding molecular adducts. The formation and monitoring of these adducts may yield a 10- to 1000-fold increase in sensitivity. Further fragmentation of ions generated by the interface can be formed by increasing electrical potentials in pre-quadrapole areas such as the entrance orifice and specifically designed chambers to fragment ions by collision reactions with an inert gas known as collision induced dissociation (CID).[6] Mass spectrometry of nonpolar phospholipids can be performed by directly infusing a low concentration solution (<1 µg/ml) of the compound dissolved in mobile phase at a low flow rate (<0.1 ml/min). Sensitivities of 0.01 pmol/µl were reported for 1,2-dipalmitoyl-*sn*-glycero-3-phosphocoline by direct infusion into the interface at a flow of 1 µl/min.[7] Maximum ionization and subsequent mass detection can be enhanced by ionizing the compound in the proper ionization mode. Zwitterionic species such as PC, phosphatidylethanolamine (PE), and sphingomyelin typically are more sensitive in the positive ionization mode. Negatively charged acidic species of phosphatidylglycerol (PG), phosphatidylinositol (PI), phosphatidylserine (PS), and phosphatidic acid (PA) are more sensitive in the negative mode. Automated processes such as injection control, HPLC flow control, changing of interface and mass spectral parameters, and data analysis allow for unattended high through-put analysis on multiple and varied phospholipid compounds.

3. Stereochemistry

Phospholipids have at least one optically active center located at the C2 carbon of the glycerol backbone. The naturally occurring stereoisomer in most biological systems is the L-isomer form, and the D-isomer form is biologically inactive.[8] For research designed to mimic biological systems or clinical applications where the lipid should be cleared by enzymatic degradation, the L-isomer form is the preferable stereoisomer and D-isomer contamination should be minimal. There are two methods for determining stereoisomers of lipids: optical rotation and enzymatic digestion by phospholipase A_2 (PLA_2).

a. Optical Rotation

Standard instrumentation exists for determining optical rotation of lipid solutions. In general, chloroform solutions of phospholipids at approximately 1g/ml are analyzed at 589, 578, 546, 436, and 365 nm and 25°C. Recovered values are ambiguous as they differ depending on the acyl composition and good reference values do not exist for all compounds. While optical rotation is the established method for determining the stereochemistry of compounds, we feel it is relatively insensitive to small amounts of D-isomer contamination and suffers from a lack of standards for comparison.

b. Phospholipase A_2 Digestion

Phospholipids incubated in the presence of PLA_2 from snake venom will be hydrolyzed to the lyso lipid, liberating free fatty acid in the process.[9] PLA_2 cleaves the fatty acid at the C2 carbon of the glycerol backbone and is entirely specific for the L-isomer form of the lipid. In fact, not only is the D-isomer inactive as a substrate, being a structural analog of the L-isomer, it competitively inhibits the enzyme by binding at the active site.[10] Therefore, the presence of D-isomer is amplified in the results. Analysis of the incubation can be performed by TLC with a suitable standard or HPLC in an appropriate mobile phase for separation of diacyl lipid, monoacyl lipid, and free fatty acid.

B. CHROMATOGRAPHIC ANALYSIS

1. Thin Layer Chromatography

TLC analysis is a fast, qualitative look at a lipid sample. It allows a great deal of flexibility in mobile phase selection to optimize separation and identification of lipid species. Using appropriate lipid standards, the major species of lipids in a sample can be identified quickly, facilitating HPLC methods development. When selecting phospholipid standards for TLC analysis, choose unsaturated synthetic lipids (C18:1) or tissue-derived lipids, as these lipids stain easily in iodine. In general, C18:1-containing lipids are preferred. C18:1-containing lipids provide adequate staining capacity and they are more stable to oxidation and degradation since they lack the polyunsaturated fatty acids which are found in tissue-derived lipids.

Mobile phase selection can be quite confusing when analyzing phospholipids. Phospholipids do not always retain their migration on the TLC plate relative to other lipids when changing mobile phase type (e.g., neutral to acidic, or neutral to basic). Within the same type of mobile phase, phospholipids behave as expected. The more polar lipids bind tighter to silica gel and migrate lower on the plate. To increase the distance these polar lipids migrate on the plate, increase the polarity of the mobile phase, thus disrupting the interaction between the lipid and silica gel. See Method 1 for general instructions on TLC analysis of phospholipids, including two mobile phase systems useful in separating the major classes of phospholipids.

Numerous detection methods have been developed to identify functional groups and assist in identification of lipid species present in the sample. The most useful detection reagents for phospholipids include iodine vapors,[11] ninhydrin reagent,[12] and molybdate reagent.[13] Iodine detection is accomplished by simply placing a developed TLC plate into a closed container containing 1 to 2 g of iodine crystals. Iodine vapors provide a general stain for many organic compounds; however, compounds which do not contain double bonds or other groups which react with iodine will not stain easily (e.g., saturated phospholipids like DPPC). Ninhydrin reagent is used for detection of lipids containing primary amines such as PS and PE. The reagent is applied by spraying the developed plate with a solution of ninhydrin reagent then heating the plate until staining occurs. Primary amine-containing lipids appear as pink spots on a white background. Molybdate reagent is also applied by spraying the developed plate and will stain any phosphorus-containing compound blue on a white background. Since the reagent is prepared as a sulfuric acid solution, continued heating of the plate will result in charring of carbon-containing material, especially lipids containing double bonds. See Method 2 for instructions on preparing ninhydrin solutions and molybdate solutions.

Quantitating results from TLC analysis is quite difficult. Methods based on densitometry of plates visualized with iodine vapors or molybdate have been tried,[14–17] however, these methods suffer from the fact that lipids stain with different densities based on their degree of unsaturation. Standardizing these methods poses significant problems. Commercial instrumentation is available for quantitative TLC including those based on densitometry analysis of developed plates. The closest to true quantitative TLC is found in an instrument using quartz rods with flame ionization detection. Lipids are applied to one end of a quartz rod, the rod is developed in an appropriate mobile phase in a standard TLC tank, and the rod is dried and placed in the flame ionization detector (FID). As the FID scans along the rod, lipids are burned off the rod and detected. Using appropriate standards, each species in the sample can be quantitated. The disadvantages of this technology are (1) the equipment cost is comparable to an HPLC system, (2) it does not support mobile phase gradients, (3) as the rods age, resolution of lipid species deteriorates and the rods must be replaced and calibrated, and (4) resolution of multiple compounds is concentration dependent, thus complicating analysis of complex mixtures or natural extracts.

2. Gas Chromatography

a. *Analysis of Fatty Acid Methyl Esters*

Capillary GC with flame ionization detection is a versatile tool for lipid analysis.[18,19] Not only can it be used to identify the fatty acyl species of phospholipids, it can also be used for qualifying fatty acid raw materials or standards, and determining residual organic volatiles in dry lipid samples. The latter topic will be discussed in detail later in this section. The procedure for generation of fatty acid methyl esters (FAME) of phospholipids involves treating the lipid with a solution of sodium methoxide. The resulting transesterification liberates the fatty acid as the methyl ester, a volatile derivative of fatty acids. Injection of the solution into the capillary GC yields separations based on hydrocarbon chain length and unsaturation. Quantitation limits of 0.1% contaminating fatty acid can be achieved. See Method 3 for information on column and GC parameters, including instructions on generation of FAME. One disadvantage of GC is that it only provides information on the purity of the acyl portion of the compound. Structural information can be obtained only when coupled with another technique. An example of this would be analysis of asymmetric or mixed-acid phospholipids (e.g., palmitoyl oleoyl phosphatidylcholine [POPC]). GC analysis of the FAME from POPC should theoretically be 50% each of palmitoyl and oleoyl. A validated system would detect <1% of contaminating fatty acids such as myristoyl or stearoyl; however, structurally different contaminants such as DPPC, dioleoyl phosphatidylcholine (DOPC), or oleoyl palmitoyl phosphatidylcholine (OPPC) would not be detected. The first two contaminants could be detected by HPLC or mass spectrometry (MS), but OPPC can only be detected by PLA_2 digestion and HPLC or GC analysis of the resulting lyso lipid, or by an MS/MS technique. Another problem associated with analyzing tissue-derived lipids is that generation of FAME from the lipid sample does not derivatize the ether lipids or plasmalogens present in the sample. Data are biased to reflect only the fatty acid composition of the ester lipids in the sample. Also, tissue-derived lipids often contain fatty acid metabolic intermediates for which retention time standards are not available.

b. *Analysis of Residual Organic Volatiles*

Phospholipid materials to be used in pharmaceutical preparations must be low in residual solvents used in the manufacture of the product. The U.S. Pharmacopeia sets the limits for common chlorinated solvents at 50 ppm. Suggested limits for other volatile organic solvents such as ethanol, methanol, acetone, hexane, cyclohexane, and toluene are approximately 100 ppm with a total residual solvent load not more than 250 ppm. Capillary gas chromatography with flame ionization detection has proven to be a sensitive and fast analytical method for the screening and quantitative analysis of residual solvents. Electron capture detection can be utilized if chlorinated hydrocarbons are specified below 50 ppm by the user. Simple extraction of the solvent from a powder can be performed into dimethylformamide, with recovery being an important parameter in calculating the final result expressed as μg of residual solvent per gram of product. GC column stationary phases such as Carbowax and Polyethylene Glycol phases are ideal for these separations but suffer from low operational temperatures (<220°C). This limits the use of the GC for other analysis at high temperatures. Moderately nonpolar bonded phases such as 5% phenylmethylsilane can be used at the low temperatures required for separation without much sacrifice in resolution and peak symmetry.

3. High Performance Liquid Chromatography

Early HPLC analysis of lipids was severely limited due to the types of detectors available. Ultraviolet (UV) and refractive index detectors were insensitive to some lipids and limited the solvents or gradients employed for separation of lipid species. Solubility in solvents used with these detectors was low for some lipid species. Lipid analysts now have two types of detectors for HPLC that not only expand the available solvents for developing isocratic and gradient separations, but are sensitive to the nanogram range. Two types of mass detectors, flame ionization detectors (FIDs) for liquid chromatography (LC) and evaporative light scattering detectors (ELSD), have emerged as powerful instruments for the HPLC system. FIDs are sensitive to approximately 500 ng, while some ELSDs can reach 50 ng and lower sensitivity. Therefore, quantitation limits of 0.1% contaminants can be achieved with only 50 to 500 μg of sample loaded on the column. Figure 2 demonstrates the separation power available with these new detectors. This simple gradient, using solvents which are compatible with all phospholipids, successfully separates all the major classes of phospholipids in a single chromatogram.[20] Method 4 outlines the parameters for this gradient system. Unlike other detectors, including the ELSD, which shuttle data to a PC-based integration package, the FID uses a Hewlett-Packard 3396a integrator, an obvious disadvantage given the power of the PC-based packages available.

C. BIOLOGICAL ANALYSIS

1. Endotoxins

Analysis of phospholipid samples for endotoxins can only be done by the gel-clot limulus amoebocyte lysate (LAL) assay. Standard sample preparation procedures do not apply to lipid samples. We and others have developed a simple method for properly preparing lipid samples for LAL quantitation by the gel-clot assay. Briefly, a dry lipid sample is hydrated with sterile water at a temperature above the phase transition temperature of the lipid for 15 min and vortexed. It is important to hydrate the sample above the transition temperature; otherwise, the lipid particles formed are not fully hydrated and heterogeous. Hydration at elevated temperatures produced a uniform lipid suspension which presented no difficulties during the subsequent assay. Following hydration, the hydrated sample can then be assayed according to the directions for the LAL assay. The sample should not be sonicated or disrupted by any means other than vortexing. Sonication can lead to endotoxins becoming trapped in the lipid membrane and not available to the clotting assay, thus giving a false negative. This method is generally applicable to typical phospholipid species (PC, PG, PE, etc.), however, some special lipid species such as phosphatidylethanolamine-N-glutaric acid require hydration with sterile saline. When these samples are hydrated with sterile water, they must be diluted 1000-fold to recover a positive control. By using saline to hydrate the sample, positive controls are obtained at dilutions low enough to ensure an endotoxin-free sample.

2. Bioburden

Dry lipid samples can be hydrated in sterile water and tested for microbial content by usual methods. Special preparation or handling is not required for lipid samples. The general specifications for microbial quality of nonsterile pharmaceutical products are total aerobic microbial count not more than 100 CFU/g, total yeasts and molds not more than 100 CFU/g, and negative for *Escherichia coli*, *Salmonella* species, *Staphylococcus aureus*, and *Pseudomonas aeruginosa*.[21]

D. MISCELLANEOUS ANALYTICAL TESTS

1. Water Analysis by Karl Fischer Titration

Excessive water present in lipid products may lead to hydrolysis of the lipid. However, some lipids bind water tightly in the headgroup region and it is rather difficult and not advisable to try and remove these water molecules. Doing so could compromise the stability and integrity of the product. For example, PC tightly binds 2 to 3 water molecules per headgroup which translates to 5 to 7% water in a dry lipid sample.[22] A typical specification for PC is <10% by weight which is roughly four water molecules per lipid. Above four water molecules, the water is bound less tightly and could potentially cause hydrolysis problems. PG, on the other hand, does not require water associated with its headgroup to stabilize the compound, presumably due to the presence of hydroxyl groups in the headgroup region. Typical water values for PG are <1% by weight; therefore, a specification on water for these compounds of 1 to 2% is generally reasonable.

2. UV Analysis of Oxidation Products

Quantitation of oxidation products in lipid samples is quite difficult.[23] GC methods have been developed; however, the harsh conditions of the GC separation may produce artifacts in results.[24] The UV method for determination, while it is rather simple, is only a close approximation.[23] The extinction coefficient used in the calculation of oxidation products is derived from a nonlipid oxidation intermediate and therefore one must assume that it accurately represents the lipid oxidation products. For synthetic

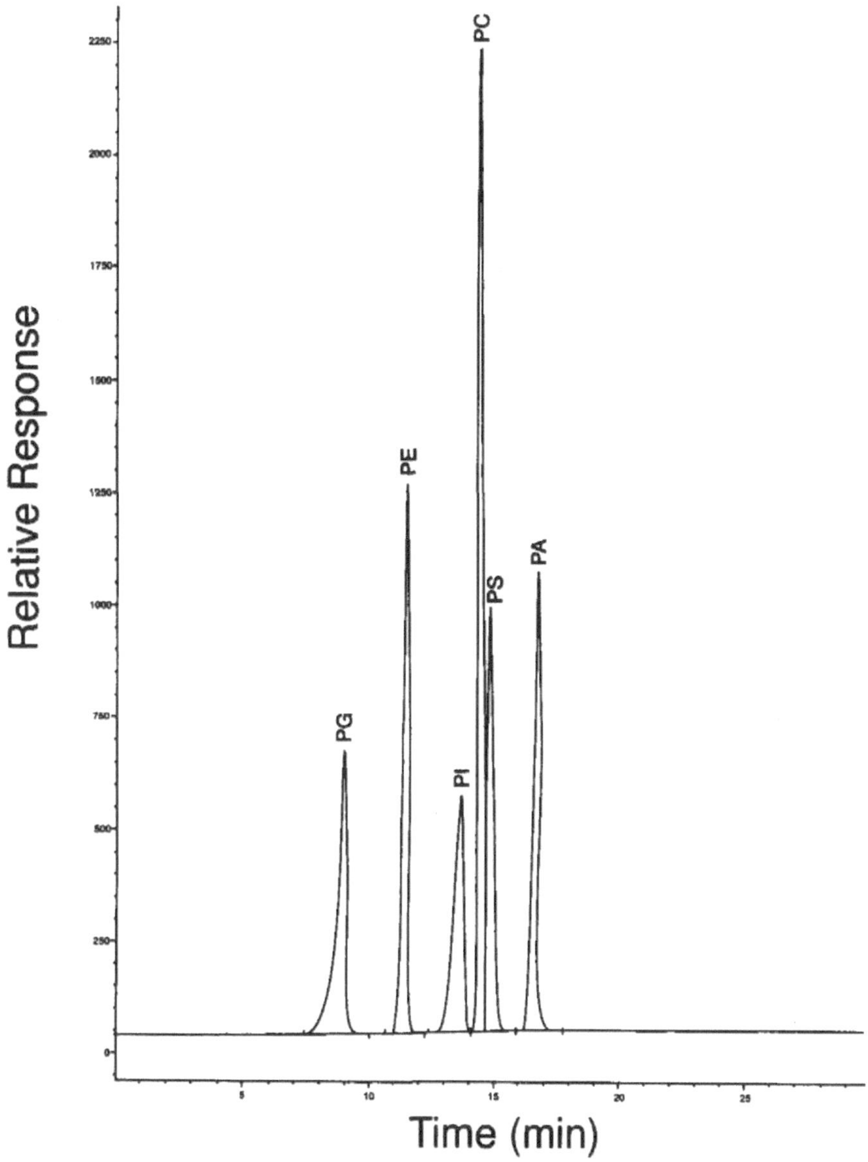

Figure 2 HPLC chromatogram of mixture of phospholipids containing DOPG, DOPE, Soy PI, DOPC, DOPS, and DOPA. Concentration of lipids is 0.5 mg/ml of each component. Column load was 20 μg of lipid. Elution was accomplished using the Becart et al.[20] gradient at 1 ml/min and detection by Sedex 55 evaporative light scattering detector.

unsaturated phospholipids, specifications of <1% peroxidized lipid as determined by the UV method are generally accepted. For tissue-derived lipids, however, while specifications of <1% is readily achieved, stability of the product determines how long after production of the lipid this oxidation level can be maintained. Use of tissue-derived products should be carefully considered for this reason and synthetic products substituted when possible.

3. Quantitation by Phosphorus Determination

Quantitation of lipid mass can be accomplished by a number of methods including gravimetric analysis, functional group titration, and elemental assays. There are numerous colorimetric assays for phosphorus requiring full digestion of the phospholipid to inorganic phosphorus using strong acids and oxidation reagents, followed by complex formation with molybdate and quantitation by UV-VIS spectroscopy.[25] Some of these assays use perchloric acid for the digestion which requires having a special hood for the assay.[26] Other assays have been revised to utilize sulfuric acid for the digestion, allowing greater flexibility in performing the assay on a routine basis. Refer to Method E for a modification of the Bartlett and Chen assays.[27,28]

III. ISSUES TO CONSIDER WHEN SELECTING LIPIDS

A. PHASE TRANSITION TEMPERATURE

The phase transition temperature is defined as the temperature required to induce a change in the lipid physical state from the ordered gel phase, where the hydrocarbon chains are fully extended and closely packed, to the disordered liquid crystalline phase, where the hydrocarbon chains are randomly oriented and fluid.[29] There are several factors which directly affect the phase transition temperature including hydrocarbon length, unsaturation, charge, and headgroup species. As the hydrocarbon length is increased, van der Waals interactions become stronger, requiring more energy to disrupt the ordered packing; thus the phase transition temperature increases. Likewise, introducing a double bond into the acyl group puts a kink in the chain which requires much lower temperatures to induce an ordered packing arrangement.

When developing a new product, procedure, or method, controlling the transition temperature of the lipid could be useful. Choosing a high transition lipid where the lipid vesicle would always be in the gel phase would provide a nonleaky packaging system. Alternatively, a lipid with a transition temperature between the starting temperature and the ending temperature of the system would provide a means of releasing packaged material as the lipid passes through its phase transition temperature and the vesicle becomes leaky.[30] Also, one should consider how the transition temperature of the lipid could impact the processing steps. Using a high transition lipid when filtration is necessary could present some technical problems.

B. STABILITY

The long-term stability or shelf life of a drug product containing lipids can be dramatically affected by the lipid species used in the formulation. Generally, the more unsaturated a compound, the easier the product is oxidized, and thus the shorter the shelf life of the product. Lipids from biological sources (e.g., egg, bovine, or soybean) typically contain significant levels of polyunsaturated fatty acids and therefore are inherently less stable than their synthetic counterparts. While saturated lipids offer the greatest stability in terms of oxidation, they also have much higher transition temperatures and thus present other difficulties in formulation. If unsaturation is a requirement, keep the degree of unsaturation as low as possible. In most cases, compounds containing oleic acid (18:1, cisΔ9) are sufficient to satisfy the need for unsaturation, and since they are monounsaturated, oleoyl-containing products are much more stable than polyunsaturated compounds.

Stability issues due to hydrolytic degradation are a general problem with lipid products. Aqueous formulations of drug products tend to be less stable since the presence of excess or bulk water leads to rapid hydrolytic degradation in lipid preparations.[31-33] This hydrolysis is dependent on several factors including pH,[31] temperature,[31,33] buffer species,[33] ionic strength, acyl chain length and headgroup,[32] and the state of aggregation.[32] A summary of the discussion on these factors can be found elsewhere.[34] Others have shown that the cause of this hydrolysis is possibly due to the penetration of water into the membrane. Simon and McIntosh[35] report water penetration depths in PE and PE:cholesterol membranes determined by X-ray diffraction and specific capacitance measurements. In PE membranes, water penetrates to near the deeper carbonyl group, while in PE membranes containing cholesterol, water only penetrates to the glycerol backbone. This indicates that cholesterol could play a role in stabilizing lipid membranes to hydrolysis.

Stabilizing membranes has been the subject of research for many years. The bulk of this research has been aimed at stabilizing intact liposomes in the dry powder form such that they retain their trapped internal contents upon reconstitution. Recently, lipid preparations have been stabilized using carbohydrates.[36,37] The possible reason for the stabilizing effect carbohydrates have on lipid membranes is that the carbobohydrate could intercalate into the headgroup region near the membrane/water interface and displace water from that region. In dry lipid formulations, this would serve to maintain a "hydrated" lipid membrane and keep the liposome structure intact. If this is true, it stands to reason that in an aqueous environment, the carbohydrate could still enter this region and displace water. This would tend to stabilize the membrane to hydrolysis from the bulk water phase.

C. CHARGE

Many biological membranes carry a net negative charge on their surface. The charge is generally imparted by the presence of anionic phospholipid species in the membrane. The major naturally occurring anionic phospholipids are PS, PI, PA, and cardiolipin. Some bacterial systems also contain PG. The charge may provide a special function for the membrane. Several steps of the blood coagulation cascade require a lipid membrane. The assembling of protein aggregates on the surface of platelets requires a negatively charged surface. For the conversion of prothrombin to thrombin, not only does it require a negative surface, the requirement is somewhat specific, limited to PS and PA.[38] Coagulation proteins bind as tightly to negatively charged surfaces containing PG and PI as they do to PS or PA membranes; however, the activity is only a fraction of that obtained with PS or PA membrane. Therefore, in some systems, not only must the charge requirement be satisfied, the system specificity for a particular species must be satisfied.

D. LIPID MIXTURES

In many cases, a single lipid species does not yield the exact physical properties needed for a particular system, or does not adequately mimic the natural system for which it is intended to replace or reproduce. For these issues, consider a complex lipid mixture composed of two or more individual lipid species, the composition being designed to create or reproduce a particular charge ratio, unsaturation ratio, phase transition temperature, or biological function. To reproduce the function of native brain tissue extracts, a blend of synthetic lipids (dioleoyl acyl composition) in the ratio 5:3:2 (wt%), PE:PS:PC, has been found to be satisfactory.[39] This represents the general phospholipid composition of most brain tissues. Also, many commercially available coagulation reagents which contained crude brain extracts in the past are being replaced by synthetic lipid blends. The advantages to this replacement system are the increased stability due to the lack of polyunsaturated fatty acids found in biological extracts, and the reproducibility of synthetic blends. Blending of multiple lipid species does not require much additional effort in sample preparation. If the quantity of lipid blend is sufficient, many times the lipid supplier will pre-blend to the user's specifications and provide a ready-to-use product.

E. CHOLESTEROL

Cholesterol is a membrane constituent widely found in biological systems which serves a unique purpose of modulating membrane fluidity, elasticity, and permeability. It literally fills in the gaps created by imperfect packing of other lipid species when proteins are embedded in the membrane. Cholesterol serves much the same purpose in model membranes. Unfortunately, cholesterol presents certain problems when used in human pharmaceuticals. High purity sources suitable for clinical applications are not widely available. Most commercially available cholesterol is derived from egg or wool grease (sheep derived). These animal sources are potentially not suitable for human pharmaceuticals due to the potential viral contamination. Also, cholesterol is readily oxidized creating a stability problem for lipid-based drug products.[40] Some of these oxidation by-products tend to be rather toxic in biological systems. The oxidation products 25-hydroxy cholesterol, 7-keto-cholesterol, 7α- and 7β-hydroxycholesterol, cholestane-3β,5α,6β-triol, and the 5- and 7-hydroperoxides, were found in a concentrate which had activity causing aortic smooth muscle cells to die.[41] This suggests that results from studies on atherosclerosis involving feeding experimental animals a diet containing cholesterol stored under adverse conditions (room temperature, open to air) could be ambiguous due to the potential presence of significant quantities of oxidized sterols.

F. SOURCE

There are two basic sources of phospholipids: synthetic and tissue-derived. Tissue-derived lipids are generally either egg-derived or bovine-derived. For clinical applications, either of these sources is not suitable due to stability problems and the possibility of viral or protein contamination. The U.S. Food and Drug Administration (FDA) issued a letter restricting the source of bovine tissue used to isolate pharmaceutical products to countries and animals certified to be free of bovine spongiform encephalopathy (BSE). Cattle in the U.S. are not certified BSE-free and cannot be used to isolate pharmaceutical products. Egg sources are not currently restricted; however, additional testing for viral contamination may be required for pharmaceutical products. Regardless of the regulatory issues, animal-derived products do not offer any advantage to synthetic lipids. They are inherently less stable due to the polyunsaturated fatty acids, and in most cases the synthetic counterpart costs the same or less than the tissue-derived product.

Synthetic lipids from different sources are not necessarily equal either. Synthetic lipids can be prepared from glycerol or glycero-3-phosphocholine (GPC) derived from a plant or animal source. The latter is sometimes referred to as semi-synthetic lipids because a portion of the molecule is derived from a natural source. Lipids derived from glycerol require the chiral center be synthetically prepared, which may lead to stereochemical impurities present in the final product. Lipids prepared using GPC obtained from an animal source may suffer from the same viral and protein contamination issues outlined above. The typical plant source for GPC is soybean lecithin.

IV. HANDLING LIPID PREPARATIONS

A. STORAGE AND HANDLING

1. Organic Solutions

Phospholipids supplied as an organic solution should be stored in a glass container layered with argon or nitrogen at below –20°C. The closure for the vial should be lined with teflon. Organic solutions should ***never*** be stored in polymer or plastic containers (polystyrene, polyethylene, polypropylene, etc.) as this will leach impurities out of the container.[42] Always use glass, stainless steel, or teflon to transfer lipids stored as organic solutions. **Do not** use plastic pipet tips to transfer organic solutions of lipids.

2. Powders

Phospholipids supplied as a powder should be handled as described below.

a. Saturated Lipids

Lipids composed of fatty acids that are completely saturated (e.g., dimyristoyl or dipalmitoyl phosphatidylcholine) are stable as powders. These lipids should be stored in a glass container with a teflon closure at below –20°C. When transferring a portion of the material out of the container, remove the container from the freezer and allow the material to reach room temperature before opening the bottle.

b. Unsaturated or Tissue-Derived Lipids

Lipids composed of fatty acids containing one or more double bonds (e.g., dioleoyl PC or egg PC) are **not** stable as powders. These lipids are extremely hygroscopic as powders and will quickly absorb moisture and become gummy upon opening the container. This could result in hydrolysis or oxidation of the material. These lipids should be dissolved in a suitable organic solvent, transferred to a glass container with a teflon closure, and stored at below –20°C.

c. Notes

Phospholipids should not be stored for long periods of time as aqueous supspensions. Storage of lipids in an excess of water results in hydrolysis of the sample.

B. SAMPLE PREPARATION

Proper sample preparation is important to achieving liposome preparations with suitable functionality and physical properties. Two procedures have been used with success. The first and simplest method is to aliquot the chloroform lipid solution to a glass container, evaporate the chloroform with nitrogen, and suspend the lipid film in an aqueous buffer. In many cases this is sufficient; however, hydration of lipid films may not produce suitable liposome suspensions for some lipid classes (e.g., phosphatidylethanolamine/cationic lipid mixtures). If difficulties are encountered, the following procedure should be used to prepare samples for hydration.

Dissolve the lipids in chloroform and aliquot the required quantity to a glass container. When using lipid mixtures or hygroscopic lipids to prepare liposome suspensions, it is necessary to first dissolve the lipids in chloroform. Hygroscopic lipids in powder form cannot be weighed accurately, and multicomponent systems must be homogeneously mixed before reconstituting in aqueous buffer.

1. Gently evaporate the bulk of the chloroform using a nitrogen stream. This should be done in a hood or well-ventilated area.
2. Resuspend the lipid in an appropriate amount of cyclohexane. This is usually 0.5 to 1.0 ml for <50 mg of lipid.
3. Freeze the cyclohexane solution using dry ice and while the solution is still frozen, place on a high vacuum lyophilization system. The sample should remain frozen when placed on the vacuum system. If the sample begins to thaw, remove from the vacuum system and refreeze. Repeat the process again.

If the sample continues to thaw, either the vacuum system is not sufficient to lyophilize the sample or too much residual chloroform was remaining in the sample.

4. Allow the sample to dry on the vacuum system for 3 to 5 h or overnight. For most phospholipids, this yields a white powder residue which easily suspends in aqueous buffers.
5. Remove the sample from the vacuum system and cap to prevent moisture uptake. The sample can now be hydrated with a suitable buffer or stored frozen until ready to use.

Note: Always hydrate the lipid mixture above the phase transition temperature of the lipids in the sample.

V. PROCEDURES AND METHODS

A. PROCEDURE FOR THIN LAYER CHROMATOGRAPHY ANALYSIS OF PHOSPHOLIPIDS

Definition: The R_f value for a phospholipid under test must fall within the limits of an appropriate standard.
Scope: Applicable to all PHOSPHOLIPIDS.

Apparatus

1. Nebulizer
2. Hot plate
3. TLC plates (Whatman K6)
4. Pasteur pipets
5. Spray chamber
6. TLC tanks
7. Iodine tank

Reagents

1. Chloroform (B&J)
2. Methanol (B&J)
3. Deionized Water
4. Ammonium Hydroxide
5. Phosphorus spray
6. Ninhydrin spray

Procedure

1. **Prepare TLC solvent mixtures.** Mix 650 ml chloroform, 250 ml methanol, and 40 ml deionized water in a 1-l graduated cylinder. Mix 650 ml chloroform, 250 ml methanol, and 40 ml ammonium hydroxide in another 1-l graduated cylinder.
2. **Developing TLC plates.** Place two clean TLC plates on the hot plate (200°C), silica side up, to reactivate silica, for approximately 3 min. Remove TLC plates from heat. Collect a sample of product (no more than 50 mg/ml) with a glass pasteur pipet. Spot sample 1/4 in. from the bottom of each TLC plate. Line the sides of the TLC tank with filter paper. Wet the filter paper and cover the bottom of one TLC tank with approximately 1 cm of the chloroform:methanol:water solvent mixture. Prepare another TLC tank as described above and cover the bottom of it with approximately 1 cm of the chloroform:methanol:ammonium hydroxide solvent mixture. Place a TLC plate in each solvent tank. Make sure the solvent does not touch product sample. Replace lids on top of tanks to ensure saturation. Remove TLC plates from tanks when they are at least three fourths saturated. Place TLC plates on hot plate to remove excess solvent. If necessary, place them in iodine tank.
3. **Spraying and interpreting TLC plates.** Place TLC plates in spray chamber. Completely spray TLC plates with ninhydrin spray. Place TLC plates on hot plate for 1 min. Read TLC plates for any ninhydrin positive material/product. Place TLC plates in spray chamber and spray with phosphorus spray. Place TLC plates on hot plate for 30 sec. Read TLC plates for phosphorus positive material. Dip plates in water and place them on the hot plate to determine if any contaminants are present. Again dip plates in water and place them on the hot plate to determine if any contaminants are present. Allow the plates to remain on the hot plate until the sample disappears or shows a charred appearance. Read TLC plates for any contaminants that appear charred.

B. DETECTION METHODS FOR THIN LAYER CHROMATOGRAPHY

1. Detection of Phosphorus

Equipment and Reagents

1. Nebulizer
2. Stirring hot plate
3. Magnetic stir bars
4. Assorted glassware
5. Molybdic anhydride
6. Molybdenum metal
7. Sulfuric acid
8. Deionized water

Procedure

Preparation of Reagent 1

1. Dissolve 40.1 g of molybdic anhydride in 1 l of concentrated sulfuric acid with heat and stirring overnight.
2. Cover the beaker or flask with plastic film (Saran® wrap) and secure with rubber band to prevent uptake of moisture.
3. Produces a clear, light blue solution.

Preparation of Reagent 2

1. Dissolve 1.78 g of molybdenum metal using 500 ml of Reagent 1 with heat and stirring overnight.
2. Cover the beaker or flask with plastic film (Saran® wrap) and secure with rubber band to prevent uptake of moisture.
3. Produces a dark blue solution.

Preparation of Phosphorus Spray

1. To 120 ml of deionized water add with stirring: 40 ml Reagent 1
   40 ml Reagent 2
2. Use **extreme caution** while adding reagents to water.
3. **Important!** Always add reagents to water, not water to reagents. Adding water to sulfuric acid solutions can be explosive.
4. Thoroughly mix the solution, allow to cool, and pour into a nebulizer.

Application of Phosphorus Spray

1. In a fume hood, spray TLC plates with molybdate solution until plate is covered with reagent.
2. Heat plate slightly until spots completely develop.
3. Phosphorus-containing compounds will appear as blue spots on white background. Additional heating will char unsaturated compounds.

2. Detection of Amines

Equipment and Reagents

1. Nebulizer
2. Assorted glassware
3. *n*-Butanol
4. Ninhydrin reagent
5. Deionized water

Procedure

Preparation of Ninhydrin Spray

1. Saturate *n*-butanol with water by shaking 200 ml of *n*-butanol with 200 ml of deionized water in a separatory funnel. Discard the water phase (lower phase).
2. Dissolve 200 mg of ninhydrin reagent in saturated *n*-butanol with stirring.
3. Produces a clear, light yellow solution.
4. Thoroughly mix the solution and pour into a nebulizer.

Application of Ninhydrin Spray

1. In a fume hood, spray TLC plates with ninhydrin solution until plate is covered with reagent.
2. Heat plate until spots completely develop.
3. Primary amine containing compounds will appear as pink spots on white background.

C. GC ANALYSIS OF FATTY ACID METHYL ESTERS (FAME)

Definition: The time at which a fatty acid methyl ester under test is observed to elute from a capillary column must fall within the limits of an appropriate standard.
Scope: Applicable to all phospholipids.

Apparatus

1. Hewlett-Packard 5890 Gas Chromatograph
2. J & W Fused Silica Capillary Column DB-225
3. IBM PC
4. Hewlett-Packard Chemstation Software
5. 10-μl syringe
6. Test tubes

Reagents

1. Hexane (B&J)
2. Methanol (B&J)
3. Reference Standard #68A (Nu-Chek-Prep, Inc.)
4. 1 N sodium methoxide

Procedure

1. **Prepare a reference standard.** Transfer the contents of 1 vial of #68A STANDARD to a small bottle with a cap. Add 4 ml of hexane to the standard. Keep the reference standard refrigerated.
2. **Prepare the sample.** Transfer 5 mg of sample into a test tube. If the sample is in powder form, dilute directly with 1 ml of hexane. If the sample is in chloroform, evaporate the solvent with nitrogen and add 1 ml of hexane. Add 3 drops of sodium methoxide to the sample in hexane. Mix. Allow the sample to stand 5 min before injecting.
3. **Prepare the gas chromatograph.** Set a temperature program which begins at 150°C, holds for 2 min, then ramps at 10°C/min up to 220°C where it holds for 30 min. The FID detector is set to 250°C. The injector is set to 220°C. The injection mode is split. Helium is the carrier gas with a flow rate of 2.5 ml/min. The pressure of the helium tank is set to 20 psi.
4. **Prepare the computer**. Set the HP Chemstation data acquisition time for the fatty acid methyl esters as follows:
   a. 3 to 12 carbons, 14 min
   b. 14 to 24 carbons, 20 min
   c. natural products, 27 min

5. **Run the sample.** Wash the syringe with methanol several times. Wash the syringe with hexane several times. Wash the syringe with the lipid sample once. Draw 1 µl of lipid sample into the syringe allowing 2 to 3 µl of head space. Inject the sample.
6. **Check GC results.** The lipid sample should be >99% of the appropriate fatty acid for symmetric phospholipids (e.g., 18:1 PC) and approximately 50 ± 5% for each fatty acid component of an asymmetric phospholipid (e.g., 16:0-18:1 PC). In order to check this, inject 1 µl of the reference standard. Compare the retention times of the reference standard to the retention times in the lipid sample. If the retention times match and there is <1% impurity, then the sample is good.
7. **Limit of Detection.** The limit of detection of any fatty acid other than the fatty acid(s) normally present in the sample is <1%.

D. HPLC ANALYSIS OF LIPIDS

Apparatus

1. Hewlett-Packard 1050 Pump Station
2. Sedex 55 Evaporative Light Scattering Detector
3. Hewlett-Packard 2-D Chemstation
4. 500- and 10-ml syringes (Hamilton Co.)
5. Astec Diol 5 µm, 80Å spherical column
6. Sonicator
7. Small glass test tubes
8. 1-l graduated cylinder
9. 1-l erlenmeyer flask

Reagents

1. Chloroform (Mallinckrodt)
2. Methanol (B&J)
3. Ammonium hydroxide (Mallinckrodt)
4. Deionized Water

Procedure

1. **Prepare Mobile Phase A:** Mix 800 ml chloroform, 195 ml methanol, and 5 ml ammonium hydroxide in a 1-l graduated cylinder. Mix well by inverting cylinder approximately 20 times. Transfer solvent mixture to a 1-l erlenmeyer flask and sonicate for 10 to 15 min to remove air bubbles.
2. **Prepare Mobile Phase B:** Mix 600 ml chloroform, 340 ml methanol, 50 ml deionized water, and 5 ml ammonium hydroxide in a 1-l graduated cylinder. Mix well by inverting cylinder approximately 20 times. Transfer solvent mixture to a 1-l erlenmeyer flask and sonicate for 10 to 15 min to remove air bubbles.
3. **Sample Preparation:** Dissolve sample in chloroform:methanol:deionized water (73:23:3 v/v/v) at a concentration of 2 mg/ml. Inject 200 µl into a 20 µl loop and load on column; elute with gradient.
4. **Gradient:**
   0 to 14 min: 100% A to 100% B linear gradient
   14 to 25 min: hold 100% B
   25 to 30 min: 100% B to 100% A linear gradient
   30 to 45 min: hold 100% A to regenerate
5. **Flow Rate:** 1 ml/min

E. PHOSPHORUS DETERMINATION

Apparatus

1. Hewlett-Packard 8452A Diode Array Spectrophotometer
2. Hewlett-Packard HP 89531A MS-DOS-UV/VIS Operating Software
3. Compatible computer
4. HELLMA UV Quartz sample cell (QS 1.000)
5. Kimwipes
6. 30-ml disposable glass tubes

7. Graduated cylinders (1-l and two 100-ml)
8. Acid-washed porous boiling chips
9. 12 marbles

Reagents

1. Deionized water
2. Concentrated H_2SO_4 (Mallinckrodt)
3. Ammonium molybdate(VI) tetrahydrate (Aldrich cat. no. 22,123-6)
4. L-Ascorbic acid (Aldrich cat. no. 25,556-4)
5. 0.65 m*M* Phosphorus standard solution (Sigma cat. no. 661-9)
6. Hydrogen peroxide (Aldrich cat. no. 21,676-3)

Procedure

Prepare the Solutions

1. **8.9 N H_2SO_4 solution.** Slowly add 247 ml of concentrated H_2SO_4 to 753 ml of deionized water, allow the heat to dissipate from the cylinder, and mix the solution well. Store in a sealed container at room temperature for up to 1 year.
2. **10% Ascorbic acid solution.** Place 10 g of ascorbic acid into a cylinder and dilute to 100 ml with deionized water. Mix the solution well. Store in an amber screw-cap bottle at 4°C for up to 1 month.
3. **2.5% Ammonium molybdate(VI) tetrahydrate solution.** Place 2.5 g of ammonium molybdate(VI) tetrahydrate into a cylinder and dilute to 100 ml with deionized water. Mix the solution well. Store in an amber screw-cap bottle at 4°C for up to 1 month.

Prepare the Sample Tubes

1. Place sample (0.1 µmol phosphorus) into the bottom of each tube. Gently remove any solvent from the tubes with N_2.

Prepare the Five Standard Tubes

1. Place the following quantities of phosphorus standard into five separate tubes: 0.0325 µmol (50 µl), 0.065 µmol (100 µl), 0.114 µmol (175 µl), 0.163 µmol (250 µl), and 0.228 µmol (350 µl).

Digestion of Organic Sample to Inorganic Phosphate

1. The samples, five standards, and three empty tubes should be labeled and placed in a vinyl-coated test tube rack.
2. Add 0.45 ml H_2SO_4 and 2 acid-washed boiling chips to each of the 12 tubes.
3. Heat all tubes in an aluminum block in the hood at 200 to 215°C for 25 min. **Important:** temperature must be above 200°C.
4. Remove tubes from the block and allow them to cool 5 min before adding 6 drops H_2O_2 to all tubes (try not to get H_2O_2 on walls of tubes).
5. Replace the tubes in the block and continue to heat for an additional 30 min. The samples should be colorless at this point (if any brown color persists, add three additional drops of H_2O_2 to all cooled tubes and continue heating tubes for 15 min). Cool tubes.
6. Add 3.9 ml deionized water to all tubes.
7. Add 0.5 ml ammonium molybdate(VI) tetrahydrate solution to all tubes and vortex each tube five times.
8. Add 0.5 ml ascorbic acid solution to all tubes and vortex each tube 5 times.
9. Place tubes in test tube rack. Put a marble on each tube to prevent evaporation. Heat all tubes in a boiling water bath for 7 min. Cool the tubes.

Spectrophotometric Analysis of Samples

1. Zero the spectrophotometer using deionized water (the reference).
2. Determine the absorbance of each of the three blanks at 820 nm.
3. Determine the absorbance of each of the five standards at 820 nm.
4. Determine the absorbance of each of the samples at 820 nm.
5. Subtract the blank absorbance from the standards and samples.
6. Generate a calibration curve using the standards and determine the concentration of phosphorus in the samples.

REFERENCES

1. **Meneses, P. and Glonek, T.,** High resolution ^{31}P NMR of extracted phospholipids, *J. Lipid Res.*, 29, 679, 1988.
2. **Berden, J. A., Barker, R. W., and Radda, G. K.,** NMR studies on phospholipid bilayers. Some factors affecting lipid distribution, *Biochim. Biophys. Acta*, 375, 186, 1975.
3. **Frost, D. J. and Gunstone, F. D.,** The PMR analysis of non-conjugated alkenoic and alkynoic acids and esters, *Chem. Phys. Lipids*, 15, 53, 1975.
4. **Henderson, T. O., Glonek, T., and Myers, T. C.,** Phosphorus-31 nuclear magnetic resonance spectroscopy of phospholipids, *Biochemistry*, 13, 623, 1974.
5. **Kates, M.,** *Techniques of Lipidology*, 2nd ed., North-Holland, Amsterdam, 1986.
6. **Murphy, R. C.,** *Handbook of Lipid Research: Mass Spectrometry of Lipids*, Vol. 7, Plenum Press, New York, 1993.
7. **Han, X. and Gross, R. W.,** Electrospray ionization of mass spectroscopic analysis of human erythrocyte plasma membrane phospholipids, *Proc. Natl. Acad. Sci., U.S.A.*, 91, 10635, 1994.
8. **Gurr, M. I. and Harwood, J. L.,** *Lipid Biochemistry: An Introduction*, 4th ed., Chapman and Hall, New York, 1991.
9. **Wells, M. A. and Hanahan, D. J.,** Phospholipase A from *Crotalus adamanteus* venom, *Meth. Enzymol.*, 14, 178, 1969.
10. **Deveer, A. M., Franken, P. A., Dijkman, R., Meeldijk, J., Egmond, M. R., Verheij, H. M., Verger, R., and de Haas, G. H.,** Competitive inhibition of lipolytic enzymes. VII. The interaction of pancreatic phospholipase A2 with micellar lipid/water interfaces of competitive inhibitors, *Biochim. Biophys. Acta*, 1125, 73, 1992.
11. **Mangold, H. K. and Malins, D. C.,** Fractionation of fats, oils, and waxes on thin layers of silici acid, *J. Am. Oil Chem. Soc.*, 37, 383, 1960.
12. **Skipski, V. P., Peterson, R. F., and Barclay, M.,** Separation of phosphatidylethanolamine, phosphatidylserine, and other phospholipids by thin-layer chromatography, *J. Lipid Res.*, 3, 467, 1962.
13. **Ellingson, J. S. and Lands, W. E. M.,** Phospholipid reactivation of plasmalogen metabolism, *Lipids*, 3, 111, 1968.
14. **Frederiks, W. M. and Broekhoven, S.,** Separation and determination of lipids by one-dimensional micro thin layer chromatography followed by densitometry, *J. Chromatogr.*, 150, 170, 1978.
15. **Macala, L. J., Yu, R. K., and Ando, S.,** Analysis of brain lipids by high performance thin layer chromatography and densitometry, *J. Lipid Res.*, 24, 1243, 1983.
16. **Schmitz, G., Assmann, G., and Bowyer, D. E.,** A quantitative densitometric method for the rapid separation and quantitation of the major tissue and lipoprotein lipids by high-performance thin layer chromatography. I. Sample preparation, chromatography, and densitometry, *J. Chromatogr.*, 307,65, 1984.
17. **Mallinger, A. G., Yao, J. K., Brown, A. S., and Dippold, C. S.,** Analysis of complex mixtures of phospholipid classes from cell membranes using two-dimensional thin layer chromatography and scanning laser densitometry, *J. Chromatogr.*, 614, 67, 1993.
18. **Evershed, R. P.,** *Lipid Analysis: A Practical Approach*, Hamilton, R. J. and Hamilton, S., Eds., IRL Press, New York, 1992, 113.
19. **Christie, W. W.,** A simple procedure for rapid transmethylation of glycerolipids and cholesteryl esters, *J. Lipid Res.*, 23, 1072, 1982.
20. **Becart, J., Chevalier, C., and Biesse, J. P.,** Quantitative analysis of phospholipids by HPLC with a light scattering evaporating detector-application to raw materials for cosmetic use, *J. High Res. Chromatogr.*, 13, 126, 1990.
21. Microbiological attributes of non-sterile pharmaceutical products, in *U.S. Pharmacopeia/National Formulary*, Vol. USP 23/NF 18, 1994, 1939.
22. **Jendrasiak, G. L. and Hasty, J. H.,** The hydration of phospholipids, *Biochim. Biophys. Acta*, 337, 79, 1974.
23. **Kim, R. S. and LaBella, F. S.,** Comparison of analytical methods for monitoring autoxidation profiles of authentic lipids, *J. Lipid Res.*, 28, 1110, 1987.
24. **Barenholz, Y. and Amselem, S.,** Quality control assays in the development and clinical use of liposome-based formulations, in *Liposome Technology: Liposome Preparation and Related Techniques*, Vol. 1, 2nd ed., Gregoriadis, G., Ed., CRC Press, Boca Raton, FL, 1993, 527.
25. **Kates, M.,** Techniques in lipidology. Isolation, analysis and identification of lipids, in *Laboratory Techniques in Biochemistry and Molecular Biology*, Vol. 3, Part 2, Work, T. S. and Work, E., Eds., Elsevier/North-Holland, Amsterdam, 1986.
26. **Sheltawy, A. and Dawson, R. M. C.,** Separation and estimation of phospholipids, in *Chromatographic and Electrophoretic Techniques*, Vol. 1, Smith, I., Ed., Heinemann, Bath, 1969, 450.
27. **Bartlett, G. R.,** Phosphorus assay in column chromatography, *J. Biol. Chem.*, 234, 466, 1959.
28. **Chen, P. S., Jr., Toribara, T. Y., and Warner, H.,** Microdetermination of phosphorus, *Anal. Chem.*, 28, 1756, 1956.
29. **Small, D. M.,** *Handbook of Lipid Research: The Physical Chemistry of Lipids, From Alkanes to Phospholipids*, Vol. 4, Plenum Press, New York, 1986.
30. **Ellens, H., Bentz, J., and Szoka, F. C.,** Destabilization of phosphatidylethanolamine liposomes at the hexagonal phase transition temperature, *Biochemistry*, 25, 285, 1986.
31. **Frøkjaer, S., Hjorth, E. L., and Wørts, O.,** Stability and storage of liposomes, in *Optimization of Drug Delivery*, Bundgaard, H., Bagger Hansen, A., and Kofod, H., Eds., Munksgaard, Copenhagen, 1982, 384.

32. **Kensil, C. R. and Dennis, E. A.,** Alkaline hydrolysis of phospholipids in model membranes and the dependence on their state of aggregation, *Biochemistry*, 20, 6079, 1981.
33. **Grit, M., de Smidt, J. H., Struijke, A., and Crommelin, D. J. A.,** Hydrolysis of phosphatidylcholine in aqueous liposome dispersions, *Int. J. Pharm.*, 50, 1, 1989.
34. **Grit, M., Zuidam, N. J., and Crommelin, D. J. A,** Analysis and hydrolysis kinetics of phospholipids in aqueous liposome dispersions, in *Liposome Technology: Liposome Preparation and Related Techniques*, Vol. 1, 2nd ed., Gregoriadis, G., Ed., CRC Press, Boca Raton, FL, 1993, 455.
35. **Simon, S. A. and McIntosh, T. J.,** Depth of water penetration into lipid bilayers, *Meth. Enzymol.*, 127, 511, 1986.
36. **Crowe, J. H. and Crowe, L. M.,** Factors affecting the stability of dry liposomes, *Biochim. Biophys. Acta*, 939, 327, 1988.
37. **Crowe, J. H., Crowe, L. M., Carpenter, J. F., and Aurell Winstrom, C.,** Stabilization of dry phospholipid bilayers and proteins by sugars, *Biochem. J.*, 242, 1 1987.
38. **Jones, M. E., Lentz, B. R., Dombrose, F. A., and Sandberg, H.,** Comparison of the abilities of synthetic and platelet-derived membranes to enhance thrombin formation, *Thromb. Res.*, 39, 711, 1985.
39. **van den Besselaar, A. M. H. P., Neuteboom, J., and Bertina, R. M.,** Effect of synthetic phospholipids on the response of the activated partial thromboplastin time to heparin, *Blood Coag. Fibrinol.*, 4, 895, 1993.
40. **Smith, L. L.,** *Cholesterol Autoxidation*, Plenum Press, New York, 1981.
41. **Taylor, C. B., Peng, S. K., Werthesen, N. T., Than, P., and Lee, K. T.,** Spontaneously occurring angiotoxic derivatives of cholesterol, *Am. J. Clin. Nutr.*, 32, 40, 1979.
42. **Pidgeon, C., Apostol, G., and Markovich, R.,** Fourier transform infrared assay of liposomal lipids, *Anal. Biochem.*, 181, 28, 1989.

Chapter 3

An Overview of Liposome Scaled-Up Production and Quality Control

Yechezkel Barenholz and Danilo D. Lasic

CONTENTS

I. INTRODUCTION

The history of liposomes as we know them today (vesicles resembling biological membranes) started in the laboratory of Bangham and co-workers 30 years ago,[1–4] although at least one medical application of small unilamellar vesicles (SUV) was performed 10 years earlier by Friedman and co-workers.[5] They had already demonstrated in 1956 that structures, which today we would call SUV, composed of phosphatidylcholine (PC) injected intravenously into rabbits and other mammals were very efficacious in treating athersclerosis.[5,6] It was not until 1990 that the first liposomal product (Ambisome™) was approved for clinical use in Europe,[7] and in 1995 Doxil™[8] and Daunoxome™ were approved in the U.S. Large-scale production of liposomes started in the mid-1980s, late in the short history of liposomes. The reason for the slow development of large-scale production is related to the time it took to resolve technological and quality control issues, including: (1) availability of high-quality lipid raw materials; (2) validated quality control assays; (3) reproducible processes for scaled-up production; (4) fool-proof methods for obtaining sterile (and for injectables, apyrogenic) liposomal products; and (5) long-term chemical and physical stability. All these issues are interrelated; for example, the lack of a complete repertoire of quality control assays slowed down the development of large-scale production and stability assessment.[9–12] The cosmetic industry, due to its need for large amounts of liposome preparations,[3] was the first to solve these issues. However, the requirements for pharmaceutical dosage forms are much higher, and therefore it took much longer to resolve these issues at the standards required.[2,3,9]

II. LARGE-SCALE PRODUCTION OF LIPOSOMES

A. GENERAL ASPECTS

The two main obstacles to large-scale production of liposomes are (a) that, despite the very large number of methods available for liposome preparation (see Tables 1 and 2), there is no agreement on the theoretical aspects of how and why liposomes are formed using these different approaches,[2,3,9,10] and (b) it is still not clear which of the liposome preparations, if any, are thermodynamically stable assemblies. (See also Chapter 4 by Leung and Lasic.) The energy input required for preparation of liposomes by all methods described in Table 1 is needed either to overcome a kinetic barrier to form a stable assembly or, alternatively, to form a kinetic trap for a metastable assembly. The fact that, at least for phospholipid vesicles, the exact structural parameters of the liposomes are determined to a large extent by the method

0-8493-4012-8/96/$0.00+$.50
© 1996 by CRC Press, Inc.

Table 1 Methods of Vesicle Preparation and Their Liposome Product[a,b]

Mechanical Methods
Vortex or hand shaking of phospholipid dispersions (MLV)[a,c]
Extrusion through polycarbonate filters (OLV or UV, at low or medium pressure, respectively)
Extrusion through a French pressure cell (mainly SUV)
"Microfluidizer" technique (mainly SUV)
High-pressure homogenization (mainly SUV)
Ultrasonic irradiation (SUV of minimal size)
Bubbling of gas (BSV)
Methods Based on Replacement of Organic Solvent(s) by Aqueous Media
Removal of organic solvent(s) (MLV, OLV, SUV)
Use of water-immiscible solvents; ether and petroleum ether infusion (solvent vaporization) (MLV, OLV, LUV)
Reverse-phase evaporation (LUV, OLV, MLV)
Methods Based on Detergent Removal by:
Gel exclusion chromatography (SUV)
"Slow" dialysis (LUV, OLV, MLV)
Fast dilution (LUV, OLV)
Other related techniques (MLV, OLV, LUV, SUV)
Methods Based on Size Transformation and Fusion
Spontaneous fusion of SUV in the gel phase (LUV, OLV, MLV)
Freeze-thawing (MLV)
Freeze-drying (MLV)
Dehydration of SUV followed by rehydration with or without sizing
Ca^{2+} ion-induced fusion (LUV, OLV, MLV)
Detergent-induced growth (LUV, OLV)
Methods Based on pH Adjustment
SUV and possibly LUV

[a] For definitions and abbreviations, see Table 2.
[b] For more details see Barenholz and Crommelin[9] and references therein.
[c] Type of liposomes is defined in parentheses.

of liposome preparation suggests that these liposomes are not at thermodynamic equilibrium. It is our opinion that, due to the fact that the formation process is in general irreversible and can yield a population with a wide size distribution, liposomes can occupy various local free energy minima and therefore the way by which these local points in the free energy function was reached determines the size and lamellarity of liposomes produced.[3,13] However, when liposomes are formed they can be very stable due to deep local free energy minima and then their thermodynamic stability may become a semantic question. Recently, it has been found that for certain amphiphile compositions, such as a mixture of two amphiphiles of opposing charges, "equilibrium" vesicles may be formed spontaneously.[14] The lack of basic knowledge on these two aspects (a and b, above) hinders development of optimal methods for liposome production and for calculating the exact structural parameters of the liposomes produced.[13] Therefore, the current method of choice for scale-up is dependent mainly on various technical considerations, as described in Table 3.

B. LIPOSOME PRODUCTION

One should distinguish between "liposome preparation" which refers to a laboratory-scale for which many methods are available (Table 1) to meet very varied needs, and "liposome production," namely, large-scale preparation of liposomes for industrial and pharmaceutical needs. For various reasons, many of the preparation methods described in Table 1 are unsuitable for scale-up as exemplified in Table 3. Liposome preparation, as well as production, consists of various steps (the number varying from method to method). However, all these methods share in common two key steps: lipid hydration and downsizing. Table 3 describes the major approaches to lipid hydration and liposome downsizing. It is important to note that regarding downsizing the two best approaches are high-pressure homogenization and medium-pressure extrusion. The currently available instrumentation of APV Gaulin[15] (Hilversum, Holland) or

Table 2 Liposome Classification

	Classification Based on Functionality	Main Features
A.	Conventional liposomes	High accessibility to interact with factors from extraliposome medium (i.e., opsonins, enzymes, low molecular weight agents)
	Sterically stabilized liposomes	Low accessibility to extraliposomal factors due to grafted steric barrier
	Activosomes	Vesicular to nonvesicular transition upon contact with medium or agents in the medium
	Classification Based on Structural Parameters[a]	**Abbreviation**
B.	Multilamellar large vesicles, >0.5 μm	MLV
	Oligolamellar vesicles, 0.1–1 μm	OLV
	Unilamellar vesicles (all size range)	UV
	Small unilamellar vesicles, 20–100 nm	SUV
	Medium-sized unilamellar vesicles	MUV
	Large unilamellar vesicles, >100 nm	LUV
	Giant unilamellar vesicles (cell-size vesicles with diameters 1 μm)	GUV
	Multivesicular vesicles (usually large >1 μm)	MVV
	Classification Based on Method of Preparation[a]	**Abbreviation**
C.	Single or oligolamellar vesicles made by reserve-phase evaporation	REV
	Multilamellar vesicles made by the reverse phase method	MLV-REV
	Stable plurilamellar vesicles	SPLV
	Frozen and thawed MLV	FAT-MLV
	Vesicles prepared by extrusion methods	VET
	Vesicles prepared by French press	FPV
	Vesicles prepared by fusion	FUV
	Dehydration–rehydration vesicles	DRV
	Bubblesomes	BSV

[a] Table based on Reference 9.

APV Rannie[16] (Albertslund, Denmark) for high-pressure homogenization enables the user to up-scale the production from a scale of 35 ml minimal volume per batch using the Rannie 8.30 Minilab one-stage homogenizer[16] having a maximum capacity of 10 l/h to a scale of many thousands of liters per hour using higher capacity, two-stage high-pressure homogenizers of APV Rannie or APV Gaulin (Weintraub, Bolotin, Bercovier, and Barenholz, unpublished). The main weak point of these homogenizers is the remote, but existing, possibility that the cooling water of the homogenizer will mix with the liposomal product. Avestin (Ottawa, ON, Canada) offers high-pressure homogenizers under the trade name EmulsiFlex® in a broad volume range, starting with 1 ml minimal volume with a maximum capacity of 20 l/h in a broad pressure range. Microfluidics Corporation (Newton, MA, U.S.) also makes high-pressure laboratory- and large-scale homogenizers called Microfluidizers® having no moving parts.[17] It is not yet clear, if the scale-up from laboratory-scale to large-scale using Microfluidizers is straightforward. The main advantage of the Avestin homogenizers and of the Microfluidizers is that there is no possibility that during their use the cooling water will mix with the product. The main limitation of the Avestin homogenizers is that, so far, this company does not have high-capacity homogenizers to approach the many-liter scale. The size distribution of vesicles obtained by high-pressure homogenization is, in general, limited to vesicles of <120 nm,[9,10,15,16] while preparation of larger and homogeneous unilamellar vesicles, according to our experience, is difficult. Often, in the case of high-pressure homogenization, the product (small vesicles) is contaminated with some giant uni- or oligolamellar liposomes. Another problem is that in the case of lipid mixtures containing a lipid with high aqueous solubility, small mixed micelles are formed and coexist with the vesicles. A post-homogenization extrusion–filtration step may improve the size distribution. Emulsiflex®-C5 (Avestin) is actually a long-awaited combination of homogenizer and extruder. Caution is also needed to minimize oxidation and hydrolysis.

Medium-pressure extrusion seems to be the "cleanest" approach to large-scale production. Its other advantage is the ability to produce liposomes of a broad spectrum of defined sizes which is determined by pore size of the filter used for extrusion, although the larger the filter pore size the greater is the size heterogeneity.[10] The main limitation of medium-pressure extrusion is that generally the instrumentation

Table 3 The Two Key Steps in Liposome Production, Hydration and Sizing and Their Upscaling Feasibilities[2,21]

Hydration		Sizing	
Method	**Upscaling Feasibility**	**Method**	**Upscaling Feasibility**
Thin lipid hydration by mechanical shaking after evaporation or lyophilization Product: MLV	Very good	Ultrasonic irradiation gives mainly unilamellar small liposomes, including those of minimal size (SUV: 20 nm diameter)	Poor
Dehydration–rehydration of SUV Product: MLV having very large trapped volume	Very good but expensive due to lyophilization step		
Organic solvent(s) replacement Products: MLV, OLV, UV		High-pressure extrusion (above 35 MPa, ca 5000 psi) gives mainly small (but not minimal size) vesicles, mostly unilamellar liposomes. Examples: French pressure cell, microfluidizer, high-pressure homogenizers	Very good for high pressure homogenizers
Solvents miscible with water, such as ethanol, methanol and isopropylalcohol	Very good		
Solvents immiscible with water, such as ether and alkyl halides	Feasible but problematic	Low- or medium-pressure extrusion (up to 14 MPa ca 2000 psi) through pores of defined sizes. Examples: Polycarbonate membranes, ceramic filters, stainless steel filters. Liposome size distribution and number of lamellae is determined by pore size.	Very good
Formation of lipid–detergent mixed micelles followed by detergent removal Products: OLV, UV, SUV	Good for OLV Reasonable for SUV		

Note: For more details on methods see References 9 and 21.

has to be developed in-house[21] and that the capacity of the filter or cartridge with large surface area (≥1 m^2) may be a bottleneck. Recently, high surface area-pleated filters mounted between two pressure vessels were successfully used in the scale-up and vesicles with tighter size distribution than the homogenized or microfluidized ones were produced (Lasic, unpublished).

III. LIPOSOME LOADING

A. GENERAL ASPECTS

The main use of liposomes is as a carrier of drugs and other agents;[10,18–20] therefore the agent has to be loaded either into the intraliposomal aqueous phase or the liposomal lipid bilayer. In general, one should distinguish between the loading during the liposome preparation, namely, loading that is performed during the lipid hydration step (referred to as passive loading) and remote loading which is done after the liposomes are produced.[9] In both cases the unloaded "free" agent has to be removed and separated from the liposomes using differences in charge, size, or other property (physical, chemical, or biological).[9,12,21]

B. PASSIVE LOADING

Passive loading is the general approach to liposome loading with any agent of choice irrespective of its molecular mass, molecule shape, charge, and balance between hydrophilicity and hydrophobicity. Passive loading can be obtained by the interaction of the encapsulated agent with the liposome bilayer (membrane association), by encapsulation in the intraliposome aqueous phase, or by a combination of both membrane association and aqueous loading.[9,22] The membrane association can be hydrophobic or electrostatic in nature and is related to the liposome-to-medium partition coefficient.[22] Aqueous loading is controlled by physical and structural parameters, such as size, shape, and number of lamellae, which control the aqueous trapped volume. The loading may also be affected (either increased or reduced) by specific interactions such as attraction or repulsion between the bilayer components and the loaded agent. In such a case the number of lamellae and the exact composition of the medium may have important contributions. The magnitude of osmotic gradient formed after lipid hydration is another important factor which determines loading efficiency, as was proven for dehydrated–rehydrated vesicles.[23] The exact composition and pH of the hydration medium, downsizing medium, and storage medium should be well controlled and defined in order to optimize liposome loading.

C. REMOTE LOADING

Remote loading of liposomes requires that the liposomes will be prepared with a "driving force" which will enable spontaneous transport of the agent from the extraliposomal medium across the lipid bilayer(s) into the intraliposomal aqueous phase. In order for the agent to be retained inside the liposomes it has to be modified in a way that prevents its release from the liposomes. For agents for which the energy of activation of release is high, stable loading can be achieved.[9,16,22] The three main methods for remote loading are based on a pH gradient,[24,25] membrane potential,[26] or an ammonium sulfate gradient.[15,27–29] The combination of ammonium sulfate gradient and liposomes composed of matrix phospholipids having a high gel-to-liquid-crystalline phase transition temperature enables attaining highly stable remote loading of doxorubicin into sterically stable liposomes.[8,16] These have been recommended for approval by the FDA for clinical use in treating cancer patients.

Another approach to remote loading is the use of liposomes composed of cationic amphiphiles to form assemblies with poly- and oligonucleotides for gene delivery and for controlling gene expression.[30,31] These liposomes are referred to by us as "activsomes" (Table 2), which means that upon interaction with the agent, a new assembly is formed which is responsible for the delivery of the agent to the site of action. The exact protocols for the application of such activosomes have still to be finalized.

IV. LIPOSOME STERILIZATION

A. GENERAL ASPECTS

One of the most important aspects in liposome preparation is how to make and keep the preparation sterile (and for injectables, apyrogenic). Most phospholipids are 100% biodegradable and serve as an excellent medium for microorganism growth. This can be prevented by working under aseptic conditions, by using a sterilization step[12,21,32] and, when permitted, using one or more agents which inhibit microorganism growth, such as parabens, benzalkonium chloride, etc. The user has to be aware that many of these agents have a rather high liposome-to-medium partition coefficient, which may reduce their efficacy and also modify the vesicle properties (such as the charge when benzalkonium chloride serves as the preservative).

B. SPECIFIC ASPECTS OF LIPOSOME STERILIZATION

Each of the main methods for liposome sterilization suffers from one or more drawbacks. Autoclaving, a very efficient method, may affect chemical and physical stability.[11,32–34] Filtration has many advantages; however, it is limited to vesicles smaller than 200 nm and it does not remove virions and small spores. Gamma-irradiation, a very effective sterilization method, seems to be destructive of phospholipids and cholesterol.[10,35,36] It is possible that in the presence of specific agents which protect the lipids and the encapsulated agent, but not the microorganisms and the virions, this approach may still be useful.[36,37] As it stands now, one has to use sterile filtration when possible or, alternatively, to try to optimize autoclaving. At times, working under aseptic conditions may be necessary.

V. QUALITY CONTROL OF LIPOSOMES

Liposome characterization and quality control was, and to some extent still is, the Achilles' tendon of liposome production. So far, there are no general protocols that are agreed upon by many users. The complexity of this issue (Table 4) is related to quality control and stability follow-up required to assay the exact composition, physical properties, toxicity, and performance. Although at the laboratory level various approaches and levels of sophistication can be used,[11,12,38,39] this is not the case for quality control of scaled-up production, where the highest level of quality control is required.[11,12]

Table 4 Summary of Quality Control Assays of Liposomal Products

	Methodology
	General Characterization Studies
pH	pH meter
Conductivity	Conductivity meter
Osmolarity	Osmometer
Phospholipid concentration (PL)	Lipid phosphorus content (Bartlett method)
Phospholipid composition	TLC (combined with the Bartlett method), HPLC
Cholesterol concentration	Cholesterol oxidase assay, HPLC (normal and/or reverse phase)
Trapped volume	Measure of osmotic intraliposomal aqueous phase
Agent concentration	Spectrophotometry, HPLC, GC, spectrofluorimetry
Residual organic solvents	NMR, GC
and heavy metals	Standard protocols
Agent/phospholipid ratio	Determination of drug and phospholipid concentrations
$[H]^+$ or ion gradient for remote loading	Fluorescent indicators, ESR indicators, ^{31}P-NMR, ^{19}F-NMR
	Physical Characterization and Stability Assays
Vesicle size distribution	
Submicron range	Photon correlation spectroscopy (PCS), gel exclusion chromatography and specific turbidity, electron microscopy (various methods)
Micron range	Coulter counter, light microscopy, light diffraction and specific turbidity
Electrical surface potential and surface pH	Use of membrane-bound electrical field probes and pH-sensitive probes
Zeta potential	Electrophoretic mobility
Thermotropic behavior, phase transition, phase separation	DSC, NMR, fluorescence methods, FTIR, Raman spectroscopy, ESR
Percentage of free drug	Gel exclusion chromatography, ion exchange chromatography, precipitation by polyelectrolyte
Liposome/aqueous phase partition coefficient (*K*p) by dilution-dependent drug release assay	Dilution effect (up to 10,000-fold) on liposomal agent/PL ratio at equilibrium
k_{off}[a]	Effect of time on agent/liposome ratio
$[H]^+$ or ion gradient	Fluorescent pH indicators, ^{31}P-NMR, ^{19}F-NMR, intraliposomal ion concentration
	Chemical Stability Assays
pH	pH meter
Phospholipid acyl chain autoxidation	Conjugated dienes, lipid peroxides, TBARS and fatty acid composition (GC)
Phospholipid hydrolysis	TLC, HPLC, total PL and/or free fatty acid concentration
Cholesterol autoxidation	TLC, HPLC
Antioxidant degradation	TLC, HPLC
Agent degradation	TLC, HPLC, spectrophotometry, spectrofluorimetry
	Biological Assays
Sterility	Aerobic and anaerobic bottle cultures
Pyrogenicity	Rabbit and/or limulus amebocyte lysate (LAL) tests
Toxicity	Related to agent and use of liposomal product
Medium-induced leakage	Gel exclusion chromatography, ion exchange chromatography, precipitation by polyelectrolyte

Note: For more details on the methodologies see Reference 12 and references listed therein.

Abbreviations: PL, phospholipids; TLC, thin layer chromatography; HPLC, high performance liquid chromatography; GC, gas liquid chromatography; NMR, nuclear magnetic resonance; PCS, photon correlation spectroscopy; DSC, differential scanning calorimetry; FTIR, Fourier transformed infra-red spectroscopy; ESR, electron spin resonance; TBARS, thiobarbituric acid reactive substances.

[a] K_{off}, constant which describes the rate of agent release of the liposomes (References 9, 16, 22, and references listed therein).

REFERENCES

1. **Bangham, A. D.,** Liposomes: the Babraham connection, *Chem. Phys. Lipids*, 64, 275, 1993.
2. **Barenholz, Y.,** Liposome production: historic aspects, in *Liposome Dermatics*, Braun-Falco, O., Korting, H. C., and Maibach, H. I., Eds., Springer-Verlag, Berlin, 1992, 69.
3. **Lasic, D. D.,** *Liposomes: From Physics to Applications*, Elsevier, Amsterdam, 1994.
4. **Barenholz Y. and Lasic D. D., Eds.,** *Handbook of Nonmedical Applications of Liposomes*, Vols. I–IV, CRC Press, Boca Raton, FL, 1996.
5. **Friedman, M., Byers, S. O., and Rosman, R. H.,** Resolution of aortic atherosclerotic infiltration in the rabbit by phosphatide infusion, *Proc. Soc. Exp. Biol. Med.*, 95, 586, 1957.
6. **Williams, K. J., Werth, V. P., and Wolf, J. A.,** Intravenously administrated lecithin liposomes: a synthetic antiatherogenic lipid particle, *Perspect. Biol. Med.*, 27, 417, 1984.
7. **Adler-Moore, J. P. and Profitt, R. T.,** Development, characterization, efficacy and mode of action of Ambisome, a unilamellar liposomal formulation of amphotericin B, *J. Liposome Res.*, 3, 429, 1993.
8. **Gabizon, A., Catane, R., Uziely, B., Kaufman, B., Safra, T., Cohen, R., Martin, F., Huang, A., and Barenholz, Y.,** Prolonged circulation time and enhanced accumulation in malignant exudates of doxorubicin encapsulated in polyethylene-glycol coated liposomes, *Cancer Res.*, 54, 987, 1994.
9. **Barenholz, Y. and Crommelin, D. J. A.,** Liposomes as pharmaceutical dosage forms, in *Encyclopedia of Pharmaceutical Technology*, Vol. 9, Swarbrick, J. and Boylan, J. C., Eds., Marcel Dekker, New York, 1994, 1.
10. **Lichtenberg, D. and Barenholz, Y.,** Liposomes: preparation, characterization and preservation, in *Methods in Biochemical Analysis*, Vol. 33, Glick, D., Ed., John Wiley & Sons, New York, 1988, 337.
11. **Barenholz, Y., Ed.,** Special Issue: Quality Control of Liposomes, *Chem. Phys. Lipids*, Vol. 64, 1993.
12. **Barenholz, Y. and Amselem, S.,** Quality control assays in the development and clinical use of liposome-based formulations, in *Liposome Technology*, Vol. I, 2nd ed., Gregoriadis, G., Ed., CRC Press, Boca Raton, FL, 1993, 527.
13. **Winterhalter, M. and Lasic, D. D.,** Liposome stability and formation: experimental theories on size distribution, *Chem. Lipids*, 64, 35, 1993.
14. **Kaller, E. W., Kaalakara, M., Rodriguez, B. E., and Zasadzinski, J. A. N.,** Spontaneous vesicle formation in aqueous mixtures of single-tailed surfactants, *Science*, 245, 1371, 1989.
15. **Masson, G.,** Advanced techniques for preparation and characterization of small unilamellar vesicles, *Food Microstr.*, 8, 11, 1989.
16. **Haran, G., Cohen, R., Bar, L. K., and Barenholz, Y.,** Transmembrane ammonium sulfate gradients in liposomes produce efficient and stable entrapment of amphipathic weak bases, *Biochim. Biophys. Acta*, 1151, 201, 1993.
17. **Mayhew, E., Mikolopoulos, G., and Siciliano, A.,** An advanced technique for the manufacture of liposomes, *Am. Biotechnol. Lab.*, (Nov/Dec issue) 36, 1985.
18. **Gregoriadis, G., Ed.,** *Liposome Technology*, Vols. I–III, CRC Press, Boca Raton, FL, 1984.
19. **Gregoriadis, G., Ed.,** *Liposomes as Drug Carriers*, John Wiley & Sons, New York, 1988.
20. **Gregoriadis, G., Ed.,** *Liposome Technology*, Vols. I–III, 2nd ed., CRC Press, Boca Raton, FL, 1993.
21. **Amselem, S., Gabizon, A., and Barenholz, Y.,** A large-scale method for the preparation of sterile and nonpyrogenic liposomal formulations of defined size distributions for clinical use, in *Liposome Technology*, Vol. I, 2nd ed., Gregoriadis, G., Ed., CRC Press, Boca Raton, FL, 1993, 501.
22. **Barenholz, Y.,** Optimization of amphiphile-based colloidal carriers, in *Handbook of Nonmedical Applications of Liposomes*, Vol. III, Barenholz, Y. and Lasic, D. D., Eds., CRC Press, Boca Raton, FL, 1996, chap. 1.
23. **Kirby, C. and Gregoriadis, G.,** Dehydration-rehydration vesicles: a simple method for high yield drug entrapment in liposomes, *Biotechnology*, 979, 1984.
24. **Nichols, J. W. and Deamer, D. W.,** Catecholamine uptake and concentration in liposomes maintaining pH gradients, *Biochim. Biophys. Acta*, 455, 269, 1976.
25. **Parr, M. J. and Cullis, P. R.,** Transbilayer transport induced by transmembrane pH gradients in liposomes: implications for biological systems, in *Handbook of Nonmedical Applications of Liposomes*, Vol. II, Barenholz, Y. and Lasic, D., Eds., CRC Press, Boca Raton, FL, 1996.
26. **Mayer, L. D., Bally, M. B., Hope, M. J., and Cullis, P. R.,** Uptake of antineoplastic agents into large unilamellar vesicles in response to a membrane potential, *Chem. Phys. Lipids*, 40, 89, 1985.
27. **Lasic, D. D., Frederik, P. M., Stuart, M. C. A., Barenholz, Y., and McIntosh, T. J:** Gelation of liposome interior. A novel method for drug encapsulation, *FEBS Lett.*, 312, 255, 1992.
28. **Ceh, B.,** Theory of loading of agents into liposomes, in *Handbook of Nonmedical Applications of Liposomes*, Vol. III, Barenholz, Y. and Lasic, D. D., Eds., CRC Press, Boca Raton, FL, 1996, chap. 5.
29. **Bolotin, E. M., Cohen, R., Bar, L. K., Emanuel, N., Ninio, S., Lasic, D. D. and Barenholz, Y.,** Ammonium sulfate gradients for efficient and stable remote loading of amphipathic weak bases into liposomes and ligandoliposomes, *J. Liposome Res.*, 4, 455, 1994.
30. **Felgner, P., Tsai, Y. J., and Felgner, J. H.,** Advances in the design and application of cytofection formulations, in *Handbook of Nonmedical Applications of Liposomes*, Vol. IV, Lasic, D. D. and Barenholz, Y., Eds., CRC Press, Boca Raton, FL, 1996, chap. 4.

31. **Remy, J., Sirlin, C., and Behr, J.**, Gene transfer with cationic amphiphiles, in *Liposomes as Tools in Basic Research and Industry*, Philippot, J. R. and Schuber, F., Eds., CRC Press, Boca Raton, FL, 1995, 159.
32. **Zuidam, N. J., Talsma, H., and Crommelin, D. J. A.**, Sterilization of liposomes, in *Handbook of Nonmedical Applications of Liposomes*, Vol. III, Barenholz, Y. and Lasic, D. D., Eds., CRC Press, Boca Raton, FL, 1996, chap. 6.
33. **Kikuchi, H., Carlsson, A., Yachi, K., and Hirota, S.**, Possibility of heat sterilization by liposomes, *Chem. Pharm. Bull.*, 39, 1018, 1991.
34. **Zuidam, N. J., Lee, S. S. L., and Crommelin, D. J. A.**, Sterilization of liposomes by heat treatment, *Pharm. Res.*, 10, 1591, 1993.
35. **Zuidam, N. J.**, Stability of Liposomes, Ph.D. thesis, University of Utrecht, the Netherlands, 1994, chaps. 4A, 4B.
36. **Samuni, A. and Barenholz, Y.**, Nitroxide antioxidants prevent γ-irradiation-induced oxidative damage to liposomal lipids, submitted.
37. **Barenholz, Y., Amselem, S., Goren, D., Cohen, R., Gelvan, D., Samuni, A., Golden, E. B., and Gabizon, A.**, Stability of liposomal-doxorubicin formulation: problems and prospects, *Medicinal Res. Rev.*, 13, 449, 1993.
38. **New, F. R. C., Ed.**, *Liposomes, A Practical Approach*, IRL Press, Oxford, 199.
39. **Burgess, S. W., Moore, J. D., and Shaw, W. A.**, Lipids as raw materials for liposomal products: properties and characterization, in *Handbook of Nonmedical Applications of Liposomes*, Vol. III, Barenholz, Y. and Lasic, D. D., Eds., CRC Press, Boca Raton, FL, 1996, chap. 2.

Chapter 4

Stability and Stabilization of Liposomes

Daisy Leung and Danilo D. Lasic

CONTENTS

I. INTRODUCTION

The stability of liposomes and efforts to improve them by physical and chemical means are extremely important issues, especially with the commercialization of liposome products. We should point out also that only improved understanding of liposome stability and interaction characteristics led to the so-called rational liposome design which resulted in some recent breakthroughs in their applications.[1,2] In this review we shall primarily concentrate on physical stability, i.e., consistency of liposome size distribution, while chemical and biological stability will be dealt with only briefly and we shall refer to other reviews.

II. FORCES IN LIPOSOME SYSTEMS

Stability of liposomes depends on intraliposomal and interliposomal forces, including intrinsic tensions, interactions with other liposomes, extrinsic particles, and agents. We shall very briefly discuss forces within a single liposome, attractive and repulsive forces between liposomes, as well as their interaction with chemical and biological agents.

A. INTRALIPOSOMAL FORCES

Liposome structure and its stability are due to several forces and energy contributions from hydration of polar heads and their electrostatic/dynamic interactions with solvents and solutes, hydrophobic interaction, and attractive London–van der Waals dispersive forces between hydrocarbon chains. These interactions in balance with energetically unfavorable interactions between polar heads (electrostatic and steric) and bending stiffness of the symmetric bilayers determine aggregate morphology and mechanical characteristics of membranes.[3,4]

We shall see later that liposomes are trapped into a higher energy state and are often referred to as "loaded springs" due to the bending into the spherical form.[1,4,5] This bending energy which is proportional to bending elasticity (K) of the membrane is counterbalanced by stretching elastic energy which opposes rupture of membrane due to cohesive dispersion forces between hydrocarbon chains. Also, in addition,

0-8493-4012-8/96/$0.00+$.50
© 1996 by CRC Press, Inc.

creation of a pore requires additional energy in order to compensate for the exposure of hydrophobic area and/or reorientation of molecules at the edge. While at larger radii of curvature, bending can be approximated to be elastic (Hookean), it contains higher order terms when curvature is high due to the stretching/compressing of polar heads in the outer/inner monolayer.

Electric and magnetic forces due to asymmetric distribution of charged and/or paramagnetic ions can be in general neglected. The influence of high surface charges or transmembrane potential on liposome morphology and stability is not yet completely understood but may cause, in addition to shape changes, spontaneous formation of unilamellar vesicles from multilamellar ones[6] due to the contributions of Gaussian curvature.

Another intraliposomal force can be due to osmotic pressure. It was found that liposomes behave as tiny osmometers.[5] However, they can sustain large osmotic imbalances before they break. Experimentally it is known that, for instance, DSPC/Chol (distearoyl phosphatidyl choline/cholesterol),(3/2 mol/mol) can sustain osmotic pressure of several dozens of atmospheres without bursting. Assuming the validity of the Law of Laplace one can estimate the osmotic pressure (Π) that vesicles of size R can sustain by using membrane tension, τ, for the surface tension and from

$$\Pi = 2\tau/R \tag{1}$$

it follows that, for instance, DPPC/Ch 1/1 vesicles with $R = 50$ nm and $\tau = 40$ dyn/cm can sustain pressure difference of 80 atm. This corresponds to concentration difference of ca 1.5 *M* NaCl. Very small vesicles contain stored additional tension due to inelastic deformation of the membrane and are therefore less stable. Because we have taken τ as phase boundary tension the Law of Laplace has a coefficient 2 and not 4, as is the case of foam bubbles which have two surfaces.[1]

In many cases, however, the vesicle membrane cannot sustain the osmotic pressure. Often it bursts and quickly reseals allowing the exchange of solutes and affecting the particle size distribution only slightly, especially in diluted solution where collisions between vesicles are less frequent.

In the hydrodynamic flow, especially when particles are small (5 to 10 times smaller than the eddies which are normally around 2 μm) and spherical, morphology of liposomes may not be affected. However, the encapsulated molecules may leak out, possibly due to the small fluctuations in shape which can create improper packing when (and where) one curvature changes into another in nonspherical particles characterized by at least two curvatures. Similar reasoning can explain also the leakage of liposomes during nebulization. In general, this effect decreases when aerosol particle size exceeds vesicle diameter severalfold and liposomes can be quite stable when aerosol particles are tenfold larger than vesicles.

B. INTERLIPOSOMAL FORCES

As in any system, the behavior of colloidal systems depends on the interparticle forces and interactions of particles with surrounding milieu, in this case solvent and solutes.

There are several attractive and repulsive forces which act between colloidal particles. Ubiquitous attractive force is van der Waals attraction which acts between all particles separated by a medium of different polarizing properties. For two spherical particles with radii r and separated by a distance d, this can be expressed as[7,8]

$$F_{vdW} = -A_H r/12/d^2 \tag{2}$$

where A_H is Hamaker constant, normally around 10^{-20} J. This gives attractive forces around pN at separations around 1 nm (for comparison, gravity force between two smaller liposomes at the same distance is ca 10^{-33} N). One of the few systems where van der Waals interactions are negligible are DNA molecules because their dielectric constants equals the one of the aqueous environment.

In contrast to this long-range force a much stronger attractive force, the hydrophobic force, which acts only at closer separations is generally recognized[9] in some systems. This is empirically expressed as[10]

$$F = -0.1359\ kT \exp(-d/14) \tag{3}$$

but its origin is still not understood (d in [Å]).

In the case of charged particles there is a strong electrostatic force. Mixing two samples of liposomes bearing opposite charges results in flocculation. Mostly, however, particles bear the same charge and this results in electrostatic repulsion which can be expressed as[11]

$$Fe = 2kT/a/d * q/tgq, \quad q * tgq = \alpha\pi z e^2\ d/kT/a \tag{4}$$

where α is the degree of ionization, e is elementary charge, a is area of molecules, and z is the valency of the charge.

The above-mentioned van der Waals attraction and electrostatic repulsion forces are the basis of the Derjaguin-Landau-Verwey-Overbeek (DLVO) theory of the stability of colloidal particles.[7] This theory states that colloidal system is stable if electrostatic repulsion exceeds van der Waals attraction. In some cases a secondary minimum is found in which the attractive potential is weak (<5 to 10 kT) and such loose aggregates can be normally (re)dispersed by gentle agitation. In many cases, however, this theory could not explain the stability of liposomes against aggregation. In the case of predicting too stable colloidal suspension, additional attractive forces such as hydrophobic attraction or ion–ion correlation attraction were introduced,[9] while for explaining the stability of uncharged liposomes and liposomes in high ionic strength media other repulsive forces were recognized.

The aggregation of noncharged liposomes is normally prevented by the repulsive hydration force which is due to the hydration of polar heads. In order to bring two bilayers together one has to remove water molecules from the polar heads. This results in the repulsive force which is normally empirically expressed as[12]

$$F_h = \text{Const } e^{-d/\lambda} \tag{5}$$

where λ is the characteristic decay length and is normally around 3 Å. Recently, protrusion repulsive force, originating in the vertical motion of polar heads was introduced.[13] The application of such repulsion to nonaggregated systems, such as DNA is not clear at present.

In soft bilayers that can thermally undulate, another repulsive force was recognized.[14] This undulation force is due to the excluded volume effect two approaching undulating membranes exhibit upon one another. In order to approach they have to rigidify. This is entropically unfavored and results in a repulsive force. This interaction can be expressed as[14]

$$F_u = 3\pi^2 (kT)^2/128/K$$

These forces can explain the stability of liposomes and provide a basis for their stabilization. Historically, mostly surface charge density, zeta potential, size distribution and homogeneity, ionic strength, and pH of the medium were found to determine colloidal stability *in vitro*. In biological systems, however, the behavior is completely different due to high salt concentration and the pressure of active liposomicidal substances.

The hydration and undulation forces fall into a more general class of steric repulsion. In a similar manner, one can therefore greatly increase the repulsive forces between the two particles by coating the surface of the particle with appropriate molecules. The additional repulsive interactions originating from the surface attached molecules give rise to the so-called steric repulsion.[15] Natural and synthetic polymers are the most common materials used. They must be inert, well solvated, and compatible with the solvent, and have polarizability close to water. Steric stabilization can be induced by surface attachment of various natural or synthetic polymers, either by adsorption, hyrdrophobic insertion, electrostatic binding, or, preferentially, by grafting via covalent bond. Nonionic, water-compatible, flexible, and well-hydrated polymers are preferred. The repulsion between surfaces with attached polymers was shown to be dependent on the grafting density and degree of polymerization. In the low grafting densities, i.e., in the region where the polymer chains cannot interact among themselves, the so-called mushroom model was defined which yields steric repulsion at distance h as[16]

$$F_{st}^{m} = (kT)/(D^2 h_c) (h_c/h)^{8/3} \tag{7}$$

where

$$h_c = N a (a/D)^{2/3} \tag{8}$$

and D is average distance between adjacent grafting points, a is the size of the segment, and N is the degree of polymerization. At higher densities where polymers start to interact this causes their extension, and the so-called brush model[6] can be applied. The repulsive force is now

$$F_{st}^{br} = (kT)/D^3 [(h_c/h)^{9/4} - (h/h_c)^{3/4}] \tag{9}$$

These equations explain the repulsion in the case of polymers which form both, the so-called adsorption and depletion layer. In the case of a depletion-layer forming polymer the conformation probably looks more like an inverse droplet or pear that alters the density profile, but leaves the scaling laws unchanged.[17] Theoretical analysis predicted extension of the chains to be 40% longer as opposed to the model above,[18] i.e., 7/5 R_F vs. R_F, where Flory radius $R_F = a\ N^{3/5}$.

The above discussed inter- and intraliposomal forces define the reactivity, interaction characteristics and stability of liposomes. Mostly they define nonspecific attraction and repulsion. In the case of specific ligands attached to the liposomes they define also specific reactivity if appropriate receptors are present.

III. STABILITY OF LIPOSOMES

In liposome research and applications, the term stability is vaguely used and is often not properly defined. In the broadest sense liposome stability involves the stability of the particle and its constituents including the entrapped molecules in any, including liposomicidal, environment. This general definition of stability contains various aspects which can be defined separately but often synergistically act together.[1]

We shall postulate that overall stability is a sum and a product of physical, chemical, and colloidal aspects. Applications of various liposome products require, in addition to minimize these inherent instability components below an acceptable and defined limit, also to define the stability of liposomes upon application, which is often in liposomicidal (liposome deleterious) environment, and the long-term stability of the product. Most of them are of a pharmaceutical nature and therefore one can define biological and shelf life stabilities, which is a cumulative property of several particular stabilities, respectively.

A. PHYSICAL STABILITY

By physical stability we define the constancy of the liposome structure and the entrapment/association of microencapsulated molecules.

The preservation of the size and encapsulation depends on the mechanical properties of the bilayers and thermodynamics of the system.

Phospholipid molecules in the vesicle membrane are in equilibrium with free molecules in solution at the critical micelle concentration (cmc). The values for phospholipids are very low and I shall use a simple example to show the virtual invariance of liposomes against dilution. From empirical equations relating the transfer free energy of alkanes from water to organic medium with the number of methyl and methylene groups one can estimate $\Delta G = -15.1$ kcal/mol for distearoyl PC. From

$$\Delta G = -RT \ln \text{cmc}, \qquad (10)$$

where T is temperature and R is universal gas constant one calculates $\text{cmc}_{DPPC} = 4\ 10^{-10}\ M$, in good agreement with the experimental value of $2\ 10^{-10}\ M$.[3] This negligible effect on the liposome stability can be illustrated by the following example: in 1 ml of 1 m*M* small unilamellar liposomes there are approximately 10^{14} vesicles. At cmc = 10^{-10} only six 10^{10} molecules are in solution and 1000-fold dilution to 1l releases additional six 10^{13} molecules into solution. This accounts for one lipid molecule released from approximately two liposomes. At higher concentrations this is even less important, for instance in the case of 100 m*M* solution one molecule would be released from 200 liposomes.

Strictly speaking, however, the low values of cmc can be treated as low solubility rather than as a dynamic exchange process[19] and I do not see a need to define parameters such as critical bilayer or vesicle concentration and the concept of Equation 10, an analogy to micelle formation, cannot be applied to explain liposome stability, structure, and size distribution rigorously.

1. Thermodynamic Stability

All single component bilayers and symmetric bilayers increase their free energy upon curving because the spontaneous curvature of the system, C_0, is zero. For large bilayered fragments one can assume that the bending is elastic and the excess energy due to curvature per unit area can be expressed as[4,20]

$$\acute{E} = 1/2\ K\ (C_1 + C_2 - C_0)^2 + k\ C_1\ C_2 \qquad (11)$$

where C_1 and C_2 are the principal curvatures, C_0 is spontaneous curvature, K is elastic bending modulus, and k is bending modulus of Gaussian curvature. In the simplest case of bending a plane into a sphere, principal curvatures are equal and reciprocal to the radius of curvature, R. Upon closure the radius of curvature equals the liposome radius and after integrating over the surface one can write

$$E = 8\pi K + 4\pi k \qquad (12)$$

Because K>>k we shall neglect the last term. Equation 2 shows the excess energy of the self-closed system due to the bending.[1,4,20] Because E exceeds thermal energy, i.e., E ~10 to 15 kT, where k is Boltzmann constant and T temperature, the entropy, which is proportional to the number of liposomes N as N lnN, cannot stabilize the system. From $8\pi K > kT$ it follows that liposomes are thermodynamically unstable.[6,20,21]

Neglecting k, Equation 2 shows that theoretically there are two possible ways to produce thermodynamically stable liposomes: entropic stabilization requires E ~ kT while E = 0 for $C_o = C_1 + C_2$ in the case of asymmetric bilayers.[22]

K is proportional to the thickness of the membrane. Membrane thinning due to the interdigitation of bilayers or action of cosurfactants, such as in microemulsions, can give rise to K ~ kT, and entropically stabilized vesicles can be formed.[22] Another way to produce thermodynamically stable liposomes is to create nonzero spontaneous curvature. This can be done by asymmetric bilayers or by asymmetric conditions across the bilayer, such as different values of pH, concentrations of counterions, or interacting (macro)molecules.[23–25] Furthermore, it is likely that some of the novel lyotropic phases,[26] such as cubic sponge phase can spontaneously disintegrate into liposome solution upon dilution due to the critical fluctuations and low values of K. Similar intermediate structures may also occur in some liposome preparation methods, such as in demulsification of particular reverse phases or in some lipid–detergent mixtures during the preparation of vesicles by detergent depletion technique. Here, however, the changes are driven by the removal of nonpolar phase.[1]

The entropically stabilized spontaneous vesicles are not interesting for most of the applications because, due to their very existence, they have low values of K which inherently reduce their barrier and permeability properties. Alternatively, their existence and stability can be based on transmembrane gradients which can dissipate. In contrast, asymmetric bilayers, including some novel amphiphiles, such as bipolar lipids with asymmetric polar heads, or poly(ethlene glycol) (PEG)-lipids, can make very stable and nonpermeable liposomes with $C_o \neq 0$.

The stability of liposomes against external perturbations such as dilution can be understood by very low cmc value of lipids and quite high energy required to open a liposome.

The energy to create a pore, break up the liposome and cause fusion or formation of liquid crystalline phase in excess solvent is[27]

$$E_o = 2\pi r\Gamma - \pi r^2\gamma \qquad (13)$$

where γ is the surface tension and Γ the edge interaction at the polar–nonpolar interface of exposed bilayer. Contamination of bilayers with single chain surfactants, which can effectively shield the interface may reduce E_o.[21] Normally, however, it has a rather large value (~20 to 50 kT) and pores, even if formed, quickly reseal. This can explain why vesicles can be relatively stable. All this indicates that liposomes can be kinetically trapped into their structures because they are formed in a nonequilibrium irreversible process in conditions of high excesses of energy[1,5,21,28] and the energy needed to open or break them can also be ~50 kT for r ~R.

The above analysis shows that part of the energy input during the liposome formation becomes stored in the conformational energy of the bilayer–bilayer curvature[29] and is responsible for the metastability.

Furthermore, a simple analysis shows the advantage of kinetically trapped systems as opposed to systems at thermodynamic equilibrium for applications such as microencapsulation. A system in equilibrium quickly adopts to the external change, such as dilution, mixing, presence of other substances, and injection into body fluids, by a phase change. For instance, micelles or microemulsion droplets quickly disintegrate upon dilution (i.e., move in phase diagram into different regions) while liposomes made from phospholipids with low values of critical vesicle concentration, are stable against dilution. This simple analysis explains many of the unexpected data of liposomal carriers.

Liposomes can be also stabilized against some phase transitions, such as freeze-thawing and freeze-drying by the addition of water soluble cryoprotectants, such as sucrose. This enables manufacturing of frozen and freeze-dried liposome formulations.

2. Mechanical Stability

According to the theory of smectic liquid crystals[30] there are three different ways to deform a thin membrane: area dilation/condensation, surface extension at constant area (shear), and bending. Each deformation is characterized by the appropriate modulus and the knowledge of material properties of the bilayer can be used to improve the behavior of membranes under stresses and prepare more stable liposomes. Deformation can be a consequence of osmotic imbalances, adsorption, hydrodynamic shears and other forces, and results in changes of vesicle shapes, membrane tension, and possibly rupture and closure with reequilibration of internal and external solutes, and topological changes, such as formation of buds, invaginations, or daughter vesicles, fission, or simply, liposome disintegration.

Obviously, mechanically stronger and more cohesive bilayers, as can be measured by critical areal strain (τ) and elastic compressibility modulus ($K_{d/c}$),[31] resist external perturbations better. Micropipette manipulation[32] studies have shown that the strongest bilayers consist of equimolar DSPC and cholesterol as measured via $K_{d/c}$ and τ. Quantitatively, this observation was explained by a theory in which cholesterol is regarded as crystal breaker of the gel phase and inducer of chain ordering in the fluid phase without rigidification of the overall phase. This gives rise to the liquid-ordered phase which has a high degree of conformational order and no resistance to the shear forces.[33]

Bilayers can be mechanically stabilized also by polymerization[34] or by using lipids with fluorocarbon chains.[35] Despite reduced van der Waais attraction among carbon chains these lipids form stronger and less permeable bilayers due to stronger hydrophobic interaction.[1] Bolaamphiphiles, two-headed surfactants, can also form very stable liposomes if asymmetrical polar heads orient within the bilayer.

Permeability of bilayers depends on the nature of permeating species, mostly on their charge, size, hydrophobicity, and hydration, and can be correlated to the phase behavior and mechanical stability of the bilayer and structural defects. In addition, it influences also the bilayer stability against chemical degradation.

While the thermodynamic behavior of bilayers and its correlation to permeability has been known for quite some time,[36] the strong correlation between mechanical properties of the bilayers and stability of liposomes was realized only recently. With several other surface characteristics this knowledge allows one to control and rationally design liposome properties.

B. CHEMICAL STABILITY

Most biomolecules undergo different degradation reactions, and lipids are one of the most sensitive ones. The most common degradation pathways are oxidation and hydrolysis. In general, polar heads are much more stable than the hydrocarbon chains. Hydrocarbon chains, especially the unsaturated ones, are subject to oxidation while hydrolysis pinches off a hydrocarbon chain from the backbone molecule. In the case of double-chain phospholipids it produces a fatty acid and lyso lipid. Ultimately, it produces two fatty acids, glycerol, and a free polar head. Because the reaction is acid and base catalyzed it exhibits strong pH dependence.[37] The protective measures include optimization of pH, normally around 6.5, and, preferentially, freezing or drying the sample in the presence of cryoprotectants.

Oxidation is a radical reaction which ultimately results in the breakage of chains or in the case of two adjacent double bonds, formation of cyclic peroxides. Protective measures which include the use of saturated lipids and addition of antioxidants, such vitamin E (α-tocopherol) or BHT (butylated hydroxi toluen), and working with the degassed solutions in the argon atmosphere result in substantial reduction of oxidation and samples can be stable, within tight specifications, for years.[38]

Chemical and physical stability are closely related. Mechanically strong and well-packed bilayers reduce the access of oxidizing or hydrolyzing agents. This in turn reduces the changes in size distribution, fusion, and mechanical properties of the bilayer due to the reduction of the concentration of the degradation products, which influence K, τ, cmc of liposome constituents (solubility), and $K_{d/c}$. Normally, these bilayers are also formed from the molecules least susceptible to the degradation and the control of pH, and addition of antioxidants can warrant prolonged chemical stability.[1]

Chemical stability becomes a much more complex issue when degradable substances are encapsulated. In these cases, the optimal stability is a compromise between various contributions and in many cases it may not be achieved even in a frozen- or freeze/spray-dried liposomal product.

Several synthetic lipids, including the ones with fatty acids linked to the glycerol backbone via ether linkage, are chemically more stable. However, they are not biodegradable by mammalian enzymes and that limits their use in parenteral drug delivery while they are often used in chemical or cosmetic applications.

C. COLLOIDAL STABILITY

Particle size distribution changes in colloidal systems mostly occur via two mechanisms: on the molecular level this can be asymmetric molecular exchange, while on the particle level this is mostly aggregation, flocculation, fusion, coacervation, or flotation (creaming)/precipitation. The first process is the so-called Ostwald ripening and the latter, in the case of liposomes, aggregation and fusion.

In addition to some theoretical considerations there are many practical aspects of liposome colloidal stability. While working experimentally one has to be aware of specific stability problems. Flotation can be another stability problem due to preparation of liposomes in dense solutions (of sugars, antibiotics) in which, due to the specific gravity, liposomes cannot disperse into the acqueous phase but float on the surface. Similarly, interactions of liposomes with monolayers can lead to monolayer-induced fusion of vesicles. The surface of every liposomes solution/dispersion is coated with at least a monolayer lipid film. Therefore, shaking, bubbling, foaming, and shipping of liposomes in liquid form in containers with large head-spaces and similar treatments can lead to substantial particle size growth.

1. Ostwald Ripening

In this process larger particles grow at the expense of smaller ones. Smaller particles have a larger surface-to-volume ratio and, additionally in the case of vesicles, higher curvature gives rise to higher chemical potential[39]

$$\mu^o_N = \mu^o_\infty + (Kv)/(2\delta R^2) \tag{14}$$

where μ^o_N is chemical potential of the lipid molecule in an aggregate of N monomers and μ^o_∞ in the flat bilayer, K is bending elasticity, v molar volume of lipid monomer, δ thickness of bilayer, and R vesicle radius. Monomer concentration in equilibrium with vesicles therefore depends on vesicle radius[39]

$$c_{eq}(R) = c_{eq}(\infty)\ [1 + \{(Kv)^{1/2}\ /\ ((2kt\delta)^{1/2}\ R^2)\}^2] \tag{15}$$

where $c_{eq}(\infty)$ is the monomer concentration above the infinite, flat bilayer. Assuming that it is equal to cmc, we see that this size disproportionation is negligible for most of phospholipids which have values of cmc $<10^{-8}$ *M*. For some synthetic surfactants, however, cmc can be much higher and this becomes the dominant size growth mechanism. In the case of sodium heptylnonyl benzenesulfonate with cmc of 5 m*M*, it was shown that vesicles grow proportionally to $t^{0.3}$, i.e., they double their size to 40 nm in 40 days.[39] Using the same equations, assuming equilibrium conditions and cmc = 5 10^{-9} *M* of lecithin, one gets the kinetics of apparent growth <5 10^{-7} nm/day.

2. Aggregation and Fusion

In phospholipid liposome preparations the increase of the particle size is mostly due to particle aggregation because the size growth due to Ostwald ripening is negligible. Aggregation is the first step in a vesicle–vesicle fusion[40] which depends on the thermodynamic stability of the sample, crystal defects, and presence of intrinsic or oxogeneous fusogenic substances or vectors. Fusion is a practically irreversible process while aggregation can be reversible: often liposome systems are only loosely aggregated and mild swirling of the sample redisperses liposomes.[41] Here we shall discuss only the origin of aggregation: we have already defined different forces acting between liposomes, commented on the DLVO theory, and below we shall introduce a more general interparticle potential. Later we shall use this concept to simulate interaction of bilayers with macromolecules. We shall not be concerned with the kinetics of aggregation and subsequent fusion.

Aggregation is a reversible process. It can be prevented by electrostatic, steric, or electrosteric stabilization. Most of the phospholipids have pK values of phosphate and amino group below 2 and above 9, respectively, and are zwitterionic in the normal pH ranges. Charge stabilization normally means inclusion of a certain fraction of charged amphiphiles into the bilayer. Here we should mention that addition of single-chain charged detergent may stabilize the system in the test tube, but is very unlikely to stabilize particles upon dilution or application. Namely, after the change in the surroundings, the equilibria (cmc of the detergent) immediately reestablish, resulting in very quick loss detergent. The stability of the system depends also on ionic strength and pH if pK of the amphiphile ranges around pH of the system. Presence of divalent counterions and oppositively-charged polyelectrolytes can cause bridge aggregation which normally results in quick aggregation followed by fusion or flocculation/precipitation. Another way to increase stability is by coating the liposome surface with appropriate macromolecules which can induce steric repulsion. These are mostly (poly)saccharides, hydrophilic polymers, and some surfactants which can be attached to the surface via adsorption, covalent binding, or hydrophobic anchoring, respectively.

The preferential method is covalent coupling to diacyl lipids because adsorption and bilayer insertion of some detergents with large polar heads generally cause leakage of liposome contents. Also, the anchoring part of the polymer-lipid has to be carefully selected because short, unsaturated chains or even cholesterol as hydrophobic anchors do not provide stable retention of these polymer-lipids in the bilayers.[42–44] Steric stabilization with nonionic polymers can be salt-concentration independent in a wide concentration range, and therefore offers a much better way to stabilize biological colloids than electrostatic stabilization as nature has proven.

Fusion is a very complex process and it mechanism is not yet understood. In order that two bilayers merge there must be a morphological change at the contact point. The first models envisioned an inverse micelle as the intermediate structure in the curvature change and merger of the bilayers, while recently most of the researchers believe that a stalk is formed as a transition structure in the fusion process. Fusion occurs in living systems very frequently (synaptic vesicles, secretion, cellular transport, vesicle trafficking) and liposomes are used as a very useful model because their size and lipid composition can be easily tailored, as well many monitoring markers can be entrapped into their interior or bound/embedded to the membrane.

D. BIOLOGICAL STABILITY

Even the most stable conventional liposomes are rather unstable in a biological milieu. In addition to interactions with various plasma components and enzymes, an active uptake of liposomes by the body's immune system takes place. It has been shown that mechanical stability can help reduce the deleterious interactions *in vitro* and reduce leakage *in vivo* but cannot slow down appreciably the uptake of liposomes by phagocytic cells upon parenteral administration.

It was assumed[42] that the predominant mechanisms of liposome clearance from blood were (a) opsonization by immune and nonimmune components in blood, such as immunoglobulin complement factors, and fibronectin which trigger subsequent macrophage uptake and (b) reaction with plasma (lipo)proteins which deplete lipid molecules from liposomes ultimately causing their disintegration.

In order for effective opsonization to take place, a macromolecule has to bind to the liposome surface. We believe that this adsorption can be electrostatic, hydrophobic, H bonding, or physisorption due to van der Waals forces, i.e.,

$$W_{tot} = W_{est} + W_{hfo} + W_{vdW} + W_{Hbond} + W_{spec} + W_{edyn} \quad (16)$$

where the total adsorption energy contains contributions from electrostatic, hydrophobic, van der Waals, hydrogen bonding, specific, and electrodynamic interactions. The latter ones mostly occur when some antibodies or haptens are attached. Obviously, an ideal drug targeting would tend to eliminate all contributions to W_{tot} with the exception of W_{spec}. Probably in some proteins the adsorption or adhesion requires conformational change to trigger subsequent macrophage uptake.

Similar conclusions may be applied to interactions with plasma proteins. Obviously more cohesive bilayers exchange lipid molecules with the surroundings more slowly. Very cohesive bilayers in the gel state, however, contain too many hydrophobic defects to be stable.[1]

This model of liposome uptake can explain the above-mentioned results. Clearly, uncharged liposomes should be more stable because any strong ionic binding is absent ($W_{est} = 0$). Hydrophobic binding, such as penetration of a segment of the macromolecule into the bilayer, should also be reduced in the presence of more mechanically cohesive bilayers (reduction of W_{hfo}). Physisorption, however, is not prevented, i.e., W_{vdW} and W_{Hbond} are <0. Later we shall show that physisorption can be greatly reduced by the presence of a steric shield of an inert, nonimmunogenic ($W_{spec} = 0$), nonionic ($W_{est} = 0$) polymer.

The absence and reduction of chemisorption, i.e., electrostatic and hydrophobic binding, does not reduce the H bonding and physical adsorption of macromolecules onto the liposome. This may explain their lower stability *in vivo* and their macrophage uptake. In contrast, the presence of a steric shield, as exhibited by a noninteractive surface-attached polymer, prevents or reduces all the different types of macromolecular interactions with liposomes. Results show that in the presence of surface-grafted polymer both fluid and less cohesive or charged liposomes can exhibit prolonged blood circulation times.[45] This property is important in designing liposomes with well-defined leakage characteristics and containing no cholesterol as well as improving their encapsulation efficiency for hydrophobic molecules.

It seems that in contrast to electrostatic stabilization, which was found to be effective in many *in vitro* applications, steric stabilization is a ubiquitous way to stabilize particles in biological systems and in media with (high and) variable salinity. Of course, sterics stabilization can be imposed on charged bilayers and in these cases *in vitro* stability may be increased while the stability is not additive in the *in vivo* case.

A combination of the two, i.e., electrostatic repulsion in the case of surface-attached polyelectrolytes, would probably increase the stability of liposomes due to thicker, more rigid, and mutually electrostatically repulsive brushes *in vitro*. However, as was discussed before, it is unlikely that it would increase biological stability because $W_{est} \neq 0$. Polyelectrolyte chains bearing opposite charge to the bilayer may, however, disrupt the bilayer integrity.

E. AN EXTENDED DLVO INTERPARTICLE POTENTIAL

The stability of colloidal particles against aggregation is normally described by the DLVO theory. This theory is based on the interplay of the attractive van der Waals and repulsive electrostatic forces, and cumulative potential determines the stability. As already mentioned, the theory cannot explain the stability of an uncharged liposome suspension. This can be corrected by invoking hydration repulsion. For the most general case, however, we propose that the total potential between the particles should include all the above-mentioned forces, including steric and

$$V_{tot} = -V_{vdW} - V_{hfo} + V_{est} + V_{hyd} + V_{und} + V_{st} \quad (17)$$

We have also included the hydrophobic attractive force despite that its existence is still controversial. A treatment, analogous to the DLVO for two potentials, should be performed to find stability of the system. In a first approximation, potentials could be treated as additive. For rigorous treatment, however, one may expect dependence between various terms. It is reasonable to expect the interdependence of electrostatic, hydration, and steric interaction in the case of charged surface and covalently bound polymers due to different water structure at the surface.

In the case of grafted polyelectrolyte chains, however, the conformation of chains themselves is a function of electrostatic interactions and surface charge and obviously detailed analyses should be performed.

F. STABILITY IN LIPOSOMICIDAL ENVIRONMENT

The biological stability previously discussed is an example of liposomicidal environment with particular properties. We shall reiterate that in circulation even the most stable liposomes, including "indisintegrable" polymerized liposomes with enzymatically nondegradable chemical bonds (ether instead of ester) are not stable due to macrophage uptake. Steric stabilization can increase this stability due to reduced adsorption/adhesion onto/of such particles as well as reduced enzymatic degradation. Stability against chemical attack (low/high pH, some ions, free radicals, oxidants, macromolecules, enzymes, etc.) can be minimized only by mechanically using the most cohesive bilayers which minimize penetration of protons/hydroxils/radicals into the bilayer. The same is true also for the stability against the detergent-induced disintegration. Normally detergents at low concentrations (few mol%) stabilize liposomes (induce spontaneous nonzero curvature due to asymmetric distribution), at higher concentrations they start to induce leakage and above 50 to 100 mol%, depending on detergent, cause liposome disintegration into mixed micelles. Some bilayers, such as DSPC/Chol 1/1 are extremely stable against detergent or ionophore-induced disintegration or leakage and may require prolonged heating in the presence of the agent to induce changes.

This reasoning potentially allows the design of liposomes which could be used for oral administration. During such applications liposomes are challenged by low values of pH in the stomach, digestive enzymes (phospholipases), and bile detergents. Briefly, one should add polymer coating for reduction of enzyme attack, use mechanically strong bilayers composed of chemically stable lipids with addition of antioxidants to minimize damage in low pH environment as well as stabilize against detergent attack. If cargo is surrounded by several bilayers this may increase its chance for the passage through the stomach.

Similar reasoning can be used to design liposomes with the highest stability in other environments. In addition to these colloidal and mechanical factors, chemical stability of the components in a particular environment has to be carefully evaluated to obtain maximal stability.

1. Shelf Life Stability

Commercial application of liposomes require also the stability of liposomes on the shelf. This can be up to 2 years at room temperature, normally with less than 5% change in any of the liposome characteristics (size distribution, concentrations, pH, encapsulation efficiency, zetapotential, osmolality). It may be obtained in liquid, frozen, or dried form. In the liquid form one has to be careful to select optimal pH, buffer, amount of antioxidant, chelators, free radical scavengers, and similar characteristics. For frozen products, addition of cryoprotectants is normally required with several sugars and peptides able to retain liposome integrity after freezing–thawing cycle(s). Drying–rehydration is normally a more difficult task and requires special cryoprotectants, most frequently, sucrose above 0.3 *M*. Both processes may depend on the rate of freezing, and in the case of freeze-drying also depends on many other lyophilization parameters (length and temperature of various steps, surface/volume ratio and mass of the sample, and other geometrical parameters). Liposomes can also be spray-dried.

IV. CONCLUSION

In conclusion, we wanted to present in this chapter a coherent view of liposome stability and ways to increase it. The multidisciplinary approach of theoretical physics, colloid and interface science, physical chemistry, biophysics, and biochemistry allowed an explanation of liposome stability and interaction characteristics, and this allowed rational design of stable liposomes for various applications. From conventional liposomes, which were characterized by nonspecific reactivity to the inert (stealth) and very reactive, polymorphic (cationic, fusogenic) liposomes were devised for respective applications, most currently in cancer chemotherapy as well as in gene therapy.[2]

REFERENCES

1. **Lasic, D. D.,** *Liposomes: From Physics to Application*, Elsevier, Amsterdam, 1993.
2. **Lasic, D. D. and Papahadjopoulos, D.,** Liposomes Revisited, *Science*, 267, 1275, 1995.
3. **Tanford, C.,** *The Hydrophobic* Effect, John Wiley & Sons, New York, 1980.
4. **Helfrich, W.,** Elasticity of bilayers, *Zeit. Naturforsch*, 28c, 693, 1973.
5. **Bangham, A. D. and Horne, R. W.,** Negative staining of phospholipids and their structured modification by surface active agents as observed in the electron microscope, *J. Mol. Biol.*, 8, 660, 1964.
6. **Winterhalter, M. and Lasic, D. D.,** Liposome stability and formation, *Chem. Phys. Lipids*, 64, 19, 1993.
7. **Verwey, E. J. and Overbeek, J. T. G,** *The Theory of the Stability of Lyophobic Colloids*, Elsevier, Amsterdam, 1948.
8. **Israelachvili, J. N.,** *Intermolecular and Surface Forces*, Academic Press, New York, 1985.
9. **Helm, C., Israelachvili, J. N., and McGuiggan, P. M.,** Molecular mechanisms and forces involved in the adhesion and fusion of amphiphilic bilayers, *Science*, 246, 919, 1989.
10. **Leon, S. I., Lee, J. H., Andrade, J. D., and deGennes, P. G.,** Protein surface interactions in the presence of PEO; simplified theory, *J. Colloid Interface Sci.*, 142, 149, 1991.
11. **Crowley, A. C., Fuller, N. L., Rand, R. P., and Parsegian, V. A.,** Measurement of repulsive forces between charged phospholipid bilayers, *Biochemistry*, 17, 3163, 1978.
12. **Lis, J. L., McAlister, M., Fuller, N., Rand, R. P., and Parsegian, V. A.,** Interaction between neutral phospholipid bilayer membranes, *Biophys. J.*, 37, 657, 1982.
13. **Israelachvili, J. N. and Wennerstrom, H.,** Entropic forces between amphiphilic surfaces in liquids, *J. Phys. Chem.*, 96, 520, 1992.
14. **Helfrich, W.,** Steric interaction of fluid membranes in multilayer systems, Z. *Naturforsch.*, 33a, 305, 1978.
15. **Naper, D. H.,** *Polymeric Stabilization of Colloidal Dispersions*, Academic Press, New York, 1983.
16. **deGennes, P. G.,** Polymers at interface; a simplified view, *Adv. Colloid Interface Sci.*, 27, 189, 1987.
17. **deGennes, P. G.,** private communication, 1992.
18. **Needham, D., Hristova, K., McIntosh, T. J., Dewhirst, M., Wu, N., and Lasic, D. D.,** Polymer grafted liposomes: physical basis for the "Stealth" property, *J. Liposome Res.*, 2, 411, 1992.
19. **Mukherjee, P.,** private communication, 1994.
20. **Helfrich, W.,** Pathways of vesiculation, *Prog. Surf. Sci.*, 95, 7, 1994.
21. **Lasic, D. D.,** A molecular model for vesicle formation, *Biochim. Biophys. Acta*, 692, 501, 1982.
22. **Lasic, D. D.,** Mixed micelles in drug delivery, *Nature*, 355, 279, 1992.
23. **Cantu, L., Corti, M., Musolino, M., and Salina, P.,** Spontaneous vesicle formation from a one component solution of a biological surfactant, *Europhys. Lett.*, 13, 1655, 1990.
24. **Gebicki, J. M. and Hicks, M.,** Ufasomes are stable particles surrounded by unsaturated fatty acid membranes, *Nature*, 243, 232, 1973.
25. **Kaler, E. W., Murthy, A. K., Rodriguez, B. E., and Zasadzinski, J. A. N.,** Spontaneous vesicle formation in aqueous mixtures of single-tailed surfactants, *Science*, 245, 1371, 1989.
26. **Lipowsky, R.,** The conformation of membranes, *Nature*, 349, 475, 1991.
27. **Klotz, K. H., Winterhalter, M., and Benz, R.,** Use of irreversible electric breakdown of lipid bilayers for the study of interactions of membranes with surface active molecules, *Biochim. Biophys. Acta*, 1147, 161, 1993.
28. **Lasic, D. D.,** The mechanism of vesicle formation, *Biochem. J.*, 252, 1, 1988.
29. **Lasic, D. D.,** The formation of membranes, *Nature*, 351, 613, 1991.
30. **deGennes, P. G.,** *The Physics of Liquid Crystal*, Oxford University Press, New York, 1974.
31. **Evans, E. and Skalak, R.,** Mechanics and thermodynamics of membranes, *Crit. Rev. Bioengineer.*, 3, 181, 1979.

32. **Needham, D. and Nunn, R.,** Elastic deformations and failure of lipid bilayer membranes containing cholesterol, *Biophys. J.*, 58, 997, 1990.
33. **Bloom, M., Mouritsen, O. and Evans, E.,** Physical properties of the fluid lipid-bilayer component of cell membrane: a perspective, *Q. Rev. Biophys.*, 24, 293, 1991.
34. **Fendler, J. H.,** *Membrane Mimetic Chemistry*, John Wiley & Sons, Chichester, 1982.
35. **Santaella, C., Vierling, P., and Riess, J. G.,** Highly stable liposomes derived from perfluoroalkyl glycero phosphoquinolones, *Angew. Chem.*, 30, 567, 1991.
36. **Papahadjopoulos, D., Vail, W. J., Newton, C., Nir, S., Post, G., and Lazo, R.,** Studies on membrane fusion, *Biochim. Biophys. Acta*, 465, 579, 1977.
37. **Grit, M. and Crommelin, D. J. A.,** Chemical stability of liposomes, *Chem. Phys. Lip.*, 64, 3, 1993.
38. **Martin, F. J.,** Pharmaceutical manufacturing of liposomes in *Specialized Drug Delivery Systems*, Tyle, E., Ed., Marcel Dekker, New York, 1990, 267.
39. **Madani, A. and Kaler, E. W.,** Aging and stability of vesicular dispersions, *Langmuir*, 6, 125, 1990.
40. **Nir, S., Bentz, J., Wilschut, J., and Duzgunes, N.,** Aggregation and fusion of liposomes, *Prog. Surf. Sci.*, 13, 1, 1983.
41. **Okhi, S., Doyce, D., Flanagan, T., Hui, S. and Mayhew, E., Eds.,** *Molecular Mechanisms of Membrane Fusion*, Plenum Press, New York, 1988.
42. **Lasic, D. D., Martin, F. J., Gabizon, A., Huang, S. K., and Papahadjopoulos, D.,** Sterically stabilized liposomes: a hypothesis on the molecular origin of the extended circulation times, *Biochim. Biophys. Acta*, 1070, 187, 1991.
43. **Lasic, D. D.,** Sterically stabilized vesicles, *Angew. Chem. Int. Ed. Engl.*, 33, 1685, 1994.
44. **Lasic, D. D. and Martin, F., Eds.,** *Stealth Liposomes*, CRC Press, Boca Raton, FL, 1995.
45. **Woodle, M. C., Newman, M., Martin, F. J., Matei, K., and Papahadjopoulos, D.,** Versatility in lipid composition showing prolonged circulation with sterically stabilized liposomes, *Biochim. Biophys. Acta*, 1105, 193, 1992.

Chapter 5

Theory of Loading of Agents into Liposomes

Boris Čeh

CONTENTS

I. INTRODUCTION

In different biological systems the driving force for many bioenergetic, signaling, and other processes are electrochemical gradients of ions across membranes. pH gradient and transmembrane potential were recognized to play an important role for accumulation of different kinds of chemical species like lipophilic ions and weak acids and bases in cell organelles, and synthetically prepared lipid vesicles. After the amines were recognized to be concentrated in the chromaffin vesicles by permeation of its neutral form,[1,2] the first similar experiment of uptake of the catecholamines into the synthetic phospholipid vesicles — liposomes — was undertaken by Nichols and Deamer in 1976.[3]

Since then, in the past two decades, liposomes have become one of the most studied colloidal drug delivery systems. A number of methods for loading ions[4] and pharmaceutical drugs into the liposomes were developed, investigated, and improved, mostly with the intention to seek the most appropriate conditions to reach the highest possible loading efficiency and highest concentration and stability of the loaded drugs.[5]

Several methods exist to improve loading of such molecules. Mostly, they are simply based on passive entrapment, taking into account special geometrical considerations. At high lipid concentrations, liposomes encapsulate up to 70% of the sample volume and upon dilution, large fractions of drug can be encapsulated. These samples are, however, difficult to size down and, if possible, an active loading method is preferable.[5]

Active loading methods consist of forced entrapment of the drug molecules upon certain concentration gradients between the liposome interior and exterior following by permeation through the liposome bilayer. In addition to the neutral forms of weak bases and acids, coexisting in the solution with their charged forms, are species which can respond to these gradients. In the presence of ionophores and

0-8493-4012-8/96/$0.00+$.50
© 1996 by CRC Press, Inc.

uncoupling agents there are also some ions which may — according to the gradients — be redistributed between both liposome compartments.[6–8]

Various gradients are often employed to actively fill some molecules into preformed colloidal particles separated from the medium by a semipermeable membrane. In the case of therapeutic applications of the liposomes there are predominantly a pH gradient[9] and ammonium sulfate gradient methods,[10,11] which are widely used to load drug molecules, being mostly weak amphipathic bases or acids. These gradients can dissipate and the encapsulated molecules leak out.[9,12]

The question arising now is if it is in general possible to simulate and model the behavior of liposome systems with the purpose of obtaining at least a semiquantitative picture of favorable final parameters (like loading efficacy, inner/outer pH values, etc.), with which we could select appropriate starting chemical parameters to perform loading experiments. Except in the case of the pH gradient loading, the influence of different mechanisms and processes affecting the loading are not always well understood. Here we present a theoretical model to treat these phenomena.

First, considering acid–base equilibria, we will develop an appropriate expression to calculate pH of any solution as well as concentration of any species in given conditions. Further, combining this expression with a generally accepted fact that the membrane is permeable only to neutral species, in the case of their chemical potential gradient, we will derive an equation-couple for calculation of the equilibrium state after permeation of an optional number of species in either direction. Additionally, similar expressions will be derived, considering also the interaction of permeated species with inner/outer bilayer compartments as well as with the previously encapsulated material resulting in the formation of the precipitate.[13]

Besides all that, there are very often the processes, which accompany the loading, and can — depending on particular conditions — to a greater or lesser extent, influence the uptake of the drugs into the vesicles. The proton movement towards a less acidic compartment, induced through the pH gradient across the membrane,[14,15] as well as efflux or influx of less permeable neutral components are some of these influences.[14] Due to the phenomena just mentioned, the redistribution (reloading and unloading) of the drugs follows.

The purpose of this contribution is to introduce the corresponding terms for physicochemical processes mentioned above, and to find an appropriate set of mathematical expressions, with which we could describe a given liposomal system, and further, calculate and simulate the influence of particular factors on properties, behavior as well as the drug uptake of the liposomes.

II. THEORY

The interior of the liposomes is separated from the surrounding medium by the boundary consisting of one or more amphipathic lipid bilayers. When liposomes, filled with a given (inner) aqueous solution, are placed in an (outer) aqueous solution of different composition, the interaction between these two systems takes place, resulting in the equalization of concentrations of different permeable species on both sides of the membrane.

Different species permeate the membrane with very different rates. Because of the amphipathic nature of lipid membranes, the permeability to many neutral species is known to be many orders of magnitude faster than the permeability of the charged species.[16] Permeation of species occurs, on the whole, in both directions and is due to the gradient of chemical potential, which represents the driving force for loading (i.e., accumulation of active molecules in the liposome interior) or leakage (release of the same) of mainly neutral species. We shall assume that the membrane is permeable only to neutral species. In other words, it means that the equilibria between both liposome compartments should be reached much faster than appreciable amounts of whichever less-permeable neutral or even ionic species will permeate the membrane.

By numerical calculations performed on the expressions obtained from the preloading starting conditions in both compartments of the liposomes, it will be almost always possible to obtain the numerical values of some valuable parameters as a consequence of the influence of various parameters and chemical gradients. These calculated values are internal/external pHs ($pH_i s/pH_o s$), the equilibrium concentration of any species in either compartment, then, in the case of bidirectional exchanges of two or more species, the influx/efflux amount ratios, and, what is very important, the entrapment efficiencies for particular drug molecules.

Before we continue with the development of mathematical formalism, we shall postulate that only neutral species can cross the bilayer and that the activity coefficients of all species at equilibrium, γ_{ij}, will be assumed to be equal to 1.

A. STATIC SOLUTION. CALCULATION OF pH IN THE "STATIONARY SYSTEMS"

By following the known rules, including equilibrium constants, mass–balance and charge–balance equations, we will firstly develop an appropriate general equation for the calculation of the pH of an optional acid–base–salt solution. Before doing it, we shall introduce several new terms, which are indispensable for understanding of basic acid–base equilibration processes.

In the polar solvent the solute either disintegrates into neutral molecules or dissociates into ions. Both kinds of particles mentioned represent sources of a series of chemically related molecules and ions, which we shall determine with the term "starting-point species." Dealing with aqueous solutions, the starting-point species may be divided into two groups. In the first one there are all those species which, excepting hydration, do not undergo any chemical reactions with water molecules, and in the second one there are the members, which, according to Brønstedt–Lowry theory, react with water molecules giving in general a series of "sister-species," differing in charge and acidity (or basicity). Each of these sister-species can act as an acid or a base, depending of the conditions in a given aqueous solution.

Before starting with the calculation of the pH in any "stationary system" (i.e., closed system where matter cannot cross the boundary), we shall assume that species present do no form precipitates and complex ions. Suppose that each starting-point species $S_{ik}^{z_{ik}}$ reacts with water molecules forming a series of (at least two) sister-species:

$$S_{i1}^{z_{i1}} \underset{-H^+}{\overset{+H^+}{\rightleftharpoons}} S_{i2}^{z_{i2}} \underset{-H^+}{\overset{+H^+}{\rightleftharpoons}} \cdots S_{ij}^{z_{ij}} \underset{-H^+}{\overset{+H^+}{\rightleftharpoons}} \cdots S_{iN_i}^{z_{iN_i}} \qquad (S_{ij}^{z_{ij}} = HS_{ij+1}^{z_{ij+1}}) \tag{1}$$

Here, $S_{ij}^{z_{ij}}$ represents a particular sister-species member having, on the whole, the charge z_{ij}. Each of these species, except the first and the last one, can act either as an acid or a base. Each couple of two neighboring species, represented in above series, is referred to as a conjugated acid–base pair. Their concentrations are related through quantitative relationships called the acidity (basicity) constants:

$$K_{i1} = \frac{[H^+][S_{i2}^{z_{i2}}]}{[S_{i1}^{z_{i1}}]}, \qquad K_{i2} = \frac{[H^+][S_{i3}^{z_{i3}}]}{[S_{i2}^{z_{i2}}]}, \quad \ldots \quad K_{i(N_i-1)} = \frac{[H^+][S_{iN_i}^{z_{iN_i}}]}{[S_{i(N_i-1)}^{z_{i(N_i-1)}}]} \tag{2}$$

Considering that reaction coefficients in all protolithic reactions in Equation 1 are equal to 1, the mass–balance relationship can be expressed as:

$$C_i = [S_{ik}^{z_{ik}}]^{st} = [S_{i1}^{z_{i1}}] + [S_{i2}^{z_{i2}}] + \ \ldots \ + [S_{ik}^{z_{ik}}] + \ \ldots \ + [S_{iN_i}^{z_{iN_i}}] \qquad S_{ij}^{z_{ij}} = HS_{ij+1}^{z_{ij+1}-1} \tag{3}$$

The relations between concentrations inside N_i-member sister-species series can always be described by $N_i - 1$ independent acidity constant and one mass balance equation — therefore, by $(N_i - 1) + 1 = N_i$ independent equations. There are always two additional species in aqueous solutions, H^+ and OH^-. Consequently, for a complete description of an aqueous solution at equilibrium, two equations are still needed; the first one is the equilibrium constant of the autoprotolysis of water ($K_w = [H^+]\cdot[OH^-]$), and the second one the charge balance equation expressing the electroneutrality of the solution:

$$\sum_{i=1}^{L}\sum_{j=1}^{N_i} z_{ij}[S_{ij}^{z_{ij}}] + \sum_{l=1}^{M} z_l c_l + \left([H^+] + (-1)\frac{K_w}{[H^+]}\right) = 0 \tag{4}$$

Each of particular three terms in above sum represents the contribution of the charges of chemically related species; the first one, the double-sum term describes that contribution of those species which undergo the protolithic reaction with water molecules, the second one describes the contribution of "inert" species, and the last one describes the contribution of H^+/OH^- ions. The behavior of the complex acid–base–salt solution containing L-member sister-species series is therefore described by $(N_1 + N_2 + \ldots + N_L) + 2$ interacted species, but also by the same number of independent equations.

The solution of this system lead us to the following two important expressions:

$$f_i([H^+]) = 1 + \sum_{j=1}^{N_{i-1}} \frac{\prod_{k=1}^{j} K_{ik}}{[\mathrm{H}^+]^j} = \sum_{j=0}^{N_{i-1}} \frac{\prod_{k=0}^{j} K_{ik}}{[\mathrm{H}^+]^j}; \qquad (K_0 = 1) \tag{5}$$

and:

$$g_i([H^+]) = z_{ij} + \sum_{j=2}^{N_{i-1}} z_{ij} \frac{\prod_{k=1}^{j-1} K_{ik}}{[\mathrm{H}^+]^{j-1}} = \sum_{j=0}^{N_{i-1}} \frac{\prod_{k=0}^{j} K_{ik}}{[\mathrm{H}^+]^j}; \qquad (K_{i0} = 1) \tag{6}$$

The first of above $[H^+]$-dependent expressions we will call the "concentration function," the second one the "charge function." From expressions just derived one can obtain the basic equation:

$$\sum_i c_i \frac{g_i([\mathrm{H}^+]\cdot)}{f_i([\mathrm{H}^+])} + \sum_l z_l c_l + \left([\mathrm{H}^+] + (-1)\frac{K_w}{[\mathrm{H}^+]}\right) = 0 \tag{7}$$

This expression, which we shall call the Master equation, allows one to calculate $[H^+]$ and p in any solution assuming the activity coefficients, $(\gamma_{ij})_\pm = 1$. Via $[H^+]$, calculated from the above equation, the concentration of any sister-species designated ij may also be obtained.

B. INTERACTION BETWEEN BOTH LIPOSOME COMPARTMENTS

1. 'Clear' Solutions. Loading in the "Exchangeable Systems"

In general, the expression "exchangeable system" represents any two compartments separated by a semipermeable membrane. Usually, the interaction between both solutions separated by the membrane takes place, resulting in crossing of different species through it.

Before loading is assumed to start, we can describe the equilibrium in both, i.e., interior and exterior liposome solutions by means of Equation 7:

$$\sum_m c_{ml} \frac{g_{ml}(x^{st})}{f_{ml}(x^{st})} + u_I(x^{st}) = 0 \tag{8}$$

$$\sum_k c_{kO} \frac{g_{kO}(y^{st})}{f_{kO}(y^{st})} + u_O(y^{st}) = 0 \tag{9}$$

The terms $u(x^{st})$ and $u(y^{st})$ represent the sum of the contributions of charges of inert species as well as H^+ and OH^- ions, and can be expressed as follows:

$$u_I(x^{st}) = \sum_l z_l c_l + \left(x^{st} - \frac{K_w}{x^{st}}\right) \quad \text{and} \quad u_O(y_{st}) = \sum_n z_n c_n + \left(y_{st} - \frac{K_w}{y^{st}}\right) \tag{10}$$

Due to the concentration gradients, we postulate that the portions of potentially permeable neutral forms of particular sister-species series cross the bilayer in either direction. For any of the outward species which cross the membrane, the following concentration relationship is valid:

$$V_O \cdot c_{kO} \cdot \alpha_k = V_I \cdot c_{kI} \tag{11}$$

A similar equation is also obtained for any species crossing the membrane in the opposite direction:

$$V_I \cdot c_{mI} \cdot \beta_m = V_O \cdot c_{mO} \tag{12}$$

On the whole there are many species which can diffuse in both directions, i.e., into and out of each of the compartments. If one can describe the state outside/inside of the liposomes by two equations, for the inner solution one can obtain:

$$\sum_m c_{mI}\left(1 - B_m\beta_m\right)\frac{g_{mI}(x)}{f_{mI}(x)} + \sum_k\left(\frac{V_O}{V_I}A_k c_{kI}\alpha_k\right)\frac{g_{kI}(x)}{f_{kI}(x)} + u_I(x) = 0 \tag{13}$$

A similar, 'symmetrical' expression can describe the state in the outer solution:

$$\sum_k c_{kO}\left(1 - A_k\alpha_k\right)\frac{g_{kO}(y)}{f_{kO}(y)} + \sum_{m,I}\left(\frac{V_I}{V_O}B_m c_{mO}\beta_m\right)\frac{g_{mO}(y)}{g_{mO}(y)} + u_O(y) = 0 \tag{14}$$

Therefore, for a particular set of α and β values, the concentrations of H^+ in both compartments, x and y, can be calculated.

At equilibrium (there is only one such state) a set of equilibrium values, α^{eq}s and β^{eq}s, exists, which can be expressed as a ratio between the amount of permeated species and sum of amounts of species present in both compartments. Thus, for a particular neutral species loaded into liposomes it is valid:

$$\alpha_k^{eq} = \frac{n_{kI}}{n_{kI} + n_{kO}} = \frac{V_I c_{kI}^{eq,t}}{V_I c_{kI}^{eq,t} + V_O c_{kO}^{eq,t}} = \frac{c_{kI}^{eq,t}}{c_{kI}^{eq,t} + K_v c_{kO}^{eq,t}} \tag{15}$$

and similar for a particular species, released in its neutral form from the inner solution:

$$\beta_m^{eq} = \frac{n_{mO}}{n_{mO} + n_{mI}} = \frac{V_O c_{mO}^{eq,t}}{V_O c_{mO}^{eq,t} + V_I c_{mI}^{eq,t}} = \frac{c_{mO}^{eq,t}}{c_{mO}^{eq,t} + K_v^{-1} c_{mI}^{eq,t}} \tag{16}$$

Now, at equilibrium the concentrations of particular permeated as well as released species on both sides of the liposome bilayer are taken to be equal:

$$\left[S_{kl_k}^0\right]_I^{eq} = \left[S_{kl_k}^0\right]_O^{eq} \quad \text{and} \quad \left[S_{ml_m}^0\right]_O^{eq} = \left[S_{ml_m}^0\right]_I^{eq} \tag{17}$$

Assuming that all the equilibrium constants in the inner and outer solutions are the same, and taking into account all the relationships previously derived, one can obtain the corresponding expression for calculation of the portion of the particular species loaded into vesicles:

$$\beta_m^{eq} = \frac{V_O y^{l_m-1} f_{mO}(y)}{V_O y^{l_m-1} f_{mO}(y) + V_I x^{l_m-1} f_{mI}(x)} = \frac{f_{mO}(y)}{f_{mO}(y) + K_v^{-1}\left(\frac{x}{y}\right)^{l_m-1} f_{mI}(x)} \tag{18}$$

and for the calculation of the portion of the particular species released from the vesicles:

$$\alpha_k^{eq} = \frac{V_I x^{l_k-1} f_{kI}(x)}{V_I x^{l_k-1} f_{kI}(x) + V_O y^{l_k-1} f_{kO}(y)} = \frac{f_{kI}(x)}{f_{kI}(x) + K_v\left(\frac{y}{x}\right)^{l_k-1} f_{iO}(y)} \tag{19}$$

Introducing Equations 18 and 19 into expressions 13 and 14, two generalized equations with two unknowns are obtained. The first one represents the equilibrium state in the outer solution:

$$\sum_k c_{kO} \frac{g_{kO}(y)}{A_k k_v \left(\frac{x}{y}\right)^{l_k - 1} f_{kI}(x) + f_{kO}(y)} + \sum_m B_m c_{mI} \frac{g_{mO}(y)}{\left(\frac{x}{y}\right)^{l_m - 1} f_{mI}(x) + K_v f_{mO}(y)} + u_O(y) = 0 \tag{20}$$

and the second one describes the state in the inner liposome compartment:

$$\sum_m c_{mI} \frac{g_{mI}(x)}{f_{mI}(x) + B_m K_v \left(\frac{y}{x}\right)^{l_m - 1} f_{mO}(y)} + \sum_k A_k c_{kO} \frac{g_{kI}(x)}{k_v f_{kI}(x) + \left(\frac{y}{x}\right)^{l_m - 1} f_{kO}(y)} + u_I(x) = 0 \tag{21}$$

The above pair of equations completely describes the equilibrium state in the inner as well as in the outer solution after permeation of any number of species in both directions.

2. Membrane Binding. Loading in the "Exchangeable Systems" in which One of the Species is Bound onto the Membrane Surface

Some species, depending on their nature, charge, and membrane composition, may interact with the inner as well as outer bilayer surface compartment. Using Cafiso–Hubbell's four compartment model,[17] we will assume that one of the charged species of the loaded drug may bind onto both membrane surface layers. In the case of symmetrical composition of the liposome bilayer, the inner as well as outer membrane-binding constant, expressed as a ratio of membrane layer concentration and inner solution compartment concentration of the bound species, are taken to be equal:

$$K_{kj}^{m_I} = K_{kj}^{m_O} = \frac{c_{kj}^{m_I}}{c_{kj}^{s_I *}} = \frac{c_{kj}^{m_O}}{c_{kj}^{s_O *}} \tag{22}$$

Volumes of the solutions of both compartments are now reduced for volumes of the membrane layers formed. In order to compare a new situation with the previous one, where no membrane binding was assumed to occur, the membrane and solution compartment concentrations of the species as well as the outer to inner volume ratio have to be recalculated on the initial outer and inner volumes, both of them reduced for volumes occupied by membrane surface layers. The relationship between initial and reduced volume ratio is now:

$$K_v^* = \frac{V_O^*}{V_I^*} = \frac{V_O - V_{mO}}{V_I - V_{mI}} = \frac{V_O(1 - V_{mO}/V_O)}{V_I(1 - V_{mI}/V_I)} = K_v \frac{1 - V_{mO}/V_O}{1 - V_{mI}/V_I} \tag{23}$$

Considering the additional presumption that, in comparison to outer solution volume, the outer membrane surface layer volume is negligible, one can obtain the following simplified equation:

$$K_v = K_v^*(1 - V_{mI}/V_I) \tag{24}$$

Now, by introducing two new terms, expressed as a ratio of the concentrations of membrane-bound species and the concentration of all sister-species remaining in the accompanying solutions,

$$R_{kjI}^m = \frac{c_{kj}^{m_I *}}{c_k^{s_I *}} \quad \text{and} \quad R_{kjO}^m = \frac{c_{kj}^{m_O *}}{c_k^{s_I *}} \tag{25}$$

a final equation can be derived which connects the x and y dependent ratio-factors in both compartments:

$$R_{kjI}^m \cdot K_m \cdot f_{kI} \cdot (x) \cdot x^{j_{kj}-1} = R_{kjO}^m \cdot K_v^* \cdot f_{kO}(y) \cdot y^{j_{kj}-1} \tag{26}$$

The relation of equilibrium concentrations of particular permeated species and membrane-bound species in all four compartments is described by the following expression:

$$V_O^{\bullet} c_{kO} \alpha_k^{eq} = V_I^{\bullet} c_{kI}^{I} = V_I^{\bullet}\left(c_k^{s_I \bullet} + c_{kj}^{m_I \bullet}\right) = V_I^{\bullet} c_k^{s_I \bullet}\left(1 + c_{kj}^{m_I \bullet} / c_k^{s_I \bullet}\right) = V_I^{\bullet} c_k^{s_I \bullet}\left(1 + R_{kjI}^{m}\right) \tag{27}$$

At equilibrium, the portion of a species permeated from the exterior into the liposome interior can also be expressed in a way similar to that in Equations 15 and 16; first, as a function of concentrations in both solutions:

$$\alpha_k^{eq} = \frac{V_I^{\bullet} c_k^{s_I \bullet}\left(1 + R_{kjI}^{m}\right)}{V_I^{\bullet} c_k^{s_I \bullet}\left(1 + R_{kjI}^{m}\right) + V_O c_k^{s_O \bullet}\left(1 + R_{kjO}^{m}\right)} \tag{28}$$

and secondly, as an x, y, and R^m dependent expression:

$$\alpha_k^{eq} = \frac{f_{kI}(x)\left(1 + R_{kjI}^{m}\right)}{f_{kI}(x)\left(1 + R_{kI}^{m}\right) + K_V^{\bullet}\left(\frac{y}{x}\right)^{l_k - 1} f_{kO}(y)\left(1 + R_{kjO}^{m}\right)} \tag{29}$$

Consequently, the complete contribution of the charges of all chemically related sister-species in the inner aqueous solution and inner membrane surface layer is now:

$$\sum z_{kj}\left[S_{kj}^{z_{kj}}\right]_I = c_{kO} \frac{g_{kI}(x) + z_{kj} R_{kjI}^{m} f_{kI}(x)}{k_v^{\bullet} f_{kI}(x)\left(1 + R_{kjI}^{m}\right) + \left(\frac{y}{x}\right)^{l_k - 1}\left(f_{kO}(y) + R_{kjI}^{m}\left(\frac{K_m}{K_v^{\bullet}}\right)\left(\frac{x}{y}\right)^{j_{kj} - 1} f_{kI}(x)\right)} \tag{30}$$

and finally, the contribution of the same sister-species remaining in the outer solution can also be expressed in a similar way:

$$\sum z_{kj}\left[S_{kj}^{z_{kj}}\right]_O = c_{kO} \frac{g_{kO}(y) + z_{kj} R_{kjI}^{m}\left(\frac{K_m}{K_v^{\bullet}}\right)\left(\frac{x}{y}\right)^{j_{kj} - 1} f_{kI}(x)}{k_v^{\bullet} f_{kI}(x)\left(1 + R_{kjI}^{m}\right)\left(\frac{x}{y}\right)^{l_k - 1} + \left(f_{kO}(y) + R_{kjI}^{m}\left(\frac{K_m}{K_v^{\bullet}}\right)\left(\frac{x}{y}\right)^{j_{kj} - 1} f_{kI}(6x)\right)} \tag{31}$$

Anyway, to describe a four-compartment liposome system at equilibrium, both of the contributions derived above can be inserted into basic Equations 20 and 21. This yields the final equation-couple, with which one can calculate x, y, and consequently all parameters we are interested in. The associated membrane binding constant for particular membrane-bound species can also be calculated by means of the following expression:

$$K_{kj} = R_{kjI}^{m} f_{kI}(x)\left(\frac{x^{j_k - 1}}{\prod_{i=0}^{j_k - 1} K_i}\right) \tag{32}$$

3. Precipitate Formation. Loading in the "Exchangeable Systems" in which Precipitate is Formed inside the Liposomes

Let us assume that inside the liposomes, after permeation of a particular number of species in both direction occurs, the precipitate is formed between one of the charged sister-species, originating from one of the permeated neutral species, the charged species being previously present in the inner compartment before loading. This process can be described with the following chemical equation:

$$a_{kj}S_{kj}^{z_{kj}} + b_{mi}S_{mi}^{z_{mi}} = \left(S_{kj}\right)_{a_{kj}}\left(S_{mi}\right)_{b_{mi}} \qquad a_{kj}\cdot z_{kj} + b_{mi}\cdot z_{mi} = 0 \tag{33}$$

To express the portion of the precipitate formed in the inner compartment, one can introduce the concentration ratio of precipitate-forming drug species and the total sum of all the sister-species remaining in the solution:

$$R_{kjI}^{p} = \frac{c_{kj}^{p_I}}{c_k^{s_I}} \tag{34}$$

The portion of the permeated species, which the precipitate-forming ion originates from, can now be expressed in a way similar to that in the Equation 29:

$$\alpha_k^{eq} = \frac{f_{kI}(x)\left(1 + R_{kjI}^{p}\right)}{f_{kI}(x)\left(1 + R_{kjI}^{p}\right) + K_v\left(\frac{y}{x}\right)^{l_k-1} f_{kO}(y)} \tag{35}$$

After precipitate formation takes place, the whole charge contribution of both precipitate-forming ions as well of their sister-species remaining in the inner compartment is now:

$$\sum_j z_{kj}\left[S_{kj}^{z_{kj}}\right]_{sI} + \sum_i z_{mi}\left[S_{mj}^{z_{mj}}\right]_{sI} = c_{mI}\frac{g_{mI}(x)}{f_{mI}(x)}$$

$$+\; c_{kO}\frac{g_{kI}(x)}{k_v f_{kI}(x)\left(1 + R_{kjI}^{p}\right) + \left(\frac{y}{x}\right)^{l_k-1} f_{kO}(y)}\left(1 - \left(\frac{b_{mi}}{a_{kj}}\right)R_{kjI}^{p}\frac{f_{kI}(x)}{g_{kI}(x)}\frac{g_{mI}(x)}{f_{mI}(x)}\right) \tag{36}$$

The final equation-couple which describes the equilibrium state in both subsystems can now be obtained; for the inner solution one gets:

$$\sum_{m\neq m^*} c_{mI}\frac{g_{mI}(x)}{f_{mI}(x) + B_m K_v\left(\frac{y}{x}\right)^{l_m-1} f_{mO}(y)} + \sum_{k\neq k^*} A_k c_{kO}\frac{g_{kI}(x)}{k_v f_{kI}(x) + \left(\frac{y}{x}\right)^{l_k-1} f_{kO}(y)}$$

$$+\; c_{k^*O}\frac{g_{k^*I}(x)}{k_v f_{k^*I}(x)\left(1 + R_{kj^*I}^{p}\right) + \left(\frac{y}{x}\right)^{l_{k^*}-1} f_{k^*O}(y)}\left(1 - \left(\frac{b_{m^*i}}{a_{k^*j}}\right)R_{kj^*I}^{p}\frac{f_{k^*I}(x)}{g_{k^*I}(x)}\frac{g_{m^*I}(x)}{f_{m^*I}(x)}\right)$$

$$+\; c_{m^*I}\frac{g_{m^*I}(x)}{f_{m^*I}(x)} + u_I(x) = 0 \tag{37}$$

and for the outer solution it follows:

$$\sum_{k\neq k^*} c_{kO}\frac{g_{kO}(y)}{A_k k_v\left(\frac{x}{y}\right)^{l_k-1} f_{kI}(x) + f_{kO}(y)} + \sum_{m\neq m^*} B_m c_{mI}\frac{g_{mO}(y)}{\left(\frac{x}{y}\right)^{l_m-1} f_{mI}(x) + K_v f_{mO}(y)}$$

$$+\; c_{k^*O}\frac{g_{k^*O}(y)}{k_v\left(\frac{x}{y}\right)^{l_{k^*}-1} f_{k^*I}(x)\left(1 + R_{kj^*I}^{p}\right) + f_{k^*O}(y)} + u_O(y) = 0 \tag{38}$$

Clearly, in addition to the precipitation (crystallization or the formation of amorphous solids), gelation, or flocculation in the liposome interior, this concept can be extended to other interactions in liposome systems. They may contain a variety of pre-encapsulated molecules such as polyions, polymers, complexing agents, chelates, dextrans, and other molecular traps onto/with which the permeated drug molecules and their charged sister-species can bind electrostatically, hydrophobically, or via dispersive forces.

To conclude this section we should mention the calculation of the x, y, and R^p dependent solubility product:

$$L_p = \left(c_{kO} \frac{1}{k_v f_{kI}(x)\left(1 + R^p_{kjI}\right) + \left(\frac{y}{x}\right)^{l_k - 1} f_{kO}(y)} \left(\frac{\prod K_{kj}}{x^{j-1} f_{kI}(x)} \right) \right)^{a_{kj}}$$

$$\left(c_{mI} - \left(\frac{b_{mi}}{a_{kj}}\right) R^p_{kjI} c_{kO} \frac{f_{kI}(x)}{k_v f_{kI}(x)\left(1 + R^p_{kjI}\right) + \left(\frac{y}{x}\right)^{l_k - 1} f_{kO}(y)} \left(\frac{\prod K_{mi}}{x^{i-1} f_{mI}(x)} \right) \right)^{b_{mi}} \tag{39}$$

4. Transmembrane Potential Gradient. Reduction of Proton Concentration by Charging of Membrane

The equilibration between the liposomes' inner and outer compartments occurs after permeation of an arbitrary number of neutral species. From now on, for simplicity's sake, the whole liposome system at equilibrium, in the above sections always described by two Master equations, will be written in a shorter form below. For the inner/outer solution the equilibrium state is now described by the following simplified expressions:

$$F_I(x, y) = 0 \tag{40}$$

$$F_O(x, y) = 0 \tag{41}$$

The two equations above express the electroneutralities of interior and exterior, respectively, both of them as a function of two unknowns — the inner and outer proton concentration. Parallel to the loading, the pH gradient, established across the membrane, can act as a driving force for one-directional movement of protons from the more acidic compartment into the less acidic environment. The charging of the bilayer, becoming substantially faster when a protonophore is added,[12,15] will now be presumed to occur in such an extent that the established transmembrane potential will be postulated to be at equilibrium with the transmembrane pH gradient. Therefore, in the equilibrium state, the electrochemical [H^+]-gradient, i.e., the sum of the transmembrane potential and the pH gradient across the vesicle membrane can be equalized to zero:

$$\Delta\mu_H = \Delta\psi + \left(2.30 \frac{RT}{F}\right) \Delta\text{pH} = 0 \tag{42}$$

Taking into account the movement of the protons into the opposite compartment, the charge of the more acidic environment will now be reduced and of opposite sign to that of the effluxed protons. If the movement of the protons occurs out of the more acidic inner/outer compartment, the following equations are obtained:

$$F_I^*(x, y) = -\frac{n(H^+)}{v_I} \quad \textit{and} \quad F_O^*(x, y) = -\frac{n(H^+)}{K_v v_I} \tag{43}$$

It is also valid:

$$n(H^+) = \frac{SC\Delta\Psi}{F} \tag{44}$$

The amount of protons effluxed from either more acidic compartment onto the surface of the bilayer(s) is membrane-capacity dependent. Combining Equations 42 to 44, and assuming the spherical shape of the vesicles, two possibilities are given. If at equilibrium $x > y$, one gets the following equation-couple:[14]

$$F_I^{*}(x,y)+\left(\frac{3RT}{F^2}\right)\frac{C}{r}\ln\frac{x}{y}=0 \qquad and \qquad F_O^{*}(x,y)=0 \tag{45}$$

and oppositely — higher acidity of exterior dictates an analogous equation-couple:

$$F_O^{*}(x,y)+\left(\frac{3RT}{F^2}\right)\frac{C}{K_v(r+d)}\ln\frac{y}{x}=0 \qquad and \qquad F_I^{*}(x,y)=0 \tag{46}$$

With Equations 46 and 47 one can now calculate better values, especially in the cases where the membrane capacity could drastically influence the pHs in either compartment.*

C. COMMENTS

To avoid possible errors due to the use of various simplifications, we derived a general equation with which one can calculate the pHs of very complex acid–base–salt solutions. Simultaneously, this equation serves as an appropriate starting point for derivation of a general model simulating the accumulation of weak acids and bases into the liposome systems.

One of the advantages of such a simulation is description of the behavior of very complex liposome loading systems. The equations can give available estimates of the values of the starting pHs, the percentages of the permeated/accumulated species, and the difference of the pHs in both solutions after the assumed equilibria are reached. The simulations of the drug loading into the liposomes can easily be accomplished when an appropriate software is established. Various calculations using a large span of values of any number of parameters could be performed, considering all possible acid–base interactions between mono- and polyprotic species, the binding of species onto the membrane, and the formations of precipitates — clearly, all without any corrections, simplifications, and omissions of "negligible" terms, which can be, in such complex systems, sources of large errors.

As pointed out in the introduction, we have used the following approximations to evaluate these systems: zero permeability to charged species across the bilayer, rapid permeability to neutral species, which equilibrate according to their chemical potential, and activity coefficients of unity. While the first two assumptions are very reasonable, the third one may be more questionable. It was shown, for example, that activity coefficients of 1:1 and 2:1 electrolytes inside the concentration range of 0 to 0.25 *M*, usually in use for loading liposomes, are not too low and have the values above 0.77 and 0.43, respectively.[18] It is also likely that the error due to the omission of activity coefficients may be partially cancelled due to similar solutions on both sides of the membrane. Similarly, but unrelated, for instance, the osmolality of 155 m*M* ammonium sulfate is around 300 mOsm/kg giving rise to an osmotic coefficient around 0.67.

Quantitative applications of these calculations might not always match experimental data. Apart from the assumptions and approximations previously mentioned, this also involves practical difficulties in performing well-defined and characterized experiments.

In our case we mostly discussed anions. In a symmetrical case, however, cations can serve as encapsulated counterions in order to improve retention of negatively charged drugs, with an example being polyvalent cations and nucleosides, nucleotides, and DNA fragments.

In summary, the simulation model presented in this article is suitable for:

- calculation of pH in any solution
- calculation of the state after permeation of one or more species occurs
- calculation of the (non)equilibrium conditions due to the induced pH, which originated by simultaneous stoichiometric exchange of cations and protons caused by the presence of ionophores and uncoupling agents
- the behavior of the system in the case of the leakage of the vesicles in a given solution
- the behavior of the liposomes upon interaction with a given solution after they were loaded with the drug

* Otherwise, when the r-distribution function of liposomes is known, the average $\left\langle \frac{1}{r} \right\rangle$-value can be used in Equations 45 and 46:

$$\left\langle \frac{1}{r} \right\rangle = \sum_i w(r_i)\frac{1}{r_i}; \qquad w(r_i) = 1$$

The model is distinguished by the following advantages:

- there is no need to introduce any simplifications and cancellations or corrections
- the possibility is always given to calculate the behavior of the systems as a function of any possible number and range of values of any parameter
- transmembrane potential correction is taken into account

The model has the following insufficiency:

- the activity coefficients are not considered (in the case of the solutions of considerable complexity and higher concentrations, as they are sometimes used in the drug delivery, it is hardly possible to calculate their real values)

Further extensions of the model could be achieved by introducing osmotic pressure correction, neutral metal drug complexes, etc. The mathematical formalism developed herein can be generalized for similar calculations in any three or more compartment systems separated by semipermeable membranes.

Before we conclude this chapter and continue with treatments of particular liposome systems, we should mention that similar accumulation of species in vesicles may be a very frequent phenomenon in living systems. For instance, gel-containing vesicles closely resemble, with respect to size, shape, and appearance, the secretory granules of eosinophils. In addition to numerous amino acids, weak bases, and other molecules which can respond to various gradients, amination and similar reactions are common in living cells. In addition to the understanding of the function of living cells these studies may also shed more light on the origin of life; it was speculated that amino acids could have concentrated into the preexisting vesicles exhibiting pH gradients.

III. GRADIENTS

From the theoretical point of view the accumulation of the drugs into the liposomes can be divided concerning the number of species crossing the membrane as well as the directions of permeation, respectively. The membrane can be crossed by one or more species only in one direction or by two or more species in both directions. We have to emphasize that from a chemical point of view there is a great difference between the two cases mentioned above: (a) unidirectional filling and (b) bidirectional exchange.

In this chapter, the simulation of the loading of various species into liposomes will be presented in the form of diagrams. Each series of several different systems treated consists of a family of curves. In this way insight is given simultaneously into two factors which influence each particular system defined by the remaining fixed parameters. The most frequently calculated and shown parameters are the percentage of the encapsulated molecules, the pH difference across the membrane, the influx/efflux amount ratio, all of them as a function of different parameters like drug concentration, the outer to inner compartment volume ratio (which depends on lipid concentration and size distribution of vesicles), acidity constants of the drugs and buffers, internal and external buffer concentrations, membrane partition term (R^{m}_{kjl}), and solubility product (L_p) of the precipitated form of the drug. These diagrams have been calculated using appropriate nonlinear equations (as indicated in the text, and parameters as specified in the figure legends), and by solving them numerically using the Newton–Raphson method. In the case of precipitate-forming liposome systems the values were calculated by the combination of Newton–Raphson and bisection methods. The solutions converge only if the starting values of pH_i/pH_o, i.e., x and y, are close enough to the right values.

In each of the two examples discussed previously, it is always possible to find a couple of Master equations by which the states of aqueous solutions inside and outside of the liposomes after one or more neutral (or, in principle, even charged) species cross the membrane in either direction is completely described.

The first of the three treated systems gives rise to the Master equation-couple 20/21 in which only acid–base equilibria are considered. In the second and third case, the possibilities of binding of one or more ions onto both sides of a symmetric membrane (Equations 30 and 31) or the formation of one precipitate inside the liposomes (Equations 37 to 39) were also considered. In all the cases mentioned, the transmembrane potential correction was also implied. After the equilibrium pHs in both compartments are numerically obtained, one can calculate the concentration of any equilibrium-state species in the system. The percentages of the influxed and effluxed species (Equations 18, 19, 29, and 35), the membrane-binding constants of particular charged species (Equation 32) as well as the solubility product of the precipitate formed inside the liposomes (Equation 39) can also be calculated, respectively.

A. ONE SPECIES–ONE DIRECTION (pH GRADIENT) LOADING

The loading of one or more species in one direction, usually known as the pH gradient loading, can be explained by a pH difference in both compartments and approximately described as a redistribution of neutral species according to the Henderson–Hasselbalch equation.[8,9,12] After crossing the membrane, the neutral form of a weak base (acid) in the solution with a lower (higher) pH binds (gives up) the proton. By protonation (deprotonation) the concentration of the neutral form diminishes (increases) continuously. The gradient of the chemical potential, perpetually reestablished by the decrease of the concentration of the neutral form, drives the species through the membrane until the equilibration of the concentrations on both sides of the membrane is reached. It is very easy to see that, in the case of the pH gradient, the effect of loading is decided by the capacities of the buffers and by the pH difference in both compartments. The loading in only one direction can be illustrated by the following chemical equilibrium equations:

$$BH_o^+ \rightleftharpoons B_o + H_o^+ \rightleftharpoons \text{mem} \rightleftharpoons B_i + H_i^+ = BH_i^+ \tag{47}$$

The underlying constituents of the pH gradient loaded liposome systems are:

- outer buffer (mono- or polyvalent acid(base)-salt mixture)
- inner buffer (mono- or polyvalent acid(base)-salt mixture)
- drug in the outer compartment

The entrapment efficacy, depending on particular circumstances, can be strongly supported also by membrane binding as well as precipitate formation. To obtain the most simple corresponding expressions, with which we can describe the equilibrium in the case of pH gradient loading, we will take only univalent acids, base, and salts. For the inner solution we therefore obtain:

$$\begin{aligned} & c_1 \frac{1 + F_m R^m \left(1 + K_{11}/x\right)}{K_v^{-1}\left(1 + F_m R^m\right)\left(1 + F_p R^p\right)\left(1 + K_{11}/x\right) + \left(\frac{y}{x}\right)\left(1 + K_{11}/y\right) + F_m R^m \left(\frac{K_m}{K_v^*}\right)\left(\frac{x}{y}\right)\left(1 + K_{11}/x\right)} + \\ & c_2 \frac{-K_{21}/x}{1 + K_{21}/x} + \left(x - \frac{K_w}{x}\right) + ac_2 + F^*\left(1 - F_c\right)\left(\frac{3RT}{F^2}\right)\frac{C}{r}\ln\frac{x}{y} = 0 \end{aligned} \tag{48}$$

and for outer solution it is valid:

$$\begin{aligned} & c_1 \frac{1 + F_m R^m \left(\frac{K_m}{K_v^*}\right)\left(\frac{x}{y}\right)\left(1 + K_{11}/y\right)}{K_v^{-1}\left(1 + F_m R^m\right)\left(1 + F_p R^p\right)\left(1 + K_{11}/x\right)\left(\frac{\chi}{\gamma}\right) + \left(1 + K_{11}/y\right) + F_m R^m \left(\frac{K_m}{K_v^*}\right)\left(\frac{x}{y}\right)\left(1 + K_{11}/x\right)} + \\ & c_3 \frac{-K_{31}/x}{1 + K_{31}/y} + \left(y - \frac{K_w}{y}\right) + bc_3 + F^* F_c \left(\frac{3RT}{F^2}\right)\frac{C}{K_v(r + d)}\ln\frac{x}{y} = 0 \end{aligned} \tag{49}$$

The solubility product can also be expressed as follows:

$$\begin{aligned} L_p = & \left(c_1 \frac{1}{K_v^{-1}\left(1 + K_{11}/x\right)\left(1 + R^p\right) + \left(\frac{y}{x}\right)\left(1 + K_{11}/y\right)}\left(\frac{1}{\left(1 + K_{11}/x\right)}\right)\right) \\ & \left(\left(c_2 - \left(\frac{1}{1}\right)R^p c_1 \frac{1 + K_{11}/x}{K_v^{-1}\left(1 + K_{11}/x\right)\left(1 + R^p\right) + \left(\frac{y}{x}\right)\left(1 + K_{11}/y\right)}\right)\left(\frac{1}{\left(1 + K_{11}/x\right)}\right)\right) \end{aligned} \tag{50}$$

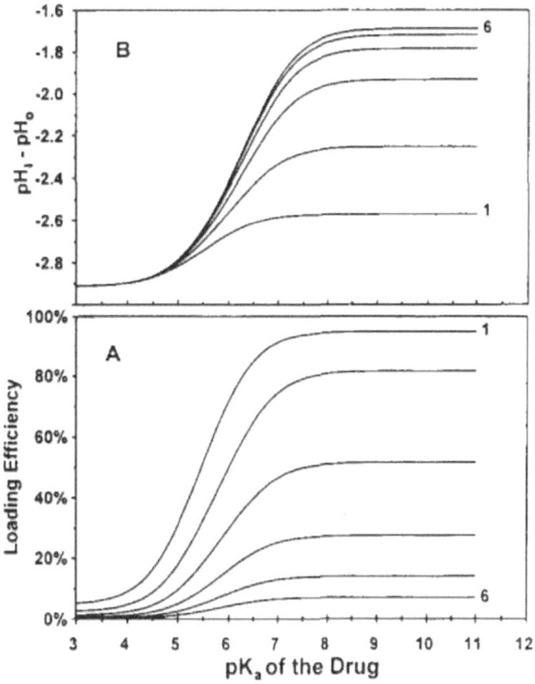

Figure 1 Inner compartment: 100 m*M* buffer with pK_a = 4.0 and [B_i^-]/[HB_i] ratio 1:1; outer compartment: 20 m*M* buffer with pK_a = 7.0 and [B_o^-]/[HB_o] ratio 1:1; the protonated drug concentration is 1.0 m*M*; K_v values of the corresponding curves are 20 (no. 1), 40, 80, 160, 320, and 640 (no. 6), respectively; C = 0 µF.

Here it has to be mentioned that in cases where $x > y$, $F_c = 0$, and inversely, if $x < y$, $F_c = 1$, respectively. Factors F_m and F_p imply either membrane-binding or precipitate formation.

To get a complete insight into the behavior of the liposome system dependence on the particular parameter, we will always present at least two related figures in the case of the pH gradient: the loading efficiency and the ΔpH as a function of the particular parameter. In all the cases of loading of the liposomes, which we will treat here, there will always be present an inner and outer buffer with pK_a values of 4 and 7, each of them with 1:1 concentration ratio of their conjugated acid–base pairs, respectively. From a theoretical point of view, the pH gradient, described by the Master equations above, represents the simplest of all possible cases of liposome loading. The results shown in the figures presented in the next section are almost always simple to explain. For where there is always only one species-one direction loading case, we can consider it as a 'one-dimensional' problem.

The drug encapsulation efficiency as a function of its pK_a for different liposome K_v values is shown in Figure 1A. For each of the calculated curves, the starting pHs of the inner and outer compartment are 4 and 7, respectively; the starting ΔpH is therefore 3. As shown in Figure 1, the percentage of the loaded drug is minimally affected by the pK_as of the drugs if their values are higher than the pHs of the buffered outer solution (>7) or lower than the pHs of the buffered inner solution (<4). For drug pK_a values between the pHs of the inner and outer compartments (4 < pK_a < 7), the portion of the loaded drug rapidly diminishes with decreasing pK_a values and is practically negligible if they are close to those of the inner buffer. This diagram also shows the quantitative dependence of entrapment efficacy on the liposome concentration. At higher liposome concentrations the drugs are more efficiently loaded. A complimentary situation can be observed at the related ΔpH values (Figure 1B). According to higher liposome concentrations, the buffer capacity is more reduced, resulting in the lower final ΔpH values. The smallest differences (curve 1) are due to the lowest liposome concentrations. For the reasons already known, the pH differences are also strongly pK_a dependent in the pH range between 3 and 7. At the pK_a values below 4 and above 8, they are practically constant.

Figure 2A represents loading as a function of a drug concentration. Inside the pK_a range between 3 and 7 the strong pK_a 'sensibility' on the loading efficacy as well as on the pH differences are observed. At lower initial outer drug concentrations the drug entrapment is higher, but not the concentration of the encapsulated drug, and, consequently, the buffer capacity is not strongly affected which results in almost unchanged starting ΔpH values (Figure 2B; curve 1).

Figures 3A and 3B underline the importance of the outer buffer concentration on drug loading. The curves are calculated for different drug concentrations. Their course, especially of that designated 1, gives us information that for a particular drug concentration there is always a minimum buffer concentration

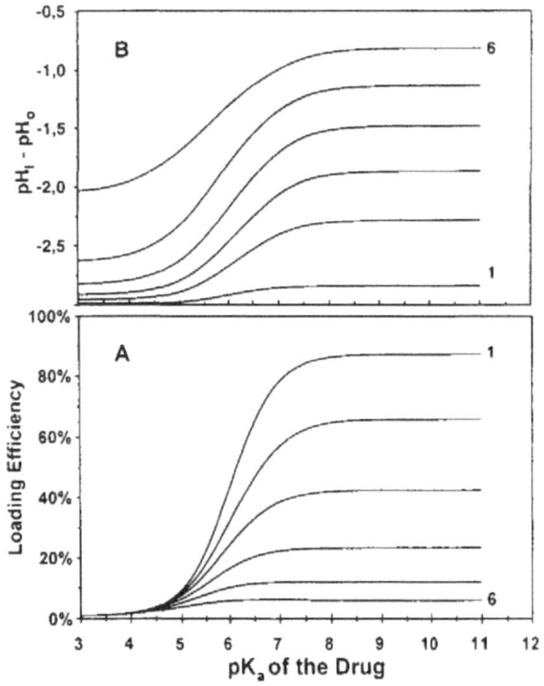

Figure 2 Inner compartment: 100 m*M* buffer with $pK_a = 4.0$ and $[B_i^-]/[HB_i]$ ratio 1:1; outer compartment: 20 m*M* buffer with $pK_a = 7.0$ and $[B_o^-]/[HB_o]$ ratio 1:1; pK_a of the protonated drug is 8.0 and the drug concentrations of the corresponding curves are 0.1 (no. 1), 0.5, 1.0, 2.0, 4.0, and 8.0 m*M* (no. 6), respectively; $K_v = 100$ and $C = 0$ μF, respectively.

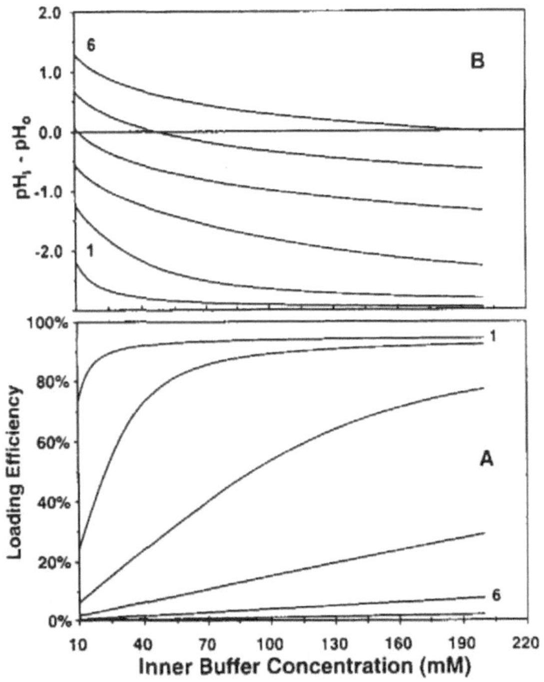

Figure 3 Inner compartment: buffer with $pK_a = 4.0$ and $[B_i^-]/[HB_i]$ ratio 1:1; outer compartment: 20 m*M* buffer with $pK_a = 7.0$ and $[B_o^-]/[HB_o]$ ratio 1:1; the protonated drug concentrations of the corresponding curves are 0.1 (no. 1), 0.5, 1.0, 2.0, 4.0, and 8.0 m*M* (no. 6), respectively; pK_a of the protonated drug is taken to be 8.0; $K_v = 50$ and $C = 0$ μF, respectively.

above which the loading efficacy does not change considerably (for example, in Figure 3A, for the curves 1 and 2 these concentrations are around 20 and 60 m*M*, respectively). In particular cases, especially at lower buffer/drug ratios, the (absolute) ΔpHs dissipate strongly and may become even positive (curves 6, and 7 in Figure 3B).

The importance of the mutual influence of the outer buffer and drug concentration is presented in Figures 4A and 4B. Also here, rapid diminution of loading efficiency as well as ΔpHs at lower outer buffer concentrations are observed. The curves lying higher (lower drug concentration) confirm the logical fact that the loading system is more effective at higher outer buffer to drug concentration ratios.

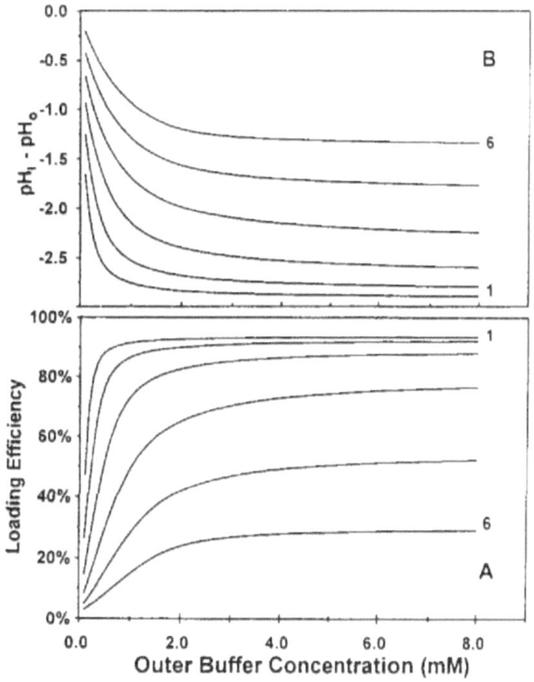

Figure 4 Inner compartment: 100 m*M* buffer with pK_a = 4.0 and [B_i^-]/[HB_i] ratio 1:1; outer compartment: 20 m*M* buffer with pK_a = 7.0 and [B_o^-]/[HB_o] ratio 1:1; the concentrations of the protonated drug of the corresponding curves are 0.1 (no. 1), 0.2, 0.4, 0.8, 1.6, and 3.2 m*M* (no. 6), respectively; pK_a of the protonated drug is taken to be 8.0; K_v = 50 and C = 0 μF, respectively.

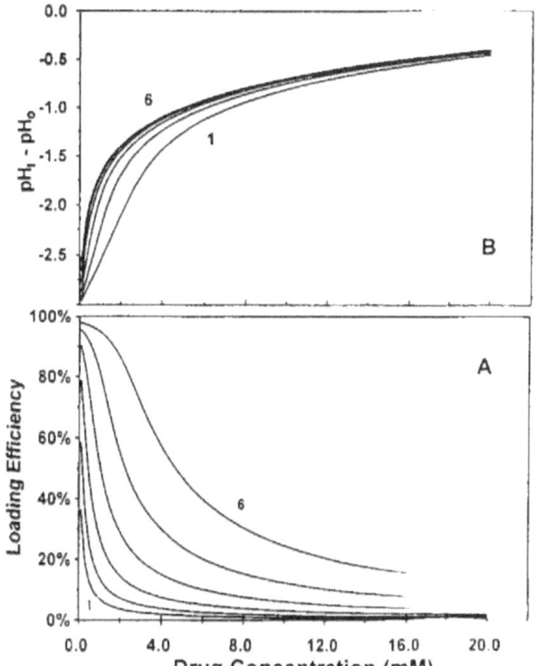

Figure 5 Inner compartment: 100 m*M* buffer with pK_a = 4.0 and [B_i^-]/[HB_i] ratio 1:1; outer compartment: 20 m*M* buffer with pK_a = 7.0 and [B_o^-]/[HB_o] ratio 1:1; the pK_a of the protonated drug is 8.0; R^m values of the corresponding curves are 20 (no. 1), 40, 80, 160, 320, and 640 (no. 6), respectively; K_v = 50 and C = 0 μF, respectively.

As was exemplified in the previous case, for each particular drug concentration in a given system, a 'reasonable' minimal outer buffer concentration can always be found above which the loading efficacy (Figure 4A) as well as ΔpH (Figure 4B) practically does not change any more.

In Figures 5A and 5B the influence of the mutual dependence of drug concentration as well as membrane-binding factor R^m onto loading efficiency is shown. Due to the Le Chatelier's principle, the proton concentration on both sides of the membrane is reduced by the membrane-bound protonated form of the drug. Because K_v values are usually much higher than 1, the reduction of protons inside the liposomes is more expressive, which results in lower ΔpH values.

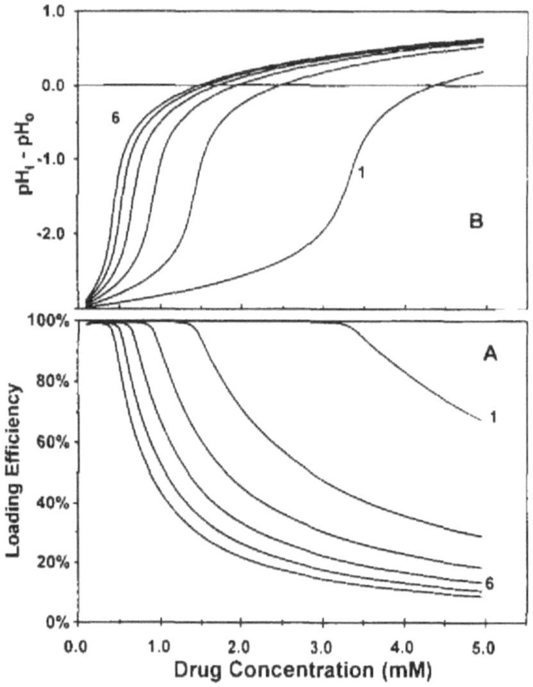

Figure 6 Inner compartment: 200 m*M* buffer with pK_a = 4.0 and $[B_i^-]/[HB_i]$ ratio 1:1; outer compartment: 100 m*M* buffer with pK_a = 7.0 and $[B_o^-]/[HB_o]$ ratio 1:1; protonated drug with pK_a of 8.2; K_v values of the corresponding curves are 30 (no. 1), 70, 110, 150, 190, and 230 (no. 6), respectively; C = 0 μF.

The influence of the solubility product of the precipitate, formed inside the liposomes between one of drug's charged species and an appropriate coion, onto the loading as well as onto the ΔpH, is shown in Figures 6A and 6B, respectively. The curves, calculated for different outer/inner volume ratios (K_vs) show us the effectiveness of the low solubility of the precipitate on loading efficiency, especially at lower liposome concentrations. From Figure 6B we can also conclude that for each of the K_v values there is always a very narrow drug concentration range where the ΔpH values rapidly dissipate. Namely, in this range the portion of the precipitated drug strongly increases, reaching a maximum at the point of inflection (calculations are not shown).

The pL_p dependence is shown in Figures 7A and 7B. One could expect that with increasing solubility product the loading efficiency should continuously grow as well. However, for the particular loading system defined, there is always a minimum pL_p at which the precipitate is formed in the inner compartment. When the concentrations are high enough, and after the precipitate formation starts (at the end of straight line of each curve), there is still slight/moderate increase in loading efficiency, but after a few units following on the pL_p-scale the efficacy practically does not change anymore. Surprisingly, the behavior of ΔpH is completely opposite (Figure 7B). At higher pL_p values and low drug concentrations the inner pHs even decrease slightly. And conversely, at higher drug concentrations the inner pH increases moderately. This unusual course of ΔpH curvatures is due to the prevailing of one of two competitive reactions upon the other. To form the precipitate, the buffer anion as well as drug cation are consumed. Because of Le Chatelier's principle, both equilibria:

$$HB \leftrightarrow H^+ + B^- \qquad \text{and} \qquad D + H^+ \leftrightarrow DH^+$$

have to be reestablished. One of the above reactions consumes protons while the other one produces them. However, in general, one of both reactions is predominant, and consecutively, depending on the particular situation, there is always surplus or insufficiency of protons into the liposome compartment — the inner solution becomes more acidic or basic.

Shown in Figures 8A and 8B are dependencies of loading efficiency and ΔpH upon the specific membrane capacity and liposome radius. The transmembrane potential is known to play an important role in charging membrane what implies the reduction of the amount of protons on either side of the bilayer. In consequence of that, the buffer capacity is diminished and due to this, the loading efficacy is diminished as well. We can see that the last term into Equation 46 affects the reduction of inner drug

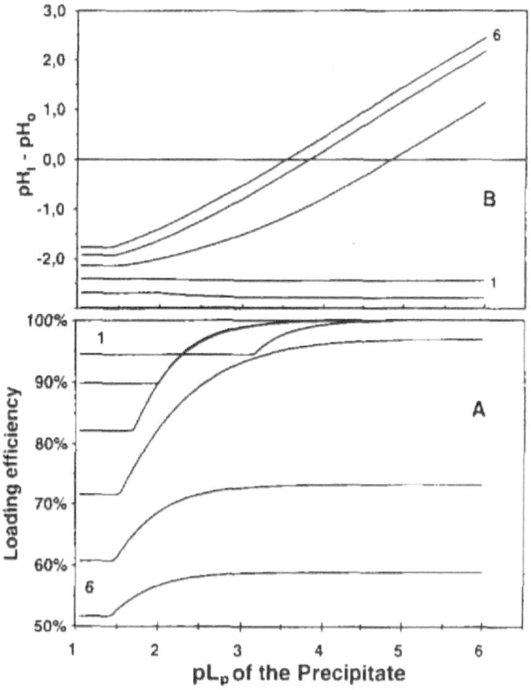

Figure 7 Inner compartment: 300 m*M* buffer with $pK_a = 4.0$ and $[B_i^-]/[HB_i]$ ratio 1:1; outer compartment: 100 m*M* buffer with $pK_a = 7.0$ and $[B_o^-]/[HB_o]$ ratio 1:1; protonated drug with pK_a of 8.0; drug concentrations of the corresponding curves are 0.1 (no. 1), 1.1, 2.1, 3.1, 4.1, and 5.1 m*M* (no. 6), respectively; $K_v = 50$; $C = 0$ μF; the precipitate 'drugH⁺buf⁻' is formed inside the liposomes.

concentration by higher specific membrane capacities as well as by lower liposome radii. The effect of both factors mentioned is observed in the figures discussed here.*

B. TWO SPECIES–TWO DIRECTIONS (AMMONIUM SULFATE) GRADIENT LOADING

The exchange loading, i.e., ammonium sulfate gradient method, is based on the exchange and/or interaction of at least two species, each of them being initially present on the opposite side of the membrane. In such a bidirectional exchange the circumstances are more complex and are based on a higher number of mutually dependent trans-membrane equilibrium reactions. The qualitative picture of the chemical occurrence in such a case is illustrated in Figure 9.

According to Le Chatelier's principle, the start of any of the reactions shown in Figure 9 triggers all the other ones. Each of these processes described schematically (also including the binding onto the membrane and formation of the precipitate), once started, supports and drives all the remaining ones until all the equilibria in both compartments are reached.

The ΔpH difference, *which can be positive or negative in the ammonium salt gradient loading* technique, is therefore a consequence and not the driving force for drug accumulation in the liposomes.

In the case of the pH gradient loaded liposomes, the underlying constituents of both interacted subsystems are:

- *co-drug species in the inner compartment*, C_2
- drug in the outer compartment having a role of being exchanged by co-drug molecules, C_1

and additionally but not necessarily:

- outer buffer (mono- or polyvalent acid(base)-salt mixture)
- inner buffer (mono- or polyvalent acid(base)-salt mixture)
- precipitate forming species in the interior of the liposomes

* It is straightforward to prove that for this case the changes for both α and ΔpH are proportional to capacity changes.

$$\frac{\left(1 + K_v \frac{y}{x}\right)^2}{K_v \frac{y}{x}} d\alpha = \alpha(\Delta \mathrm{pH})$$

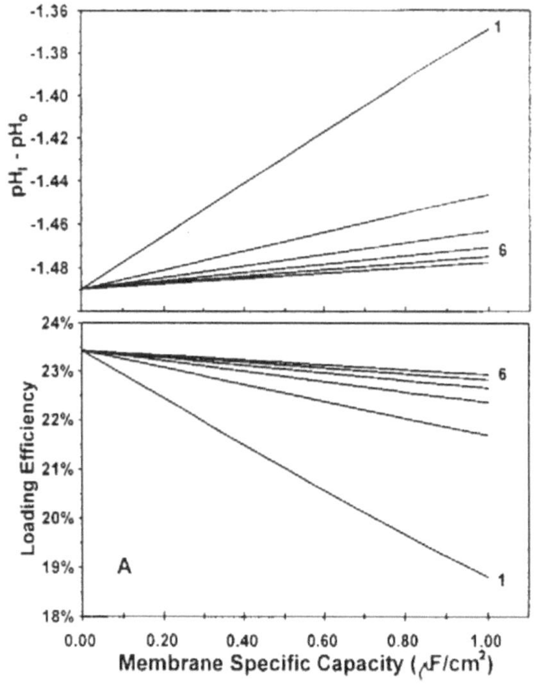

Figure 8 Inner compartment: 50 m*M* buffer with $pK_a = 4.0$ and $[B_i^-]/[HB_i]$ ratio 1:1; outer compartment: 20 m*M* buffer with $pK_a = 7.0$ and $[B_o^-]/[HB_o]$ ratio 1:1; 1.0 m*M* protonated drug with pK_a 9.0; $K_v = 100$; the inner liposomal radii of the corresponding curves are 50 (no. 1), 140, 230, 320, 410, and 500 Å (no. 6), respectively; $K_v = 50$ and $C = 0$ μF, respectively.

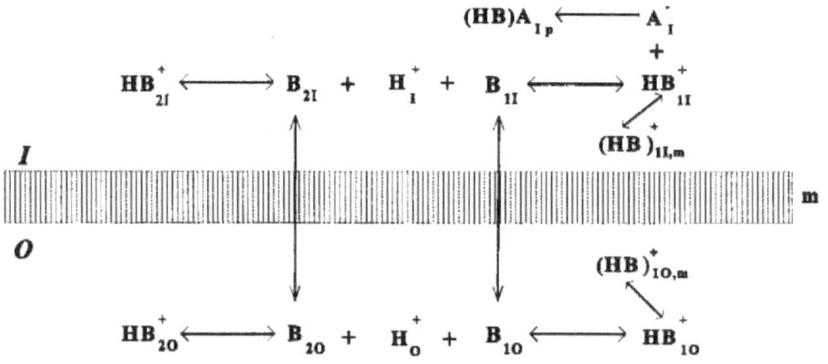

Figure 9 The mutually dependent chemical equilibria across liposome membrane at two species-two directions gradient loading.

As already mentioned, the entrapment efficacy can be, depending on particular circumstances, strongly supported also by membrane binding as well as precipitate formation. Therefore, by analogy with Equations 48 to 50, we can obtain comprehensive basic equations with which the entire drug loading system at the equilibrium state may be described (because of simplicity the membrane binding will not be considered in the equations below). For the inner solution we can write:

$$c_1 \frac{1}{K_v^{-1}(1+K_{11}/x)(1+F_pR^p)+\left(\frac{y}{x}\right)(1+K_{11}/y)} \cdot \left(1-\left(\frac{1}{2}\right)R^p \frac{(1+K_{11}/x)(-1-2K_{42}/x)}{(1+K_{11}/x)(1+K_{42}/x)}\right)+$$

$$c_2 \frac{-1-2K_{42}/x}{1+K_{42}/x} + 2c_2 \frac{1}{(1+K_{31}/x)+K_v\left(\frac{y}{x}\right)(1+K_{31}/y)} + \left(x-\frac{K_w}{x}\right)+ \tag{51}$$

$$F^*(1-F_e)\left(\frac{3RT}{F^2}\right)\frac{C}{r}\ln\frac{x}{y}=0$$

and for the outer solution it follows:

$$c_1 \frac{1}{K_v^{-1}\left(\frac{x}{y}\right)\left(1+F_pR^p\right)\left(1+K_{11}/x\right)+\left(1+K_{11}/y\right)} + 2c_2 \frac{1}{\left(\frac{x}{y}\right)\left(1+K_{31}/x\right)+K_v\left(1+K_{31}/y\right)}$$
$$+\left(y-\frac{K_w}{y}\right)+F^*F_c\left(\frac{3RT}{F^2}\right)\frac{C}{K_v(r+d)}\ln\frac{y}{x}=0 \tag{52}$$

and finally one gets the corresponding expression for the solubility product:

$$L_p = \left[\frac{c_1}{K_v^{-1}\left(1+K_{11}/x+\right)\left(1+R^p\right)+\left(\frac{y}{x}\right)\left(1+K_{11}/y\right)}\right]^2$$
$$\left[\left(c_2-\left(\frac{1}{2}\right)c_1R^p\frac{\left(1+K_{11}/x\right)}{K_v^{-1}\left(1+K_{11}/x\right)\left(1+R^p\right)+\left(\frac{y}{x}\right)\left(1+K_{11}/y\right)}\right)\left(\frac{K_{42}}{K_{42}+x}\right)\right] \tag{53}$$

In the system described above the buffer is present neither in the inner nor in the outer compartment. As already mentioned, in contrast to the 'unidirectional' pH gradient loading, in the case of exchange of more than one species, there are several additional mutual interactions which have to be taken into account. Therefore, it is hardly possible to explain or even to predict all the results obtained by the simulations. Here, besides the α^{eq} and the ΔpH, the "exchange-ratio" (the amount of species permeated into inner solution divided by the amount of species released from the inner into the outer solution — $n_{KO}/n_{ml} = K_v(c_{kO}\alpha_k/c_{ml}\beta_m)$) could be of interest and will also be presented. In all the cases calculated here the ammonium sulfate was taken to be present inside the liposomes ($pK_a(NH_3) = pK_{31} = 9.26$; $pK_{a1}(H_2SO_4) = pK_{41} = -10$; $pK_{a2}(H_2SO_4) = pK_{42} = 1.92$).

Shown in Figure 10A are loading efficacy, ΔpH, and exchange-ratio, all of them as a function of drug concentration at different K_v values. The facts of general validity like the diminishing loading efficiency with increasing K_v values are confirmed also in this particular case. In contrast to the previous case, ΔpHs are strongly dependent on ammonium sulfate concentration (Figure 10B). Using the pK_a value of the drug to be 9, the exchange-ratios observed here are all very close to 1, except those at extremely low concentrations (Figure 10C).

From Figures 11A to 11C, all of them showing drug concentration and K_v dependence, the most interesting is maybe a complete K_v independence of the ΔpH values.

One could expect that at very high membrane-binding constants the loading efficacy close to 100% should be reached. As shown in Figure 12A, the loading efficiency can be completely R^m independent, but only to some extent, which strongly depends upon particular drug concentrations. For the liposome system treated here, it is straightforward to see that at (relatively) high drug concentrations the efficacy increases minimally even at relatively high membrane-binding constants (curve 6 in Figure 12A).*

In contrast to this, at higher drug concentrations, the ΔpH values constantly decrease with increasing R^m factor (Figure 12B). For the system treated in this case, all exchange-ratios are close to 1 (Figure 12C).

Behavior very similar to that in the previous case (Figure 7) is observed in Figures 13A to 13C. However, also here, above a particular pL_p value for each of the drug concentrations used in the calculations, the increase of entrapment efficiency is no longer observed. As is seen from Figure 13A, at lower drug concentrations, the liposomes are 'saturated' already at relatively low pL_p values.

* Due to the fact that the relationship $pK_a >> pH_o$ is always valid in this case (the pK_a value of the drug is taken to be 9) and taking into account Equation 32, $R^m = K^m$.

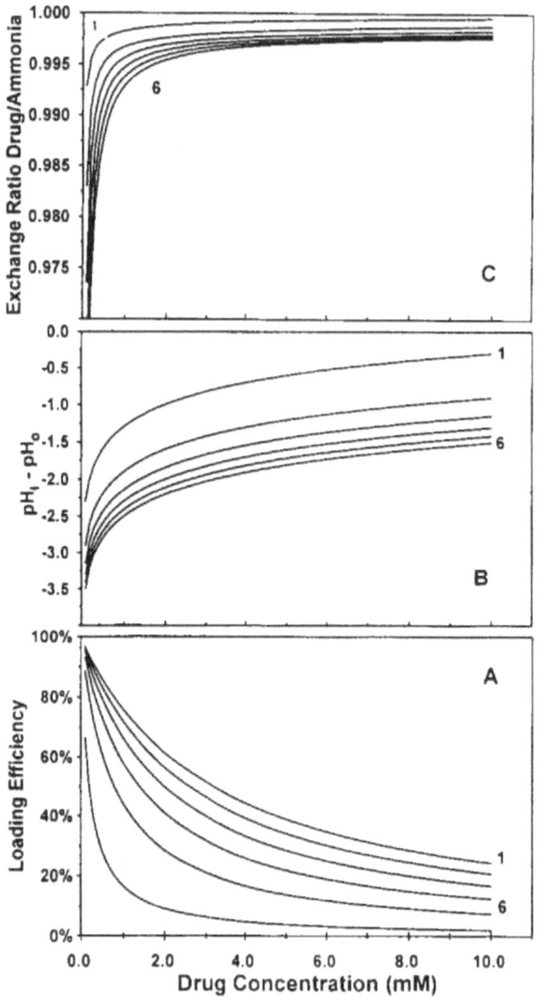

Figure 10 Inner compartment: ammonium sulfate with pK_a (NH_4^+) = 9.3, and pK_{a2} (H_2SO_4) = 1.92, respectively; the ammonium sulfate concentrations of the corresponding curves are 10 (no. 1), 40, 70, 100, 130, and 160 m*M* (no. 6), respectively. Outer compartment: protonated drug with pK_a value of 9.0; K_v = 100; C = 0 µF.

C. BIDIRECTIONAL EXCHANGE OF THREE OR MORE SPECIES

More complex systems include simultaneous efflux of at least one species, a neutral influx of one or more different kinds of molecules. In such cases there are many mutually dependent equilibria, exhibiting influence in either compartment as well as through the bilayer. Because of that, the behavior of such systems, even in approximate lines, is usually completely unpredictable. To get insight into the mutual dependence of particular parameters of interest, the calculations are necessary. To spare the space, we have omitted the extensive equations used to get the results discussed below.

The simulation of loading of the liposomes, suspended in a buffered drug containing medium and filled with a buffered NH_4HCO_3 aqueous solution, are presented in Figures 14A and B. The simultaneous release of NH_3 and CO_2 as well as uptake of the neutral form of the drug were assumed to occur.

From Figure 14A the independence of drug loading efficiency on NH_4HCO_3 concentration at higher K_v values is observable. In all cases calculated here, the initial values of a ΔpH close to 3 drop down to values between 0.4 and 1. Because of using the rather high outer buffer concentration of 100 m*M* with the pH value of 8, the larger outer volume as well as its more basic contents are responsible for almost complete unloading of CO_2 from the inner liposome compartment.

D. LEAKAGE OF THE LIPOSOMES

In all previous calculations the rapid uptake of only neutral species across the membrane in either direction was the basic presumption. Once filled with drugs and maintained in the same solution for a long time, liposomeses can be a subject of release of the drug due to the possible leakage of some species (for example, buffer components) from the inner liposome compartment.[12] Depending on the circum-

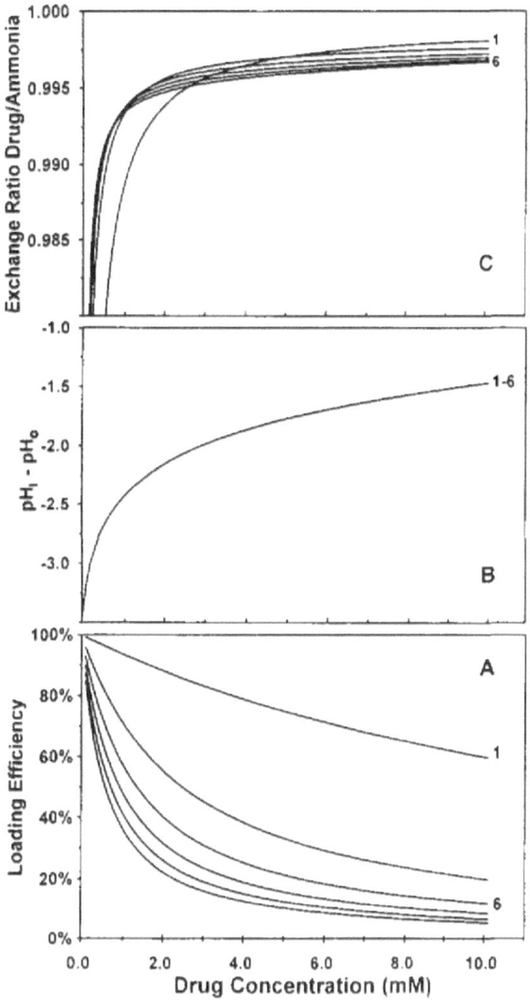

Figure 11 Inner compartment: 150 m*M* ammonium sulfate with pK_a (NH_4^+) = 9.3, and pK_{a2} (H_2SO_4) = 1.92, respectively; outer compartment: protonated drug with pK_a value of 9.0; the K_v values of the corresponding curves are taken to be 20 (no. 1), 120, 220, 320, 420, and 520 (no. 6), respectively; C = 0 µF.

stances, even influx of the drug can occur due to irruption of some species into the liposome interior. As a consequence of such a process(es), first, a renewed reestablishment of all protolytic equilibria including that of the loaded drug and subsequently a redistribution of the permeable species occur. To simulate these processes on a real-time scale, a much faster reequilibration of the drug molecules in comparison with leaked species will be assumed. Expressions will again be modified in a way very similar to that in Equations 40 and 41. In both terms, all charged species, except those taking part 'in leakage out from/loading into liposomes,' will be commonly involved under the functions designated $F_I^{+}(x, y)$ and $F_O^{*}(x, y)$. Before leakage of one or more components starts, the state in both compartments can be described as follows:

$$F_I^{+}(x, y) + \sum_r c_{rI} \frac{g_{rI}(y)}{f_{rI}(y)} = 0 \quad \text{and} \quad F_O^{+}(x, y) + \sum_s c_{sO} \frac{g_{sO}(x)}{f_{sO}(x)} = 0 \tag{54}$$

and for the state after one or more species leak from the interior or erupt in it, the reestablished equilibria are given by an equation-couple given below. The expression by which the equilibrium state in the inner solution is given, has the following form and similar is valid also for the outer compartment:

$$F_I^{+*}(x, y) + \sum_r c_{rI}(1 - \gamma_r) \frac{g_{rI}(x)}{f_{rI}(x)} + \left(\frac{V_O}{V_I}\right) \sum_s c_{sO} \delta_s \frac{g_{sI}(x)}{f_{sI}(x)} = 0 \tag{55}$$

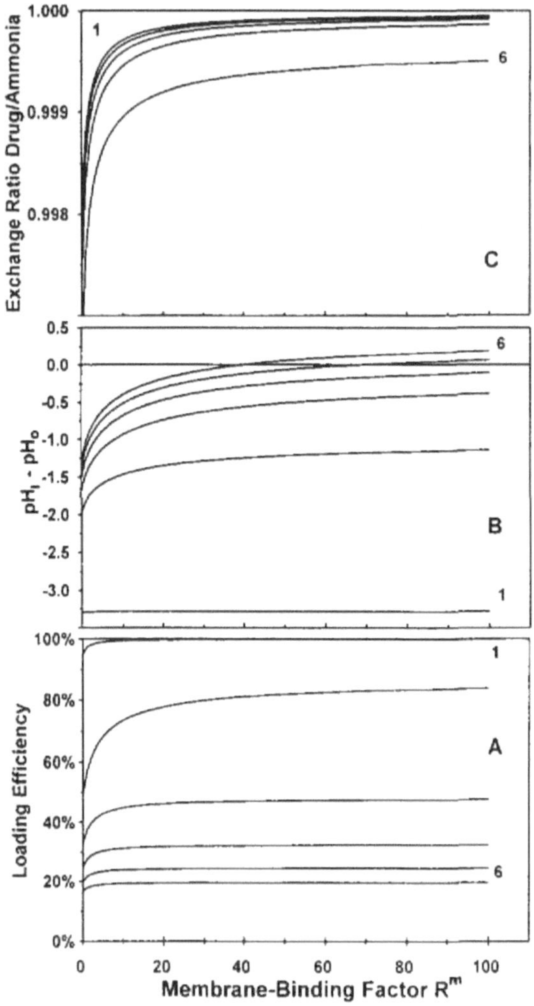

Figure 12 Inner compartment: 100 m*M* ammonium sulfate with pK_a $(NH_4^+) = 9.3$, and pK_{a2} $(H_2SO_4) = 1.92$, respectively; outer compartment: protonated drug with pK_a value of 9.0; the drug concentrations of the corresponding curves are taken to be 0.1 (no. 1), 2.1, 4.1, 6.1, 8.1, and 10.1 m*M* (no. 6), respectively; $K_v = 100$; $C = 0$ μF.

$$F_O^{+*}(x, y) + \sum_s c_{sO}(1 - \delta_s)\frac{g_{sO}(y)}{f_{sO}(y)} + \left(\frac{V_I}{V_O}\right)\sum_r c_{rI}\gamma_r \frac{g_{rI}(y)}{f_{rI}(y)} = 0 \tag{56}$$

From Equations 56 and 57, both representing a liposome system in the state of equilibrium after a leakage/eruption of slowly permeable species, for each set of values γ_s/δ_r, x and y values can be calculated as well as percentage of the loaded drug and other parameters of possible interest.

In the concrete example discussed below we will simulate the leakage of the inner buffer (in its neutral form) out from the liposomes which results in diminishing of loading efficiency as well as ΔpH values. The adequate equations are those shown in Equations 48 and 49, but slightly modified in terms including c_2, which now describe possible release of the inner buffer:

$$c_2 \frac{-K_{21}/X}{1 + K_{21}/x}\left(1 - \gamma_2\right) \qquad K_v^{-1} c_2 \frac{-K_{21}/X}{1 + K_{21}/y}\gamma_2 \tag{57}$$

For this particular case and concentrations higher than 1 m*M*, from Figure 15A it is straightforward to derive an equation, with which one may almost precisely calculate the diminished loading efficacy as a function of the portion of the buffer released: $\alpha = \alpha_0(1 - 2\gamma)$ (here α is diminished percentage of loading and γ percentage of the released buffer). The fact that at 50%-release of the buffer the portion of the loaded drug for all initial drug concentrations is close to zero is difficult to explain. At relative higher drug

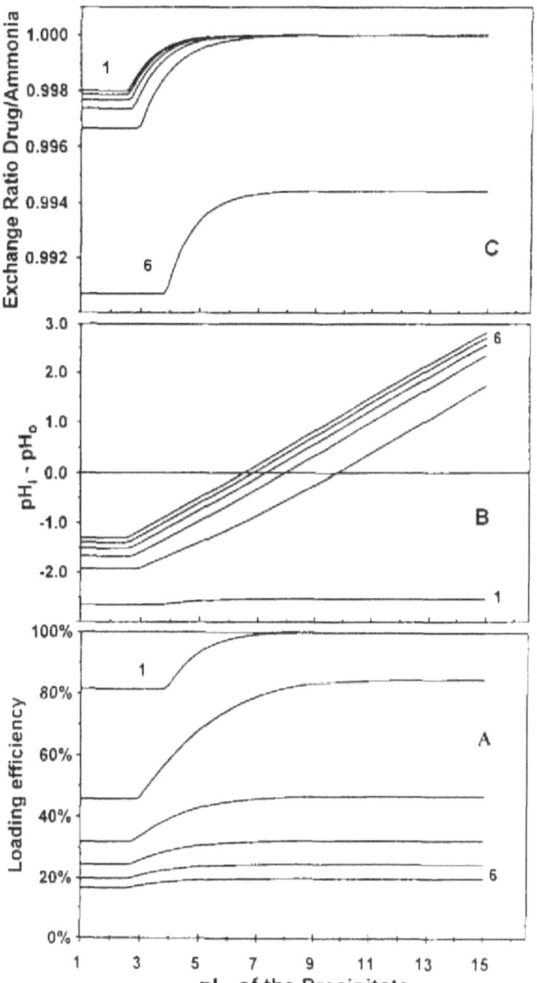

Figure 13 Inner compartment: 100 m*M* ammonium sulfate with pK_a (NH_4^+) = 9.3, and pK_{a2} (H_2SO_4) = 1.9, respectively; outer compartment: protonated drug with pK_a value of 9.0, and corresponding drug concentrations are taken to be 0.45 (no. 1), 2.40, 4.35, 6.25, 8.15, and 10.0 m*M* (no. 6), respectively; K_v = 100; C = 0 μF. The precipitate '$(drugH)_2SO_4$' is formed inside the liposomes.

concentrations (Figure 15B, curves 2–6) the inner compartment becomes even more basic in comparison with outer aqueous solution.

E. INDUCED TRANSMEMBRANE pH GRADIENT LOADING

During preparation of the liposomes the inner compartment solution will be assumed to be contaminated with an M^{n+}-salt where M^{n+} are usually alkali or earth-alkali metal cations or even well-permeable lipophilic cations. If, in a suspension of liposomes, the pH values in both compartments are equal, the pH gradient across the membrane is no longer present. After addition of an ionophore (and protonophore) into the exterior, an antiport stoichiometric exchange of M^{n+}-ions and protons may be assumed to occur. For theoretical treatment of such a bidirectional cation/proton exchange, Equations 40 and 41 will be slightly modified, similar to the previous case. Before loading, the charge concentration function of all species will be expressed as follows; for both compartments the following expressions are valid:

$$F_I^*(x, y) + \sum_l z_l c_{lO} = 0 \quad \text{and} \quad F_O^*(x, y) + \sum_n z_n c_{nO} = 0 \tag{58}$$

Because the amount of counter cations M^{n+}/H^+, unloaded/loaded from/into the inner/outer solution has to be equal, one can write:

$$V_I c_{lI} \lambda_l = V_O c_{lO} \quad \text{and} \quad V_O c_{nO} v_n = V_I c_{nI} \tag{59}$$

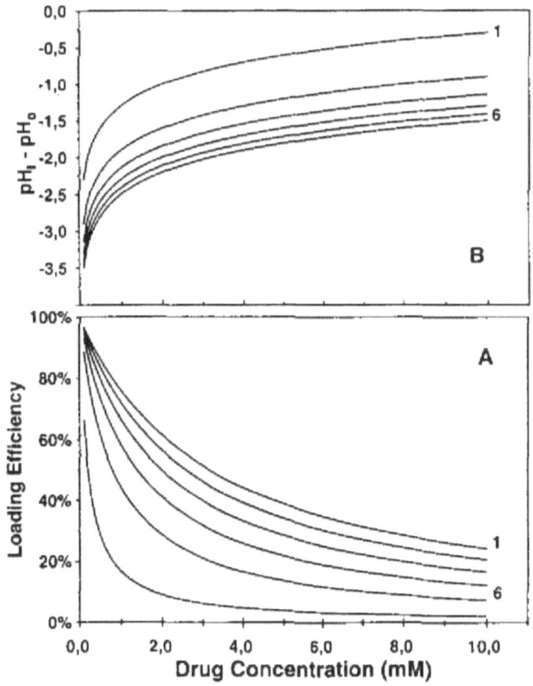

Figure 14 Inner compartment: 100 m*M* buffer with $pK_a = 5.0$ and $[B_I^-]/[HB_I]$ ratio 1:1, and NH_4HCO_3 with pK_a (NH_4^+) = 9.3, and pK_{a1} ('H_2CO_3') = 6.4, and pK_{a2} ('H_2CO_3') = 10.2; outer compartment: 100 m*M* buffer with $pK_a = 8.0$ and $[B_o^-]/[HB_o]$ ratio 1:1, 10 m*M* protonated drug with pK_a of 8.0; the corresponding values of NH_4HCO_3 concentrations are 1.0 (curve no. 1), 41, 81, 121, 161, and 201 m*M* (curve no. 6), respectively; $K_v = 50$; $C = 0$ μF.

Hence, assuming a stoichiometric antiport of protons and M^{n+} ions, acidity (or basicity) in both compartments is now changed and can be calculated from the following equations:

$$F_I^{*+}(x,y) + \sum_l z_l c_{lI}(1-\lambda_l) + \left(\frac{V_O}{V_I}\right)\sum_n z_n c_n v_{nO} = 0 \tag{60}$$

$$F_O^{*+}(x,y) + \sum_n z_n c_{nO}(1-v_n) + \left(\frac{V_I}{V_O}\right)\sum_l z_l c_l \lambda_{lI} = 0 \tag{61}$$

Now, through x and y values, calculated from the equations above, it is possible to get loading percentages and other values as a function of the portion of the permeated cations.

Similarly, as in the case of 'leakage of liposomes' the results of calculations will be based on Equations 48 and 49, slightly modified on two terms, each of them implying the counter-crossing of M^+/H^+ through the bilayer:

$$ac_2(1-\lambda_2) \quad \text{and} \quad ac_2K_v^{-1}\lambda 2 \tag{62}$$

The results of simulation are present in Figures 16A and 16B.

IV. CONCLUSION

The simulation model presented in this chapter combines a general approach of an overall treatment of the acid–base–salt equilibria in the aqueous solution with the interaction of two liposomal aqueous subsystems, implying permeation and equilibration of permeated neutral species on both sides of the liposome membrane. Simultaneously, it considers most of the possible accompanying physicochemical phenomena such as precipitate formation, membrane-binding, charging of the membrane, leakage of the liposomes, and induced pH gradient. The mathematical description of the liposome system usually results in two or three nonlinear equations which can be numerically solved by use of the Newton–Raphson method. A broad span of different parameters combined with a wide spectrum of the potentially

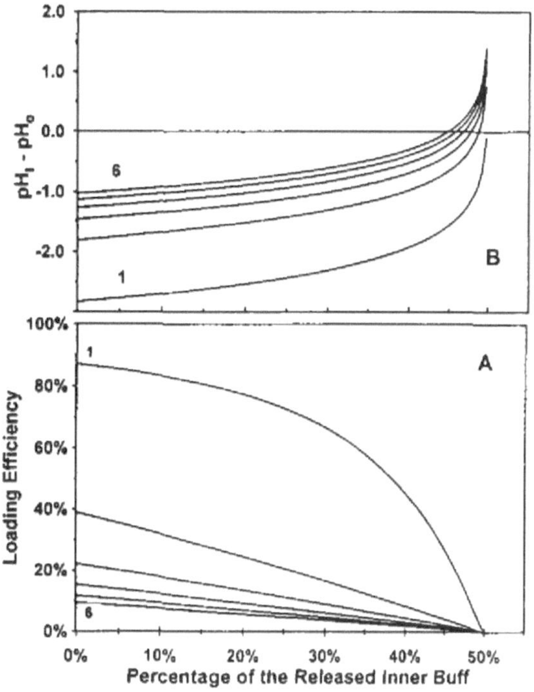

Figure 15 Inner compartment: 100 buffer with pK_a = 4.0 and [B_i^-]/[HB_i] ratio 1:1; outer compartment: 10 m*M* buffer with pK_a = 7.0 and [B_o^-]/[HB_o] ratio 1:1; the protonated drug concentrations of the corresponding curves are 0.1 (no. 1), 1.1, 2.1, 3.1, 4.1, and 5.1 m*M* (no. 6), respectively; pK_a of the protonated drug is taken to be 9.0; K_v = 100; *C* = 0 μF.

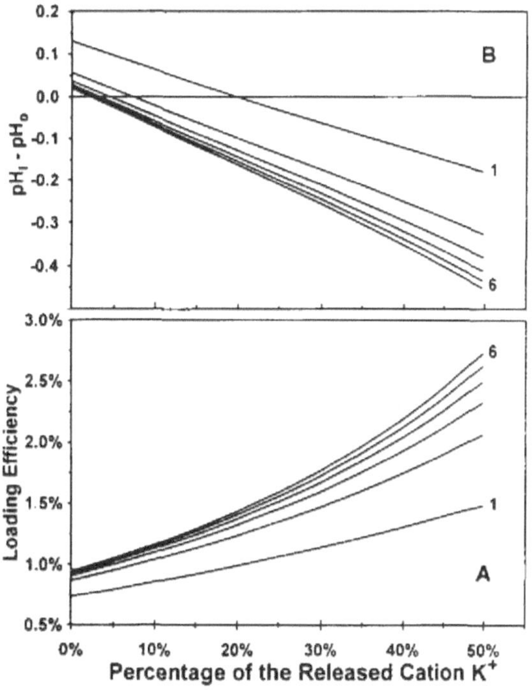

Figure 16 Inner compartment: buffer with pK_a = 4.0 and [B_i^-]/[HB_i] ratio 1:1, the corresponding inner buffer concentrations are 10 (curve no. 1), 18, 46, 64, 82, and 100 m*M* (curve no. 6), respectively. Outer compartment: 10 m*M* buffer with pK_a = 7.0 and [B_o^-]/[HB_o] ratio 1:1, and 1.0 m*M* protonated drug with pK_a of 9.0. K_v = 100; *C* = 0 μF.

possible situations in loading liposomes can be simulated. The obtained results, calculated with an appropriate computer software program can be either evaluated or compared with those of the experiments.[19] The calculated numerical values — most favorably if presented in the form of diagrams — could give us insight into the circumstances at equilibrium, and thereby make it possible to find out or to sort out such loading-defined parameter ranges that if an experiment is accomplished with them, could give better results.

V. GLOSSARY OF SYMBOLS, INDEXES, AND LABELS

A_k — Factor which 'allows' the neutral species designated kI to load into the inner solution (= 1) or not (= 0)

a_{kj} — Reaction coefficient of one of the sister-species kj which forms the precipitate after permeation into the inner compartment

B_m — Factor which 'allows' the species mI to unload from the liposomes (= 1) or not (= 0)

b_{mi} — Reaction coefficient of one of the inner sister-species mi which forms precipitate with the species designated kj

c_i — Total concentration of all sister-species of the corresponding starting-point species ik

c_{kI}, c_{mO} — Total concentrations of the species k, m in both compartments after crossing the membrane

c_{kO}, c_{mI} — Concentration of the starting-point species designated k/m in the outer/inner solution before loading/unloading

$c_{kj}^{m_I}, c_{kj}^{m_O}$ — Concentration of the charged species bound onto the inner/outer membrane surface layer

$c_{kj}^{m_I *}, c_{kj}^{m_O *}$ — Recalculated concentration of the charged species bound onto the inner/outer membrane surface layer

$c_{kj}^{s_I *}, c_{kj}^{s_O *}$ — Concentration of the charged species remaining in the inner/outer solution after binding onto membrane

$c_k^{s_I}, c_m^{s_I}$ — Total concentration of the drug sister-species kj and mi remaining in the inner/solution after precipitation occurs

$c_k^{\dot{s}_I}, c_k^{\dot{s}_O}$ — Concentration of all sister-species which remain in both solutions after membrane binding of species kj onto both sides of the bilayer

c_{kI}^{I} — Total concentration of the membrane-bound species in the inner compartment

$k{*}O$, $m{*}I$ — Designation of the precipitate-forming starting-point species k, m

$c_{kj}^{p_I}$ — Concentration of the precipitated species kj in the inner compartment

$c_{kI}^{eq,t}, c_{kO}^{eq,t}$ — Total concentration of the species k in the inner/outer solution after the equilibrium is reached (before loading, species designated k is only in the outer solution)

c_l, c_m — Start concentration of lth/mth 'inert' species (does not react with water) in the inner/outer compartment

c_{lO}, c_{mI} — Concentration of lth/mth 'inert' species in the outer/inner compartment after crossing the membrane

$c_{mI}^{eq,t}, c_{mO}^{eq,t}$ — Total concentration of the species m in the outer/inner solution after the equilibrium is reached (before loading, m is only in the outer solution)

c_{rI}, c_{sO} — Initial concentrations of the species r/s which leaks through bilayer

C — Specific membrane capacity

d — Thickness of the membrane surface layer

f_i ([H$^+$]) — H$^+$-dependent concentration function

$f_{kI}, f_{kO}, f_{mI}, f_{mO}$ — Concentration function of the species k/m, in both compartments after loading/unloading occurs

$f_{rI}, f_{rO}, f_{sI}, f_{sO}$ — Concentration function of the species r/s in both compartments after leakage in either direction occurs

F — Faraday constant

F^*, F_c, F_m, F_p — Factors (F^* 'allows' reduction of the proton concentration in either compartment (= 1) or not (= 0); then $F_c = 1$, if $pH_O < 7$, and $F_c = 0$, if $pH_O > 7$; F_m 'allows' binding of protonated drug onto both sides of the bilayer (= 1) or not (= 0); F_p 'allows' formation of precipitate inside the liposomes (= 1) or not (= 0))

$g_{kI}, g_{kO}, g_{mI}, g_{mO}$ — Charge function of the species k/m in both compartments after loading/unloading occurs

$g_{rI}, g_{rO}, g_{sI}, g_{sO}$ — Charge function of the species r/s in both compartments after leakage occurs

j_{kj} — Running number of the sister-species kj in the starting-point series k (Ex.: if NH_4^+ is bound onto membrane, in the series NH_4^+/NH_3 there is NH_4^+ occupying the first place in this two-member series — therefore, $j_{kj} = 1$)

K_v — Ratio between outer and inner compartment volume

K_v^* — Reduced ratio between outer and inner compartment volume

k_v — $= K_v^{-1}$

K_m	Ratio of the binding surface volumes on both sides of the membrane
L	Number of starting-point species placed in a given solution (Ex.: $(NH_4)_2HPO_4$; HPO_4^{2-} is starting-point species $i3$, giving sister-species H_3PO_4 ($i1, j = 1$), $H_2PO_4^-$ ($i2, j = 2$), HPO_4^{2-} ($i3, j = 3$), PO_4^{3-} ($i4, j = 4$); z_{ij} values are: 0, −1, −2, and −3; $i4$= 1, 2, 3, and 4, respectively)
l_k, l_m	Running number of the neutral species $S^0_{kl_k}/S^0_{ml_m}$ in a given ordered series (Ex.: in the series NH_4^+/NH_3, NH_3 is on the second place — therefore, for this two-member series, $l_k = 2$)
n_{kI}, n_{kO}, n_{mI}, n_{mO}	Total amount of species k/m in both compartments after equilibration
r	Inner liposome radius
R	Gas constant
R^m_{kjI}, R^m_{kjO}, R^m_{miI}, R^m_{miO}	Ratio between the concentration of the ions kj/mi, bound onto inner/outer surface membrane layer and concentration of all sister-species remained in both solutions
$[S^0_{kl_k}]^{eq}_I$, $[S^0_{kl_k}]^{eq}_O$	Inside/outside equilibrium concentration of the neutral species designated l_k after loading occurs
T	Absolute temperature
v_l	Volume of the liposome vesicle
V_I, V_O	Inner and outer compartment volume of the liposome system
V_I^*, V_O^*	Inner and outer 'reduced' and recalculated volume of the liposome system
V_{mI}, V_{mO}	Volume of the inner/outer membrane equals binding layer
x, y	H^+ equilibria concentration in both compartments
x^{st}, y^{st}	Initial H^+ concentration in both compartments before loading
z_{kj}, z_{mi}	Charge of the species kj/mi in the outer/inner solution
z_l, z_n	Charge of the lth/nth 'inert' species
α_k	Percentage of the species k which is loaded into the liposomes
β_m	Percentage of the unloaded species m
γ_s, δ_r	Percentage of the species s/r which leaks into outer/inner compartment
λ_l, ν_n	Percentage of the 'inert' influxed/effluxed species l/n
$\Delta\mu_H$	Electrochemical transmembrane H^+ gradient
$\Delta\Psi$	Transmembrane potential

REFERENCES

1. **Johnson, R. G. and Scarpa, A.,** Internal pH of isolated chromaffin vesicles. Relation to ammonium ion uptake, *J. Biol. Chem.*, 251, 2189, 1976.
2. **Crofts, A. R.,** Amine uncoupling of energy transfer in chloroplasts, *J. Biol. Chem.*, 242, 3352, 1967.
3. **Nichols, J. W. and Deamer, D. W.,** Catecholamine uptake and concentration by liposomes maintaining pH gradients, *Biochim. Biophys. Acta*, 455, 269, 1976.
4. **Veiro, J. A. and Cullis, P. R.,** A novel method for the efficient entrapment of calcium in large unilamellar phospholipid vesicles, *Biochim. Biophys. Acta*, 1025, 109, 1990.
5. **Lasic, D. D.,** *Liposomes: From Physics to Applications*, Elsevier, Amsterdam, 1993, chap. 11.
6. **Deamer, D. W., Prince, R. C., and Crofts, A. R.,** The response of fluorescent amines to pH gradients across liposome membranes, *Biochim. Biophys. Acta*, 274, 323, 1972.
7. **Lee, H. C. and Forte, J. G.,** A study of H⁺ transport in gastric microsomal vesicles using fluorescent probes, *Biochim. Biophys. Acta*, 508, 339, 1978.
8. **Mayer, L. D., Tai, L. C. L., Bally, M. B., Mitilenes, G. N., Ginsberg, R. S., and Cullis, P. R.,** Characterization of liposomal systems containing doxorubicin entrapped in response to pH gradients, *Biochim. Biophys. Acta*, 1025, 143, 1990.
9. **Madden, T. D., Harrigan, P. R., Tai, L. C. L., Bally, M. B., Mayer, L. D., Redelmeier, T. E., Loughrey, H. C., Tilcock, C. P. S., Reinish, L. W., and Cullis, P. R.,** The accumulation of drugs within large unilamellar vesicles exhibiting a proton gradient: a survey, *Chem. Phys. Lipids*, 53, 37, 1990.
10. **Lasic, D. D., Frederik, P. M., Stuart, M. C. A., Barenholz, Y., and McIntosh, T. J.,** Gelation of liposome interior. A novel method for drug encapsulation, *FEBS Lett.*, 312, 255, 1992.
11. **Haran, G., Cohen, R., Bar, L. K., and Barenholz, Y.,** Transmembrane ammonium sulfate gradients in liposomes produce efficient and stable entrapment of amphipathic weak bases, *Biochim. Biophys. Acta.*, 1151, 201, 1993.
12. **Harrigan, P. R., Hope, M. J., Redelmeier, T. E., and Cullis, P. R.,** Determination of transmembrane pH gradients and membrane potentials in liposomes, *Biophys. J.*, 63, 1336, 1992.
13. **Čeh, B. and Lasic, D. D.,** A rigorous theory of drug loading into liposomes, *Langmuir*, 1995, in press.

14. **Čeh, B. and Lasic, D. D.,** A rigorous theory of remote loading of drugs into liposomes. Transmembrane potential induced pH gradient caused loading and leakage of the liposomes, *J. Colloid Interface Sci.*, submitted.
15. **Sharon, J. A. and Cafiso, D. S.,** Voltage-dependent conductance for alamethicin in phospholipid vesicles. A test for the mechanism of gating, *Biophys. J.*, 60, 380, 1991.
16. **Lasic, D. D.,** *Liposomes: From Physics to Applications*, Elsevier, Amsterdam, 1993, chap. 2.
17. **Cafiso, D. S. and Hubbell, W. L.,** Estimation of transmembrane pH gradients from phase equilibria of spin labelled amines, *Biochemistry*, 17, 3871, 1978.
18. **Outhwaite, C. W., Molero, M., and Bhuiyan, L. B.,** Primitive model electrolytes in the modified Poisson–Boltzmann theory, *J. Chem. Soc., Faraday Trans.*, 89, 1315, 1993.
19. **Lasic, D. D., Čeh, B., Guo, L., Frederik, P. M., Stuart, M. C. A., and Barenholz, Y.,** Transmembrane gradient induced phase transitions in vesicles: lesson for drug delivery, *Biophys. Biochim. Acta*, 1995, in press.

Chapter 6

Sterilization of Liposomes

Nicolaas J. Zuidam, Herre Talsma, and Daan J. A. Crommelin

CONTENTS

I. INTRODUCTION

A. AIM AND LAYOUT

One of the problems during the production of liposome-containing pharmaceuticals, for example, for parenteral use, is to achieve sterility and apyrogenicity. Contaminated liposome dispersions might be toxic for the patient, especially upon parenteral use, and could also affect the quality of the liposome dispersion upon storage. Production of liposomes under aseptic conditions is possible,[1,2] but complex and expensive. Thus, there is a need for a widely applicable sterilization method. The aim of this chapter is to assess the possibilities and limitations of several techniques which can be applied to achieve sterility. At the end we will discuss the sterilization method of choice for liposomes under different circumstances and monitor future developments.

B. WHAT IS STERILE?

Sterile pharmaceutical products should comply with the tests as described, for instance, in the monographs of the European Pharmacopoeia[3] or the U.S. Pharmacopeia (USP).[4] At least three definitions for the term 'sterile' are in use; these are:

1. free from living microorganisms
2. a product that meets the requirements as stipulated in a special monograph in a *Pharmacopeia*
3. if the chance of finding a unit that is contaminated with living microorganisms is less than 1 in 10^6 sterilized units of the final product.[3,4]

The first definition stipulates what is desired: a product free of living microorganisms. This definition is not very useful in practice, since it cannot be experimentally assessed. (What is free?) The second definition is generally used in the case of a dispute concerning sterility. However, it should be emphasized that in general even products that are severely contaminated might easily pass the tests described in the *Pharmacopeia*, unless large numbers of units are tested. This is illustrated in Table 1.[5] Absolute sterility cannot be practically demonstrated without complete destruction of all samples. Therefore, in practice sterility can not be guaranteed by testing of the final product alone, but it is assured by following strictly defined production procedures and by process control as well. Process validation (Good Manufacturing Practice, GMP) in combination with strict control over quality of raw material offers a better guarantee than the (destructive) end testing of the product. The standard approach

0-8493-4012-8/96/$0.00+$.50
© 1996 by CRC Press, Inc.

Table 1 The Relationship Between the Degree of Contamination and the Number of Samples Taken at Random from a Batch Containing More than 10,000 Units[5]

	Probability of Accepting a Product Which is Still Contaminated[a]		
Number of Samples	**0.1% Contaminated**	**1.0% Contaminated**	**5.0% Contaminated**
1	99.9	99.0	95.0
5	99.5	95.1	77.4
10	99.0	90.4	59.9
20	98.0	81.8	35.8
50	95.1	60.5	7.7
100	90.5	36.6	0.6

[a] The probability that all units are sterile is equal to $(1-p)^n \times 100\%$, where p is the fraction of units contaminated ($0 < p < 1$) and n is the number of units tested ($n > 10{,}000$ units). The calculation is based on the assumptions that the units were taken out randomly and that the unit is rejected upon finding one contaminated unit.

is to combine a low degree of contamination at the start of the sterilization process with an effective, fully-validated sterilization step. This should result in a degree of contamination of the production units that complies with the third definition.

A low contamination degree is defined as 100 or less colony forming units (cfu)/ml. An effective sterilization process gives at least 10^{12}-fold reduction of the initial degree of contamination of a process-related test organism. Microbiological tests are, in general, less reliable and accurate than analytical determinations. This is partly due to the large variety of microorganisms that might contaminate the product and that cannot all be detected upon 'sterility' testing. Therefore, sterilization methods are validated by using microorganisms known to be highly resistant to that particular sterilization method (worst-case assumption). For example, for autoclaving, thermoresistant spores of *Bacillus stearothermophilus* are used to test the effectivity of the sterilization process.[3,4]

II. STERILIZATION PROCESSES

A. PRECONDITIONS OF STERILIZATION METHODS

Methods for sterilizing liposomes should be destructive for microorganisms and should not affect the liposome dispersion. However, in most sterilization methods one has to compromise between inactivation of the contaminating microorganisms and degradation of the pharmaceutical product. Ideally, the physical and chemical characteristics of the liposome dispersions should remain unchanged in the sterilization process. To assess whether changes occur, liposome characteristics such as particle size, chemical composition, retention of encapsulated drug, and bilayer melting properties (if present) should be evaluated before and after the sterilization process. An evaluation of all liposome characteristics should clarify the suitability of a certain sterilization technique. It is hard to set limits for acceptance for each parameter. In the decision-making process to accept or reject a sterilization method, one should consider factors such as route of administration, influence of the changes on the pharmocokinetics of an encapsulated drug, possible alternatives to prepare sterile liposomes, toxicity of the degradation products (if present), etc. As a rule of thumb, pharmaceutical companies use 10% degradation as the acceptance limit in the case of an excipient if the degradation products are nontoxic. The value of 10% indicates that upon sterilization the percentage degradation should not exceed the 2 to 5% range under the conditions mentioned above, so that a margin for further degradation upon storage is still available.

B. STERILIZATION METHODS

Several sterilization methods for liposomes described in the literature will be discussed below: autoclaving, high pressure sterilization, ethylene oxide treatment, γ-irradiation, and filtration. Especially, the chemical and physical consequences for liposomes upon sterilization will be discussed here. In the last part of this chapter, we will discuss the method of choice for several liposome dispersions.

1. Autoclaving

Autoclaving is a sterilization process based on an established time–temperature combination and is carried out in a chamber (autoclave) filled with saturated steam under pressure, e.g., 134°C for a minimum holding time of 3 min or, as used most of the time, 121°C for a minimum holding time of 15 min. Several articles have appeared over the years demonstrating that autoclaving is a suitable sterilization technique for liposomes under proper conditions.[6–8] Upon a 10- to 15-min heat treatment at 121°C of a gentamicin (an antibiotic) containing egg phosphatidylcholine (EPC)-liposome dispersion, Cherian et al.[6] found an overall drug leakage of about 5%. They claimed that a slow cool-down step is the preferred way to prevent drug leakage. However, they did not provide convincing experimental support for this claim; only an indication is given in their patent. The heat treatment caused only a small increase of 2 to 5% in the content of lysophosphatidylcholine (LPC, one of the hydrolysis products of phosphatidylcholine). No information was provided about drug stability, aggregation or fusion of liposomes, or oxidation of the phospholipids. The efficacy of the heat sterilization was tested by Cherian et al.[6] by seeding liposomes (without gentamicin) with spores of *B. stearothermophilus*. They claimed that the microbiological activity had been completely eliminated upon heating at 121°C for about 15 min.[6]

More extensive studies were published by Kikuchi et al.[7] and by Zuidam et al.[8] Kikuchi et al.[7] heat-treated several types of liposomes at 121°C for 20 min. They observed aggregation of uncharged and charged liposomes in 0.9% sodium chloride solutions. In the presence of 5% glucose or 2% propylene glycol, no aggregation was observed. They claimed that the absence of electrolytes was the reason for the stabilizing effect of the glucose and propylene glycol. However, in the study of Zuidam et al.[8] no aggregation of charged, 0.2 μm diameter liposomes in a buffer-salt medium was observed. The reason for this discrepancy is unclear. Upon autoclaving Kikuchi et al.[7] found signs for oxidation in liposomes composed of EPC/egg phosphatidylglycerol (EPG) with a high peroxide value; the liposome dispersion turned yellowish. This could be prevented by using EPC with a low peroxide value, phospholipids with saturated bonds, or by the use of nitrogen. Zuidam et al.[8] also found no indication for an increase in the degree of oxidation upon autoclaving liposomes composed of EPC/EPG 10/1 with low peroxide values; no increase in oxidation index and no formation of thiobarbituric acid (TBA)-reactive products were found. The only chemical damage to the phospholipids by the autoclaving process observed by them was an increase in the percentage of hydrolyzed phospholipids after autoclaving. At pH 7.4 this increase was a minor one (<1%). At pH 4.0 the increase was 3 to 7%.[8] The somewhat larger increase in percentage hydrolysis of the phospholipids in liposomes at pH 4.0 is caused by the reported proton- and hydroxyl-catalyzed hydrolysis of phospholipids.[9,10] In a medium with constant pH hydrolysis of liposomal phospholipids follows Arrhenius kinetics.[11–13] It was demonstrated by Zuidam et al.[8] that the observed percentage of hydrolyzed liposomal phospholipids found after autoclaving at pH 4 and pH 7.4 corresponds well with the predicted percentages by using Arrhenius kinetics.

As reported by Zuidam et al.,[8] minor differences in the melting properties of dipalmitoylphosphatidylcholine/dipalmitoylphosphatidylglycerol (DPPC/DPPG) 10/1-liposomes were found by differential scanning calorimetry (DSC) upon autoclaving. At pH 7.4 a small decrease (20 ± 9%) in the enthalpy (ΔH) of the pretransition was found and a shoulder appeared at the left-hand side of the main transition (see Figure 1B) after the heat treatment. At pH 4.0 the differences after autoclaving were more pronounced; the pretransition was not detectable anymore and a minor change of 16 ± 11% in the ΔH of the main-transition was found (see Figure 1A). The exact nature of this phenomenon is still under investigation. The appearance of the shoulders at different sides of the main transition might be explained by differences in degree of ionization of the hydrolysis product palmitic acid (noncharged at pH 4.0 and partly negatively charged at pH 7.4).

An additional problem may be encountered when autoclaving a liposome dispersion with an encapsulated drug. Calcein, a highly water soluble compound, and doxorubicin (DXR), with an amphiphilic character leak out of the liposomes, but the lipophilic agents N-trifluoroacetyldoxorubicin-14-valerate (AD-32) and α-tocopherol remained encapsulated after autoclaving (see Table 2).[8] Of course, the chemical stability of the drug upon autoclaving should be assessed. Both DXR and AD-32 showed substantial degradation after autoclaving (see Table 2).[8]

From the previously mentioned studies it can be concluded that under neutral pH conditions, liposomes without encapsulated agents or with heat-stable, bilayer interacting (lipophilic) agents can be sterilized by autoclaving. Both saturated and unsaturated phospholipids (with low peroxide values) showed similar behavior as hydrolysis, oxidation, and leakage were negligible. For other pH conditions hydrolysis might be a problem. Autoclaving of liposomes with water-soluble, nonbilayer interacting agents is an option: (1) if the agent is encapsulated after the autoclaving process (for instance with 'remote loading' techniques under aseptic conditions), (2) if the free drug is removed after autoclaving (under aseptic conditions), or (3) if the presence of nonencapsulated material is not interfering with the desired therapeutic effect.

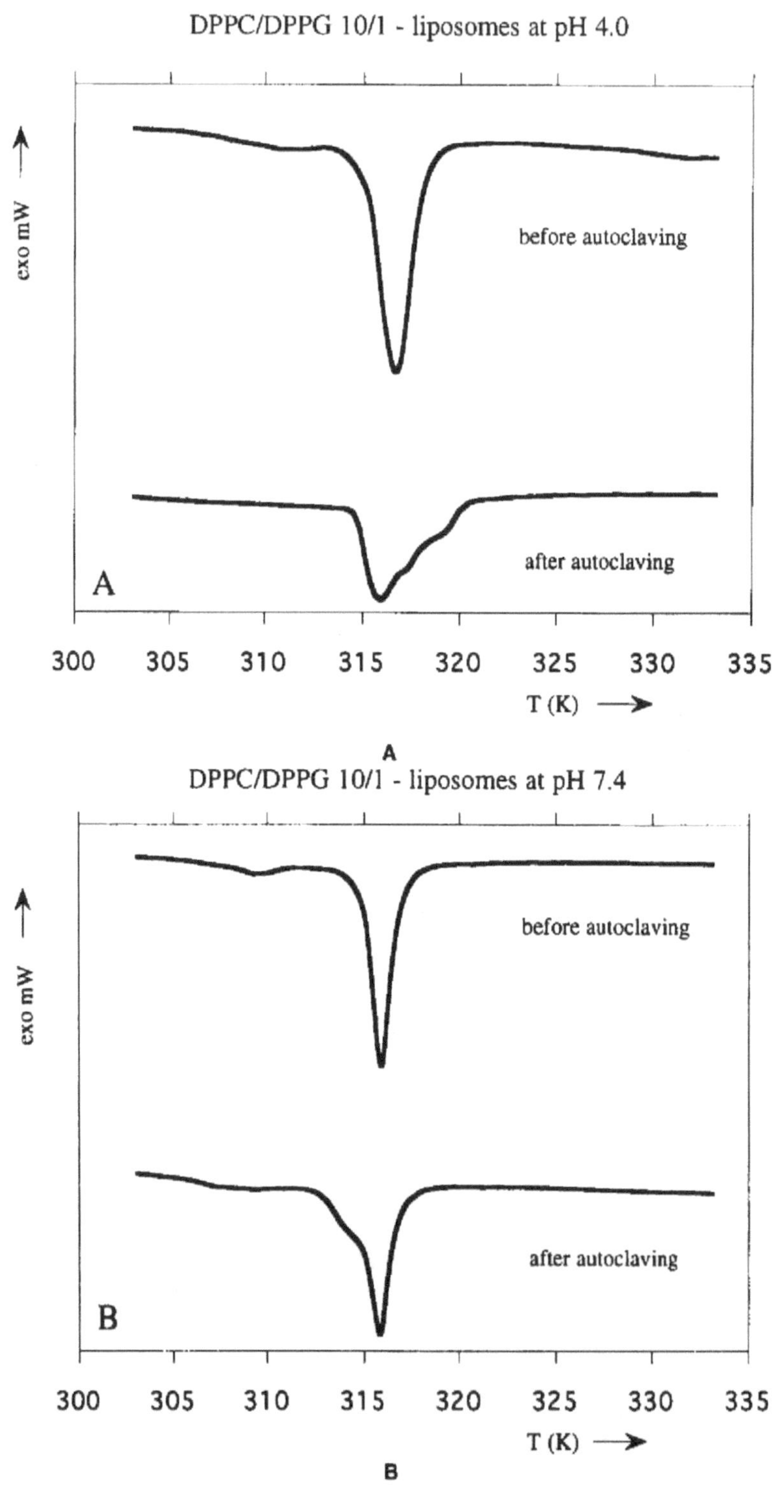

Figure 1 Typical examples of calorimetric scans of 0.2 μm DPPC/DPPG 10/1-liposomes in pH 4.0 (A) and pH 7.4 (B) before (top) and after (bottom) autoclaving. (From Zuidam, N. J., Lee, S. S. L., and Crommelin, D. J. A., *Pharm. Res.*, 10(11), 1591, 1993. With permission.)

2. High-Pressure Sterilization

Sterilization of liposomes in a 10-m*M* phosphate-saline buffer was performed by Mentrup et al.[14] by using protocols based on a selected time–temperature–high-pressure combination, e.g., 5 h at 60°C and 2500 bar or 21 h at 40°C and 2000 bar. Mentrup et al.[14] showed in their article that this treatment could effectively

Table 2 The Leakage and Degradation Percentage of Encapsulated Agents After Autoclaving[a]

Lipid Composition	pH	Encapsulated Model Compound	Leakage[b] (%)	Degradation (%)
DPPC/DPPG/CHOL 10/1/4	4.0	DXR	20 ± 5	26 ± 8
EPC/EPG 10/1	4.0	AD-32	—[c,d]	39 ± 4
DPPC/DPPG/CHOL 10/1/4	7.4	calcein	44 ± 4	—[d]
DPPC/DPPG/CHOL 10/1/4	7.4	α-tocopherol	—[d]	—[d]

[a] All experiments were done in triplicate.

[b] For DXR and calcein leakage was defined as % free compound/total compound. For the lipophilic compounds AD-32 and α-tocopherol the % disappearance was deduced by comparing the ratio mg or m*M* agent/mol P before and after autoclaving after 1 min or centrifugation at 2000 g in a table centrifuge.

[c] The degradation product of AD-32, doxorubicinone-14-valerate, remained also in the bilayer. A disappearance of 4 ± 4% was calculated.

[d] No significant degradation or disappearance was observed.

From Zuidam, N. J., Lee, S. S. L., and Crommelin, D. J. A., *Pharm. Res.*, 10(11), 1591, 1993. With permission.

reduce the degree of contamination of several types of bacteria, but the reduction factor for spores of *B. stearothermophilus* was only 10^3. The stability of liposomes under the chosen conditions depended both on the chosen time, temperature, and pressure conditions, and on the characteristics of the liposomes used. For instance, liposomes composed of hydrogenated soybean PC or dimyristoylphosphatidylcholine (DMPC) and cholesterol (in equal molar ratio) were stable in size and in retention of carboxyfluorescein (<5%) after 5 h at 60°C and 2500 bar. However, sterilization of liposomes composed of only hydrogenated soybean PC or composed of unsaturated soybean PC and cholesterol (in equal molar ratio) showed a pronounced change in vesicle size and high leakage of carboxyfluorescein (≤45%) under these conditions. Liposomes composed of unsaturated soybean PC and cholesterol (in equal molar ratio) could be efficiently treated for only 21 h at 40°C and 2000 bar. In their article, Mentrup et al.[14] did not provide information on degradation of encapsulated agent of phospholipids. Also, no experiments were done with bilayer-interacting encapsulated drug(s) or with charged liposomes to prevent size changes.

3. Ethylene Oxide

Ethylene oxide is a chemical agent which kills microorganisms by replacing labile hydrogen atoms with hydroxylethyl groups, thus blocking many reactive groups needed in essential metabolic reactions. Ethylene oxide is commonly used for sterilizing medical disposable devices. Ratz et al.[15] used ethylene oxide to sterilize freeze-dried liposomes. Liposomes composed of soybean PC and cholesterol (in an equal molar ratio) were prepared in a 50-m*M* phosphate buffer (pH 7.4) containing Isovist (a contrast medium). After removal of the free agent, trehalose was added to act as a cryoprotectant (4 g trehalose per 1 g lipid). The liposome dispersion was freeze-dried and treated with ethylene oxide for 2.5 h. After this treatment the liposome dispersions were claimed to be sterile. However, the sterilizing capacity of this method was not validated in this study as no challenge test was performed. Therefore, it was not checked if, for example, the penetrating of the gas into the freeze-dried cakes was efficient. As pointed out by Ratz et al.[15], ethylene oxide is carcinogenic and mutagenic (and also flammable and explosive). Residues of ethylene oxide and/or contaminants caused by ethylene oxide might stay behind after the sterilization treatment. Another drawback of the treatment of ethylene oxide is that the use of the method is limited to only (freeze-)dried liposome dispersions.

4. γ-Irradiation

γ-Irradiation becomes more and more accepted as a sterilization method of pharmaceuticals.[16,17] γ-Irradiation exposes substances to high-energy gamma photons, emitted from γ-irradiation sources such as cobalt 60 or cesium 137. The major concern in γ-irradiation sterilization is product damage (e.g., disruption of chemical bonds). Historically, a dose of 25 kGy (=2.5 Mrad) has generally been accepted as a suitable sterilization dose.[3,4,17] The choice of this dose was based on the radiation response of radiation resistant bacterial spores of the *B. pumilus*.[17]

Several publications appeared over the years demonstrating that γ–irradiation might be a useful sterilization technique for 'dry' phospholipids[18] and for liposomes.[19–22] Albertini and Rustichelli[23] have summarized the physical consequences for liposomes upon γ–irradition. Their results were mostly based on irradiation experiments with liposomes composed of saturated phospholipids. Upon γ–irradiation of these liposomes, changes were observed by NMR-, DSC-, and X-ray diffraction experiments (e.g., see Figure 2 for the effect of γ–irradiation on DPPC-liposomes as determined by DSC). However, chemical degradation of liposomal phospholipids after γ–irradiation was hardly monitored in an appropriate way at that time. γ–Irradiation may affect phospholipids both directly via rupture of the chemical bonds (in particular rupture of carbon–carbon bonds and hydrogen–carbon bonds) and indirectly via reactions of species induced by γ–irradiation (in particular radicals).[23] Recently, Tinsley and Maerker[24] chemically analyzed the effect of a 9.66 kGy irradiation dose on liposomes composed of several phospholipids in water at 0 to 4°C, under air (see Table 3). Of the phospholipids studied DPPG was the most labile upon γ–irradiation. About half of the disappearance of the phospholipids was accounted for by the formation of dipalmitoylphosphatidic acid (DPPA) and lysophospholipid (see Table 3). Tinsley and Maerker[25] also found traces of palmitoylphosphocholinepropanediol upon γ–irradiation of DPPC-liposomes (not quantitated by them). Degradation routes for this and other possible products were published by Handel and Nawar[26] and by Sevilla et al.[27] Both groups performed experiments on dry PE. We obtained similar results as those of Tinsley and Maerker. Upon γ–irradiation of liposomes composed of saturated phospholipids we found no influence of atmosphere (air/nitrogen) or effect of the presence of 10-m*M* phosphate buffer (pH 7.4) or 0.13 *M* sodium chloride. However, we found a pronounced effect of the concentration of liposomes and vesicle size.

Not only phospholipids but also cholesterol degrades upon γ–irradiation. Maerker and Jones[28] found several degradation products upon γ–irradiation of cholesterol containing liposomes with doses ≤4.83 kGy under air. However, the quantity of degraded cholesterol is not clear from their study.

In Russian literature it has been reported that the most effective way to protect drug solutions or liposome dispersions from irradiation damage is freezing.[29–31] This method is called cryoradiation. With cryoradiation the capacity to inactivate viruses or microorganisms upon exposure to a dose of 25 kGy is

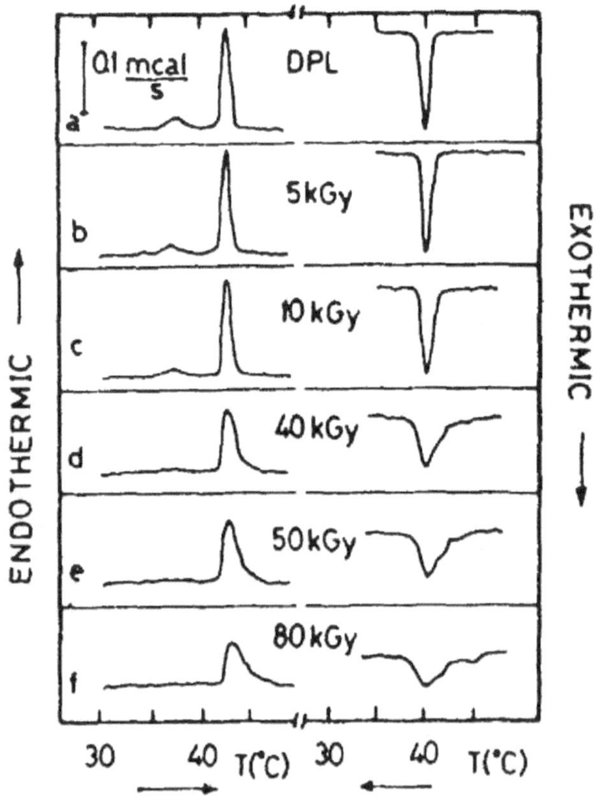

Figure 2 Typical examples of calorimetric scans of DPPC (=DPL)-liposomes in water, before (a) and after γ-irradiation at indicated doses (left heating curves, right cooling curves). (From Albertini, G. and Rustichelli, F., *Liposome Technology*, Gregoriadis, G., Ed., CRC Press, Boca Raton, FL, 1993, chap. 23. With permission.)

Table 3 The Effect of γ–Irradiation on Liposomal Phospholipids in Water After a Dose of 9.66 kGy at 0 to 4°C Under Air

Lipid Composition[a]	Degradation[b] (%)	mol % DPPA Formed After Irradiation	mol % Lysophospholipid Formed After Irradiation
DPPC	3	0.4	0.6
POPC	3	—	—
DOPC	7.5	—	—
DPPE	2	0.2	0.1
POPE	2	—	—
DOPE	–0.3	—	—
DPPG (Na^+-salt)	29	20	—
POPG (Na^+-salt)	23	—	—
DOPG (Na^+-salt)	21	—	—
DPPS	3	—	—
DPPS (Na^+-salt)	11	—	—
POPG (Na^+-salt)	19	—	—
DOPS (Na^+-salt)	27	—	—

[a] POPC, 1-palmitoyl-2-oleoyl-phosphatidylcholine; DOPC, dioleoylphosphatidylcholine; DPPE, dipalmitoylphosphatidylethanolamine; POPE, 1-palmitoyl-2-oleoyl-phosphatidylethanolamine; DOPE, dioleoylphosphatidylethanolamine; POPG, 1-palmityol-2-oleoyl-phosphatidylglycerol; DOPG, dioleoylphosphatidylglycerol; DPPS, dipalmitoylphosphatidylserine; POPS, 1-palmitoyl-2-oleoyl-phosphatidylserine; DOPS, dioleoylphosphatidylserine.

[b] Experiments were done two to four times. The concentration of the nonsized liposomes was 5 m*M*.

Adjusted from Tinsley, P. W. and Maerker, G., *JAOCS*, 70, 187, 1993. With permission.

maintained. By freezing, the damaging effect by indirect action is suppressed. Most of the irradiation damage is not caused by direct 'hits' on the bilayer structure by the protons, but it is caused indirectly: by γ–irradiation generated reactive agents outside the bilayer. Kapanin et al.[29] found no increase in oxidation and no decrease in insulin content after γ–irradiation of liposomes containing insulin when the solution was frozen at –196°C while using 10% choline chloride as a cryoprotectant. No further information on the chemical and physical characteristics of the cryoradiated liposomes was provided (e.g., phospholipid content or leakage). This does not allow us to properly evaluate this method.

We also did some experiments to prevent irradiation damage of liposomes, not only by freezing but also by freeze-drying (results to be published). As a cryoprotectant, 10% trehalose was used. Trehalose reacted with species induced by γ–irradiation as demonstrated by large drops in pH. Interestingly, we found a substantial decrease in the irradiation damage of the liposomes in the presence of trehalose. However, no information is available on the toxicity of the degradation products of trehalose formed. Even if these products are nontoxic, γ–irradiating liposomes in the presence of trehalose is far from elegant because of its instability. To prevent physical changes by the freeze-drying process itself, a cryoprotectant has to be added to the dispersion.[32,33] The cryoprotectants used are all organic compounds and they all will probably be affected by the γ–irradiation process. Until the discovery of a cryoprotectant that is stable upon γ–irradiation, freeze-drying is an unlikely option for γ–sterilization of liposomes. γ–Sterilization of liposome dispersions in combination with freezing and/or a powerful antioxidant (butylhydroxytoluene (BHT) was not efficient enough) is an option that should be pursued.

In conclusion, the potential of γ–irradiation as a sterilization method for liposomes is limited because of degradation of liposomal phospholipids. Future experiments are necessary to establish whether γ–sterilization of liposomes in the presence of powerful antioxidants in combination with freezing will provide an acceptable sterilization procedure.

5. Filtration

Filtration is widely used to minimize the content of microorganisms in dispersions with small liposomes.[34,35] To remove contaminating microorganisms, filtration must be performed through filters with a pore size of ≤0.22 μm.[3,4] This requirement limits the use of this technique, because only liposomes with a size of 0.2 μm or smaller can be treated. A major advantage of filtration is that it is not a destructive technique for liposomes smaller than 0.2 μm. A potential problem with this technique may be clogging of the filter upon treating a large volume of liposomes. Besides, as the filtration step is not the last step in the production process, contamination of the liposomes might occur after the filtration step. This calls for carefully validated production protocols and well-trained personnel. According to the USP, the probability of nonproduct-related contamination may be in the order of 10^{-3} during an aseptic operation,[4] much more than the contamination level required after sterilization (10^{-6}; see Section I.B). Therefore, filtration is not accepted by everyone as a sterilization technique.

III. CONCLUSIONS

Sterility cannot be guaranteed by testing of the final product, but is assured by validated well-defined procedures (see Section I.B). Sterility will be achieved if a low degree of contamination (<100 cfu/ml) is combined with an effective sterilization step immediately after finishing the preparation procedure. To achieve this low degree of contamination of the end product, one should check the raw materials and equipment to be used for microorganism burden.

The limitations and efficacy of the different sterilization techniques discussed in Section II.B are summarized in Table 4. High-pressure sterilization has not been used extensively and is not a sterilization technique of the end product (see Section II.B.2). Its limited efficacy against the spores of *B. stearothermophilus* probably limits the use of this technique, which would have been of interest to sterilize temperature-sensitive and high-pressure resistant drugs such as proteins. The use of ethylene oxide is limited by the possibility that toxic residues may remain and its sterilizing capacity for freeze-fried liposome dispersions has not been proven yet (see Section II.B.3). If possible, we suggest autoclaving as the method of choice. This method is simple and validated extensively. Another advantage is that autoclaving can be applied to the end product. However, autoclaving may not be acceptable when using a nonneutral pH, or when using a water-soluble or a heat-labile drug (see Section II.B.1).

The use of γ–irradiation as a sterilization technique is still under debate. This technique is of interest to sterilize temperature-sensitive drugs, such as proteins. After γ–irradiation of aqueous liposome dispersions with a dose of 25 kGy, too much degradation of the liposomal phospholipids has been found (see Section II.B.4). More studies are necessary to evaluate this method in combination with freezing and/or a powerful antioxidant.

Although filtration is not a sterilization technique to be considered as a first option, it is widely used to produce parenteral products that could not be produced with other sterilization techniques. The major advantage of filtration is that it is not destructive for small liposomes. Therefore, to reduce initial contamination we recommend filtration of small liposome dispersions through filters with a pore size of ≤0.22 μm always, even when autoclaving is used as a final sterilization technique.

Table 4 Limitations and Efficacy of the Different Sterilization Methods for Liposomes

Sterilization Method	Efficient Reduction of Microorganism	Advantages	Drawbacks	Ref.
Autoclaving	Yes	Simple technique Sterilization in final container	Limited to neutral pH Leakage of water-soluble drugs Limited to aqueous dispersions	6–8
High-pressure sterilization	Yes/?	No size changes Hardly any water-soluble drug leakage Sterilization in final container	Not an extensively used technique Reduction factor for spores of *B. stearothermophilus* only 10^3	14
Ethylene oxide	?	Not destructive for liposome dispersions	Remaining residues of gas might be toxic Limited to freeze-dried dispersions	15
γ–Irradiation	Yes	Sterilization in final container	Not for freeze-dried liposomes	19–25, 28–31
Filtration	Yes	Simple technique Not destructive for small liposomes	Not for large liposomes Sterilization not in final container	34, 35

Not much attention has been paid to combining different approaches to reach acceptable reduction factors that are not possible with one approach alone. For instance, a low dose of γ–irradiation (±5 to 10 kGy) in combination with filtration might be a suitable sterilization method for those dispersions which cannot be sterilized by autoclaving. It is clear that more work is necessary to validate this and other combined sterilization methods.

Finally, more work is also necessary to evaluate the use of preservatives. Nothing is known about the effect of preservatives on the sterilization capacities of the different sterilization techniques for liposomes. In other pharmaceutical products preservatives can have a positive effect on the reduction factor of the sterilization technique.[3]

ACKNOWLEDGMENT

We gratefully acknowledge the financial support provided by Bayer A. G. (Leverkusen, FRG).

REFERENCES

1. **Amselem, S., Gabizon, A., and Barenholz, Y.**, A large-scale method for the preparation of sterile and nonpyrogenic liposomal formulations of defined size distributions for clinical use, in *Liposomal Technology*, Vol. I, 2nd ed., Gregoriadis, G., Ed., CRC Press, Boca Raton, FL, 1993, 501.
2. **Martin, F. J.**, Pharmaceutical manufacturing of liposomes, in *Specialized Drug Delivery Systems: Manufacturing and Technology*, Vol. 41, Tyle, P., Ed., Marcel Dekker, New York, 1990, 267.
3. **European Pharmacopoeia,** 2nd ed., Part II, European Treaty Series No. 50, Maissonneuve, S. A., France, 1992.
4. **The United States Pharmacopeia (USP) XXII,** U.S. Pharmacopeial Convention, Rockville, MD, 1989.
5. Military Standard 105D: Sampling procedures and tables for inspection attributes, U.S. Government Printing Office, Washington, D.C., 1963.
6. **Cherian, M., Lenk, R. P., and Jedrusiak, J. A.**, Heat treating liposomes, PCT Int. Appl. Patent No. WO 90/03808, 1990.
7. **Kikuchi, H., Carlsson, A., Yachi, K., and Hirota, S.**, Possibility of heat sterilization of liposomes, *Chem. Pharm. Bull.*, 39(4), 1018, 1991.
8. **Zuidam, N. J., Lee, S. S. L., and Crommelin, D. J. A.**, Sterilization of liposomes by heat treatment, *Pharm. Res.*, 10(11), 1591, 1993.
9. **Grit, M., Zuidam, N. J., Crommelin, D. J. A.**, Analysis and hydrolysis kinetics of phospholipids in aqueous liposome dispersions, in *Liposome Technology*, Vol. I, 2nd ed., Gregoriadis, G., Ed., CRC Press, Boca Raton, FL, 1993, chap. 26.
10. **Grit, M. and Crommelin, D. J. A.**, Chemical stability of liposomes: implications for their physical stability, *Chem. Phys. Lipids*, 64, 3, 1993.
11. **Grit, M., De Smidt, J. H., Stuijke, A., and Crommelin, D. J. A.**, Hydrolysis of phosphatidylcholine in aqueous liposome dispersions, *Int. J. Pharm.*, 50, 1, 1989.
12. **Grit, M., Underberg, W. J. M., and Crommelin, D. J. A.**, Hydrolysis of saturated soybean phosphatidylcholine in aqueous liposome dispersions, *J. Pharm. Sci.*, 82(4), 362, 1993.
13. **Grit, M., Zuidam, N. J., Underberg, W. J. M., and Crommelin, D. J. A.**, Hydrolysis of partially saturated egg phosphatidylcholine in aqueous liposome dispersions and the effect of cholesterol incorporation on hydrolysis kinetics, *J. Pharm. Pharmacol.*, 45, 490, 1993.
14. **Mentrup, E., Butz, P., Stricker, H., and Ludwig, H.**, Hochdrucksterilization von Lipsomen, *Pharm. Ind.*, 50(3), 363, 1988.
15. **Ratz, H., Freise, J., Magerstedt, P., Schaper, A., Preugschat, W., and Keyser, D.**, Sterilization of contrast media (Isovist) containing liposomes by ethylene oxide, *J. Microencapsulation*, 6(4), 485, 1989.
16. **Bögl, W.**, Radiation sterilization of pharmaceuticals — chemical changes and consequences, *Radiat. Phys. Chem.*, 25, 425, 1985.
17. **Jacobs, G. P.**, Radiation in the sterilization of pharmaceuticals, in *Sterile Pharmaceutical Manufacturing. Applications for the 1990's*, Vol. I, Groves, M. J., Olson, W. P., and Anisfeld, M. H., Eds., Interpharm Press, Inc., Buffalo Grove, IL, 1991, chap. 4.
18. **Anderson, P. M., Hanson, D. C., Hasz, D. E., Halet, M. R., Blazar, B. R., and Ochoa, A. C.**, Cytokines in liposomes: preliminary studies with IL-1, IL-2, IL-6, GM-CSF and interferon-γ, *Cytokine*, 6(1), 92, 1994.
19. **Gregoriadis, G.**, Medical applications of liposome-entrapped enzymes, *Meth. Enzym.*, 44, 698, 1976.
20. **Brassinne, C., Atassi, G., Frühling, J., Penasse, W., Coune, A., Hildebrand, J., Ruysschaert, J. M., and Laduron, C.**, Antitumor activity of a water-insoluble compound entrapped in liposomes on L1210 Leukemia in mice, *JNCI*, 70(6), 1081, 1983.

21. **Coune, A., Sculier, J. P., Frühling, J., Stryckmans, P., Brassinne, C., Ghanem, G., Laduron, C., Atassi, G., Ruysschaert, J. M., and Hildebrand, J.,** IV administration of a water-insoluble antimitotic compound entrapped in liposomes. Preliminary report on infusion of large volumes of liposomes to man, *Cancer Treat. Rep.*, 67(11), 1031, 1983.
22. **Sculier, J. P., Acoune, A., Brassinne, C., Laduron, C., Atassi, G., Ruysschaert, J. M., and Frühling, J.,** Intravenous infusion of high doses of liposomes containing NSC 251635, a water-insoluble agent. A pilot study with pharmacokinetic data, *J. Clin. Oncol.*, 4(5), 789, 1986.
23. **Albertini, G. and Rustichelli, F.,** Effects of gamma irradiation on the liposomal structure, in *Liposomal Technology*, Vol. I, 2nd ed., Gregoriadis, G., Ed., CRC Press, Boca Raton, FL, 1993, chap. 23.
24. **Tinsley, P. W. and Maerker, G.,** Effect of low-dose γ-radiation on individual phospholipids in aqueous suspension, *JAOCS*, 70, 187, 1993.
25. **Tinsley, P. W. and Maerker, G.,** Isolation and identification of palmitoylphosphocholinepropanediol from γ-irradiated dipalmitoylphosphatidylcholine, *JAOCS*, 70(8), 815, 1993.
26. **Handel, A. P. and Nawar, W. W.,** Radiolysis of saturated phospholipids, *Radiat. Res.*, 86, 437, 1981.
27. **Sevilla, C. L., Swarts, S., and Sevilla, M. D.,** An ESR study of radical intermediates formed by γ-radiolysis of tripalmitin and dipalmitoyl phosphatidylethanolamine, *JAOCS*, 60(5), 950, 1983.
28. **Maerker, G. and Jones, K. C.,** A-ring oxidation products from γ-irradiation of cholesterol in liposomes, *JAOCS*, 70(3), 255, 1993.
29. **Kapanin, P. V., Soboleva, N. N., and Trofimov, V. I.,** Feasibility of liposome cryoradiation sterilisation, *Khim.-Farm. Zh.*, 22(4), 479, 1988.
30. **Trofimov, V. I. and Nisnevich, M. M.,** Production of sterile drug-containing liposomes and control over their properties, *Vestn. Akad. Med. Nauk. SSSR*, 6, 28, 1990.
31. **Talrose, V. L. and Trofimov, V. I.,** The most effective way to protect drugs from ionizing radiation injury in water containing media. Publication of Institute of Energy Problems of Chemical Physics of Russia, A.S., 38/2 Leninsky prospects, Moscow.
32. **Crommelin, D. J. A., Grit, M., Talsma, H., and Zuidam, N. J.,** Physical stability on long-term storage, in *Phospholipids Handbook*, Cevc, G., Ed., Marcel Dekker, New York, 1993, 335.
33. **Özer, Y., Talsma, H., Crommelin, D. J. A., and Hincal, A. A.,** Influence of freezing and freeze-drying on the stability of liposomes dispersed in aqueous media, *Acta Pharm. Technol.*, 34(3), 129, 1988.
34. **Freise, J.,** The preparation of sterile drug-containing liposomes, in *Liposome Technology*, Vol. I, Gregoriadis, G., Ed., CRC Press, Boca Raton, FL, 1984, 131.
35. **Barenholz, Y. and Amselem, S.,** Quality control assays in the development and clinical use of liposome based formulations, in *Liposome Technology*, Vol. I, 2nd ed., Gregoriadis, G., Ed., CRC Press, Boca Raton, FL, 1993, 527.

Chapter 7

Environmental Effects of Phospholipids Used in Liposomes: Relevance for Safety

Michael J. Parnham, Miklós Ghyczy, and Armin Wendel

CONTENTS

I. INTRODUCTION

The major constituents of most liposomal products are phospholipids. Traditionally, liposomes have been prepared using naturally occurring phospholipids, though increasing sophistication, particularly in the formulation of liposomes for medical uses, has led to expanding inclusion of modified or synthetic phospholipids in liposomal products.

Naturally occurring phospholipids, such as soya bean lecithin, have been used in a wide variety of pharmaceutical, nutritional, cosmetic, and technical applications for many years. Control and regulation of the quality, safety, and efficacy of these phospholipids for both medical and nonmedical purposes, including liposomes, is therefore highly developed.[1] Generally speaking, modified or synthetic phospholipids have been treated in the same way as natural phospholipids, even though their physical properties and biological activities may differ.[2] Attention has recently been drawn to the possibility that the safety of these modified phospholipids may not always be identical to that of the naturally occurring compounds.[3]

Large-scale production of phospholipids, particularly as liposomes, is associated with specific challenges regarding stability, purity, and reproducibility and may be associated with the formation of side-products.[4] These may affect the safety of the final commercial product.

0-8493-4012-8/96/$0.00+$.50
© 1996 by CRC Press, Inc.

The widespread use of phospholipids, as such and as liposomes, inevitably raises the question as to their toxicity and environmental impact. Naturally occurring phospholipids are remarkably nontoxic and have been employed in plant protection and waste disposal, as well as for veterinary and medical purposes. Phospholipids can be found in most living organisms in which they serve specific biological functions.

In this chapter, we shall briefly review the distribution of phospholipids in the environment together with their pathways of degradation and their interactions with agents used in environmental protection. We also indicate approaches that may be taken in assessing the environmental impact of modified or synthetic phospholipids and related compounds used in liposomal preparations.

II. ENVIRONMENTAL DISTRIBUTION

Phospholipids are the major lipid components of most cell membranes, be they mammalian, fish, plant, invertebrate, bacterial, or fungal.[5]

A. MICROORGANISMS

In microorganisms the membrane lipid content varies considerably, mainly due to the effects of environment, light, temperature, pH, and so on.[6] The phospholipid content of several organisms is shown in Table 1, the differences probably reflecting the various conditions under which the measurements were made.[6]

The major bacterial phospholipid classes, found on the inner leaflet of cell membranes, are PE (constituting 50 to 90% of the phospholipids), PG, and cardiolipin.[5,6] Other phospholipids can also be detected in small amounts. The distribution of specific phospholipids is further subject to variation depending on environmental factors.

In fungi, the main phospholipids are PC, PE, and PS; PC being present on the outer leaflet of the membrane bilayer, the other two on the inner leaflet.[7,8] Examples of the distribution of phospholipid classes in fungi are given in Table 2.

A major fatty acid present in the phospholipids of microorganisms is palmitic acid (16:0), linoleic acid (18:2 n-6) also occurring in high amounts in fungi.[5–7] Interestingly, the degree of unsaturation of phospholipid fatty acid constituents decreases with increasing environmental temperature.[6] The distribution of phospholipid fatty acid constituents has recently been used as a biomarker for microorganisms in the sediments of polluted coastlines in Japan. The presence of linoleic acid as the only unsaturated fatty acid, together with some fatty acids only found in bacteria, such as *cis*-vaccenic acid (18:1) was used to assess the contamination of the sediment.[9]

B. ALGAE AND PLANTS

In most algae, phospholipids represent a major component of the total lipids, varying between 15 to 50%.[6] PC, PG, or PE are frequently the major classes, though PS, PI, and cardiolipin may also occur in large amounts depending on the species.

Table 1 Phospholipid Content of Several Microorganisms[6]

Microorganism	Phospholipid Content (% w/w of Major Lipids)
Bacterium	
Arthrobacter AK19	6.7
Yeasts	
Cryptococcus terricolus	1.8
Endomycopsis vernalis	4.0
Lipomyces lipofer	14.0
Rhodotorula glutinis	2.0
Molds	
Cladosporium herbarum	7.4
Fusarium oxysporum	28.0
Penicillium lilacinum	6.4
Pythium irregulare	17.8
Rhizopus oryzae	11.8

Table 2 Phospholipid Composition of Fungi[10]

Fungal species	PC	PE	PI	DPG	PS
	Phospholipid[a] (% of Total)				
Saccharomyces cerevisiae	43	23	21	3	7
Endomycopsis selenospora	46	23	17	—	14
Hansenula anomala	53	23	9	—	15
Bullera alba	54	30	8	—	8
Candida macedoniensis	41	26	16	—	17
Brettanomyces truxellensis	41	21	22	—	16
Saccharomyces marrianus	42	22	17	—	19
Claviceps purpurea	27	19	28	—	13

[a] PC, phosphatidylcholine; PE, phosphatidylethanolamine; PI, phosphatidylinositol; DPG, diphosphatidylglycerol; PS, phosphatidylserine.

Table 3 Phospholipid Composition of Plant Tissue[5]

	Total PL (μmol/g Fresh wt.)	Phospholipid Composition (%)[a]				
		PC	PE	PG	PI	DPG
Leaves						
Corn	1.7	26	14	28	7	14
Lettuce	0.8	38	26	12	7	16
Tomato	2.2	51	21	20	5	5
Poplar	4.6	40	18	24	11	8
Alfalfa	2.8	27	18	24	10	21
White clover	3.9	36	22	29	7	7
Runner bean	2.7	45	17	22	9	—
Rose	3.1	48	18	19	9	6
Spinach	2.3	47	17	26	9	1
Sugar beet	4.7	47	23	19	11	—
Pine	2.3	35	12	24	16	13
Root						
Sugar beet	3.5	35	17	6	5	2
Parsnip	0.9	38	20	8	13	22
Fruit						
Potato	0.7	55	28	2	12	1
Apple	0.5	46	31	7	13	2
Seed						
Castor bean	1.5[b]	13	83	—	1	—
Soybean	1.2[b]	45	15	—	25	4
Groundnut	0.7[b]	44	18	—	24	—
Pine	1.2[b]	54	11	—	30	—
Cotton	—	33	22	—	37	—
Sunflower	—	51	23	—	22	—
Cocoa	4.0[b]	40	17	—	29	—

[a] PC, phosphatidylcholine; PE, phosphatidylethanolamine; PG, phosphatidylglycerol; PI, phosphatidylinositol; DPG, diphosphatidylglycerol.

[b] Molar concentration calculated from weight concentration.

This large variety in phospholipid content is also reflected in plants generally, in which phospholipids are primary constituents of lipoprotein membrane structures.[5,6] An indication of this variability is provided by the data shown in Table 3. The major factors determining the distribution of phospholipids in plant membranes are temperature and light, though no general trends are detectable.[5,6] Rather, individual higher plant species show different responses to temperature stress.

On the other hand, the fatty acid composition of the phospholipids of higher plants shows much greater consistency, palmitic (16:0), linoleic (18:2, n-6), and linolenic (18:3) acids being the major constituents.[5]

Most of the natural lecithin used commercially is extracted from mature soya beans. In keeping with other plants, the major constituent of soya lecithin is PC with a high content of linoleic acid in both *sn*-1 and *sn*-2 positions.[11] However, depending on the environmental conditions, the phospholipid content of soya beans also varies. For instance, immature soya beans contain high concentrations of N-acylated PE which reportedly changes during germination.[12,13]

C. INVERTEBRATES

1. Protozoa

Protozoa, such as *Trypanosoma*, *Plasmodium*, and *Paramecium* have a phospholipid distribution which resembles that of mammals in that they contain large amounts of PC and PE, with smaller amounts of PS, PI, and sphingomyelin.[6] Several phospholipids have been detected in protozoa, though, which are not detectable in mammals, such as ethanolamine sphingo-phosphonolipids.[6]

2. Insects

There are more insect species than of any other living creatures. An extensive review of the tissue phospholipid compositions of many insect species has been provided by Fast.[14] Generally speaking, the major phospholipid class in insects is PC, though in flies (*Diptera*) and aphids ethanolamine phospholipids dominate. In the honeybee (*Apis mellifera*), at least, the fatty acids in the phospholipid fraction are mainly of the long-chain saturated variety.[6]

3. Molluscs

Molluscs differ from other organisms in that the largest percentage of their total polar lipids are plasmalogens.[15] The relative distribution of ethanolamine, choline, and serine plasmalogens varies considerably with the season and temperature. Mollusc lipids are of particular interest because of their very high content of n-3 polyunsaturated fatty acids.[15] The beneficial effects of these fatty acids on inflammatory and cardiovascular diseases in humans has drawn the attention of nutritionists to marine invertebrates and fish in recent years.[16] In this respect, shrimps, which are now bred almost exclusively under artificial conditions for culinary purposes, have been studied very intensively.[17] Shrimp oil contains high levels of 20:5 n-3 and 22:6 n-3 fatty acids and consists of 35 to 50% phospholipids.[17] In order to maintain this normal balance, soya bean lecithin, which contains a high proportion of polyunsaturated fatty acids, is frequently added to feeds in shrimp cultures.

D. FISH

Like molluscs, fish have a very high content of n-3 polyunsaturated fatty acids.[18,19] In other respects, the phospholipid distribution is similar to that of mammals, with large amounts of PC and PE.[5] Although the distribution varies from tissue to tissue, the effects of temperature on this distribution are not as great in fish as they are on aqueous invertebrates.[6]

E. MAMMALS

The distribution of phospholipids throughout the various tissues, cellular and subcellular organelles of a wide variety of mammals has been the subject of extensive investigation and lies beyond the scope of this chapter. A recent review[5] provides an overview of the field. As for fish and several other organisms, the predominant phospholipids in mammals are PC and PE, with cholesterol playing an important role in the membranes of a number of tissues. In mammalian phospholipids the usual distribution of the fatty acid constituents is a saturated fatty acid at *sn*-1 position and a polyunsaturated fatty acid at the *sn*-2 position. A clear membrane asymmetry in phospholipid species is detectable, the aminophosphatides, PS, PI, and PE generally being found in the inner leaflet of the membrane and the choline phosphatides, PC, and sphingomyelin generally in the outer leaflet.

From this brief survey, it is clear that the major classes of phospholipids are found in almost all biological organisms, and both terrestrial and marine application of naturally occurring phospholipids does not represent an insult to the environment. This is particularly so since all these organisms possess effective phospholipid catabolic pathways.

III. BIOLOGICAL DEGRADATION OF PHOSPHOLIPIDS

The catabolic pathways for phospholipids have been textbook knowledge for many years. Any effects which phospholipids may have on the environment, however, must take into consideration the fact that

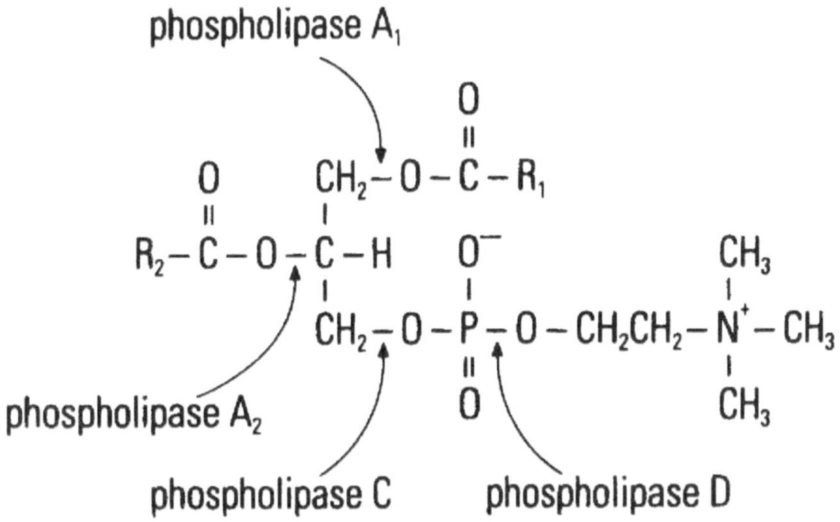

Figure 1 Metabolism of phosphatidylcholine by phospholipases (R_1 and R_2 are fatty acid residues).

both natural and synthetic phospholipids will be taken up, metabolized, and probably incorporated into the tissues of the exposed organism. For the sake of completeness, it is, therefore, important to briefly consider the major pathways for the metabolism of phospholipids in different organisms.

The main catabolic pathways for phospholipids in all organisms involve phospholipases, as shown in Figure 1. The relative importance of each individual phospholipase varies according to the organism concerned. The hydrolysis products formed are available for further degradation or for reincorporation into the same or other lipids. Thus, phospholipases A release fatty acids which can be degraded by beta-oxidation or reincorporation into phospholipids, triglycerides, sphingomyelin, etc. Phospholipase C forms diacylglyerol which can act as a mediator of cell signal transduction and/or may be further metabolized by diacylglycerol lipase to generate free fatty acids. PA, which is initially generated by the action of phospholipase D, is also active in cell signal transduction and can be further metabolized to diacylglycerol.[20]

A. MICROORGANISMS

Phospholipase A_1 activity is widely distributed throughout bacteria in which it is normally associated with the outer membrane.[21] Phospholipase A_2, on the other hand, is less common. Lysophospholipases, which catalyze the further breakdown of phospholipids are also widely distributed in microorganisms and may protect the outer membranes from the lytic activity of the surface active lysophospholipids generated by the action of phospholipase A_1.[22]

Both phospholipases C and D are also found ubiquitously in bacteria,[22] phospholipase C being released extracellularly. It can act as a bacterial toxin, as is the case with the alpha-toxin of *Clostridium perringens* type A.[23] While being an integral part of the phospholipid catabolic process in microorganisms, it is worth noting that phospholipase C can be inhibited by phospholipids which also occur naturally in bacterial cell membranes.[24] In other words, even some naturally occurring lipids are resistant to degradation by the biological environment, a fact which is likely to be true also for some synthetic phospholipids.

The final degradation products of phospholipids in microorganisms are carbon dioxide and methane. For instance, in marine sediments under aerobic conditions, carbon dioxide is the major product and under anaerobic conditions both carbon dioxide and methane are formed, though in reduced amounts in comparison to those under aerobic conditions.[25]

The extent of phospholipid degradation in such sediments is directly related to their content of organic matter; the greater the content of organic matter, the slower the rate of phospholipid degradation by the microorganisms. Thus phospholipid degradation by microorganisms is greater in beach or mangrove sediments than in organic matter-rich marsh sediments.

B. PLANTS

In addition to the presence of other phospholipases, plants are particularly characterized by a marked activity of phospholipase D. The enzyme in plant leaves is closely associated with chromatophores such as chlorophyll and carotene and has selectivity for the intact phospholipid molecule.[26] In seeds, phospholipase D is associated with the proplastiol. The hydrolysis of phospholipid by phospholipase D to form phosphatidic acid is generally rapid. The rate of turnover of phospholipids in plants, however, varies according to the tissue compartment investigated.[27]

The degradation of exogenous phospholipids by the leaves of plants is dependent on their ability to cross the hydrophobic cuticle. Phospholipids generally are surface active and effective wetting agents and this will be considered later with regard to the interactions between phospholipids and herbicides.

C. ANIMALS

The latest developments in our understanding of the metabolic pathways of phospholipids in animal cells have recently been reviewed.[20] In all animal species investigated, phospholipids are rapidly absorbed and metabolized by the well-known pathways. As a result, naturally occurring phospholipids, particularly those from plants, can be ingested in large quantities by animals and humans without any untoward effects.[3] The extent to which novel, synthetic phospholipids, such as those which may be prepared in large quantities for liposomal preparations, are degradable by known catabolic pathways requires investigation in each individual case.

IV. LEGAL PROVISIONS FOR COMMERCIAL LECITHIN

A. SPECIFIC STATUS OF LECITHIN

On January 19, 1981, a request submitted by the Decco Tilbelt Agricultural Division, Monrovia, California to the U.S. Environmental Protection Agency (EPA) was announced in which an exemption from the requirement of a tolerance for lecithin under 40 CFR 180.1001(c) was proposed. No requests for referral to an advisory committee were received and an official exemption from tolerance testing for lecithin as an ingredient in pesticide formulations applied to growing crops or to raw agricultural commodities after harvest was approved, effective from March 24, 1981.[28] Soya lecithin is now widely used as an adjuvant to enhance the efficacy of various pesticides.

Subsequently, in 1982, the FDA completed a comprehensive review of the use of lecithin obtained from plant sources as a human food ingredient. A total of 3092 literature citations were considered and since nonbleached and peroxide-bleached lecithin had been in use in the U.S. as food ingredients since 1930, these two products were evaluated as a single ingredient. The average daily intake of lecithin as an added food ingredient in the U.S., based on a 1970 study, was considered by the FDA to be approximately 1.5 mg/kg adult body weight. No adverse reactions were found in any of the reports.

The agency also concluded that the specifications on the limits of residual unreacted peroxide, adopted in the Food Chemicals Codex, 3rd edition, 1981, were adequate to assure the continued safe use of lecithin as a food ingredient. The FDA consequently issued in the Federal Register new regulations (21 CRF 184.1400) affirming the "generally recognized as safe" GRAS status of commercial, plant-derived lecithin containing up to 35 triglycerides, for direct and indirect use as a food ingredient, provided that the current good manufacturing practice (GMP) conditions were adhered to.[29]

On the basis of these two regulations, commercial lecithin from vegetable sources (mainly soya beans) is legally recognized in the U.S. as being without untoward effect on food crops.

B. GENERAL ECOTOXICITY REQUIREMENTS

Few countries have a legal framework for the testing of ecotoxic effects of industrial chemicals. With the growing public concern about environmental pollution, however, there is an increased awareness of the need for legal standards for ecotoxicity testing. Generally, the Guidelines of the OECD for Testing of Chemicals tend to be followed by manufacturers. These guidelines (adopted between 1981 and 1982) cover a wide variety of tests for chemicals including physical–chemical properties and effects on biotic systems. An even larger number of draft guidelines are currently under consideration which cover, in addition to properties and effects, degradation and accumulation and health effects of chemicals. Tests adopted by the Organization for Economic Cooperation and Development (OECD) include toxicity tests on algae, water fleas, fish, birds, earthworms, plant growth, sludge respiration, and genetic toxicity. Copies of these guidelines can be obtained from OECD, Environmental Health and Safety Division, 2 Rue André Pascal, F-75775 Paris Cedex 16, France.

In the U.S., the environmental hazard of an industrial chemical is currently considered on a case-by-case basis and the potential hazard is estimated from structure–activity relationships. The close relationship between a new chemical and a known dangerous chemical may lead to the request by the regulatory authorities for extensive ecotoxicity and animal toxicity testing.

In the European Union (EU), a directive was recently published[30] and following implementation in Germany in May 1994, it is to be implemented in all EU countries by the end of 1995. This directive, which provides criteria for substances dangerous to the environment, defines specific tests which must be carried out to assess the acute risk of chemicals to the aquatic environment. These are the 96 h LC_{50} (lethal concentration for 50% of the population) for a relevant freshwater fish, the 48 h EC_{50} for *Daphnia maqua* (water flea) and the 72 h IC_{50} (concentration inhibiting growth by 50%) for the growth of unicellular green algae. Compounds with values for these criteria of <100 mg/l are considered harmful, provided the substances are also not readily degradable.

In this respect, a substance is considered readily degradable if in 28-day biodegradation studies: 70% degradation is achieved in tests based upon dissolved organic carbon; 60% degradation is achieved in tests based on oxygen depletion or carbon dioxide generation; these figures are achieved within 10 days from the time at which 10% of the substance has been degraded. Alternatively, convincing scientific evidence must be provided that the substance can be degraded (biotically and/or abiotically) in the aquatic environment to a level of >70% within a 28-day period.

Clearly, by these criteria, most phospholipids are readily degraded, as discussed in the previous section and cannot be considered in any way harmful to the aquatic environment. Employment of less-easily degradable phospholipid liposomes for agricultural use, however, could conceivably increase their ecotoxicological relevance.

Currently, no specific test criteria have been adopted for nonaquatic environmental toxicity, but the criteria are under assessment by the OECD and the EU and are likely to be published in the near future.

V. INTERACTIONS WITH ENVIRONMENTAL PROTECTANTS AND POLLUTANTS

The inclusion of lecithin as an adjuvant in liposomal pesticide formulations is considered in another chapter. Here the relevance for safety of this interaction together with the interactions of phospholipids with environmental pollutants will be considered.

The interactions of lecithin with a wide variety of animal and human poisons have been known for nearly a century. Already in 1941 Kunze[31] was able to review a series of studies demonstrating that lecithin enhances the poisonous activity of compounds as diverse as strychnine, curare, diphtheria toxin, and phosphorus. On the other hand, at different dosages, lecithin was also protective against several poisons. Presumably in some cases the lecithin was acting as a carrier, similar to liposomes, thereby redistributing and concentrating the poison. In other cases, the lecithin was exerting a direct biological action to repair tissue damage, particularly that to the liver. In fact, highly purified soya lecithin is still widely used as a treatment for viral or toxic damage to the liver.[32]

Because of their surface-active properties, phospholipids may also be able to facilitate the entry of chemicals into plants and microorganisms.

A. HERBICIDES

Liposomal formulations of agrochemical pesticides are discussed elsewhere in this publication.[33] Here we are concerned with the biological effects of the combination of a herbicide with a phospholipid. Soya lecithin has been shown to be an effective dispersant and wetting agent which helps to spread aqueous material, such as herbicides, over the surface of the leaves of plants. By reducing the size of drops used in pesticide sprays, the lecithin also potentially reduces drift and the hazards of inhalation.[34]

Soya lecithin itself has no phytotoxic effect and increases the content of nutrient phosphate in field-grown soya bean plants.[35] The authors of this study describe a simple test for phytotoxicity involving the measurement of the lowering of the surface tension induced by the phospholipid at the leaf surface and a scoring system for leaf burning 4 days after treatment. This test is easily adaptable to the testing of other formulations of phospholipids with herbicides and can be used as a screening system for new (liposomal) phospholipids.

Several studies have been performed on the effects of the combination of soya lecithin with herbicides on various plants. These are summarized in Table 4. Generally, soya lecithin causes cuticle wetting and increased uptake of the herbicide into the plants. In a comparative study on hemp dogbane, which has a very waxy cuticle, and common milkweed, which has little wax in the cuticle, soya lecithin exerted a similar effect on both plants. The authors proposed that the main site of action of the phospholipid in

Table 4 Effects of Herbicides in Combination with Soya Lecithin

Herbicides	Phospholipid	Plant	Effect	Ref.
2,4-D or 2,4,5-T[a]	Soya lecithin	Potato parenchymal tissue	Increased herbicide uptake	36
Glyphosphate	Centromix (soya phosphatides)	Common milkweed; hemp dogbane	Slight increases in phytotoxicity	37
Glyphosphate	Natipide 6008 (soya lecithin product)	Winter wheat seedlings	Little effect	38
Isoproturon + various herbicides	Natipides	Cereals, sugar beet, maize	More rapid onset of action, 40–50% reduction in herbicide requirement	39
Ethephon + mepiquat Cl	Acidified soya lecithin	Winter barley	Facilitated inhibition of stem extension and crop yield	40

[a] 2,4-D = (2,4-dichlorophenoxy)acetic acid; 2,4,5-T = (2,4,5-trichlorophenoxy)acetic acid.

increasing herbicide uptake is to break the barrier function of the plasmalemma. However, in the two studies on glyphosphate, soya lecithin was not very effective at enhancing herbicide phytotoxicity, suggesting that this compound may interact in a different manner than other herbicides with phospholipids.

An important study with regard to environmental safety is that of Maas.[39] He was able to demonstrate that as a result of the increased uptake of 10 different herbicides in combination with various soya phospholipids, the amount of herbicide sprayed could be reduced by 40 to 50% to provide the same effect. This represents a considerable reduction in the general burden on the surrounding soil and biosphere.

In one of the few studies of its kind, the effect of the herbicide cycloate (Ro-Neet™) alone or in combination with soya lecithin (Glycerid-P™) has been studied on the activity of soil microorganisms for up to 20 weeks.[41] While the herbicide inhibited soil dehydrogenase activity, possibly as a result of increased mineralization, combination with phospholipid had little additional effect. Merely an increase in soil respiration after 20 weeks was observed, which was attributed to enhanced phospholipid degradation.

Combination of herbicides with (liposomal) phospholipids, therefore, not only increases targeting of the herbicides to the plants, but by reducing the dose required also reduces the side effects on soil microorganisms. The use of nonnatural phospholipids in liposomal formulations must take into consideration possible deviant effects to those of soya lecithin on both phytotoxicity and the biological activities of soil.

B. FUNGICIDES

An emulsifiable preparation containing 42% soya lecithin (Lecithinon™) has been available for nearly 20 years in Japan for the treatment of powdery mildew (*Sphaerotheca fuliginea*) on cucumbers and strawberries.[42] The activity of this product is based on the fungistatic properties of soya lecithin, which when sprayed on *S. fuliginea*, stunts the growth of germ tubes, forms a membrane over hyphal tips and causes conidiospores to stick to conidiophores rather than being released.[43] A fungicidal action is not observed. Similar products containing soya lecithin are also available in Germany (Lipomel™, Bio Blatt™ Spritzmittel, Substral-Mehltaufrei) for the treatment of mildew on roses. In the U.K., LI-700, containing 750g/l soya phospholipids, is sold as an adjuvant for a variety of pesticide formulations used in crop sprays.

In contrast, when added to cultures of certain fungi, different soya lecithin fractions are able to support the growth of the fungi. The lecithin fractions (NAT30, NAT70, NAT100) stimulated mycelial growth and virulence of the endoparasitic fungi *Verticillium balanoides* and *Drechmeria coniospora* but not of *Harposporium anguillulae*.[44] This growth-promoting action can be considered beneficial, since *D. coniospora* is employed for the biological control of nematode infestations of plants and artificial growth in culture is necessary for large scale production. A similar mycelial growth-promoting action with different phospholipid mixtures mainly containing negatively charged phospholipids (NAT50, NAT70, NAT89, NAT106) from soya lecithin when added to *Phanerochaete chrysosporium* INA-12 in culture.[45] These effects, however, have only been obtained in artificial conditions for industrial application, whereas the effects of lecithin on mildew are obtained in the field.

Direct interactions of phospholipids with fungicides have also been observed under *in vitro* conditions. Thus, a variety of lipids were tested for their interactions with several fungicides in cultures of *Candida albicans* and *Ustilago avenae*.[46,47] Sterols, unsaturated fatty acids, and several phospholipids (PI, PS, PC) all counteracted the inhibitory actions of triadimefon, triadimenol, nuarimol, and imazalil on *U. avenae*.[46]

Phospholipids, but only *cis*-saturated fatty acids, also counteracted the inhibition of *C. albicans* by clotrimazole and miconazole.[47]

Direct binding of fungicides, which is likely to be dependent on the degree of saturation, is clearly a possible environmental effect of phospholipids. As with herbicides, such an interaction is likely to decrease the nonspecific dispersion of sprayed materials. In this context, it is worth mentioning that the inclusion of lecithin in a commercial fungicidal product containing a mixture of fatty acids has been claimed to reduce significantly the phytotoxic side effects of the product.[48] Sprayed on plants (*Phaseolus vulgaris*) infested with the fungus, the combination product killed the fungus with much reduced lesioning of the leaves of the plant. This is an important consideration in view of the fact that soya lecithin can enhance chemical protectant uptake into both plants and fungi. Any new or synthetic compound developed as a liposomal carrier for agrochemicals will inevitably require testing for the specificity of uptake. For instance a liposomal fungicide should not also be taken up in increasing amounts into plants, which would increase phytotoxicity.

Botrytis cinerea is a particular problem for winegrowers since it infects grapevines and resistant strains are on the increase. The results of several studies[49–51] on the use of phospholipids in combination with fungicides in the treatment of *Botrytis* infection of grapevines have shown that a liposomal mixture of soya phospholipids (Natipid 6008) enhances the potency of the same dose of fungicide (mainly Euparen™ or Rovral™) by up to 40%. The effects of the admixture of the phospholipids were dependent on the dosage of fungicide (Figure 2), achieving a maximum and declining at higher doses.[52] This was attributed to maximal incorporation of phospholipids into fungal cell membranes. Any possible adverse ecotoxicological effects of the mixture of phospholipid and fungicide would be attributable to the overzealous spraying of the fungicide rather than to an effect of the phospholipid. Certainly, the use of the phospholipid had no effect on the taste of the subsequent wine produced, indicating that uptake of the fungicide into the grapes was not promoted.[52]

As with herbicides, the spraying of fungicides with phospholipids appears to be associated with increased targeting of the fungus and a consequent reduction in ecotoxic hazard.

C. BACTERICIDES

In agreement with its growth-promoting activity for other cell types, lecithin also promotes the growth of bacterial colonies (*Staphylococcus aureus, Streptococcus viridans, Neisseria catarrhalia*) *in vitro*.[53] This only appears to be the case for Gram-positive organisms, since the growth of Gram-negative bacteria

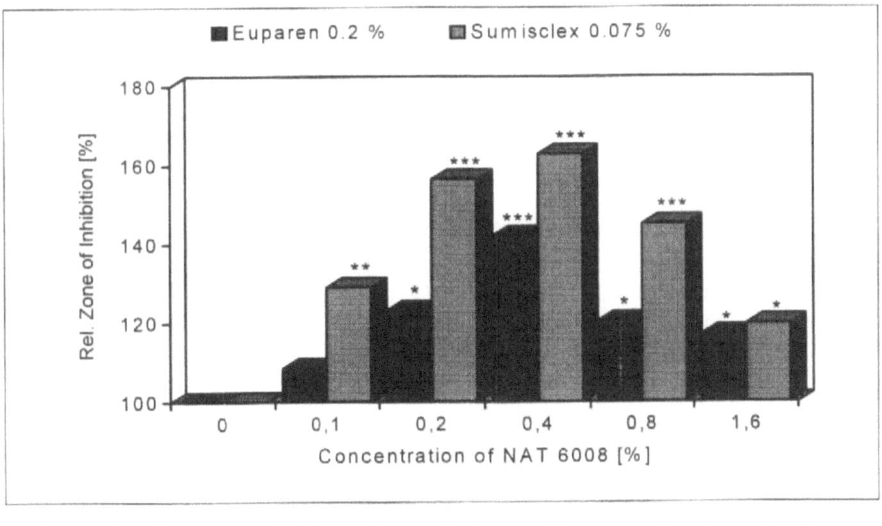

Influence of applied amount of NAT 6008 in combination with Euparen and Sumisclex in CDT.

* significant compared to Euparen alone
** highly significant compared to Euparen alone
*** very highly significant compared to Euparen alone

Figure 2 Influence of the amount of NAT 6008 applied in combination with Euparen™ and Sumisclex on the inhibition of *Botrytis cinerea* in the cuticular diffusion test in *Vicia faba*. (From Ghyczy, M., et al., *Med. Fac. Landbouw. Rijksuniv. Gent*, 56, 843, 1991. With permission.)

was not affected. This growth-promoting activity of lecithin has recently been found ecologically of benefit in the microbiological breakdown of industrial lubricants. Particularly, a phospholipid mixture with a high proportion of negatively-charged soya phospholipids (Phospholipon 5) was found to facilitate the microbiological breakdown process.[54]

Little has been published on the interactions between phospholipids and bactericidal agents. Lecithin was found in an early study to antagonize the bactericidal action of detergents *in vitro*,[55] probably by the well-known protection of cell membranes.[32] However, large amounts of phospholipids would be required to observe such an effect in field studies. Wherever bacterial growth is likely, such as in anaerobic waste effluent or stagnant water, the presence of unusually large amounts of phospholipids must be considered as a factor likely to alter the ecological balance toward increased bacterial growth. For industrial application of phospholipids, this possibility must be borne in mind.

D. INSECTICIDES

Lecithin by itself has been used as an effective means of controlling mosquito populations. Sprayed as a monolayer on the surface of the water, the surface tension is reduced, the lecithin film spreads and forms a stable, soap bubble-like film. As a result, the respiratory horns of the anopheles pupae are covered by the phospholipid film, the horn interior becomes wetted, the ventral air space is flooded and the pupae are asphyxiated.[56–58] The lowered surface tension also prevents the mosquito larvae from surfacing, causing drowning.[58] A number of field studies have confirmed the usefulness of this technique.

As discussed above for herbicides and fungicides, lecithin is able to interact directly with insecticides. Dipalmitoyl PC forms a complex with compounds of the dichlorodiphenyltrichloroethane (DDT) type, an interaction which has been proposed to explain DDT concentration in and toxicity to the central nervous system.[59,60] From a toxicological point of view, this binding of DDT to phospholipids could theoretically contribute to accumulation and persistence of insecticides in animal and human tissues. However, it has been shown that DDT and pyrethroids preferentially partition into triglycerides rather than into phospholipids.[61] This may offer an explanation for the earlier finding that resistance of boll weevils to toxaphene plus DDT was directly correlated with the concentration of triglycerides but not that of phospholipids in the weevils.[62] Thus, phospholipids appear to assist in targeting insecticides to their central nervous site of action without facilitating their persistence.

As a corollary to these findings in insects, lecithin has been shown to facilitate the metabolism of the pyrethroid, lindane,[63] and the clearance of DDT metabolites in the rat;[64] DDT metabolites were conjugated to fatty acids. In keeping with its general protective action against liver damage,[32] lecithin also decreased the accumulation of hepatic and skeletal muscle glycogen induced by a variety of insecticides.[65] Therefore, while a direct interaction of phospholipids with insecticides occurs, this is associated with detoxifying effects in organisms other than insects. In view of the differing effects of naturally occurring phospholipids and triglycerides on insecticide partition, though, it should be borne in mind that synthetic alteration of phospholipids could affect the partitioning of pesticides with which they may be formulated, thereby causing ecotoxicity.

E. WASTES AND POLLUTANTS

The use of liposomes in the bioreclamation of soil contaminated with hydrophobic waste is discussed elsewhere in this volume. In addition, phospholipid liposomes supply nutrients and stimulate the growth of microorganisms which are able to break up oil spills on water.[65] Other ecologically "friendly" effects of phospholipids include use as a nontoxic dispersing agent in paints[67] and as a wetting agent to control dust generation.[68] It has been suggested that the use of synthetic liposomes may be more effective for the control of environmental contamination.[69] However, such synthetic products would need to undergo ecotoxicological testing on potential target organisms, as described in previous sections.

VI. IMMUNOADJUVANTICITY

Since naturally occurring soya lecithin, the major commercially available phospholipid, has the GRAS status, it is usually considered that all phospholipids are essentially nontoxic to humans. While this may be true for natural phospholipids, some toxicological limitations do exist, particularly when phospholipids or modified phospholipids are combined with other compounds in liposomal formulations.[3,70] Liposomes, as particulate material, are rapidly taken up and cleared by cells of the mononuclear phagocyte system, both on intravenous and aerosol administration.[71,72] For this reason, liposomes have been found to be effective immunoadjuvants, targeting antigens to the antigen-presenting phagocytic cells and thereby enhancing the immune response to antigen. Such immunoadjuvant effects of liposomes have been found, for instance, with antigens administered intravenously and orally.[73–75] Formulation with phospholipid liposomes has been shown to increase drug targeting to lung macrophages on aerosol administration[72] and even to facilitate absorption of peptides through the skin.[76]

Environmental spraying with liposomal phospholipid formulations of a variety of chemicals always carries the danger of accidental contact with the skin, inhalation, or ingestion. The liposomal phospholipids may not only facilitate the absorption of harmful chemicals, but may also act as immunoadjuvants to increase the allergenic potential of the transported chemicals. Fortunately, reports of the possible adjuvant effect of phospholipids on allergenic responses are very rare, two recent reports being limited to drug allergies in sick people. In one case, a hypersensitivity response to the soya lecithin contained in an aerosol administered to an asthmatic patient was thought to have arisen,[77] probably in association with an effect of the phospholipid on the lung surfactant layer. In the other case, an allergy to AL721 was reported in an AIDS patient.[78] This product is a mixture of egg yolk phospholipids and it is possible that an allergenic protein might be present. Consequently, natural phospholipids do not appear to significantly enhance the allergenic potential of drugs or environmental chemicals, but the immunoadjuvanticity of novel liposomal formulations must be given some consideration. Several tests for the toxicity screening of liposomes have been proposed recently[79] and should be complemented by studies of immunogenicity.[73,80]

VII. CONCLUSIONS

Phospholipids are widely distributed throughout all natural organisms and the extensive environmental application of such naturally occurring phospholipids is not associated with ecotoxic effects. To the contrary, spraying natural phospholipids facilitates the breakdown of chemical pollutants by microorganisms, traps oil and dust, and is effective in the control of plant mildew and mosquito pupae and larvae. Furthermore, formulation with phospholipids improves targeting of pesticides, reduces drift on spraying and enhances penetration, thereby promoting a reduction in the required dose of pesticides. In this regard, the phospholipid micelles formed on mixture with aqueous solutions can be considered as harmless as the more complex liposomal preparations of phospholipids.

Since natural phospholipids applied in this way are rapidly degraded, they do not represent an environmental hazard. However, the effects reviewed here may be subject to change when the formulations applied contain synthetic or semisynthetic phospholipids. In such cases it is important, and increasingly a legal necessity to carry out tests for potential excotoxicity. Liposomal pesticides will need to be tested as free and liposomal products, while liposomal constituents to be used *per se* will need to be tested for environmental degradation and submitted, at least, to tests for aquatic ecotoxicological tests and the limits of toxicity. Within the next couple of years further guidelines will necessitate vigilance on the part of manufacturers to meet the expanding and justifiable limitations on environmental application of chemicals.

REFERENCES

1. **Ostro, M. J.,** Industrial application of liposomes: what does that mean? in *Liposomes as Drug Carriers*, Gregoriadis, G., Ed., John Wiley & Sons, Chichester, 1988, chap. 61.
2. **Blume, A.,** Phospholipids as basic ingredients, in *Liposome Dermatics*, Braun-Folco, O., Korting, H. C., and Maibach, H. C., Eds., Springer-Verlag, Berlin, 1992, 29.
3. **Parnham, M. J.,** Liposome phospholipids: toxicological and environmental advantages, in *Liposome Dermatics*, Braun-Folco, O., Korting, H. C., and Maibach, H. C., Eds., Springer-Verlag, Berlin, 1992, 57.
4. **Klimchak, R. J. and Lenk, R. P.,** Scale-up of liposome products, *BioPharm*, February, 18, 1988.
5. **Yorek, M. A.,** Biological distribution, in *Phospholipids Handbook*, Cevc, G., Ed., Marcel Dekker, New York, 1993, chap. 21.
6. **Neidleman, S. L.,** Occurrence and response to environmental stresses in nonmammalian organisms, in *Phospholipids Handbook*, Cevc, G., Ed., Marcel Dekker, New York, 1993, chap. 2.
7. **Chopra, A. and Khuller, G. K.,** Lipids of pathogenic fungi, *Progr. Lipid Res.*, 22, 189, 1983.
8. **Rothman, J. E. and Leonard, J.,** Membrane asymmetry, *Science*, 195, 743, 1977.
9. **Rajendran, N., Suwa, Y., and Urushigawa, Y.,** Microbial community structures in sediments of a polluted bay as determined by phospholipid ester-linked fatty acids, *Mar. Pollut. Bull.*, 24, 305, 1992.
10. **Wassef, M. K.,** Fungal lipids, *Adv. Lipid Res.*, 15, 159, 1977.

11. **Cherry, J. P. and Kramer, W. H.,** Plant sources of lecithin, in *Lecithins. Sources, Manufacture and Uses*, Szuhaj, B. F., Ed., AOCS, Champaign, IL, 1989, 16.
12. **Combe, N.,** Etude analytique des lecithines en vue de leur utilisation dans l'agro-alimentaire, *Ind. Aliment. Agric.*, 108, 369, 1991.
13. **Chapman, K. D. and Moore, T. S.,** N-Acylphosphatidylethanolamine synthesis in plants: occurence, molecular composition, and phospholipid origin, *Arch. Biochem. Biophys.*, 301, 21, 1993.
14. **Fast, P. G.,** Insect lipids, *Progr. Chem. Fats*, 11, 179, 1972.
15. **Joseph, J. D.,** Lipid composition of marine and estriarine invertebrates, *Progr. Lipid Res.*, 21, 109, 1982.
16. **Horrobin, D. F.,** Low prevalences of coronary heart disease (CHD), psoriasis, asthma and rheumatoid arthritis in eskimos: are they caused by high dietary intake of eicosapentaenoic acid (EPA), a genetic variation of essential fatty acid (EFA) metabolism or a combination of both? *Med. Hypoth.*, 22, 421, 1987.
17. **Akiyama, D. M., Dominy, W. G., and Lawrence, A. L.,** Penaeid shrimp nutrition, in *Marine Shrimp Culture: Principles and Practices*, Fast, A. W. and Lester, L. J., Eds., Elsevier, Amsterdam, 1992, chap. 25.
18. **Henderson, J. and Tocher, D. R.,** The lipid composition and biochemistry of freshwater fish, *Progr. Lipid Res.*, 26, 281, 1987.
19. **Malins, D. C. and Wekell, J. C.,** The lipid biochemistry of marine organisms, *Progr. Chem. Fats*, 10, 337, 1967.
20. **Scherphof, G. L.,** Phospholipid metabolism in animal cells, in *Phospholipids Handbook*, Cevc, G., Ed., Marcel Dekker, New York, 1993, chap. 22.
21. **Finnerty, W. R.,** Microbial lipid metabolism, in *Microbial Lipids*, Vol. 2, Ratledge, C. and Wilkinson, S. G., Eds., Academic Press, New York, 1989, 115.
22. **Mackie, R. I., White, B. A., and Bryant, M. P.,** Lipid metabolism in anaerobic ecosystems, *CRC Crit. Rev. Microbiol.*, 17, 449, 1991.
23. **MacFarlane, M. G. and Knight, B. C. J. G.,** The biochemistry of bacterial toxins. I. The lecithinase activity of Clostridium welchii toxins, *Biochem. J.*, 35, 884, 1941.
24. **Pagano, R. E. and Sleight, R. G.,** Emerging problems in the cell biology of lipids, *Trends Biochem. Sci.*, 10, 421, 1985.
25. **Harvey, R. H., Fallon, R. D., and Patton, J. S.,** The effect of organic matter and oxygen on the degradation of bacterial membrane lipids in marine sediments, *Geochim. Cosmochim. Acta*, 50, 795, 1986.
26. **Pardun, H.,** Die Pflanzenlecithine. Gewinnung, Eigenschaften, Verarbeitung und Anwendung pflanzlicher Phosphatidpräparate, H. Ziolkowsky, Augsburg, 1988, 143.
27. **Mudd, J. B.,** Phospholipid biosynthesis, in *The Biochemistry of Plants*, Vol. 4, Stumpf, P. K., Ed., Academic Press, New York, 1980, chap. 9.
28. **Johnson, E. L.,** Lecithin: tolerances and exemptions from tolerances for pesticide chemicals in or on raw agricultural commodities, *Fed. Regist.*, 46, 1831, 1981.
29. **Randolph, W. F.,** GRAS status of lecithin, *Fed. Regist.*, 48, 51149, 1983.
30. **European Commission Directive 91/325/EEC,** *Off. J. Eur. Commun.*, 34, 1, 1991.
31. **Kunze, R.,** Lecithin, Verlag Rosenmeier & Dr. Saenger, Berlin, 1941, 142.
32. **Kuntz, E.,** The "essential" phospholipids in hepatology — 50 years of experimental and clinical experiences, in *50th Anniversary of Phospholipid Research (EPL)*, Gundermann, K.-J. and Schumacher, R., Eds., wbn-Verlag, Bingen, 1990, 49.
33. **Quinn, P. J.,** Liposomal formulations of agricultural pesticides, in *Nonmedical Applications of Liposomes*, Barenholz, Y. and Lasic, D. D., Eds., CRC Press, Boca Raton, FL, 1993.
34. **Quinn, P. J., Perrett, S. F., and Arnold, A. C.,** An evaluation of soya lecithin in crop spray performance, *Atom. Spray Technol.*, 2, 235, 1986.
35. **Stein, L. A. and Storey, J. B.,** Influence of adjuvants on foliar absorption of nitrogen and phosphorus by soybeans, *J. Am. Soc. Hort. Sci.*, 111, 829, 1986.
36. **Smith, A. E.,** Accumulation of (2,4-dichlorophenoxy)acetic acid and (2,4,5-trichlorophenoxy)acetic acid by parenchyma tissue as influenced by metabolic inhibitors and lecithin, *Physiol. Plant.*, 27, 338, 1972.
37. **Wyrill, J. B. and Burnside, O. C.,** Glyphosphate toxicity in common milkweed and hemp dogbane as influenced by surfactants, *Weed Sci.*, 25, 275, 1977.
38. **Meinen, E., de Ruiter, H., and Uffing, A. J. M.,** Influence of biodegradable adjuvants on the toxicity of glyphosphate on winter wheat seedlings, Nattermann Phospholipids Report No. 1204.
39. **Maas, G.,** Pesticide dose rate reduction by combining with phospholipids of plant origin, *Nachtrichtenbl. Deut. Pflanzenschutzd.*, 34, 1982.
40. **Green, C. F., Chalmers, I. F., and Packe-Drury-Lowe, S. J.,** Enhancing the performance of ethephon with mepiquat chloride on barley (Hordeum distichon cv. Panda) using an adjuvant comprising acidified soyal phospholipid, *Ann. Appl. Biol.*, 113, 177, 1988.
41. **Malkomes, H.-P. and Halstrick, S.,** Einfluß des Herbizids Ro-Neet und dessen kombination mit einem Phospholipid auf bodenbiologische Aktivitäten unter Laborbedingungen, *Zbl. Mikrobiol.*, 140, 381, 1985.
42. **Misato, T.,** Lecithinon, an agricultural chemical obtained from natural substance — its characteristics and effective utilization, *Jpn. Pest. Inf.*, 35, 32, 1978.

43. **Homma, Y., Takahashi, H., Mizumo, H., and Misato, T.**, Effect of soybean lecithin on cucumber powdery mildew, Spaerotheca fuliginea, at various growth stages, *J. Pest. Sci.*, 2, 33, 1977.
44. **Lohmann, U., Sikora, R. A., and Höfer, M.**, Influence of phospholipids on growth, sporulation and virulence of the endoparasitic fungi Drechmeria coniospora, Verticillium balanoides and Harposporium anguillulae in liquid culture, *J. Phytopathol.*, 125, 139, 1989.
45. **Capdevila, C., Moukha, S., Ghyczy, M., Theilleux, J., Gelie, B., Delattre, M., Corrieu, G., and Asther, M.**, Characterization of peroxidase secretion and subcellular organization of Phanerochaete chrysosporium INA-12 in the presence of various soybean phospholipid fractions, *Appl. Environ. Microbiol.*, 56, 3811, 1990.
46. **Buchenauer, H.**, Interaction of different lipid components with various fungicides, *Z. Pflanzenkrankh. Pflanzensch.*, 87, 335, 1980.
47. **Yamaguchi, H.**, Antagonistic action of lipid components of membranes from Candida albicans and various other lipids on two antimycotics, clotrimazole and miconazole, *Antimicrob. Chemother.*, 12, 16, 1977.
48. **Kulenkampff, H. E.**, U.S. Patent 5,093,124, 1992.
49. **Müller, F.**, Einfluß von Phospholipiden auf die Wirksamkeit von Fungiziden gegenüber Botrytis cinera, Doctoral Thesis, University of Giessen, Germany, 1986.
50. **Holz, B.**, Phospholipide verbessern die Wirksamkeit von Botryziden im Weinbau, *Mitt. BBA Berlin*, 266, 216, 1990.
51. **Wohlfahrt, P.**, Phospholipids — Mittel zur Wirkungssteigerung bei Spezialbotrytiziden, *Der Deutsche Weinbau*, 10, 488, 1988.
52. **Ghyczy, M., Hager, J., and Böhm, R.**, The influence of liposomes from soybean lecithin on the efficacy of fungicides, *Med. Fac. Landbouw. Rijksuniv. Gent*, 56, 843, 1991.
53. **Gershenfeld, L. and Johnson, G. C.**, Effect of certain surface active agents upon bacterial growth, *J. Am. Pharm. Soc.*, 42, 187, 1953.
54. **Lautenschläger, H.**, German Patent 4223945.1, 1992.
55. **Binder, T.**, Zur Frage der chemischen Spezifität der antagonistischen Hemmung der bakteriziden Wirkung von Invertseifen, *Schweiz. Z. Pathol. Bakt.*, 12, 85, 1949.
56. **McMullen, A. E. and Hill, M. N.**, Anoxia in mosquito pupae under insoluble monolayers, *Nature*, 234, 51, 1971.
57. **Reiter, P. and McMullen, A. I.**, The action of lecithin monolayers on mosquitoes. I. General observations, *Ann. Trop. Med. Parasitol.*, 72, 163, 1978.
58. **Reiter, P.**, The action of lecithin monolayers on mosquitoes. II. Action on the respiratory structures, *Ann. Trop. Med. Parasitol.*, 72, 169, 1978.
59. **Tinsley, I. J. Haque, R., and Schmedding, D.**, Binding of DDT to lecithin, *Science*, 174, 145, 1971.
60. **Haque, R., Tinsley, I. J., and Schmedding, D.**, Lipid binding and mode of action of compounds of the dichlorodiphenyltri-chloroethane type: a proton magnetic resonance study, *Mol. Pharmacol.*, 9, 17, 1973.
61. **Sandermann, H.**, Triglyceride/phospholipid partitioning and persistence of environmental chemicals, *Chemosphere*, 8, 499, 1979.
62. **Moore, R. F. and Taft, H. M.**, Relationship between phospholipids and triglyceriders in the boll weevil and susceptibility to toxaphene + DDT, *J. Econ. Entomol.*, 65, 1733, 1972.
63. **Chadwick, R. W., Simmons, W. S., Bryden, C. C., Chuang, L. T., Key, L. M., and Chadwick, C. J.**, Effect of dietary lipid on lindane metabolism, *Toxicol. Appl. Pharmacol.*, 37, 113, 1976.
64. **Leighty, E. G.**, Decreased retention of fatty acid conjugated DDT metabolites in rats given injections of heparin, bile salts or lecithin, *Res. Commun. Chem. Pathol. Pharmacol.*, 31, 69, 1981.
65. **El-Beih, N. M.**, Protective effect of phospholipids and other chemicals on plasma glucose and glycogen content of liver and skeletal muscle in rats treated with chemical insecticides, *Ann. Zool. (India)*, 16, 185, 1980.
66. **Gatt, S., Bercovier, J. H., and Barenholz, Y.**, Use of liposomes to combat oil spills and their potential application to bioreclamation, in *On-Site Bioreclamation*, Hinchee, R. E. and Olfenbuttel, R. F., Eds., Butterworth, Stoneham, MA, 1991, 293.
67. **Hajak, J.**, New aspects in the formulation of environmentally friendly primers, *Farbe Lack*, 99, 109, 1993.
68. **Glanville, J. O., Halay, L. H., and Wightman, J. P.**, Coal dust phenomena and control technology, *Surfact. Sci. Ser.*, 32, 379, 1989.
69. **Lasic, D. D.**, Liposomes in ecology and other applications, in *Liposomes: From Physics to Applications*, Elsevier, Amsterdam, 1993, chap. 23.
70. **Parnham, M. J.**, Do phospholipids have toxicological limitations? in *Phospholipids — Characterization, Metabolism and Novel Biological Applications*, Cevc, G. and Paltauf, F., Eds., Am. Oil Chem. Soc., Champaign, IL, 1995, 56.
71. **Allen, T. M.**, Toxicity and systemic effects of phospholipids, in *Phospholipids Handbook*, Cevc, G., Ed., Marcel Dekker, New York, 1993, 801.
72. **Schreier, H., Gonzalez-Rothi, R. J., and Stecenko, A. A.**, Pulmonary delivery of liposomes, *J. Contr. Release*, 24, 209, 1993.
73. **Gregoriadis, G.**, Immunological adjuvants: a role for liposomes, *Immunol. Today*, 11, 89, 1990.
74. **Fries, L. F., Gordon, D. M., Richards, R. L., Egan, J. E., Hollingdale, M. R., Gross, M., Silverman, C., and Alving, C. R.**, Liposomal malaria vaccine in humans: a safe and potent adjuvant strategy, *Proc. Natl. Acad. Sci. U.S.A.*, 89, 358, 1992.

75. **Fujii, Y., Aramaki, Y., Hara, T., Yachi, K., Kikuchi, H., and Tsuchiya, S.,** Enhancement of systemic and mucosal immune responses following oral administration of liposomes, *Immunol. Lett.*, 36, 65, 1993.
76. **Artmann, C., Röding, J., Ghyczy, M., and Pratzel, H. G.,** Liposomes from soya phospholipids as percutaneous drug carriers, *Arzneim. Forsch./Drug Res.*, 40, 1363, 1990.
77. **Fine, S. R.,** Possible reactions to soya lecithin in aerosols, *J. Allergy Clin. Immunol.*, 87, 600, 1991.
78. **Wolf, R., Ophir, J., and Yust, I.,** Atopic dermatitis provoked by AL721 in a patient with acquired immunodeficiency syndrome, *Ann. Allergy*, 66, 421, 1991.
79. **Parnham, M. J. and Wetzig, H.,** Toxicity screening of liposomes, *Chem. Phys. Lipids*, 64, 263, 1993.
80. **Fogler, W. E.,** Immunologic properties and activities of phospholipids, in *Phospholipids Handbook*, Cevc, G., Ed., Marcel Dekker, New York, 1993, 817.

NONLIPID LIPOSOMES AND NONLIPOSOMAL LIPID PARTICLES

In this short introductory to the following chapters we shall introduce other molecules which can also make liposomes and some other colloidal particles which can be used for similar applications as liposomes.

Other Lipids: Most liposomes are prepared from phospholipids and sterols, almost exclusively cholesterol. Other phospholipids, such as sphingomyelins are also occasinally used. Glycolipids, including glycero–glyco lipids and sphingo–glyco lipids are sometimes utilized. Some other diacyl lipids are also commercially available. The formation of liposomes from lipids like diacyl glycerol may cause problems because of the tendency of such lipids to form inverted hexagonal phases and often mixing with other lipids is required to formulate them into bilayers and liposomes. This can often be observed during dry lipid film preparation and hydration: uneven film, presence of crystallites, difficult hydration, and the inability to quantitatively empty the round bottom glass flask may indicate a hydration problem and formation of nonbilayered phases.

In general, in addition to natural lipids, fatty acids, lysolecithins, and sterols, several other polar synthetic lipids are used in the preparation of liposomes. The most used are nonionic surfactants which alone, or in mixtures with cholesterol, form nonionic liposomes (niosomes) which are extensively used in cosmetics due to their increased chemical stability, ease of large scale liposome production, and cost. We won't discuss all the possible lipid and nonlipid systems which can be used to prepare liposomes [detergents, acylated crown ethers and heterocycles, acylated amino acids and ammonium salts, polymerizable lipids, bolaamphiphiles, polymer-lipids (PEGylated lipids, other polymers), block copolymers, mixtures of sterols, silicon containing lipids, and others], but we shall mention that hydrocarbon chains can be (partially) replaced by fluorocarbon chains resulting in an extremely rich phase behavior of the aqueous systems of these surfactants. Due to increased hydrophobic interaction and despite reduced attractive van der Waals interactions among chains, liposomes with different properties (higher stability, lower leakage, etc.) can be prepared.

Other particles: Most of these lipids can be used for preparation of other lipid colloidal particles. In aqueous suspensions some of these form micelles. The ones whith bulkier hydrophobic parts and/or smaller polar heads cannot form micelles but can form mixed micelles.

In mixtures with oils, emulsions and microemulsions can be made. In most of the cases we are interested more in oil-in-water (o/w) than in water-in-oil (w/o) systems. While microemulsions form spontaneously, emulsions are systems — like liposomes — which require some energy dissipation such as sonication, homogenization, or extrusion to be formed. Then, however, they can be stable due to the fact that they are in a kinetically trapped state.

There is, however, an important distinction between micelles and microemulsions on one side and liposomes on the other. In contrast to the former ones, the latter one is not a system at thermodynamic equilibrium. This has a profound effect on their properties and directly on their preparation and application. While thermodynamically stable systems are relatively easy to prepare, stable, and batches are reproducible, the kinetically stabilized systems are more difficult to prepare reproducibly and may have shelf-life stability problems. However, after application, micelles or microemulsion droplets rapidly change their phase while liposomes are stable due to the fact that they are kinetically trapped into their state and therefore an energy is needed to open or break them. For example, micelles dissolve and microemulsion droplets aggregate upon dilution, while liposomes are stable and practically do not change. This stability upon application clearly shows their superior potential for microencapsulation and delivery of active substances.

Extrusion, homogenization, or sonication of lyotropic liquid crystalline phases can result in the formation of colloidal dispersions of droplets of different liquid crystalline systems, resulting in hexasomes, cubosomes, and similar. Several groups are exploiting these particles for various applications.

D. D. Lasic
Y. Barenholz

Chapter 8

Membranes, Vesicles, and Other Supramolecular Systems Made from Fluorinated Amphiphiles

Jean G. Riess, Frédéric Frézard, Jacques Greiner, Marie Pierre Krafft, Catherine Santaella, Pierre Vierling, and Leila Zarif

CONTENTS

I. INTRODUCTION

Bilayer membranes and vesicles formed from self-aggregating amphiphiles are primarily known for the synthetic model membranes they provide[1] and as drug carrier and delivery systems.[2,3] Since the discovery in 1965 that liposomes are formed upon dispersion of phospholipids in water,[4] much effort has been devoted to the synthesis of amphiphiles different from natural lipids but which also self-organize into supramolecular systems.[5–8] The first examples of synthetic bilayer membranes and vesicles were reported in the mid-1970s.[9–12] The formation of stable single- and multi-walled vesicles from dispersions of didodecyldimethylammonium bromide was then recognized.[9] Bilayer and/or vesicle formation has since

0-8493-4012-8/96/$0.00+$.50
© 1996 by CRC Press, Inc.

been observed for a large variety of synthetic amphiphiles which contain one to four hydrophobic alkyl tails, various nonionic, cationic, anionic or zwitterionic polar heads, and sometimes also polymerizable units located either in the hydrophobic chains or in the polar head. These membranes generally present the advantage of being better defined and more easily modulable than biomembranes.

Since the early 1980s the array of well-defined synthetic self-assembling amphiphiles has been enriched by the development of many perfluoroalkylated amphiphiles.[6,7,13–16] These amphiphiles contain one to three highly fluorinated tails and display characteristics that combine the self-organizing behavior of their hydrocarbon counterparts with some of the unique characteristics of fluorocarbons, among which one should emphasize their low affinity for both hydrophilic and lipophilic phases. Fluorinated chains are considerably more hydrophobic than are hydrocarbon ones. Their presence greatly intensifies the surfactant's amphiphilic character, resulting in enhanced hydrophobic interactions and surface activity, hence the fluorinated amphiphiles' stronger tendency to collect at interfaces and to self-associate into a variety of supramolecular assemblies.[6,7,17,18] Fluorinated amphiphiles being also lipophobic display limited miscibility with hydrocarbon amphiphiles whether in micelles[19–22] or in bilayers and vesicles,[23–25] in monolayers or in Langmuir–Blodgett films.[25,26] Advantage of this peculiar lipophobic character was taken to prepare bilayer assemblies in organic media or even in fluorocarbons.[27]

A large diversity of fluorinated amphiphiles has recently been synthesized[13–17] as part of intensive research aiming at providing new, highly surface-active though biocompatible surfactants for the pharmaceutical industry and, among other things, at improving and controlling the characteristics of fluorocarbon emulsions to be used as injectable oxygen carriers, contrast agents, and drug delivery systems.[28–30] The availability of series of pure and well-defined fluorinated amphiphiles has made possible the investigation of their colloidal properties, of the relationships of these properties to molecular structure, and of the impact of the perfluoroalkyl (or *F*-alkyl*) chains. Interestingly, the preliminary assessment of the biological tolerance of fluorinated amphiphiles definitely indicates lower hemolytic activity compared to their hydrocarbon analogs and somewhat reduced acute toxicity.[15,18,31] This is of particular interest when their use in the biomedical, pharmaceutical, cosmetic, and household areas is contemplated.

This review will be primarily concerned with fluorinated amphiphiles, their phase behavior (lyotropism, thermotropism), and the structure and properties of the membranes, vesicles, and other supramolecular aggregates they form in water and in organic media. The properties that will be discussed include shelf-stability, lipophilicity, fluidity/rigidity, dynamics, permeability, encapsulation of diverse materials, and stability with respect to their release, whether the amphiphiles are used alone or in conjunction with other fluorinated or nonfluorinated amphiphiles (limited miscibility, phase-separated liposomes). Some background literature on fluorinated monolayers, Langmuir–Blodgett multilayers, and interfacial films in emulsions will also be presented. The specific properties induced by the highly fluorinated chains should have potential in nonmedical applications where the formulation, encapsulation, protection, controlled release of material, or template effect needed, examples of which are found in reaction control and catalysis, photosynthesis, surface coating, and the engineering of electronic devices. This review will also highlight the impact of the introduction of *F*-alkyl tails on phase behavior and colloidal properties, as well as the dependence of the observed effects on the number, length, and relative weight of these tails with respect to the total hydrophobic chains, i.e., the degree of "fluorination" of the membrane. The reader is also referred to the excellent overviews on self-organization of amphiphiles (including some fluorinated ones) that have been published by Fendler,[1] Ringsdorf,[6] Kunitake,[7] Ostro,[32] Friberg and Lindman,[33] Murakami,[34] and many others.

II. SUPRAMOLECULAR ORGANIZATION OF FLUORINATED AMPHIPHILES

A. SPECIFIC FEATURES OF FLUORINATED AMPHIPHILES

Significant structural differences exist between fluorocarbon and hydrocarbon surfactants.[16,17,35] Because of the larger size of the fluorine atom (van der Waals radius estimated as ca. 1.47 Å) compared to the hydrogen atom (1.20 Å),[36] fluorinated chains are more bulky than hydrogenated ones, with cross-sections of ca. 30 $Å^2$ vs. 20 $Å^2$, respectively.[37,38] The average volumes of CF_2 and CF_3 groups are estimated as 38 $Å^3$ and 92 $Å^3$, respectively, compared to 27 $Å^3$ and 54 $Å^3$ for CH_2 and CH_3 groups.[37,39–41] Another consequence of the larger size of the fluorine atom is the greater stiffness of fluorocarbon chains[42–44] which is related to the loss of the gauche/trans conformational freedom. The gauche/trans energy

* The symbol "*F*" means that all hydrogen atoms attached to the carbon atoms of the prefixed compound or radical are replaced by fluorine atoms; i.e., *F*-hexyl means C_6F_{13}.

difference is of ca. 2.0 kJ mol^{-1} and 4.6 kJ mol^{-1} for hydrocarbon and fluorocarbon chains, respectively.[45,46] Finally, in order to minimize steric hindrance, fluorinated chains usually adopt a helical conformation.[47,48] As a result of lesser flexibility of the fluorinated chains, fluorinated surfactants have higher melting points and higher Krafft points than the corresponding hydrocarbon surfactants.[49]

The C–F bond is the strongest encountered in organic chemistry (450 kJ mol^{-1} in CH_3F). Its strength is further increased when several fluorine atoms are present on a same carbon atom (485 kJ mol^{-1} in CF_4) and was reported to reach 531 kJ mol^{-1} in terminal CF_3 groups.[50] Furthermore, the high electron density of the fluorine atoms results in a compact electron shield which protects the molecule against the approach of chemical reagents by a kind of "scotchguard effect."[51]

The presence of fluorinated chains can endow amphiphilic molecules with certain of the unique characteristics of perfluorocarbons. The larger surface presented by the fluorinated chains, in conjunction with the low polarizability of the fluorine atoms, results in enhanced hydrophobicity. The "hydrophobic" effect of the chain is roughly proportional to its area in contact with water.[39] The strong hydrophobic interactions developed by fluorinated chains increase dramatically the tendency of fluorinated amphiphiles to self-assemble when dispersed in water and also to collect at interfaces, hence to display strong surface activity. This is illustrated by the fact that the incremental changes in the free energies of adsorption, $\Delta\Delta G°$, for the transfer of a CF_2 group from water to the air/water interface are almost twice those of a CH_2 group, i.e., –5.1 vs. –2.6 kJ mol^{-1}.[52,53] For a CF_2 group added to the hydrophobic chain of a fluorinated amphiphile, the $\Delta\Delta G°$ of transfer from water to a perfluorohexane/water interface (–5.35 kJ) is more negative, but only by 0.25 kJ, than for its transfer to the air/water interface (–5.10 kJ). This is in contrast with the values found for a CH_2 group added to the chain of hydrocarbon surfactant (–3.42 kJ for the hexane/water interface and –2.60 kJ for the air/water interface: a difference of 0.84 kJ).[52,53] This means that the interaction of a CF_2 group of a surfactant with perfluorohexane is significantly weaker than that of a CH_2 group with hexane. In polymers, replacing part of a hydrocarbon side-chain by an *F*-alkyl one increases the hydrophobic character but appears to reduce chain–chain interactions.[54] These results are consistent with the expected weaker interactions between fluorocarbon molecules in comparison to hydrocarbon molecules.[52]

The increase in hydrophobic interactions resulting from the replacement of hydrogenated chains by fluorinated ones appears to be large enough to compensate for weaker interactions between fluorinated chains and promote self-aggregation, provided these chains have more than four carbon atoms.[18]

Perfluoroalkylated surfactants have lower critical micellar concentration (cmc) values than their hydrocarbon counterparts, the cmc values for the former being roughly equivalent to those of hydrocarbon surfactants with a 50% longer chain.[49,55] It was shown that the contribution of a CF_2 group toward micellization for metal salts of carboxylic acids or sulfonates is roughly equivalent to 1.6 times that of a CH_2 group.[49] Such a simple equivalency rule is, however, not valid for all types of surfactants and for all their properties. For example, the cloud points of fluorinated poly(oxyethylene) surfactants are lower than those of hydrogenated ones even twice as long.[56] Moreover, a mere examination of binary and ternary phase diagrams shows that an oxyethylene group in the polar head does not have identical hydrophilic properties when the hydrophobic chain is fluorinated or hydrogenated.[56,57]

Above the cmc, solutions of fluorinated surfactants have much lower surface tensions than those of hydrocarbon analogs.[15,58,59] Small structural changes in the perfluoroalkyl chain can, however, dramatically affect certain physicochemical properties. The substitution in $F(CF_2)_6CH_2O(C_2H_4O)_3CH_3$ of one fluorine atom in terminal position by a hydrogen atom provoked, for example, a strong decrease in surface activity.[60]

While fluorinated chains convey higher surface activities than hydrogenated ones, they can also be qualified as less detergent toward membranes, as shown by their ability to reduce or suppress hemolytic activity.[18,31] In line with this, poorer protein solubilization potency was found for compound $C_{10}F_{21}C_2H_4S[CH_2CH(R)]_4H$ with $R = C(O)NHC(CH_2OH)_3$ compared to its hydrocarbon analogs and to Triton® X-100, which indicates a lower tendency to disrupt phospholipid membranes.[61] Advantage was taken of this smoother action by using the *F*-alkylated phosphocholine derivative $C_8F_{17}(CH_2)_{11}OP(O)^-{}_2O(CH_2)_2N^+(CH_3)_3$ to extract a flavoprotein-cytochrome b_{558} complex, apparently without denaturation.[62]

As a consequence of their strong self-aggregating tendency, fluorinated chains have a significant impact on the morphology of micelles. While micelles made from hydrocarbon surfactants are usually spherical, those obtained from fluorinated amphiphiles have a strong tendency to be rod-shaped. This was predicted by calculations based on the geometry of fluorinated amphiphiles,[37] and recently supported by experimental work on fluorinated polyethoxylated amides.[63,64] Aqueous dispersions of the latter compounds were investigated using ^{19}F NMR. Above the cmc, the ^{19}F NMR spectra showed separate signals

Table 1 Fluorinated Single-Chain Amphiphiles Forming Bilayers and Vesicles in Aqueous Solutions

Basic structure	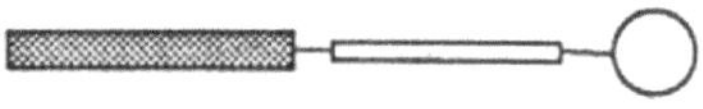		ref.
Cationic	$CF_3(CF_2)_7CH_2CH_2OC(=O)-C_6H_4-N{=}CH-C_6H_4-O(CH_2)_{10}-\overset{+}{N}(CH_3)_3$		93
Zwiterionic	$CF_3(CF_2)_nCH_2CH_2OP(=O)(O^-)OCH_2CH_2\overset{+}{N}(CH_3)_3$	n = 5 **1a** n = 7 **1b**	89
Nonionic	$C_nF_{2n+1}CH_2(OC_2H_4)_xOH$	n = 6, 7, 8; x = 3, 4, 5, 6	84
	$C_nF_{2n+1}CH_2C(O)N[(C_2H_4O)_3CH_3]_2$	n = 6, 8, 10	87
	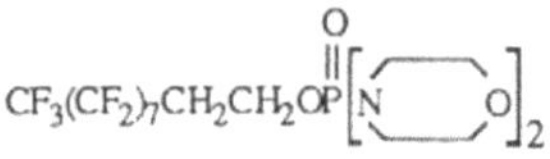	**2**	102
	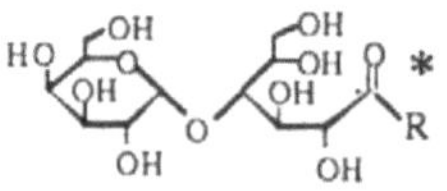	R = $-NH(CH_2)_2C_8F_{17}$ **3a** $-NH(CH_2)_9CH_3$ $-NHCH_2C(O)NH(CH_2)_2C_8F_{17}$ **3b**	107 107 91,106
Basic structure	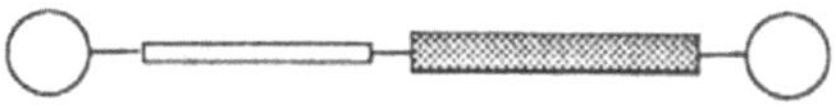		
Anionic	$NaO_2C(CH_2)_{10}(CF_2)_8O(CF_2)_2SO_3Na$	**4**	171

* Those form also supramolecular assemblies other than vesicles.

due to the simultaneous presence of monomers and micelles, indicating slower exchange of the amphiphiles between these two species than in any other previously reported micellar system.[63] These dispersions exhibit non-Newtonian pseudoplastic behavior with viscosities which increase very rapidlv with increasing concentration. Large aggregates (100 to 200 nm) were also reported to form. It was suggested that these aggregates may exist in a cubic phase. On the other hand, some of the less hydrophobic fluorocarbon surfactants and the hydrocarbon analogs have normal rheological properties (Newtonian with low viscosities). The non-Newtonian behavior was shown to be due to the formation of long rod-like micelles (80 to 200 nm in length, 4 to 5 nm in diameter) as evidenced by freeze-fracture electron microscopy.[63] No such rod-like micelles were observed in the case of the hydrocarbon analogs.

Tables 1 to 3 collect examples of single-, double-, and triple-chain fluorinated amphiphiles known to form lamellar phases and vesicles. Many of these amphiphiles belong to families of homologous compounds with a modular design which allows incremental structural variations, thus facilitating the establishment of structure/properties relationships.[13–15] These surfactants present a wide structural diversity, can be nonionic, anionic, cationic, or zwitterionic, and include models and close analogs of natural phospholipids, but also of glycolipids, glycophospholipids, and derivatives of polyols or of tris(hydroxymethyl)aminomethane, ammonium or phosphate groups, aminoacids, and other polar groups. They usually comprise a combination of one to three hydrophobic tails of variable length and fluorine content, diverse linkages, a variety of connecting units, and the polar head groups listed above.

Table 2 Fluorinated Double-Chain Amphiphiles Forming Bilayers and Vesicles in Aqueous Solutions

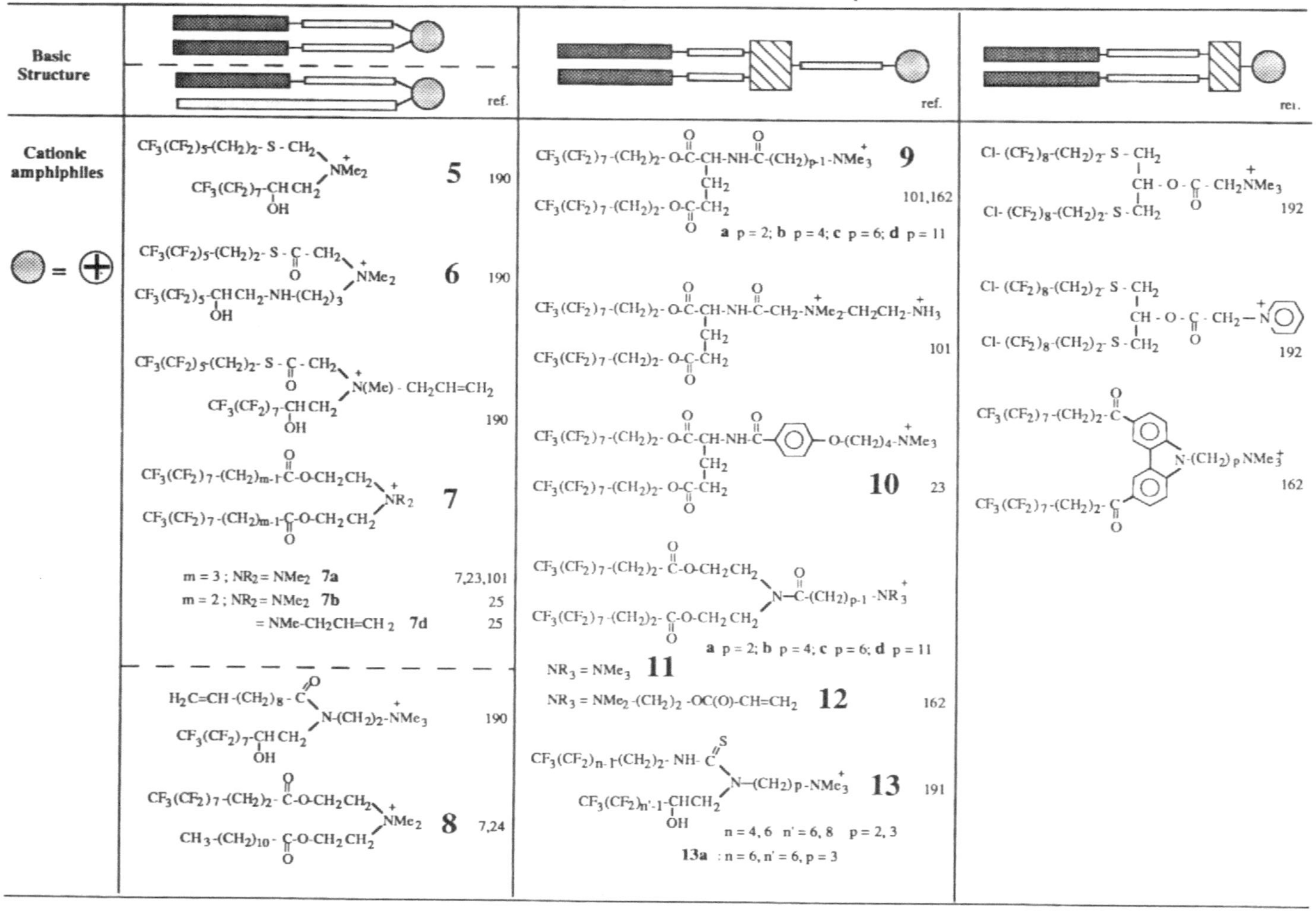

Table 2 (continued)

Basic Structure	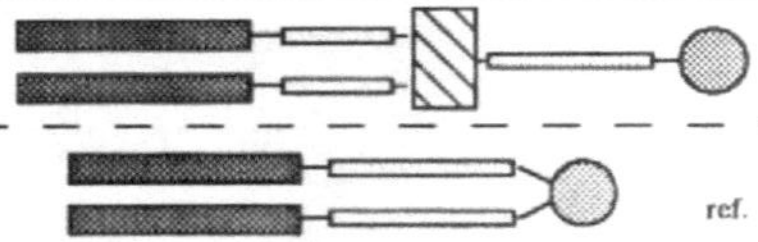	ref.	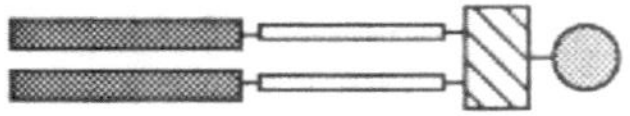	ref.	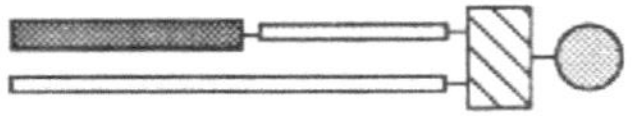	ref.
Anionic amphiphiles 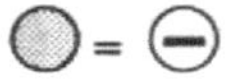	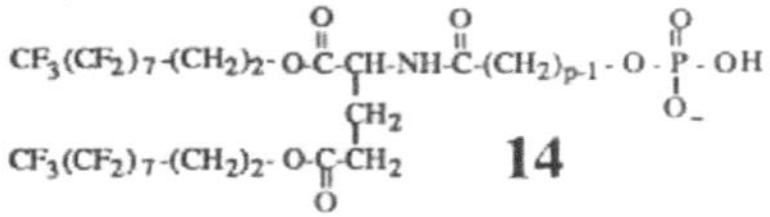 **14** a p = 2; b p = 4; c p = 6; d p = 11	162	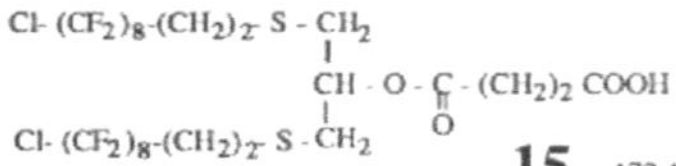 **15**	172,192	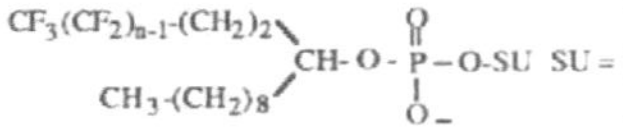 SU = - Gal6 **18**, - Glc3 **19**, - Glc6 **20**, - Man6 **21**; a n = 6; b n = 8	131,193
	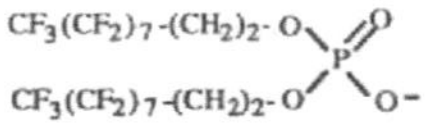	24	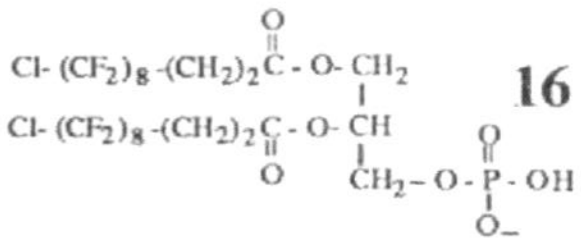 **16**	192	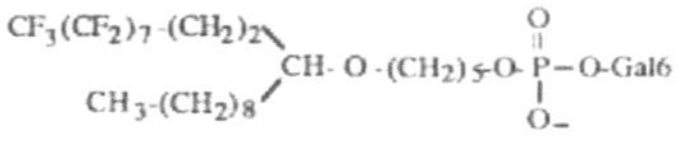 **22**	131,178
	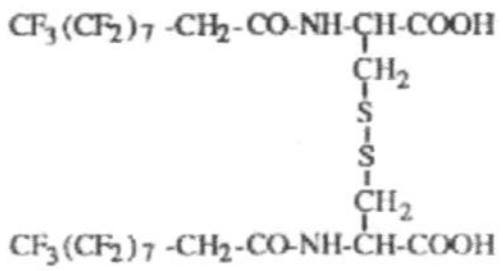	6	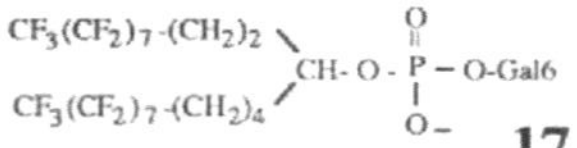 **17**	178		
Non-ionic amphiphiles	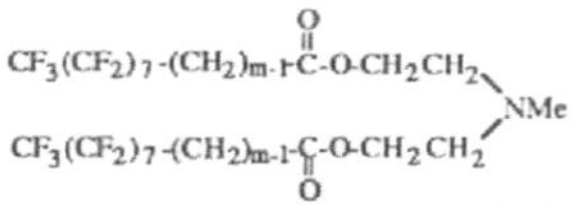 **7** m = 2 **7c**	25	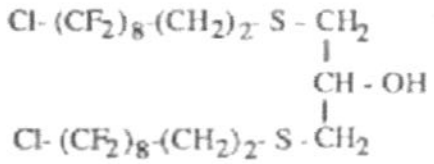	192	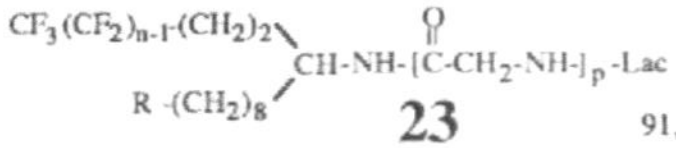 **23** R = CH_3- ; n = 6, p = 1 **23a**; n = 8, p = 1 **23b**; $CH_2=CH-$; n = 6, p = 1 **23c**; n = 8, p = 1 **23d**; CH_3- ; n = 6, p = 2 **23e**	91,112,113
					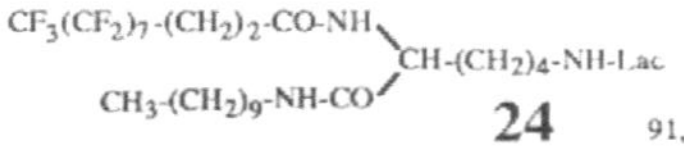 **24**	91,112,113
					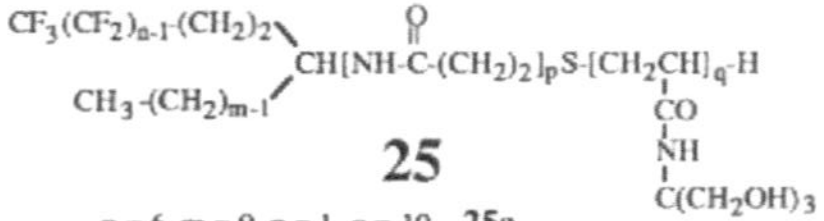 **25** n = 6, m = 9, p = 1, q = 10 **25a**; n = 6, m = 11, p = 1, q = 11 **25b**; n = 8, m = 11, p = 0, q = 20 **25c**; n = 8, m = 9, p = 1, q = 14 **25d**	114,115

Basic Structure	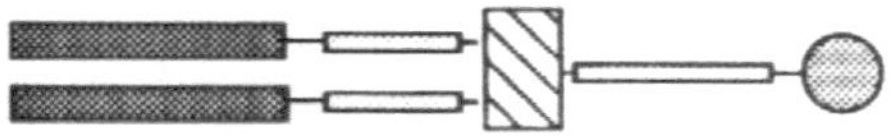 ref.	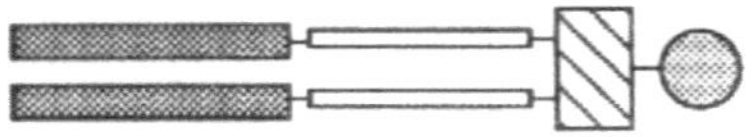 ref.	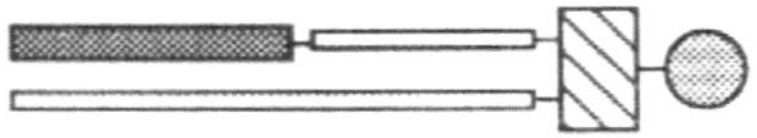 ref.
Zwitterionic amphiphiles 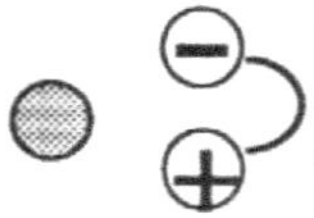	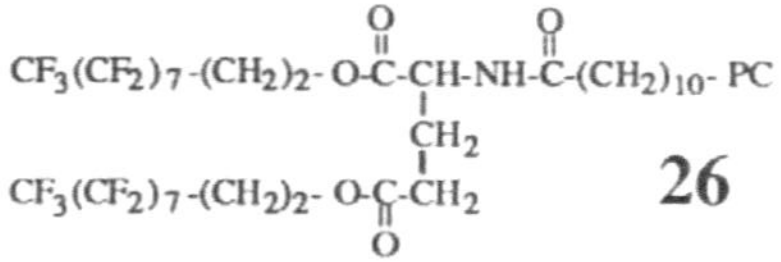 **26** 162 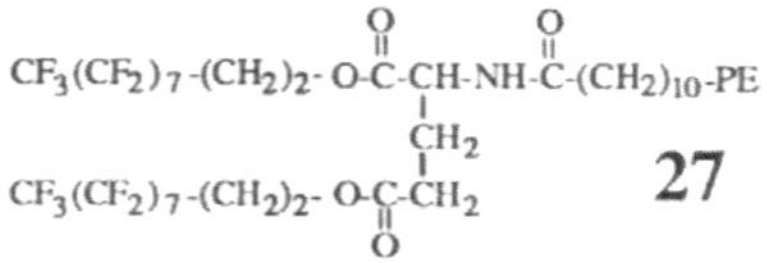 **27** 162 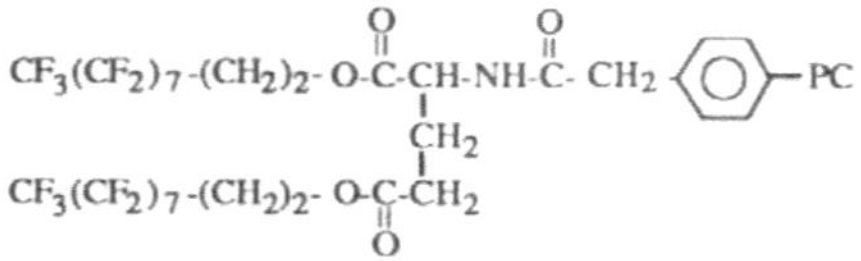 194	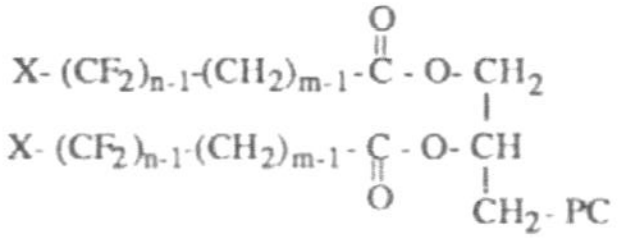 X = Cl n = 9 m = 3 **28** 195 X = CF_3 **DFnCmPC** **29** 111,157 **a** n = 4, m = 5; **b** n = 4, m = 11 **c** n = 6, m = 5; **d** n = 6, m = 7 **e** n = 6, m = 11; **f** n = 8, m = 5 **g** n = 8, m = 11 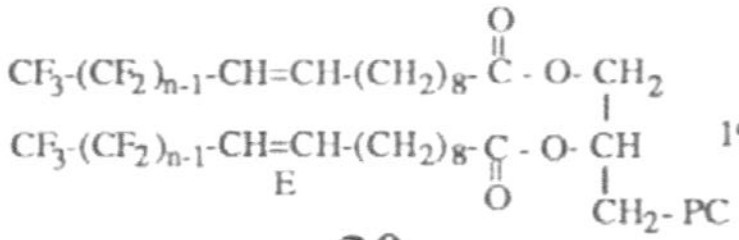 196 **a** n = 6; **b** n = 8 **30**	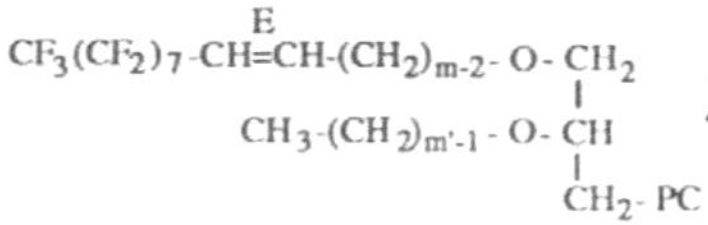 **34** **a** m = 5, m' = 14 **b** m = 11, m' = 16 159 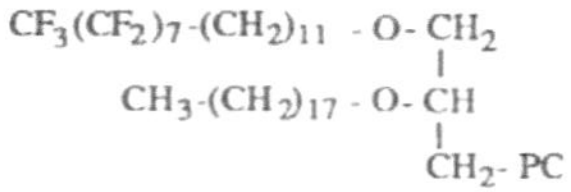 **35** 159

Table 2 (continued)

Basic Structure	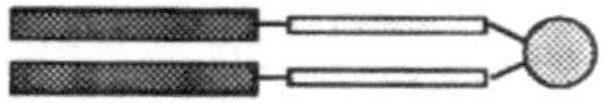 ref.	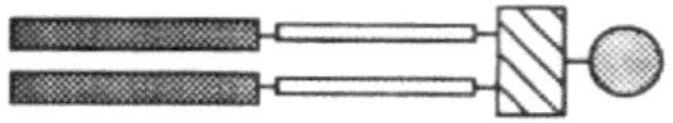 ref.	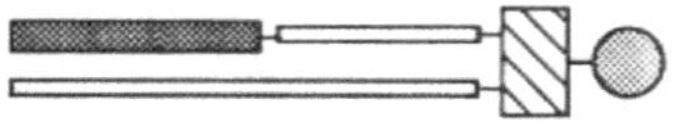 ref.
Zwitterionic amphiphiles 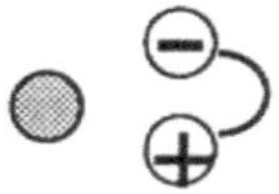	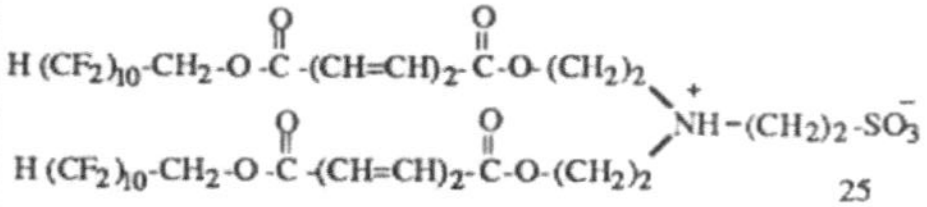 **47** 25	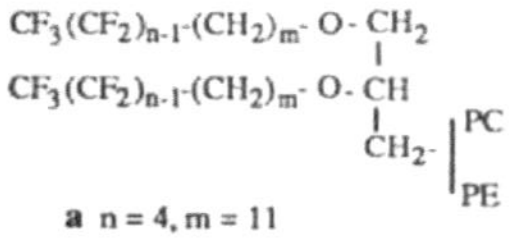 **31** (PC) **32** (PE) **a** n = 4, m = 11 **b** n = 8, m = 11 159 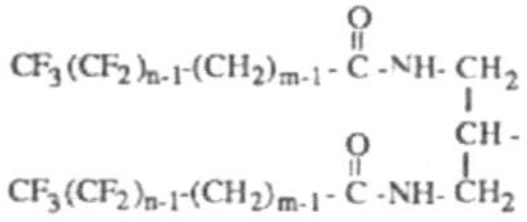 **33** **a** n = 4, m = 11 **b** n = 6, m = 11 161	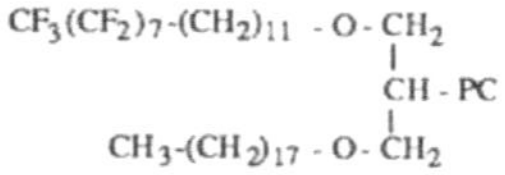 **36** 159

Note:

 : **fluorocarbon tail**

 : **hydrocarbon spacer or tail**

 : **connecting unit**

 : **polar head**

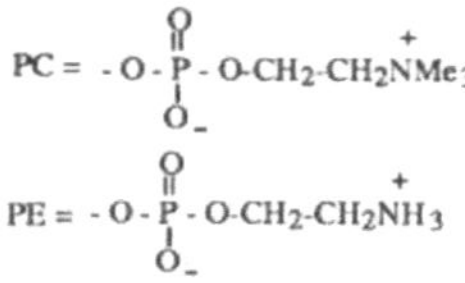

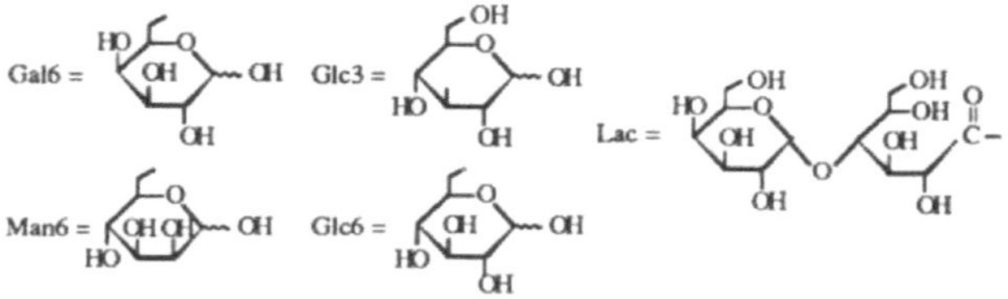

Table 3 Fluorinated Triple-Chain Amphiphiles Forming Bilayers and Vesicles in Aqueous Solution

Basic Structure	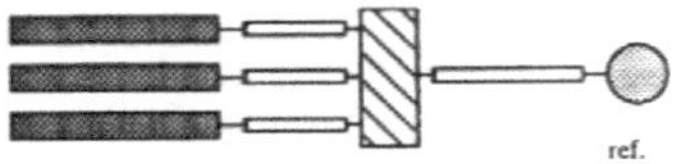 ref.	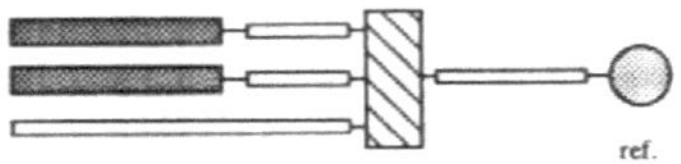 ref.	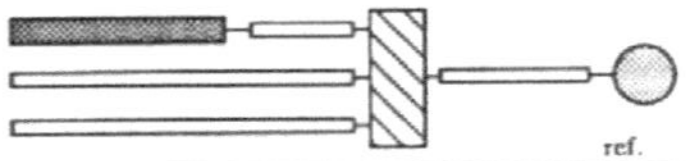 ref.
Cationic amphiphiles 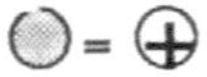	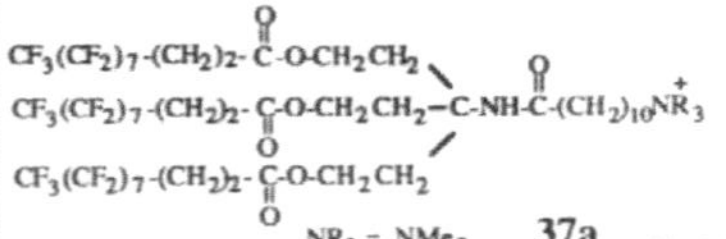 NR_3 = NMe_3 **37a**; = NMe_2Ac* **37b** 158,164	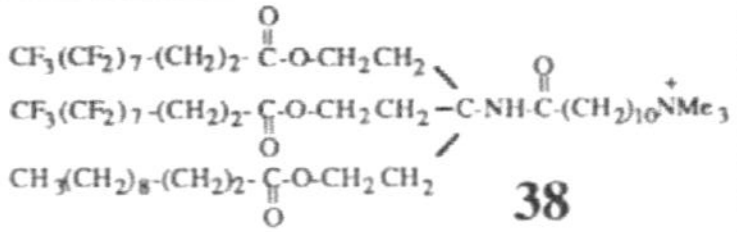 **38** 158,164	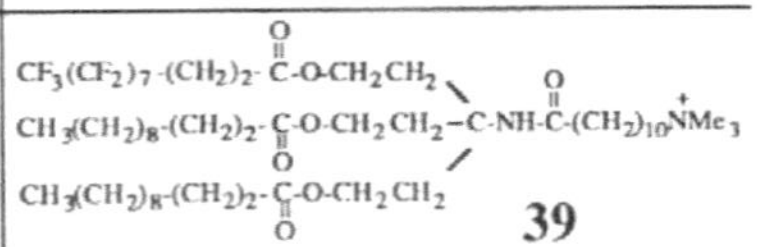 **39** 158,164

* Ac = $-(CH_2)_2-OC(O)-CH=CH_2$

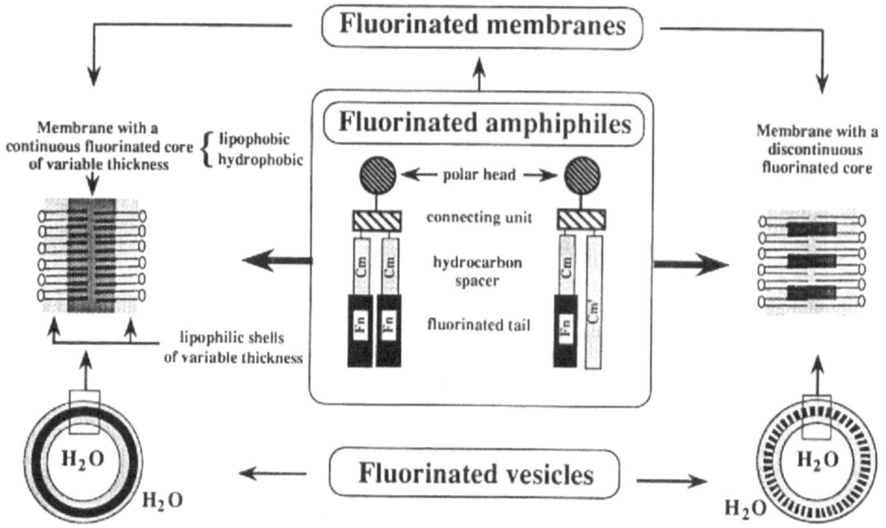

Figure 1 Structure of double-tailed fluorinated amphiphiles, membranes, and vesicles.

Table 4 Nonconventional Fluorinated Amphiphiles Forming Bilayers and Vesicles in Hydrocarbon and Fluorocarbon Media

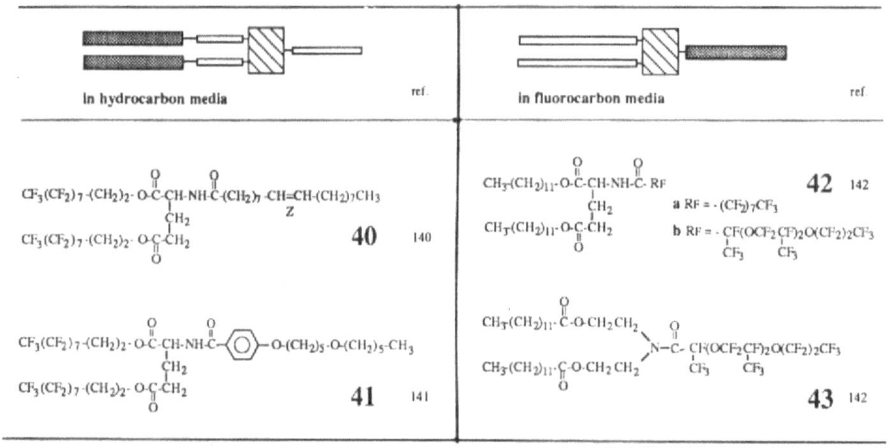

These structural features are expected to play a role in determining the hydrophilic/lipophilic/lipophobic balance, chemical and biological stability, and *in vivo* fate of the surfactants, as well as the physicochemical and biological properties of the membranes, vesicles, and other aggregates of which they are part. In particular, it would be expected that the presence of a fluorinated core inside a bilayer membrane (Figure 1) would modify its shelf-stability, fluidity/rigidity, permeability, drug encapsulation, and release parameters, stability in biological fluids, *in vivo* recognition, intravascular persistence, adsorption, distribution, metabolism and excretion characteristics, and biocompatibility. Certain features, such as the presence of sugar moieties, can be built into the structure of the amphiphiles in view of achieving drug targeting.[14,65]

Table 4 collects the few nonconventional fluorinated amphiphiles reported to form bilayers, vesicles, and other self-assemblies in fluorocarbon and hydrocarbon solvents.

B. MONOLAYERS AND LANGMUIR–BLODGETT (LB) FILMS FROM FLUORINATED AMPHIPHILES

Numerous Langmuir and Langmuir–Blodgett (LB) films of fluorinated amphiphiles have been prepared and studied in recent years. Langmuir monolayers are useful models for investigating the structure and functions of biological membranes and their interactions with lipids, steroids, sugars, proteins, ions, etc.

They have also been used to induce nucleation of diverse organic or inorganic three-dimensional crystals from molecules or ions present in the sub-phase. This further allowed the study of the interactions between crystal surface and solvents. Langmuir monolayers can be transferred onto solid supports for the formation of multilayer (LB) films with potential as piezoelectric sensors, nonlinear optical devices, for the passivation of surfaces, etc.

1. Formation and Stability

The *F*-alkylated double-chain amine **7c**, and ammonium salts **7b,d** with short 9-carbon atom hydrophobic chains, were shown to form at the air–water interface monolayers that are significantly more stable than those formed by their hydrocarbon counterparts, as evidenced by collapse pressures that are remarkably high (50 to 60 mN m^{-1}) for such short chains.[25] No significant enhancement of the thermal and mechanical integrity was observed for their LB films.[54] Polymerization of fluorinated amphiphiles containing appropriate groups,[66] or the introduction of polymeric properties by the formation of a complex between a fluorinated amphiphile and a polyelectrolyte counterion,[67] led to LB films of increased stability. The replacement of hydrocarbon by fluorocarbon chains in polymers also enabled the formation of thinner LB multilayers with improved thermal stability.[54,68]

The high chemical stability of perfluorinated compounds allowed the formation of films on the surface of concentrated corrosive acids. Thus, the *F*-alkylpolyether $CF_3[CF(CF_3)CF_2O]_{27(average)}CF_2CF_3$ gave stable, insoluble monolayers when spread on concentrated sulfuric, nitric, and phosphoric acids.[69] Surface pressure-area measurements indicated that, at monolayer coverage, the *F*-alkylethers preferentially pack lying down on sulfuric acid. It was suggested that spreading was driven by hydrogen bonding between the acid and the ether oxygen or fluorine atoms of the perfluorocarbon.

More recently, it has been reported that a totally nonpolar linear fluorocarbon, $C_{20}F_{42}$, can form ordered, stable Langmuir monolayers on water with the molecules aligned vertically, even though they have no amphiphilic character.[70] It was suggested that the interactions within the layer were sufficiently strong to maintain the layer packing when one face of the layer is in contact with water, and that the van der Waals forces between this layer and the water surface were sufficient to stabilize the system as a Langmuir monolayer.

2. Structure and Properties

The replacement of hydrocarbon by fluorocarbon chains in the double-chain amine and ammonium salts **7**, led to larger limiting cross-sectional molecular areas occupied by the amphiphiles.[25] A strong influence of the polar head was also demonstrated. Moreover when the uncharged head group in **7c** was replaced by a charged one, as in **7b,d**, a decrease in collapse area was observed in the case of the fluorocarbon amphiphiles, while the opposite is normally observed for lipids with hydrocarbon chains. This effect certainly warrants further investigations.[25] Similarly, an increase in collapse area was found when the C_4H_9 or C_6H_{13} terminations were replaced by fluorinated ones in DMPC or DPPC, as in **29d** (DF4C11PC), and **29e** (DF6C11PC).[71] By contrast, introducing a C_8F_{17} tail in DMPC, as in **29f** (DF8C5PC), appeared to decrease (by ca. 5%) the limiting cross-sectional molecular area. DF8C5PC thus displayed a lower collapse area than its longer and less fluorinated homologs DF6C11PC and DF4C11PC.

The development of surface specific X-ray diffraction and fluorescence techniques have recently revolutionized the study of Langmuir monolayers. These methods allow the determination of the in-plane and vertical structure of monolayers with a resolution approaching the atomic level.[72] Langmuir monolayers of fluorinated molecules are characterized by a higher crystallinity than those formed by their hydrocarbon counterparts. Grazing incidence diffraction data on fluorinated monolayers of 4-[2-(*F*-decyl)ethyl] aspartate, $C_{10}F_{21}C_2H_4OC(O)CH_2CH(NH_3^+)CO_2^-$, and (*F*-decyl)acetic acid, $C_{10}F_{21}CH_2CO_2H$, suggested the coexistence of ordered condensed islands and of a dilute disordered phase.[72,73] The molecular tilt in the ordered phase was very small (near 0°) and remained unchanged throughout the coexistence region. In the hydrocarbon monolayers, on the other hand, the ordered phase were compressible with a variable tilt angle ranging from 30° to 0° at closest packing.[73] The differences were attributed to the enhanced stiffness of the *F*-alkyl chain. These experimental data are in qualitative agreement with recent molecular dynamics simulations.[74,75] Such simulations also showed that the concentration of gauche defects in the hydrogenated monolayer ($CH_3(CH_2)_{10}COOH$) is much higher (2.16%) than in the fluorinated one ($CF_3(CF_2)_{10}COOH$, 0.11%).[75] It is interesting to note that the defects are primarily located near the polar head for the fluorinated surfactants and near the tail end for the hydrogenated ones.

3. Applications

Langmuir monolayers formed from functionalized amphiphiles at the air–water interface may serve as powerful, selective tools for the extraction and 2D crystallization of proteins[76] and for molecular recognition via hydrogen bonding or ionic pairing.[77] Recently, monolayers of double fluorocarbon chain steroidal amphiphiles bearing cellubiose or maltose as the hydrophilic head groups were shown to exhibit pressure-area behaviors which were related, respectively, to specific and nonspecific interaction with the water-soluble enzyme cellulase (which is known to interact specifically with cellubiose and nonspecifically with maltose).[78]

Another potential development of fluorocarbon multibilayers lies in their immobilization by casting in the form of transparent films which are expected to exhibit gas permeability and permselectivity properties.[79,80] Highly ordered self-supporting thin films conserving the bilayer structure were obtained by casting of aqueous fluorocarbon bilayers made from double-chain ammonium amphiphiles in the presence (or not) of poly-vinyl alcohol. The surface structure of the composite films based on a fluorinated amphiphile and a polymer was analyzed by wide-angle X-ray diffraction and X-ray photoelectron spectroscopy.[81,82] Such films exhibited enhanced gas permeation and were shown to permeate oxygen in preference to nitrogen, while films obtained from analogous hydrocarbon bilayers exhibited lower gas permeation and lower oxygen-enrichment capacities.[79,80]

By observing the frequency change of a multibilayer-coated piezoelectric crystal in an aqueous solution, selective molecular adsorption and penetration of long-chain hydrocarbon and fluorocarbon alcohols into synthetic hydrocarbon and fluorocarbon multibilayer matrices, respectively, were detected.[83] This appears to be the first example of selective molecular recognition of synthetic bilayer membranes. It should also be mentioned that the fluorinated alcohol investigated, $C_8F_{17}C_2H_4OH$, displayed very large partition coefficients for all multibilayers, irrespective of their nature, the highest value being, however, found for the fluorocarbon multibilayer. This was attributed to the higher hydrophobicity of the fluorocarbon alcohol which favors its penetration even into hydrocarbon multibilayers.

C. BILAYERS AND VESICLES FORMED FROM FLUORINATED SINGLE-CHAIN AMPHIPHILES

The first phase diagram for nonionic fluorinated amphiphiles was published only in 1987.[84] Lamellar phases are strongly favored for aqueous dispersions of both ionic and nonionic *F*-alkylated amphiphiles even when dilute.[37,85,86] The phase equilibria have been determined between 0 and 100°C for a number of $C_nF_{2n+1}CH_2(OC_2H_4)_xOH$ compounds (n = 6–8, x = 3–6). The most striking result is the existence for these surfactants of very large domains of lamellar Lα phase (roughly for surfactant contents in the 50 to 90% w/w range) and the absence of the usual L_1 micellar phase (i.e., the surfactant appears almost "water insoluble"). Typically, the Lα phase melts at about 70 to 80°C, a noticeably higher temperature than for the parent hydrogenated compounds. In the water-rich domain the Lα phase is dispersed in the water phase. Below 35°C even the short-chain fluorinated surfactant $C_{10}F_{21}CH_2C(O)N[(C_2H_4O)CH_3]_2$ forms gel phases in water. These gel phases are closely related to the lamellar phases, both having a layer structure of alternating hydrophobic and water regions. These phases are typical of systems which involve highly hydrophobic single- and/or double-chain amphiphiles.[87] The hydrophobic chains can be interdigitated and/or crystallized. The formation of vesicles has been suggested for low surfactant concentrations (0 to 5%). The substitution in $C_nF_{2n+1}CH_2(OC_2H_4)_xOH$ of one single fluorine atom in terminal position by a hydrogen atom was shown to modify the phase diagram drastically.[87] Following the initial observation of a lamellar phase formed by ammonium perfluorooctanoate,[88] it has been shown that other ionic derivatives give hexagonal, intermediate, and lamellar phases.[37,86]

Replacement of a hydrocarbon chain by a fluorinated one can drastically affect the aggregation behavior of a surfactant. This is the case for example for some *F*-alkylated derivatives of phosphocholine and glycolipids which form small unilamellar vesicles (SUV), multilamellar vesicles (MLV), and other organized systems, including fibers and tubules rather than micelles.[89–91]

Hydrogenated single-chain amphiphiles at concentrations higher than their critical micellar concentration tend normally to aggregate into micelles which may be spherical, cylindrical, or plate-shaped.[1,6,92] They usually do not form stable vesicles unless some additional structural features or interactions are brought into play. These interactions are needed to promote the alignment of molecules and maintain tight packing. They can be of various natures and may include steric effects, electrostatic effects, hydrogen bonding, or polymerization. Thus, the insertion into hydrogenated single-chain amphiphiles of a rigid, rod-like segment between the polar head and the hydrophobic tail, e.g., a diphenyl or diacetylenic unit, has led to vesicles.[93] The formation of vesicles from "hyperextended" highly hydrophobic polyammonium amphiphiles, i.e., amphiphiles with a very long hydrocarbon tail (24, 28, or 35 carbons), has recently been reported.[94] The authors suggest that such long chains either strongly interdigitate or fold in their center to form "pseudo-double-chain" entities. Another approach consists of incorporating cholesterol in dispersions of nonionic amphiphiles; this allowed, for example, the preparation of the "niosomes" used in cosmetics.[95,96] The enabling intermolecular associations can also result from electrostatic forces. For example, the equimolar association of cationic and anionic amphiphiles led to the formation of very stable vesicles;[97,98] the ion-pairing of the polar heads then again mimics a double-tail situation. pH-Dependent vesicles, in which self-association results primarily from hydrogen bonding, were prepared by titration of an alkaline dispersion of a fatty acid by a dilute acid solution.[11,12] Finally, membranes formed by amphiphiles bearing a diacetylenic or a vinylic unit can be polymerized, which assures their stability.[6] The use of a polymeric counterion (polyacrylate) as a matrix to induce bilayer formation from an ammonium surfactant was also reported.[100]

Single-chain fluorinated amphiphiles bearing a rigid rod-shaped segment in their hydrophobic chain have been synthesized (see Table 1). Their general aggregation behavior was closely related to that of their hydrogenated counterparts.[7,24,101] Various ordered structures which display thermotropic liquid-crystalline behavior were observed to form. The formation of vesicles from fluorinated ethoxylated alcohols has also been reported.[85,87]

The question as to whether vesicles and/or other organized supramolecular systems could be obtained from a single-chain amphiphile by principally enhancing the hydrophobic interactions through the use of a highly fluorinated tail was investigated.[89,102,103] It was found possible to obtain vesicles from aqueous dispersions of *F*-alkylated single-chain amphiphiles derived from phosphocholine (e.g., **1b**) or from dimorpholinophosphoramide (e.g., **2,** Table 1).[102–105] The preparation of these vesicles includes first the formation of a gel phase (ca. 10^{-1} *M*) by hand shaking, followed by dilution with water (ca. 10^{-2} *M*) and brief sonication at 30°C. The average size of these vesicles was established by laser light scattering (LLS) measurements. The vesicles were characterized by freeze-fracture electron microscopy. Their shelf-stability depends on the amphiphile's structure. Those prepared from **1b** (ca. 70 ± 40 nm) were very stable, with no significant variation in average size for at least 3 months at 25 or 40°C. They were also stable enough to withstand heat sterilization under standard conditions (121°C, 15 min, 10^5 N m^{-2}) without noticeable degradation (Figure 2). On the other hand, vesicles prepared from **2** were stable for a few hours only.

The strong tendency to aggregation demonstrated for the *F*-alkylated single-chain phosphocholine **1b** is in sharp contrast with the behavior of its hydrocarbon analog, decylphosphocholine.[102] For the latter, no liquid crystal formation was visualized by polarization microscopy, and no aggregates were seen by transmission electron microscopy either. LLS measurements showed the absence of aggregates larger than micelles. Altogether the results indicated that the dispersions of the hydrocarbon amphiphile consist only in fluid micelles, even at concentrations up to 30% w/v.

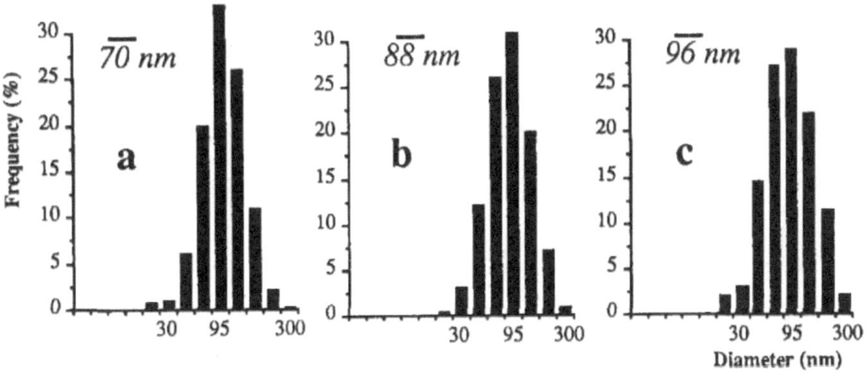

Figure 2 Particle size distribution histograms of vesicles made from single-chain perfluoroalkylated phosphocholine derivative **1b** measured by laser light scattering spectroscopy (a) immediately after preparation, (b) after heat sterilization, and (c) after 3 months at 40°C. (From Riess, J.G. and Krafft, M. P., *Chem. Phys. Lipids,* 75, 1, 1995. With permission.)

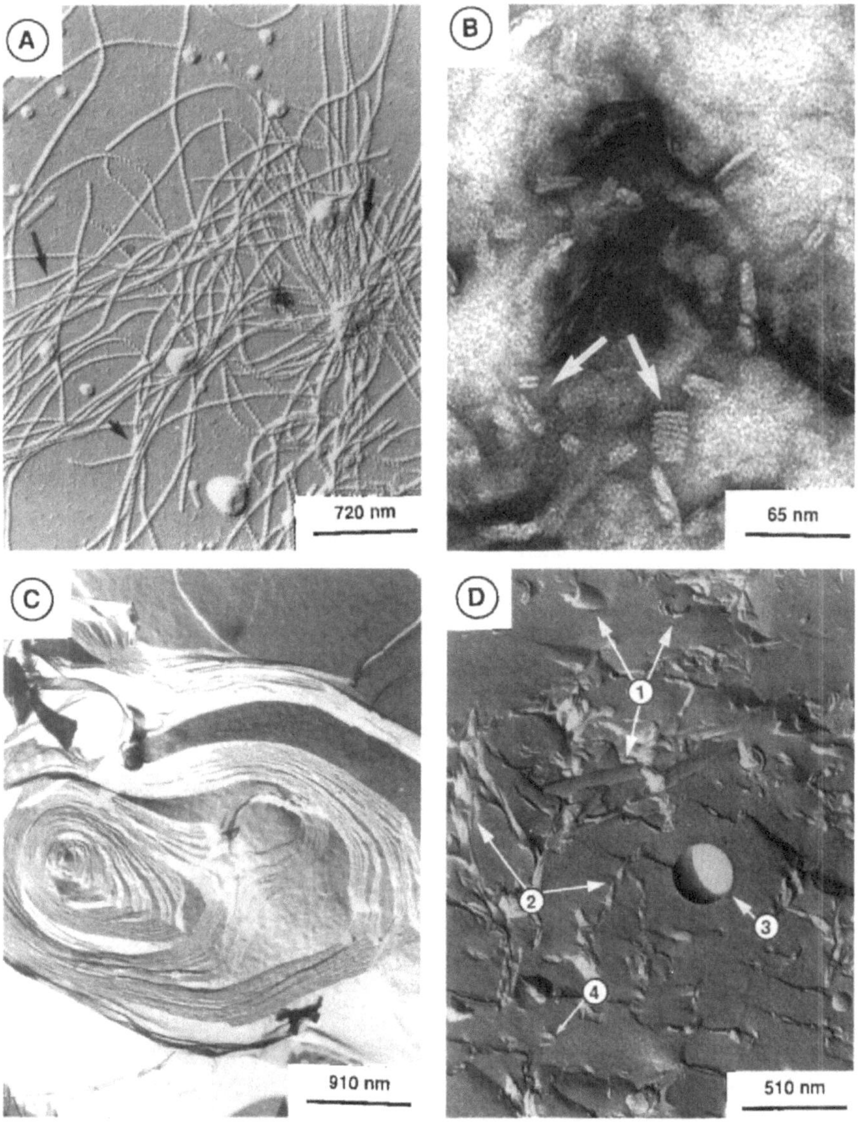

Figure 3 Phase behavior of glycolipid-derived amphiphiles in aqueous media. (A) electron micrograph of an aqueous dispersion of the single-chain fluorinated glycolipid **3a** after shadowing with C-Pt; the arrows point to a bundle of helixes. (B) electron micrograph of an aqueous dispersion of the single-chain glycolipid **3b** after negative staining by phosphotungstic acid; the arrows point to stacked disks. (C) freeze-fracture electron micrograph of an aqueous dispersion of the mixed double-chain glycolipid **23d**; note the stacked bilayers. (D) freeze-fracture electron micrograph of an aqueous dispersion of glycolipid **24**; arrows point to (1) tubules, (2) helices, (3) liposomes, (4) disks.[91,107]

The glycolipid **3a** (Table 1, 100% D-configuration) forms long, left-handed helices, several microns in length.[91,106] The helical structures are shown in Figure 3A which represents an electron micrograph of a dispersion of this glycolipid after C/Pt shadowing.[107] The fluorinated glycolipid **3b**, bearing a glycine spacer, was shown to self-organize spontaneously in disk-like assemblies, which are likely to be precursors of vesicles. The disks were visualized by transmission electron microscopy after negative staining (Figure 3B).

For hydrocarbon amphiphiles, Israelachvili's concept of "optimal surface area" per polar head, for which the total free energy per amphiphile molecule is minimum, allows the prediction of the formation, type, and size of aggregates (spherical or rod-like direct micelles, planar bilayers, vesicles, and inverted

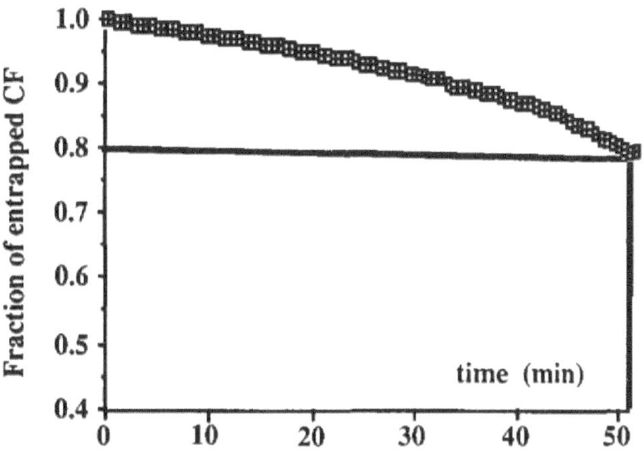

Figure 4 Release of 5(6)-carboxyfluorescein from vesicles made from equimolar amounts of 2-(perfluorooctyl)ethylphosphocholine **1b** and (perfluorooctyl)ethane. (From Trevino, L., et al., *Proc. Int. Symp. Control. Rel. Bioact. Mater.*, 21, 606, 1994. With permission.)

micelles).[40] This theory is only applicable to a certain extent to *F*-alkylated single-chain amphiphiles. The pertinent surface-activity data and geometrical characteristics needed to calculate the packing parameter (P) were determined for **1b** and **2**, and for their hydrogenated analogs.[108] The structures predicted from these data were lamellar phases and vesicles in the case of the *F*-alkylated amphiphiles (P ca. 0.7) and micelles in the case of their hydrocarbon analogs (P ca. 0.5); such systems were indeed observed to form.[89,102] The model did not, however, predict the highly versatile behavior of **1b** and **2** and the formation of tubules and fibers (Section II.F.).

The formation of stable multi- and unilamellar vesicles from an *F*-alkylated phosphocholine, when compared to the inability of its hydrocarbon analog to form anything more organized than micelles, establishes that a single *F*-alkyl chain can suffice to induce the alignment of amphiphilic molecules in bilayers. One major component of this driving force is likely to derive from the strongly hydrophobic forces developed by the presence of *F*-alkyl chains; these forces are of course also related to the volume and stiffness of these chains.[89,103]

The fluorinated vesicles obtained from **1b** were highly permeable to entrapped 5(6)-carboxyfluorescein (CF) in a phosphate-buffered saline solution (PBS) (half-life <1 min). This is probably due to a lack of stability toward the dehydrating electrolytes present in the buffer, and can be corrected to a certain extent by using mixtures of the single chain *F*-alkylated amphiphile **1b** with the mixed fluorinated/hydrogenated compound $C_8F_{17}C_2H_5$ (F8H2).[109] The stability in saline (NaCl 0.1 *M*) of the fluorinated vesicles prepared from an equimolar complementary mixture of **1b** and F8H2 was significantly increased and the encapsulation of CF in a PBS solution then became possible (Figure 4); 80% of the entrapped probe was still present in the vesicles after 52 min.[110] This association of a single-chain phosphocholine amphiphile with a mixed fluorocarbon/hydrocarbon compound of appropriate structure probably leads to the reconstitution of a pseudo double-tail amphiphile, the cohesion between the hydrophilic-fluorophilic component, **1b**, and the entirely hydrophobic second component F8H2 being provided by hydrophobic interactions. Tighter packing of the *F*-alkylated tails within the membrane was expected to reduce its permeability, and this was borne out by the experimental data.

Much attention has been focused on the influence of entropic forces on interactions in both natural and synthetic membranes; there is, however, a scarcity of useful model membranes for studying these effects, which may have retarded the understanding of aggregation behavior. The single-chain systems reported above and the unique aggregation properties induced by fluorinated end groups may help provide insight into interactions between amphiphilic surfaces.

D. BILAYERS AND VESICLES MADE FROM FLUORINATED MULTI-CHAIN AMPHIPHILES

Tables 2 and 3 collect the various fluorinated double- and triple-chain amphiphiles for which bilayer, uni-, and/or multilamellar vesicle formation was reported. These amphiphiles are composed of a large variety of nonionic, cationic, anionic, or zwitterionic head groups and at least one fluorinated chain. The hydrophobic chains are linked to the polar head through various bonds (ester, thioester, ether, thioether, amide, or C–C) to connecting units derived from glycerol, thioglycerol, glutamic acid, diethanolamine, diaminopropanol, glycine, glycyl-glycine, lysine, tris(hydroxymethyl)aminomethane, polyol, sugar, etc.

The first examples of fluorinated bilayers and liposomes were reported in 1982 by Kunitake et al.[101] for amphiphiles of the series **9** and **11** (Table 2). The fluorinated phospholipid analogs **29** (DFnCmPC) were shown to form stable uni- or multilamellar vesicles when the total length (Fn + Cm) of hydrophobic chains ranges from 9 to 17 atoms, whatever the length of the Fn tail.[111] In comparison, conventional hydrocarbon phosphatidylcholines form uni- or multilamellar vesicles for a chain length of 12 to 24 carbon atoms and primarily micelles when the chain length is shorter than 10 carbon atoms. The hydrophobic interactions and geometrical characteristics of the DFnCmPCs are thus seen to promote a bilayer structure in water even for chains as short as $C_4F_9(CH_2)_4C(O)$, as in **29a**.

Unilamellar and/or multilamellar vesicles were also obtained from the mixed fluorinated/hydrogenated double-tailed glycolipids **23** to **24** (Table 2).[91,112,113] The intercalation of a glycyl–glycine spacer between the hydrophobic chain and the polar head results in the immediate formation of SUVs by simple hand shaking.

Vesicles were likewise formed, along with numerous micelles, from the mixed fluorinated/hydrogenated double-tail tris(hydroxymethyl)aminomethane telomers **25**.[114–116] The very large, very hydrophilic polar head present in these amphiphiles may explain that micelles and not only vesicles are present in this case.

E. FLUORINATED MEMBRANES AND LIPOSOMES FROM PHOSPHOLIPIDS AND MIXED FLUOROCARBON/HYDROCARBON AMPHIPHILES

An alternative approach to fluorinated liposomes has recently been devised which consists in combining standard phospholipids with linear mixed fluorocarbon/hydrocarbon compounds such as (perfluoroalkyl)alkanes ($C_nF_{2n+1}C_mH_{2m+1}$, FnHm) or (perfluoroalkyl)alkenes ($C_nF_{2n+1}CH{=}CHC_mH_{2m+1}$, FnHmE). The co-organization of the two complementary types of amphiphiles (one hydrophilic-lipophilic, the other lipophilic-fluorophilic) results indeed in liposomes with an internally fluorinated core within their bilayer membrane.[103,117,118] The same surfactant system had previously been highly successful for stabilizing fluorocarbon emulsions[119,120] and controlling their particle sizes.[121] The incorporation of the FnHm compounds was then shown by the authors to result in tighter packing of the phospholipidic surfactant film.[122] Side by side comparison of FnHm compounds and nonamphiphilic higher boiling point fluorocarbons, both used as stabilizers of fluorocarbon/phospholipid emulsions, indicated that the stabilization mechanism is different and that the former compounds are incorporated in the surfactant film.[123] It was also shown that the adsorption of sodium laurate at a fluorocarbon/water interface is increased when FnHm molecules are present.[124]

The case of the liposomes made from the DMPC/F4H10E combination has been investigated in detail.[117] For a 1:2 molar ratio of the two components a single population of small (ca. 20 nm) SUVs was obtained. The incorporation of (*F*-alkyl)alkenes resulted in marked changes of properties. The physical stability of the mixed DMPC/FnHmE liposomes was significantly increased with respect to DMPC-only liposomes (Figure 5). A slight lowering of the crystal-to-liquid crystal phase transition temperature was also observed. Black lipid membrane (BLM) experiments indicate that membranes formed from equimolar amounts of egg yolk phospholipids and FnHm have higher capacitance values and can resist to higher applied voltages than those prepared without the mixed hydrocarbon/fluorocarbon amphiphile.[125]

Measurements of the liposome membrane's permeability to 5(6)-carboxyfluorescein (CF) and calcein (CA) were made in an aqueous buffer and in human serum for various FnHm compounds.[117,118] A significant increase in encapsulation stability of the entrapped molecules was found with respect to DMPC alone. The largest effect was observed with F4H10E with a leakage half-time in a HEPES/Na_2SO_4 buffer of 15 h, i.e., 16 times longer than for the DMPC-alone reference liposomes (Figure 6). The presence of a double bond between the Fn and Hm segments had a beneficial impact on encapsulation stability. Thus, with F4H10E a half-life of ca. 15 h was found for CF in buffer, compared to 4.3 h for F4H12. Stability in human serum was, however, very low, although superior to that found for DMPC alone. The incorporation of FnHm and FnHmE compounds in DMPC liposomes was also shown to reduce the leakage rate of CA in serum,[118] and the release of adriamycin[110] in buffer (Figure 7). That the presence of the mixed fluorocarbon/hydrocarbon compound in the DMPC dispersion does strongly influence the half-life of encapsulated material is further evidence that it is incorporated into the liposomal membrane.

An interesting finding is that the incorporation of FnHm components in the membrane of liposomes made of DMPC or DPPC can result in a significant reduction in the rate of hydrolysis of the phospholipids by porcine pancreatic phospholipase A_2.[126] It was found that for a given phospholipid, hence for a given fatty acid chain length, there is a precise minimal Hm chain length for this hydrolysis rate reduction to occur (Figure 8). This chain length depends directly on the chain length of the phospholipid utilized; Hm has to have at least 10 carbons for DMPC, 12 for DPPC. The fluorinated Fn segment is also indispensable for the effect to occur, indicating that it plays a major role in structuring the liposomal membrane.

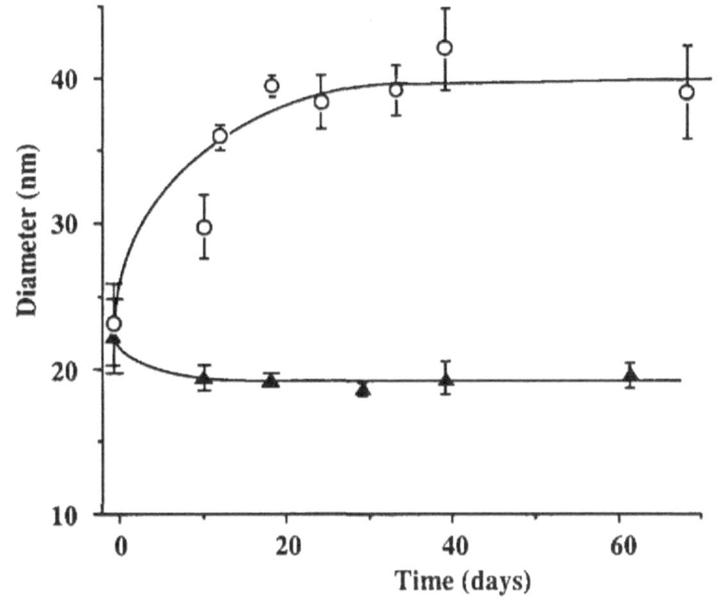

Figure 5 Coarsening of DMPC and DMPC/F4H10E ($C_4F_9CH{=}CHC_{10}H_{21}$) (1:2) liposomes as a function of time: O, DMPC liposomes; ▲, F4H10E/DMPC. (From Trevino, L., *Colloids Surf.*, 84, 223, 1994. With permission.)

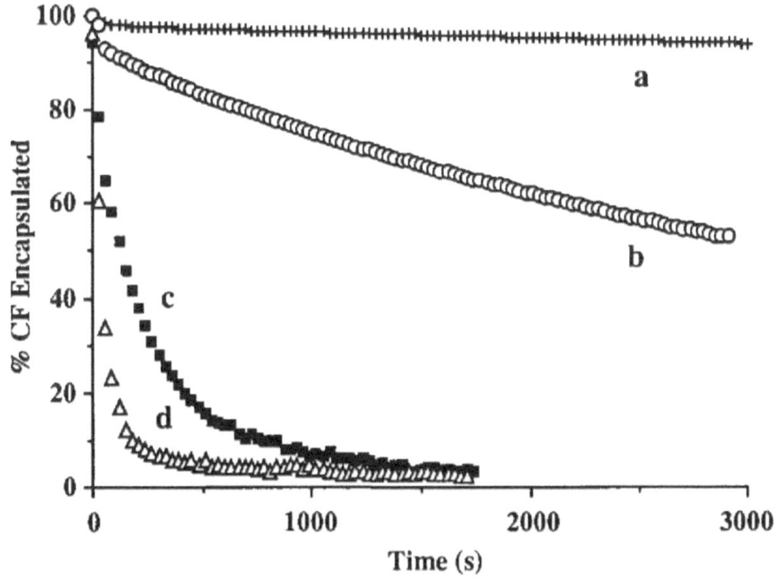

Figure 6 Release of carboxyfluorescein (CF) from SUVs composed of DMPC/F4H10E ($C_4F_9CH{=}CHC_{10}H_{21}$) (curves a and c) and DMPC alone (curves b and d) in HEPES/sodium sulfate buffer (curves a and b) and in human serum (curves c and d). (From Trevino, L., et al., *J. Liposome Res.*, 4, 1017, 1994. With permission.)

The incorporation of mixed fluorocarbon/hydrocarbon components in classical liposomes may thus provide a new means of controlling diverse physical and biochemical characteristics of a liposomal system and deserves further investigation. One of its advantages is that it combines standard components such as phospholipids, which are well accepted in the pharmaceutical, cosmetic, and food industries, with low cost materials which are chemically and biologically inert as compared to the usual surfactants.[103] When injected intravenously as an emulsion, F6H10E was shown not to be metabolized and no sign of toxicity was seen at a 3.6 g/kg body weight in mice.[127]

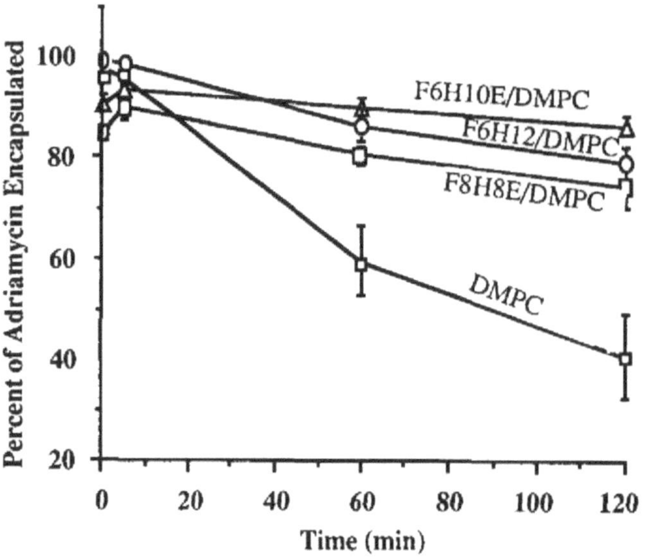

Figure 7 Release of adriamycin from FnHm (or FnHmE)/DMPC vesicles in buffer at 37°C. (From Trevino, L., et al., *Proc. Int. Symp. Control. Rel. Bioact. Mater.*, 21, 606, 1994. With permission.)

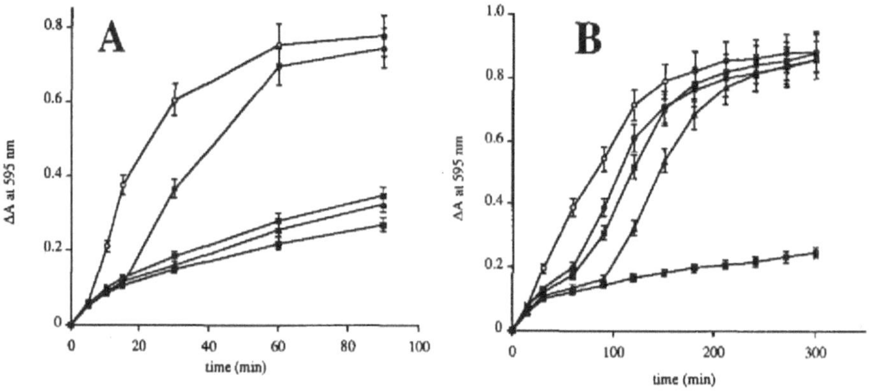

Figure 8 Hindering of the enzymatic hydrolysis (porcine pancreatic phospholipase A_2) of phospholipids by incorporation of (perfluoroalkyl)alkanes, FnHm, into the liposomal membrane. (A) DMPC at 23°C; (B) DPPC at 37°C. ΔA: variation of the absorption of phenolsulfonephtalein, utilized as a pH indicator. (From Privitera, N., et al., *Biochim. Biophys. Acta*, 1254, 1, 1994. With permission.)

F. OTHER AGGREGATES FROM FLUORINATED AMPHIPHILES

An outstanding feature of fluorinated amphiphiles is the readiness with which they engender a wealth of colloidal behaviors.[18] Thread-like structures, most probably consisting of parallel alignments of rod-like micelles rather than lamellar sheets, were shown to form from nonionic *F*-alkylated *N*-polyethoxylated amides.[63] Rigid tubule-like cylindrical aggregates, up to 50 μm in length, were observed to grow spontaneously at room temperature from concentrated (6 to 12% w/v) aqueous dispersions of [(F-alkyl)alkyl] dimorpholinophosphoramide, for example, **2**.[90] The presence of multiple bilayers wrapped around an aqueous core was seen by freeze-fracture electron microscopy. These tubules, which are formed without the help of hydrogen bonds between the polar heads, chiral effect or rigid rod-like segments, or any stabilizing additive,[8,128] did not show any significant change in morphology for at least 6 months at 20°C when observed by phase contrast optical microscopy. When heated above the phase transition temperature they are reversibly converted into giant vesicles. Such tubules can also be obtained in solvents such as ethanol or dimethylformamide.[129] The mechanism of formation of these tubules included the aggregation of sonicated SUVs, their fusion into large flat multibilayer sheets and their subsequent rolling up at edges.

The colloidal behavior of [2-(*F*-octyl)ethyl]phosphocholine, **1b**, is particularly versatile, with the formation, depending on concentration, energy input, temperature, and time, of stacked bilayers, SUVs,

MLVs, fibers, globules, and giant vesicles.[89,102] Long flexible fibers were, for example, found to grow in unsonicated, 6% w/v concentrated dispersions of **1b** upon standing.[90] After 4 months at 20°C they could reach 50 to 500 μm in length (1 to 5 μm in diameter) and then became visible by phase contrast and Nomarsky microscopy. These fibers reversibly transform into large vesicles when temperature is increased above 50°C. The rate of formation of the long flexible fibers was significantly augmented when equimolar mixtures of **1b** and **2** were utilized. Large fibers, 50 μm in length, visible by optical microscopy were then obtained after ca. two weeks.[90] The addition of the fluorinated alcohol $C_8F_{17}C_2H_4OH$ to vesicles made from **1b**, depending on the alcohol/**1b** ratio, induced the formation of flexible fibers, tubules, or crystallized platelets.[129a]

Fluorinated glycolipids also display remarkable versatility in their phase behavior since, depending on molecular structure, not only vesicles but also disks, tubules, or helical aggregates were obtained.[91] The single-chain fluorinated glycolipid **3a** (Table 1) (with no peptide spacer), when dispersed in water by simple shaking, was found to arrange into helical assemblies (Figure 3A) while its hydrocarbon analog did not show any evidence of large size supramolecular arrangement. When a glycine spacer was present, as in **3b**, the fluorinated single-chain glycolipid arranged into disks (Figure 3B). With the same polar head (galactose) and peptide spacer (glycine) as in **3b**, the double-tailed fluorocarbon/hydrocarbon glycolipid **23d** (Table 2) gave, after sonication, typical multilayered vesicles (Figure 3C). The lysine-containing amphiphile **24**, upon dispersion in water by sonication, also yielded a variety of microstructures which included stacked disk-like assemblies, tubules, and twisted helical structures, along with vesicles (Figure 3D). Transparent, long fiber-containing gels have been obtained from formamide solutions of *N*-fluoroalkylglucoamides.[130] Certain of the fluorinated glycophospholipids with polar heads derived from glucose were also found to form tubules which converted reversibly into vesicles upon heating.[131,132]

Tubules and fibers have recently aroused considerable interest.[7,8,128] Specific examples of applications include the manufacture of electroactive composites, including cathodes for vacuum field emission, and use as microvials for long-term release applications. Applications in catalysis, separation systems, and "smart" materials were also suggested.[128] So far, however, in the hydrocarbon series, the hydrophobic interactions that develop between the surfactant and water had to be supplemented by hydrogen bonding, the presence of rigid rod-like segments, polymerization or metal-coating, the addition of divalent ions, and generally also by the use of chiral molecules, in order to achieve the stabilization of such elongated aggregates.[8,133-136] The formation of tubules and fibers from simple nonchiral, fluorinated amphiphiles[90] indicates that combination of the intense hydrophobic effect developed by fluorinated chains and of the appropriate overall geometry of the surfactant can suffice to allow the formation of microstructures more stable than vesicles. The opposite situation is found with certain bolaamphiphiles (the arborols) which also form fibers, but for which it is the interactions between a multitude of amide groups which bind the polar heads tightly together, to the point that the hydrophobic effect is considered negligible.[137]

An original phase behavior was found in the DFnCmPC series **29**, for DF6C11PC, **29e**, and DF8C11PC, **29g**, which, as evidenced by freeze-fracture electron microscopy and X-ray diffraction, display a ribbon-like phase (two-dimensional centered rectangular lattice) at room temperature (Figure 9) and a lamellar phase above their gel-to-fluid phase transition temperatures.[111] Upon cooling, the lamellar gel phase converted very slowly back into the ribbon-like phase.

G. SELF-ORGANIZATION OF FLUORINATED AMPHIPHILES IN HYDROCARBON AND FLUOROCARBON MEDIA

A major driving force for self-organization of amphiphiles into membranes in water consists in hydrophobic interactions. These are lost in hydrophobic media. Semifluorinated *n*-alkanes, $C_nF_{2n+1}C_mH_{2m+1}$ (FnHm) were reported by Rabolt et al. to present a bilayer-type crystal structure[138] and to form micellar aggregates in toluene and in fluorinated solvents.[139] Kunitake and co-workers recently took advantage of the immiscibility of fluorocarbons with hydrocarbons in order to prepare bilayers in conventional organic solvents such as cyclohexane or benzene[27,140,141] as well as in fluorocarbons.[27,142] Amphiphiles which contain hydrocarbon and fluorocarbon moieties (Table 4) were used for this purpose. Hydrophobic interactions are then replaced by the lipophobic and fluorophobic interactions needed for the formation of bilayers in hydrocarbons and fluorocarbons, respectively. For instance, as a result of aggregation and alignment of its fluorocarbon chains, compound **40** gave lamellar and rod-like systems in benzene,[140] and **41** gave mainly vesicles accompanied by twisted tapes and fiber-like aggregates in cyclohexane.[141] Conversely, amphiphiles **42** and **43** gave stable bilayers in various fluorocarbons due to aggregation and alignment of their hydrocarbon chains.[142]

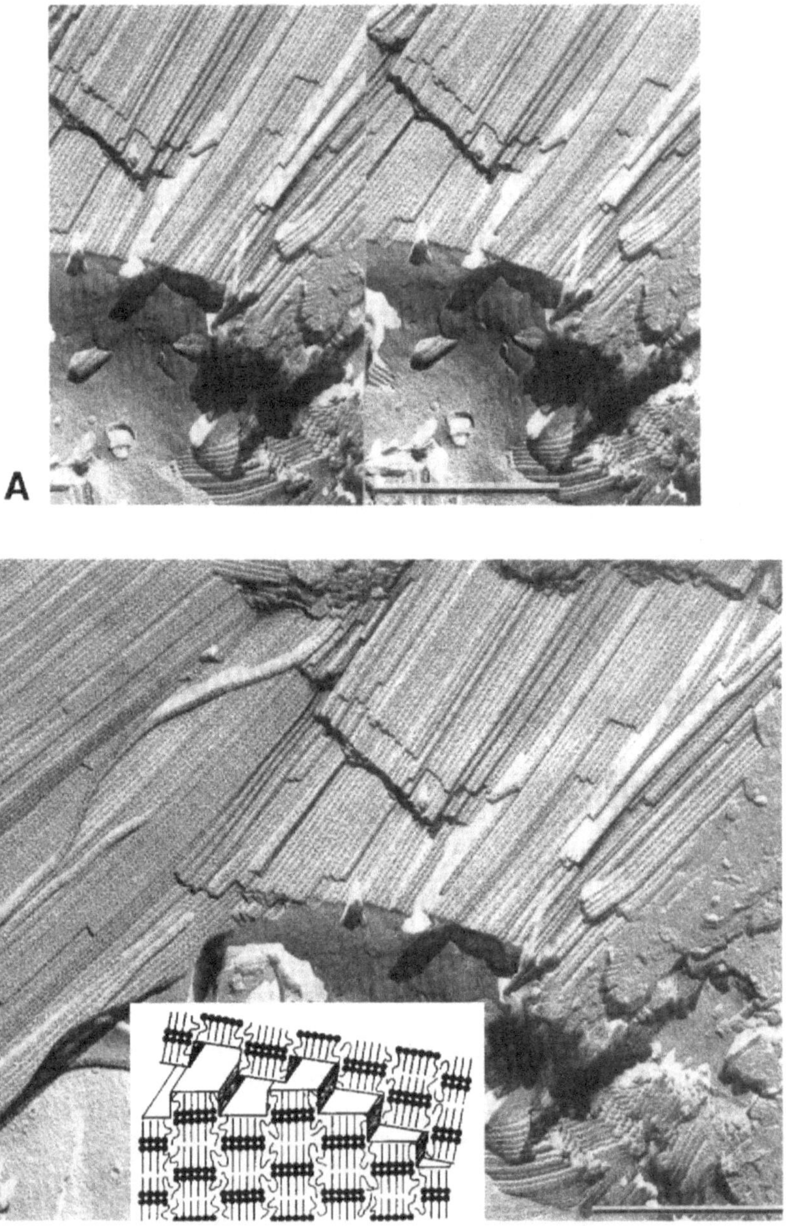

Figure 9 Freeze-fracture images of a DF6C11PC, **29e**, centrifugation pellet obtained from a 3% w/v aqueous dispersion. (A) Stereo-pair images; (B) Enlarged view of the same image. Note the roughness of the fractured surfaces and the presence of ribbon-like elements. (From Santaella, C., et al., *Biochim. Biophys. Acta*, 1190, 25, 1994. With permission.)

The concept of bilayer assembly has been generalized to protic media, aprotic hydrocarbon media, and aprotic fluorocarbon media.[27] The conclusion was that "organized molecular assemblies are produced in any medium if the amphiphilic/amphiphobic nature of the solute molecules is properly designed." The driving force of the molecular assembly is the solute solvent immiscibility (enthalpic force) that arises from differences in cohesive energy between the solute (amphiphile) and solvent.

H. FLUORINATED AMPHIPHILES AT A WATER/FLUOROCARBON INTERFACE

Due to their aptitude to strongly reduce fluorocarbon/water interfacial tensions and their strong tendency to organize at interfaces, *F*-alkylated surfactants can be utilized, alone or in conjunction with other surfactants, to stabilize and control the characteristics of fluorocarbon emulsions.[13,17,103,120,143] Reverse water-in-fluorocarbon emulsions, stable for at least 6 months at 25°C, were obtained, using small amounts of *F*-alkylated phosphomorpholidate $C_8F_{17}(CH_2)_{11}OP(O)[N(CH_2CH_2)_2O]_2$ as the surfactant.[144] Such emulsions can carry a variety of drugs and have potential for delivery through the pulmonary route. Fluorinated surfactants also allowed the preparations of fluorocarbon microemulsions.[56,145–148]

When small amounts (typically 1% w/v) of a *F*-alkylated amine oxide, $C_7F_{15}C(O)NH(CH_2)_3N(O)(CH_3)_2$, were dispersed in fluorocarbons with the help of a small amount of water, the formation of gels was observed.[149] These gels, which are heat sterilizable and stable, were shown to consist in a matrix of "polyaphrons," i.e., polyhedral fluorocarbon domains surrounded by a thin membrane of hydrated surfactant. The latter may consist in an inverse bi- or multilayer membrane. These gels can be diluted with water to produce stable individualized "aphrons"; the polyaphron matrix tends to form again upon standing.[150]

At the other end of the concentration scale aqueous gels have been reported which contain 50 to 99% v/v of water as the internal phase, the continuous phase being a hydrocarbon[151,152] or a fluorocarbon.[153,154] In the latter case the surfactant was a *F*-alkylated ethoxylated alcohol and the "continuous phase" was shown to be a w/o microemulsion. Recently the relationship between the structure of these "gel-emulsions" and their viscoelastic properties was elucidated.[155]

III. PROPERTIES OF FLUORINATED MEMBRANES AND VESICLES

A. SHELF-STABILITY OF FLUORINATED VESICLES

The impact of the fluorinated chains on the thermal resistance and shelf stability of vesicles is most strikingly illustrated by the obtaining of stable, heat-sterilizable vesicles from single-chain phosphocholine derivatives (Figure 2).[89,103]

The vesicles based on the fluorinated phosphatidylcholines DFnCmPCs, **29**,[156,157] glycolipids **23** and on the fluorinated galactosyl- and mannosyl-phosphates **17**, **18**, **21**[131,177,178] were also found to exhibit remarkable stability with respect to particle size and particle size distribution. The morphology and average sizes of the fluorinated vesicles were almost unaffected by heat sterilization and upon aging for 10 months at room temperature. This is in strong contrast with the poor stability of vesicles made from pure hydrocarbon analogs.[18] For glycophospholipids the nature of the sugar has a significant impact on stability.[131] The galactose- (as in **17**, **18ab** and **22**) or mannose-derived polar heads (as in **21a,b**) led to higher vesicle stability than the glucose-derived ones, whether the latter was linked to the phosphate group in position 3 or 6 (as in **19a,b** or **20a**, respectively). Thus, the vesicles made from the fluorinated galactosyl- and mannosyl-phosphates remained essentially unchanged (30 to 65 nm) after 10 months at 25°C while those made from the hydrocarbon analog of **21b** precipitated after 12 h, and those made from their 3- or 6-glucosyl phosphate analogs **19a,b** precipitated after 10 to 20 days, respectively, or, in the case of **20b**, turned into a gel after 1 day. This gel was found to contain tubules.[132]

B. PHASE TRANSITION/THERMOTROPISM

Bilayer membranes of hydrocarbon amphiphiles usually display a crystal (or gel)-to-liquid crystal (or fluid) phase transition which is detected most directly by differential scanning calorimetry (DSC). Bilayers formed from fluorocarbon amphiphiles also display such a phase transition. Their thermotropic phase behavior has usually been investigated by DSC, and in some cases also through turbidity measurements, the use of fluorescent and paramagnetic probes, or by ^{19}F or ^{31}P NMR. Table 5 collects the phase transition temperatures, Tc, ΔH, and ΔS values.

The gel-to-fluid phase transition of lipid bilayers is an energetic event reflecting a cooperative transition of the lipid chains from an ordered to a disordered state. The relative energetic contents and the packing modes of the lipid chains in two different systems can thus be compared on the basis of the thermodynamic parameters associated with their phase transitions. Differences in Tc, ΔH, and ΔS values for fluorocarbon vs. hydrocarbon bilayers may be assigned to differences in packing characteristics due to differences in average conformation of the chains, lateral van der Waals chain–chain interactions and hydrophobic interactions in the lamellar gel state.

The results collected for double- and triple-chain amphiphiles in Table 5 show that for similar hydrophobic chain length L (L = n + m = number of carbon atoms) the introduction of a C_8F_{17} tail in all the chains caused a significant increase of Tc[111] while introducing the same tail in only one of the chains had the opposite effect.[158,159] As an example, in the *F*-alkylated phosphatidylcholine series, the Tc of DF8C5PC, **29f** (~69°C for L = 13) is higher than that of DMPC (~23°C, L = 14) and even DSPC (~55°C, L = 18). Introducing a C_8F_{17} tail in only one of the chains as in the mixed **34b** (Tc = 5°C) caused a drastic decrease in Tc as compared to DSPC or DPPC (~41°C, L = 16).

Table 5 Crystal-to-Liquid Crystal Phase Transition Temperatures (Tc) of Fluorocarbon Bilayers and of Their Hydrocarbon Analogs and Their Associated Enthalpies (ΔH) and Entropies (ΔS)

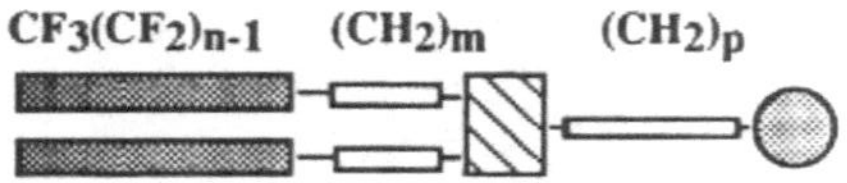

COMPOUNDS*	Fluorocarbon amphiphiles				Hydrocarbon amphiphiles (Fn = H)				
	Tc [a]	ΔH [b]	ΔS [c]	Ref.		Tc [a]	ΔH [b]	ΔS [c]	Ref.
9a: n = 8, m = 2, p = 2	95.0	3.3	9	162	m = 12 p = 2	no transition detected between 0 - 95°C			162
9b: n = 8, m = 2, p = 4	92.0	8.8	24	162	m = 12 p = 4	4	18.4	66	162
9c: n = 8, m = 2, p = 6	90.0	4.6	13	162	m = 12 p = 6	12	28.0	98	162
9d: n = 8, m = 2, p = 11	84.0	7.1	20	162	m = 12 p = 11	33	45.3	148	162
10: n = 8, m = 2, p = 4	no transition detected between 15 - 50°C			23	m = 12 p = 4	31	44.3	146	162
11a: n = 8, m = 3, p = 2	75.0	2.2	6	162	m = 12 p = 2	21.3	53.5	182	162
11b: n = 8, m = 3, p = 4	75.0	2.1	6	162	m = 12 p = 4	20.0			162
11c: n = 8, m = 3, p = 6	71.0	1.7	5	162	m = 12 p = 6	14.6	39.4	137	162
11d: n = 8, m = 3, p = 11	53.0	3.1	9	162	m = 12 p = 11	12.0	30.6	107	162
12d: n = 8, m = 3, p = 11	44.8	3.8	12	162					
13a: n = 6, n' = 8, p = 3	51.7	2.2	6.8	191					
14d: n = 8, m = 2, p = 11	76.0 (pH4)	3.7	10	162	m = 12 p = 11	59.5 (pH 6.5)	41.5	125	162
26: n = 8, m = 2, p = 11	74.0	8.9	26	162					
27: n = 8, m = 2, p = 11	77.8 (pH 7)	8.0	23	162	m = 12 p = 11	63 (pH 8.5)	47.3	141	162

COMPOUNDS*		Fluorocarbon amphiphiles T_c[a]	ΔH[b]	ΔS[c]	Ref.	Hydrocarbon amphiphiles (Fn = H)	T_c[a]	ΔH[b]	ΔS[c]	Ref
$CF_3(CF_2)_{n-1}$ $(CH_2)_m$ 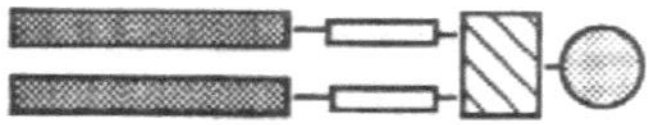										
15: n = 9, m = 2 (d)		71.8			192					
16: n = 9, m = 3 (d)		no transition detected between 30 - 80°C			195					
28: n = 9, m = 3 (d)		53			195					
29a: n = 4, m = 5	**DF4C5PC**	< -20			111	m = 12	-1	16	59	160
29b: n = 4, m = 11	**DF4C11PC**	18.6	7.5	26	111	m = 13	14	18	63	160
29c: n = 6, m = 5	**DF6C5PC**	< -20			111	m = 14	23.5	26	88	160
29d: n = 6, m = 7	**DF6C7PC**	4.4	6.6	24	111	m = 15	33.5	32	104	160
29e: n = 6, m = 11	**DF6C11PC**	48	18.4	57	111	m = 16	41.5	36.5	116	160
29f: n = 8, m = 5	**DF8C5PC**	69.3	10.0	29	111	m = 17	48	37	115	160
29g: n = 8, m = 11	**DF8C11PC**	> 95			111	m = 18	55.5	46	140	160
						m = 19	61.5	45	134	160
						m = 20	66	50	147	160
30a: n = 6, m = 11		54.6	8.7	26	196					
30b: n = 8, m = 11		no transition detected between 5 - 95 °C			196					
31a: n = 4, m = 11		17.0	11.1	38	159	m = 14	28.5			160
31b: n = 8, m = 11		85.7	28.0	78	159	m = 16	43.5			160
32a: n = 4, m = 11		30.8	20.6	68	159	m = 14	55			160
						m = 16	68.5			160
33a: n = 4, m = 11		14.1			161					
33b: n = 6, m = 11		41.6	8.7	28	161					

Table 5 (continued)

CODE NAME*	Fluorocarbon amphiphiles Tc [a]	ΔH [b]	ΔS [c]	Ref.	Hydrocarbon amphiphiles (Fn = H)	Tc [a]	ΔH [b]	ΔS [c]	Ref.
$CF_3(CF_2)_{n-1}$ $(CH_2)_m$ 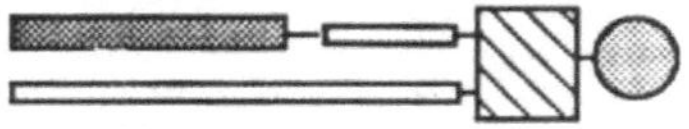 $CH_3(CH_2)_{m'-1}$									
20a: n = 6, m' = 8 **21a:** n = 6, m' = 8	no transition detected between - 30 to 90°C			131,178					
18b: n = 8, m' = 8 **19b:** n = 8, m' = 8 **20b:** n = 8, m' = 8 **21b:** n = 8, m' = 8	no transition detected between -30 to 90°C			131,178					
22 : n = 8, m' = 8	no transition detected between -30 to 90°C			131,178					
23a: n = 6, m = 2, m' = 9, p = 1	no transition detected between -10 to 90°C			91,112, 113					
23b: n = 8, m = 2, m' = 9, p = 1	no transition detected between -10 to 90°C			91,112, 113	m = 11, p = 1 m' = 12	14.0	32.3	112.5	91,112, 113
23c: n = 6, m = 2, m' = 9, p = 1	no transition detected between -10 to 90°C			91,112, 113					
23d: n = 8, m = 2, m' 9, p = 1	no transition detected between -10 to 90°C			91,112, 113	m = 11, p = 1 m' = 10	22.0	34.9	118.3	91,112, 113
23e: n = 6, m = 2, m' = 9, p = 2	no transition detected between -10 to 90°C			91,112, 113					
25b: n = 6, m = 11, p = 1, q = 11 **25c:** n = 8, m = 11, p = 0, q = 20	no transition detected between -10 to 90°C			115					
34a: n = 8, m = 5, m' = 14	< -25			159					
34b: n = 8, m = 11, m' = 16	5.1	6.3	23	159					
36: n = 8, m = 11, m' = 18	46.4	16.2	51	159	m = 18 m' = 18	59			160

COMPOUNDS*	Fluorocarbon amphiphiles				Hydrocarbon amphiphiles (Fn = H)				
	Tc[a]	ΔH[b]	ΔS[c]	Ref.		Tc[a]	ΔH[b]	ΔS[c]	Ref.
$CF_3(CF_2)_{n-1}$ $(CH_2)_m$ 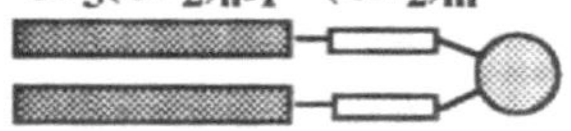 **5**	71.5			190					
6: n = 6, m = 2	69.5			190					
7a: n = 8, m = 3	no transition detected between 0 - 70°C			23	m = 12	29.0	38.5	127	23
7d: n = 8, m = 2	80			25	m = 16	49			25
8 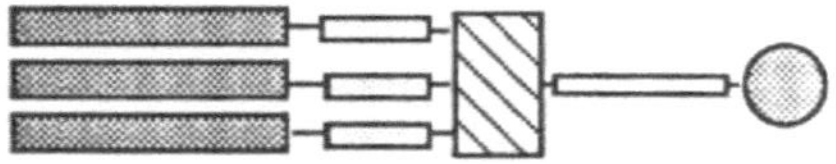	no transition detected between 0 - 70°C								
37a: n = 8, m = 3	53.7	5.9	18	158					
37b: n = 8, m = 3	54.8	13	40	158					
38: n = 8, m = 3, m' = 12	46	8.8	28	158					
39: n = 8, m = 3, m' = 12 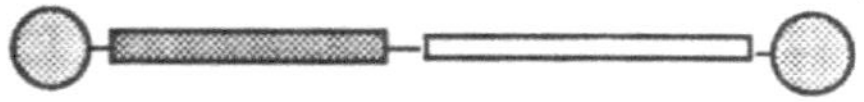	36.1	20.9	68	158	m = 12, m' = 12	41.5	31.8	101	158
4: n = 8, m = 10 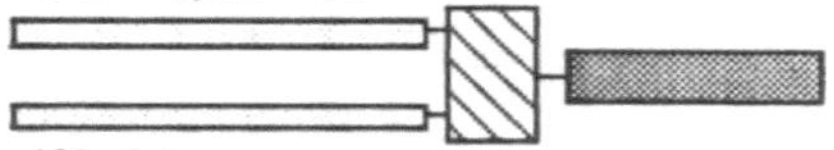	32.5			171					
42b (e)	2.3	42	152	142					
43 (e)	-12.6	19.5	75	142					

* : for the exact structure see tables 1 to 4; when present, C(O) and CH=CH are included in the chain carbon count; a) Tc in °C; b) ΔH in kJ mol^{-1}; c) ΔS in J mol^{-1} K^{-1}; d) Cl replaces CF_3; e) measured in methyl perfluoro-(2,5,8)-trimethyl-3,6,9-trioxadecanoate.

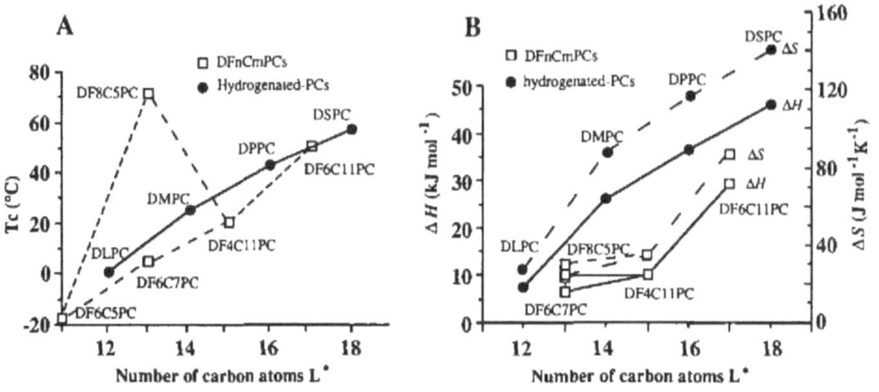

Figure 10 (A) Gel-to-fluid transition temperature, Tc, and (B) corresponding ΔH and ΔS variations *vs* length (number of carbon atoms, L) of the hydrophobic chain for the DFnCmPCs **29**, and hydrocarbon-PCs. The Tc, ΔH, and ΔS values of the hydrocarbon PCs were taken from References 111 and 160. For the DFnCmPCs, L = n + m; for the hydrocarbon-PCs, L (= p) is the number of carbon atoms of the $CH_3(CH_2)_{p-2}C(O)$-chain.

Introducing a C_6F_{13} segment had almost no effect on Tc: the Tc of DF6C11PC, **29e**, (48°C, L = 17) is indeed between those of DPPC and DSPC.[111] The introduction of a short C_4F_9 tail, as in DF4C11PC, **29b**, resulted on the contrary in a decrease in Tc (~19°C), as compared to DMPC, which appears to indicate that the hydrophobic contribution is not sufficient to compensate for the lower van der Waals attraction between fluorinated chain as compared to hydrogenated ones.

The gel-to-fluid phase transition temperature of the fluorinated phosphatidylcholines is related to the total length of their hydrophobic chains and, more markedly, to the length of the fluorinated tails (Figure 10). Tc rises (a) when, for a given Cm spacer, the Fn tail's length increases; or (b) when, for a given Fn tail, the Cm spacer's length increases.[111] In the hydrogenated phosphatidylcholine series, a chain lengthening of two CH_2s is accompanied by a nearly monotonic Tc increase of about 20°C (Figure 10A, Table 5).[160] By contrast, the addition of two CF_2s in the fluorinated series resulted in an increase in Tc of at least 90°C from DF6C5PC (**29c**) to DF8C5PC (**29f**), and 30°C from DF4C11PC (**29b**) to DF6C11PC (**29e**).[111] Figure 10A shows that Tc is primarily modulated by the length of the Fn tail rather than by the total chain length. The nature of the chemical junction (ester, ether, amide) between the hydrophobic chains and the connecting unit in the glycerol or diaminopropanol series has little impact on Tc.[159,161]

Interestingly, it was found that increasing the length of the hydrocarbon spacer in the glutamate compounds **9** and in the diethanolamine compounds **11** (Table 2), resulted in a decrease in Tc.[162] The same trend was observed for hydrocarbon analogs of the diethanolamine series while the opposite was found for the hydrocarbon analogs of the glutamate series (Table 5).[162] Larger spacers appear to increase the disorder in the crystalline state of the fluorinated bilayers. On the other hand, longer spacers in the hydrocarbon series promote regular packing in the crystalline state and give rise to larger structural changes during the phase transition, as reflected by larger Tc and ΔH and ΔS values. The Tc values were also found to be little affected by the nature of the polar head in the series **26**, **27** and **14** (Table 5).

The ΔH and ΔS values measured for the gel-to-fluid phase transition of fluorocarbon bilayers are much smaller than those found for their hydrocarbon analogs (Figure 10B, Table 5). This was taken to indicate that the observed phase transition is mainly due to the reorganization of the hydrocarbon part of the chains.[162] The temperature dependence of the linewidth of the ^{19}F NMR resonances showed that the phase transition is preceded by the melting of the Fn tail and that each portion of this tail, and the hydrocarbon spacer, melt separately.[111,163] Figure 11B shows that the melting of the C_8F_{17} chain of DF8C5PC, **29f**, starts with the terminal CF_3 group at approximately 30°C, continues at 35°C with the CF_2 in α of the CF_3 group, and finishes with the CF_2 closest to the Cm spacer at nearly 50°C, a temperature which is still far below the Tc of the main phase transition (69°C). That the melting of the fluorocarbon chains in this compound occurs over such a large temperature range (almost 20°C) is in marked contrast with what is observed for the hydrocarbon phosphatidylcholines for which chain melting, due to high cooperativity, occurs within a very short temperature range (1°C) at their phase transition temperature. When only a short portion of the hydrophobic chains is replaced by a fluorocarbon segment, i.e., C_4H_9 by C_4F_9, the cooperative effects are maintained. The terminal CF_3 and $CF_2(CH_2)$ groups of the C_4F_9 tail both melt

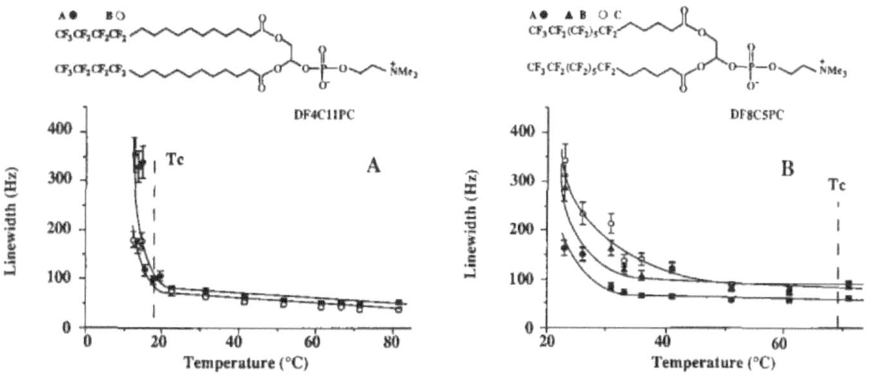

Figure 11 Temperature dependence of the linewidths of the ^{19}F NMR resonances of (A) DF4C11PC, **29b**, and (B) DF8C5PC, **29f**. (From Santaella, C., et al., *Biochim. Biophys. Acta*, 1190, 25, 1994. With permission.)

within 2 to 4°C at a temperature close to the Tc (Figure 11A).[111] Noncooperative melting of the fluorinated chain was also observed for the single-chain phosphocholine **1b**, the CF_2 groups recovering their freedom one after the other over a still larger temperature range of ca. 30°C.[89]

The differences observed between fluorocarbon and hydrocarbon bilayers concerning Tc, ΔH, and ΔS were accounted for by recognizing that the introduction of a *F*-alkylated chain can influence the packing characteristics of the membranes in two opposite directions: (a) enhancement of packing disorder and fluidity, resulting in lower Tc, ΔH, and ΔS values due mainly to weaker lateral inter- and intramolecular van der Waals interactions and increased steric repulsions between the fluorinated chains as compared to hydrocarbon chains,[52] and (b) increase in order or rigidity, as expressed by an augmentation of Tc, related to an increase in hydrophobic interactions. In the phosphatidylcholine series, a short C_4F_9 tail appears to be insufficient to counterbalance the former effect, resulting in membrane fluidification. For C_6F_{13}, the antagonist contributions seem to be equilibrated, while for C_8F_{17} the hydrophobic interactions predominate, resulting in enhanced membrane ordering and rigidity.[111]

The fluidity of fluorocarbon vs. hydrocarbon membranes made from triple-chain amphiphiles (Table 3) was evaluated using fluorescence depolarization.[164] In their gel state, the fluidity of the membranes near their surface and in their bulk was virtually the same, irrespective of the presence of fluorocarbon chains. By contrast, in the liquid-crystalline state, fluidity increased near the membrane's surface whereas it decreased in their bulk with increasing number of fluorocarbon chains.

C. MEMBRANE MISCIBILITY AND APPLICATIONS

1. Formation of Phase-Separated Vesicles or Domains in Membranes

Fluorocarbons display a very low affinity for hydrocarbons. This is exemplified by the formation of heterogeneous micelles from fluorocarbon and hydrocarbon surfactants,[19,20] a property that has been used to control phase separation in membranes, membrane permeation, and catalysis. Fluorocarbon and hydrocarbon amphiphiles show indeed limited miscibilities and tend to form separate domains within monolayers and membranes.[25] For instance, freeze-fracture electron micrographs revealed, in the case of liposomes made from a mixture of DMPC (95 mol %) and the fluorinated double-chain amine **7c** (5 mol %), that phase-separated liposomes with distinct domains constituted by the individual lipids are obtained rather than a stochastic distribution of the two lipids within one type of liposomes or two different liposome populations.[25]

Limited miscibility was also demonstrated in the case of various other double- and triple-chain fluorocarbon amphiphiles.[23,24,164,165] For example and as illustrated in Figure 12A, it was found by DSC that even the least fluorinated of the phosphatidylcholines, DF4C11PC, **29b**, was not miscible with DPPC.[165] It was also found that DF4C11PC, **29b**, and DF6C11PC, **29e**, which differ only by two difluoromethylenes, display a nonideal mixing behavior in the fluid phase and partial miscibility in the gel phase (Figure 12B) while their hydrocarbon analogs DMPC and DPPC, which also differ by two methylenes, exhibit ideal mixing irrespective of the physical state.[166] The low miscibility of fluorocarbon and hydrocarbon amphiphiles can, however, be improved. Thus, in the case of triple-chain amphiphiles **37** and **39**, the compounds with only fluorocarbon and only hydrocarbon chains become more miscible upon addition of a compound that possesses both hydrocarbon and fluorocarbon chains.[23,24,164]

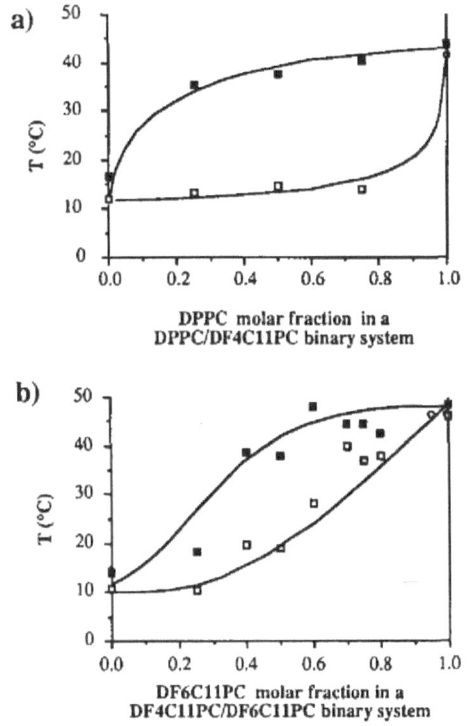

Figure 12 Fluid-solid equilibrium phase diagrams for aqueous dispersions of: (A) a mixture of two fluorinated phospholipids DF6C11PC, **29e** and DF4C11PC, **29b**; (B) a mixture of a fluorinated phospholipid DF4C11PC, **29b** and a hydrogenated one, DPPC.[165]

2. Selective Opening of Phase-Separated Vesicles

Phase-separated liposomes were used by Ringsdorf et al.[6] to simulate the hole formation process which occurs when an activated macrophage attacks and kills a tumor cell.[6] In order to avoid complete destruction when the holes were created by "uncorking" specific phase-separated domains, these liposomes were composed of a polymerized matrix formed from the polymerizable fluorocarbon amphiphile **47** which was surrounding labile domains made of the cleavable nonpolymerizable disulfide hydrocarbon compound **46**. Phase-separated liposomes were also formed from the polymerizable hydrocarbon amphiphile **45** and the disulfide fluorocarbon compound **44** (Figure 13). Evidence for the formation and opening of the labile "cork" domains in these mixed liposomes using dithiothreitol as a "corkscrew" (the water-insoluble disulfide is reductively cleaved into the more soluble single-chain thiol) was provided by freeze-fracture electron microscopy as well as by the release of an entrapped dye.

3. Permeability Control

Formation of phase-separated fluorocarbon domains within hydrocarbon membranes was also used to control membrane permeability.[167,168] Ion (K^+ and Cs^+) permeation through a ternary composite membrane composed of a polycarbonate, a liquid crystal-forming derivative (*N*-(4-ethoxybenzylidene)-4′-butylaniline), and a crown ether fitted with a fluorocarbon chain (which forms microheterogeneous phase-separated aggregates within the membrane), was completely thermocontrolled. Ion permeation was fast above the crystal-nematic liquid crystal phase transition temperature, T_{KN}, and completely suppressed below. This behavior contrasts strongly with that of the membrane formed with the hydrocarbon crown ether: permeation of K^+ was lower above T_{KN} and was not suppressed below; Cs^+ permeation did not occur at all either above or below T_{KN}. The faster transport of K^+ across the *F*-alkylated crown ether-containing membrane above T_{KN} was assigned to faster motion of this crown ether within the membrane. The latter crown ether should be relatively more desolvated than its hydrocarbon analog owing to the more pronounced lipophobic character of the fluorocarbon chain. Complete suppression of K^+ permeation below T_{KN} was attributed to the phase-separated dispersion state of the fluorocarbon crown ether: in the frozen state of the membrane ions are transported according to a site-to-site jump mechanism which can be operative only when the carrier sites are dispersed homogeneously, i.e., in the case of the hydrocarbon crown ether. However, when the carriers form separated aggregates, as the fluorocarbon crown ether does, aggregate-to-aggregate jump of ions would be almost impossible. The difference in Cs^+ permeation was rationalized in terms of the formation of 1:2 metal-crown ether sandwich complexes which facilitates

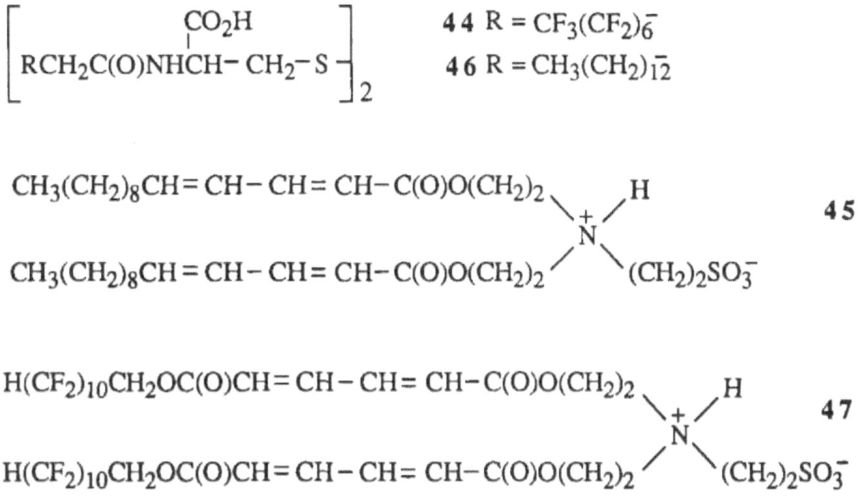

Figure 13 Examples of molecules used for selective opening of phase-separated liposomes.[6]

the ion's transmembrane diffusion. Formation of such complexes can occur more easily with the aggregated fluorocarbon crown ether, while it is almost prohibited with its hydrocarbon analog which is homogenously distributed within the membrane.

4. Reaction Control

Alteration of reaction rates in micelles and bilayer matrices[1] has also been recognized for fluorinated systems.[24,169,170] Catalytic reactions supported on bilayer matrices are not only affected by the physical state (gel or fluid) of the membrane but also by the mode of distribution of the reacting species within the membrane. Phase separation of fluorocarbon components embedded in hydrocarbon bilayers was shown to have a drastic effect on the hydrolysis of fluorocarbon- vs. hydrocarbon-containing phenylesters catalyzed by a fluorocarbon-containing histidine derivative: it was found that the hydrolysis of the fluorocarbon substrate, owing to closer proximity with the fluorocarbon catalyst in the fluorocarbon domain, was considerably faster than that of the hydrocarbon substrate which is homogeneously distributed within the hydrocarbon bilayer matrix.[24]

D. MEMBRANE DYNAMICS (FUSION, COMPONENT EXCHANGES, FLIP-FLOP)

When fluorescent vesicles and quencher vesicles were mixed, it was found that fluorinated vesicles did not undergo fusion and/or component exchange with other vesicles, whether hydrogenated or fluorinated.[101]

Using the sensitivity to membrane fluidity and component distribution of the excimer emission of the benzene chromophore, it was found, however, that fusion and/or component exchange between two distinct populations of fluorocarbon vesicles can occur.[23] Indeed, upon mixing vesicles made from fluorinated double-chain amphiphiles containing this chromophore with unlabeled vesicles, the fluorinated probe existed as monomers within several fluorinated vesicles and even within some hydrocarbon ones (the extent of fusion and/or component exchange was evaluated by the decrease of the ratio of the excimer/monomer emission fluorescence intensities).

Mutual phobicity and the thicker fluorocarbon segment attached to the larger sulfonate group in the single-chain hybrid bolaamphiphile $HO_2C(CH_2)_m(CF_2)_nSO_3H$, **4**, favored the formation of vesicles in which the outer and inner surfaces consist in fluorocarbon and hydrocarbon semimonolayers, respectively.[171] It was shown that hydrocarbon and fluorocarbon spin labels were located in their own domain in these hybrid vesicles and that the hydrocarbon spin labels undergo a flip-flop process from the inward to the outward surface, hence into the fluorocarbon domain. These results indicate that, although fluorocarbon and hydrocarbon compounds display limited miscibility, kinetic processes such as the translocation flip-flop, can still occur.

Biphasic kinetics and flip-flop behavior within vesicles of *F*-alkylated amphiphiles with a 1,3-disubstituted glycerol structure, **15**, have been reported.[172] It was shown that a fluorocarbon bilayer made

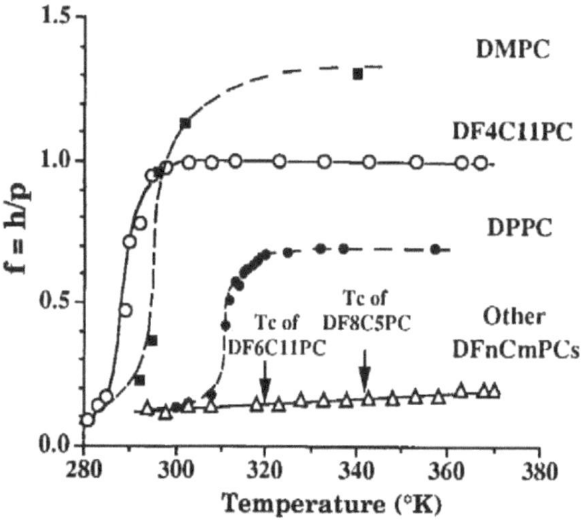

Figure 14 ESR peak height ratio (f = h/p) as a function of temperature for TEMPO partitioning between the hydrophilic and hydrophobic regions of vesicles made from DFnCmPCs, **29**, DMPC, and DPPC. (From Santaella, C., et al., *Biochim. Biophys. Acta*, 1190, 25, 1994. With permission.)

from an amphiphile with an ammonium head group was an "insulator" for the permeation of hydroxide anions. Hydrolysis of an amphiphilic probe carrying a cleavable *p*-nitrophenyl ester group showed a biphasic kinetic pattern in which the fast and slow reactions were attributed to the hydrolysis of the probe molecules at the outer and inner vesicular surfaces, respectively. It was suggested that the slow process is an outward flip-flop process of probe molecules within the fluorocarbon domain.

E. MEMBRANE LIPOPHILICITY

The presence of a fluorinated core within a bilayer membrane decreases its lipophilicity, hence the solubility of lipophilic compounds. This was demonstrated, in particular, by examining the partitioning of the lipophilic/hydrophilic paramagnetic probe TEMPO between the aqueous and lipidic phases present in dispersions of fluorinated amphiphiles using ESR.[111] In the fluorinated DFnCmPC phosphatidylcholine series **29** an increase in fluorophilicity results indeed in a dramatic decrease of the membrane's permeability. Thus, DF4C11PC, **29b**, the least fluorophilic and most lipophilic (lowest Fn/Cm ratio) of the DFnCmPCs investigated, exhibited a TEMPO distribution pattern comparable to that of DMPC and DPPC, indicating that the probe penetrates the membrane (Figure 14). By contrast, the spin probe was found to be located only in the aqueous phase in the presence of vesicles based on the other DFnCmPCs, which have longer fluorinated chains and higher Fn/Cm ratio. This most likely arises from an increase of the membranes' fluorophilic/lipophobic character. It could also, to some extent, arise from an increased order within the fluorinated membranes in their fluid state. These characteristics are expected to modify their interactions with hydrophobic enzymes, proteins and peptides, and cells, and may therefore have potential in membrane/liposome-supported reactions and in drug delivery systems.

F. MEMBRANE PERMEABILITY (ENCAPSULATION, RELEASE IN BUFFER)

The elaboration of liposomes allowing stable encapsulation of a variety of materials is a challenging objective. It has been achieved to some extent for drugs by increasing the rigidity or order of the membrane by including cholesterol (CH) or by using phospholipids which tend to self-organize into a rigid gel lamellar phase.[173,174] A hydrocarbon membrane constitutes, however, only a tenuous barrier, especially for sustaining water-soluble lipophilic species, and the modulation of release from vesicles with rigid membranes is difficult to achieve.

Endowing the liposomal membrane with some of the unique properties of fluorocarbons, i.e., high hydrophobicity and high lipophobicity, hence low affinity for both inorganic and organic species, appeared to be an interesting approach to modifying membrane permeability and acting on liposome stability with respect to the release of encapsulated materials.[18] The modular molecular structure of the diverse families of fluorinated amphiphiles now available (Tables 1 to 4) allows the formation of membranes with a continuous or discontinuous internal lipophobic fluorinated core. It also allows stepwise adjustment of the thickness of the lipophilic shells and fluorinated core (Figure 1), hence of the physicochemical properties (stability, gel-to-fluid phase transition temperature, fluidity/rigidity) of fluorinated membranes and liposomes. It permits a detailed analysis of the impact of some of these parameters on the permeability of fluorinated membranes and on the release of various probes encapsulated into fluorinated liposomes, as compared to conventional ones.

1. Permeability to Hydrophilic Organic Probes

The impact on membrane permeability of the fluorinated lipophobic core, of the physical state (gel or fluid) of the liposomal membrane, and of the molecular structure of the fluorinated amphiphiles, was determined from release experiments of entrapped CF or CA from unilamellar liposomes made from *F*-alkylated glycerophosphocholines[159,175,176] and glycophospholipids.[131,177,178] Table 6 collects the half-leakage times of CF and CA from such fluorinated liposomes and from various hydrogenated ones. Table 7 lists the permeability coefficients, P, and the activation energies, Ea, of CF permeation across fluorinated and hydrogenated phosphatidylcholine-based membranes.[175]

It was found that the fluorinated core within the liposomal membrane constitutes in the fluid state of the membrane a very efficient barrier to the permeation of CF. The "fluid" fluorinated vesicles made from DFnCmPC, **29**, retain entrapped CF much more efficiently (from 1 to at least 7500 times) and exhibit much lower membrane permeability coefficients (from 1.5 to over 2500 times) than any of their "fluid" hydrocarbon counterparts.[175] The fluorinated core more than compensates for the increase in leakage which would be expected from reducing membrane thickness, for example, when going from DSPC to DF6C5PC, **29c**, which among the phospholipids investigated possesses the shortest chains (Table 7). A higher capacity to retain entrapped CF as compared to DSPC was also found for "fluid" liposomes made from the *F*-alkylated glycophospholipid **17d**, which also has short but highly fluorinated hydrophobic chains.[178] The efficiency of the fluorinated chains in reducing permeation is further illustrated by the high CF half-leakage time found for the mixed "fluid" **34b**-based liposomes which have a discontinuous fluorinated core (Figure 1), owing to the presence of one fluorocarbon chain and one hydrocarbon chain in **34b**.[159] Some of the "fluid" fluorinated liposomes were even less leaky than "gel" or "gel-like" ones. This is the case at 37°C for the "fluid" DF6C7PC, **29d**, liposomes as compared to the "gel" DF6C11PC, **29e**, or DSPC liposomes, or the EPC/CH ones, which are in a highly ordered fluid state (Tables 6, 7).[175,176] In the gel state the fluorinated core appears to have almost no effect on CF permeation (Table 6).

Permeation across a membrane is controlled, among others, by the solubility and diffusion of the permeant species in the hydrophobic phase of this membrane and requires dehydration of the permeant species. It was suggested that the latter step determines the activation energy of permeation.[179,180] For a given species, almost identical activation energies were generally found for its permeation across various membranes, in spite of widely different permeability coefficients, indicating that the selectivity of membrane permeation is entropy-controlled. This seems to hold for diffusion across fluorinated membranes: the activation energies calculated for the diffusion of CF across these membranes in their fluid state are indeed in the same range as those found for its diffusion across "fluid" hydrogenated ones (Table 7).[175] However, the fact that a change in phospholipid or in physical state of the bilayer is accompanied by significant changes in permeability indicates the importance of the hydrophobic and lipophobic regions of the membrane in the permeation process. Since a membrane is more structured than the surrounding water, the transfer of CF into this membrane should be accompanied by an unfavorable entropy change. The lower permeability found for fluorinated membranes in their fluid state was therefore attributed to a reduction of the probe's solubility and diffusion in these membranes. Furthermore, the fluorinated tails can impose on the hydrocarbon spacer, hence on the lipophilic shells (Figure 1), a more rigid and more ordered arrangement, which might also contribute to reducing the solubility and/or diffusion of the dye in the membrane. These differences would result in a change in permeability without change in activation energy. The fact that the physical state — gel or fluid — of the membrane is much less important for the fluorinated phosphatidylcholine-based membranes than for conventional ones was attributed to differences in packing of the lipids between the fluid and gel states.[175] These differences appear to be more tenuous for the fluorinated membranes as a result of the rigidifying and ordering effect induced by the fluorinated tails. The small differences in permeability between fluorinated and hydrogenated membranes in their gel state, may indicate that penetration is almost as unfavored in both cases and that the decrease in entropy which accompanies the permeation of CF is comparable.

In terms of molecular structure vs. permeability (or CF leakage) relationships (Figure 15), permeability of the "fluid" fluorinated membranes decreases, as expected, when either the fluorinated tail or the hydrogenated spacer of the fluorinated phospholipid is lengthened.[175] The fact that permeation either remains unmodified from DF4C11PC, **29b**, to DF6C5PC, **29c**, or decreases from DF6C11PC, **29e**, to

Table 6 Half-Leakage Times, $t_{1/2}$, for 5(6)-Carboxyfluorescein or Calcein Leakage from Small Unilamellar Vesicles Made from Various Fluorocarbon and Hydrocarbon Amphiphiles at Different Temperatures

Encapsulated Probe		5(6)-Carboxyfluorescein			5(6)-Carboxyfluorescein 37°C			Calcein 37°C		
Temperature		7°C[a] $t_{1/2}$ h (±)	37°C[a] $t_{1/2}$ h (±)	75°C[a] $t_{1/2}$ h (±)	a = $t_{1/2}$ h (±) (NaCl Buffer)	b = $t_{1/2}$ h (±) (Human Serum)	b/a	$t_{1/2}$ h (±)	$t_{1/2}$ h (±) (Serum)	Ref.
Fluorinated liposomes										
29c	**DF6C5PC**	*253 (160)*	*53 (10)*	*1.8 (0.1)*	*56 (5)*	*0.66 (0.06)*	*0.012*		*1.2 (0.2)*	175
29f	**DF8C5PC**	160 (20)	27 (5)	*204 (33)*	24 (4)	25 (3)		too slow[c]	43 (2)	175
						77 (4)[b]	3			
29d	**DF6C7PC**		*140 (20)*	*9 (2)*	*74 (20)*	*14 (1)*	*0.2*			175
29b	**DF4C11PC**	203 (51)	*83 (24)*	*1.9 (0.2)*	*83 (12)*	*3.6 (0.1)*	*0.04*	*too slow*[c]	*36 (2)*	175
29e	**DF6C11PC**	405 (85)	85 (25)	*31 (5)*	60 (12)	>160				175
					66 (10)[b]	>600[b]	>10			
30a			22 (3)		5 (1)	41 (4)	8			175
30b			*67 (26)*		*84 (15)*	*11.2 (0.3)*	*0.1*			175
34a			*21 (7)*		*15 (4)*	2.9 (0.5)	0.2			175
31a/30b 1/1			62 (16)		57 (1)	63 (3)	1.1			175
33a					*7.9 (0.5)*	*0.74 (0.01)*	*0.1*			175
17			*82.5 (13.5)*			*0.3 (0.1)*				175
18b			*0.7 (0.1)*			*fast*[d]		*16.4 (4.5)*[a]	*fast*[d]	178
17/18b 1/2			*1.1 (0.1)*			*fast*[d]				
18b/CH[e] 2/1								*1.9 (0.6)*		178
19b			*0.3 (0.1)*					*0.8 (0.1)*[a]	*0.20 (0.05)*	178
21b			*0.8 (0.1)*					*3.8 (0.3)*[a]	*fast*[d]	178
18a			*0.3 (0.1)*					*1.6 (0.2)*[a]	*fast*[d]	178
21b/H21b[f] 2/1								*0.3 (0.1)*[a]		178
21a			*0.70 (0.01)*					*1.4 (0.2)*[a]	*fast*[d]	178
22			18.2 (0.4)			*fast*[d]				178
22/DSPC 1/4			5.3 (0.4)			0.7 (0.2)	0.1			178
22/CH[e] 2/1			22.4 (0.8)			0.60 (0.05)	0.04			178

Conventional liposomes									
DLPC	fast[d]	*fast*[d]							175
DMPC	130 (20)	*1.1 (0.2)*	*0.027 (0.003)*	*0.33 (0.05)*	*fast*[d]		*9.4 (1.1)*[a]	*fast*[d]	175
DPPC	110 (3)	21 (9)	*0.19 (0.03)*						175
DSPC	35 (8)	14 (3)	*1.4 (0.3)*	10 (2)	19 (1)				
					40 (5)[b]	4			175
EPC[e]				*13 (1)*	*0.12 (0.02)*	*0.01*		*2.0 (0.5)*	175
EPC/CH[e]				33 (9)	35 (1)		too slow[c]	too slow	175
					27 (2)[b]	0.8			

Note: Values in *italics* indicate that the liposomal membranes are in the fluid state at the given temperature. The $t_{1/2}$ values given correspond to the mean (± S.D.) for 3 to 6 independent experiments. Mean size from 20 to 80 nm (>95% of the population).

[a] experiments run in the Na_2SO_4 containing buffer.
[b] values calculated from 24 h kinetic experiments (otherwise 2 h).
[c] values calculated from experiments run in the NaCl containing buffer.
[d] the CF release is too fast to be measured in our conditions, indicating that the $t_{1/2}$ lie below 0.017 h.
[e] **CH** and **EPC** mean cholesterol and egg phosphatidylcholine, respectively.
[f] ***H21b*** is the hydrocarbon analog of **21b**.

Table 7 Permeability Coefficients, P, and Activation Energies, Ea, of 5(6)-Carboxyfluorescein Diffusion Across Membranes Made from DFnCmPCs and Hydrocarbon-PCs at Different Temperatures

	DF6C5PC 29c	DF8C5PC 29f	DF6C7PC 29d	DF4C11PC 29b	DF6C11PC 29e	DMPC	DPPC	DSPC
7°C								
P (10^{12}) cm/s (±)	*0.49 (0.23)*	1.24 (0.17)		0.44 (0.14)	0.57 (0.13)	0.42 (0.10)	0.74 (0.31)	3.3 (0.7)
37°C								
P (10^{12}) cm/s (±)	*1.75 (0.15)*	8 (1)	*1.1 (0.2)*	*1.2 (0.4)*	3.2 (0.8)	*53 (7)*	4 (2)	9 (2)
75°C								
P (10^{12}) cm/s (±)	*56 (1)*	*0.84 (0.13)*	*17 (3)*	*48 (4)*	*7.1 (1.2)*	*2150 (300)*	*305 (34)*	*85 (3)*
Ea kJ/mole (±)	84.2 (5.5)	137 (57)	129 (19)	120.7(6.5)	85.4(3.3)	84.5(3.8)	108.2(10.8)	117.2(5.1)

Note: Values in *italics* indicate that the liposomal membranes are in the fluid state at the given temperature.

From Frézard, F., Santaella, C., Vierling, P., and Riess, J. G., *Biochim. Biophys. Acta*, 1192, 61, 1994. With permission.

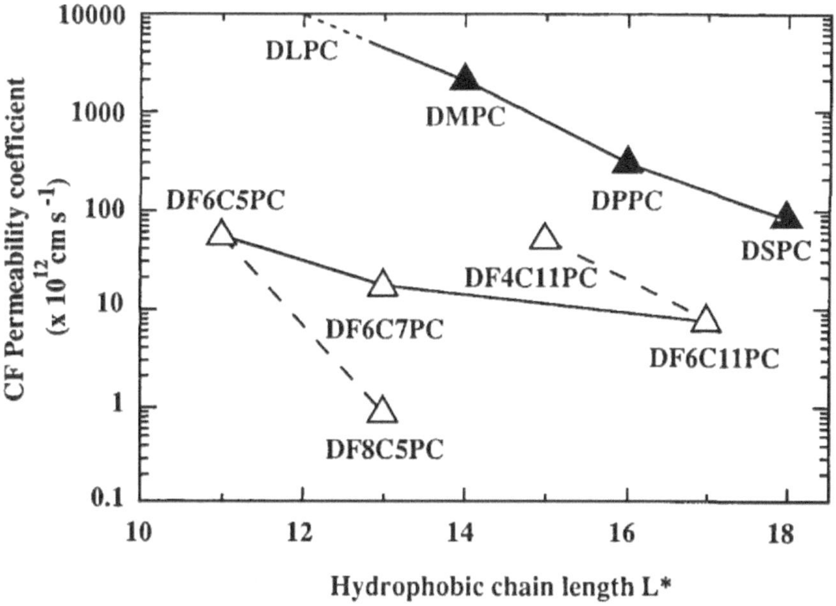

Figure 15 Variation of the permeability coefficient of carboxyfluorescein (CF) at 75°C for the DFnCmPC, **29**, and hydrocarbon-PC-based membranes as a function of the length (number of carbon atoms) of the phospholipid hydrophobic chain; Δ, fluorinated membranes; ▲, hydrogenated membranes. Error bars are smaller than the signs. *For the DFnCmPCs, **29**, L = n + m; for the hydrocarbon-PCs, L (= p) is the number of carbon atoms of the $CH_3(CH_2)_{p-2}C(O)$-chain. (From Frézard, F., et al., *Biochim. Biophys. Acta*, 1192, 61, 1994. With permission.)

DF8C5PC, **29f**, shows that the effect on permeation of two CF_2s compensates or outweighs that of six CH_2s. The thicker the fluorinated lipophobic core, the more efficient the barrier to CF permeation, as further illustrated with compounds DF6C7PC, **29d**, and DF8C5PC, **29f**, (same hydrophobic chain length but different degrees of fluorination): in the fluid state liposomes made of the latter compound are almost 20 times less leaky and less permeable than those made from the former.

A similar trend was found in the fluorinated glycophospholipid series **17–22** where a lengthening of the fluorinated tail or of the hydrocarbon spacer resulted in a decrease of CF or CA release.[131,178] The presence of two fluorinated chains, as in **17**, rather than one fluorinated and one hydrogenated one, as in **18b**, hence the passage from a discontinuous to a continuous fluorinated core, also reduced CF release very efficiently. The same trend was, however, not found in the fluorinated double-chain glycerophosphocholine series, where the liposomes based on the mixed fluorocarbon/hydrocarbon compound **34b** showed as little leakage as those made from the double fluorocarbon-tailed DFnCmPC, **29**.[159] Where the nature of the sugar moiety in the glycophospholipid series is concerned, the galactosyl or mannosyl derivatives gave rise to more stable liposomes than their glucosyl analogs (Table 6).[131]

The mixed fluorocarbon/hydrocarbon-tailed amphiphiles were synthesized and investigated in order to facilitate homogeneous dispersion of fluorinated amphiphiles in standard lipid bilayers, impart to them some of the properties of the fluorinated compounds, and help control the properties of the resulting bilayers. However, when compound **20b** was mixed with its hydrocarbon counterpart, or **17b** with CH, calcein release was observed to be significantly faster than when the fluorinated amphiphile was utilized alone.[131] The low miscibility of the two types of amphiphiles and the formation of separate domains may account for this behavior.

2. Permeability to Amphipathic Probes

Doxorubicin (DOX), an efficient anticancer drug, is a weak amphiphilic base whose encapsulation into liposomes can be efficiently mediated using transmembrane ammonium sulfate or pH gradients. It can also be considered as a model for lipophilic/hydrophilic drugs. Its active encapsulation into liposomes made from fluorinated DFnCmPCs, **29**, and the stability of these liposomes with respect to DOX release as compared to conventional liposomal DOX formulations, have been investigated.[181] It was found that the fluorinated core within the membrane was not a permeation barrier for the remote loading of DOX

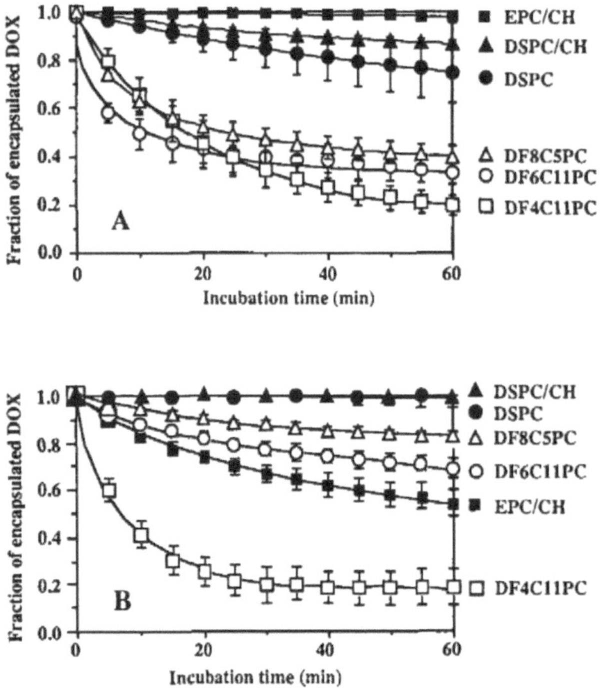

Figure 16 Time-course of doxorubicin (DOX) release at 37°C from fluorinated and conventional liposomes incubated in (A) PBS, and (B) in 50% v/v human serum/PBS solutions. The encapsulation of DOX (2 m*M* for a 1/5 DOX/lipid molar ratio) into the liposomes was performed using an ammonium sulfate gradient. (From Frézard, F., et al., *Biochim. Biophys. Acta*, 1194, 61,1994. With permission.)

into the liposomes made from DFnCmPC, **29**. Efficient (>90% for a DOX/phospholipid molar ratio as high as 0.2) and stable loading (less than 5% of release after one week) of DOX into the fluorinated liposomes' inner aqueous phase could indeed be achieved using a transmembrane ammonium sulfate or even, in the absence of Na^+, a transmembrane pH gradient (see next section).

When incubated in a sodium-containing physiological buffer (PBS), the DOX-loaded fluorinated liposomes were, however, significantly less stable than some conventional DOX-loaded liposomes (Figure 16A). This was attributed to the higher permeability of the fluorinated membranes to Na^+ (see next section). The presence of Na^+ in the outer-liposomal aqueous phase results in an increase of the internal pH, after a H^+/Na^+ exchange, hence in an increase in deprotonated neutral DOX which can diffuse more easily across membranes than protonated DOX. This further indicates that the lipophobic character of fluorinated membranes, although it strongly reduces the solubility of lipophilic/hydrophilic compounds,[18,111] does not hinder significantly the diffusion of a lipophilic/hydrophilic drug across these membranes when driven by a concentration gradient.

Reduced permeation through fluorinated bilayers was also noted for fluorescamine.[101] This compound reacts with an amine to give a highly fluorescent product. When mixed with amine-labeled unilamellar vesicles, rapid reaction with the amine groups of the outer surface was observed, which was followed by a slower process consisting of the reaction of the amino groups of the inner surface of the vesicle, a process which is controlled by the inward diffusion of fluorescamine across the membrane. This second step was much slower for fluorinated vesicles than for a hydrogenated analog.

3. Permeability to Inorganic Ions ($H^+/Na^+/OH^-$)

The active encapsulation of DOX into the fluorinated DFnCmPC (**29**)-based liposomes using a transmembrane pH gradient was effective only when the extraliposomal aqueous phase did not contain Na^+.[181] The low level of DOX entrapment and the reduced stability observed in the presence of Na^+ were assigned to rapid dissipation of the transmembrane pH gradient, and revealed high proton permeability and higher Na^+ permeability for the fluorinated membranes as compared to conventional ones. This was confirmed by the rapid decrease of the fluorescence of entrapped pyranine, used as a pH indicator of the intraliposomal aqueous phase (Figure 17).[181] Altogether these data indicate that membrane permeability to Na^+ increased along the sequence DSPC, EPC/CH (1/1), DMPC/CH (1/1) << DF4C11PC (**29b**) < DMPC << DF6C11PC (**29e**), DF8C5PC (**29f**). These results differ significantly from those concerning CF permeability. The higher Na^+ permeability of the membranes based on DF6C11PC, **29e**, or DF8C5PC, **29f**, was attributed to the presence of defects arising from their semi-fluid semi-gel like state, owing to partial fusion of the

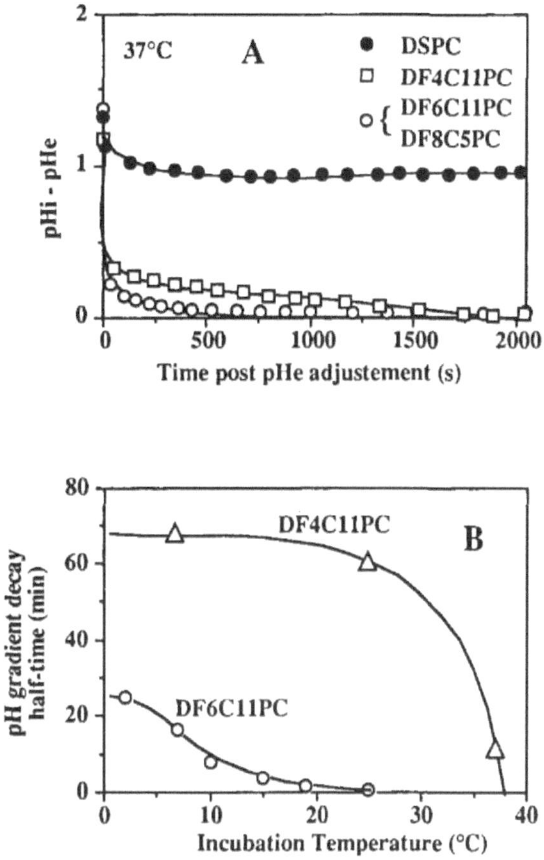

Figure 17 (A) Stability of the transmembrane pH gradient for DF4C11PC, **29b**, DF6C11PC, **29e** and DSPC liposomes prepared in a 2.5 m*M* pyranine solution (pH 8.2). After removal of unencapsulated pyranine, the liposomes were incubated at 37°C and pHe was then lowered to 6.7. The dissipation of the pH gradient (pHi(t)-pHe) as a function of time was monitored using the fluorescence decrease of entrapped pyranine. (B) Half-times ($t_{1/2}$) of transmembrane pH gradient decay as a function of the incubation temperature for DF4C11PC, **29b**, and DF6C11PC, **29e**, liposomes. (From Frézard, F., et al., *Biochim. Biophys. Acta*, 1194, 61, 1994. With permission.)

fluorinated tails.[111] These defects would facilitate permeation of smaller ions such as Na^+ rather than that of larger ones, such as CF.

Lower permeability of fluorinated membranes to OH^- has been evidenced using entrapped riboflavin (pKa ca. 10) which is nonfluorescent at high pH values.[24,182] The quenching of its fluorescence, which depends on OH^- permeation across the membrane, was complete after 30 min with a 1/200 *N* NaOH solution in the case of the vesicles made from the hydrocarbon analog of **10**. It required much more concentrated NaOH solutions (1/10 N) and longer incubation times (1 h) in the case of the corresponding fluorinated vesicles made from **10**. When vesicles were made from the fluorinated compound to which 10 mol % of the hydrogenated one was added, the quenching rate was seen to increase, suggesting that the hydrogenated component formed separate domains, more permeable to OH^-, within the fluorinated bilayer matrix.

G. FLUORINATED AMPHIPHILES AND VESICLES IN BIOLOGICAL ENVIRONMENTS

The highly hydrophobic and lipophobic character of a fluorinated liposomal membrane was expected to have an impact on its interactions with serum components and other biological compounds. This could modify the stability of such liposomes in biological fluids with respect to entrapped-drug leakage, as well as their *in vivo* behavior, including recognition by the mononuclear phagocytic system (MPS), hence intravascular persistence and biodistribution. While the exploration of the aptitude of fluorinated amphiphiles to form membranes, liposomes and other phases, is well advanced, the assessment of their biological characteristics is only beginning. A few studies have nevertheless been devoted to the *in vitro* action (on cell cultures, hemolytic activity, stability with respect to dye release in human serum) and *in vivo* behavior (acute toxicity in mice, intravascular persistence) of fluorinated amphiphiles and liposomes.[15,18,112,183,184] Their adsorption, biobistribution, metabolism, and excretion characteristics still remain to be established. The acceptable level of toxicity is obviously much lower for the average nonmedical than for medical applications, and in the latter case depends on application, dose, and route of administration.

1. *In Vitro* Behavior

a. *Effect on Cell Cultures and Hemolytic Activity*

The replacement of hydrocarbon chains by fluorocarbon ones in an amphiphile was generally found to enhance its *in vitro* biological tolerance.[15,18,31] The hemolytic action of fluorinated amphiphiles on human red blood cells is consistently and considerably lower than for their hydrocarbon analogs, and often totally suppressed.[31]

b. *Stability in Biological Fluids*

The stability of some fluorinated liposomes with respect to the release of entrapped CF and CA, when incubated at 37°C in human serum, has been investigated (Table 6). As expressed by the $t_{1/2}$ ratio collected in Table 6, the fluorinated DFnCmPC **29**-based liposomes whose membranes were in the fluid state at 37°C were destabilized by serum components but to a considerably lesser extent than "fluid" conventional ones.[174–176] This indicates that the fluorinated core inside the liposomal membrane protects the vesicles to some degree, possibly by reducing their interactions with the serum components, which would limit their adsorption at the liposome's surface and/or hinder their penetration into the bilayer, or by reducing phospholipid exchange between the liposomes and the high-density lipoproteins present in serum.

On the other hand, fluorinated liposomes whose membranes were in the gel state (i.e., DF6C11PC, **29e**) and, surprisingly, also those which were in a semi-fluid semi-gel state (i.e., DF8C5PC, **29f**) displayed a significantly greater stability in serum than in a physiological buffer, and were even noticeably more stabilized in serum than conventional "gel" (i.e., DSPC) or "gel-like" (i.e., EPC/CH) liposomes.[175,176] These results were attributed to the suppression of packing defects, generally present in gel or gel-like membranes, by serum components. They correlate well with the study in which DOX was used as a probe of liposome stability.[181] It was shown (Figure 16), that for the fluorinated liposomes whose membranes were in a gel- (i.e., DF6C11PC, **29e**) or in a semi-fluid semi-gel state (i.e., DF8C5PC, **29f**) at 37°C, the retention of DOX was increased when going from buffer (Figure 16A) to human serum (Figure 16B). The stability of the DOX-loaded fluorinated phosphatidylcholine-based liposomes in human serum was nevertheless lower than that of the conventional DSPC and DSPC/CH 1/1 ones, albeit more stable than the EPC/CH 1/1 ones.[181]

Human serum had also a strong destabilization effect on the "fluid" vesicles formulated with fluorinated glycophospholipids.[131,178] Except for the vesicles made from **17c** and **18b,** CF and CA were released too rapidly to allow $t_{1/2}$ determination (Table 6). Using CF as a probe for vesicles stability, it appeared that the "fluid" fluorinated vesicles made from **17c**, which were almost as stable in buffer as the "fluid" ones made of fluorinated phosphatidylcholines, were destabilized by serum to a considerably larger extent than the latter ones. This could reflect a higher affinity of some serum components for a glycosyl-coated surface compared to a phosphocholine-coated one.

2. *In Vivo* Behavior

a. *Acute Toxicity*

The available acute toxicity data (lethal dose, LD_{50}, or maximum tolerated dose, MTD, i.v. in mice) show wide differences depending on the amphiphile's structure, with no specific rule emerging yet.[15,18,183] Some of the *F*-alkylated compounds have rather low acute toxicity, with LD_{50} or MTD values in the gram range, both in the single- and in the double-chain series. Typical examples of such low toxicity compounds are the phosphatidylcholine derivative DF4C11PC, **29b**, (LD_{50} ~7700 mg/kg body weight in mice) and the tris(hydroxymethyl)acrylamidomethane telomer **25d** (MTD ~3750 mg/kg).

b. In Vivo *Blood Circulation Time*

In the last few years much effort has focused on the development of vesicles which evade the MPS and exhibit prolonged circulation times in the blood stream. Significant advances were achieved in this respect by including some specific natural glycolipids (such as the monosialylganglioside G_{M1})[185,186] and more recently by using synthetic phospholipids with hydrophilic, bulky and mobile poly(ethylene glycol) (PEG) groups.[187]

The clearance from circulation of fluorinated liposomes made from **29e** and **29f**, FnCmPCs, was investigated in mice using liposome-entrapped CF.[184] Using this probe, circulation half-lives were observed that can reach 8.6 h, and were 6 to 13 times and 3 to 6 times larger than those of similarly sized DSPC and DSPC/CH liposomes, respectively. Furthermore, their blood clearance was dose-independent in the 3.3 to 330 μmol/kg body weight dose range investigated. Their behavior, in terms of high circulation half-times and dose independence thus came within reach of those of some of the PEG-labeled "stealth" liposomes.

Liposome clearance from blood stream occurs as a result of interactions with blood proteins and other plasma components, leading to liposome opsonization followed by macrophage uptake, and/or to liposome disintegration by lipid exchange and depletion.[2,188,189] In the case of PEG-labeled "stealth" liposomes, steric stabilization, and surface mobility of the long polymeric PEG chains are thought to decrease opsonization and, consequently, recognition and uptake by the MPS.[187] In the case of the fluorinated liposomes, it is likely that the lipophobic fluorinated core inside the membrane could reduce the adsorption and anchoring of opsonizing proteins onto their membrane. The enhanced hydrophobic character of the fluorinated membrane may also hinder their disintegration and depletion through lipid exchange induced by plasma proteins, in line with the high *in vitro* stability observed for the fluorinated liposomes both in serum and in whole blood.[184]

These results further show that not only direct modification of the liposomes' surface but also more subtle and *a priori* "invisible" changes introduced inside the membrane are of importance for them to evade *in vivo* recognition and MPS uptake.[18]

ACKNOWLEDGMENTS

We wish to thank our co-workers whose names are listed in the references, Mme V. Briet for preparing the manuscript and figures, the Centre National de la Recherche Scientifique, and Applications et Transferts de Technologies Avancées (ATTA) for financial support.

REFERENCES

1. **Fendler, J. H.,** *Membrane Mimetic Chemistry*, Wiley-Interscience, New York, 1983.
2. **Gregoriadis, G., Ed.,** *Liposome Technology*, Vols. I–III, 2nd ed., CRC Press, Boca Raton, FL, 1993.
3. **Lasic D. D.,** *Liposomes: From Physics to Applications*, Elsevier, Amsterdam, 1993.
4. **Bangham, A. D., Standish, M. M., and Watkins, J. C.,** Diffusion of univalent ions across the lamellae of swollen phospholipids, *J. Mol. Biol.*, 13, 238, 1965.
5. **Eibl, H.,** Phospholipids as functional constituents of biomembranes, *Angew. Chem., Int. Ed. Engl.*, 23, 257, 1984.
6. **Ringsdorf, H., Schlarb, B., and Venzmer, J.,** Molecular architecture and function of polymeric oriented systems: models for the study of organization, surface recognition, and dynamics of biomembranes, *Angew. Chem., Int. Ed. Engl.*, 27, 113, 1988.
7. **Kunitake, T.,** Synthetic bilayer membranes: molecular design, self-organization, and application, *Angew. Chem., Int. Ed. Engl.*, 31, 709, 1992.
8. **Fuhrhop, J. H. and Helfrich, W.,** Fluid and solid fibers made of lipid molecular bilayers, *Chem. Rev.*, 93, 1565, 1993.
9. **Kunitake, T.,** A totally synthetic bilayer membrane, *J. Am. Chem. Soc.*, 99, 3860, 1977.
10. **Tran, C. D., Klahn, P. L., Romero, A., and Fendler, J. H.,** Characterization of surfactant vesicles as potential membrane models. Effect of electrolytes, substrates, and fluorescence probes, *J. Am. Chem. Soc.*, 100, 1622, 1978.
11. **Gebicki, J. M. and Hicks, M.,** Ufasomes are stable particles surrounded by unsaturated fatty acid membranes, *Nature*, 243, 232, 1973.
12. **Hargreaves, W. R. and Deamer, D. W.,** Liposomes from ionic, single-chain amphiphiles, *Biochemistry*, 17, 3759, 1978.
13. **Riess, J. G., Arlen, C., Greiner, J., Le Blanc, M., Manfredi, A., Pace, S., Varescon, C., and Zarif, L.,** Design, synthesis and evaluation of fluorocarbons and surfactants for *in vivo* applications: new perfluoroalkylated polyhydroxylated surfactants, in *Blood Substitutes*, Chang, T. M. S. and Geyer, R. P., Dekker, New York, 1989, 421.
14. **Riess, J. G. and Greiner, J.,** Perfluoroalkylated sugar derivatives as surfactants and co-surfactants for biomedical uses, in *Carbohydrates as Organic Raw Materials II*, Descotes, G., Ed., VCH, Weinheim, 1993, 209.
15. **Greiner, J., Riess, J. G., and Vierling, P.,** Fluorinated surfactants intended for biomedical uses, in *Organofluorine Compounds in Medicinal Chemistry and Biomedical Applications*, Filler, R., Kobayashi, Y., and Yagupolskii, L. M., Eds., Elsevier, Amsterdam, 1993, 339.
16. **Kissa, E.,** *Fluorinated Surfactants. Synthesis, Properties, Applications*, Marcel Dekker, New York, 1994.
17. **Riess, J. G.,** Highly fluorinated systems for oxygen carriage, diagnosis and drug delivery, *Colloids Surf.*, 84, 33, 1994; **Riess, J. G.,** Fluorine in our arteries, *New Chem.*, 19, 893, 1995.
18. **Riess, J. G.,** Fluorinated vesicles, *J. Drug Target.*, 2, 455, 1994; **Riess, J. G.,** Introducing a new element — fluorine — into the liposomal membrane, *J. Liposome Res.*, 5, 413, 1995.
19. **Mukerjee, P. and Yang, A. Y. S.,** Nonideality of mixing micelles of fluorocarbon and hydrocarbon surfactants and evidence of partial miscibility from differential conductance data, *J. Phys. Chem.*, 80, 1388, 1976.
20. **Funasaki, N. and Hada, S.,** Existence of critical demicellization concentration, *J. Colloid Interface Sci.*, 73, 425, 1980.
21. **Mysels, K. J.,** Critical demicellization concentration?, *J. Colloid Interface Sci.*, 66, 331, 1978.
22. **Asakawa, T., Miyagishi, S., and Nishida, M.,** Coexistence of two kinds of mixed micelles revealed by gel filtration, *Langmuir*, 3, 821, 1987.

23. **Kunitake, T., Tawaki, S. I., and Nakashima, N.,** Excimer formation and phase separation of hydrocarbon and fluorocarbon bilayer membranes, *Bull. Chem. Soc. Jpn.*, 56, 3235, 1983.
24. **Kunitake, T. and Higashi, N.,** Bilayer membranes of fluorocarbon amphiphiles, *Makromol. Chem., Suppl.*, 14, 81, 1985.
25. **Elbert, R., Folda, T., and Ringsdorf, H.,** Saturated and polymerizable amphiphiles with fluorocarbon chains. Investigation in monolayers and liposomes, *J. Am. Chem. Soc.*, 106, 7687, 1984.
26. **Overney, R. M., Meyer, E., Frommer, J., Brodbeck, D., Lüthi, R., Howald, L., Güntherodt, H. J., Fujihira, M., Takano, H., and Gotoh, Y.,** Friction measurements on phase-separated thin films with a modified atomic force microscope, *Nature*, 359, 133, 1992.
27. **Ishikawa, Y., Kuwahara, H., and Kunitake, T.,** Self-assembly of bilayers from double-chain fluorocarbon amphiphiles in aprotic solvents: thermodynamic origin and generalization of the bilayer assembly, *J. Am. Chem. Soc.*, 116, 5579, 1994.
28. **Riess, J. G., Ed.,** Blood Substitutes and Related Products: The fluorocarbon Approach, Proc. Int. Symp. Blood Substitutes, Vol. III, *Art. Cells, Blood Subst., Immob. Biotech.*, 22, 1994.
29. **Riess, J. G.,** Overview on progress in the fluorocarbon approach to *in vivo* oxygen delivery, in *Blood Substitutes and Oxygen Carriers*, Chang, T. M. S., Ed., Marcel Dekker, New York, 1993, 24.
30. **Zuck, T. F. and Riess, J. G.,** Current status of injectable oxygen carriers, *Crit. Rev. Clin. Lab. Sci.*, 31, 295, 1994.
31. **Riess, J. G., Pace, S., and Zarif, L.,** Highly effective surfactants with low hemolytic activity, *Adv. Mater.*, 3, 249, 1991.
32. **Ostro, M. J., Ed.,** *Liposomes: From Biophysics to Therapeutics*, Marcel Dekker, New York, 1987.
33. **Friberg, S. and Lindman, B., Eds.,** *Organized Solutions. Surfactants in Science and Technology*, Marcel Dekker, New York, 1992.
34. **Murakami, Y.,** *Supramolecular Assemblies*, Mita Press, Tokyo, 1990.
35. **Rosen, M. J.,** *Surfactants and Interfacial Phenomena*, John Wiley & Sons, New York, 1978.
36. **Bondi, A.,** Van der Waals volumes and radii, *J. Phys. Chem.*, 68, 441, 1964.
37. **Tiddy, G. J. T.,** Concentrated surfactant systems, in *Modern Trends of Colloid Science in Chemistry and Biology*, Eicke, E., Ed., Birkäuser Verlag, Basel, 1985, 148.
38. **Rosen, M. J., Cohen, A. W., Dahanayake, M., and Xi-Tuan Hua, J.,** Relationship of structure to properties in surfactants. X. Surface and thermodynamic properties of 2-dodecyloxypoly(ethenoxyethanol)s, $C_{12}H_{25}(OC_2H_4)_xOH$, in aqueous solution, *J. Phys. Chem.*, 86, 541, 1982.
39. **Tanford, C.,** *The Hydrophobic Effect*, 2nd ed., John Wiley & Sons, New York, 1980.
40. **Israelachvili, J. N., Mitchell, D. J., and Ninham, B. W.,** Theory of self-assembly of lipid bilayers and vesicles, *Biochim. Biophys. Acta*, 470, 185, 1977.
41. **Weast, R. C. and Astle, M. J., Eds.,** *CRC Handbook of Chemistry and Physics*, 63rd ed., CRC Press, Boca Raton, FL, 1982.
42. **Asakawa, T., Mouri, M., Miyagishi, S., and Nishida, M.,** Probe methods for the second cmc of fluorocarbon and hydrocarbon surfactant mixtures, *Langmuir*, 5, 343, 1989.
43. **Hoffmann, H., Kalus, J., and Thurn, H.,** Small angle neutron scattering measurements on micellar solutions of perfluor detergents, *Colloid Polym. Sci.*, 261, 1043, 1983.
44. **Eaton, D. F. and Smart, B. E.,** Are fluorocarbon chains "stiffer" than hydrocarbon chains? Dynamics of end-to-end cyclization in a C_8F_{16} segment monitored by fluorescence, *J. Am. Chem. Soc.*, 112, 2821, 1990.
45. **Gruen, D. W. R.,** A model for the chains in amphiphilic aggregates. II. Thermodynamic and experimental comparisons for aggregates of different shape and size, *J. Phys. Chem.*, 89, 153, 1985.
46. **Flory, P. J.,** *Statistical Mechanics of Chain Molecules*, Interscience Publishers, New York, 1969, 153.
47. **Rigby, H. A. and Bunn, C. W.,** A room temperature transition in polytetrafluoroethylene, *Nature*, 164, 583, 1949.
48. **Furukawa, G. T., McCoskey, R. E., and King, G. J.,** Calorimetric properties of polytetrafluoroethylene (Teflon), *J. Res. Nat. Bur. Stand.*, 49, 273, 1952.
49. **Shinoda, K., Hato, M., and Hayashi, T.,** The physicochemical properties of aqueous solutions of fluorinated surfactants, *J. Phys. Chem.*, 76, 909, 1972.
50. **Smart, B. E.,** in *The Chemistry of Functional Groups, Suppl. D. The Chemistry of Halides, Pseudo Halides and Azides*, Part 2, Patai, S. and Rappoport, Z., Eds., John Wiley & Sons, Chichester, 1983, 603.
51. **Riess, J. G.,** Perfluorochemical emulsions for intravascular use: principles, materials and methods, in *Fluorine in Medicine*, Banks, R. E. and Lowe, K. C., Eds., Rapra Technology Ltd, Shawbury, 1994, paper 20.
52. **Mukerjee, P. and Handa, T.,** Adsorption of fluorocarbon and hydrocarbon surfactants to air-water, hexane-water, and perfluorohexane-water interfaces. Relative affinities and fluorocarbon-hydrocarbon nonideality effects, *J. Phys. Chem.*, 85, 2298, 1981.
53. **Mukerjee, P.,** Fluorocarbon-hydrocarbon interactions in micelles and other lipid assemblies. at interfaces and in solutions, *Colloids Surf.*, 84, 1, 1994.
54. **Schneider, J., Erdelen, C., Ringsdorf, H., and Rabolt, J. F.,** Structural studies of polymers with hydrophilic spacer groups. II. Infrared spectroscopy of Langmuir-Blodgett multilayers of polymers with fluorocarbon side chains at ambient and elevated temperature, *Macromolecules*, 22, 3475, 1989.
55. **Mukerjee, P. and Mysels, K. J.,** Critical micelle concentrations of aqueous surfactant systems, NSRDS-NBS-36, U.S. Government Printing Office, Washington, D.C., 1971.

56. **Mathis, G., Leempoel, P., Ravey, J. C., Selve, C., and Delpuech, J. J.,** A novel class of nonionic microemulsions: fluorocarbons in aqueous solutions of fluorinated poly(oxyethylene) surfactants, *J. Am. Chem. Soc.*, 106, 6162, 1984.
57. **Ravey, J. C., Gherbi, A., and Stébé, M. J.,** Comparative study of fluorinated and hydrogenated nonionic surfactants. I. Surface activity properties and critical concentrations, *Prog. Colloid Polym. Sci.*, 76, 234, 1988.
58. **Kunieda, H. and Shinoda, K.,** Krafft points, critical micelle concentrations, surface tension, and solubilizing power of aqueous solutions of fluorinated surfactants, *J. Phys. Chem.*, 80, 2468, 1976.
59. **Riess, J. G., Greiner, J., Abouhilale, S., and Milius, A.,** Stabilization of fluorocarbon emulsions by sugar-derived perfluoroalkylated surfactants and co-surfactants, *Prog. Colloid Polym. Sci.*, 88, 123, 1992.
60. **Selve, C. and Achilefu, S.,** Synthesis of monodisperse perfluoroalkyl-oxyethylene surfactants with methoxy capping: surfactants of high chemical inertness, *J. Chem. Soc., Chem. Commun*, 911, 1990.
61. **Maurizis, J. C., Pavia, A. A., and Pucci, B.,** Efficiency of nonionic telomeric surfactants for the solubilization of subcellular fractions proteins, *Bioorg. Med. Chem. Lett.*, 3, 161, 1993.
62. **Dermardirossian, C., Krafft, M. P., and Lederer, F.,** Utilisation d'un détergent fluoré pour la solubilisation du cytochrome b_{558}, composant membranaire de la NADPH oxydase de neutrophiles humains, *Eur. J. Clin. Invest.*, 25(Suppl. 2), Abstr. 141, 1995.
63. **Fung, B. M., Mamrosh, D. L., O'Rear, E. A., Fresh, C. B., and Afzal, J.,** Unusual micellar properties of a new class of fluorinated nonionic surfactants, *J. Phys. Chem.*, 92, 4405, 1988.
64. **Guo, W., Brown, T. A., and Fung, B. M.,** Micelles and aggregates of fluorinated surfactants, *J. Phys. Chem.*, 95, 1829, 1991.
65. **Jones, M. N.,** Carbohydrate-mediated liposomal targeting and drug delivery, *Adv. Drug Delivery Rev.*, 13, 215, 1994.
66. **Laschewsky, A., Ringsdorf, H., and Schmidt, G.,** Polymerization of hydrocarbon and fluorocarbon amphiphiles in Langmuir–Blodgett multilayers, *Thin Solid Films*, 134, 153, 1985.
67. **Higashi, N. and Kunitake, T.,** Stabilization and facilitated deposition of surface monolayers of fluorocarbon amphiphiles through polyion complex formation, *Chem. Lett.*, 105, 1986.
68. **Erdelen, C., Laschewsky, A., Ringsdorf, H., Schneider, J., and Schuster, A.,** Thermal behaviour of polymeric Langmuir–Blodgett multilayers, *Thin Solid Films*, 180, 153, 1989.
69. **Klassen, J. K., Mitchell, M. B., Govoni, S. T., and Nathanson, G. M.,** Insoluble perfluorinated monolayers and thin films on the surfaces of corrosive acids, *J. Phys. Chem.*, 97, 10166, 1993.
70. **Li, M., Acero, A. A., Huang, Z., and Rice, S. A.,** Formation of an ordered Langmuir monolayer by a non-polar chain molecule, *Nature*, 367, 151, 1994.
71. **Santaella, C. and Vierling, P.,** unpublished results, 1994.
72. **Jacquemain, D., Grayer Wolf, S., Leveiller, F., Deutsch, M., Kjaer, K., Als-Nielsen, J., Lahav, M., and Leiserowitz, L.,** Two-dimensional crystallography of amphiphilic molecules at the air-water interfaces, *Angew. Chem., Int. Ed. Engl.*, 31, 130, 1992.
73. **Barton, S. W., Goudot, A., Bouloussa, O., Rondelez, F., Lin, B., Novak, F., Acero, A., and Rice, S. A.,** Structural transitions in a monolayer of fluorinated amphiphile molecules, *J. Chem. Phys.*, 96, 1343, 1992.
74. **Shin, S., Collazo, N., and Rice, S. A.,** A molecular dynamics study of the packing structures in monolayers of partially fluorinated amphiphiles, *J. Chem. Phys.*, 96, 1352, 1992.
75. **Collazo, N., Shin, S., and Rice, S. A.,** Molecular-dynamics studies of the structure and properties of monolayers of perfluorinated amphiphiles, *J. Chem. Phys.*, 96, 4735, 1992.
76. **Lebeau, L., Schultz, P., Célia, H., Mésini, P., Nuss, S., Keinger, C., Ollard, S., Ondet, P., and Nioskowski, C.,** Specifically designed lipid assemblies as tools for two dimensional crystallization of soluble biological macromolecules, in *Handbook of Nonmedical Applications of Liposomes*, Vol. II, Barenholz, Y. and Lasic, D. D., Eds., CRC Press, Boca Raton, FL, 1996, in press.
77. **Sasaki, D. Y., Kurihara, K., and Kunitake, T.,** Self-assembled multifunctional receptors for nucleotides at the air-water interface, *J. Am. Chem. Soc.*, 114, 10994, 1992.
78. **Tian, Y., Lian, B., Li, Z., Liang, Y., Lu, L., and Huang, W.,** Specific and non-specific interaction between double perfluoroalkyl-chain steroidal glycoside monolayers and cellulase at the air-water interfaces, *J. Chem. Soc., Chem. Commun.*, 31, 1993.
79. **Kunitake, T., Higashi, N., and Kajiyama, T.,** Oxygen enrichment by fluorocarbon bilayer films, *Chem. Lett.*, 717, 1984.
80. **Higashi, N., Kunitake, T., and Kajiyama, T.,** Surface structure and oxygen permeation in mixed multibilayer films of hydrocarbon and fluorocarbon amphiphiles, *Macromolecules*, 19, 1362, 1986.
81. **Takahara, A., Higashi, N., Kunitake, T., and Kajiyama, T.,** State of aggregation and surface chemical composition of composite thin films composed of poly(vinyl alcohol) and fluorocarbon amphiphile, *Macromolecules*, 21, 2443, 1988.
82. **Takahara, A., Morotomi, N., Hiraoka, S., Higashi, N., Kunitake, T., and Kajiyama, T.,** Analysis of the surface structure of a built-up film of fluorocarbon amphiphile and polymer/(fluorocarbon amphiphile) composite thin film by means of X-ray photoelectron spectroscopy, *Macromolecules*, 22, 617, 1989.
83. **Okahata, Y., Ebato, H., and Ye, X.,** Molecular selective adsorption on a multibilayer-coated piezoelectric crystal, *J. Chem. Soc., Chem. Commun.*, 1037, 1988.

84. **Ravey, J. C. and Stébé, M. J.**, Phase and structure behaviour of fluorinated nonionic surfactant systems, *Prog. Colloid Polym. Sci.*, 73, 127, 1987.
85. **Ravey, J. C. and Stébé, M. J.**, Properties of fluorinated non-ionic surfactant-based systems and comparison with non-fluorinated systems, *Colloids Surf.*, 84, 11, 1994.
86. **Fontell, K. and Lindman, B.**, Fluorocarbon surfactants. Phase equilibrium and phase structures in aqueous systems of a totally fluorinated fatty acid and some of its salt, *J. Phys. Chem.*, 87, 3289, 1983.
87. **El Moujahid, C.**, Propriétés physicochimiques de systèmes aqueux de tensioactifs non ioniques fluorés, *Thèse de Doctorat*, Université de Nancy, France, 1993.
88. **Tiddy, G. J. T.**, Fluorocarbon surfactant-water mesomorphic phases. I. Study of the ammonium perfluorooctanoate-water system by optical microscopy and low angle X-ray diffraction, *J. Chem. Soc., Faraday Trans. 1*, 68, 608, 1972.
89. **Krafft, M. P., Giulieri, F., and Riess, J. G.**, Can single-chain perfluoroalkylated amphiphiles alone form vesicles and other organized supramolecular systems?, *Angew. Chem., Int. Ed. Engl.*, 32, 741, 1993.
90. **Giulieri, F., Krafft, M. P., and Riess, J. G.**, Stable flexible fibers and rigid tubules made from single-chain perfluoroalkylated amphiphiles, *Angew. Chem., Int. Ed. Engl.*, 33, 1514, 1994.
91. **Zarif, L., Gulik-Krzywicki, T., Riess, J. G., Pucci, B., Guedj, C., and Pavia, A. A.**, Alkyl and perfluoralkyl glycolipid-based supramolecular assemblies, *Colloids Surf.*, 84, 107, 1994.
92. **Bleasdale, T. A. and Tiddy, G. J. T.**, Surfactant liquid crystals, in *The Structure, Dynamics and Equilibrium Properties of Colloidal Systems*, Bloor, D. M. and Wyn-Jones, E., Eds., Kluwer Academic, Dordrecht, The Netherlands, 1990, 397.
93. **Kunitake, T., Okahata, Y., Shimomura, M., Yasunami, S. I., and Takarabe, K.**, Formation of stable bilayer assemblies in water from single-chain amphiphiles. Relationship between the amphiphile structure and the aggregate morphology, *J. Am. Chem. Soc.*, 103, 5401, 1981.
94. **Menger, F. M. and Yamasaki, Y.**, Hyperextended amphiphiles. Bilayer formation from single-tailed compounds, *J. Am. Chem. Soc.*, 115, 3840, 1993.
95. **Handjani-Vila, R. M., Ribier, A., and Vanlerberghe, G.**, Les niosomes, in *Les Liposomes — Applications Thérapeutiques*, Puisieux, F., Delattre, J., Eds., Lavoisier, Paris, 1985, 297.
96. **Bouwstra, J. A., Hofland, H. E. J., Gooris, G. S., and Junginger, H. E.**, Characterization of niosomes, in *Emulsions — A Fundamental and Practical Approach*, Sjöblom, J., Ed., Kluwer Academic, Dordrecht, The Netherlands, 1992, 227.
97. **Kaler, E. W., Murthy, A. K., Rodriguez, B. E., and Zasadzinski, J. A. N.**, Spontaneous vesicle formation in aqueous mixtures of single-tailed surfactants, *Science*, 245, 1371, 1989.
98. **Fukuda, H., Kawata, K., Okuda, H., and Regen, S. L.**, Bilayer-forming ion-pair amphiphiles from single-chain surfactants, *J. Am. Chem. Soc.*, 112, 1635, 1990.
99. **Benz, R., Prass, W., and Ringsdorf, H.**, Black lipid membranes from polymerizable lipids, *Angew. Chem., Int. Ed. Engl.*, 21, 368, 1982.
100. **Wakita, M., Edwards, K. A., Regen, S. L., Turner, D., and Gruner, S. M.**, Use of a polymeric counterion to induce bilayer formation from a single-chain surfactant, *J. Am. Chem. Soc.*, 110, 5221, 1988.
101. **Kunitake, T., Okahata, Y., and Yasunami, S.**, Formation and enhanced stability of fluoroalkyl bilayer membranes, *J. Am. Chem. Soc.*, 104, 5547, 1982.
102. **Krafft, M. P., Giulieri, F., and Riess, J. G.**, Supramolecular assemblies from single-chain perfluoroalkylated phosphorylated amphiphiles, *Colloids Surf.*, 84, 113, 1994.
103. **Riess, J. G. and Krafft, M. P.**, Fluorinated phosphocholine-based amphiphiles as components of fluorocarbon emulsions and fluorinated vesicles, *Chem. Phys. Lipids*, 75, 1, 1995.
104. **Krafft, M. P., Rolland, J. P., Vierling, P., and Riess, J. G.**, New perfluoroalkylated phosphocholines. Effects on particle size and stability of fluorocarbon emulsions, *New J. Chem.*, 14, 869, 1990.
105. **Krafft, M. P., Vierling, P., and Riess, J. G.**, Synthesis and preliminary data on biocompatibility and emulsifying properties of perfluoroalkylated phosphoramidates as injectable surfactants, *Eur. J. Med. Chem.*, 26, 545, 1991.
106. **Zarif, L., Guedj, C., Pucci, B., Pavia, A. A., and Riess, J. G.**, Modular glycolipids with perfluoroalkylated, hydrogenated, or mixed perfluoroalkylated/hydrogenated hydrophobic tails as components for fluorinated vesicles and other supramolecular assemblies, *Abstract Int. Symp. Liposomes in Drug Delivery: The Nineties and Beyond*, London, 1993, 126.
107. **Zarif, L., Gulik-Krzywicki, T., Riess, J. G., Pucci, B., Guedj, C., and Pavia, A. A.**, unpublished results, 1994.
108. **Giulieri, F. and Krafft, M. P.**, Can the formation of vesicles from single-chain perfluoroalkylated amphiphiles be predicted?, *Colloids Surf.*, 84, 121, 1994.
109. **Giulieri, F., Krafft, M. P., and Riess, J. G.**, Fluorinated vesicles from single-chain perfluoroalkylated amphiphiles, *Abstract Int. Symp. Liposomes in Drug Delivery: The Nineties and Beyond*, London, 1993, 78.
110. **Trevino, L., Krafft, M. P., Frézard, F., Giulieri, F., and Riess, J. G.**, Fluorinated liposomes. Use of (*F*-alkyl)alkyl components to impart impermeability to the liposomal membrane, *Proc. Int. Symp. Control. Rel. Bioact. Mater.*, 21, 606, 1994.
111. **Santaella, C., Vierling, P., Riess, J. G., Gulik-Krzywicki, T., Gulik, A., and Monasse, B.**, Polymorphic phase behavior of perfluoroalkylated phosphatidylcholines, *Biochim. Biophys. Acta*, 1190, 25, 1994.
112. **Zarif, L., Riess, J. G., Pucci, B., Guedj, C., and Pavia A. A.**, Double-tailed perfluoroalkylated glycolipids as components for drug delivery and targeting systems. Preliminary biocompatibility results, *Art. Cells, Blood Subst., Immob. Biotech.*, 22, 1485, 1994.

113. **Guedj, C., Pucci, B., Zarif, L., Coulomb, C., Riess, J. G., and Pavia, A. A,** Vesicles and other supramolecular systems from biocompatible synthetic glycolipids with hydrocarbon and/or fluorocarbon chains, *Chem. Phys. Lipids*, 72, 153, 1994.
114. **Myrtil, E., Zarif, L., Greiner, J., Riess, J. G. Pucci, B., and Pavia, A. A.,** Double-tailed perfluoroalkyl telomeric surfactants derived from tris(hydroxymethyl)-acrylamidomethane for medical applications, *Macromol. Chem. Phys.*, 195, 1289, 1994.
115. **Myrtil, E., Zarif, L., Greiner, J., Riess, J. G. Pucci, B., and Pavia, A. A.,** unpublished results, 1994.
116. **Sternberg, B., Zarif, L., Greiner, J., and Riess, J. G.,** unpublished results, 1994.
117. **Trevino, L., Frézard, F., Rolland, J. P., Postel, M., and Riess, J. G.,** Novel liposome systems based on the incorporation of (perfluoroalkyl)alkenes (FmHnE) into the bilayer of phospholipid liposomes, *Colloids Surf.*, 84, 223, 1994.
118. **Trevino, L., Frézard, F., Postel, M., and Riess, J. G.,** Incorporation of a (perfluoroalkyl)alkane into the phospholipid bilayer of DMPC liposomes results in greater encapsulation stability, *J. Liposome Res.*, 4, 1017, 1994.
119. **Riess, J. G., Solé-Violan, L., and Postel, M.,** A new concept in the stabilization of injectable fluorocarbon emulsions: the use of mixed fluorocarbon-hydrocarbon dowels, *J. Dispersion Sci. Technol.*, 13, 349, 1992.
120. **Riess, J. G. and Postel, M.,** Stability and stabilization of fluorocarbon emulsions destined for injection, *Biomat., Art. Cells, Immob. Biotech.*, 20, 819, 1992.
121. **Cornélus, C., Krafft, M. P., and Riess, J. G.,** Improved control over particles sizes and stability of concentrated fluorocarbon emulsions by using mixed fluorocarbon/hydrocarbon molecular dowels, *Art. Cells, Blood Subst., Immob. Biotech.*, 22, 1183, 1994.
122. **Cornélus, C., Krafft, M. P., and Riess, J. G.,** About the mechanism of stabilization of fluorocarbon emulsions by mixed fluorocarbon/hydrocarbon additives, *J. Colloid Interface Sci.*, 163, 391, 1994.
123. **Cornélus, C., Krafft, M. P., and Riess, J. G.,** The dowel effect — mechanism of stabilization of fluorocarbon emulsions by mixed fluorocarbon/hydrocarbon amphiphiles, *Abstract, 14th Int. Symp. Fluorine Chemistry*, Yokohama, Japan, 1994, 49.
124. **Trevino, L., Postel, M., and Riess, J. G.,** The dowel effect of (perfluoroalkyl)alkenes at the fluorocarbon/water interface: increased adsorption of sodium laurate, *J. Colloid Interface Sci.*, 166, 414, 1994.
125. **Cohen, J. A. and Krafft, M. P.,** unpublished results, 1994.
126. **Privitera, N., Naon, R., and Riess, J. G.,** Hydrolysis of DMPC or DPPC by pancreatic phospholipase A_2 is slowed down when (perfluoroalkyl)alkanes are incorporated into the liposomal membrane, *Biochim. Biophys. Acta*, 1254, 1, 1994.
127. **Zarif, L., Postel, M., Septe, B., Trevino, L., Riess, J. G., Mahé, A. M., and Follana, R.,** Biodistribution and excretion of mixed fluorocarbon-hydrocarbon dowel molecules used as stabilizers of fluorocarbon emulsions. A quantitative study by fluorine NMR, *Pharm. Res.*, 11, 122, 1994.
128. **Schnur, J. M.,** Lipid tubules: a paradigm for molecularly engineered structures, *Science*, 262, 1669, 1993.
129. **Krafft, M. P., Giulieri, F., and Riess, J. G.,** Enhanced proclivity to self-aggregation of phosphorous-based amphiphiles when fluorinated: the (perfluoroakyl)alkyl dimorpholinophosphates, in *Phosphorous, Sulfur, Silicon*, in press.
129a. **Giulieri, F., Krafft, M. P., and Riess, J. G.,** About the self-organization of single chain fluorinated amphiphiles with fluorinated alcohols, *Thin Solid Films*, in press.
130. **El Ghoul, M., Rico, I., and Lattes, A.,** *Colloids Surf.*, in press.
131. **Guillod, F.,** Synthèse et évaluation physico-chimique de glycophospholipides perfluoroalkylés composants de liposomes pour la vectorisation de principes actifs, *Thèse de Doctorat,* Université de Nice, France, 1994.
132. **Guillod, F., Giulieri, F., Greiner, J., Krafft, M. P., and Riess, J. G.,** unpublished results, 1994.
133. **Papahadjopoulos, D., Wail, W. J., Jacobson, K., and Poste, G.,** Cochleate lipid cylinders: formation by fusion of unilamellar vesicles, *Biochim. Biophys. Acta*, 394, 483, 1975.
134. **Fuhrhop, J. H., Schnieder, P., Rosenberg, J., and Boekema, E.,** The chiral bilayer effect stabilizes micellar fibers, *J. Am. Chem. Soc.*, 109, 3387, 1987.
135. **Fuhrhop, J. H., Demoulin, C., Rosenberg, J., and Boettcher, C.,** Cloth-like aggregates of micellar fibers made of *N*-dodecyltartaric acid monoamides, *J. Am. Chem. Soc.*, 112, 2827, 1988.
136. **Imae, T., Takahashi, Y., and Muramatsu, H.,** Formation of fibrous molecular assemblies by amino acids surfactants in water, *J. Am. Chem. Soc.*, 114, 3414, 1992.
137. **Newkome, G. R., Baker, G. R., Arai, S., Saunders, M. J., Russo, P. S., Theriot, K. J., Moorefield, C. N., Rogers, L. E., Miller, J. E., Lieux, T. R., Murray, M. E., Phillips, B., and Pascal, L.,** Synthesis and characterization of two-directional Cascade molecules and formation of aqueous gels, *J. Am. Chem. Soc.*, 112, 8458, 1990.
138. **Russel, T. P., Rabolt, J. F., Twigg, R. T., Siemens, R. L., and Farmer, B. L.,** Structural characterization of semi-fluorinated *n*-alkanes. II. Solid-solid transition behavior, *Macromolecules*, 19, 1135, 1986.
139. **Turberg, M. P. and Brady, J. E.,** Semifluorinated hydrocarbons: primitive surfactant molecules, *J. Am. Chem. Soc.*, 110, 7797, 1988.
140. **Ishikawa, Y., Kuwahara, H., and Kunitake, T.,** Self-organization of solvophobic double-chained fluorocarbon derivatives in organic media, *Chem. Lett.*, 1737, 1989.
141. **Ishikawa, Y., Kuwahara, H., and Kunitake, T.,** Self-assembly of bilayer membranes in organic solvents by novel "amphiphilic" compounds, *J. Am. Chem. Soc.*, 111, 8530, 1989.

142. **Kuwahara, H., Hamada, M., Ishikawa, Y., and Kunitake, T.,** Self-organization of bilayer assemblies in a fluorocarbon medium, *J. Am. Chem. Soc.*, 115, 3002, 1993.
143. **Riess, J. G., Greiner, J., Abouhilale, S., and Milius, A.,** Stabilization of fluorocarbon emulsions by sugar-derived perfluoroalkylated surfactants and co-surfactants, *Progr. Colloid Polym. Sci.*, 88, 123, 1992.
144. **Stadtler, V., Krafft, M. P., and Riess, J. G.,** Water-in-fluorocarbon emulsions for drug delivery, *Proc. Int. Symp. Control. Rel. Bioactive Mater.*, 22, 464, 1995.
145. **Rosano, H. L. and Gerbacia, W. E.,** Fluorocarbon microemulsions, U.S. Patent 2,452,513, 1975.
146. **Robert, A. and Tondre, C.,** Solubilization of water in binary mixtures of fluorocarbon and nonionic fluorinated surfactants: existence domains of reverse microemulsions, *J. Colloid Interface Sci.*, 98, 515, 1984.
147. **Lattes, A. and Rico-Lattes, I.,** Microemulsions of perfluorinated and semi-fluorinated compounds, *Art. Cells, Blood Subst., Immob. Biotech.*, 22, 1007, 1994.
148. **Schubert, K. V. and Kaler, E. W.,** Microemulsifying fluorinated oils with mixtures of fluorinated and hydrogenated surfactants, *Colloids Surf.*, 84, 97, 1994.
149. **Krafft, M. P. and Riess, J. G.,** Stable highly concentrated fluorocarbon gels, *Angew. Chem., Int. Ed. Engl.*, 33, 1100, 1994.
150. **Krafft, M. P. and Riess, J. G.,** Fluorocarbon polyaphrons, *Abstracts 14th Int. Symp. on Fluorine Chemistry*, Yokohama, Japan, 1994, 255.
151. **Solans, C., Dominguez, J. G., Parra, J. L., Heuser, J., and Friberg, S. S.,** Gelled emulsions with a high water content, *Colloid Polym. Sci.*, 266, 570, 1988.
152. **Kunieda, H., Solans, C., Shida, N., and Parra, J. L.,** The formation of gel-emulsions in a water/nonionic surfactant/oil system, *Colloids Surf.*, 24, 225, 1987.
153. **Ravey, J. C. and Stébé, M. J.,** Structure of inverse micelles and emulsions-gels with fluorinated nonionic surfactants. A small-angle neutron scattering study, *Prog. Colloid. Polym. Sci.*, 82, 218, 1990.
154. **Ravey, J. C. and Stébé, M. J.,** Small angle neutron scattering studies of aqueous gels with fluorinated nonionic surfactants, *Physica B*, 156–157, 394, 1989.
155. **Pons, R., Erra, P., Solans, C., Ravey, J. C., and Stébé, M. J.,** Viscoelastic properties of gel-emulsions: their relationship with structure and equilibrium properties, *J. Phys. Chem.*, 97, 12320, 1993.
156. **Riess, J. G., Santaella, C., and Vierling, P.,** New perfluoroalkylated phosphatidylcholines as surfactants for biomedical applications, *Proc. XXII CED Meeting on Surfactants, Palma de Mallorca, Comité Espagnol Detergencia, Tensioactivos y Afines*, Ed., Barcelona, 1991, 157.
157. **Santaella, C., Vierling, P., and Riess, J. G.,** Highly stable liposomes derived from perfluoroalkylated glycerophophocholines, *Angew. Chem., Int. Ed. Engl.*, 30, 567, 1991.
158. **Kunitake, T., Ando, R., and Ishikawa, Y.,** DSC studies of the phase transition behavior of synthetic bilayer membranes. II. Bilayer membranes of single-chain and triple-chain amphiphiles, *Mem. Fac. Eng., Kyushu Univ.*, 46, 245, 1986.
159. **Ravily, V., Gaenztler, S., Santaella, C., and Vierling, P.,** unpublished results, 1994.
160. *Phospholipids Handbook*, Cevc, G., Ed., Marcel Dekker, New York, 1993.
161. **Clary, L., Santaella, C., and Vierling, P.,** unpublished results, 1994.
162. **Kunitake, T., Ando, R., and Ishikawa, Y.,** DSC studies of the phase transition behavior of synthetic bilayer membranes. I. Bilayer membranes of double-chain amphiphiles, *Mem. Fac. Eng., Kyushu Univ.*, 46, 221, 1986.
163. **Higashi, N., Matsumoto, T., and Niwa, M.,** Specific phase transition behavior of fluorocarbon bilayer membrane on the basis of ^{19}F-NMR, *Sci. Eng. Rev., Doshisha Univ.*, 29, 24, 1988.
164. **Kunitake, T. and Higashi, N.,** Bilayer membranes of triple-chain, fluorocarbon amphiphiles, *J. Am. Chem. Soc.*, 107, 692, 1985.
165. **Rolland, P., Santaella, C., and Vierling, P.,** unpublished results, 1994.
166. **Mabray, S. and Sturtevant, J. M.,** Investigation of phase transitions of lipids and lipid mixtures by high sensitivity differential scanning calorimetry, *Proc. Natl. Acad. Sci. U.S.A.*, 73, 3862, 1976.
167. **Shinkai, S., Torigoe, K., Manabe, O., and Kajiyama, T.,** Complete thermocontrol of ion permeation through ternary composite membranes composed of polymer-liquid crystal-fluorocarbon-containing crown ethers, *J. Chem. Soc., Chem. Commun.*, 933, 1986.
168. **Shinkai, S., Torigoe, K., Manabe, O., and Kajiyama, T.,** Temperature regulation of crown-mediated ion transport through polymer/liquid crystal composite membranes. Remarkable transport ability of fluorocarbon-containing crown ethers, *J. Am. Chem. Soc.*, 109, 4458, 1987.
169. **Ihara, H., Hashiguchi, Y., and Kunitake, T.,** Functional fluorocarbon micelles. Phase separation and reactivity change of hydroxamate nucleophiles in mixed micelles of hydrocarbon and fluorocarbon surfactants, *Chem. Lett.*, 733, 1983.
170. **Kunitake, T., Ihara, H., and Hashiguchi, Y.,** Functional fluorocarbon micelles. Specific rate enhancement in the catalytic hydrolysis of phenyl esters due to selective binding of fluorocarbon and hydrocarbon substrates, *J. Am. Chem. Soc.*, 106, 1156, 1984.
171. **Liang, K. and Hui, Y.,** Vesicle of a hybrid bolaamphiphile: flip-flop behavior of spin labels, *J. Am. Chem. Soc.*, 114, 6588, 1992.

172. **Liang, K. and Hui, Y.,** Biphasic kinetics and flip-flop behaviour of vesicles of fluorocarbon amphiphiles with a 1,3-disubstituted glycerol structure, *J. Phys. Org. Chem.*, 5, 715, 1992.
173. **Senior, J. and Gregoriadis, G.,** Stability of small unilamellar liposomes in serum and clearance from the circulation: the effect of the phospholipid and cholesterol components, *Life Sci.*, 30, 2123, 1982.
174. **Kirby, C., Clarke, J., and Gregoriadis, G.,** Effect of the cholesterol content of small unilamellar liposomes on their stability *in vivo* and *in vitro*, *Biochem. J.*, 186, 591, 1980.
175. **Frézard, F., Santaella, C., Vierling, P., and Riess, J. G.,** Permeability and stability in buffer and in human serum of fluorinated phospholipid-based liposomes, *Biochim. Biophys. Acta*, 1192, 61, 1994.
176. **Frézard, F., Santaella, C., Vierling, P., and Riess, J. G.,** unpublished results, 1992.
177. **Guillod, F., Greiner, J., Milius, A., and Riess, J. G.,** Amphiphilic sugar phosphates with single or double perfluoroalkylated hydrophobic chains for use in oxygen and drug delivery systems, *Art. Cells, Blood Subst., Immob. Biotech.*, 22, 1273, 1994.
178. **Guillod, F., Greiner, J., and Riess, J. G.,** unpublished results, 1994.
179. **Papahadjopoulos, D., Nir, S., and Ohki, S.,** Permeability properties of phospholipid membranes: effect of cholesterol and temperature, *Biochim. Biophys. Acta*, 266, 561, 1971.
180. **Bresselers, J. M., Goderis, H. L., and and Tobback, P. P.,** Measurement of the glucose permeation rate across phospholipid bilayers using small unilamellar vesicles, *Biochim. Biophys. Acta*, 772, 374, 1984.
181. **Frézard, F., Santaella, C., Vierling, P., and Riess, J. G.,** Fluorinated phosphatidylcholine-based liposomes: H^+/Na^+ permeability, active doxorubicin encapsulation and stability in human serum, *Biochim. Biophys. Acta*, 1194, 61, 1994.
182. **Kunitake, T., Okahata, Y., and Yasunami, S.,** Substrate entrapment and formation of pH gradient in ammonium bilayer vesicles, *Chem. Lett.*, 1397, 1981.
183. **Mahé, A. M., Manoux, J., Valla, A., Follana, R., Zarif, L., Greiner, J., Vierling, P., and Riess, J. G.,** Perfluoroalkylated surfactants: relationships between structure and acute toxicity in mice, *Biomat., Art. Cells, Immob. Biotech.*, 20, 839, 1992.
184. **Santaella, C., Frézard, F., Vierling, P., and Riess, J. G.,** Extended *in vivo* blood circulation time of fluorinated liposomes, *FEBS Lett.*, 336, 481, 1993.
185. **Allen, T. M. and Chonn, A.,** Large unilamellar liposomes with low uptake into the reticuloendothelial system, *FEBS Lett.*, 223, 42, 1987.
186. **Gabizon, A. and Papahadjopoulos, D.,** Liposome formulations with prolonged circulation time in blood and enhanced uptake by tumors, *Proc. Natl. Acad. Sci. U.S.A.*, 85, 6949, 1988.
187. **Lasic, D. D.,** Sterically stabilized vesicles, *Angew. Chem., Int. Ed. Engl.*, 33, 1685, 1994.
188. **Gregoriadis, G.,** *Liposomes as Drug Carriers: Recent Trends and Progress*, John Wiley & Sons, Chichester, 1988.
189. **Scherphof, G., Damen, J., and Hoekstra, D.,** Interactions of liposomes with plasma proteins and components of the immune systems, in *Liposomes: From Structure to Therapeutic Applications*, Knight, C. G., Ed., Elsevier, Amsterdam, 1981, 299.
190. **Szönyi, S. and Cambon, A.,** New synthetic strategy to vesicle-forming perfluoroalkylated amphiphiles, *New. J. Chem.*, 17, 425, 1993.
191. **Trabelsi, H., Szönyi, S., Gaysinski, M., and Cambon, A.,** Synthesis and vesicle formation of cationic surfactants based on trisubstituted perfluoroalkylated thiourea, *Langmuir*, 9, 1201, 1993.
192. **Liang, K. and Hui, Y.,** Characterization and phase transition study for vesicles of fluorocarbon amphiphiles with 1,3-disubstituted glycerol structure, *Chin. J. Chem.*, 10, 396, 1992.
193. **Guillod, F., Greiner, J., and Riess, J. G.,** Synthesis of double-tailed (perfluoroalkyl)alkyl phosphosugars: new components for drug-carrying and -targeting systems, *Carbohydr. Res.*, 261, 37, 1994.
194. **Higashi N., Adachi T., and Niwa M.,** Molecular weight control of photopolymerization at an oriented bilayer surface using phase separation of fluorocarbon- and hydrocarbon-amphiphiles, *J. Chem. Soc., Chem. Commun.*, 1573, 1988.
195. **Gu, J. H., Luo, S. M., Kong, H. K., and Hui, Y. Z.,** Synthesis of fluorocarbon phospholipids and the formation of their liposomes, *Acta Chim. Sin.*, 46, 913, 1988.
196. **Montisci, M. J. M., Clary, L., Ravily, V., Santaella, C., and Vierling, P.,** unpublished results, 1994.

Chapter 9

Delivery of Microparticulated Liquid Systems in Food

Nissim Garti

CONTENTS

I. INTRODUCTION

Sustained, slow, controlled, and targeted release formulations for pharmaceutical and medical applications have been subject to extensive research.[1-4] The modern food industry has shown an increasing interest in controlled release formulation, mostly to improve rheological and physical properties of the product, to upgrade its quality and to improve its stability and appearance.[5-7]

Tremendous progress has been made in the last decade in the understanding of the surface and colloid properties of foods, and in developing advanced techniques to microparticulate and disperse solid particles and liquid droplets in various continuous phases.

Many techniques for reducing particles' sizes and for stabilizing the dispersed matter from aggregating, flocculating, and coalescing (liquids) have been developed. These include formation of liposomes and vesicles, swollen micelles and microemulsions, emulsions and double emulsions (for liquid dispersions) and formation of micro- and nanoparticles, nano- and microspheres, macrocapsules and coascervates, and a combination of the above (for solid particles).

Food scientists and technologists use the term "food colloids" for "dispersed systems," including dispersed solids and liquids as both exist in natural foods (milk, eggs, meat, etc.) and in processed foods (margarines, dressings, ice creams) even when the dispersed particles are not in the colloid range of sizes.[8] This chapter will deal only with systems in which the particles are liquid and in the colloidal range or obey the basic rules of colloidal science. It is most important to state that the increasing awareness of scientists

0-8493-4012-8/96/$0.00+$.50
© 1996 by CRC Press, Inc.

and technologists, of the relevance of the principles of colloid and surface science in solving many of the problems in the food industry, opened new horizons in that area.

A manufactured food product is typically a complex, multi-component, multiphase colloidal system, and its structural and textural properties are determined by the number (and strength) of interactions between the various types of particles and macromolecules making up the system. In many cases technologists have only a limited understanding of processes such as adsorption, aggregation, crystallization, wetting, gelation, etc.; therefore, much of their work is in the art-side.

Recently, progress has been made by theoreticians and experimentalists in explaining "steric" and "mechanical" stabilization in dispersed systems, facilitated by formulating synthetic dispersions with required properties. Furthermore, sophisticated customers, demanding products of better quality, with improved rheological properties, better palatability, more taste and color, and the presence of essential additives (such as vitamins, minerals, amino acids, etc.), have triggered a strong demand for artificial dispersed systems.

Liquid or solid dispersions can encapsulate, solubilize, entrap, and immobilize various substances, and deliver them at proper temperature, pressure, time, physical, or chemical conditions (pH, active compounds, UV, etc.) to the continuous phase.

Dispersed systems have high surface areas that can be utilized as "microreactors" for sophisticated surface and interfacial reactions (mostly enzymatic processes) that will be enhanced and catalyzed at the interface by a slow and controlled release of the active ingredients.[9]

This chapter will not discuss naturally occurring food colloids such as milk, meat, eggs, etc. For a comprehensive review of food colloids the reader is referred to books such as *Food Colloids*,[8,10] *Food Emulsions*,[11] *Food Emulsions and Foams*,[12] *Emulgatoren fur Lebensmittel*,[13] *Advances in Food Emulsions and Foams*,[14] *Colloids in Food*,[15] *Food Polymers, Gels, and Colloids*,[16] and the many reviews related to this subject.

This chapter will not analyze and discuss structural aspects of any industrial (processed) food dispersions like ice cream, beverages, chocolate, baked goods, etc. The aim of this review, the first of its kind, is to present the concepts related to dispersed food systems, in which the idea of controlled release (delivery) of active surface is expressed, and to bring examples of basic formulations in which the new concepts have been exercised. All examples are of man-made systems, often used in foods to improve texture, palatability, color, shelf-life, stability, texture, etc., or systems used to achieve special properties not granted by the basic food formulation.

A close examination of food colloids and emulsions reveals that there is a need in modern food industry for controlled release of additives that will improve the quality, flavor, texture and appearance of the final product, and will upgrade its value on the market.[11] For example, low-fat, low-calorie, no-cholesterol margarines will replace butter only if the added "butter-flavor" is dispersed homogeneously and the flavor will be controlled-released in our mouth. Colors, spices, flavorants, vitamins, and other active ingredients will improve food products if released slowly and at a required temperature.

Technologists can think of hundreds of products in which the use of controlled, encapsulated (liquid or solid) formulations, when incorporated in a product, will improve food attractively to the customer. The food industry is very reluctant in using additives at "added price." However, if the added value is significantly pronounced, and the added cost is tolerable, chances are that the formulation will be adopted.

One must not forget that foods are tightly controlled by the health authorities and only very seldom is the use of new emulsifiers or stabilizers permitted.[6] In many cases approved additives used in one product will not be permitted in another.

The following review will be, in many aspects, an introduction to the basic concepts of dispersion technology, and will act, mostly, as fundamental and applied guidelines for food scientists or technologists rather than a descriptive book of the structural aspects of existing products.

An attempt is made to discuss the possible delivery systems, conceptually and mechanistically, from the colloid scientist's point of view.

The particulated systems are grouped into two categories: (1) liquid–liquid dispersions (one liquid in another immiscible liquid to be termed "emulsions") and (2) solid dispersions in liquids (suspensions) and solid dispersions in solid matrices (solids in polymers). The chapter will concentrate on the first category of colloids.

II. MICELLES

Amphiphiles, consisting of a hydrophilic "heads" and a lipophilic "tails," tend to concentrate as a monolayer at the water interface and reduce the surface tension of the air–water interface. Any excess

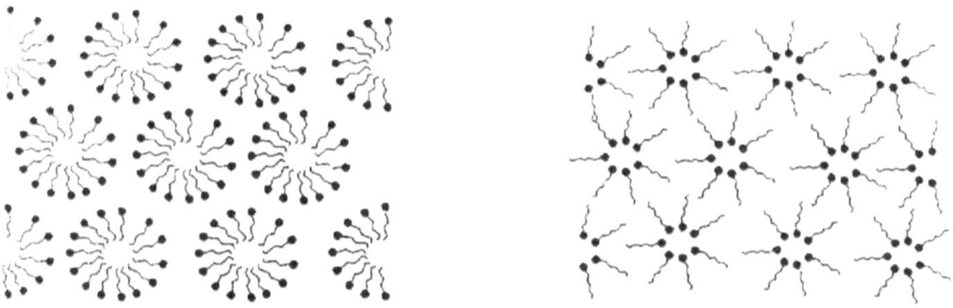

Figure 1 Schematic illustration of direct and reversed micelles.

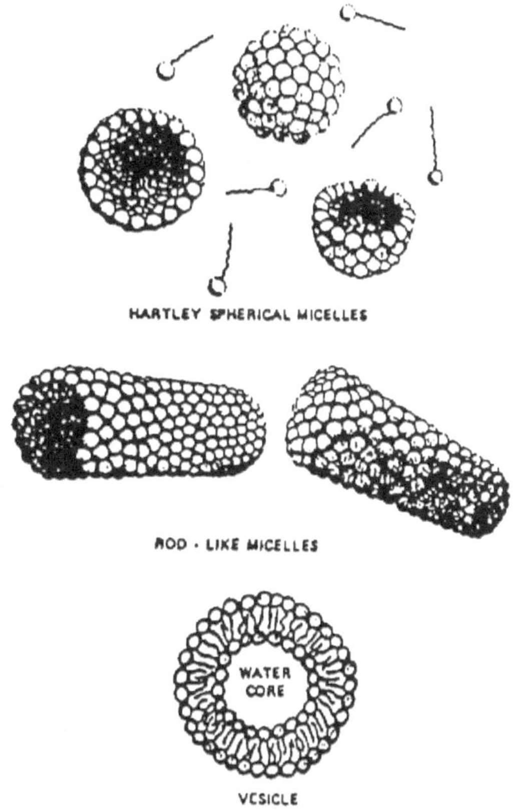

Figure 2 Illustration of various types of aggregates: Hartley spherical micelles, rod-like micelles, vesicles.

surfactant that is not accommodated at the interface migrates to the bulk and forms aggregates termed "micelles" when the bulk is water or "reverse micelles" when the bulk is organic solvent or oil (Figure 1).[17,18]

The nucleation of micelles is spontaneous. Many amphiphiles, sometimes over a couple of hundred, aggregate together, forming a structure that can vary between spherical and cylindrical, depending on the molecular organization of the amphiphiles, the composition of the solution, and temperature (Figure 2).[19,20]

The arrangement of the direct micelles is such that the hydrophobic "tails" collect together within the interior of the micelle while the "head" groups are in contact with the water. The interior of the micelle is "liquid-like" and mostly hydrophobic, capable of dissolving apolar compounds, preferentially within the core region.

It is important to note that micellar aggregates are mobile, the interfaces are flexible and there is a rapid exchange of molecules between the neighboring region and the aggregate (Figure 3).[21,22] At higher concentrations, more extended structures form (Figures 3 and 4). Micelles cannot be regarded as persistent entities, having well-defined geometrical shapes.[23,24] Thus, both micelles (termed S_1-phase) and reverse micelles (S_2-phase) represent isotropic solutions containing aggregates, which are dynamic

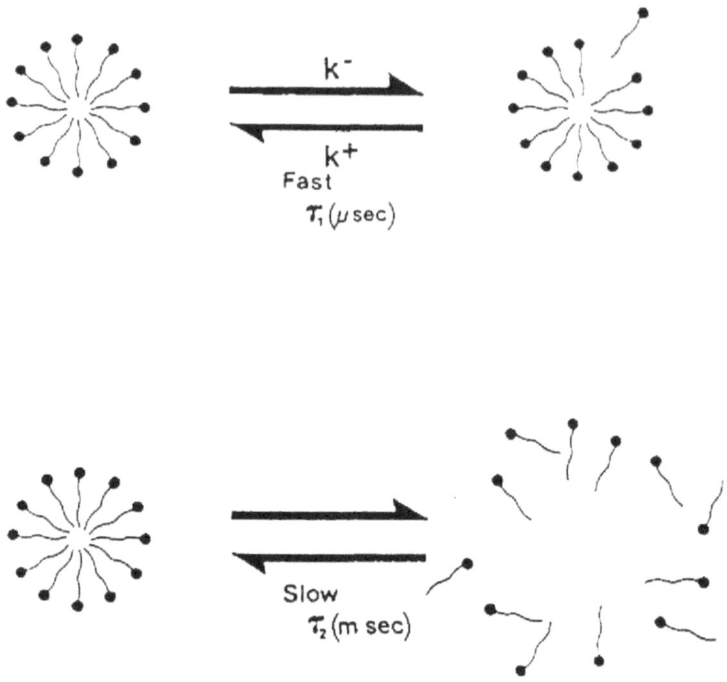

Figure 3 Two relaxation mechanisms proposed for dilute micellar solutions of pure surfactants. (The slow process is thought to take place through a sequence of increasingly smaller aggregates and not suddenly as shown.[24])

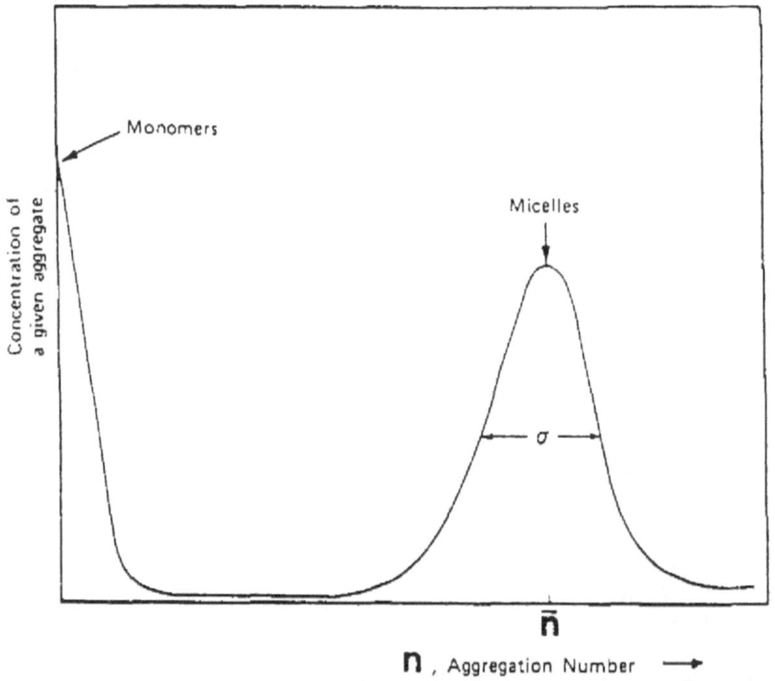

Figure 4 The distribution curve for aggregates responsible for the existence of two relaxation times: T_1 (monomers) and T_2 (micelles).

Table 1 Selected Critical Micelle Concentrations of Various Anionic, Cationic, Zwitterionic, and Nonionic Surfactants

Structure	Temperature (°C)	CMC (m*M*)
$C_8H_{17}SO_3^-Na^+$	40	160
$C_8H_{17}SO_4^-Na^+$	40	140
$C_{10}H_{21}SO_3^-Na^+$	40	41
$C_{10}H_{21}SO_4^-Na^+$	40	33
$C_{12}H_{25}SO_3^-Na^+$	40	9.7
$C_{14}H_{29}SO_3^-Na^+$	40	2.5
$C_{12}H_{25}SO_4^-Na^+$	25	8.2
$C_{12}H_{25}SO_4^-Li^+$	25	8.9
$C_{12}H_{25}SO_4^-K^+$	40	7.8
$(C_{12}H_{25}SO_4^-)_2Ca^{2+}$	70	3.4
$C_{12}H_{25}SO_4^-N(CH_3)_4^+$	25	5.5
$C_{12}H_{25}SO_4^-N(C_2H_5)_4^+$	30	4.5
$C_{16}H_{33}SO_4^-Na^+$	40	5.8×10^{-1}
$C_{12}H_{25}CH(SO_4^-Na^+)C_3H_7$	40	1.7
$C_{10}H_{21}CH(SO_4^-Na^+)C_5H_{11}$	40	2.4
$C_{13}H_{27}CH(CH_3)CH_2SO_4^-Na^+$	40	8×10^{-1}
$C_{12}H_{25}CH(C_2H_5)CH_2SO_4^-Na^+$	40	9×10^{-1}
$C_{11}H_{23}CH(C_3H_7)CH_2SO_4Na^+$	40	1.1
	Cationics	
$C_8H_{17}N(CH_3)_3^+Br^-$	25	140
$C_{10}H_{21}N(CH_3)_3^+Br^-$	25	68
$C_{10}H_{21}N(CH_3)_3^+Cl^-$	25	61
$C_{12}H_{25}N(CH_3)_3^+Br^-$	25	16
$C_{12}H_{25}N(CH_3)_3^+Cl^-$	25	20
	Zwitterionics	
$C_8H_{17}N^+(CH_3)_2CH_2COO^-$	27	250
$C_8H_{17}CH(COO^-)N^+(CH_3)_3$	60	86
$C_{10}H_{21}CH(COO^-)N^+(CH_3)_3$	27	13
$C_{12}H_{25}CH(COO^-)N^+(CH_3)_3$	27	13
	Nonionics	
n-$C_4H_9(OC_2H_4)_6OH$	20	800
n-$C_4H_9(OC_2H_4)_6OH$	40	710
$(CH_3)_2CHCH_2(OC_2H_4)_6OH$	20	910
n-$C_6H_{13}(OC_2H_4)_6OH$	20	74
$(C_2H_5)_2CHCH_2(OC_2H_4)_6OH$	20	100
$(C_4H_9)_2CHCH_2(OC_2H_4)_6OH$	20	3.1
$(C_4H_9)_2CHCH_2(OC_2H_4)_9OH$	20	3.2
n-$C_{12}H_{25}(OC_2H_4)_4OH$	25	4×10^{-2}
n-$C_{12}H_{25}(OC_2H_4)_{14}OH$	25	5.5×10^{-2}
n-$C_{12}H_{25}(OC_2H_4)_{23}OH$	25	6.0×10^{-2}
$C_{16}H_{33}O(C_2H_4O)_9H$	25	2.1×10^{-3}

entities, uniformly distributed and rapidly exchanging molecules with the surrounding continuous region (Figure 3).

Solutions containing micelles are thermodynamically stable, but because each individual aggregate consists of a relatively small number of amphiphilic molecules, a particular aggregate has a finite life span.

The micelles are composed of nonionic, anionic, cationic, and/or zwitterionic surfactants. The critical concentration of a surfactant required to form a micelle (critical micellar concentration, cmc) depends on the amphiphile structure (length of tail chains, size of head), the type and size of the counter-ion, added electrolyte and organic additives, and on an external factors such as temperature and pressure.

Table 2 Micellar Size Distribution (Aggregation Numbers) of Surfactants at Various Temperatures

Surfactant	Temperature (°C)	Aggregation Number
$C_{10}H_{21}SO_3^-Na^+$	30	40
$C_{12}H_{25}SO_3^-Na^+$	40	54
$(C_{12}H_{25}SO_3^-)_2Mg^{2+}$	60	107
$C_{12}H_{25}SO_4^-Na^+$	23	71
$C_{14}H_{29}SO_3^-Na^+$	60	80
$C_{12}H_{25}N(CH_3)_3^+Br^-$	23	50
$C_8H_{17}O(CH_2CH_2O)_6H$	30	41
$C_{10}H_{21}O(CH_2CH_2O)_6H$	35	260
$C_{12}H_{25}O(CH_2CH_2O)_6H$	15	140
$C_{12}H_{25}O(CH_2CH_2O)_6H$	25	400
$C_{12}H_{25}O(CH_2CH_2O)_6H$	35	1400
$C_{14}H_{29}O(CH_2CH_2O)_6H$	35	7500

The micelles are dynamic structures, but can be viewed as aggregated entities at an equilibrium, consisting of a number of surfactant molecules. Essentially, two dominant "species" exist in a solution, unaggregated surfactant and a micellar structure with a given aggregation number n. The aggregation numbers are small (up to a hundred) for anionic and cationic surfactants, but can be very large for nonionic surfactants (Table 2).[22]

Much has been written on the shape, size, fluidity, and thermodynamics of micelles (enthalpy of micellization and the standard free energies) of various surfactants with relation to structure of the surfactant, temperature, electrolytes, etc. The reader is referred to a comprehensive review in *Microemulsions and Related Systems*, Chapter 2.[22]

III. MIXED MICELLES

When two or more different types of surfactant molecules are dissolved in water, the micelles that form are mixed; that is, an aggregate containing representatives of each type of amphiphile present in the aqueous solution (Figure 5).[24,25]

Furthermore, the proportions of each type of surfactant in a given micelle can be very different from its proportions in the solution. Thus, in such a mixture the micelle will contain a larger proportion of long-chain molecules than in the solution. Monomer composition changes continuously upon increasing surfactant concentration.

Certain surfactants form ideal mixed-micelles; that is, the mixture properties can be predicted, assuming that the micellar pseudo phase is an ideal mixture of surfactants.[26] Such properties are manifested in variety of properties, such as solubilization power, surface tension etc. (Figure 6). However,

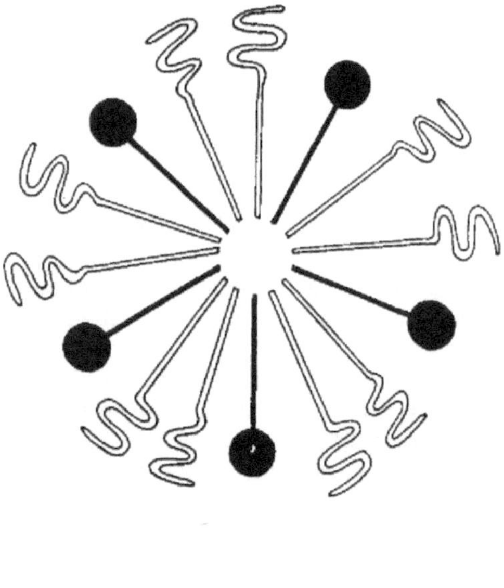

Figure 5 Schematic illustration of a mixed micelle.

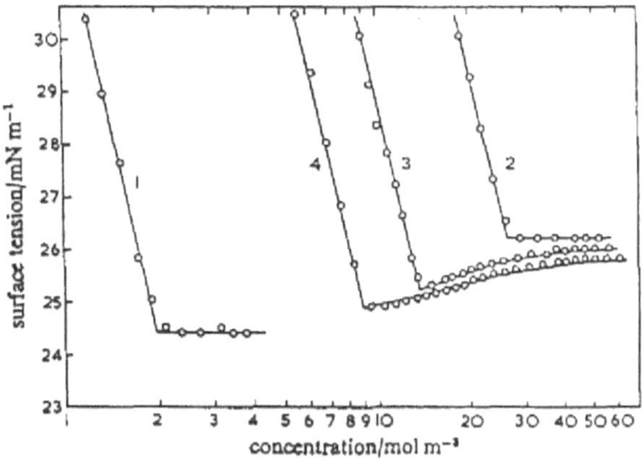

Figure 6 The surface tension of pure *n*-decyl methyl sulphoxide (curve 1) and pure *n*-octyl methyl sulphoxide (curve 2) and for two mixtures, all shown as function of concentration; curve 3, mole fraction of *n*-decyl component of 0.075; curve 4, mole fraction of *n*-decyl component of 0.156.

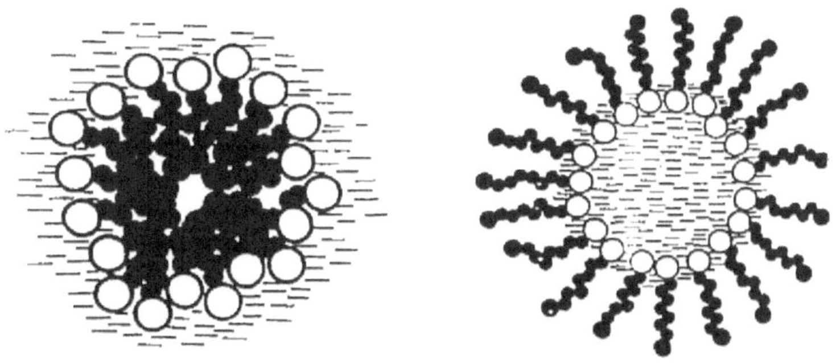

Figure 7 Schematic illustration of swollen micelles (direct and reversed microemulsions).

there are surfactant mixtures that do not obey the definition of ideally. For example, mixtures of nonionic and anionic surfactants are non-ideal systems.[27,28]

IV. SWOLLEN MICELLES AND MICROEMULSIONS

The micelles can accommodate organic immiscible-with-water molecules in their hydrophobic core or may associate them with the amphiphile within the interfacial layer. Therefore, the micelle can solubilize some immiscible matter. The increased solubility of the immiscible matter was termed by McBain "solubilization."[29]

The micelles that swell ("swollen micelles") because of the solubilized matter will, in many cases, have a spherical structure (Figure 7). The systems are amorphous, isotropic, transparent, and translucent and in essence represent organized solutions (mixtures) of water, oil, and surfactant(s). There is much debate regarding the terminology of these "mixtures." Some will call them "swollen micelles," while others will use the term "microemulsions." An attempt has been made by scientists to distinguish between the two terms. Microemulsions are, according to this terminology, systems in which the solubilized matter is concentrated at the core of the micelle, as "reservoir of liquid" matter. We will use the term "solubilized micellar solutions," "microemulsions" and "swollen micelles" without any structural preference.

The word "microemulsion" was originally proposed by Hoar and Schulman[30] in 1943 (for systems containing water, benzene, hexanol, and potassium oleate). Their definition for microemulsion was based on the use of a fourth component (in addition to the water, oil, and amphiphile) termed "coemulsifier" or "cosolvent," being usually an alcohol, to "clear" the emulsion and form, spontaneously, a transparent system. They proposed that the coemulsifier, when titrated with the emulsion, adsorbs to the surface and reduces the water–oil interface up to a "negative value." The "negative interfacial tension" was proved at a later stage to be zero or very low tension, obtained during redistribution of the cosurfactant on the interfacial film. Microemulsions are thermodynamically stable, form almost spontaneously, and have submicronal droplets size. More detailed studies revealed that the structures are not necessarily always spherical and depend on the system internal composition and the nature of the amphiphile (single or double tailed). Other, very sophisticated organizations and structures have been characterized. The details structures will be further discussed. Several books and reviews have been written on microemulsions.[34-39] The reader will find a fascinating new world of structures and properties and very complex relationships between the structure and the microemulsions solubilization capacity.

In connection with solvent properties of aqueous solutions of members of homologous series of amphiphilic salts, a distinction is often made between "solubilization" and "hydrotropy." With amphiphilic salts of short-chain length (e.g., C_3–C_8), the solubilization effect becomes marked only with their rather concentrated aqueous solutions ("hydrotropy").[31,32] This distinction is artificial and misleading. According to Hartley,[33] the difference between a system where alcohol or acetone is the cosolvent, and one where paraffin-chain salt is the cosolvent, is quantitative only. The terms, however, have a certain customary significance, apart from any mechanistic implications and will be used by some scientists to describe different phenomena of dissolving organic matter.

Microemulsions have been a subject of much fundamental research focusing on noningestible systems.[34-41] Only very few attempts were made to use microemulsions in food systems,[42-44] and practically nothing was done with relation to controlled release of active matter from microdroplets. More detailed discussion on microemulsions will be given in the section related to ternary systems.

V. BINARY MESOPHASES (LYOTROPIC LIQUID CRYSTALS)

Micelles are formed in very dilute surfactant solutions. The micelles are essentially noninteracting systems. However, at intermediate to high surfactant concentrations in solution (with or without the presence of organic solvent), additional conjugated liquid phases are formed ("mesophases," intermediate structures between isotropic solutions to crystalline phases). The mesophases are better known as "lyotropic phases" to denote the fact that the solvent molecules are not uniformly distributed throughout the structure, but are more numerous in some regions than in others (Figure 8).

In this text, systems consisting of amphiphile in aqueous or organic solutions (binary systems) and systems composed of amphiphile, solvent and water (ternary systems) will be discussed.

Binary systems, composed of high concentrations of amphiphiles in water, can form a wide variety of structures[42,43] (Figure 9). Designation of some of the common lyotropic phases is given in Table 3.[19,22,45]

The structures follow each other in a definite sequence related to the proportions of water and amphiphile. The first mesophase to appear from a swollen rod-like micelle is the "middle phase" known also as the "hexagonal I phase" or the "M_1-phase." If the amphiphile is dispersed in an organic phase the middle phase will be termed "reverse hexagonal II" and will be defined as the "M_2-phase." This phase consists of infinitely long rods in hexagonal array, as depicted in Figure 10.[46,47] X-ray diffraction in the low angle region is one of the most important methods for identification of lyotropic mesophases (Luzatti and Hussan[48]). The hexagonal system is periodic in two dimensions and consists of cylindrical aggregates of which there are two types. In "hexagonal I" (also called the "middle phase", M_1), the lipophilic chains

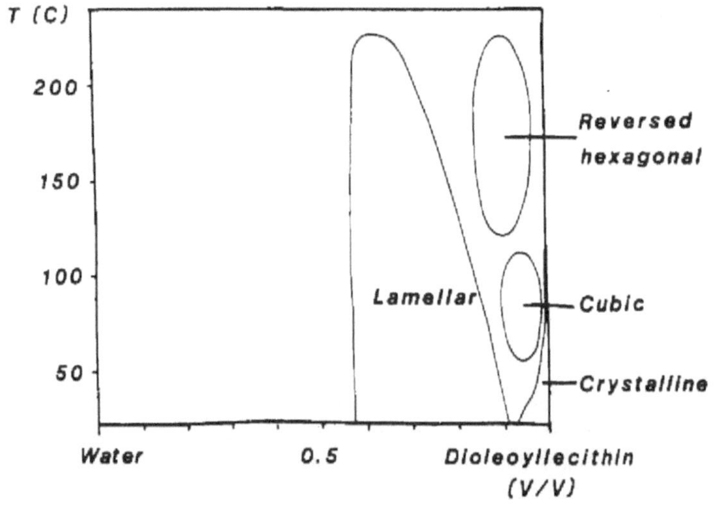

Figure 8 Phase diagram of dioleyl-lecithin/water system.[45]

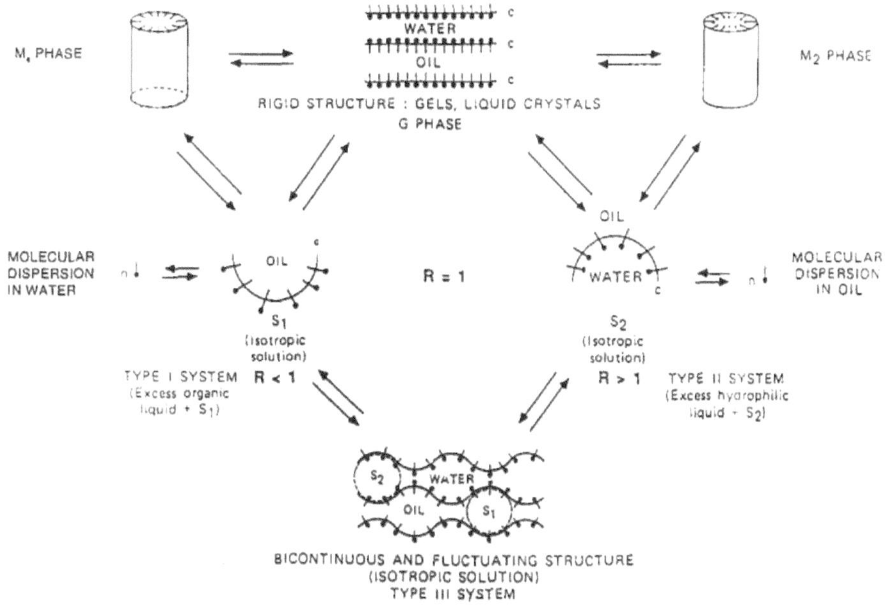

Figure 9 Intermicellar equilibrium and associated phase changes shown by certain series of amphiphilic solutions. Phase symbols are defined in text.

Table 3 Designation of Some of the Common Lyotropic Phases[45,19]

	Basic Structure	Designation
1. Micellar solutions (optically isotropic)	More or less spherical, swollen, micelles containing solubilized, organic liquids	S_1
2. Middle phase, normal (anisotropic)	Indefinitely long, mutually parallel rods in hexagonal array. The rods consist of more or less radically arranged amphiphiles. The hydrophiles are in contact with the surrounding continuous aqueous phase.	M_1
3. Neat phase (anisotropic)	Coherent double layers of amphiphilic molecules with the hydrophiles in the interfaces with intervening layers of water.	G
4. Middle phase, reversed (anisotropic)	Indefinitely long, mutually parallel rods in hexagonal array. The lipophiles are arranged so that the surrounding continuous organic phase is in contact with the lipophile.	M_2
5. Inverted micellar solutions (optically isotropic)	More or less spherical inverted micelles containing solubilized water.	S_2, L_2

of the amphiphile occupy the linear core of the cylinders, and the polar groups are arranged on the outer side in contact with the continuous water phase. In "hexagonal II" phase, the inner core of the cylinders consists of water surrounded by a shell of polar groups of the surfactant, and the hydrocarbon chains of the amphiphile make up the continuous outer phase between the cylinders. At polarized light under the microscope, the appearance of the "hexagonal I" and the "lamellar" phases is easily identified. The distinction between hexagonal I and II is not possible by this method. The dilution with water will reveal the differences. Type I can be diluted, since spherical micelles are formed. Dilution with water is not possible for type II due to the lipophilic nature of the amphiphile.[46,47]

At higher amphiphile concentrations the "neat phase" with lamellar structure may appear.[49-51] This structure is designated as the "G-phase" and its structure is depicted in Figure 11.

Between the M_1 and G phases other mesophases sometimes occur with unidentified structures. At very high amphiphile concentrations and within the M_1 and G phases a most viscous phase sometimes forms.

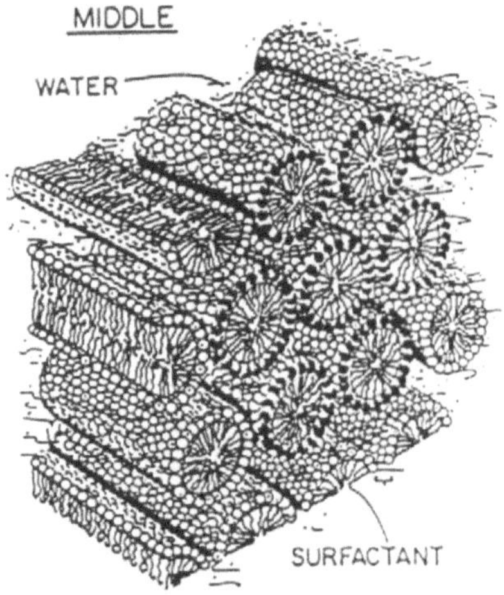

Figure 10 "M_1 phase"–"Hexagonal I" mesophase (also known as "middle phase"). Rod-like micelles of indefinite length and arranged in hexagonal array.

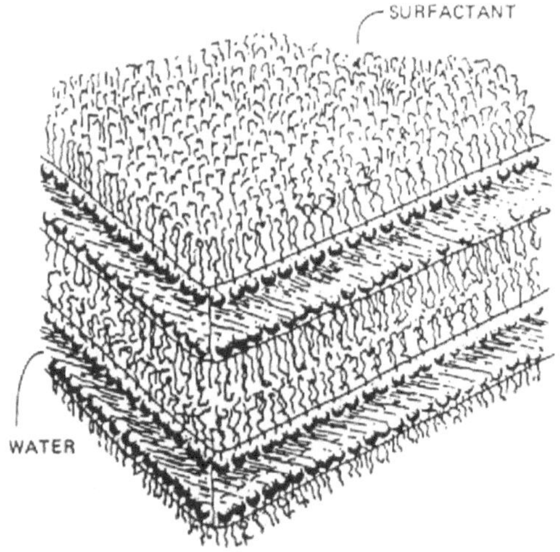

Figure 11 "G phase"–"Lamellar structures" — known also as the "neat phase."

This phase is termed the "V-form" or the "viscous-phase." Recently, the term "cubic phase" was adopted.[49-51] The structure of the cubic phase has been much discussed and is still debatable[48-51] (Figure 12).

Liquid crystalline structures are mesophases between crystals and isotropic solutions. Some investigators treat such systems from this perspective. Such approach is essential when dealing with food emulsions. A common practice is to add "self-emulsifying" or "aqueous emulsion" of monodiglycerides of fatty acids (MDG, glycerol mono- and distearate, 42% -mono) into food products. The MDG is premixed with boiling water under vigorous stirring. The mixture forms an hydrated type of amphiphile. However, when this procedure is repeated with distilled monoglycerides of fatty acids (GMS, min. 90% alpha-mono), "gums" or "gels" types of structures are formed. Out of practice it is known that the dispersion of monoglyceride must be very carefully heated and treated since various liquid crystalline mesomorphic phases in water are formed, depending on the temperature and the concentration (Figures 13, 14). It is, therefore, very important to outline the full phase diagram (phase behavior) of the amphiphile in aqueous systems as a function of temperature (Figure 15). In a crystalline form, the amphiphiles are oriented with the polar groups against each other, and the lipophilic hydrocarbon chains are parallel and densely packed, as shown in Figure 15a (Larsson and Friberg, *Food Emulsions*[11]).When the emulsifier crystals are in contact with water and the temperature is raised to the point where the Kraft temperature (Tc) is reached, the hydrocarbon chains will transform into a disordered, liquid state due to the thermal energy. At the same time, water will penetrate between the polar layer as shown in Figures 14 and 15. In practice, this is how the liquid crystalline mesophase of the lamellar type will look.

When the lyotropic lamellar liquid crystalline mixture of the water and GMS is cooled down, again below the Kraft point, the hydrocarbon chains will crystallize and arrange themselves in a regular lattice. The water may still be present between the polar groups. In this case a so-called "gel phase" is formed

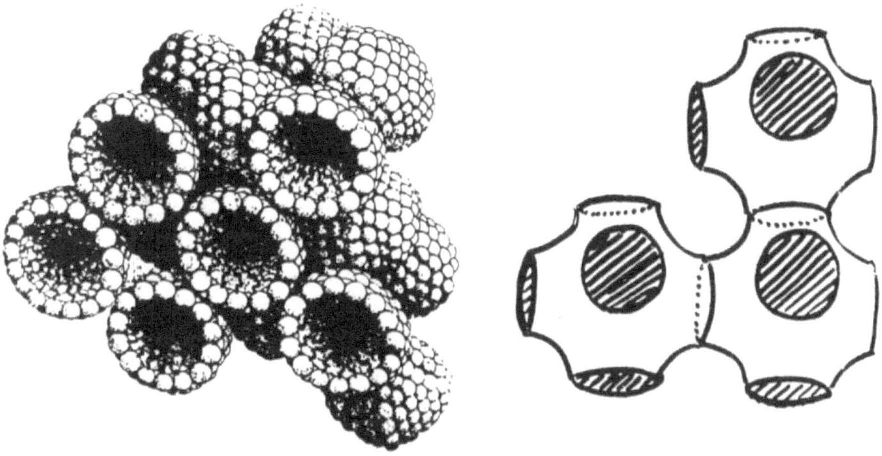

Figure 12 Schematic illustration of some of the proposed structures of the cubic phase. The most acceptable structure is the C structure (on the left).[50,51]

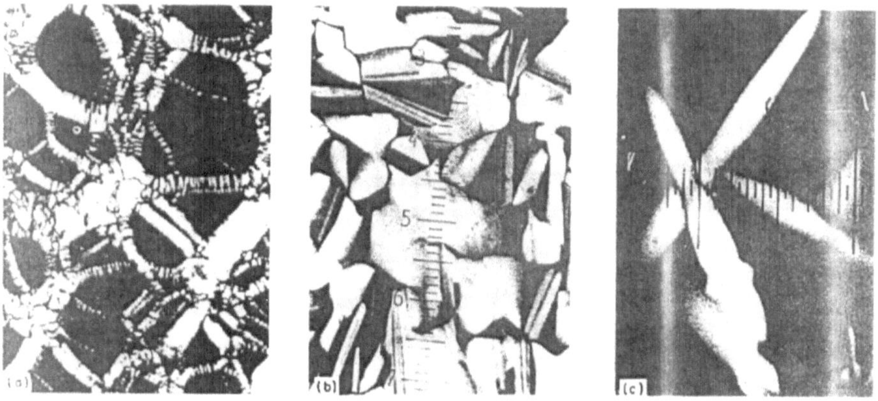

Figure 13 Photo micrographs of liquid crystalline phases; (a) lamellar, (b) hexagonal type II, (c) batonnets. Magnification X 300, polarized light.

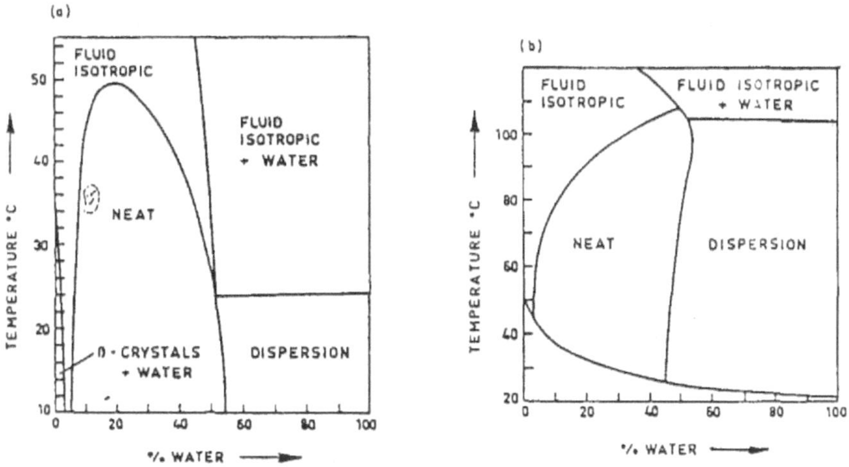

Figure 14 Phase diagrams of various pure 1-monoglycerides (saturated and unsaturated)/water systems.[11]

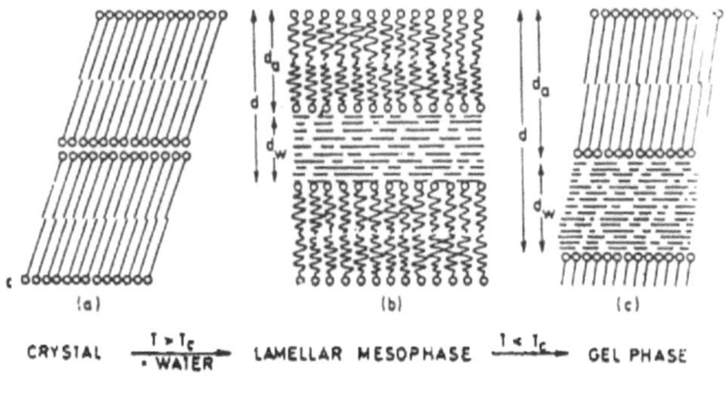

Figure 15 Schematic model showing the formation of a liquid crystalline, lamellar mesophase, and the alpha crystalline gel phase.

as shown schematically in Figure 13c. Such gels are normally metastable and will eventually change due to a decrease in the water layer thickness (d_w). When all the water is expelled, the gel phase is transformed into a microcrystalline suspension ("coagel") of the emulsifier in water (Figures 14, 15).[52]

Phase diagrams of unsaturated monoglycerides like monoolein and monolinolein in water (as first described by Lutton[53]) are also shown in Figure 14. The dominating mesophase in aqueous systems of unsaturated monoglycerides is the viscous isotropic phase. At room temperature unsaturated monoglycerides form a lamellar phase with a very small content of water, whereas at high temperatures, a viscous phase is formed if the water content is increased to approximately 20%. This phase is optically isotropic, and X-ray diffraction shows that it has a cubic structure. Above 70°C, the cubic structure is transformed into a hexagonal II structure, and on further heating the mixture becomes fluid and isotropic. Due to this phase behavior, unsaturated monoglycerides cannot be dispersed in water in diluted concentrations. They can only absorb up to about 40% water by weight in the cubic or hexagonal II mesophase. When more water is present, the viscous isotropic mesophase will separate out as gelatinous lumps. Monoglycerides made from liquid oils, such as sunflower, cottonseed, or soya bean oil are commercially available in molecular distilled form with a monoester content of 90% or more. They are often used in w/o emulsions, such as low-calorie spreads.

The phase diagram of distilled monoglycerides made from sunflower oil is quite similar to the phase diagram for pure monolinolein. The phase diagram of monoelaidin differs substantially from that of monoolein, and it corresponds more to the diagram of monostearin, except for the fact that the temperature limits are approximately 30°C lower. Monoerucin exhibits only one mesophase, neat, in a small region containing 10 to 20% water.

Phase diagrams of 1-monocaprylin/water and 1-monocaprin/water are shown in Figure 16 (Larsson[54]). The lamellar neat mesophase is the only existing phase in aqueous systems of monocaprylin and monocaprin. From Figure 16 it appears, also, that the neat phase of the monocaprin/water system exists in the temperature interval from 30°C to above 100°C at 40% water. In both phase diagrams the neat phase is present at concentrations from 5% to approximately 50% water.

The phase behavior of saturated, distilled monoglycerides, made from fully hardened lard was examined in detail for the first time by Krog and Larsson[55] (see Figure 17). The mixed fatty acid composition in the commercial product caused only small changes in the phase diagram compared to a diagram of pure monopalmitin.[56]

Diglycerides and triglycerides of unsaturated fatty acids do not exhibit lyotropic mesomorphism as do monoglycerides. Thus, it was not until the introduction of "pure" monoglycerides (manufactured on a commercial scale by means of molecular distillation) that industrial products became sufficiently pure to form well-defined mesophases of various types.

VI. TERNARY SYSTEMS — MICROEMULSIONS

Binary amphiphile solutions (water or oil) exhibit, as described, a series of intermediate phases with liquid crystalline structures. Similar intermediate mesophases exist also when substantial amounts of both oil and/or water are added to the surfactant solutions. These mesophases have similar structures, as described previously for the binary systems, except that the mesophases solubilize simultaneously both water, within the head groups, and oil, within the tail groups. The ternary systems offer very significant practical and scientific elements, and contribute to the understanding of the stability/instability mechanisms of macroemulsions.

In practice, oil, water, and surfactant are blended together and allowed to equilibrate. At certain compositions one-phase transparent systems will be seen (microemulsions), in other compositions

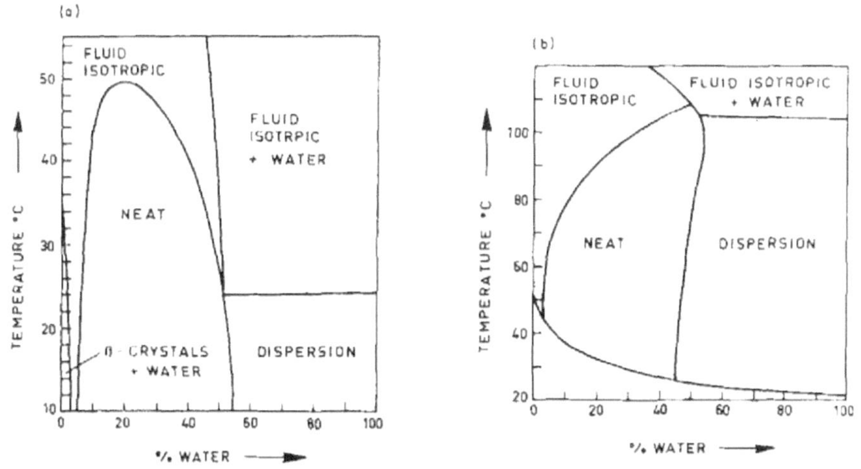

Figure 16 Phase diagrams of (a) 1-monocaprylin in water and (b) 1-monocaprin in water.

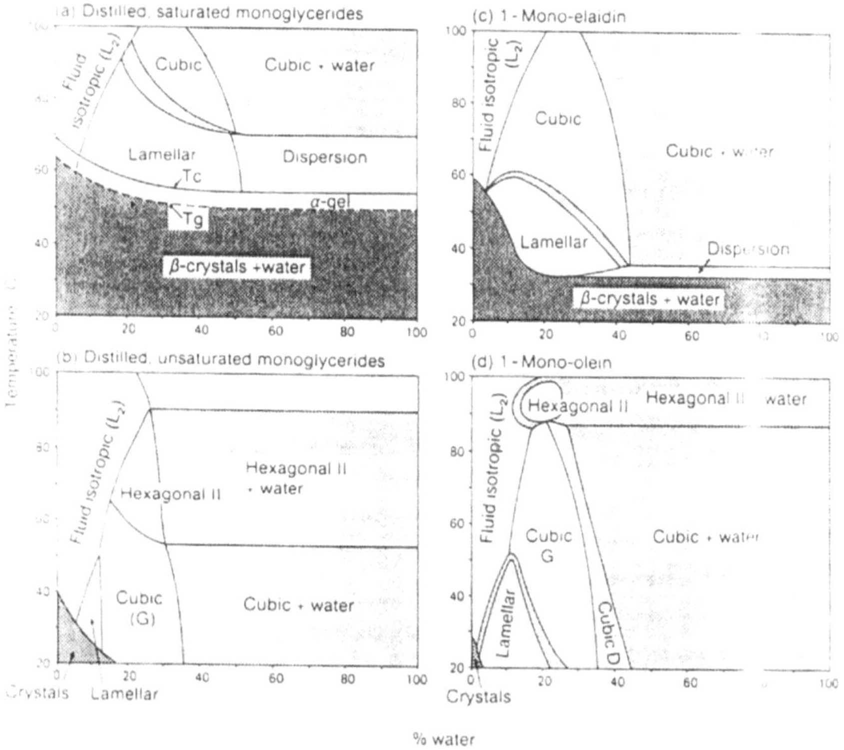

Figure 17 Binary phase diagram of pure various 1-monoglycerides in water: (a) monopalmitin, (b) monoelaidin and (c) monoolein.

viscous transparent gel-like systems will be detected, and in others two or more phases (up to five) may appear. In many cases almost the entire amount of the surfactant will reside in one of the phases, together with various proportions of oil and water. That phase, containing the bulk of the surfactant, is often called the "micellar phase," the "surfactant phase" and will have a similar structure to the above-described microemulsion.

The phase behavior of three-component systems can, at fixed pressure and temperature, be best represented by a ternary diagram. These diagrams provide a simple perspective of phase behavior that is difficult to capture in any other way (Figure 18).

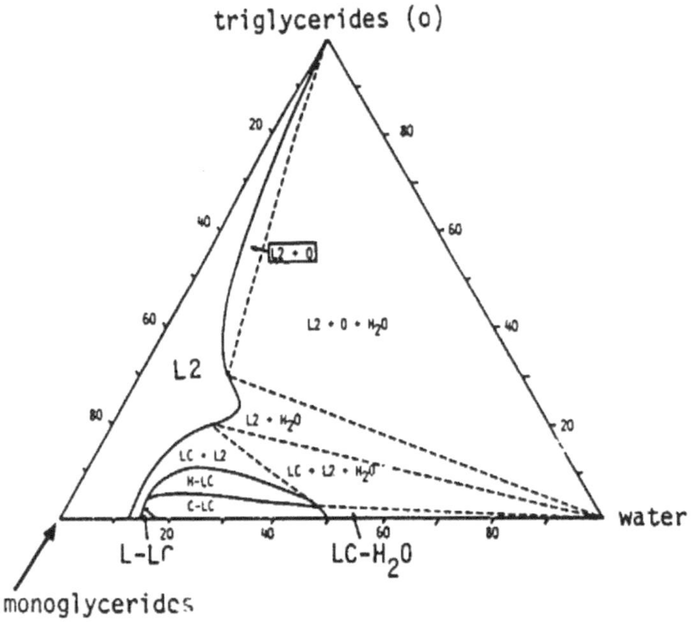

Figure 18 Ternary phase diagram of soybean oil/sunflower oil monoglycerides-water at 25°C; LC: liquid crystalline phase; L2: an inverse micellar phase.[55]

According to a phase rule, a three-component system (oil-water-amphiphile) which, at an equilibrium of a certain internal composition (point O) shows only one single phase, is bivariant, while the same system, at an equilibrium of a different internal composition (point L) can separate into two phases or three phases (Figure 19).[57] The two phases will be nonvariant. Any system whose overall composition lies within the two phases region, will exist as two phases whose composition are represented by the ends of the tie-lines. At point P (the plait point) the compositions of the two phases will be the same. If three phases coexist, the system will be invariant, meaning that three phases have compositions represented by the corners of the tie-triangle.[58] The three phase system is always bordered by three two-phase systems (Figure 20).

In practice, phase ternary diagrams can have two types of one-phase regions. The one called type I (or Winsor I) consisting of two phases, the one rich in surfactant (the micellar-phase) separated from the other containing mostly oil. The other is termed type II (Winsor II) consisting also, of two phases, the one rich in surfactant (the micellar phase) separated from the other, containing mostly water (Figure 21). Type I system can transform into type II system through an intermediate three-phase system to be termed also Winsor type III system. Therefore, type III system exhibits a tie-triangle having two micellar solutions (Figures 21 to 23).[59,60]

The micellar or surfactant phases are of great importance since they have practical applications. It is the goal of the food technologists to find the proper oils and the right amphiphiles that will give a one-phase region in the phase diagram, with maximum area. Large isotropic areas mean, in practice, thermodynamically stable surfactant phase, rich in both oil and water. If, for example, one is interested in solubilizing substantial amounts of water (ca. 30 wt%) it would be important to find the surfactant or a combination of surfactants that will exhibit large L_2 areas (Figure 24). A larger L_2 area in the phase diagram means more solubilized water (Figure 25). The three-phase tie-triangle is also very important since it was found both experimentally and by theoretical calculations that any macroemulsion prepared with the composition represented by the corners of the tie-triangle will have better thermodynamic stability than any other emulsion prepared with a composition of any two-phase regions in the diagram. Typical phase diagrams of soybean oil/water and various amphiphiles are shown in Figure 26, demonstrating the differences in the areas of the isotropic phases as a function of the type of the amphiphile.

For further reading, *Microemulsions and Related Systems*, by Bourrel and Schechter[52] is highly recommended.

Ternary and four-component phase diagrams are common practice in many labs dealing with emulsion or microemulsion science, including food systems. A typical phase diagram of triglycerides, monoglycerides, and water is depicted in Figures 27 and 28.[51]

One must also note that the microemulsion regions are in equilibrium with the lamellar liquid crystalline regions. This is one of the reasons that in practice we add cosurfactant, with medium chain

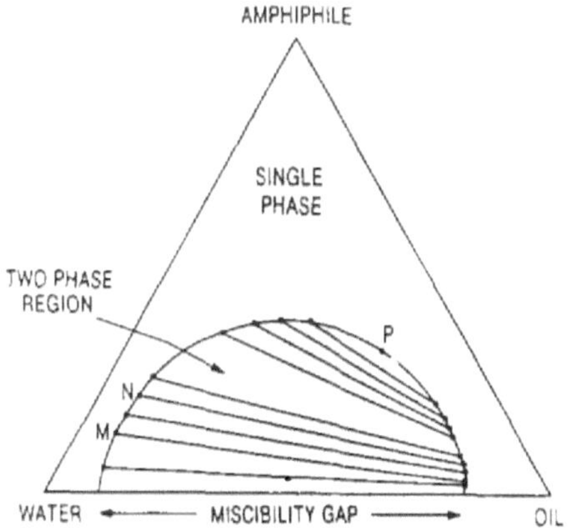

Figure 19 Ternary diagram representation of two-phase region. The sloping lines are tie lines connecting conjugated phases.[52]

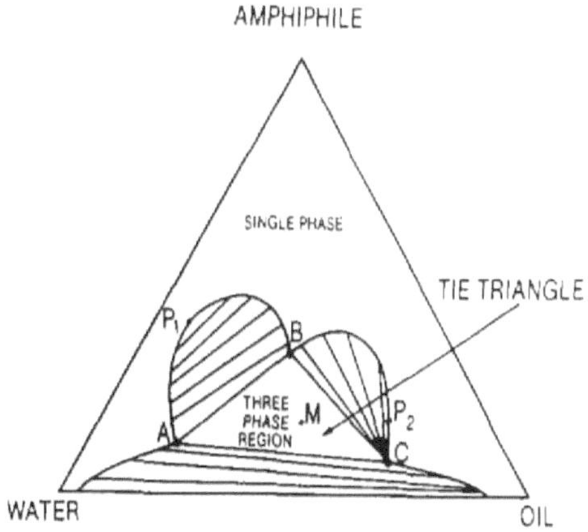

Figure 20 Ternary phase diagram showing a three invariant triangular region.

length (C_5-alcohol) to the lamellar liquid crystalline structure in order to destabilize the mesophase and transform it into microemulsion. The C_5–OH will "upset" the lamellar packing and will facilitate formation of very small spherical droplets with high curvature. The change in curvature (from flat or almost flat to highly curved) leads to a pronounced change in the free energy, (e.g., the bending component of the surface free energy) and leads to thermodynamic stability (Figure 29).

VII. EMULSIONS

A. DEFINITION AND CHARACTERIZATION

The three- and four-component phase diagrams show at equilibrium large areas of phase separation into two or three phases. If one does not wait for equilibrium and applies at the same time mechanical or shear forces to the system, the oil (or the water) will form dispersed droplets in the continuous phase. The droplets will eventually reseparate into two or three phases since they do not exhibit thermodynamic stability, have large surface curvatures, and will gain free energy upon separating into the original phases.

Macroemulsions or emulsions (Figures 30, 31) are, therefore, defined as liquid dispersions, in droplet form, of one immiscible phase into another in the presence of amphiphile (termed "emulsifier").[2,6,7,11,12,61]

It should be noted (Table 4)[37] that while for microemulsions it is essential (in many cases) to add a small quantity of cosolvent/cosurfactant, like medium chain alcohol, to stabilize the curvature and to destabilize the liquid crystalline phases, for emulsions, it is essential to add coemulsifier with chain length similar to the emulsifier, so that the huge radius with almost parallel packing of surfactant/cosurfactant will be optimal. It was well established by several investigators that the set of the two emulsifiers, to be selected for optimal stabilization, must have internal "structure fitness"; one amphiphile being hydrophilic in nature with good "dangling" groups in the water and the other being hydrophobic with good stabilizing (oil-soluble) chains. The two emulsifiers should form a structural complex between its functional groups to allow strong viscoelastic film (network) on each droplet. Such network system is known as "Sherman complex"[63] (Figure 32).

It should also be noted that emulsions can be stabilized kinetically (sometimes for very long periods of time) if the droplets are reduced in size (increase in curvature and gain of free energy) or stabilized by layers of emulsifiers with full coverage, so that attraction interaction between the droplets (as they approach each other due to Brownian forces) will be minimized and repulsive forces (electrostatic or steric) will be maximized. The formation of layered lamellar liquid crystalline structures around the oil droplets is very beneficial to such a stabilization mainly in food applications. Figure 33 represents photomicrographs of the lamellar liquid crystalline structures as viewed under the polarized light microscope and the scanning electron microscope.

B. EMULSION FORMATION

An emulsion is thermodynamically unstable and the tendency of the two liquids to disrupt and separate is a natural and favorable process. The success of emulsion technologists lies in keeping the system in

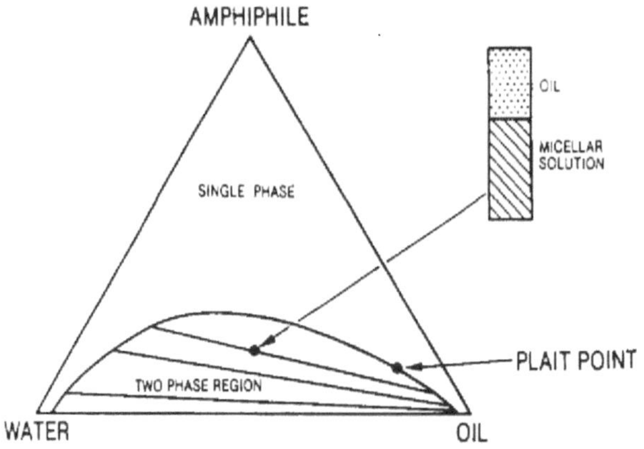

Figure 21 A typical S1 system, also known as Winsor I type system.

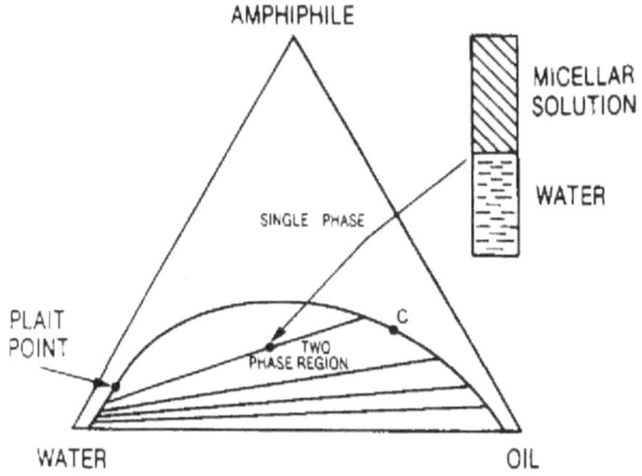

Figure 22 A typical Winsor II type system.

a state by opposing the mutual approach of the droplets. In practice, emulsion technology deals with two separate processes: the formation of the emulsion and the destruction of the droplets, meaning a constant "battle" to keep the emulsion stable and preventing it from merging into a bulk liquid.

Emulsion formation[61] is a "fast" two-step process: the deformation of droplets, (τ_{def}) and the transport of surfactant molecules to the newly created surface where they are adsorbed, (τ_{ads}). The entire process takes place in milliseconds (τ_{def} in a laminar flow is 0.1 ms, and in a turbulence flows is ca. 0.01 ms; τ_{ads} are 0.1 to 6 ms, respectively), and is achieved by applying mechanical energy. First, the interface between the two phases is deformed, forming a liquid film between the two liquids, which is followed by film deformation to such an extent that droplets are formed (Figure 34).

These droplets are mostly far too large and are subsequently broken up or distributed into smaller ones. Hence, the disruption of droplets is a critical step in emulsification. The deformation of the droplets is opposed by the Laplace pressure which is the difference in the pressure at the concave side of a curved interface, with interfacial tension γ, and that at the convex side of the droplets by an amount of $\Delta P = \frac{2\gamma}{R}$, (where R is the radius of a spherical droplet). Any droplet deformation leads to an increase in ΔP. The interface at the deformed droplet requires additional energy known as the "viscous forces" (velocity gradients) exercised by surrounding liquids and a "viscous stress" that has to be overcome. The forces are of the same magnitude as Laplace pressure. The pressure gradients or the velocity gradients needed (3 to 5 KJ/m^{-3}) are mostly supplied by agitation and other external shear forces.

C. EMULSIFIERS

The amphiphilic molecules that are capable of adsorbing onto the droplets surfaces and changing their nature, which consequently reduce their interfacial tensions and lower the gradient viscosity pressure, are called emulsifiers (Figure 35).

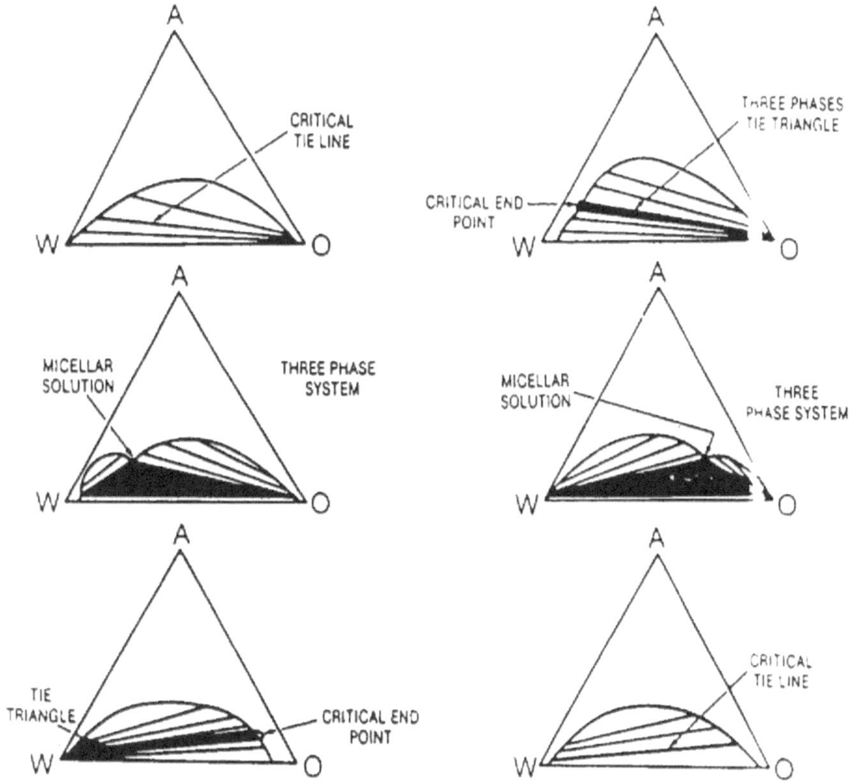

Figure 23 A sequence of ternary systems which are continuously transferred from type I to type III to type II. The existence of two critical tie lines is most notable.

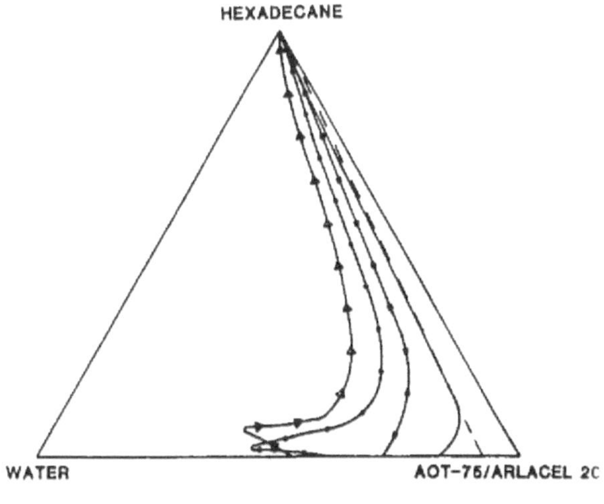

Figure 24 Pseudo-ternary phase diagram at 24°C for the water/hexadecane/Arlacel 20 (sorbitan monooleate)/AOT-75 diisoactyl sodium sulfosuccinate (75% active matter) system. The AOT-75 : Arlacel 20 (sorbitan monooleate) weight ratio is 40:60 (▷ ▷), 30:70 (● ●), 20:80 (■ ■), 10:90 (—), and 100% (---).

The emulsifier lowers interfacial tensions, γ_i (e.g., from 40 to 5 mN/m^{-1}) and thereby lowers the Laplace pressure. The emulsifier also forms a film on the dispersed phase allowing a flow of the continuous phase between the droplets (which is essential for the separation of one droplet from the other). The adsorption and the surface film formation is greatly speeded by the intense agitation and depends greatly on the nature and concentration of the surfactant. Adding a suitable surfactant may considerably reduce the agitation energy (factor of 10) needed to obtain a certain droplet size since the

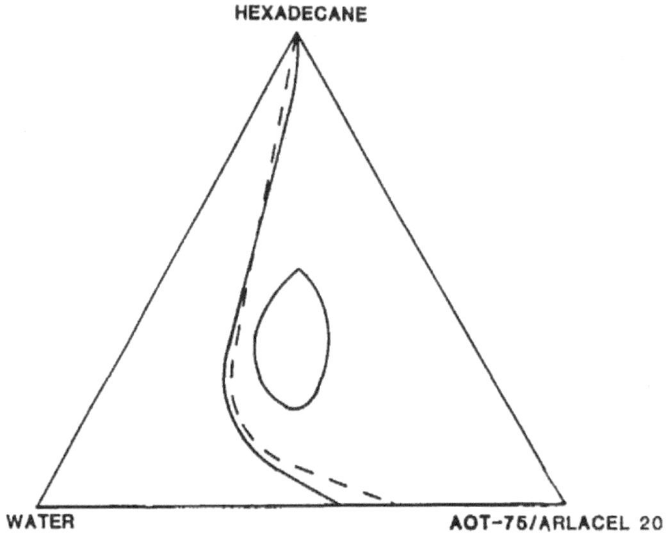

Figure 25 Pseudo-ternary phase diagram at 25°C for the water/hexadecane /AOT-75/Arlacel 20 system. The ratio of AOT-75 to Arlacel 20 are as follows: (– –) for 57/43 wt ratio, while the solid line (—) is for the 54/ 46 AOT-75/ Arlacel 20 ratio. Note the two phase region within the single phase region for the 54/46 AOT-75/Arlacel 20 system.

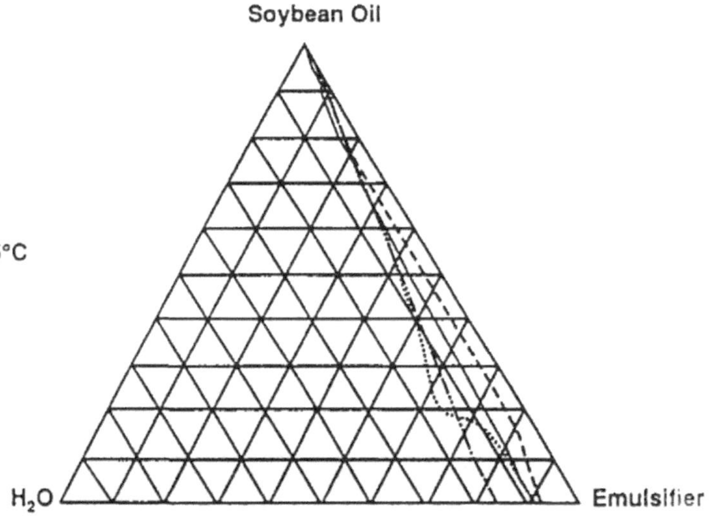

Figure 26 Complete phase diagram of soybean oil/water/emulsifier: (······) polyglycerol linoleate, (–––) polyglycerol oleate, (–––) monoglyceride stearate, and (‚‚‚‚‚‚) Tween 80 (ethoxylated-20-sorbitan monooleate).

surfactant reduces the gradient viscosity pressures. One must realize that much of the surfactant is depleted from the interface during the emulsification process due to the large increase in the interfacial surface area. Emulsion formation immediately causes the droplets to seek ways to recombine and coalesce again. Luckily, the coagulation process is comparatively "slow" and in practical systems it takes minutes or months before the net process results in the formation of metastable dispersed liquid droplets in the continuous phase.

An alternative way of making metastable emulsion, with a large number of droplets, is to start from a very tiny nuclei and then allow them to grow to the required size (the "condensation method"). The method is relatively easy with solid particles dispersed in liquid, but is difficult for liquid in liquid dispersions.

The role of the surfactant in the formation of emulsion is critical since it largely determines which phase is going to be the continuous one: it was experimentally established, using Bancroft's rule, that the continuous phase will be the one in which the surfactant is more soluble.

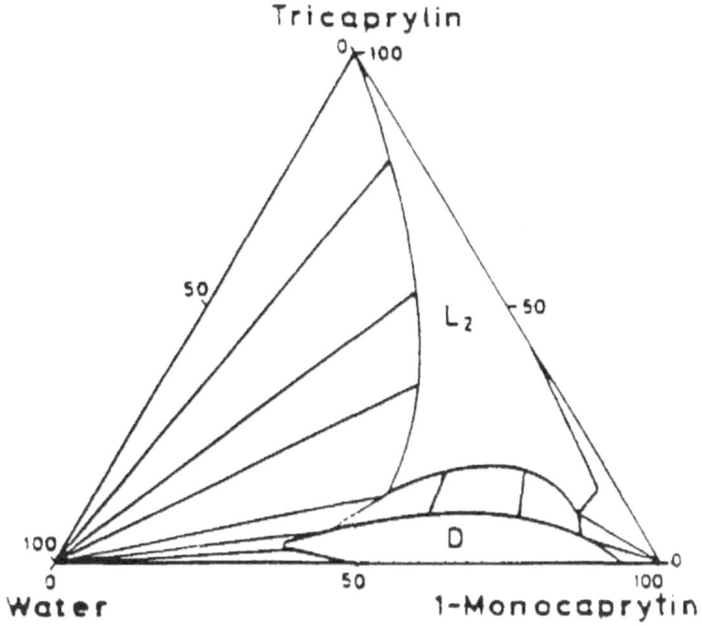

Figure 27 The w/o microemulsion region (L_2) in the system water/1-monocaprylin and tricaprylin.

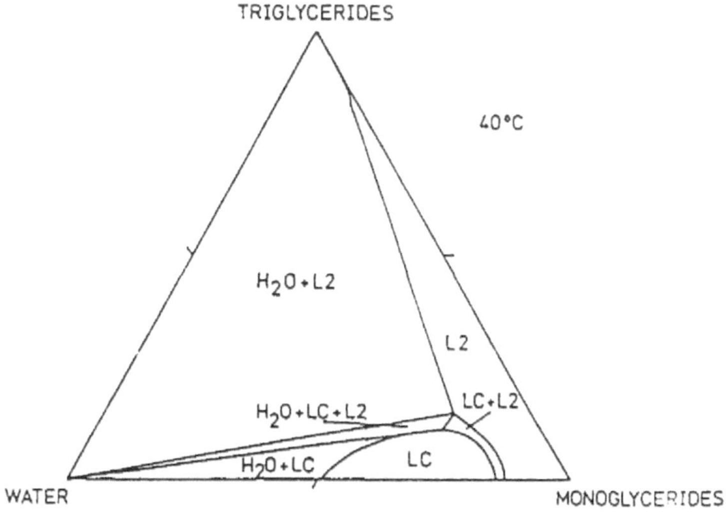

Figure 28 Phase diagram of a ternary system monoglyceride/triglyceride/water at 40°C.

Emulsions in which the water is the internal dispersed phase are termed water-in-oil emulsions (w/o), while emulsions in which oil is the dispersed phase and the water is the continuous phase are known as w/o emulsions. More complex systems, in which one emulsion is further dispersed into another continuous phase, will be called "double emulsions" or "multiple emulsions" (w/o/w or o/w/o).

D. EMULSIFICATION TECHNIQUES AND PARAMETERS

Many devices have been designed to produce emulsions.[64,65] The principles of some of these emulsifying machines are based on applying high shear forces or cavitation on the two liquids to allow rupturing of the dispersed phase into small discrete droplets. Many types of equipment have been designed and tested but only a few are in practical use in labs and in industrial units. Some of the machines are based on construction of baffles on which the liquid can impinge (laminar and/or turbulent pipe-flow); injection of one liquid into the other by a jet stream; rotor-stator machines for stirring; colloid mills; ball mills and

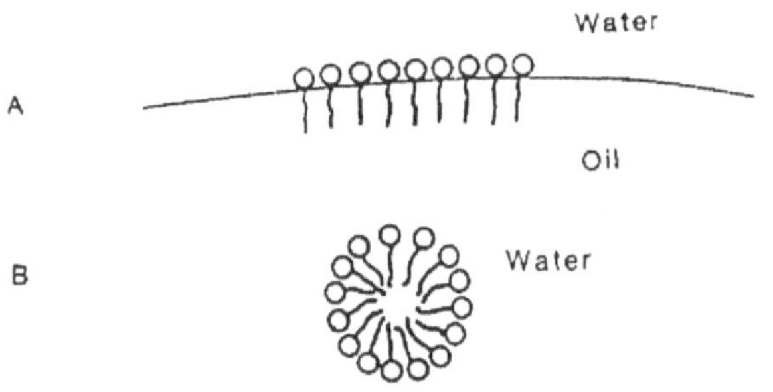

Figure 29 The curvature in an emulsion droplet (A) is extremely small and the bending component of the surface energy is not significant. A change in curvature does not lead to a change in the free energy. In the microemulsion droplet (B), on the other hand, a change in free energy in curvature leads to a pronounced change in free energy, e.g., the bending component of the surface energy is pronounced.

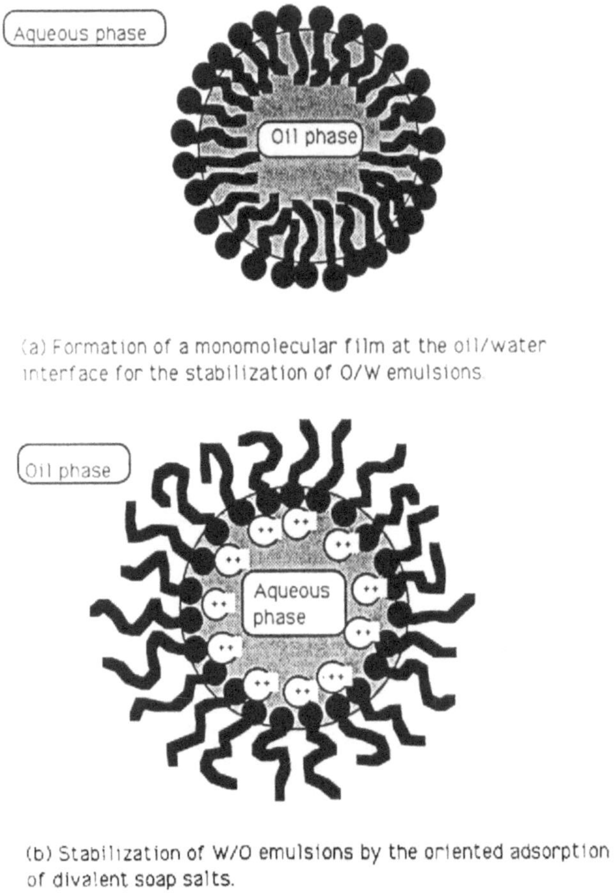

Figure 30 Schematic representation of typical o/w emulsion, the orientation and packing of surfactant molecules at emulsion droplet interface.

other mills; high pressure homogenizers; ultrasonic techniques; aerosol to liquids addition; and electrical and condensation apparatus. It is beyond the scope of this review to elaborate on the various methods. Figure 36 and Table 5 summarize the principals and the concepts of the main methods in use.

The way the substances used in the emulsion preparation process are added is important. Usually, the surfactant which should be most soluble in the continuous phase is indeed dissolved in that phase, but it should also have some solubility in the dispersed phase, in which case it can result in the formation of smaller droplets. *In situ* formation of an emulsifier is also possible (the so-called nascent-soap method). Slow addition of the dispersed phase to the continuous phase during emulsification is often advantageous: it may enable the addition of the surfactant to the dispersed phase. If both phases are mixed in bulk, the one containing the surfactants usually becomes the continuous phase. Therefore, it is essential to consider the temperature of the emulsification prior to any emulsifying action. One must also remember that the

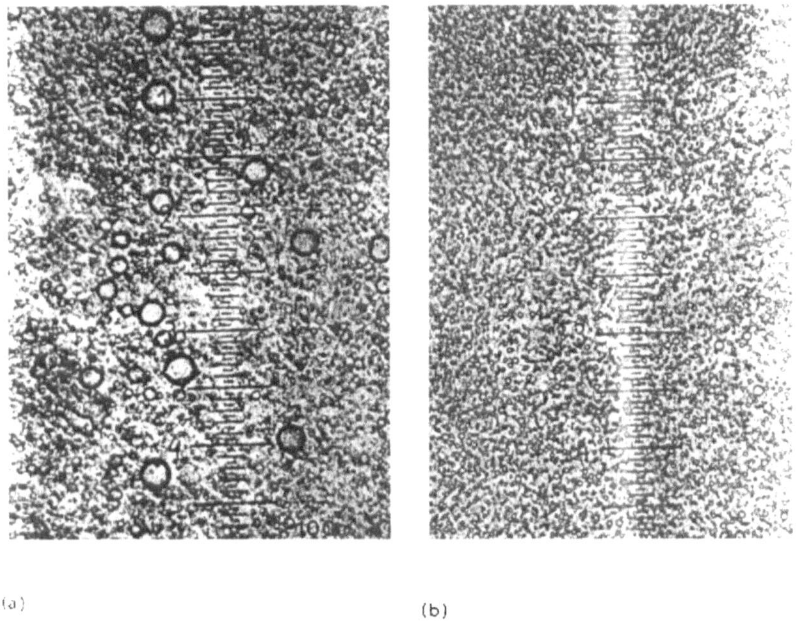

Figure 31 Water droplets size distribution in: (a) a margarine emulsion, and (b) the finished margarine.[62]

Table 4 Characteristics of Emulsions and Microemulsions

	Emulsion	Microemulsion
Appearance	Turbid	Transparent
Droplet size, μm radius	0.15–100	0.0015–0.15
Formation	Mechanical or chemical energy added	Spontaneous
Thermodynamic stability	No	Yes (No)

emulsification processes produce energy (the dissipation of the viscosity gradient energy), which may affect many variables.

The emulsification process can be evaluated in terms of emulsion capacity (the maximum amount of dispersed phase that can be emulsified under specified conditions), the emulsion stability (the amount of phase separation that takes place with time under specified conditions), and the droplets' size distribution (which seems to be the most practical way to study emulsification).

Much has been written about the different stages of emulsion formation. An excellent discussion can be found in the Walstra chapter's on "Formation of Emulsions" in the *Encyclopedia of Emulsion Technology*.[61] The main conclusions stressed there include: any emulsification process first goes through a film formation between the liquids. The film tends to drain rapidly under the influence of gravity. Interfacial forces tend to straighten the interface. The surfactant stabilizes the film and it allows an interfacial tension to exist, causing a considerable resistance against the drainage of the film. As a result, it implies slow stirring at the beginning of the emulsification (lower drainage) and high surfactant concentrations. At the second stage the plane (film) interface is disrupted into droplets. The plane disruption is accelerated by turbulence.

E. EMULSION STABILITY

Emulsions consisting of microscopic droplets of 0.1 to 100 μm, dispersed in a continuous phase, are thermodynamically unstable and tend to coalesce so that there is a net reduction in the interfacial area. Hence, coalescence is a thermodynamic spontaneous process. Nevertheless, there are numerous examples of spontaneous, naturally occurring emulsions that are stable. Milk is a good example. Stable emulsions

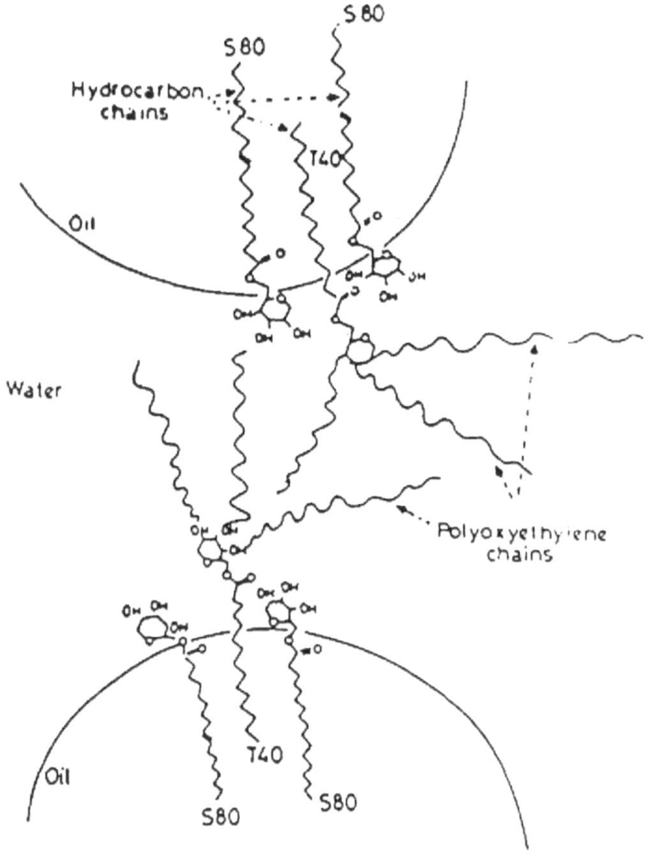

Figure 32 Schematic illustration of the "Sherman complex" composed of Tween 40® (ethoxylated sorbitan monopalmitate and Span 80® (sorbitan monooleate) at the o/w interface.

exist in some areas of application; in foods (mayonnaise), pharmaceutics (intravenous fat emulsions), cosmetics (some skin creams), agriculture (some pesticide emulsions), etc. Much has been done in the last decades to understand the factors affecting such unusual stabilities. In order to better understand the concepts of stability/instability, it is essential to study the steps and mechanisms that lead to phase separation of the liquids. Three to four main "events" can take place either in parallel or in sequence in each emulsion: "creaming," "flocculation," "coalescence," and in some cases also "Ostwald ripening" (Figure 37). All the processes lead, eventually to an "emulsion breaking" and phase separation. It is essential to distinguish between the processes and to learn to control them to some extent.

"Creaming" is the rise of dispersed droplets under the action of gravity, the droplets remaining separated when they touch. Creaming takes place in any dispersion if the phases are not exactly equal in density and if the dispersion medium is truly fluid.

It is well known that for small spherical undistorted droplets creaming will depend strongly on the radius of the droplets, the difference in the density of the two phases, and the viscosity of the continuous phase; the rate of creaming is governed by the Stoke's equation:

$$\upsilon = 2gr^2(\rho_c - \rho_d)/9\eta$$

where υ is the creaming rate, $(\rho_c - \rho_d)$ is the density difference between the two phases, η is the viscosity of the continuos phase, and r is the radius of the droplets.

Stoke's law assumes that the following conditions are valid:[66] (i) a spherical shape; (ii) no deformation; (iii) a single particle (dilute conditions); (iv) Newtonian fluid of the continuous with continuous properties and; (v) no diffusion. Emulsion droplets under rest are spherical, but they can be deformed in the shear field caused by the creaming. With large droplets or droplets stabilized by solid interfacial particles, distortion of the droplets can occur owing to changes in pressure on the different sides of the droplets (top vs. bottom) which will lead to an increase in the surface area and result in variation from the Stoke equation. The deformation effect takes place (as can be easily calculated) only for very large droplets (>1000 nm) and therefore it is not a severe drawback. However, Stoke's law has an additional limitation since it is good only for non-interacting droplets, which means it is good only for diluted systems. For more concentrated systems a term known as "mass rate of creaming" has been adopted

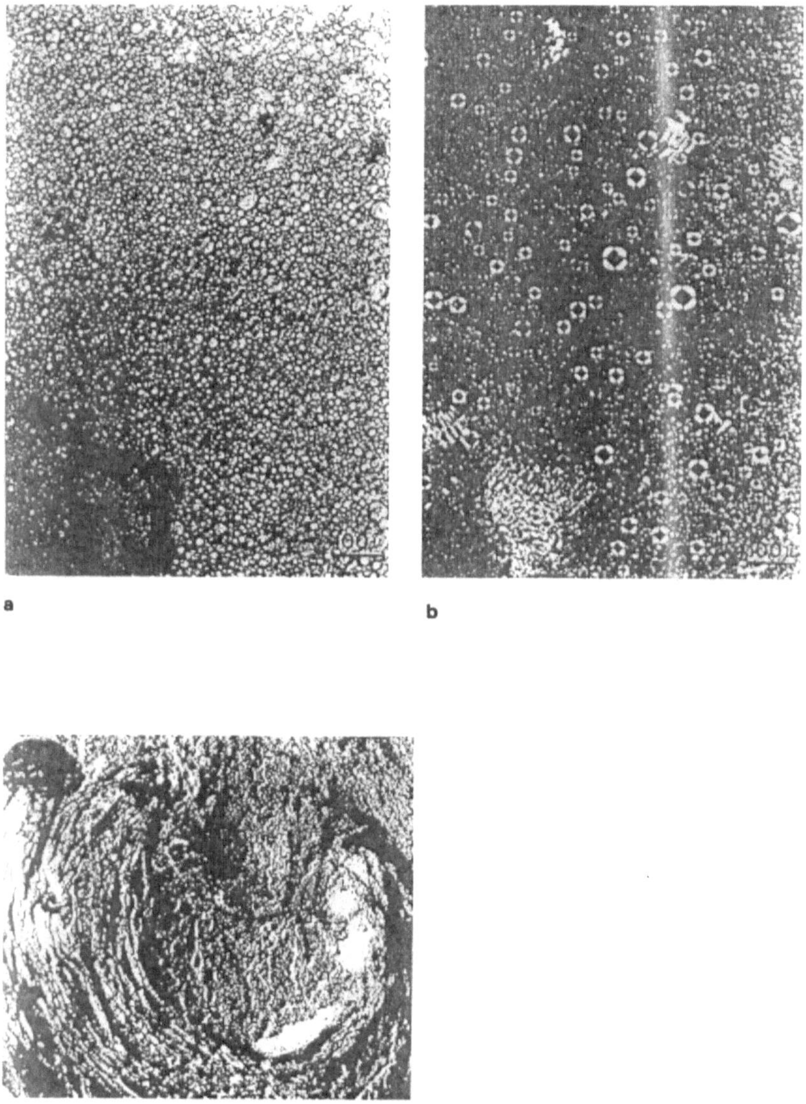

a b c

Figure 33 (a) Optical microscopy pattern of a lamellar liquid crystal, when viewed between crosspolarizers; (b) a typical view of emulsion droplets stabilized by liquid crystals; (c) electron microscopy of fractured sample showing the individual layers of the adsorbed liquid crystals.

(i.e., the motion of the center of the mass of the dispersed phase.[67] Additional complexation is related to the fact that the emulsion droplets vary in size (polydispersity) which also affects the creaming rate and, finally, parallel phenomena such as flocculation and coalescence should not be neglected when creaming rates are measured or evaluated.

"Creaming" can be hastened by applying (for charged emulsions) an external electrostatic field across the end of the vessel, centrifugation, or dilution. Creaming rates can be slowed down by decreasing the emulsion droplet sizes, equalizing the densities of the two phases, and by increasing the viscosity of the continuous phase (by adding viscosity builders, with high specific gravity, such as polysaccharides).

"Flocculation" is the act of sticking together of droplets and the formation of three dimensional clusters. In other words, "flocculation" is a process in which an aggregation of droplets takes place due to collisions, in combination with adhesive interdroplets forces. It is clear that since in both, creaming and flocculation, there is no flow of liquid from one dispersed droplet to another, these two phenomena are reversible.

"Flocculation" is a complex phenomenon caused by Brownian gravity and shear motions of the droplets, and is a very difficult phenomenon for quantitative evaluation. The free energy change of

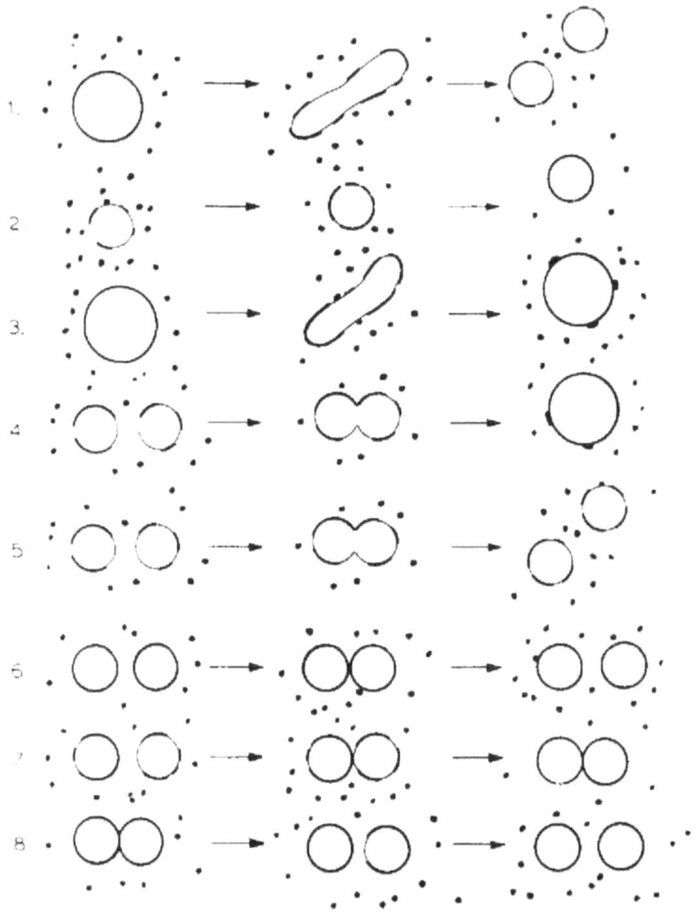

Figure 34 Processes taking place during the later stage of emulsification. Surfactant is depicted by heavy lines and dots. Not to scale and highly schematic. (After Walstra, P., *Encyclopedia of Emulsion Technology*, Vol. 1, Becher, P., Ed., Marcel Dekker, New York, 1983, chap. 2.)

flocculation, for weakly interacting droplets ΔG_{floc}, can be positive or negative depending on the droplets' concentration and thus, one can assume a critical flocculation concentration below which the emulsion is thermodynamically stable with respect to flocculation, but above which reversible flocculation to an equilibrium extent occurs. For strongly interacting droplets, this critical droplets concentration is so low as to be practically insignificant, and the flocculation may be classified as being irreversible. The flocculation process is then sometimes referred to as "coagulation." In thermodynamic, flocculated unstable systems, flocculation may be effectively prevented if a large enough free energy barrier exists between the flocculated droplets. The free energy barrier is derived from the interfacial forces such as long-range van der Waals forces and electrostatic forces. The net force of the two sets of interactions will determine the behavior of the emulsion droplets to flocculate and further coalesce or to stay stable in the kinetic sense (Figures 38 to 40).

A quantitative treatment of the three different fundamental mechanisms which may induce flocculation (Brownian, shear-induced, and gravity-induced) was recently reviewed by Bergenstahl.[66] The Brownian flocculation was first introduced by Smoluchowski[68] and described as the reaction (interaction) between two particles colliding and producing new aggregate. The flocculation rates depend on the number of particles present in the dispersion (concentration) and on the interaction time. Bergenstahl[66] found two additional factors that affect the Brownian flocculation: the surface forces (surface potential or surface charges) and the hydrodynamic interactions. Figure 41 illustrates the two effects on the emulsion stability to flocculate. The two factors are very significant when Brownian flocculation takes place and must be strongly addressed if such flocculation has to be minimized.[69]

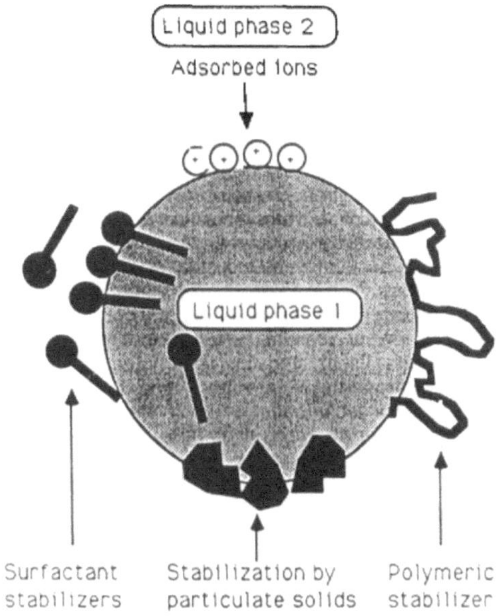

Figure 35 *Schematic view of an o/w droplet stabilized by different types of surface active molecules.*

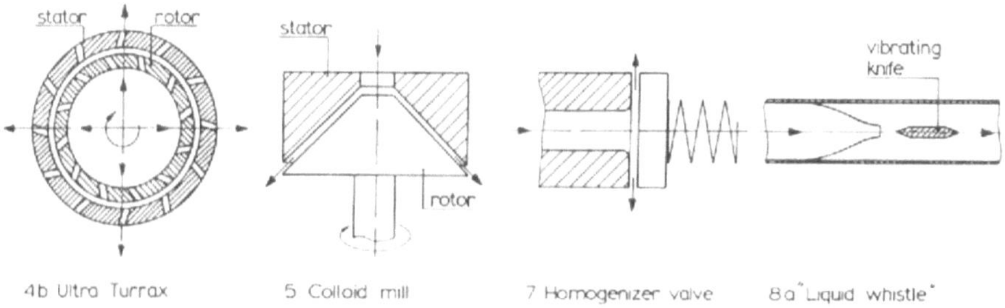

Figure 36 Principles of some emulsifying machines. The numbers refer to Table 5. The slit width in the homogenizer valve is very much exaggerated. Arrows indicate the directions of flow and rotation.[61,65]

Bergenstahl[66] also finds that the shear-induced flocculation (also known as the orthorhombic flocculation) as described by Smoluchowski[70] has similar limitations to those of the Brownian flocculation; that is, surface forces were not included, hydrodynamic interactions were not considered, and deflocculation was neglected. Gravity-induced flocculation, due to settling of particles, was first estimated by Saffman and Turner[71] assuming that every gravity-induced collision leads to attachment of the particles. Bergenstahl[66] added to the basic Stoke's law for settling of particle, the hydrodynamic interactions, the surface interactions (as described by Melik and Fogler[72,73]) and was able to predict stability areas for gravity-induced sedimentation as a function of the droplet size, surface potential, and the density differences between the particles. The mathematical expressions are very complicated but allow better evaluation of the possible mechanisms leading to flocculation, better predictions of the factors affecting the emulsions, and good evaluation of the stability to fluctuation that any emulsion can have.

"Coalescence" is joining of small droplets in an emulsion to form large droplets, or the process in which two droplets form one larger droplet leading ultimately to the formation of two separate liquid layers. The coalescence process, examined from the point of view of the individual droplets in an emulsion, contains four basic steps:

1. The approach of two droplets toward each others to achieve an adhesive contact
2. The drainage of the film between droplets in contact. The rate of the drainage process determines how rapid the critical thickness of the film is reached
3. The rupture of the film is next. This is a stochastic process, the probability of which is determined by the thickness of the film
4. Then follows the merging of the droplets. If the viscosity is low, the process is rapid.

Table 5 Methods and Apparatus to Produce Emulsions[61]

Method	Related to Method	Drop Formation	Drops Mainly Disrupted by[a]	Energy Density[b]	Mode of Operation[c]	Restrictions[d]
1. Shaking	4a	+	(T)	L	B	N
2. Pipe flow						
a. laminar	5	(+)	V	L–M	C	V
b. turbulent	4a	+	T	L–M	C	N
3. Injection	10a	+	—	L	C	
4. Stirring						
a. simple stirrer	1,2b	+	T, V	L	B, C	
b. rotor-stator	(5)	+	T, V	M–H	B, C	V
c. scraper	5	+	V	L–M	B, C	N
d. vibrator	8a	+	?	L	B, C	V
5. Colloid mill	2a, 4c, 6	(+)	V	M–H	C	V
6. Ball and roller mills	5	+	V	M	B (C)	V
7. High-press. homogenizer	(2b)	—	T, C, V	H	C	N
8. Ultrasonic						
a. vibrating knife	4d	+	C, T	M–H	C	W
b. magneto-striction		+	C	M–H	B, C	W
9. Electrical	10b	—	Electrical charge	M	B (C)	Several
10. Aerosol to liquid						
a. mechanical	3	+	—	L–M	B, C	
b. electrical	9	+	—	M	B, C	Several
11. Foaming or boiling		(+)	Spreading	L–M		(W)
12. Condensation		+	—			Several

(a) V = Viscous forces in laminar flow, T = turbulence, C = cavitation.
(b) L = low, M = medium, H = high.
(c) B = batch, c = continuous.
(d) The continous phase should be V = viscous, N = not too viscous, W = aqueous.

However, emulsions with high viscosity will merge very slowly and may even be disrupted during the process. Table 6 summarizes the factors influencing time (rate) of coalescence. It can be seen that the droplets' size and the interfacial tensions are key factors in the coalescence times. The coalescence process is very difficult to explain in simple mechanical events and the literature is still quite confusing. The main treatments are based on the kinetics of the individual droplets as was first suggested by Van den Tempel[74] and later elaborated by others. The main steps include creaming, consolidation (flocculation) (Figure 42), film drainage, and film rupture. Bergenstahl,[66] in his recent close examination and critical review of the coalescence process, brought a set of principal effects that can influence the coalescence process (Table 7). The large number of parameters affecting the coalescence emphasize the complexity of the process that takes place in the course of the collision of two droplets.

F. EMULSIFIERS AND THE HLB CONCEPT

Surfactants play a significant positive role in minimizing the destabilization phenomena (flocculation and coalescence) by forming adsorbed films at the interface, which contribute to the repulsive forces and the London dispersion forces. Mixed surfactants are particularly more efficient than single surfactants, with respect to the coalescence rate. The enhanced stability is attributed to the formation of intermolecular complexes at the oil–water interface, forming a densely packed layer and reducing the interfacial tension to very low values of ca. 0.1 mN/m^{-1}. In addition, the interfacial complexes are capable of increasing the interfacial viscosity (although each one of the component exhibits low viscosity). Some investigators[57] claim that the synergism is derived from the enhanced adsorption kinetics of the surfactant when two such surfactants mixtures are used. The coherent strong interfacial film should act as a barrier, preventing coalescence by virtue of its rheological properties, e.g., its high dilational viscoelasticity. As a result, most technologists are using mixtures made of two surfactants in most of their emulsion applications.

The selection of surfactants is still quite difficult and requires much experience and experimental work. Griffin[75,76] introduced the empirical concept of HLB (hydrophile–lypophile balance) to select or blend surfactants. Surfactants with low HLB will form w/o emulsions and those with high HLB will form o/w emulsions. Several methods[77] have been proposed to determine the HLB of nonionic surfactants, and

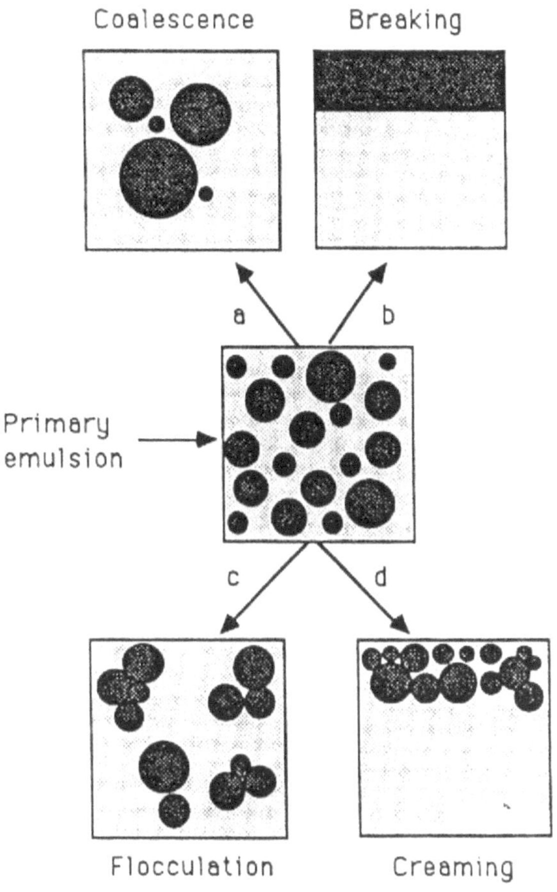

Figure 37 Schematic representation of the instability processes occurring in emulsions.

tables with the HLB values are available in any surfactant science textbook. Moore and Bell[78] suggested the H/L numbers.

Greenwald[79] suggested a water titration procedure (the water numbers) of organic solution of surfactants, but the method has not been widely used.

Davis[77] has developed a method by which to calculate the HLB values of surfactants directly from their chemical formula, using empirically determined group numbers:

$$\text{HLB} = 7 + \Sigma \text{ hydrophilic group numbers} - \Sigma \text{ lipophilic group numbers}$$

He gave the values of the group numbers in a simple table. The match between his method of evaluation of HLB and the empirical work is quite good. The experimental determination of HLB of a surfactant is also not an easy task.

One must realize that the HLB concept is based on a simplified picture of the coalescence and formation process, and does not take into account the important role of liquid-film thinning between droplets and the importance of shear and dilatation properties of the adsorbed layer of the surfactant around the droplets. Therefore, it is no surprise that the literature is full of examples of emulsifications that do not obey the HLB rules and concepts. Many authors have cast doubt on the validity of the concept. In spite of the criticism, the HLB index can be very useful in the practical formulation of emulsions. In practice, in formulating an emulsion one must determine, with a set of known emulsifiers (special kit) the "required HLB" of a given oil used for the preparation of the emulsion (by finding the minimum oil separation at given conditions and with constant concentration of a set of known emulsifiers). Many of the common oils have required HLB values given in tables. For any new emulsion to be prepared, the best set of emulsifiers will be determined on the basis of the "required HLB" and the minimum oil separation after storage. The HLB values are additive and can be easily calculated for any set of two nonionic emulsifiers.

Shinoda and his co-workers[80–82] devoted years of research to finding a better concept to describe the relationship of the surfactant structure, its solubility, and the emulsification parameters. Concepts such as phase inversion temperature (PIT), HLB numbers, and the emulsion inversion point (EIP) were introduced. Phase diagrams were constructed. The Shinoda concepts have sound a thermodynamic grounding and explain many of the phenomena attributed to the nonionic emulsifiers.

G. POLYMERIC EMULSIFIERS — AMPHIPHILIC MACROMOLECULES

The presence of polymers in the continuous phase and/or on the emulsion droplet surface has a very significant stabilizing effect since it strongly affects the surface forces between the droplets. These effects were studied in great detail in view of the DLVO theory and the steric stabilization. Polymers and mainly proteins are essential for food emulsions stability. A waste majority of food emulsions are stabilized by an adsorbed layer of proteins at the oil–water or air–water interface.[12,14–16] The physical properties of interfacial films in protein-based foams and emulsions are determined by the structural properties of the adsorbed proteins. Some of the structural features are segmental flexibility, surface hydrophobicity, reduced net charge, etc. A fairly compact folded polymer is strongly adsorbed to the oil droplets. As the proteins are large molecules, the entropy of mixing is small compared to the sum of all the interactions at the interface and contribute to the stability. Only those proteins that are (i) highly hydrophobic, (ii) have the ability to unfold at the interface, and (iii) have significant solubility in the continuous phase will adsorb strongly.[86–92] If the desorption is minor and if the protein achieves full coverage, the emulsion will have satisfactory stability. Many of the food proteins have the proper surface hydrophobicity (see Table 8) and the proper adsorbed concentrations (ca. 3 mg/m^{-2} for S_1-casein, 1.6 for BSA, and 3.0 for lactoglobulin) and therefore will contribute to the stability.[83–87]

In order to understand better the topics related to protein adsorption on oil–water interfaces and its effect on the stabilization of emulsions, the chapter of Dickinson, Murray, and Stainsby is a good reference for further reading.[88]

Emulsions can also be stabilized by naturally occurring macromolecular substances such as polysaccharides or gums. Such stabilization is common practice in food and pharmaceutical emulsions. In the past, stabilization by macromolecules was attributed only to proteins since they form thick "films" at the interface. The adsorption of macromolecules at the liquid–liquid interface is characteristic of amphiphilic polymers like proteins (casein, BSA, soy proteins, etc.), and strongly depends on the molar mass distribution of the adsorbed molecule. The molecules form an adsorbed layer comprised of "loops" and "tails" that are anchored in the continuous phase. However, polysaccharides are water soluble and very hydrophilic hydrocolloids, and do not have either the flexibility nor the hydrophobicity required from polymeric emulsifier, and therefore, were considered to act via only their viscosity effect on the continuous phase. Such stabilization is known as "depletion stabilization" or "depletion flocculation" and is characteristic of polysaccharides such as guar gum, carrageenan, etc.[93–97]

In a set of experiments, Bergenstahl was able to show that certain polysaccharides adsorb to polystyrene particles and may stabilize them sterically. Bergenstahl concluded that hydrocolloids are not surface active enough to generate or support the emulsification themselves and are usually used in combination with emulsifiers. He suggested a model for the adsorbed layer of hydrocolloid on top of the emulsifier.[98]

However, based on recent research, it is our belief that certain hydrocolloids, in addition to gum Arabic[97] (proven to adsorbs via its portentous sites), can adsorb directly onto the oil–water interface and

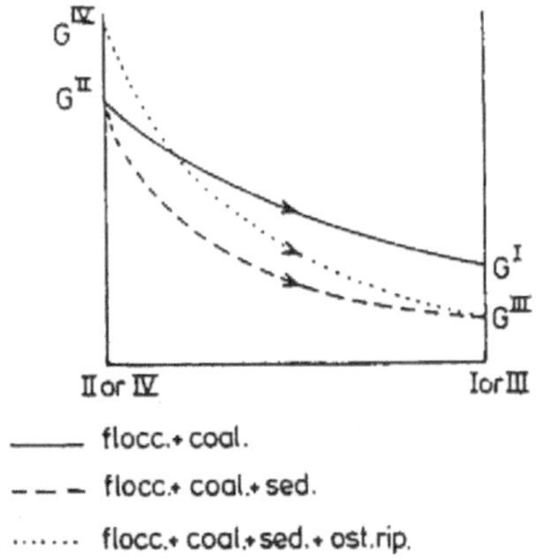

Figure 38 Free energy path in emulsion breakdown.

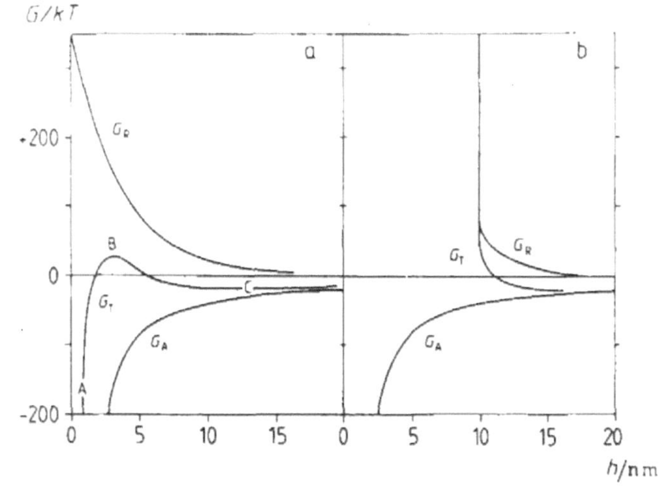

Figure 39 Calculated examples of repulsive (G_R), attractive (G_A), and total interaction free energy (G_T) as a function of surface separation distance h of two identical spheres: (a) electrostatic repulsion and van der Waals attraction (DLVO); (b) steric repulsion and van der Waals attractions.

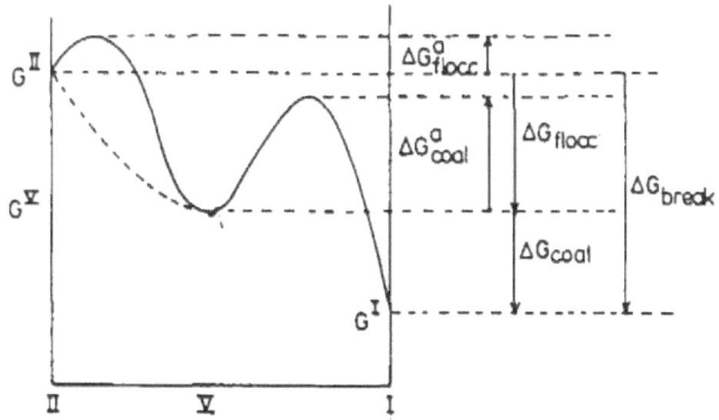

Figure 40 Free-energy path for breakdown (flocculation and coalescence) for droplets having both attractive and repulsive (electrostatic and steric) interactions.

stabilize the emulsions. It was demonstrated that guar gum, fenugreek gum, and other galactomannans adsorb weakly, form ordered birefringent layers, and stabilize some oil-in-water emulsions.[99,100]

VIII. MULTIPLE EMULSIONS

A. DEFINITION AND FORMATION

Multiple emulsions are complex systems, termed "emulsions of emulsions"; the droplets of the dispersed phase themselves contain even smaller dispersed droplets of the dispersed phase. Each dispersed globule in the double emulsion forms a vesicular structure with single or multiple aqueous compartments separated from the aqueous phase by a layer of oil phase compartments.[101-104]

Multiple emulsions were first described by Seifriz[105] in 1925, but it is only in the past 20 years that they are being studied in more detail. The two major types of multiple emulsions are the water-in-oil-in-water (w/o/w) and oil-water-oil (o/w/o).

A schematic representation of a w/o/w multiple emulsion droplet is shown in Figure 44, showing the inner and outer aqueous phases separated by an internal hydrophobic emulsifier interface (E1) and an external hydrophilic emulsifier interface (E2). Figure 45 is a microscopic picture of a typical double emulsion consisting of multiple compartments water droplets. Some double emulsions are of single or few internal water compartments (Figure 46). The number of compartments depends on the nature of the emulsifiers, nature of oil, and methods of preparation.

Multiple emulsions were prepared in two main modes: one-step emulsification and two-step emulsification. Several reports have been made on the one-step emulsification for the preparation of w/o/w double emulsion, which included strong mechanical agitation of the water phase containing a hydrophilic

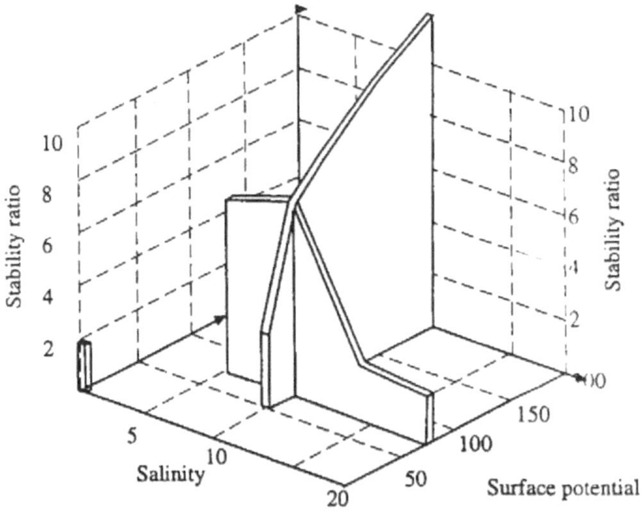

Figure 41 The stability to flocculation factor including viscous interactions vs. the surface potential and the salinity (surface charge) of the emulsion.[66,69]

Table 6 Factors Influencing Time for Coalescence of Emulsions (Measured as Time for Separation Rate of Droplets' Size Increase or as Droplets Life Time)

Parameter	Effects
Droplet size	Exponential increase with increasing radius Exponential decrease with increasing radius
Interfacial tension	Increase with increasing interfacial tension Decrease with increasing interfacial tension

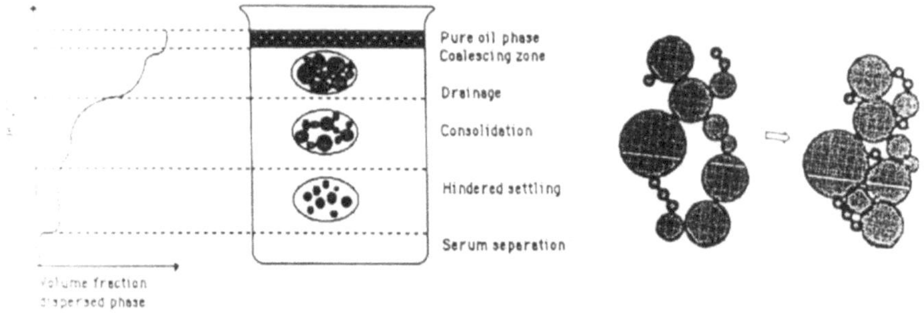

Figure 42 Various instability processes at different positions in the container in a settling, flocculating, and coalescing emulsion, and schematic illustration of the consolidation process.

emulsifier, and an oil phase containing large amounts of hydrophobic surfactant. The w/o emulsion tends to invert, and forms a w/o/w double emulsion[106] (Figure 47). In addition, double emulsions can be prepared by forming a w/o emulsion with a large excess of relatively hydrophobic emulsifier and a small amount of hydrophilic emulsifier followed by heat-treating the emulsion until, at least in part, it will invert. At a proper temperature, and with the right emulsifier's HLB, emulsion can be found in the system. However, there is usually little chance of reproducing these "accidental" preparations.

In most of the recent studies, double emulsions are prepared in a two-step emulsification process by two sets of emulsifiers: an hydrophobic surfactant, emulsifier I (E_1) (for the water-in-oil emulsion) and a hydrophilic surfactant — emulsifier II (E_2) (for the oil-in-water emulsion) as shown in Figure 48.

The primary w/o emulsion is prepared under high shear conditions (ultrasonification, homogenization), while the secondary emulsification step is carried out without any severe mixing (an excess of mixing can rupture the drops, resulting in a simple oil-in-water emulsion). A typical droplet size distribution of double emulsions is illustrated in Figure 49.

Table 7 Principal Effects of Different Variables on the Various Process Steps Involved in the Phase Separation

Variable	Creaming	Rupture	Consolidation	Drainage
Droplet viscosity	0	0	0	+
Continuous phase viscosity	+	0	+	0
Droplet size	–	+	+	–
Volume fraction of dispersed phase	+	–	–	0
Density difference	+	+	+	0
Surface tension	0	0	–	+
Surface viscosity	0	0	0	+
Surface elasticity	0	0	+	+
Depth of attractive minima	0	+	–	0
Repulsive interaction	0	–	0	+
Adsorbed layer thickness	0	–	+	+
Solubility of emulsifier in the dispersed phase	0	0	–	0
HLB (hydrophilic–lipophilic balance of the emulsifier)	0	–	+	+

Note: The symbol 0 stand for no effect; + stand for positive effects; – stands for negative effects.

The composition of the multiple emulsion is of significant importance since the different surfactants, along with the nature and concentration of the oil phase, will affect the stability of the double emulsion.[102,107]

Much work has been done on the nature of oils and their influence on manufacturing conditions, as well as on the stability of double emulsion.[107]

Ionic and nonionic surfactants have been used for different applications. It was, however, well established that combinations of emulsifiers at the outer phase have a beneficial effect on stability, and that the inner hydrophobic emulsifier (E_1) must be used in great excess ($E_1/E_2 = 10$, E_1 is 10 to 30 wt% of the inner emulsion, Figure 50), while the hydrophilic emulsifier E_2 must be used in low concentration (0.5 to 5 wt%). The inner emulsifier was found to migrate, in part, to the outer interface and influence the outer emulsifiers.[108,109] The HLB of the outer emulsion was found to be a "weighted HLB" of the contribution of the two types of emulsifiers.

In addition, parameters such as phase volume and nature of entrapped materials in the inner phase have been discussed and optimized.[109,110]

Multiple emulsions have shown significant promise in many technologies, particularly in pharmaceutical and medical applications. Potential biopharmaceutical applications (as a consequence of the dispersal of one phase inside the droplets of another) include uses such as adjuvant vaccines, prolonged drug delivery systems, sorbing reservoirs of drug overdose treatments, and immobilization of enzymes.[110–119] Multiple emulsions have also been formulated as cosmetics and as household products such as wax polish and foods.

B. STABILITY AND RELEASE

Many review articles have been written on the potential practical applications of multiple emulsions, and on the main problem associated with this technology — their inherent thermodynamic instability.[120]

The complex nature of double emulsions has caused significant difficulties in the assessment of stability and in detecting rupture and coalescence phenomena. The main technique is based on the measurement of the number and size of the double emulsion drops over a period of time. This technique produces only limited information on the emulsion stability, since no information on the coalescence of the inner droplets can be deduced. Similar information is obtained from time-dependent photomicrography.

It was, and still is, very difficult to determine if the internal droplets coalesce, aggregate, or tend to rupture. Several other techniques have been applied, including freeze-etching, viscosity measurements, and quantitative estimation of addenda transported from the inner phase to the outer phase and vice versa. In addition, engulfment or shrinkage of the double emulsion in the presence of water migration in or out of the droplets, has been studied and interpreted in terms of stability (Figure 52).

Table 8 Adsorbed Amounts in Emulsifiers as Measured by Serum Depletion and on Macroscopic Surfaces from Ellipsometry[83–87]

Source	Composition		Molecular Weight	Surface Hydrophobicity
Casein	α_{s1} casein	38%	19,000	4.89
	α_{s2} casein	10%	24,000	4.64
	β casein	36%		5.58
	κ casein	13%		5.12
Plasma proteins	BSA	60%	65,000	4.3
Whey proteins	β lactoglobulin	59%	18,600	4.41[c]
	α lactoalbumin	22%	14,400	1.66[c]
	BSA		66,300[d]	10.33[c]
Egg protein[b,c]	Ovoalbumin		45,000	1.33

[a] According to Kato and Nakai.[84]
[b] See Townsend and Nakai.[85]
[c] See Keshavarz and Shuryo.[86]
[d] See Wahlgren, et al.[87b]

It was also found that surfactants with high HLB values, used as the secondary hydrophilic emulsifier that were used in order to improve stability to coalescence, increased the release rates, primarily by increasing the rate of diffusion of methotrexate (MTX) through the nonaqueous liquid membrane, as shown in Figure 51. These findings caused significant difficulties to the formulators and opened a dilemma that retarded the development of new double emulsions. As we increase the emulsion stability to coalescence we also increase the release rates of the entrapped materials.

One can think of several approaches to overcome the instability and release problems:

1. Stabilizing the inner emulsion by reducing the inner droplet size; forming microemulsions or microspheres; increasing the water viscosity of the inner phase.
2. Modifying the nature of the oil phase by increasing its viscosity; adding carriers and/or complexing agents.
3. Stabilizing the inner and/or the outer emulsion interfaces by using polymeric emulsifiers or colloidal solid particles to form a strong and flexible film at the interface.

The stability of the double emulsions can be improved (as explained above) by forming a polymeric film or macromolecular complex across the oil–water interfaces. Omotosho et al.[121–123] have suggested using macromolecules to form such stabilizing complexes through the interfacial interaction between macromolecules such as albumin and non-ionic surfactants.

Release rates of MTX encapsulated in the internal phase of w/o/w emulsions, stabilized by a film formed as a result of an interfacial interaction between albumin and sorbitan monooleate (Span 80), were measured as functions of two formulation variables — the oil phase and the secondary emulsifier composition. The release rate was significantly affected by the nature of the oil phase and decreased in the following order: isopropylmyristate > octadecane > hexadecane > dodecane > octane, which is a reflection of the increasing internal droplet size of the emulsion, as shown in Figure 53.

The formation of multiple w/o/w emulsions with improved stability due to the formation of an interfacial complex film between gum Acacia, gelatin, polyvinylpyrrolidone (PVP) and sorbitan monooleate, have been described.[121,123] The long-term stability as assessed by microscopy showed no significant changes in droplet size with time in w/o/w double emulsions prepared with Acacia, gelatin, or PVP in the internal phase, indicating good stability in these systems, as shown in Figure 54.

Oza andFrank[115,124,125] have suggested the use of colloidal microcrystalline cellulose (CMCC) together with various monomeric surfactants to stabilize double emulsions via a "mechanical stabilization" mechanism. It was demonstrated (monitored by microscopy) that the double emulsions were stable over a period of one month, much longer than the emulsions prepared with monomeric emulsifiers, and slow release of certain drugs was achieved by gelling the oil phase.

Recently, the authors have used BSA (bovine serum albumin) as a polymeric emulsifier added both to the inner and to the outer interfaces in the absence and presence of conventional monomeric emulsifiers such as Span 80® and Tween 80®, and have tried to evaluate in more detail the stabilizing effect of the blend (complex) of BSA and Tween 80®.[126] Since it was believed that the release mechanism is associated with a micellar transport which is diffusion controlled, attempts were made to obtain stable, double

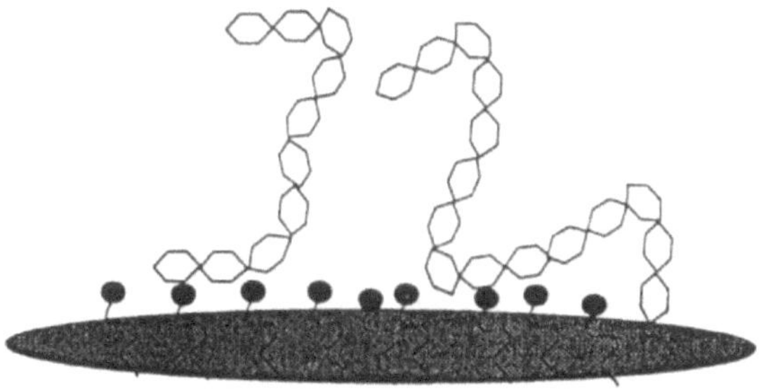

Figure 43 (a) Emulsion droplets stabilized by adsorbed proteins or other polymer amphiphile via a steric stabilization mechanism; (b) emulsion stabilized by hydrocolloids (gum, polysaccharides) via a "depletion stabilization" mechanism; (c) emulsion droplets stabilized by an emulsifier and hydrocolloid combination (according to Bergenstahl et al.[98]).

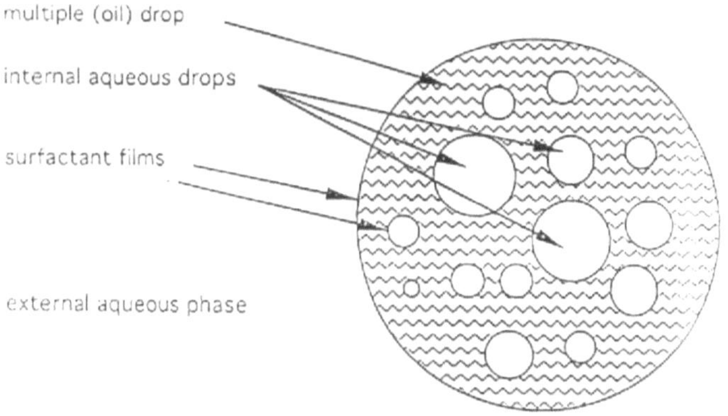

Figure 44 Schematic presentation of the two interfaces, the w/o, and the o/w, in a typical double emulsion of w/o/w.

emulsions with relatively small double emulsion droplets, and minimum oil micellization capacity. Double emulsions were prepared with 10 wt% Span 80, 0 to 0.5 wt% BSA and 2 wt% NaCl in the inner aqueous phase. The outer interface was stabilized with 5 wt% of Span 80®–Tween 80®. The emulsion droplet size did not change significantly even after 25 h of aging. Plot of the percent release of NaCl vs. wt% of BSA present in the inner phase, reveals that BSA retards the release of NaCl, as shown in Figure 55. Its maximum effect was obtained at 0.2 wt% BSA.

It seems that BSA adsorbs at the interface, together with Span 80® and contributes both to the stability of the double emulsion and to the release rates.

In accordance with the previous studies, it is suggested that the polymeric surfactant forms a complex with a monomeric lipophilic surfactant. The complex is a thick gel and forms a strong film that imparts elasticity and resistance to the rupture of the inner droplets. The film improves the mechanical and steric stability of the double emulsions and slows the coalescence rates. In addition, it appears that it depresses the formation of reverse micelles in the oil phase and slows the transport of NaCl (BSA is reducing the number of reverse micelles that are formed by the monomeric emulsifiers and that are responsible to the transport mechanism). However, the monomeric emulsifiers covering the outer interface (E2) do not prevent the coalescence of the outer droplets in the double emulsions and in fact, after 1 week of aging, a significant change in the droplet size distribution, as well as in the flocculation rates, was detected. Therefore, BSA was also added to the outer aqueous phase during the second step of the emulsification, in addition to the monomeric blend of Span 80®–Tween 80®. The Coulter counter measurements, as shown in Figure 56 clearly indicate that BSA in the outer phase contributes to the stability of the double emulsions. Double emulsions with smaller droplets are obtained in the presence of the BSA at any of the emulsifier combinations used in the inner phase.

The photomicrographs and the Coulter counter measurements of the droplets size distribution after 6 weeks of aging show practically no change in size distribution (Figure 57a–c).

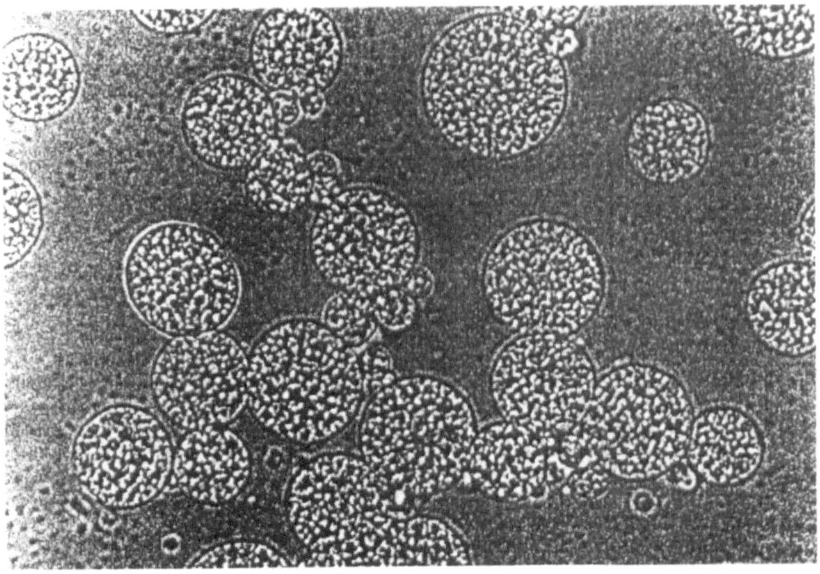

Figure 45 Photomicrograph of a typical multiple compartment double emulsion.

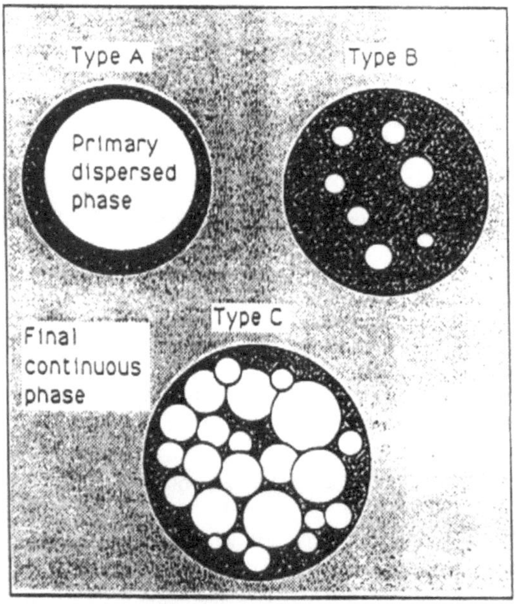

Figure 46 Droplet classification in multiple emulsions. Type A is a single compartment double emulsion; type B is few compartment double emulsion; type C is a multiple compartment emulsion.

C. RELEASE MECHANISMS

Several possible pathways and mechanisms, by which materials may be transported across the oil layer of the w/o/w systems, have been proposed and discussed.[102,105,108] The possible pathways include: (i) coalescence of the inner droplets until reaching a critical size followed by migration through the external interface to the outer aqueous phase (the rupture pathway) (Figure 52, internal drop growth); (ii) coalescence of the oil droplets leading to phase separation and "spill" of the entrapped matter (the engulfment mechanism) (Figure 52, external drop coalescence); (iii) release of the entire water through the external interface via engulfment (droplets release mechanism) (see Figures 58 and 57b–c); (iv) release of the internal water soluble matter through the oil membrane without flocculation or coalescence (molecular release mechanism) (Figures 59 and 60).

The rupture (coalescence) mechanism was discussed in the paragraph dealing with emulsions. The engulfment pathway is schematically presented in Figure 58. It can be clearly seen under the microscope in certain emulsions (Figure 57b–c).

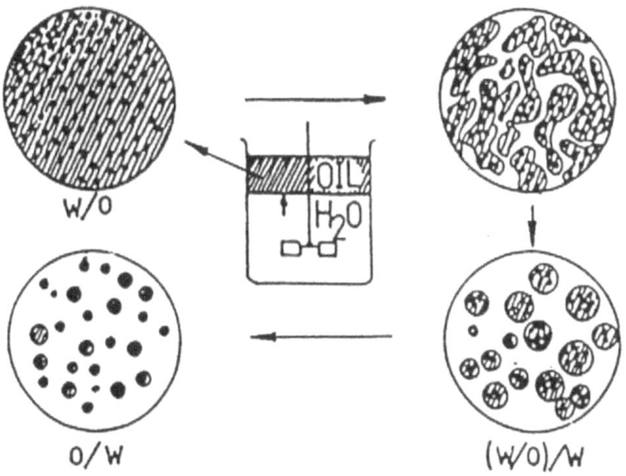

Figure 47 A single step emulsification for the preparation of a w/o/w double emulsion. The oil phase contains large amounts of hydrophobic emulsifier while the aqueous phase consists of small concentrations of hydrophilic emulsifier. The double emulsion is formed in the stage of phase inversion.

A common molecular release is the diffusion controlled mechanism for: (i) ionized lipid-insoluble materials at given pH, that turn lipid-soluble at different pH conditions (Figure 59), and (ii) nonionized lipid-insoluble materials at the storage or release conditions (the working conditions). The release will be dependent on the nature of the entrapped material (including its dissociation constant), the nature of the oil, as well as on the pH of the aqueous phase. Case (i) can take place in both directions, for addenda dissolved in the outer phase and migrating to the inner phase (concentrating the active material) or for materials dissolved in the inner phase and transported to the outer aqueous phase (sustained and controlled release), for example, in the treatment of overdose of acidic drugs such as aspirin and barbiturates.[127] At low pH values the barbiturate (present at the outer continuous phase) would exist almost exclusively as the unionized form and so would be readily soluble in the oil phase.[127-129] The drug could, therefore, easily pass across the oil layer to the internal aqueous phase containing a basic buffer which would ionize the drop that becomes now insoluble in the oil phase, and would become trapped within the internal aqueous phase. Similar transport (in the opposite direction) can take place for drugs entrapped in the inner phase and migrating to the outer phase. The drug would then be carried out with the emulsion to the gastrointestinal tract, as shown in Figure 59. In other cases, designed for drug delivery, the internal entrapped material is either ionized or totally nonionic at the working conditions and is completely insoluble in the oil phase. In spite of its insolubility in the oil membrane, the material is transferred to the aqueous outer phase. The mechanism describing such release is diffusion controlled and takes place in reverse micelles (Figure 60).

An important application of the double emulsion is related to the controlled and sustained release of water-soluble drugs in pharmaceutical applications. A controlled release of the drug is expected only in the intestinal tract. It has been demonstrated that water molecules, electrolytes, and both ionized and nonionized water-soluble drugs can easily migrate through the oil membrane without affecting the double emulsion stability.[129-131]

The mechanism based on micellar transport from one phase to the other has been extensively studied and described in some detail (Figure 60). It was also demonstrated that one can alter the diffusion coefficients through the oil membrane by controlling the nature and the amount of the reverse micelles. It also suggests that the diffusion of the addenda through the oil phase is the rate determining step that must be fully controlled, and that the inner water phase does not have any effect on the determination of the release rates. Kita et al.[132] have suggested two possible diffusion controlled mechanisms for the permeation of water and water soluble materials through the oil phase, the first being via the reverse micellar transport (Figure 60a), and the second by diffusion across a very thin lamellae of surfactant formed where the oil layer is very thin, as shown in Figure 60b.

Our detailed studies[129-131] on the release of electrolytes from the inner aqueous phase to the outer aqueous phase have indicated that the osmotic pressure gradient between the two aqueous phases is a strong driving force for migration of addenda and/or water from one phase to the other by both mechanisms. However, it was clearly demonstrated that even if the osmotic pressure of the two phases was equilibrated and no coalescence took place (neither of the inner phase droplets nor of the outer phase droplets), electrolytes tend to be transported out through a reverse micellar mechanism controlled by the viscosity and the nature of the oil membrane. The mechanism is similar to the one described by Higuchi and others for release from polymeric matrices — "slab into the sink."[128,132,133]

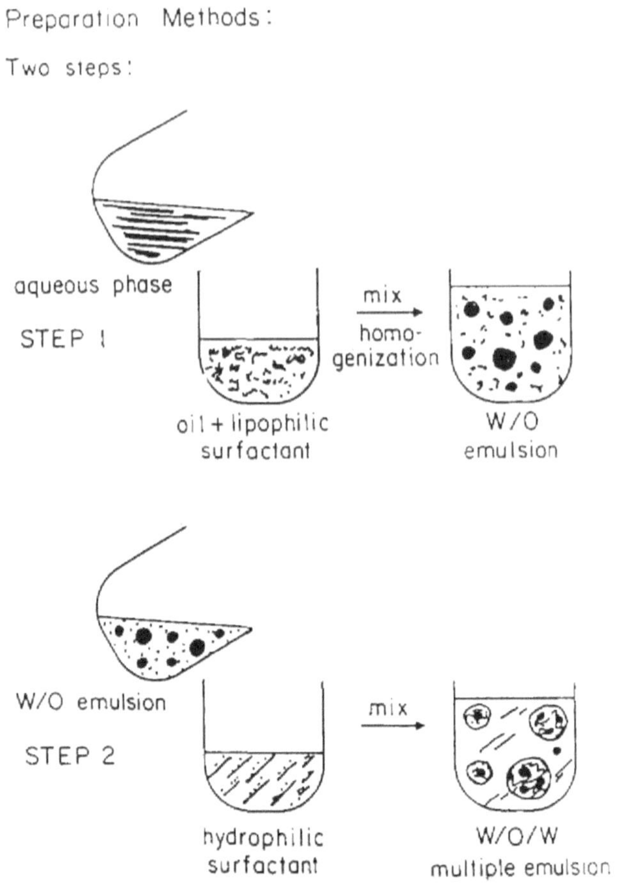

Figure 48 Two-step emulsification process for the preparation of double emulsion. In the first step, the w/o emulsion is formed. In the second step, the double emulsion is obtained.

A modified diffusion controlled equation was adapted to explain the release results. It has been demonstrated that the release factor B[129-131]:

$$\left(B = 3/2\left[1 - (1 - F)^{2/3}\right] - F = \frac{3Det}{r_0^{\,2}Co}\right)$$

where F is the release fraction of the marker soluble in the inner phase; De — effective diffusion coefficient; r_0 — radius of the droplets; t — time of release; Co — initial concentration of marker, can be plotted against 1/C and t with excellent correlation coefficients (Figure 61a–b). Such excellent fit suggests a "diffusion controlled" release mechanism of reverse micellar transport."[129,131] The effective, as well as the real diffusion coefficient, could, therefore, be calculated.

Although the release mechanism was clarified to some extent, it is still very difficult to slow or control the rate of release mainly due to the fact that the monomeric emulsifiers that have been used serve both as stabilizing moieties and as transport species. In addition, the fast exchange that those emulsifiers undergo between the two interfaces, and the fact that shear should be avoided in the second emulsification stage, lead to the formation of relatively large double emulsion droplets with very limited thermodynamic stability.

Efforts have been made during the last 10 years to replace the monomeric emulsifiers by macromolecular amphiphilic molecules both synthetic or natural. Florence et al.,[133,134] Dickinson et al.,[135] and Garti et al.[129] have used BSA as polymeric emulsifier both in the inner and in the outer phase and found that the BSA has a very beneficial effect on the emulsion stability when added mainly to the outer aqueous phase (Figure 62). The BSA was very helpful in slowing the release rates of the electrolytes and formed

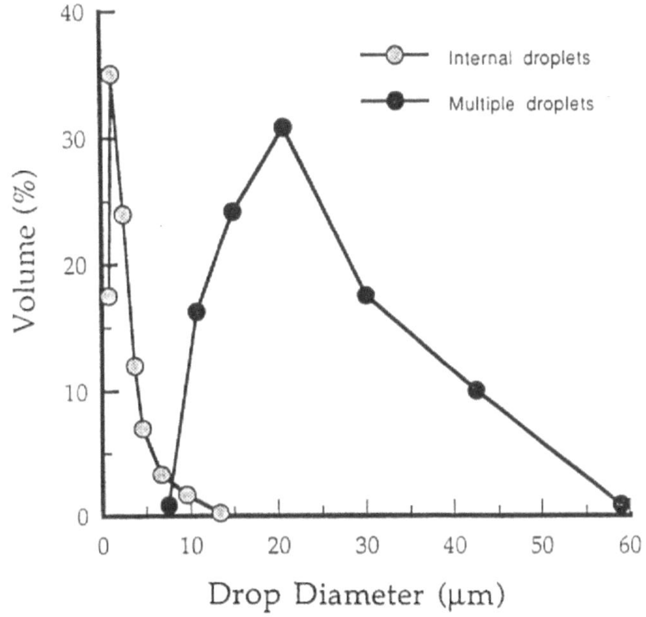

Figure 49 A typical droplet size distribution of the inner (w/o emulsion) and the outer (w/o/w emulsion) droplets in a typical multiple compartment double emulsion.

complex with the E_1 monomeric emulsifier that had better interfacial sealing effect. The release mechanisms were studied in view of the diffusion controlled reverse micellar release mechanism (Figure 62).

The use of polymeric emulsifiers opened a new era in the field of double emulsions and raised new hopes that the technique will eventually have food and pharmaceutical applications. The emulsions have longer shelf life, better stability, smaller droplets, better resistance to shear, and more controlled release rates. An example of double emulsion stabilized with polymeric emulsifiers in which the droplets are almost monodispersed and very small (3 to 5 μm) are shown in Figure 63.

IX. EMULSIFIED MICROEMULSIONS AND EMULSIFIED MICROSPHERES

Nanoparticles are solid colloidal particles ranging in size from 10 to 1000 nm. They consist mostly of macromolecular materials in which the active material is dissolved, entrapped, or encapsulated, and/or to which the active material is adsorbed or attached.[136] This definition includes "nanopellets," "nanocapsules" with a shell-like wall, as well as "microspheres," if they are below 10 nm in size. It is beyond the scope of this chapter to discus these important families of technologies. We will mention only those formulations in which solid particles, microspheres, have been used to improve the stability of the liquid formulations and mainly in order to improve the stability of the double emulsions.

Only limited efforts[137–140] have been made to stabilize double emulsions by encapsulating solid particles within the oil phase — emulsified microspheres (s/o emulsions).[137] The only study dealing with this technology does not give much detail regarding the experimental part, and therefore, it is difficult to evaluate the approach. More work is required in this area (Table 9).

X. FOOD APPLICATIONS

A. MICROEMULSIONS

In 1991, El-Nokaly and Cornell[37] attempted to compile and review the studies on microemulsions in food applications. In his introductory remarks, Becher (the editor) reports that a computer search based on the key words "food microemulsions," yielded only four references (as compared to 475 for "food emulsions"), indicating that microemulsions have not yet been extensively explored as a potential technology for food applications, and that the potential of these systems in food is limited. In fact, none of the books (except El-Nokaly and Cornell's) on food emulsions, including the most recent ones, contain any references to food microemulsions.

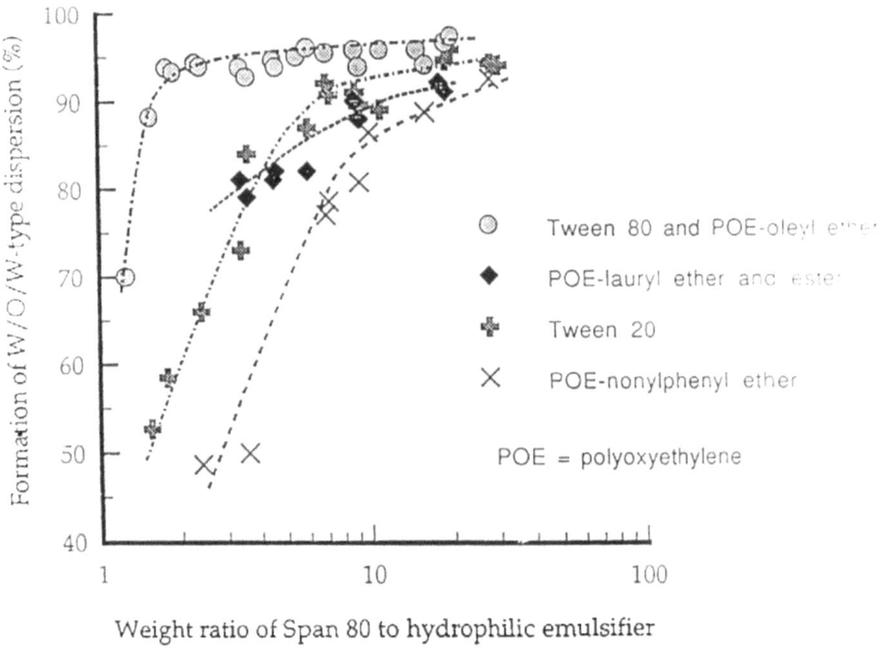

Figure 50 Yield of preparation (percent of w/o/w formation) of a double emulsion as a function of weight ratio of E1/E2 (weight ratio of Span 80® (Sorbitan monooleate) to various hydrophilic emulsifiers).

El-Nokaly and Cornell[37] compiled an earlier scientific work on the use of microemulsions of w/o for food application, mostly to entrap water soluble nutrients and flavors (see Table 10).

Even though much has been accomplished in recent years in the general field of microemulsions, there are very few available industrial examples of indigestible w/o or o/w microemulsion systems.

The scarcity of work available on water solubilization in triglycerides for food applications is obviously due to difficulties inherent in the structure of oil, in finding appropriate co-surfactants, and the need to use minimum amounts of food-approved surfactants (emulsifiers) not to adversely affect the oil properties.

Food microemulsions have been examined mostly from the descriptive point of view in the form of phase diagrams (Figure 64).[140] One can easily see that water in significant amounts can be solubilized in the oil phase in the presence of large amounts of amphiphiles.

These studies clearly indicate that food microemulsions are not difficult to formulate even with the limited number of food emulsifiers, and it is only a matter of needs and additional work in order before real food microemulsions systems in which an active matter will be entrapped in the microemulsion and delivered in a controlled way, will be offered to the marketplace. Few of the examples available in the literature are:

O/W microemulsions. (1) milk fortified with vitamin A solubilized in a microemulsion.[141](2) Dissolution of essential oils in water with and without an alcohol as co-surfactant for aromatization of beverages (Figure 65) or pharmaceutical formulations.[141–144]

Soybean oil (10%) was solubilized in water with a 30/70 surfactant mixture of Tween 20 and G1045 respectively (ICI, Inc.). This mixture had an HLB of 13.1.[145]

W/O systems. The literature available on w/o, the oil being triglycerides, microemulsion, is reviewed in Table 10. Medium chain alcohols or acids were needed as co-surfactant to form the microemulsion (with the exception of the work done by Larsson[146] and Trepow[144]). Such co-surfactants are usually not acceptable for taste, safety, or performance reasons in indigestible oil formulations and therefore can not be used.

Larsson[146] claimed to emulsify reversed micellar phases (L_2 or microemulsion) and cubic phases of lipids. The L_2 phase will then give a w/o/w structures. His early work will be further discussed together with the section on double emulsions.

El-Nokaly and Cornell[37] reported on the use of soybean oil and Crisco® oil, and a large number of food grade emulsifiers to form microemulsions. Some of the emulsifiers are

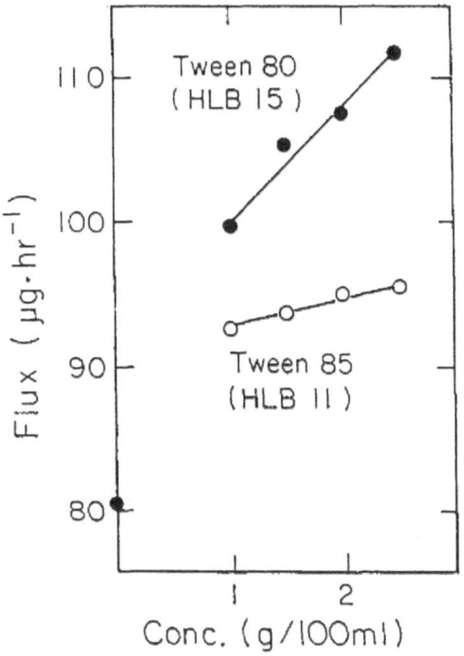

Figure 51 The effect of hydrophilicity of the external emulsifier and its concentration on the flux of additives from the inner phase to the outer phase.

- Polyglycerol oleate based on esterification of glycerol with oleic acid (min 92%).
- Polyglycerol linoleate based on esterification of polyglycerol with sunflower oil fatty acids. The average composition of the polyglycerol moiety in these esters is 18% di-, 18% tri-, 31% tetra-, 13% penta-, 10% hexa-, 2% hepta-, and the remaining 8% is glycerol and higher glycerols. The fatty composition is 36% mono- and 63% diester.
- Monoglyceride ester with fully unsaturated fatty acids known also as Dimodan LS.
- Polyoxyethylene sorbitol oleate (Atlas G1186, ICI).

The preparation of the w/o microemulsions is carried out by mixing the Crisco® oil, the water, and the polyglycerol polyricinoleate (PGPR) and constructing phase diagram. The diagram reveals that only minor amounts of water can be solubilized in any of the combinations of Crisco® oil and PGPR, thus the efficiency of such microemulsion is very limited.

Similar results were obtained with soybean oil and various food grade emulsifiers. Attempts were made to study the structure of these w/o microemulsions, but the results were inconclusive as to the presence of well-defined spherical aggregates.

El-Nokaly and Cornell[37]added water-soluble flavors and flavor enhancers to the microemulsions in very low levels (ppm's) (Table 11). The microemulsion had good flavor, low foaming, and was nonspattering during frying. It solubilized various water-soluble flavors and nutrients which could be delivered to a variety of food products. The authors claim that some flavor was released upon storage. No further details were given.

Osborne et al.[147] studied the SDS-sorbitan monolaurate phase behavior in relation to the microemulsion formation and stated that "while the microemulsion systems studied were intended for topical drug delivery vehicles, the AOT–Span 80® surfactant combination may be of interest for the food microemulsions." There are no commercial indications that such microemulsion is in use in any real food system.

Some additional microemulsions that have been mentioned as possible foods systems are

1. Wax microemulsions as citrus coatings[148]
2. Solubilization of vitamin[149]
3. Continuous glycerolisis of olive oil by lipase immobilizes in reversed micelles[150]
4. Microemulsions as flavoring materials using edible oils[151]

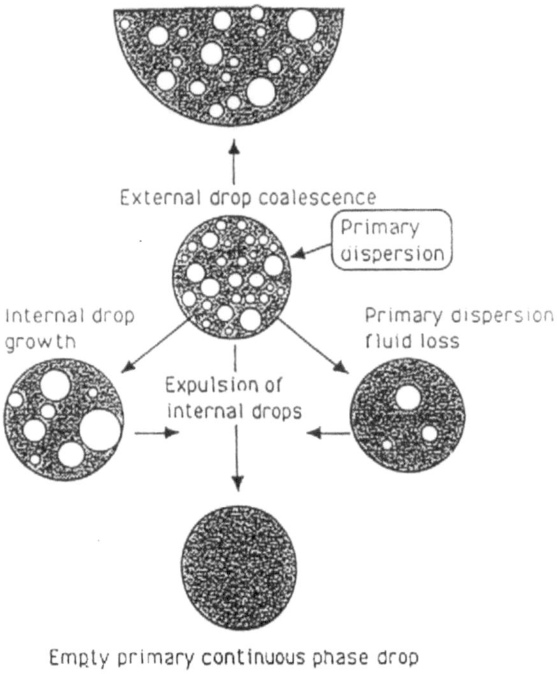

Figure 52 Possible pathways and mechanisms of release in double emulsions.

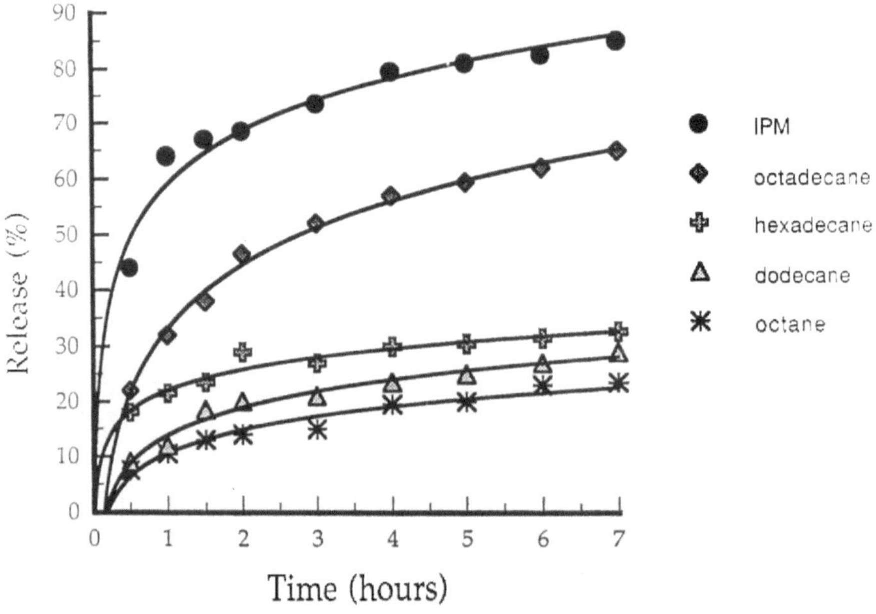

Figure 53 Rates of release of MTX from double emulsions prepared with BSA/Span 80® in the inner phase and various types of oils as the oil membrane (phase). The external emulsifier was Tween 80®.

5. Solubilization of essential oils for beverages[152]
6. Oxidation retardation of fish oils when stored in microemulsions[153]
7. Use of reverse micelles for extraction of oil and proteins from vegetable meal[154]

There are obvious advantages to having one stable phase which can solve hydrophilic as well as lipophilic substances, for example, in connection with flavoring. Larsson[155] was particularly interested in the use of L_2 phases in order to protect lipids against oxidation.

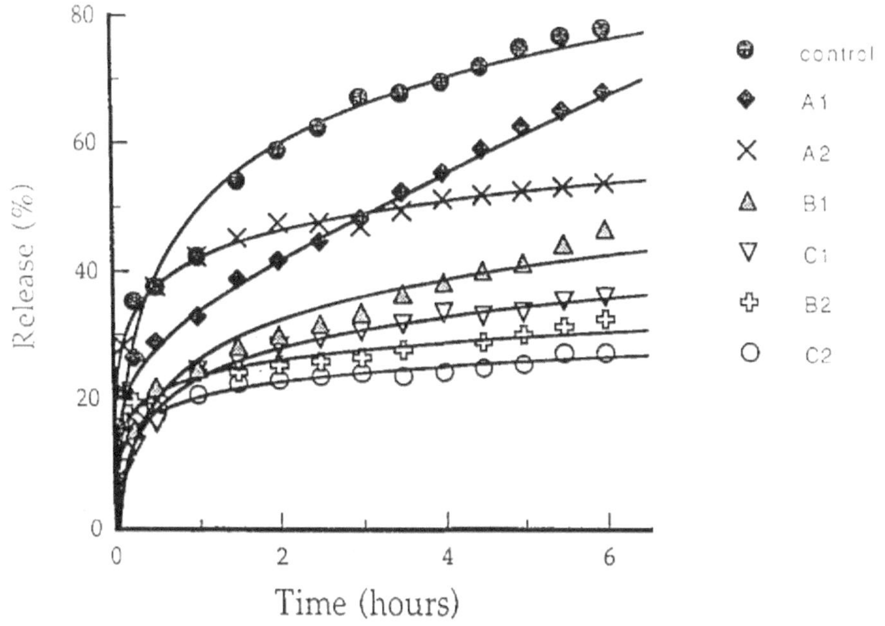

Figure 54 Double emulsions stabilized with different natural occurring macromolecules.

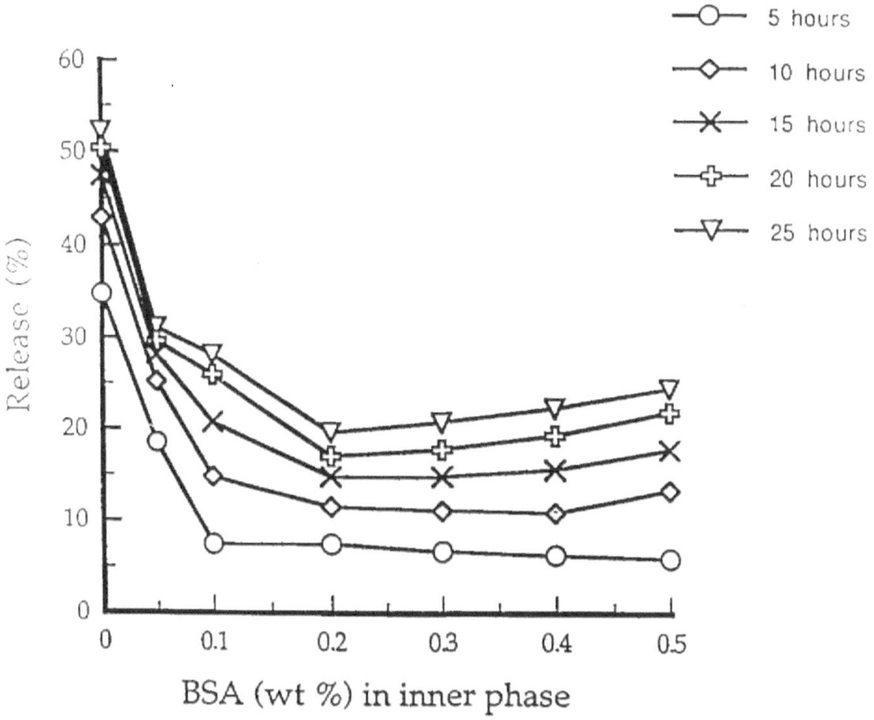

Figure 55 Plot of the percent release of NaCl vs. wt% of BSA, from double emulsion prepared with Span 80® and BSA in the inner phase and Tween 80® in the outer interface.

In order to simulate the lipid oxidation protection in a living system such as the cell membrane, it is necessary to keep tocopherol and ascorbic acid in molecular contact. The L_2 phase provides such a possibility. In our studies we used an L_2 phase able to solubilize 1% (w/w) of water, and in that water 5% (w/w) of ascorbic acid was solved. In the oil, a concentration of 200 ppm tocopherol was used.

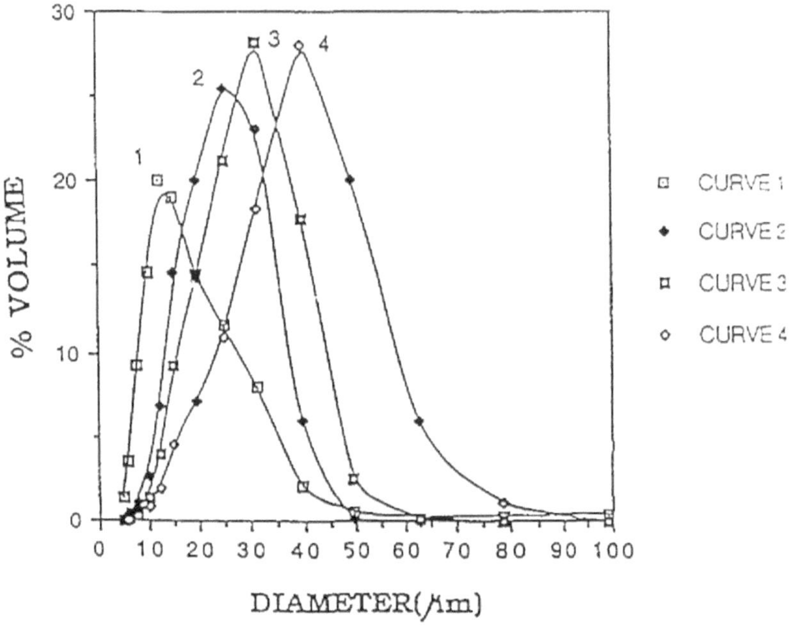

Figure 56 Percent release of NaCl as a function of BSA concentration from emulsions stabilized with Span 80® in the inner interface and Tween 80® and BSA in the outer interface.

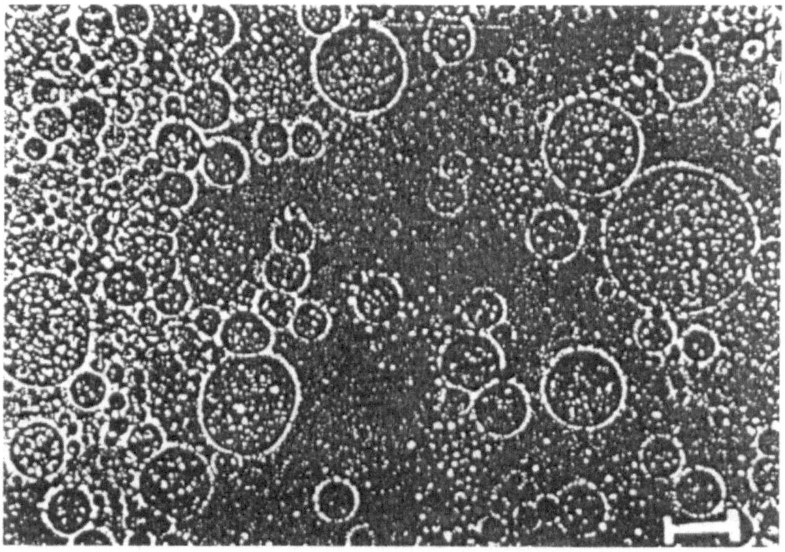

(a)

Figure 57 Photomicrographs of double emulsions stabilized with BSA and Span 80® in the inner interface and BSA and Tween 80® in the outer interface (a) immediately after preparation, (b) after one week, and (c) after 6 weeks.

Larsson[155] also mentions some possible applications for a dispersed cubic phase, which turns into liposomal stable structures.

Another application possibility is to introduce glucose oxidase into the phase in order to remove oxygen, for example, in the form of a surface layer on a food within a gas-proof packing material. Engström[156] has found that this enzyme keeps its activity for months when it is incorporated into this phase.

1. Microemulsions as Microreactors

Although microemulsions are thermodynamically stable, they are limited in their applications to industrialized foods, mostly because in practice, foods are very complicated systems, composed of many

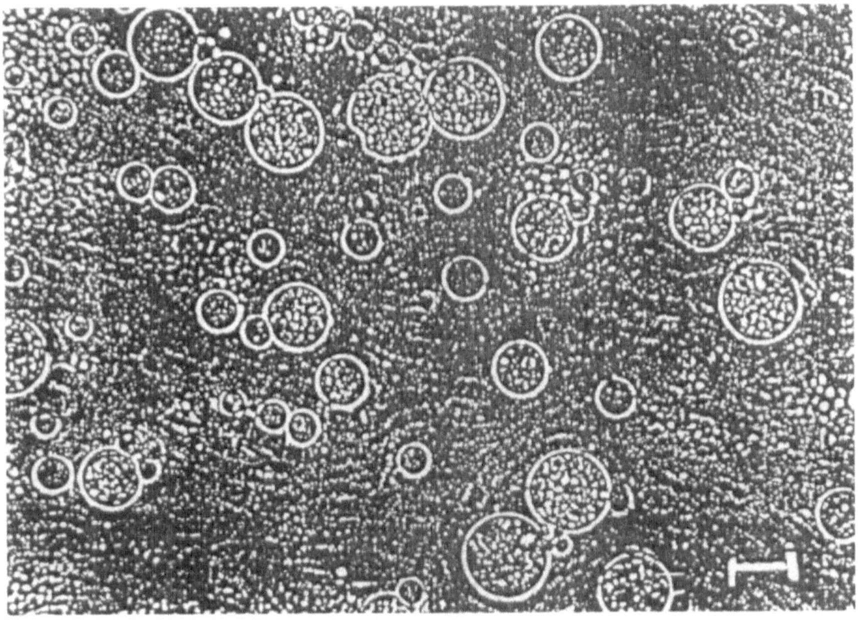

Figure 57(b)

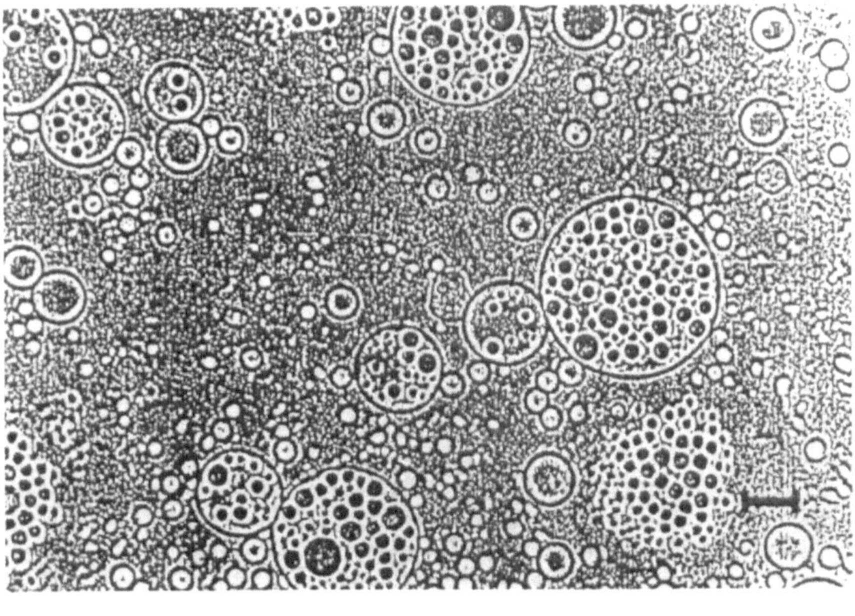

Figure 57(c)

ingredients that will destabilize the droplets and unbalance the equilibrium in which the oil-water-amphiphile must coexist.

However, microemulsions can serve as very attractive media for many enzymatic interfacial reactions to produce food ingredients, mostly by converting fats and oils into emulsifiers.[157–161]

Lecithin is a very inexpensive amphiphile with a variety of food applications. However, the complexity of its internal composition, the bitter taste that it imparts to many foods, and its limited hydrophilicity have limited its use in certain food applications. Efforts have been made to hydrolyze lecithin into lysolecithin[158] which is more hydrophilic and can thus serve as an emulsifier for o/w citrus emulsions and in baked goods (as an anti-stalling agent). Most of the past reactions were carried out in two phases either chemically or enzymatically. The processes were very inefficient and only low conversions were achieved. Recently,[158,162] we used the microemulsion composed of lecithin, vegetable

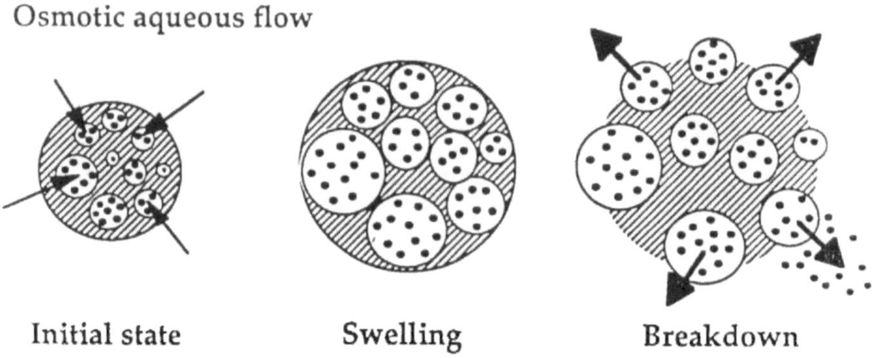

Figure 58 Diagram of the release mechanism by swelling–breakdown of a hydrosoluble molecule entrapped in the internal aqueous phase of the multiple emulsion.

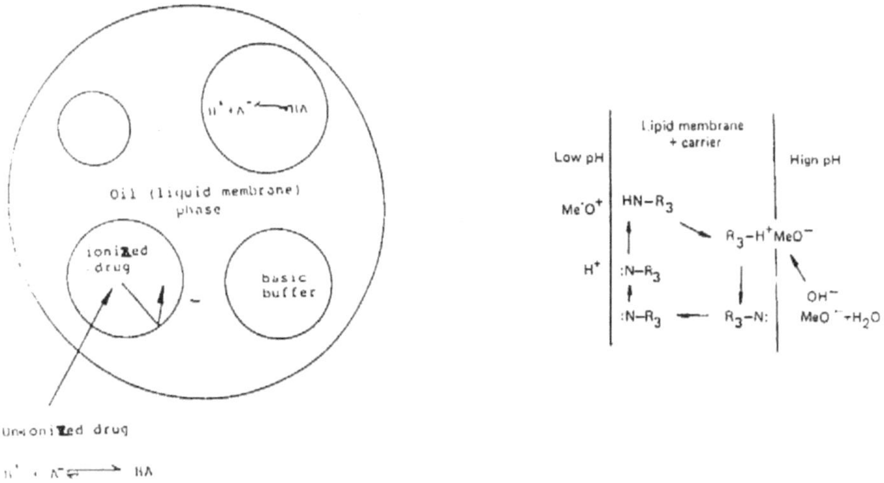

Figure 59 Schematic diagram for the facilitated transport of methyl orange (MeO) across a lipid barrier. The carrier is a long-chain tertiary alkyl amine (R_3N). Diffusion controlled mechanisms of ionized matter at given pH changing into unionized lipid soluble material at other pH. (The entrapping mechanism is also known as the liquid membrane mechanism.)

oil, and water as a microreactor and as a substrate. We managed to obtain full conversion of the lecithin to lysolecithin at high rates and with very high selectivity in the presence of specific PLA_2 enzyme solubilized in the inner phase of the water-in-oil microemulsion (see Figures 66 and 67). The reaction is influenced by the enzyme concentration, calcium concentration, type of the lecithin, and the nature of the microemulsion structure. Much work was done on the relationship between the rate of the emulsion and the enzyme reactivity.

Similar enzymatic processes are presently being studied for obtaining other food ingredients at high conversion rates and high selectivity.[162]

B. EMULSIONS

Processed foods are often colloidal systems such as dispersions, emulsions, and foams. Foods such as milk, cream, butter, ice cream, margarine, salad dressings, and mayonnaise are typical examples of foods in the form of emulsions. The emulsion can be also an intermediate step in food processing such as in cake mixes or coffee whiteners (dried emulsions). Krog and Lauridsen[47] summarized the typical food emulsions as shown in Table 12.

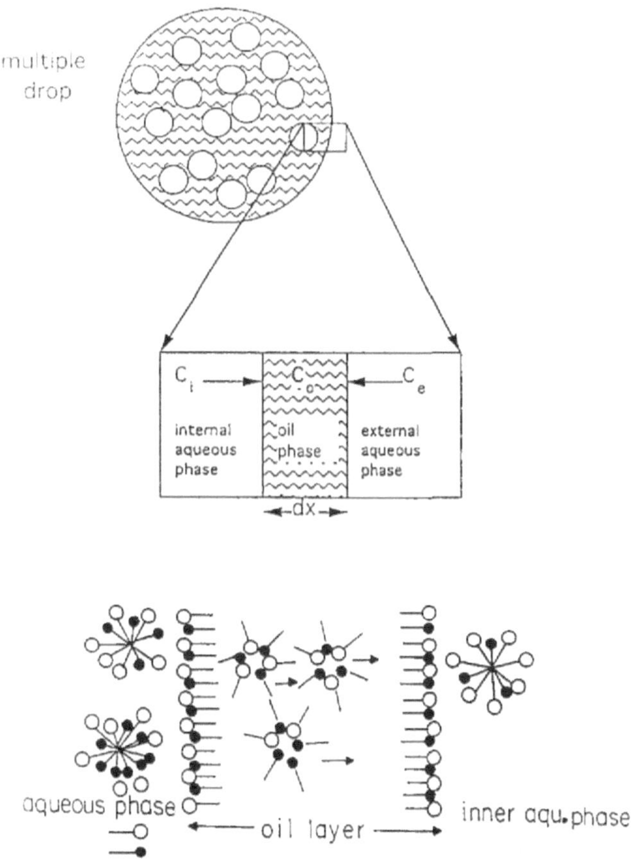

Figure 60 Diffusion controlled mechanisms of nonionized lipid-insoluble materials transferred to the external phase via: (a) reverse micelles, (b) through the external interface via a lamellar diffusion.

Some examples of man-made emulsions considered major achievements at the time they first appeared were the first butter substitute, oleomargarine, salad dressing and mayonnaise, reconstituted sausage, cake batters, and cake mixes.

It should be noted that most food emulsions are o/w and only a few are of w/o emulsions. Most emulsions are very complex systems and generally composed of fat, carbohydrates, and/or proteins. In addition, the system consists of various amounts of cellulose, gums, and other active ingredients that complicate the ability to carry out detailed scientific studies. In spite of the complexity scientists, by simulating the core of the systems, managed to learn much of the factors affecting the emulsions stability and managed to recommend the use of various emulsifiers (monomeric or proteins) to improve stability and to obtain the food with the required properties.

Recent studies[163] concentrate on: (1) the competitive adsorption of emulsifiers on oil/water interfaces; (2) the role of hydrocolloids in stabilizing the emulsions; (3) the use of various fat replacers in food systems; (4) the role of solid particles such as fats and sugars in stabilizing emulsions; (5) the formation of liquid crystalline phases in foods and man-made foods; (6) the effects of structure formation on food rheology and stability; and (7) methods to release addenda in a sustained and controlled manner.

Much work is done with attempts to replace many of the synthetic emulsifiers with natural occurring materials. New emulsifiers such as lysolecithin, glycolipids, saponins, etc., have been suggested for use in foods.

For more detailed information the reader is referred to more specific reviews such as: *Colloids in Food* (Dickinson and Stainsby[15]); *Advances in Food Emulsions and Foams* (Dickinson and Stainsby[14]);

Emulsions and Emulsion Technology, Part 1 (Finch and Griffin[164]); *Food Colloids and Polymers: Stability and Mechanical Properties* (Dickinson et al.[170]).

C. DOUBLE EMULSIONS

Food double emulsions are not yet completely commercial. Only a few products can be found on the shelves. However, many such products have been suggested in patents. Below are some of the most interesting suggested applications:

1. Flavor release from a protein stabilized double emulsion[165]
2. Controlled release of L-tryptophan and vitamin B_2[166]
3. Preparation of fine protein-stabilized w/o/w emulsion[167]
4. Oil-water-oil double emulsions composite as margarine[168]
5. Double emulsion spread preparation[169]
6. Double emulsion controlled nutrient release[170]

D. EMULSIFIED MICROEMULSIONS

Friberg[37] in his chapter on "surfactants, association structures, microemulsions and emulsions in food," discussed the problem related to the use of microemulsions in foods and suggested: "... the use of microemulsions as the inner phase for double emulsions, which can add to the stability of the double emulsions ..." because the w/o part, which is the difficult part to stabilize, would now be thermodynamically stable.

However, stabilization using surfactant combinations, as previously discussed, has fundamental difficulties. The surfactant combination for the inner microemulsion will rapidly exchange with the one for the external emulsion, which leads to destabilization of both the emulsion and the microemulsion.

The emulsion stabilizers that favor a lamellar packing of the surfactant at the interface are, hence, not suitable for the microemulsion; they will reduce the maximum solubilized water in it drastically, forming instead an unstable w/o emulsion. In addition, the microemulsions in combination will destroy the packing of surfactant/cosurfactant around the emulsion droplets; the layer will become more disordered and mobile and o/w emulsion stability will be lost also.

This dilemma has been resolved in an elegant manner by Larsson.[155] He used a surfactant combination to stabilize the w/o microemulsion but avoided the problem of the emulsion part by using a polymer as its stabilizing agent. The polymer, being water soluble, is virtually insoluble in the oil part of the microemulsion and its dimensions prevent its inclusion into the w/o droplets. It will, hence, not interfere with the w/o microemulsions stabilization system.

In addition, the polymer strongly adsorbs to an interface, thereby eliminating the competition about the emulsion interface from the surfactant combination, and hence, a very elegant solution to a difficult problem.

Close examination of Larsson's work on emulsified reversed micellar phases and aqueous dispersions of cubic phases of lipids reveals that the idea of emulsifying microemulsions for stability and controlled release properties was generated in Larsson's labs some 12 years ago.

Larsson explains that the technology of emulsification of oils has reached a high level of sophistication with regard to processing techniques as well as the use of various emulsifiers. Less is known concerning dispersion of the lamellar liquid-crystalline phase (L_α) in excess water into a liposomal dispersion. One obvious requirement is that the lamellar phase is not allowed to exhibit "infinite" swelling, and therefore anionic or cationic lipids cannot be used. It should be pointed out that liposomes in this context are defined as multilayer particles contrary to vesicles, which are closed unilayers.

There are three other types of lipid phases (in addition to the crystalline state), which can coexist in equilibrium with an excess of water and thus can form aqueous dispersions. Two are liquid-crystalline; the reversed hexagonal (H_{II}) and the cubic types. The third is the reversed micellar phase, sometimes microemulsion type, which is termed the L_2 phase. Two of these phases are particularly interesting from an application point of view, the dispersion of cubic phases and emulsions of L_2 phases.

In order to produce stable emulsions, various proteins were tried. The best result was obtained by sodium caseinate. Even if the additional reduction in surface tension was not very high, the final value was about 1 mN/m^{-1}, the emulsions obtained were quite stable. Obviously this protein is irreversibly adsorbed at the interface with many segments. With a concentration of 1% protein in the outside water medium, quite stable emulsions of about 10% (w/w) L_2 phase in water was obtained. Such an emulsion can also be described as a w/o/w emulsion.

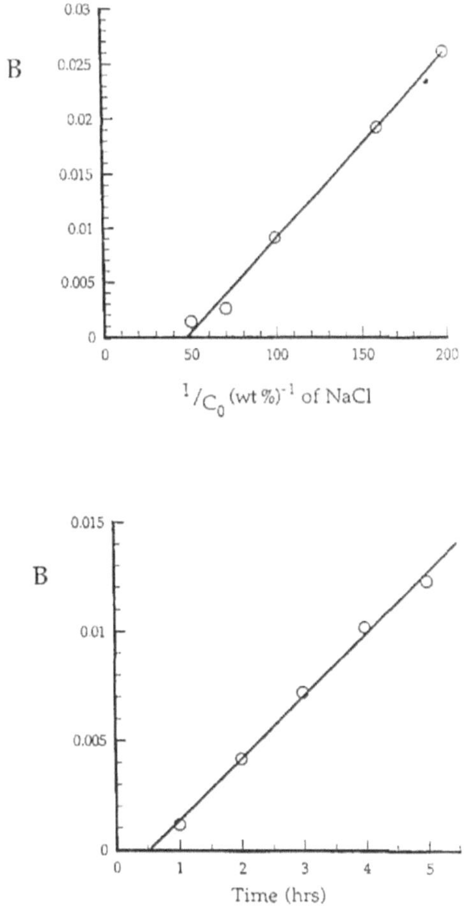

Figure 61 Rate of release of electrolytic addenda (factor B, see text) as a function of: (a) $1/C_0$ and (b) time (t).

Figure 62 Schematic illustration of possible organization and the stabilization mechanism of BSA and monomeric emulsifiers (Span 80) at the two interfaces of the double emulsion.

In a similar way, cubic phases of monoglycerides in water (note that no oil was used) were prepared.

Cubic phases are found in aqueous systems of most types of polar lipids, such as monoglycerides, phospholipids, and galactolipids. They have a remarkable structure, which best can be described as infinite periodic minimal surfaces. Such a surface has zero average curvature everywhere, i.e., it is as convex as it

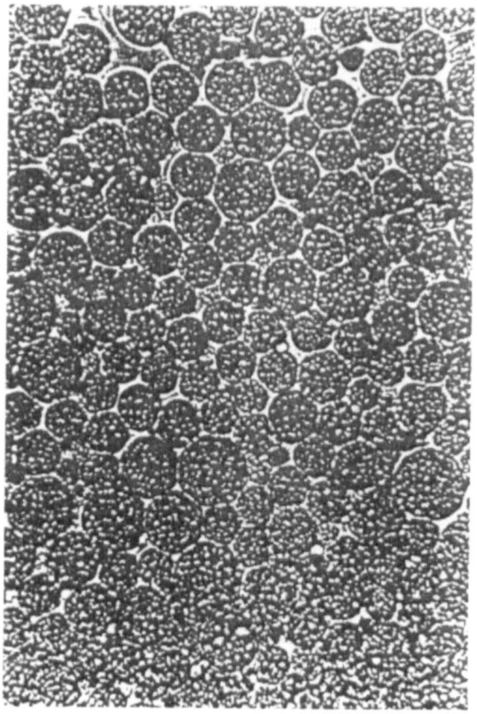

Figure 63 Photomicrographs of w/o/w double emulsion stabilized with polymeric emulsifiers.

Table 9 Composition of W/O Emulsion and S/O Emulsion

W/O Emulsion		S/O Emulsion (microspheres in oil)	
Water phase (anticancer agent)	0.35 ml	Water phase (anticancer agent)	0.35 ml
Distilled water	q.s.	Gelatin	70 mg
Oil phase	2.0 ml	Distilled water	q.s.
Sesame oil	1.84 ml	Oil phase	2.0 ml
SO-15	0.13 ml	Sesame oil	1.84 ml
HCO-60	0.03 ml	SO-15	0.13 ml
		HCO-60	0.03 ml
Total volume	2.35 ml	Total volume	2.35 ml

Note: SO-15 is soybean-oil; HCO is hydrogenated castor oil.

is concave in all positions on the surface. Such periodic surfaces without intersections were described in mathematics during the last century. From X-ray space-group symmetry, electron microscopy data, NMR diffusion data, etc., it is now clear that the cubic phases of lipids are ordered in just this way.

If the cubic phase is stirred in excess water, a kinetically stable dispersion is obtained under conditions when there is an equilibrium between the cubic phase, the lamellar phase, and water. Such an equilibrium can be obtained in the ternary system of bile salts/monoolein/water or in phosphatidylcholine/monoolein/water. The stability of these dispersions is comparable to liposomal dispersions, and they can also be obtained in a similar size distribution, related to the mechanical energy input.

From X-ray and optical studies of such dispersions it is clear that the core has a cubic structure with its outer bilayer fused with the spherical inner bilayer of a liposome.[155] These particles were, therefore, termed "cubosomes."[155] In the extreme case there might be only one such outer layer. Processing aspects on the dispersion technology is further discussed below.

XI. CONCLUSIONS

Only a few studies on microemulsions, double emulsions, emulsified microemulsions, and dispersed lyotropic liquid crystals in real or man-made foods are available in the reviewed journals and only very few commercial systems are available in the marketplace, indicating that although the systems may have

Table 10 Examples of Water-in-Triglycerides Microemulsions

Oil	Surfactant	Co-Surfactant
Soybean oil	POE sorbitol oleate	—
Tricaprylin	Alkyl aryl polyglycol ether	Pentanol
Trioctanoin	Sodium octanoate (soap)	Octanoic acid
Soybean oil	Monoglyceride	—
1,2,3-[Tris(2 ethyl-hexanoyloxy)] propane	Tetraethylene glycol dodecyl ether ($R_{12}EO_4$)	Hexadecanol
Canola oil	Monoglyceride acetic acid ester of monoglyceride	Tert-butanol isopropanol 1-hexanol
Triglyceride	None[a]	Ethanol, 1-butanol
Soybean oil	0-alkyl-3 D-glucose	Ethanol

[a] Detergentless microemulsion.

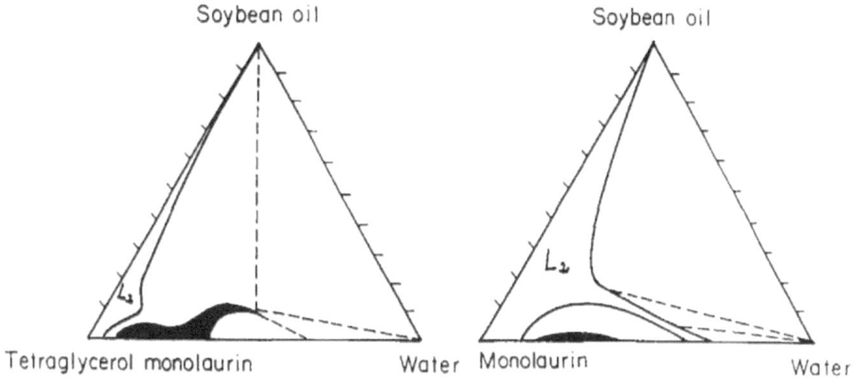

Figure 64 Ternary-phase diagram of soybean oil/water and (a) tetraglycerol monolaurin, (b) glycerol monolaurate at 25°C. L2 is the microemulsion area.

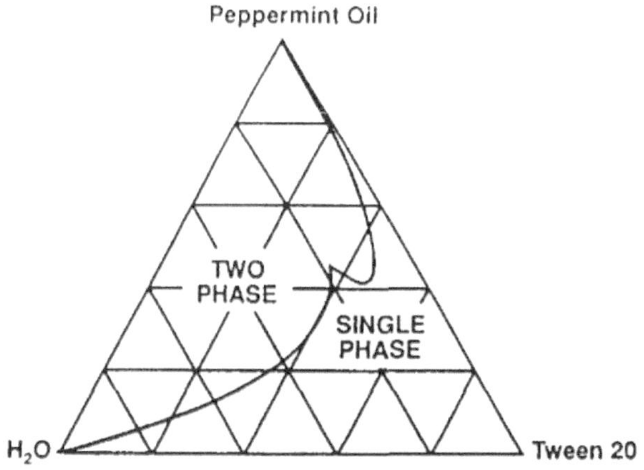

Figure 65 Phase diagram of a peppermint oil/Tween 20/water system at 250°C showing the single microemulsion phase region.

potential in food application to serve as delivery systems, actual use is so far very limited. Microemulsions are thermodynamically stable, but exchange the solubilized matter with the continuous phase very quickly and no significant control can be obtained. In addition, the solubilization capacity is still quite limited.

Emulsions, on the other hand, have been extensively studied and used in a variety of foods as delivery systems, but are found only in very limited applications as controlled delivery systems because of the fast exchange of the emulsifiers and thermodynamic instability.

Emulsified microemulsions and dispersed lyotropic mesophases have significant potential, but have not been fully explored and much work remains to be done prior to using them in real food systems.

Double emulsions are excellent systems for controlled and slow-release of active matter from the inner phase, through the oil phase, to the outer continuous phase. Double emulsions have also high dissolution

Table 11 Water-Soluble Flavors Delivered by Crisco® Oil/Water/Polyglycerol Linoleate Microemulsion

Water Soluble Flavor	Amount in Oil (ppm)
Ribotide	1000
Furaneol	10
Cyclotene	100
Methionine sulfoxide	100
Caramel furanone	100
S-methyl methionine	100
3-hydroxy-4,5-dimethyl-2(5H)-furanone	100
Natural butter flavor	1000
Honey flavor	1000
Fructose	5000
Others	
Enzymes, sugars, salts, citric, or ascorbic acids	

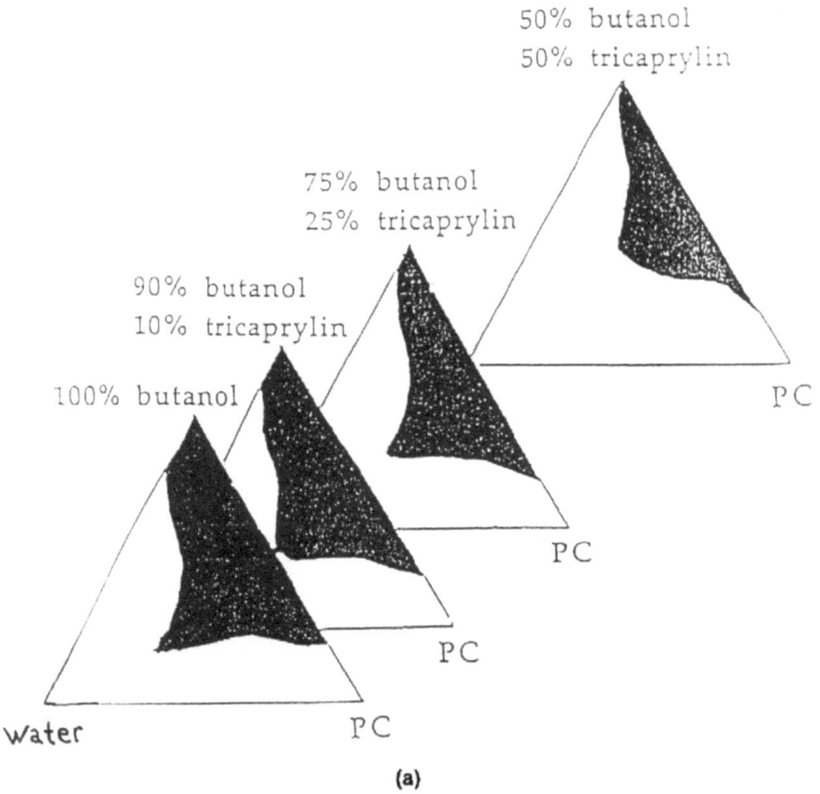

(a)

Figure 66 (a) Phase diagram of phosphatidylcholine, tricaprylin, butanol, and water. The L_2 area is the w/o microemulsion in which the hydrolysis reaction was carried out. (b) The compositions of the reaction mixtures in which hydrolysis was carried out. The reaction was carried out at constant PC concentration and with increased water concentrations and decreased oil phase concentrations.

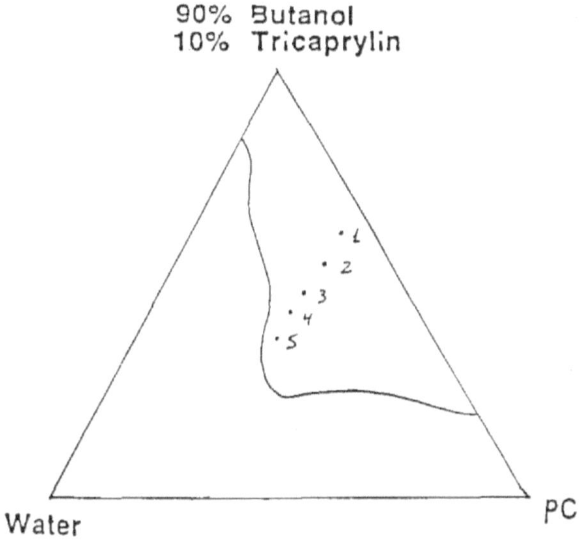

Figure 66(b)

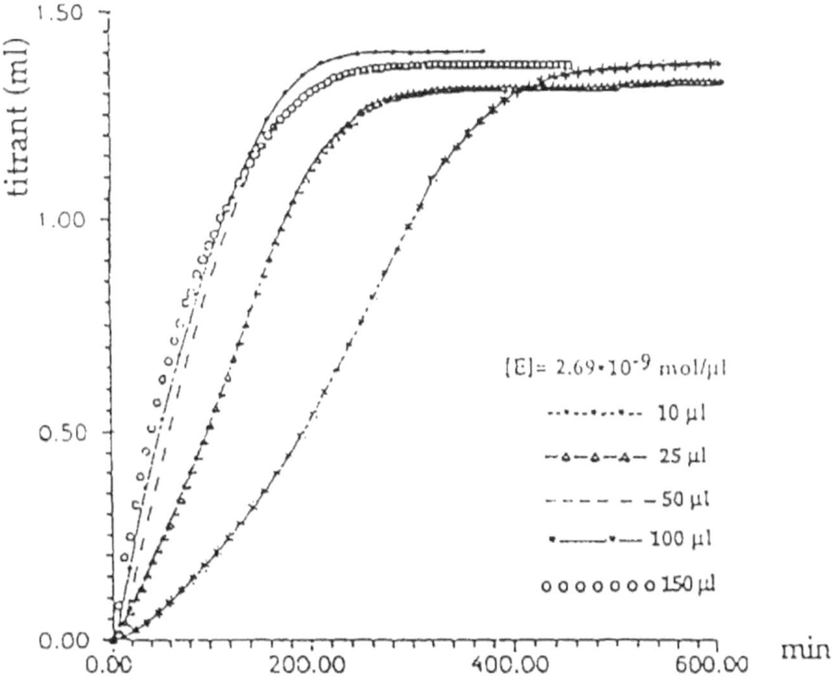

Figure 67 Microemulsion as microreactor; the hydrolysis rate of lecithin converted to lysolecithin in the L_2 phase.

Table 12 Typical Food Emulsions[47]

Product	Emulsion Type	Composition of Dispersed Phase	Composition of Continuous Phase	Stabilization Factors, etc.
Milk, cream	O/W	Butterfat triglycerides, partially crystalline and liquid oils Droplet size: 1–10 µm Volume fraction: Milk: 3–4%; Cream: 10–30%	Aqueous solution of milk proteins, salts, minerals, etc.	Lipoprotein membrane, phospholipids, and adsorbed casein
Ice cream	O/W (aerated to a foam)	Butterfat (cream) or vegetable, partially crystallized fat Volume fraction of air phase: approx. 50%	Water and ice crystals, milk proteins, carboxydrates (sucrose, corn syrup) Approx. 85% of the water content is frozen at −20°C	The foam structure is stabilized by agglomerated fat globules forming the surface of air cells Added surfactants act as "destabilizers" controlling fat agglomeration. Semisolid frozen continuous phase
Butter	W/O	Buttermilk: milk proteins, phospholipids, salts Volume fraction: 16%	Butterfat triglycerides, partially crystallized and liquid oils; genuine milk fat globules are also present	Water droplets distributed in semisolid, plastic continuous fat phase
Imitation cream (to be aerated)	O/W	Vegetable oils and fats Droplet size: 1–5 µm Volume fraction: 10–30%	Aqueous solution of proteins (casein), sucrose, salts, hydrocolloids	Before aeration: adsorbed protein film After aeration: the foam structure is stabilized by aggregated fat globules, forming a network around air cells; added lipophilic surfactants promote the needed fat globule aggregation
Coffee whiteners	O/W	Vegetable oils and fats Droplet size: 1–5 µm Volume fraction: 10–15%	Aqueous solution of proteins (sodium caseinate), carbohydrates (maltodextrin, corn syrup, etc.), salts, and hydrocolloids	Blends of nonionic and anionic surfactants together with adsorbed proteins
Margarine and related products (low-calorie spread)	W/O	Water phase, may contain cultured milk, salts, flavors Droplet size: 1–20 µm Volume fraction: 16–50%	Edible fats and oils, partially hydrogenated, of animal or vegetable origin Colors, flavor, vitamins	The dispersed water droplets are fixed in a semisolid matrix of fat crystals; surfactants added to reduce surface tension/promote emulsification during processing
Mayonaise	O/W	Vegetable oil Droplet size: 1–5 µm Volume fraction: Minimum 65% (U.S. food standard)	Aqueous solution of egg yolk, salt, flavors, seasonings, ingredients, etc. pH: 4.0–4.5	Egg yolk proteins and phosphatides
Salad dressing	O/W	Vegetable oil Droplet size: 1–5 µm Volume fractions: Minimum 30% (U.S. food standard)	Aqueous solutions of egg yolk, sugar, salt, starch flavors, seasonings, hydrocolloids, and acidifying ingredients pH: 3.5–4.0	Egg yolk proteins and phosphatides combined with hydrocolloids and surfactants, where permitted by local food law

capacity, but the systems suffer from thermodynamic instability, large droplets, and lack of sufficient steric stabilizers (polymeric materials) to control the release. Encapsulated microspheres and nanoparticles have a better chance to achieve controlled release, but have the disadvantage of being, in part, solids.

It is our belief that double emulsions are the future systems for delivery of active matter in the most sophisticated future food products. The use of food proteins to stabilize the double emulsion is a significant contribution toward achieving this goal.

REFERENCES

1. **Davis, S. S., Hadgraft, J., and Palin, K. J.,** Medical and pharmaceutical applications of emulsions, in *Encyclopedia of Emulsion Technology*, Vol. 2, Becher, P., Ed., Marcel Dekker, New York, 1985, 159.
2. **Tadros, Th. F. and Vincent, B.,** Emulsion stability, in *Encyclopedia of Emulsion Technology*, Vol. 1, Becher, P., Ed., Marcel Dekker, New York, 1983, 129.
3. **Davis, S. S.,** *Optimization of Drug Delivery*, Bundgaard, H., Bagger-Hansen, A. and Kofod, A., Eds., Munksgaard, Copenhagen, 1982, 198.
4. **Bierman, E. L.,** *Fat Emulsions in Parenteral Nutrition*, Meng, H. C. and Wilmore, D. W., Eds., American Medical Association, Chicago, 1976, 1.
5. **Tornberg, E., Olsson, A., and Persson, K.,** The structural and interfacial properties of food proteins in relation to their function in emulsions, in *Food Emulsions*, Larsson, K. and Friberg, S. E., Eds., Marcel Dekker, New York, 1990, 247.
6. **Krog, N. J., Riisom, T. H., and Larsson, K.,** Applications in the food industry: I, in *Encyclopedia of Emulsion Technology*, Vol. 2, Becher, P., Ed., Marcel Dekker, 1985, 321.
7. **Jaynes, E. N.,** Applications in the food industry: II, in *Encyclopedia of Emulsion Technology*, Vol. 2, Becher, P., Ed., Marcel Dekker, 1985, 367.
8. **Graham, H. D.,** *Food Colloids*, Avi Publishers, Westport, CT, 1977.
9. **Silberstein, T.,** The hydrolysis of phosphatidyl choline by phospholipase A_2 in microemulsion as microreactors, Ph.D. Thesis, the Hebrew University of Jerusalem, 1995.
10. **Bee, R. D., Richmond, P., and Mingins, J., Eds.,** *Food Colloids*, Royal Society Chemistry, Bristol, U.K., 1989.
11. **Larsson, K. and Friberg, S. E., Eds.,** *Food Emulsions*, Marcel Dekker, New York, 1990.
12. **Dickinson, E., Ed.,** *Food Emulsions and Foams*, Royal Society Chemistry, London, 1987.
13. **Schuster, G., Ed.,** *Emulgatoren fur Lebensmittel*, Springer-Verlag, Berlin, 1985.
14. **Dickinson, E. and Stainsby, G., Eds.,** *Advances in Food Emulsions and Foams*, Elsevier Applied Science, London, 1988.
15. **Dickinson, E. and Stainsby, G., Eds.,** *Colloids in Food*, Applied Science Publishers, London, 1982.
16. **Dickinson, E., Ed.,** *Food Polymers, Gels and Colloids*, Royal Society Chemistry, London, 1991.
17. **Myers, D.,** Surfaces, *Interfaces and Colloids*, VCH, New York, 1991.
18. **Friberg, S. E. and Lindman, B., Eds.,** *Organized Solutions*, Vol. 44, Surfactant Science Series, Marcel Dekker, New York, 1992.
19. **Laughlin, R. G.,** Phase equilibria and mesophases in surfactant systems, in *Surfactants*, Tadros, Th. F., Ed., Academic Press, London, 1984, 53.
20. **Hunter, R. J.,** *Foundations of Colloid Science*, Vol. 1, Clarendon Press, Oxford, 1989, 564.
21. **Rosen, M. J.,** *Surfactants and Interfacial Phenomena*, John Wiley & Sons, New York, 1989.
22. **Bourrel, M. and Schechter, R. S.,** *Microemulsions and Related Systems*, Vol. 30, Surfactant Science Series, Marcel Dekker, New York, 1988, 31.
23. **Winsor, P. A.,** *Chem. Rev.*, 68, 1, 1968.
24. **Mallioris, A., Lang, J., and Zana, R.,** *J. Chem. Phys.*, 90, 655, 1986.
25. **Ogino, K. and Abe, M., Eds.,** *Mixed Surfactant Systems*, Vol. 46, Surfactant Science Series, Marcel Dekker, New York, 1992.
26. **Clint, J. H.,** *J. Chem. Soc. Faraday Trans. I*, 71, 1327, 1975.
27. **Lange, H. and Beck, K.,** *Kolloid Z. Z. Polym.*, 251, 424, 1973.
28. **Rubingh, D. N.,** Mixed micelle solutions, in *Solution Chemistry of Surfactants*, Vol. 1, Mittal, K. L., Ed., Plenum Press, New York, 1979, 337.
29. **McBain, J. W.,** *Colloid Science*, D. C. Heath, Boston, 1950, 59.
30. **Hoar, T. P. and Schulman, J. H.,** *Nature*, 152, 102, 1943.
31. **Neuberg, C.,** *Biochem. Z.*, 76, 107, 1916.
32. **Shinoda, K. and Friberg, S. E.,** *Emulsions and Solubilization*, John Wiley & Sons, New York, 1986.
33. **Hartley, G. S.,** *Rep. Prog. Appl. Chem.*, 45, 33, 1948.
34. **Rosano, H. L. and Clausse, M., Eds.,** *Microemulsion Systems*, Vol. 24, Surfactant Science Series, Marcel Dekker, New York, 1987.
35. **Bourrel, M. and Schechter, R. S.,** *Microemulsions and Related Systems, Formulation, Solvency and Physical Properties*, Vol. 30, Surfactant Science Series, Marcel Dekker, New York, 1988, 127.

36. **Holt, S. L.,** *J. Dispersion Sci. Technol.*, 1, 423, 1980.
37. **El-Nokaly, M. and Cornell, D., Eds.,** *Microemulsions and Emulsions in Foods*, American Chemical Society, Washington, D.C., 1991.
38. **Friberg, S. E. and Venable, R. L.,** Microemulsions, in *Encyclopedia of Emulsion Technology*, Vol. 1., Becher, P., Ed., Marcel Dekker, New York, 1983, 287.
39. **Zana, R., Ed.,** *Surfactant Solutions: New Methods of Investigation*, Vol. 22, Surfactant Science Series, Marcel Dekker, New York, 1987.
40. **Prince, L. M., Ed.,** *Microemulsions: Theory and Practice*, Academic Press, New York, 1977.
41. **Robb, I. D., Ed.,** *Microemulsions*, Plenum Press, New York, 1982.
42. **El-Nokaly, M., Hiler, G., Sr., and McGrady, J.,** Solubilization of water and water-soluble compounds in triglycerides, in *Microemulsions and Emulsions in Foods*, El-Nokaly, M. and Cornell, D., Eds., American Chemical Society, Washington, D.C., 1991, 26; Mazur, A. W., Hiler, G. D., Sr., and El-Nokaly, M., Preparation of 2-monoglycerides, in *Microemulsions and Emulsions in Foods*, El-Nokaly, M. and Cornell, D., Eds., American Chemical Society, Washington, D.C., 1991, 51.
43. **Larsson, K.,** Emulsions of reversed micellar phases and aqueous dispersions of cubic phases of lipids: some food aspects, in *Microemulsions and Emulsions in Foods*, El-Nokaly, M. and Cornell, D., Eds., American Chemical Society, Washington, D.C., 1991, 44.
44. **Larsson, K.,** *J. Dispersion Sci. Technol.*, 1, 267, 1980.
45. **Ekwall, P.,** Composition, properties and structures of liquid crystalline phases in systems of amphiphilic compounds, in *Advances in Liquid Crystals*, Vol. 1, Brown, G. H., Ed., Academic Press, New York, 1975, 1.
46. **Brown, G. H. and Wolken, J. J.,** *Liquid Crystals and Biological Structures*, Academic Press, London, 1979.
47. **Krog, N. and Lauridsen, J. B.,** Food emulsifiers and their associations with water, in *Food Emulsions*, Friberg, S. E., Ed., Marcel Dekker, New York, 1976, 67.
48. **Luzzatti, V. and Husson, F.,** *J. Cell. Biol.*, 12, 207, 1962.
49. **Larsson, K.,** *J. Phys. Chem.*, 73, 7304, 1989.
50. **Larsson, K.,** *J. Colloid Interface Sci.*, 113, 299, 1986.
51. **Larsson, K. and Dejmek, P.,** Crystal and liquid crystal structures of lipids, in *Food Emulsions*, Larsson, K. and Friberg, S. E., Eds., Marcel Dekker, New York, 1990, 97.
52. **Bourrel, M. and Schechter, R. S.,** *Microemulsions and Related Systems*, Vol. 30, Marcel Dekker, New York, 1988, 102.
53. **Lutton, E. S.,** *J. Am. Oil Chem. Soc.*, 42, 1068, 1965.
54. **Larsson, K.,** *Zeit. Phys. Chem. Neue Folge*, 56, 173, 1967.
55. **Krog, N. and Larsson, K.,** *Chem. Phys. Lipids*, 2, 129, 1968.
56. **Friberg, S. E. and Mandrell, L.,** *J. Am. Oil Chem. Soc.*, 47, 149, 1970.
57. **Bourrel, M. and Schechter, R. S.,** *Microemulsions and Related Systems*, Vol. 30, Surfactant Science Series, Marcel Dekker, New York, 1988, 127.
58. **Polatnik, L. S. and Landau, A. I.,** *Phase Equilibria Multicomponent Systems*, Holt, Rinehart and Winston, New York, 1964, 247.
59. **Winsor, P. A.,** *Trans. Faraday Soc.*, 44, 376, 1984.
60. **Winsor, P. A.,** *Solvent Properties of Amphiphilic Compounds*, Butterworth, London, 1954.
61. **Walstra, P.,** Formation of emulsions, in *Encyclopedia of Emulsion Technology*, Vol. 1, Becher, P., Ed., Marcel Dekker, New York, 1983, 57.
62. **Torenberg, T.,** *Functionality and Protein Structure*, ACS Symp. Ser. 92, American Chemical Society, Washington, D.C., 1979, 105.
63. **Boyd, J., Parkinson, C., and Sherman, P.,** *J. Colloid Interface Sci.*, 41, 359, 1972.
64. **Clausse, D.,** Research techniques utilizing emulsions, in *Encyclopedia of Emulsion Technology*, Vol. 2, Becher, P., Ed., Marcel Dekker, New York, 1985, 77.
65. **Becher, P.,** *Emulsions: Theory and Practice*, Krieger Publications, New York, 1977, 267.
66. **Bergenstahl, B.,** *Topics in Food Emulsions*, Doctoral thesis, University of Lund, Lund, 1994.
67. **Greenwald, H. L.,** *J. Soc. Cosmetic Chem.*, 6, 164, 1955.
68. **Smoluchowski, M.,** *Phys. Z.*, 17, 593, 1916.
69. **Spielman, L. A.,** *J. Colliod Interface Sci.*, 33, 562, 1970.
70. **Smoluchowski, M.,** *Z. Phys. Chem.*, 92, 129, 1917.
71. **Saffman, P. F. and Turner, J. S.,** *J. Fluid Mech.*, 1, 16, 1956.
72. **Melik, D. H. and Fogler, H. S.,** *J. Colloid Interface Sci.*, 101, 72, 1984.
73. **Melik, D. H. and Fogler, H. S.,** *Encyclopedia of Emulsion Technology*, Vol. 3, P. Becher, Ed., Marcel Dekker, 1988, 3.
74. **Van den Tempel, M.,** *Rec. Trav. Chim.*, 72, 433, 1953.
75. **Griffin, W. C.,** *J. Soc. Cosmetic Chemist*, 5, 311, 1949.
76. **Griffin, W. C.,** *J. Soc. Cosmetic Chemist*, 5, 249, 1954.
77. **Davis, S. S.,** *Interfacial Phenomena*, Academic Press, New York, 1963, 371.
78. **Moore, C. D. and Bell, M.,** *Soap, Perfum. Cosmet.*, 29, 893, 1956.

79. **Greenwald, H. L., Brown G. L., and Fineman, M. N.,** *Anal. Chem.*, 28, 1693, 1956.
80. **Shinoda, K. and Arai, H.,** *J. Phys. Chem.*, 68, 3485, 1964.
81. **Shinoda, K. and Shaito, H.,** *J. Colloid Interface. Sci.*, 26, 70, 1968.
82. **Shinoda, K. and Kunieda, H.,** Phase properties of emulsions, in *Encyclopedia of Emulsion Technology*, Vol. 1, Becher, P., Ed., Marcel Dekker, New York, 1983, 337.
83. **Tornberg, E., Olsson, A., and Persson, K.,** The stuctural and interfacial properties of food proteins in relation to their function in emulsions, in *Food Emulsions*, Larsson, K. and Friberg, S. E., Ed., Marcel Dekker, New York, 1990, 247.
84. **Kato, A. and Nakai, S.,** *Biochem. Biophys. Acta*, 624, 13, 1980.
85. **Townsend, A. and Nakai, S. J.,** *Food Sci.*, 48, 588, 1983.
86. **Keshavarz, E. and Shuryo, N.,** *Biochem. Biophys. Acta*, 576, 269, 1979.
87a. **Peters, T. J.,** *Adv. Protein Chem.*, 37, 161, 1985.
87b. **Wahlgren M., Paulsson, M., and Arnebrant, T.,** *Colloids Surfaces*, 70, 139, 1993.
88. **Dickinson, E., Murray, B. S., Stainsby, G.,** Protein adsorption at air-water and oil-water interfaces in, *Advances in Food Emulsions and Foams*, Dickinson, E. and Stainsby, G., Eds., Elsevier Applied Science, London, 1988 123.
89. **Dickinson, E.,** *An Introduction to Food Colloids*, Oxford Science Publishers, Oxford, 1992.
90. **Walstra, P.,** The role of proteins in the stabilisation of emulsions, in *Gums and Stabilisers for the Food Industry*, Vol. 4, Phillips, G. O., Williams, P. A., and Wedlock, D. J., Eds., IRL Press, Oxford, 1988, 233.
91. **Dickinson, E., Porgson, D. J., Robson, E. W., and Stainsby, G.,** *Colloids Surf.*, 14, 135, 1985.
92. **Walstra, P.,** *Food Colloids and Polymers, Stability and Mechanical Properties*, Dickinson, E. and Walstra, P., Eds., Special Publication 113, Royal Society Chemistry, Cambridge, 1993, 3.
93. **van den Vliet, T., Kokelar, A. M., and Jansen, A. M.,** *Food Colloids and Polymers, Stability and Mechanical Properties*, Dickinson E. and Walstra, P., Eds., Special Publication 113, Royal Society Chemistry, Cambridge, 1993.
94. **Sanderson, G. R.,** *Food Technology*, 50, 1981.
95. **Glicksman, M.,** *Gum Technology in the Food Industry*, Academic Press, New York, 1969.
96. **Sandford, P. A. and Baird, J.,** *The Polysaccharides*, Vol. 2, Aspinall, G. O., Ed., Academic Press, New York, 1983.
97. **Williams, P. A., Phillips, G. O., and Randall, R. C.,** Structure-function relationship in gum arabic, in *Gums and Stabilisers for the Food Industry*, Vol. 3, Phillips, G. O., Williams, P. A., and Wedlock, D. J., Eds., IRL Press, Oxford, 1990, 25.
98. **Bergenstahl, B. S., Fogler, S., and Stenius, P.,** The influence of gums on the stability of dispersions, in *Gums and Stabilisers for the Food Industry*, Vol. 5, Phillips, G. O., Williams, P. A., and Wedlock, D. J., Eds., IRL Press, Oxford, 1988, 283.
99. **Garti, N. and Reichman, D.,** *Food Hydrocolloids*, 8, 155, 1994.
100. **Reichman, D.,** Ph.D. thesis, Characterization of hydrocolloids surface activity and interaction with monomeric emulsifiers and surfaces, The Hebrew University of Jerusalem, Jerusalem, 1994.
101. **Becher, P.,** *Emulsion Theory and Practice*, 2nd ed., Reinhold, New York, 1965, 2.
102. **Davis, S. S., Hadgraft, J., and Palin, J. K.,** *Encyclopedia of Emulsion Technology*, Vol. 2, Becher, P., Ed., Marcel Dekker, New York, 1983, 159.
103. **Florence, A. T. and Whitehill, J.,** *J. Int. Pharm.*, II, 277, 1982.
104. **Whitehill, D.,** *Chemist Druggist*, 213, 130, 1980.
105. **Seifriz, W.,** *J. Phys. Chem.*, 29, 738, 1925.
106. **Lin, T. J., Kurihara, H., and Ohta, H.,** *J. Soc. Cosmet. Chem.*, 26, 121, 1975.
107. **Florence, A. T. and Whitehill, D.,** *J. Colloid Interface Sci.*, 79, 243, 1981.
108. **Matsumoto, S. and Kang, W. W.,** *J. Dispersion Sci. Technol.*, 10, 455, 1989.
109. **Matsumoto, S., Kita, Y., and Yonezawa, D.,** *J. Colloid Interface Sci.*, 57, 353, 1976.
110. **Florence, A. T. and Whitehill, D.,** *J. Colloid Interface Sci.*, 79, 243, 1981.
111. **Garti, N., Frenkel, M., and Shwartz, R.,** *J. Dispersion Sci. Technol.*, 4, 237, 1983.
112. **Morimoto, Y., Sugibayashi, K., Yamaguchi, Y., and Kato, Y.,** *Chem. Pharm. Bull.*, 27, 3188, 1979.
113. **Zheng, S., Beissinger, R. L., and Wasan, D. T.,** *J. Colloid Interface Sci.*, 144, 72, 1991.
114. **Omotosho, J. A.,** *Int. J. Pharm.*, 62, 81, 1990.
115. **Oza, K. P. and Frank, S. G.,** *J. Dispersion Sci. Technol.*, 10, 163, 1989.
116. **Taylor, P. J., Miller, C. L., Pollock, T. M., Perkins, F. T., and Westwook, M. A.,** *Hygiene*, Vol. 67, Cambridge, 1969, 485.
117. **Elson, L. A., Mitchley, B. C. V., Collings, A. J., and Schneider, R.,** *Rev. Europ. Etudes Clin. Biol.*, 15, 87, 1970.
118. **Collings, A. J.,** British Patent #1,235,667, 1971.
119. **Benoy, C. H., Elson, L. A., and Schneider, R.,** *Rev. Europ. Etudes Clin. Biol.*, 45, 135, 1972.
120. **Brodin, A. F. and Frank, S. G.,** *Acta Pharm. Suec.*, 15, 1, 1978.
121. **Omotosho, J. A., Law, T. K., Whateley, T. L., and Florence, A. T.,** *Colloids Surf.*, 2, 133, 1986.
122. **Omotosho, J. A., Law, T. K., Whateley, T. L., and Florence, A. T.,** *J. Pharm. Pharmacol.*, 38, 865, 1986.
123. **Omotosho, J. A., Law, T. K., Whateley, T. L., and Florence, A. T.,** *J. Microencapsulation*, 6, 183, 1989.
124. **Oza, K. P. and Frank, S. G.,** *J. Dispersion Sci. Technol.*, 7, 543, 1986.
125. **Oza, K. P. and Frank, S. G.,** *J. Dispersion Sci. Technol.*, 10, 187, 1989.
126. **Garti, N., Aserin, A., and Cohen, Y.,** *J. Controlled Release*, 29, 49, 1994.

127. **Chaing, C., Fuller, G. C., Frankenfeld, J. W., and Rhodes, C. T.,** *J. Pharm. Sci.*, 63, 1987.
128. **Higuchi, T.,** *J. Pharm. Sci.*, 52, 1145, 1963.
129. **Magdassi, S. and Garti, N.,** *J. Controlled Release*, 3, 273, 1986.
130. **Magdassi, S., Frenkel, M., and Garti, N.,** *J. Dispersion Sci. Technol.*, 5, 49, 1984.
131. **Magdassi, S., Frenkel, M., and Garti, M.,** *Drug Ind. Pharm.*, 11, 791, 1985.
132. **Kita, Y., Matsumoto, S., and Yonezawa, D.,** *J. Colloid Interface Sci.*, 62, 87, 1977.
133. **Law, T. K., Florence, A. T., and Whateley, T. L.,** *J. Pharm. Pharmacol.*, 36, 50, 1984.
134. **Law, T. K., Whateley, T. L., and Florence, A. T.,** *Int. J. Pharm.*, 21, 277, 1984.
135. **Dickinson E., Evison, J., and Owusu, R. K.,** Preparation of fine protein-stabilized water-in-oil-in-water emulsions, *Food Hydrocolloids*, 5, 481, 1991.
136. **Kreuter, J.,** Nanoparticles, *Colloidal Drug Delivery Systems*, Kreuter, J., Ed., Marcel Dekker, New York, 1994, 219.
137. **Change, T. M. S.,** *Microencapsulation*, Nixon, J. R., Ed., Marcel Dekker, New York, 1976, 57.
138. **Myers, D.,** *Surfactant Science and Technology*, VCH Publishers, New York, 1988, 235.
139. **Pilman, E., Torenberg, E., and Larsson, K.,** *J. Disp. Sci. Technol.*, 3, 335, 1982.
140. **Shah, D. O., Walker, R. D., Hsieh, W. C., Shah, N. J., Dwivedi, S., Nelander, J., Pepinsky, R., and Deamer, D. W.,** *Some Structural Aspects of Microemulsions and Co-solubilized Systems*, SPE 5815, paper presented in Tulsa, OK, March 22–24, 1976.
141. **Duxbury, D. D.,** *Food Processing*, May 1988, 62–64.
142. **Wolf, P. A. and Havekotte, M. J.,** U.S. Patent 4,835,002, 1989.
143. **Thoma, K. and Pfaff, G.,** *Perfumer and Flavorist*, 2, 27, 1978.
144. **Treptow, R. S.,** *Research and Development Report*, Procter & Gamble, Cincinnati, OH, June 1, 1971.
145. **Gulik-Krzywicki, T. and Larsson, K.,** *Chem. Phys. Lipids*, 35, 127, 1984.
146. **Vesala, A. M., Rosenholm, J. B., and Laiho, S.,** *J. Am. Oil Chem. Soc.*, 62, 1379, 1985.
147. **Osborne, D. W., Pesheck, C. V., O'Neill, K. J., Ward, A. J. I., and El-Nokaly, M.,** *Cosmetics Toiletries*, 103, 53, 1988.
148. **Hagenmaier, R. D. and Baker, R. A.,** Wax microemulsions and emulsions as citrus coatings, *J. Agric. Food Chem.*, 42, 899, 1994.
149. **Chiu, Y. C. and Yang, W.L.,** *Colloids Surf.*, 63, 311, 1992.
150. **Chang, P. S., Rhee, J. S., and Kim, J. J.,** Continuous glycerolisis of olive oil by lipase immobilizes in reversed micelles, *Biotechnol. Bioeng.*, 38, 1159, 1991.
151. **Wolf, P. A. and Havekotte, M. J.,** Stable microemulsions using edible oils as flavoring materials, U.S. Patent 4,835,002, 1989.
152. **Slocum, S. A.,** Solubilization of essential oil components and orange oil in microemulsions, Ph.D. thesis, Pennsylvania State University, University Park, 1990.
153. **Sivik, B., Gruvmark, J., and Jakobsson, M.,** Oxidation studies on fish oil in microemulsions and microcapsules, *Swed. Proc. Scand. Symp. Lipids*, 147, 1991.
154. **Leser, M. E., Luisi, P. L., and Palmieri, S.,** Use of reverse micelles for extraction of oil and proteins from vegetable meal, *Biotechnol. Bioeng.*, 34, 1140, 1989.
155. **Larsson, K.,** *J. Phys. Chem.*, 73, 7304, 1989.
156. **Engström, L.,** *J. Dispersion Sci. Technol.*, 11, 479, 1990.
157. **Holmberg, K.,** *Adv. Colloid Interf. Sci.*, 51, 137, 1994.
158. **Garti, N. and Silberstein, T.,** *Food Microstructure*, in press.
159. **Almog, Y. and Lichtenberg, D.,** *Biochemistry*, 27, 873,1988.
160. **Martinik, K., Levashov, A. V., and Klyachko, N.,** *Eur. J. Biochem.*, 155, 453, 1988.
161. **Jain, K. M. and De Haas, G. H.,** *Chem. Biophysica Acta*, 736, 157, 1983.
162. **Garti, N., Lichtenberg, D., and Silberstein, T.,** Enzymatic reactions in microemulsions as microreactors, *J. Dispersion Sci. Technol.*, submitted.
163. **Dickinson, E.,** Protein-polysaccharide interactions in food colloids, in *Food Colloids and Polymers: Stability and Mechanical Properties*, Dickinson, E. and Walstra, P., Eds., Royal Society of Chemistry, Wiltshire, 1993, 77.
164. **Finch, W. M. and Griffin W. C.,** Food Emulsions, in *Emulsions and Emulsion Technology*, Lissant, K. J., Ed., Marcel Dekker, New York, 1974.
165. **Dickinson, E., Evison, J., Gramsaw, J. W., and Schwoe, D.,** *Food Hydrocolloids*, 8, 63, 1994.
166. **Owusu, R. K., Zhu, K., and Dickinson, E.,** Controlled release of L-tryptophane and vitamin B2 from model water/oil/water multiple emulsions, *Food Hydrocolloid*, 6, 443, 1992.
167. **Dickinson, E., Evisson, J., Gramshow, J. W., and Schwope, D.,** Flavor release from a protein-stabilized W/O/W emulsion, *Food Hydrocolloids*, 8, 63, 1994.
168. Kao, Corp., Oil-water-oil double emulsion composit as margarine, JP Kokai Tokkyo JP 59216628, 1984.
169. Morinaga Milk Industry Co., Double emulsion spread preparation, JP 06007985, 1994.
170. **Dickinson, E., Evisson, J., Owusu, R. K., and Zhu, Q.,** Studies of W/O/W multiple emulsions: stabilization and controlled nutrient release, *Food Colloids and Polymers: Stability and Mechanical Properties*, Special Publication 13, Royal Society of Chemistry, Cambridge, 1993, 276.

Chapter 10

Nonionic Surfactant Vesicle-in-Water-in-Oil (v/w/o) Systems

Alexander T. Florence and Toshimitsu Yoshioka

CONTENTS

I. INTRODUCTION

Nonionic surfactants of a wide variety of structural types have been found to be useful alternatives to phospholipids in the fabrication of vesicular systems.[1] In spite of the chemical differences in the monomer units which form unilamellar or multilamellar vesicles, these niosomes (NSVs or nonionic surfactant vesicles) possess physical properties which are similar to liposomes formed from phospholipids. It has, in fact, been difficult to discern differences of a substantial nature in their performance as drug delivery vehicles, although the surface characteristics of the niosomes, with polyoxyethylene, polyglycerol, sorbitan, or sugar head groups,[1] are un-ionized and might form a steric barrier to adsorption of proteins, or at least offer a different environment for the adsorption and interactions with smaller molecules *in vitro* and *in vivo*. Both niosomes and liposomes have been used in cosmetics,[2] and have been investigated as topical vehicles where they may act as both reservoirs and delivery vectors[3] and have potential in diagnostic devices.[4]

Generally, vesicular systems are used as aqueous suspensions, but there are potential uses for systems in which the external phase is nonaqueous. Inverse vesicles have been reported but, as with inverse micelles, the structural constraints on formation of both amphiphile and oil phase are such that this approach is perhaps limited at this time. We have formulated and describe here a system in which aqueous suspensions of vesicles are dispersed in a continuous oil phase, both for a study of their potential as a drug delivery system and immunological adjuvant, and also as an intrinsically interesting colloidal system. The system is an emulsion prepared from a dispersion of niosomes in water, reemulsified in an oil using a surfactant mixture of low HLB to achieve a stable w/o emulsion. The resulting vesicle-in-water-in-oil (v/w/o) system (Figure 1) is thus a close analog of o/w/o emulsions and similar in concept to microsphere(s) s/o/w dispersions[5] or w/o/w multiple emulsions.[6] The formulation of a v/w/o system has a dual advantage, as not only can water-insoluble and -soluble agents be dispersed in the internal vesicle phase,* but there is an external oil barrier at the w/o interface and particularly an external nonaqueous barrier through which the encapsulated solute must diffuse to be released from stable systems. In unstable systems release is also accomplished through coalescence of the internal aqueous droplets as in w/o/w emulsions.[7]

* The terminology used in this chapter describes the niosome phase of the emulsion as the internal niosome phase which is dispersed in the internal aqueous phase; the external (continuous) phase comprises the oil.

0-8493-4012-8/96/$0.00+$.50
© 1996 by CRC Press, Inc.

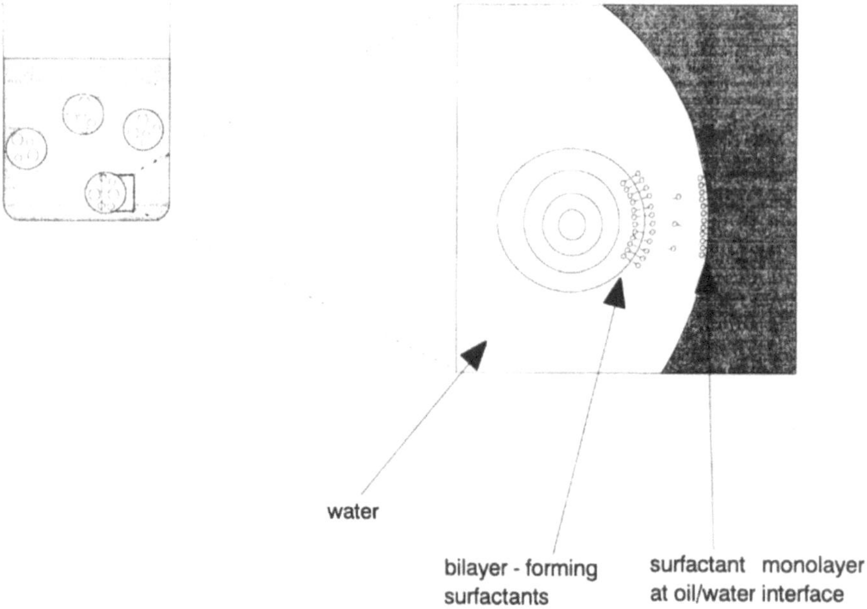

Figure 1 Diagrammatic representation of the v/w/o formulation.

Experience with multiple w/o/w systems has also shown that one aspect of instability is brought about by the migration of stabilizing surfactant from the oil/water interface to the water/oil interface and vice versa, changing the initial concentration of surfactant stabilizing both the co-exisiting o/w and the w/o emulsions.[8] There is the potential, immediately post-formation of the system, of exchange of surfactant molecules from the vesicles to the interfacial film. Thus, one critical feature of the v/w/o formulation is the choice of surfactants: the nonionic surfactant used to form the vesicles is also a component of the stabilizing system for the w/o emulsion.

Surfactants of the sorbitan monoester (Span) class, widely used in foodstuffs and pharmaceuticals and generally regarded as safe, were employed in our initial studies, and phospholipid vesicles have also been substituted for niosomes. The formation and characteristics of sorbitan-monoester vesicles (Span 20, 40, 60, and 80) have been studied both by Chandraprakesh et al.[9] and by the present authors.[10]

II. PREPARATION

The preparation of a specimen system is described below.[11] Many alternatives no doubt exist. In general terms, conventional niosome in water dispersions are prepared and these are then emulsified in a continuous oil phase (Figure 2) using the vesicle-forming surfactant as at least one of the components of the w/o stabilizing system. The niosomes used were not sonicated to allow the use of large vesicles which could be visualized microscopically, but sonicated vesicles can be incorporated into the final system.

The nonionic surfactants were sorbitan monolaurate (Span 20), sorbitan monopalmitate (Span 40), sorbitan monostearate (Span 60), and sorbitan monooleate (Span 80). Dicetyl phosphate (DCP) and cholesterol were key components. 5(6)-Carboxyfluorescein (CF) was used as a model solute after purification. Octane, hexadecane, and isopropyl myristate were chosen as the model external oil phases.

A. PREPARATION OF THE INTERNAL PHASE OF NONIONIC SURFACTANT VESICLES (NIOSOMES)

Multilamellar nonionic surfactant vesicles (niosomes) were prepared by a hand-shaking method. Three hundred μmol of surfactant (Span 20, 40, 60, or 80), cholesterol, and dicetyl phosphate (giving a molar ratio of 47.5:47.5:5.0) were the lipid ingredients dissolved in 15 ml of chloroform in a 100 ml round-

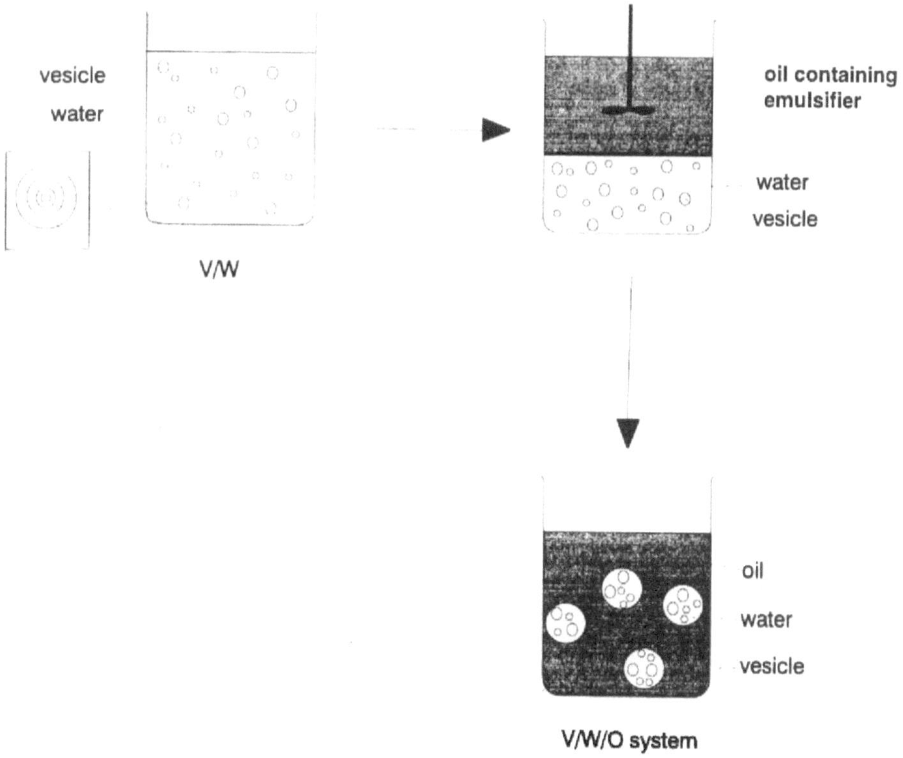

Figure 2 The steps in the preparation of the v/w/o system.

bottomed flask. The organic solvent was evaporated under reduced pressure at a temperature of 60°C using a rotary evaporator to form a thin film on the flask wall. The excess organic solvent was then removed with oxygen-free nitrogen for 10 min and the film was dried completely. The dried lipid film was hydrated with 6 ml of CF solution (or other solutes for encapsulation in the vesicles) by shaking with a mechanical shaker in a water bath at 60°C for 1 h. It was essential to prepare the vesicles at a temperature above the gel–liquid transition temperature of the nonionic surfactants. As Span 60 had the highest phase transition temperature of about 50°C, all preparations were carried out at 60°C, and the resulting multilamellar nonionic surfactant vesicle dispersion was then left to cool. The vesicles containing trapped solute were separated from unentrapped material by gel chromatography. Table 1 summarizes some of the properties of the surfactants and the niosomes formed from them, Figure 3 showing freeze-fracture electron micrographs of typical vesicles.

B. PREPARATION OF THE v/w/o SYSTEM

The same surfactant as used to form vesicles was added to the oil and dissolved completely at 60°C, and then one part of the niosome suspension was added to four parts of an oil containing 5% of the surfactant

Table 1 Properties of Span Niosomes Prepared by the Hand-Shaking Method

Span	Type	HLB	%CF	Size (μm)
20	Sorbitan monolaurate	8.6	11.4	3.4
40	Sorbitan monopalmitate	6.7	16.2	1.1
60	Sorbitan monostearate	4.7	16.4	0.96
80	Sorbitan monooleate	4.3	10.9	0.9
85	Sorbitan trioleate	1.8	9.1	0.65

Figure 3 Freeze-fracture photomicrographs of the Span surfactant vesicles used to prepare the v/w/o system.

at the same temperature. Emulsification was carried out using a vortex mixer to produce the v/w/o emulsion. Excessive shear was avoided to prevent micronisation of the aqueous phase.

C. STRUCTURE OF THE v/w/o EMULSION

Freeze-fracture electron microscopy of the v/w/o systems was unsuccessful in fracturing across the aqueous droplets, but micrographs such as Figure 4 show clearly the emulsion structure, and the complex structure is clear from optical photomicrographs (Figure 5).

As can be seen, the v/w/o emulsion comprised vesicles dispersed in large water droplets of around 5 to 25 μm. Although all the v/w/o emulsions had vesicles in the aqueous disperse phase, their size is a reflection of the Span used and the mode of preparation of the vesicles. Vesicle size is dependent on the HLB of the Span used when hand-shaking is used without sonication.

III. RELEASE OF SOLUTES

Possible routes and mechanisms of drug or solute release from stable v/w/o systems are shown in Figure 6, these being in this order: initial release of solutes from the vesicles, diffusion through the aqueous phase,

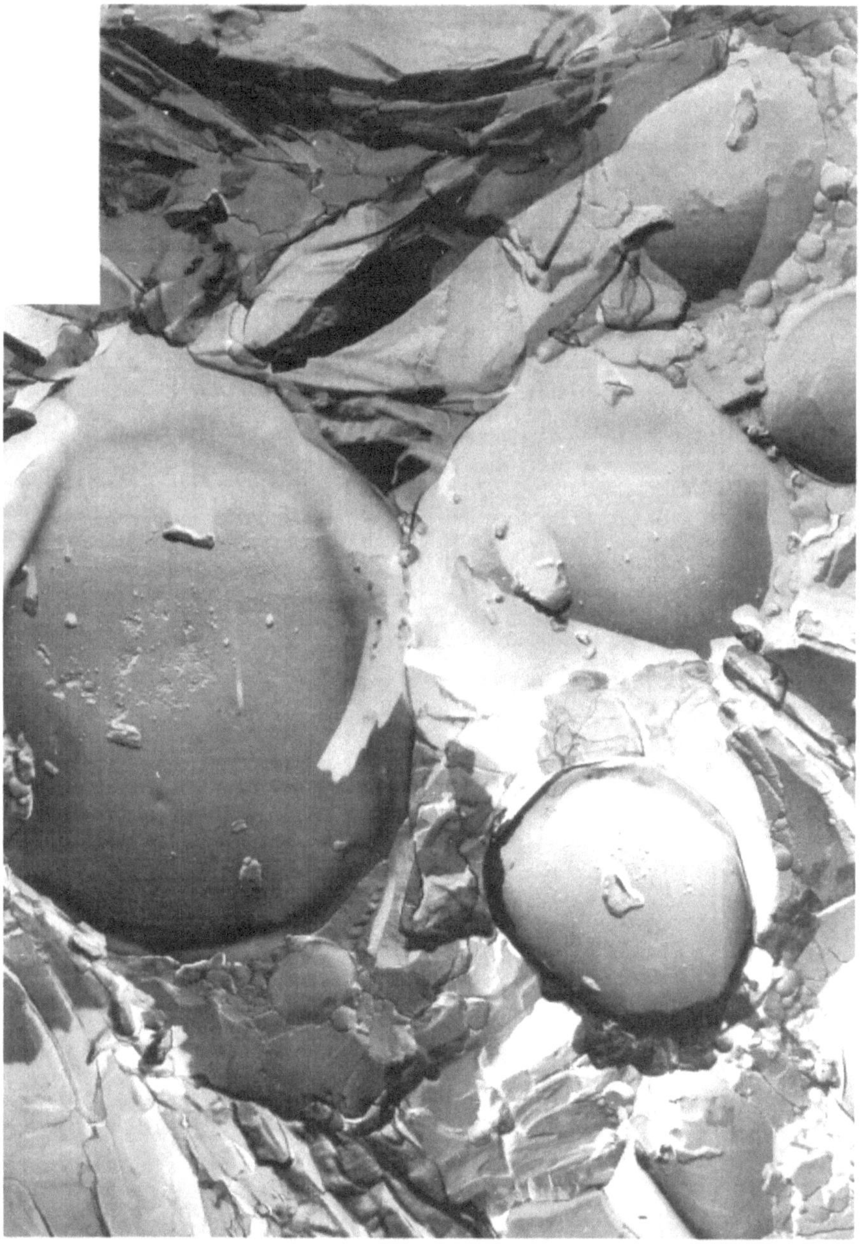

Figure 4 Freeze-fracture electron micrograph of the v/w/o system showing the internal water droplets disappear in (crystallized) oil phase.

transport across the o/w interface, and finally diffusion through the oil, a process in which w/o/w emulsions have been found to be mediated by the inverse micellar phase of the stabilizing surfactants.[12]

Some of the factors affecting *in vitro* release from the v/w/o formulations have been investigated using a simple dialysis system to contain the emulsion. Cuprophan tubing size 5 (Visking 20/32 in) was washed several times with distilled water and left soaking in distilled water overnight before use. Five ml each of CF solution, the relevant vesicle suspension (v/w), w/o emulsion, or v/w/o emulsion were pipetted into a bag made of Cuprophan tubing and sealed. Each of the samples contained the same concentrations of CF. The Cuprophan bag was placed in 250 ml of buffer solution (pH 5.0) in a 300-ml conical flask with constant shaking at 25 or 37°C. At various time intervals, the

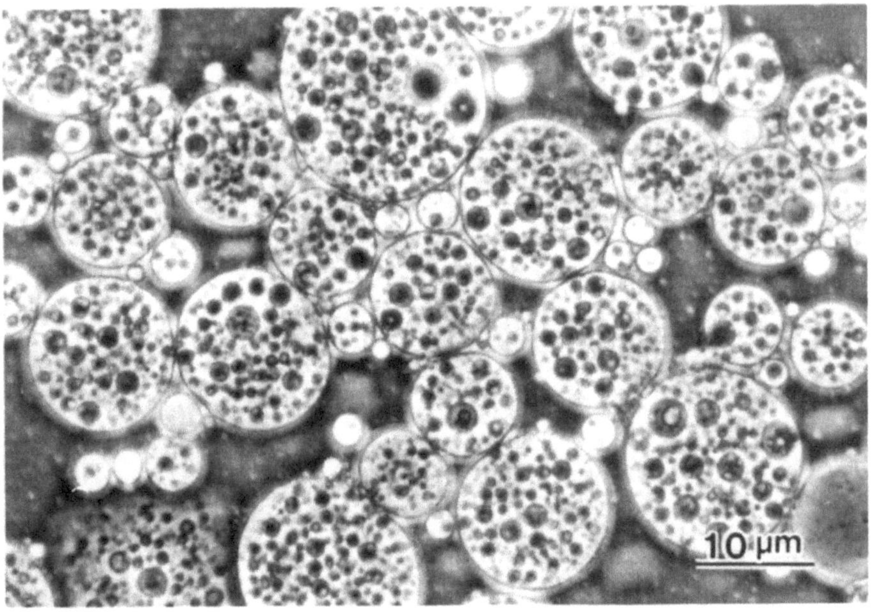

Figure 5 Photomicrograph of a v/w/o emulsion formulated with large niosomes prepared with Span 80, 5% of Span 80 in the oil phase, and octane.

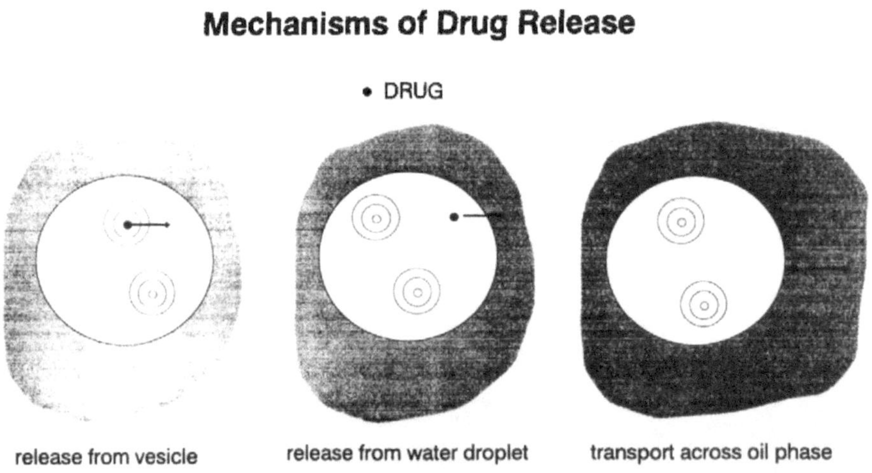

Figure 6 Mode of release of solutes from the formulation.

fluorescence of the buffer solution was measured at an excitation wavelength of 486 nm and emission wavelength of 514 nm.

A. FACTORS AFFECTING RELEASE OF SOLUTES FROM THE INTERNAL PHASE

The results of dialysis experiments are shown in Figure 7 together with those for free CF solutions, v/w suspensions, and w/o emulsions, for comparison. The release of CF was the slowest from the v/w/o emulsion. The release of CF decreased in the order: free solution > vesicles suspension > w/o emulsion > v/w/o emulsion, as anticipated, regardless of the Span used, although sometimes the differences between the systems is not large. Preliminary experiments with another solute, 5-fluorouracil did not always display the same advantage for the v/w/o system over simpler comparators, and it is clear that further work to explore the relationship between solute physical chemistry and behavior in the formulations is required.

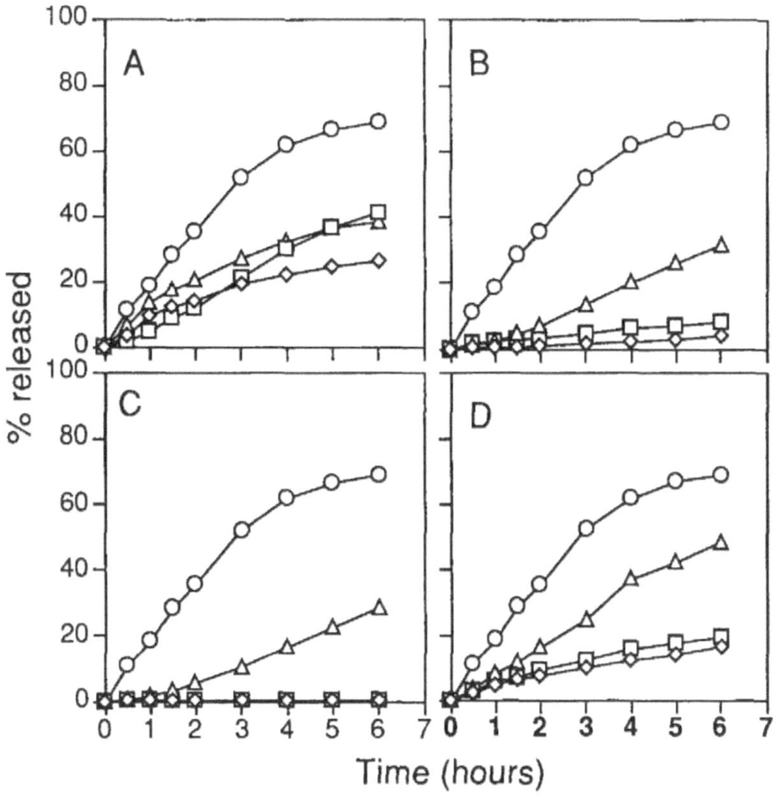

Figure 7 *In vitro* release profile of CF from various formulations prepared with Span 20® (A), Span 40® (B), Span 60® (C), and Span 80® (D). O: CF solution; △: vesicle suspension; □: w/o emulsion; ◇: v/w/o emulsion.

1. Effect of Surfactant on Release

The effect of the particular Span surfactant on the release of CF was investigated by preparing niosomes and their equivalent w/o dispersions using the Span series. The rate of release of CF entrapped into the vesicles in the v/w/o system was found to change depending on the nature of the surfactants used both in niosome preparation and as the secondary emulsifier. Among the components of the w/o emulsions and v/w/o emulsions, the Span surfactant is common to both and is most likely to be responsible for the different release patterns. The degree of hydrophobicity of the Span surfactant had a significant influence on the release rate. In the Span 60 formulation, the release rate was the slowest, the most likely explanation being that sorbitan monostearate and, to a lesser extent, Span 40 (sorbitan monopalmitate) caused the gelation of the oil phase at 25°C. The gelation was thought to to be the result of the crystallization and solidification of the excess nonionic surfactant in the oil phase, but recent work has demonstrated surfactant fibril formation as a cause (S. Murdan and A. T. Florence, unpublished).

From the semi-logarithmic linear relationship of the percent solute remaining as a function of time, the values of a release rate constant, K, were extracted for the systems and plotted vs. surfactant HLB (Figure 8). The release rate was the lowest at HLB values around 4.7. Span 20 and Span 80 are liquids, but Span 40 and Span 60 are solids at room temperature and, as stated above, caused gelation of the oil phase of the v/w/o emulsions. K values for the vesicle suspensions were almost the same, independent of the surfactants, but those for the w/o emulsions and the v/w/o emulsion were different. This suggests that the rate-limiting step was not transported through the vesicle membrane but transported through the oil layer, as suggested by Omotosho et al.[13]

2. Effect of the Oil Phase

Using all the Spans in turn, the effect of the oil phase on release rate was examined with three oils: octane, hexadecane, and isopropyl myristate (IPM). The results are shown in Figure 9. In Span 40 and Span 60

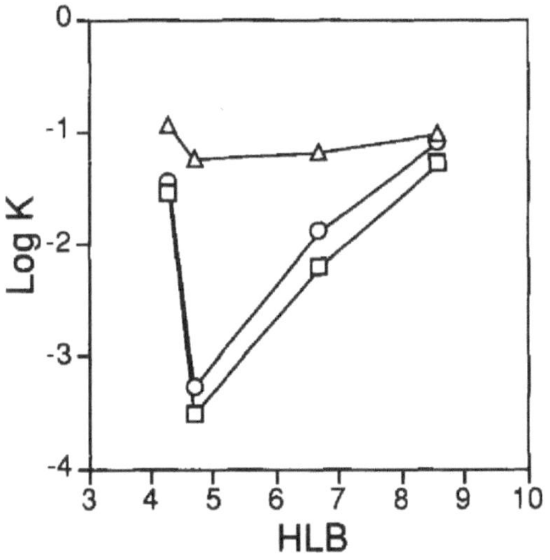

Figure 8 First-order release rate constants of various formulations prepared with different surfactants plotted as a function of surfactant HLB. ○: w/o emulsion; △: vesicle suspension; □: v/w/o emulsion.

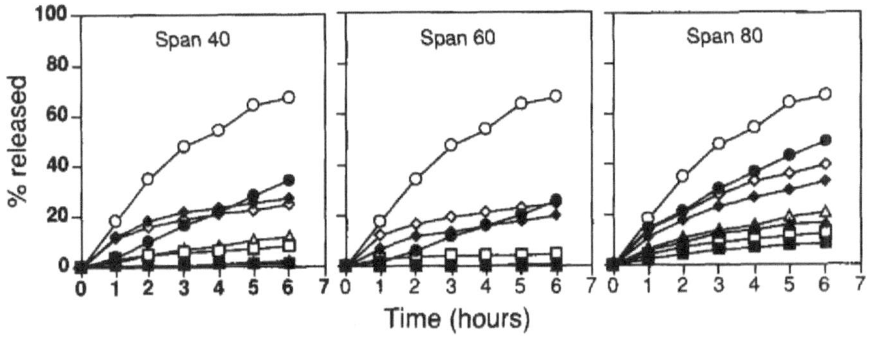

Figure 9 *In vitro* release profile of CF from various Span vesicle and emulsion formulations prepared with different oils. ○: CF solution; ●: vesicle suspension; △: w/o emulsion (octane); □: w/o emulsion (hexadecadeE); ◇: w/o emulsion (IPM); ▲: v/w/o emulsion (octane); ■: v/w/o emulsion (hexadecade); ◆: v/w/o emulsion (IPM).

v/w/o emulsions prepared with octane and hexadecane, CF was slowly released, because these formulations were viscous gels. From the semi-logarithmic linear relationship of the percent drug remaining as a function of time, the K values were extracted and plotted vs. HLB (Figure 10) as before. Isopropyl myristate as the oil phase provides the fastest release rates virtually invariant in all systems. In Span 40 and Span 60 systems, the release rate constants of v/w/o emulsions prepared with hexadecane were almost the same, but those prepared with octane varied with the surfactant. Omotosho et al.[13] examined the effect of the nature of the oil phase and the release of solutes from multiple w/o/w emulsions and demonstrated that the main factor in determining the differences in rates of release from the hydrocarbon w/o/w emulsions was the droplet size of the internal aqueous phase; the degree of solute leakage from the internal phase of the multiple emulsion being greatest with systems of low particle size and largest surface area. It is also reasonable to suppose that, as with conventional emulsions, the nature of the oil phase in the v/w/o systems can markedly affect the behavior of the system; the stability of the oil membrane against leakage of the entrapped material will depend, among other factors, on the nature of that oil. Also, for a given internal (w) phase volume, a decrease in droplet size would increase the surface area, resulting in an increased release rate, which should explain the large differences in rates of release of CF from v/w/o emulsions prepared with octane, hexadecane, and isopropyl myristate. However, the overriding differences in release rate constants appear to be due to gel formation.

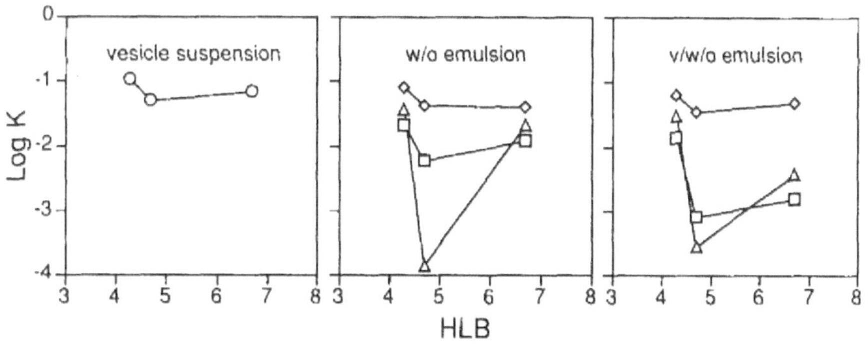

Figure 10 First-order release rate constants for a vesicle suspension (a), w/o emulsion (b), and v/w/o emulsion (c) prepared with different oils.

3. Effect of Temperature

The effect of temperature on the release rate of CF from various formulations was evaluated at 25 and 37°C. Increasing the temperature resulted in increased release rates in Span 40 and Span 80 systems, but in the Span 60 system the release rate did not increase due to the gelation of the oil phase.These results suggest that the release rate of CF is largely dependent on the physical state of the external oil phase.

IV. POTENTIAL USES

Our work has concentrated so far on potential pharmaceutical uses of the system as there are the possibilities for using it, or its variants, as an oral or a topical delivery formulation. If biodegradable or biodispersible oils are used, the v/w system can be effectively delivered in a nonaqueous phase as the oil phase degrades or disperses, such as after intramuscular injection, which is the original concept of the approach. Such approaches delay the exposure of the vesicle to the biophase. Cosmetic applications might involve application of appropriate v/w/o formulations to the skin, where application might release components from the sheared vesicle phase. Agricultural applications could be envisaged if the oil phase was volatile, although the shear-dependency of stability would require investigation because of the size of the water globules and internal vesicles, if spray applications were employed. There may be applications in the food industry in which flavors could be dispersed in the vesicular phase of oil-based foodstuffs.

Preliminary work has been carried out using the system as a vaccine adjuvant.[14] Tetanus toxoid was used as the experimental antigen and free toxoid compound against toxoid delivered in niosomes, w/o, and v/w/o emulsions. Results have shown enhanced responses (Figure 11) which might be due to the delayed release from the systems, although the true nature of adjuvant effects of surfactant systems is not clear.[15]

V. CONCLUSIONS

In conclusion the system described here, *inter alia*,

1. allows the delivery of aqueous suspensions of vesicles in a nonaqueous vehicle;
2. might form the basis of a useful oral, intramuscular, or topical pharmaceutical delivery system;
3. has potential as a vaccine adjuvant with improved effects over existing systems;
4. deserves further study to explore applications outside the pharmaceutical arena.

Since the work was completed a patent by Wallach and co-workers[16] has described a very similar system.

ACKNOWLEDGMENTS

The work described above was largely carried out during the tenure of a research fellowship by Toshimitsu Yoshioka on leave from Snow Brand Milk Products, Japan, who is thanked for providing the fellowship. The authors wish to thank M. Sagara of the Technical Research Institute and Snow Brand Milk Products Co., Ltd. for the microphotographs of the v/w/o emulsion.

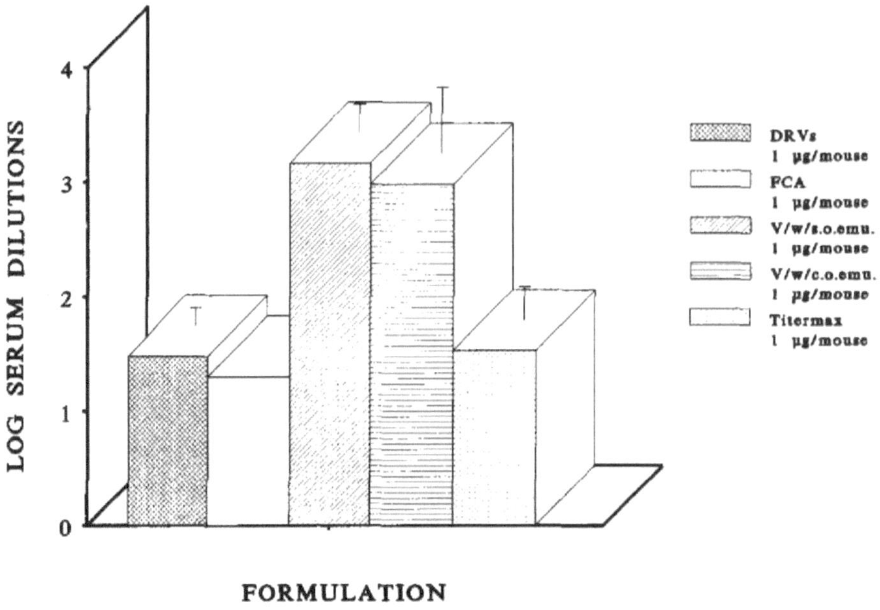

Figure 11 Secondary responses (IgG_1) to tetanus toxoid in liposomes (DRVs), Freund's complete adjuvant, two v/w/o systems (SO = sesame oil, CO = cottonseed oil) and Titermax®.

REFERENCES

1. **Florence, A. T.,** Non-ionic surfactant vesicles: preparation and characterization, in *Liposome Technology,* Vol. II, 2nd ed., Gregoriadis, G., Ed., CRC Press, Boca Raton, FL, 1993, 157.
2. **Handjani-Vila, R. M., Ribier, A., and Vanlerberghe, G.,** Liposomes in the cosmetic industry, in *Liposome Technology,* Vol. II, 2nd ed., Gregoriadis, G., Ed., CRC Press, Boca Raton, FL, 1993, 210.
3. **Mezei, M.,** Liposomes in topical applications of drugs: a review, in *Liposomes as Drug Carriers, Recent Trends and Progress,* Gregoriadis, G., Ed., John Wiley & Sons, Chichester, 1988; Mezei, M., Liposomes and the skin, in *Liposomes in Drug Delivery,* Gregoriadis, G., Florence, A. T., and Patel, H. M., Eds., Harwood, Amsterdam, 1993, 125.
4. **Hillis, L. R., Handly, J. D., and Babbitt, B. P.,** Utilization of contact sensitive liposome formulations in membrane lytic homogeneous immunoassays, in *Liposome Technology,* Vol. III, 2nd ed., Gregoriadis, G., Ed., CRC Press, Boca Raton, FL, 1993, 301.
5. **Hashida, M., Liao, M. H., Muranishi, S., and Sezaki, H.,** Dosage form characteristics of microsphere-in-oil-in water multiple emulsions. II. Examination of some factors affecting lymphotropy, *Chem. Pharm. Bull.,* 28, 1659, 1980.
6. **Florence, A. T., Omotosho, J. A., and Whateley, T. L.,** Multiple w/o/w emulsions as drug vehicles, in *Controlled Release from Drug Polymers and Aggregate Systems,* Rosoff, M., Ed., VCH, New York, 1989.
7. **Florence, A. T. and Whitehill, D.,** Some features of breakdown in water-in-oil-in-water multiple emulsions, *J. Colloid. Interface Sci.,* 79, 243, 1981.
8. **Omotosho, J. A., Law, T. K., Whateley, T. L., and Florence, A. T.,** The stabilization of w/o/w emulsions by interfacial interacton between albumin and non-ionic surfactants, *Colloids Surf.,* 20, 133, 1986.
9. **Chandraprakesh, K. S., Udupa, N., Umadevi, P., and Pillai, G. K.,** Pharmacokinetic evaluation of surfactant vesicle entrapped methotrexate in tumor bearing mice, *Int. J. Pharm.,* 61, R1-R3, 1990.
10. **Yoshioka, T., Sternberg, B., and Florence, A. T.,** Preparation and properties of vesicles (niosomes) of sorbitan monoesters (Span 20, 40, 60 and 80) and a sorbitan triester (Span 85), *Int. J. Pharm.,* 105, 1, 1994.
11. **Yoshioka, T. and Florence, A. T.,** Vesicle (niosome)-in-water-in-oil (v/w/o) emulsions: an *in vitro* study, *Int. J. Pharm.,* 108, 117, 1994.
12. **Omotosho, J., Law, T. K., Whateley, T. L., and Florence, A. T.,** Surfactant mediated diffusion of water through liquid membranes, in *Surfactants in Solution,* Vol. 10, Mittal, K., Ed., Plenum, New York, 1989, 333.
13. **Omotosho, J. A., Whateley, T. L., Law, T. K., and Florence, A. T.,** The nature of the oil phase and the release of solutes from multiple w/o/w emulsions, *J. Pharm. Pharmacol.,* 38, 865, 1986.
14. **Yoshioka, T., Skalko, N., Gursel, M., Gregoriadis, G., and Florence,** *J. Drug Targeting,* 2, 533, 1995.
15. **Gregoriadis, G.,** Immunological adjuvants: a role for liposomes, *Immunology Today,* 11, 89, 1990.
16. **Albert, E. C., Mathur, R., and Wallach, D. F. H.,** W09217179, U.S. Patent 4,911,928, 1992.

Chapter 11

Emulsomes™, A New Type of Lipid Assembly

Shimon Amselem, Eduardo Zawoznik, Ayala Yogev, and Doron Friedman

CONTENTS

I. INTRODUCTION

Emulsions are defined as heterogeneous systems of one liquid dispersed in another in the form of droplets usually exceeding 1 μm in diameter. The two liquids are immiscible and chemically unreactive or slowly reactive. An emulsion is a thermodynamically unstable dispersed system. Instability is a result of the system's tendency to reduce its free energy by separating the dispersed droplets into two liquid phases. Instability of an emulsion during storage is evinced by creaming, flocculation (reversible aggregation), and/or coalescence (irreversible aggregation).

One major drawback of standard o/w emulsions as drug carriers is limited drug loading capacity and that they are suitable only for oil-soluble compounds. When drug encapsulation above 1% is required,

0-8493-4012-8/96/$0.00+$.50
© 1996 by CRC Press, Inc.

a correspondingly larger oil phase (10 to 20%) is required to dissolve the compound. However, high oil content reduces emulsion stability, and the addition of a synthetic surfactant is necessary.

Liposomes are stable lipoidal vehicles prepared in the absence of synthetic ionic or nonionic surfactants. In liposomes, drugs can be either encapsulated in the enclosed aqueous space or intercalated into the lipid bilayer. They are the optimal carrier for trapping water-soluble materials in their internal aqueous compartments. However, due to their relatively low content of lipid molecules, the loading capacity of liposomes for lipophilic compounds is low and limited to the phospholipid bilayer domain.

This work was aimed to develop a carrier system which could provide high drug-trapping efficiencies mainly for hydrophobic compounds but in the absence of toxic surfactants. Our approach was to design a new lipid particle comprising an internal lipophilic core as in o/w emulsions surrounded not by a monolayer of phospholipids but by phospholipid bilayers as in liposomes, thus stabilizing the particles. We succeeded in developing this novel drug delivery technology and assigned the term "Emulsome™" to the new type of lipid assembly since they can be considered as intermediate particles or "hybrid" systems between o/w emulsions and liposomes.[1]

Although the Emulsome technology was developed mainly as a drug delivery system,[2] Emulsome particles can be very useful for nonmedical applications in agriculture and in cosmetic and food industries in areas where stable submicron fat particles are needed.

II. DESCRIPTION OF EMULSOME™ TECHNOLOGY

The Emulsome technology represents a new entity as a lipoidal vehicle and its successful development was achieved by the incorporation of a relatively high lecithin content (5 to 10%) compared to standard emulsions (0.5 to 2%), the use of fats or triglycerides which are solid at room temperature instead of oils, and the utilization of adequate emulsification technique. The combination of the specific lipid composition and manufacturing technology resulted in stable lipid particles in the nano-size range.

Emulsome particles comprise a hydrophobic core, as in standard o/w emulsions, but composed of solid fats instead of oils, surrounded and stabilized by one or more phospholipid bilayers as in liposomes (Figure 1). The particle size distribution of emulsomes is in the range of 50 to 250 nm in diameter.

There are several distinctive features which distinguish Emulsomes from standard oil-in-water emulsions and liposomes and they are detailed in Table 1 and schematically illustrated in Figure 1.

In contrast to standard o/w emulsions which are obtained by dispersing one liquid into another, Emulsomes are prepared by dispersing a solid in a liquid. No oils are used in the preparation of Emulsomes, but solid fats or triglycerides are used. A key feature of Emulsome particles is that the hydrophobic core is composed of fats which are in a solid or liquid–crystalline phase at room temperature. Due to the high phospholipid content of Emulsomes, one or more bilayers or envelopes of phospholipid molecules are formed around the particles stabilizing the internal hydrophobic core. Emulsomes are formulated in the absence of any synthetic ionic or nonionic surfactant or cosurfactant. Emulsomes described here differ from liposomes since their internal core is a hydrophobic fat compartment, while in liposomes the internal core is an aqueous compartment.

Having partially the characteristics of both liposomes and emulsions, Emulsomes represent a novel kind of lipoidal carrier with the advantages of high hydrophobic loading in the internal fat core and the ability to encapsulate water-soluble compounds in the aqueous compartments of the surrounding phospholipid bilayers.

Table 1 Major Differences Among a Submicron o/w Emulsion, an Emulsome, and a Liposome

	o/w Emulsion	Emulsome	Liposome
Definition	Dispersion of an oil in water	Dispersion of a solid fat in water	Dispersion of phospholipids in water
Internal core	Oil	Solid fat	Water
Phospholipid content (w/v)	0.5–2%	5–10%	0.1–10%
Nonnatural surfactant	Present	Usually absent	Usually absent
Cosurfactant	Present	Usually absent	Usually absent
Lipophilic loading	Up to 10 mg/ml	Up to 100 mg/ml	Up to 10 mg/ml
PC/total lipid (mol/mol)	0.01–0.1	0.1–0.5	0.6–0.1

Note: Based on data taken from emulsion and liposome literatures.

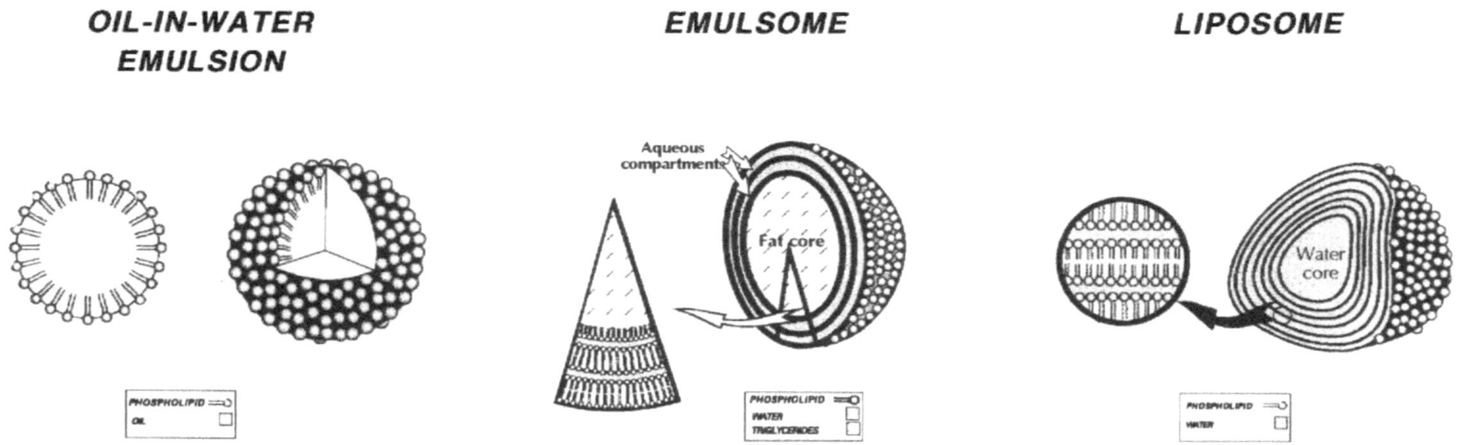

Figure 1 Schematic illustration of a liposome, an o/w submicron emulsion droplet, and proposed structure for an Emulsome particle.

III. CHARACTERIZATION OF EMULSOME™ PARTICLES

The parameters most frequently used to characterize lipid particles are morphology, particle size and size distribution, lamellarity, and trapped volume. Emulsome structure was demonstrated and characterized by several techniques including electron microscopy, photon correlation spectroscopy, and NMR.

A. PARTICLE SIZE DISTRIBUTION

The particle size distribution of Emulsome formulations were determined by photon correlation spectroscopy (PCS), based on measuring laser scattered-light fluctuations, using a N4MD Coulter submicron particle size analyzer (Coulter Electronics, England) working at the differential weight % operation mode of the instrument.[3]

The particle size distribution of Emulsomes, determined using the N4MD Coulter submicron particle analyzer, indicated the existence of single homogeneous population of emulsomes with a mean particle diameter in the range of 50 to 250 nm and in most cases around 100 nm.

B. ELECTRON MICROSCOPY

Negative stain electron microscopy (EM) has been used to characterize the morphology, size, and shape of the lipid particles present in phospholipid dispersions. EM gives also a gross estimate of lamellarity since stain penetrates interbilayer spaces and allows lamellae to be resolved. Figure 2 is an electron micrograph of an Emulsome preparation showing spheric particles having a diameter in the range of 80 to 130 nm surrounded by well-defined phospholipid bilayers enveloping the internal hydrophobic core.

C. ^{31}P-NMR MEASUREMENTS

One of the most accurate and straightforward procedures for quantitatively determining the lamellarity of phospholipid dispersions is to use NMR spectroscopy and especially a phosphorus nuclear magnetic resonance (^{31}P-NMR) signal to monitor the phospholipid phosphorous signal intensity.[4] In particular, adding an impermeable paramagnetic shift reagent to the external medium will decrease the intensity of the initial ^{31}P-NMR signal by an amount proportional to the fraction of lipid exposed to the external medium.

Phosphorous nuclear magnetic resonance (^{31}P-NMR) spectra of Emulsome preparations were recorded on a Brucker AM400 instrument employing a 50 kHz sweep width, 3 s interpulse delay, and broadening proton decoupling. Phosphoric acid in D_2O was used as standard for 0 ppm chemical shift.

Figure 3 shows ^{31}P-NMR spectra of liposomes, Emulsomes, and submicron o/w emulsion (SME) recorded before and after the addition of the lanthanide ion Pr^{+3}. Emulsomes (10% total lipids) were prepared as described in Section IV.D but suspended in D_2O as the aqueous phase. SME were prepared using a proprietary technology.[5–7] Liposomes were prepared in D_2O by high pressure homogenization using a Micron Lab 70 high pressure homogenizer (APV Gaulin International SA, Holland).

For the SME, after the addition of the paramagnetic ion, the whole ^{31}P-NMR signal was shifted downfield (Figure 3, B_1). This result is expected since in the SME all the phospholipid molecules are located in the surface of the emulsion oil droplets as a monolayer, thus the Pr^{+3} ions interact with all the exposed phospholipid molecules. On the other hand for Emulsomes, the ^{31}P-NMR signal was split into two peaks after Pr^{+3} addition (Figure 3, B_2), indicating that only a portion of the phospholipid molecules (those in the outer monolayer exposed to the aqueous medium) interacted with the Pr^{+3} ions, demonstrating the existence of bilayer structures. The same picture was obtained for small unilamellar liposomes (Figure 3, B_3) used as control, where the existence of phospholipid bilayer domains have been well documented.

D. QUANTITATIVE DETERMINATION OF LAMELLARITY BY TNBS-LABELING OF OUTER AMINOPHOSPHOLIPIDS

The existence of lamellae or phospholipid bilayers in the Emulsome particles was further demonstrated by a quantitative approach. The average number of phospholipid bilayers was quantified by determining the exposure extent of aminolipids after chemical labeling of phosphatidylethanolamine (PE) with the membrane impermeable hydrophilic reagent trinitrobenzenesulfonate (TNBS) to give tinitrophenyl-PE (TNP-PE).[7] The amount of TNP-PE in outer monolayer and total PE were quantified by separation on TLC plates, lipid extraction, and determination by modified Bartlett assay.[3,9]

For bilayer vesicles, the mol% of exposed PE should reflect the lipid mass distribution between the two leaflets, which is dependent on vesicle size.[10] For large unilamellar vesicles, the ratio of intact to total labeled bilayer vesicle should approach a minimal value of 0.5 indicating that 50% of the PE molecules are oriented toward the internal aqueous core where no contact with TNBS could occur. Reducing the

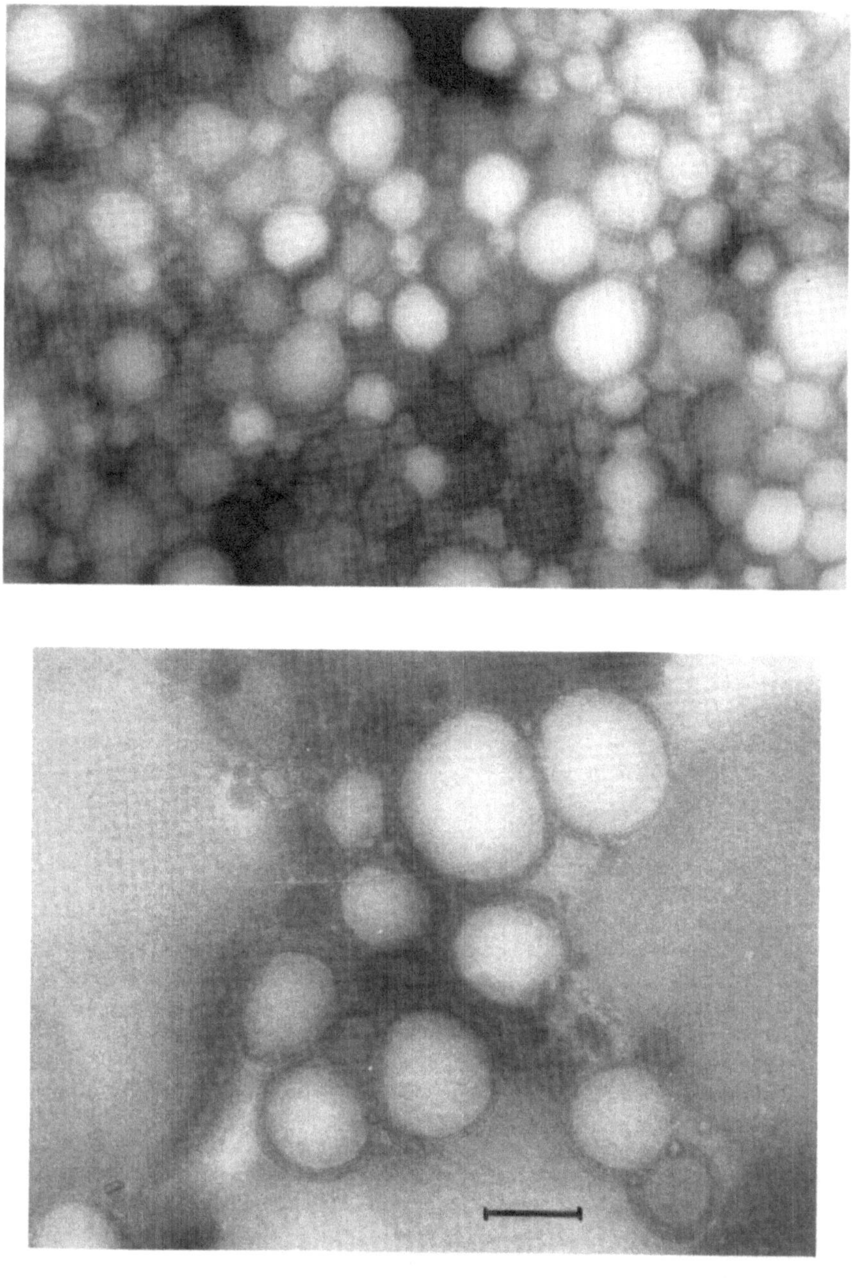

Figure 2 Transmission electron micrographs of Emulsome preparation negatively stained with 1% phosphotungstic acid solution at two different magnifications. Bar = 100 nm.

vesicle size increases the curvature and therefore increases the ratio of lipid in the outer leaflet over the lipid in the inner leaflet (out/in ratio), until it reaches a value close to 2 (66%/33%).[8,10] Increasing the number of lamellae reduces the % PE exposed to below 50% (i.e., for 2 lamellae per vesicle in average, only 25% of PE should be labeled by TNBS). The average number of lamellae is estimated from the mol% of PE reacting with TNBS to give TNP-PE.

The lipid composition of the Emulsome formulation was adjusted to include about 10 mol% PE of total phospholipid. They were prepared by high pressure homogenization as described in Section IV.D. Trinitrophenylation of the PE molecules was carried out using freshly prepared 5% TNBS solution (Sigma Chemical Co., St. Louis). TNBS labeling of intact emulsomes and total labeling of solubilized Emulsome were performed separately as described below.

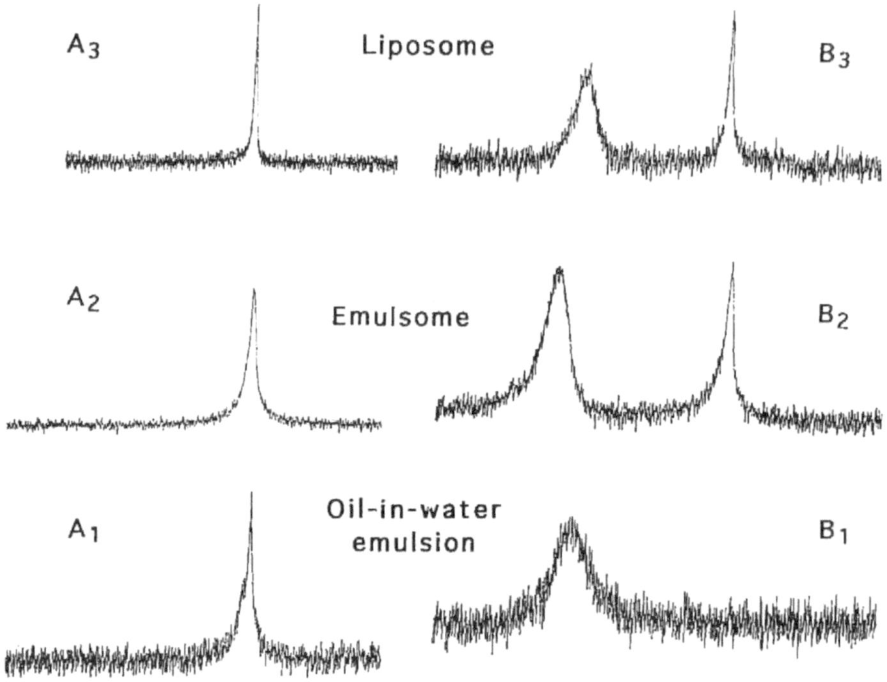

Figure 3 ^{31}P-NMR spectra of submicron o/w emulsion, Emulsomes, and liposomes recorded in the absence (A) and presence (B) of $PrCl_3$ (30 m*M*).

1. Labeling of Intact Emulsomes™

Forty microliters of Emulsome formulation were diluted with 125 µl of water in Corex® tubes and 0.8 ml of a 0.8 N solution of $NaHCO_3$ were added to each tube. To each sample (triplicates), 80 µl of 5% TNBS solution was added and the mixture was vortexed and allowed to incubate in the dark for 1 h. After complete trinitrophenylation was achieved (usually 1 h incubation), as determined from the lack of ninhydrin-positive aminophospholipids on TLC,[3] the reaction was terminated by adding water (0.5 ml), chloroform (1.5 ml), and methanol (1.5 ml) and the tubes were shaken and centrifuged at 2000 × g for 5 min at room temperature. Two phases were formed after vortexing and centrifugation. The upper phase was discarded and the lower chloroform phase was washed using a synthetic upper phase composed of chloroform:methanol:water 6:94:96 (v/v), until unreacted TNBS was completely removed as indicated by the absence of yellow color in the upper aqueous phase. The washed lower phase was transferred to conical tubes and the solvents were evaporated by a stream of nitrogen.

2. Total Labeling of Solubilized Emulsomes™

Forty microliters of Emulsome formulation were solubilized in 1 ml of solvent mixture containing isopropanol:heptane:0.8M $NaHCO_3$ (40:10:1). Then 120 µl of a 0.8 N solution of $NaHCO_3$ and 80 µl of 5% TNBS solution were added to each tube (triplicates), and the mixture was vortexed and allowed to incubate in the dark for 1 h. At the end of incubation, 0.5 ml heptane and 0.5 ml water were added and the mixture was vortexed and centrifuged at 2000 × g for 5 min at room temperature. The upper heptane phase was discarded and 1.5 ml chloroform, 0.8 ml water, and 0.7 ml isopropanol were added to the lower phase followed by mixing and centrifugation at 2000 × g for 5 min at room temperature. The aqueous upper phase was discarded and the lower chloroformic phase was washed using a synthetic upper phase composed of chloroform:methanol:water 6:94:96 (v/v). The washed lower phase was transferred to conical tubes and the solvents were evaporated by a stream of nitrogen.

3. Separation and Quantification of Lipids by TLC

For TLC analysis, the resulting yellow dry samples were dissolved in 50 µl chloroform:methanol 2:1 (v/v), the test tubes were vortexed and centrifuged in a desk centrifuge to remove lipids from test tube walls, and the supernatant applied to TLC glass plates (silica gel G 250 µm, 20 × 20 cm, Analtech, Newark, DE)

Table 2 Comparison of Structural Parameters of a Submicron o/w Emulsion Droplet, an Emulsome, and a Liposome

	o/w Emulsion (SME)	Liposome (SUV)	Emulsome
Mean size	140 ± 63 nm	73 ± 29 nm	101 ± 27 nm
% External phospholipids	98 ± 3	50.4 ± 4.7	36 ± 15
Estimated average number of bilayers	0	1	1–3
Trapped aqueous volume (µl/µmol phospholipid)	0	0.3 ± 0.2	1.1 ± 0.6

which have a low phosphorus level (around 0.5 nmol/ 2 cm^2 spot). The test tubes were then washed with chloroform:methanol 2:1 (v/v) and the supernatant applied to the TLC plate; this was repeated until no yellow color (TNP derivatives) remained in the test tubes. The plate was developed using the solvent mixture chloroform:acetone:methanol:glacial acetic acid: water, 6:8:2:2:1 (v/v). Spots were detected by their fluorescence after primuline spray (sensitivity limit 0.5 µg phospholipids per spot.[3,9] TNP derivatives (quenched spots in the primuline detection) were detected by their yellow color. Spots were identified using authentic markers of PE and TNP-PE. Spots were scraped and analyzed directly for their phosphorus content by a modified (higher sensitivity) Bartlett procedure (sensitivity limit 5 nmols phosphorus per spot).[3] The separation on TLC is described by the Rf values of the various spots, Rf = 0.78 for PE and 0.96 for TNP-PE. Percent PE exposure was calculated based on the following formula:

% PE exposed = [TNP-PE in intact Emulsomes/total TNP-PE in solubilized Emulsomes] × 100

Small unilamellar liposomes (SUV) and SME were used as control. They were prepared in 0.9% NaCl solution by high pressure homogenization as described in Section III.C and labeled using TNBS in the same way as for Emulsomes.

The results in Table 2 show that for the SUV liposomes, the TNBS method resulted in labeling of 50% of external phospholipids indicating an estimated average number of bilayers of 1 as expected for small unilamellar vesicles. For SME, 98% of all external phospholipids were labeled indicating the absence of bilayers, as it is well known for o/w emulsions where the oil droplets are surrounded by a monolayer of phospholipid molecules. For emulsomes, the % of external phospholipids labeled by TNBS is in the range of 15 to 50 (36 ± 15 average value, Table 2), indicating the existence of 1 to 3 bilayers, which is in good agreement with the lamellarity seen in the EM pictures (Figure 2) and with the results obtained by ^{31}P-NMR.

E. DETERMINATION OF TRAPPED VOLUME

The volume of the aqueous compartments of lipid vesicles (trapped volume) is usually defined in microliters per micromol lipid. The volume captured in lamellar phospholipid particles such as liposomes is a function of the vesicle shape and size. The trapped volume increases with increasing vesicle outer diameter. For multilamellar and oligolamellar vesicles, the trapped volume is also affected by the number of phospholipid bilayers. Size, lamellarity, and captured volume are dependent variables. As the lipid concentration increases, there are more structures of higher lamellarity and as a consequence the trapped volume decreases.[11]

The captured volume of liposomes also varies very significantly with the preparation method. Liposomes prepared using the dehydration–rehydration technique (DRV) or reverse phase evaporation method (REV) have larger trapped volume than SUV prepared by sonication or extrusion techniques. Freeze-thaw of multilamellar vesicles (MLV) increases capture volume by decreasing lamellarity.[11]

Numerous methods are available for measuring the captured volume of liposomes.[11] We have used the method of Oku et al.[12] to determine the trapped aqueous volume of Emulsomes. This procedure utilizes the fluorescent marker calcein, a chelating derivative of fluorescein. The fraction of the total aqueous volume that is within the phospholipid bilayers of Emulsomes was obtained as the fraction of the fluorescence that remains after adding cobalt (II) ions which, when chelated by calcein, quench its fluorescence. Separation of the Emulsomes from the medium is not required.[12]

Emulsomes containing calcein were prepared as described in Section IV.D, but in the presence of calcein in the aqueous phase. Fluorescence intensity was determined using a Perkin-Elmer LS-5 spectrofluorimeter. Samples were excited at 490 nm and the emission spectrum was recorded. Fluorescent liposomes and SME prepared in the presence of calcein were used as controls.

Table 2 shows the trapped volumes obtained for Emulsomes, liposomes (SUV), and submicron emulsion (SME). As expected for SME, since no phospholipid bilayers are present in SME droplets, no captured volume was detected. For SUV liposomes, a trapped volume of 0.3 ± 0.2 μl/μmol phospholipid was obtained. This value is in good agreement with the trapped volume reported for several SUV preparations.[11,13] In the case of Emulsomes, a trapped volume of 1.1 ± 0.6 μl/μmol phospholipid was obtained. This result is reasonable taking in consideration that they are oligolamellar and that they lack the internal aqueous core of liposomes.

IV. PREPARATION METHODS OF EMULSOMES™

A. COMPOSITION OF THE LIPID PHASE

An essential characteristic of Emulsomes is an internal hydrophobic core comprising a fat or lipid which exhibits solid or liquid crystal or mixed solid and liquid crystal phases at room temperature. The lipid may be a single compound or a mixture.

Lipids which form the lipid core of Emulsomes are composed almost exclusively of nonpolar moieties which therefore do not exhibit a preference for the lipid–water interface. The solid fats or lipids suitable for the preparation of Emulsomes enveloping an internal fat core are triglycerides, waxes, solid monoesters, and cholesterol or cholesteryl esterssters.

1. Triglycerides

Triglycerides are a preferred material from which the fat core of Emulsomes may be prepared. The triglyceride core may be composed of a single pure triglyceride, usually available as a synthetic triglyceride, or may be a mixture of several triglycerides. Fats isolated from natural sources usually are available only as mixtures of triglycerides. Such natural mixtures are suitable for preparation of Emulsomes, provided that the melting characteristics of the mixture are such that they exhibit a solid or liquid-crystal phase at room temperature. Detailed summaries of the phase behavior of various pure and mixed triglycerides are available.[14,15]

Many triglycerides which are solid at room temperature have fully saturated fatty acid chains. Saturated fatty acids are advantageous because they are incapable of undergoing peroxidation reactions, which lessen the acceptable storage life of o/w emulsions.

Examples of solid triglycerides suitable for the preparation of Emulsomes are triglycerides composed of natural, even-numbered and unbranched fatty acids with chain lengths in the C10 to C18 range, or microcrystalline glycerol triesters of saturated, even-numbered, and unbranched fatty acids of natural origin such as tricaprin, trilaurin, trimyristin, tripalmitin, and tristearin.

Triglycerides containing polyunsaturated fatty acids may be present in small amounts in the lipid core of Emulsomes, provided that the resulting triglyceride mixture is in the solid or liquid-crystal phase at room temperature.

2. Monoesters

Other preferred lipid core components of Emulsomes are esters of monounsaturated fatty acids. Although monounsaturated fatty acids are capable of undergoing peroxidation, they are less reactive than typical polyunsaturated fatty acids. Natural monounsaturated fatty acids have the *cis* configuration. In general, these are lower melting than completely saturated fatty acid esters. Usually, therefore, monounsaturated fatty acid esters will be most useful in a mixture with higher melting saturated fatty acid esters.

The lipid or hydrophobic core of Emulsomes also may be composed of waxes. In general, waxes are long-chain fatty alcohol esters of fatty acids. Many waxes have suitable melting characteristics for use in Emulsomes, since they are solid at room temperature. Examples include the esters from beeswax and spermaceti, such as cetyl palmitate. Preferred waxes are made from saturated or monounsaturated fatty acids and saturated or unsaturated fatty alcohols. An example of the latter is provided by arachidyl oleate.

Other satisfactory monoesters include solid monoglycerides such as glyceryl monostearate, and fatty acid esters of short-chain alcohols such as ethyl stearate.

3. Cholesteryl Esters and Cholesterol

Cholesterol and cholesteryl esters optionally may be incorporated into the lipid core or the surrounding phospholipid envelope of Emulsome particles. Cholesterol and its esters change the packing structure of lipids, and in high concentrations they induce the formation of a liquid-crystal phase. A liquid-crystal phase may coexist with a solid phase under some conditions.

Preferred cholesteryl esters are those of saturated or monounsaturated long-chain fatty acids, such as palmitoyl or oleoyl, respectively. Cholesteryl esters may be present in levels up to 50 mol% relative to the triglyceride or other solid lipid core component.

Squalene (an all trans isoprenoid containing six isoprene units, intermediate in biosynthesis of cholesterol) and squalane (obtained by complete hydrogenation of squalene) can be also added to the lipid phase of Emulsomes. They tend to increase stability of lipoidal vehicles and to increase loading capacity of lipid-soluble compounds.

4. Phospholipids

The preferred phospholipids which constitute the surrounding envelopes of Emulsomes are lecithins such as soybean lecithin or egg lecithin, pure phospholipids like phosphatidylcholine, synthetic phospholipids such as dimyristoyl or dipalmitoylphosphatidylcholine, and hydrogenated or partially hydrogenated lecithins and phospholipids.

Many phospholipid-based particles undergo a well-defined transition from gel to liquid-crystalline phase, which can be recorded using a differential scanning calorimeter. The phase transition temperature, T_m, primarily depends on the particle composition; however, it also depends on particle size and curvature.[10] T_m is also strongly dependent on phospholipid chain length and saturation. The phospholipid components of Emulsomes may be either saturated or unsaturated, and may have a gel to fluid phase transition temperature either above or below room temperature. Egg or soy phosphatidylcholines (egg or soy PC) are examples of phospholipids with transition temperatures well below room temperature (about –10°C for egg PC). Dimyristoyl phosphatidylcholine (DMPC) has a transition temperature close to room temperature (21 to 24°C). Dipalmitoylphosphatidylcholine (DPPC) and distearoylphosphatidylcholine (DSPC) are examples of phospholipids with transition temperatures well above room temperature and in fact even above physiological body temperature (T_m = 41 and 58°C respectively). Acceptable Emulsomes may be made with these and many other phospholipids. In general, Emulsomes prepared with phospholipids which are in the gel phase at 37°C are expected to have more rigid bilayer envelopes.

Emulsomes may be prepared with molar ratios of phospholipid to total lipid in the range of 0.1 to 0.5 (10 to 50 mol%).

5. Negatively Charged Lipids

Negatively charged lipid molecules such as oleic acid, or negatively charged phospholipids such as phosphatidylglycerol (PG) can be added to the lipid phase of Emulsomes to increase the zeta potential of the composition, thus stabilizing the particles. Additionally, the incorporation of these negatively charged lipid compounds in Emulsomes results in the formation of phospholipid bilayers with opposing charges, thus increasing the loading of water-soluble molecules in the aqueous compartments formed by the phospholipid bilayers surrounding the lipid core. This effect results from the larger aqueous spaces between the bilayers caused by the electrostatic repulsion between them. Another beneficial role of the inclusion of negatively charged lipid molecules in emulsomes is to reduce the likelihood of particle aggregation, which minimizes destabilizing processes such as coalescence, flocculation, or fusion. Aggregation is prevented by the repulsive forces between the approaching particles.

Negatively charged phospholipids such as phosphatidylglycerol have been incorporated into liposomal formulations used in human clinical studies.[16,17] The significance of zeta potential in analyzing and predicting the properties of phospholipid bilayers is discussed elsewhere.[18] Both lipoidal particle size and particle stability vary as a function of zeta potential. For liposomes, zeta potential and particle size increase in proportion to the content of negatively charged phospholipid, up to 50 weight % of negatively charged phospholipid.

The preferred range of negatively charged lipid in Emulsome particles is 0 to 30 mol% relative to total phospholipid and charged lipid.

6. Antioxidants

The lipid core of Emulsome particles optionally may contain one or more antioxidants. A preferred antioxidant is α-tocopherol or its derivatives, which are members of the vitamin E family. Other antioxidants include butylated hydroxytoluene (BHT) and butylated hydroxyanisol (BHA). Usually the concentration of lipid antioxidant used is in the range of 0.005 to 0.02%.

Antioxidants lessen the formation of oxidative degradation products of unsaturated lipids, such as peroxides. The need for antioxidants may be lessened by preparing the lipid core from saturated fatty acids.

B. COMPOSITION OF THE CONTINUOUS AQUEOUS PHASE

Emulsomes are suspended in an aqueous medium as the continous phase which may be water, saline, or any other suitable aqueous solution which can yield an isotonic- and pH-controlled preparation.

In addition, the continuous aqueous phase of Emulsome formulations may also comprise conventional additives such as preservatives, osmotic agents, or pressure regulators and antioxidants. Typical preservatives include sorbic acid, chlorobutanol, benzalkonium chloride, parabens, etc. Typical osmotic pressure regulators include glycerol and mannitol, with glycerol being preferred. The aqueous phase may also include antioxidants such as ethylene diamino tetraacetic acid (EDTA), which acts as chelating agent of metal ions, avoiding the formation of radical ions responsible for chemical oxidation reactions.

C. INCORPORATION OF ACTIVE MATERIALS

1. Lipophilic Compounds

Water-insoluble compounds can be incorporated in Emulsomes by dissolving them in a suitable organic solvent together with the lipid ingredients of the composition, phospholipids, and solid fats, evaporating the solvent until complete dryness, and hydrating the drug–lipid mixture with the aqueous phase utilizing mechanical shaking. The resultant lipid dispersion is consequently homogenized by a high-shear homogenizer for several minutes to get micronrange particles and finally the formulation is submitted to high pressure homogenization (800 bar) to reduce the particle size to the nano-size range (see Section IV.D).

2. Hydrophilic Compounds

Water-soluble materials or active ingredients may be encapsulated in Emulsomes by dissolving them in the aqueous medium, hydrating the dry fat–phospholipid mixture with the aqueous phase utilizing mechanical shaking, and sizing the resultant dispersion by high pressure homogenization to the desired final size range as described in more detail in Section IV.D.

D. MANUFACTURING PROCESS

Figure 4 describes the different steps involved in the preparation of Emulsomes.

1. Dissolution of Lipids

All the hydrophobic components, solid fat, phospholipids, fat-soluble antioxidant, and lipophilic active material are dissolved in a volatile and chemically unreactive organic solvent such as dichloromethane and transferred to a round-bottomed flask. The solvent is removed, typically under reduced pressure in a rotary evaporator (Heidolph, Germany) or under a stream of inert gas until complete dryness is achieved.

2. Hydration of Lipid Phase

The resulting lipid film is hydrated and dispersed by covering and shaking with the aqueous solution. The hydrophilic active material is dissolved in the aqueous hydration solution. Hydration is performed by

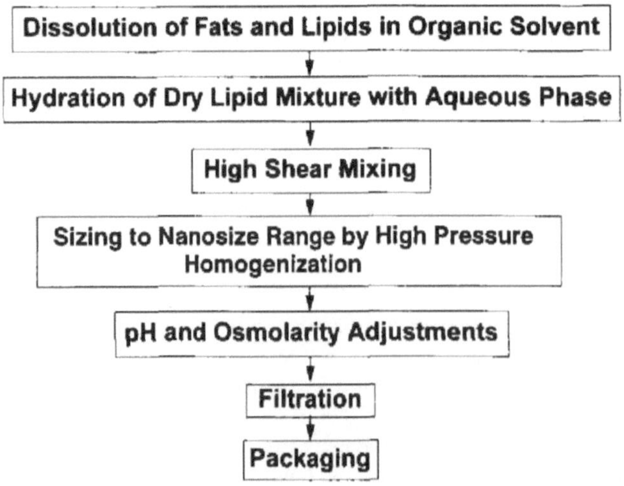

Figure 4 Flow chart of manufacturing process of Emulsomes.

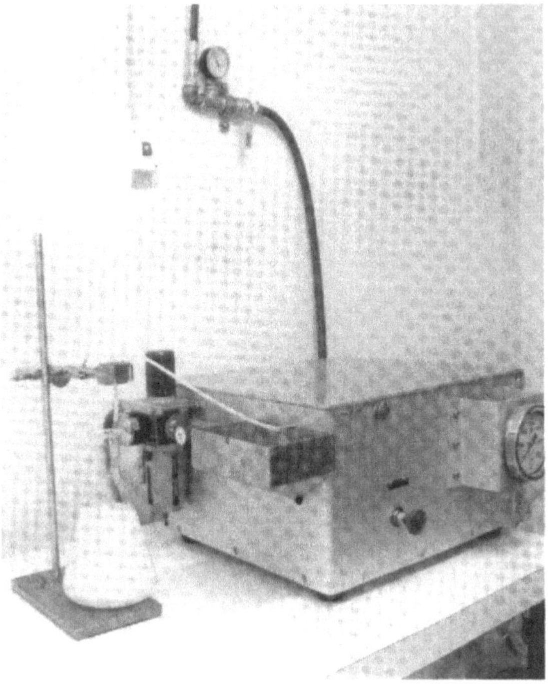

Figure 5 EmulsiFlex™ C-30 high pressure homogenizer (Avestin Inc., Ottawa, Canada).

mechanical shaking at room temperature for 1 to 2 h using a Multi-Wrist shaker (Lab-Line Instruments, Inc., IL)[19] until all the lipids are homogeneously dispersed in the aqueous phase.

3. High Shear Mixing

The dispersion is homogenized usually for 5 min at 15,000 rpm using a Polytron PT 3000 high shear homogenizer (Kinematica, AG, Switzerland). The speed can be varied depending on the exact composition of the Emulsome formulation. Thus, for preparations containing protein material, it should be slow to avoid foaming and denaturation. At this step the particle size of the resultant Emulsome dispersion is in the micron range.

4. Sizing to Nano-Size Range

The lipid suspension or dispersion is sized to submicron range, typically by high pressure homogenization at pressures up to 800 bar in a EmulsiFlex™ C-30 (Figure 5) or Gaulin Micron Lab 70 (Figure 6) high pressure homogenizers. The use of high pressure Gaulin homogenization was recently described in detail by Brandl et al.[20] The EmulsiFlex™ C-30 is a new high pressure homogenizer developed by Avestin, Inc. (Canada) for continuous processing of sample volumes ranging from 8 ml and up to several liters, at adjustable pressures up to 30,000 psi (ca. 2000 bar = 200 MPa) and maximum flow rate of 12,000 l/h. This homogenizer is also equipped with a stainless steel heat exchanger which allows cooling or heating during homogenization. The entire instrument including the homogenizing valve is easily accessible for cleaning. The instrument is also sterilizable by autoclaving, allowing the processing of sterile samples. Alternative high pressure homogenizers such as Rannie (APV Rannie AS, Denmark), Nanojet® (Nanojet Engineering GmbH, Germany), and Niro-Soavi (Niro Atomizer Co., Hudson, WI) homogenizers are also suitable for the preparation of Emulsomes.

Usually the particle size distribution of the resultant Emulsome formulations, determined using a N4MD Coulter particle size analyzer (Coulter Electronics, England) using the differential weight % mode of the instrument, indicates the existence of single homogeneous populations of Emulsome particles with a mean particle diameter in the range of 50 to 250 nm.

Emulsomes may also be prepared by high pressure extrusion through polycarbonate membranes. In this procedure, the sizing step of the lipid dispersion is performed using a pressure extruder, such as the stainless steel GH76-400 Extruder or Pressure Cell (Nucleopore, U.S.A.), rather than a high pressure homogenizer. A modification of the extrusion technique using the GH76-400 pressure extruder with additional accessories for liposome preparation is described in detail by Amselem et al.[19]

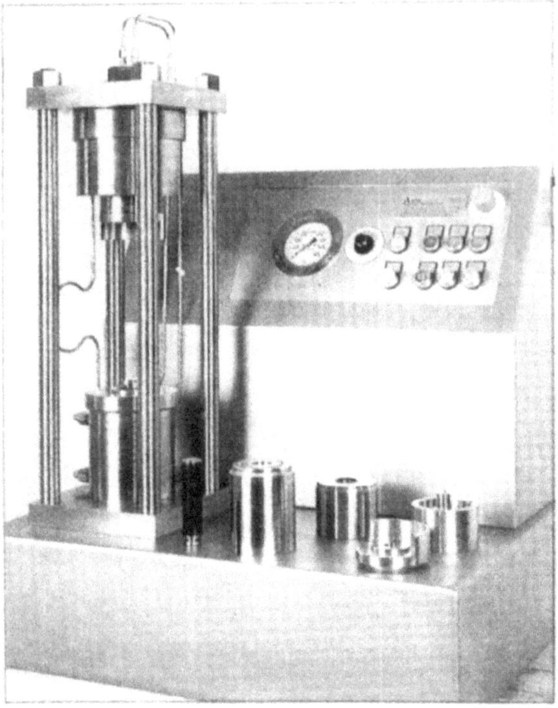

Figure 6 Gaulin Micron Lab 70 high pressure homogenizer (APV International SA, Holland).

5. pH and Osmolarity Adjustments

Since changes in pH and osmolarity can take place during the manufacturing process, it is recommended after the sizing step to readjust the osmolarity of the resultant Emulsome formulation to the required values by adding glycerol and the pH by titration with NaOH or HCl solutions.

6. Final Filtration

Due to the small size of Emulsomes, they can be sterilized by final sterile filtration through 0.2 μm filter membranes made of cellulose acetate, polycarbonate, polysulfone, nylon, or PTFE materials. We used a disposable 115 ml filter system with low extractable sterilized nylon membrane (Corning, Inc., NY).

E. POLYMERIC COATING OF EMULSOMES™

Biodegradable polymers can be incorporated in Emulsomes. Polymeric Emulsomes can be prepared containing biodegradable polymers such as polyesters of lactic and glycolic acids, polyanhydrides, polycaprolactones, polyphosphazenes, polyorthoesters, or natural polymers like gelatin, albumin, and collagen. The advantages of polymeric Emulsomes is in the area of controlled release for the delivery of drugs and biological compounds in a sustained dosage form.

The structure, selection, and use of degradable polymers in drug delivery vehicles have been reviewed in recent publications by Domb and Amselem.[21-25] In general, the ratio of polymer to lipid core material (e.g., triglyceride) may be up to 50% (w/w).

For most synthetic polymers, which are soluble in organic solvents, the polymer may be codissolved with the solid fat and phospholipid prior to the evaporation step. For natural polymers which are soluble in aqueous solution, the polymer may be dissolved in the water phase used.

V. APPLICATIONS

In the therapeutic area, the use of parenteral emulsions as drug delivery systems is still comparatively rare because of the necessity of achieving stable microdoplets of less than 1 μm to prevent emboli in the blood vessels. Thus, formulation options are severely restricted by the very limited selection of stabilizers and emulsifiers approved for parenteral injection. Due to their detergent characteristics, most of them are hemolytic agents acting as membrane solubilizers and causing toxic side effects. Since Emulsomes are formulated in the absence of any synthetic ionic or nonionic surfactant such as poloxamers, polysorbates, or deoxycholate, their use as a drug delivery system has demonstrable advantages, particularly for parenteral administration of poorly water-soluble lipophilic drugs which either could not be administered parenterally in the absence of cosolvents or toxic surfactants or, if so administered, would cause undesirable side effects. Preliminary results obtained in our lab have shown Emulsome formulations to provide effective pharmaceutical delivery for a broad variety of lipophilic and hydrophilic drugs with minimal local or systemic toxicity.

In veterinary medicine, Emulsomes have been shown in animal preclinical studies to serve as vaccine carriers and adjuvants, enhancing the immunogenicity of several antigens.[26–29] The ability of the Emulsome delivery system to enhance immunogenicity is related to the physicochemical characteristics of Emulsome particles. Vaccines formulated in Emulsomes have demonstrated enhanced antibody production and the ability to evoke strong immune responses for a broad variety of both lipophilic and hydrophilic immunogens with minimal local or systemic toxicity. Emulsomes can also be formulated to act as a depot for slow release of antigens, avoiding the need for repeated vaccinations.

Mucosal immunity plays an important role in the early defense against viral infections and other diseases. Preclinical studies have shown induction of mucosal and systemic immunogenicity in mice vaccinated intranasally with mucoadhesive Emulsome vaccines.[26–29] Immunization against biologic threat agents, such as staphylococcus enterotoxin B (SEB) requires the development of vaccines that can protect against respiratory challenge. Intranasal immunization of mice with SEB-toxoid Emulsome vaccine conferred intranasal immunogenicity.[26] Mice immunized with SEB-toxoid Emulsome vaccine were also protected against systemic challenge with lethal doses of SEB.[29]

The enhanced immunogenicity obtained with the Emulsome vaccine carrier system are encouraging for the development of safe, cost-effective, and immunogenic veterinary vaccines to protect against bacterial, viral, and parasitic infections.

Emulsions play a critical role in the cosmetic field.[30] They offer many advantages to the cosmetic chemist, allowing compounding immiscible ingredients into single formulations. Liposomes have also gained major interest in cosmetics in recent years.[31–33] Marketed products containing liposomes and tested by dermatologists revealed a whole series of positive effects including improvement of skin texture, better cutaneous hydration, increase in skin glow, decrease of wrinkles depth, and decrease of eye puffiness.

Since Emulsomes are considered as an intermediate system between emulsions and liposomes, having part of the characteristics of both lipoidal vehicles, we believe that topical fomulations of Emulsomes may have potential applications in cosmetics. Generally, the best way of applying Emulsomes as natural moisture regulators is adding them to any kind of gel-like lotion or cream preparations. Emollients and lubricants can be easily incorporated in Emulsome formulations. Emollients and lubricants are used in cosmetics to improve consumer acceptance of the product by providing skincare preparations with the appropriate slip, tactile feel, and rub-in properties to encourage the consumer to use the product more liberally and more frequently. Emulsomes can be also considered for use in topical preparations to replace alcohols used as solvents and coolants (e.g., aftershaves and antiseptic solutions). Other potential cosmetic uses of interest for Emulsomes might include hair lotions, sunscreens, after-sun gels, and encapsulation of fragrances and perfumes. Water-insoluble substances used in personal care products as antioxidants such as tocopherols (vitamin E) and retinoids (vitamin A) have been successfully incorporated in Emulsome formulations.

Emulsions are of widespread importance in the food industry.[34,35] Dispersion of one medium in another is required in many food products for control of texture, appearance, and flavor. The use of emulsions in agriculture is also very broad.[36] Emulsomes may be useful in agriculture as a vehicle for improved delivery of fertilizers, pesticides, and fungicides and in the food industry as a vehicle for the encapsulation of flavors or to improve the stability of products with high lipid content.

REFERENCES

1. Patent applications.
2. **Amselem, S., Yogev, A., Zawoznik, E., and Friedman, D.,** Emulsomes, a novel drug delivery technology, *Proc. Int. Symp. Control. Rel. Bioacti. Mater.*, Nice, 1994, 21.
3. **Barenholz, Y. and Amselem, S.,** Quality control assays in the development and clinical use of liposome-based formulations, in *Liposome Technology*, Vol. 1, Gregoriadis, G., Ed., 2nd ed., CRC Press, FL, 1993, 527.
4. **Lichtenberg, D., Amselem, S., and Tamir, I.,** *Biochemistry*, 18, 4169, 1979.
5. **Amselem, S., Schwarz, Y., Cohen, A., Wellner, E., Neumann, R., Beilin, M., Garti, N., and Friedman, D.,** Submicron emulsion as ocular delivery system for adaprolol maleate, a soft beta blocker, *Pharm. Res.*, 10, S205, 1993.

6. **Amselem, S., Schwarz, Y., Cohen, A., Wellner, E., Baru, H., Beilin, M., Bar-Ilan, A., Neumann, R., and Friedman, D.**, Formulation in submicron emulsion of HU-211, a novel non-psychotropic cannabinoid with ocular hypotensive activity, *Pharm. Res.*, 10, S236, 1993.
7. **Bar-Ilan, A., Baru, H., Beilin, M., Friedman, D., Amselem, S., and Neumann, R.**, Extended activity and increased bioavailability of indomethacin formulated in submicron emulsion, compared to commercially available formulation, *Reg. Immunol.*, 6, 166, 1994.
8. **Y. Barenholz, Gibbs, D., Litman, B., Gall, J., and Thompson, T. E.**, A simple method for the preparation of homogeneous phospholipid vesicles, *Biochemistry*, 16, 2806, 1977.
9. **Amselem, S., Cohen, R., and Barenholz, Y.**, In vitro tests to predict in vivo performance of liposomal dosage forms, *Chem. Phys. Lipids*, 64, 219, 1993.
10. **Lichtenberg, D. and Barenholz, Y.**, Liposomes: preparation, characterization and preservation, in *Methods in Biochemical Analysis*, Vol. 33, Glick, D., Ed., John Wiley & Sons, New York, 1988.
11. **Perkins, W. R., Minchey, S. R., Ahl, P. L., and Janoff, A. S.**, The determination of liposome captured volume, *Chem. Phys. Lipids*, 64, 197, 1994.
12. **Oku, N., Kendall, D. A., and McDonald, R. C.**, A simple procedure for the determination of the trapped volume of liposomes, *Biochim. Biophys. Acta*, 691, 332, 1982.
13. **Goren, D., Gabizon, A., and Barenholz, Y.**, Correlation between physical characteristics and pharmacological behaviour of doxorubicin-containing liposomes, *Biochim. Biophys. Acta*, 1029, 285, 1990.
14. **Small, D.**, Glycerides, in *The Physical Chemistry of Lipids from Alkanes to Phospholipids*, Plenum Press, New York, 1985, chap. 10.
15. **Kates, M.**, *Techniques of Lipidology*, Elsevier, New York, 1972, chap. 1.
16. **Amselem, S., Gabizon, A., and Barenholz, Y.**, Optimization and upscaling of doxorubicin-containing liposomes for clinical use, *J. Pharm. Sci.*, 79, 1045, 1990.
17. **Amselem, S., Cohen, R., Druckmann, S., Gabizon, A., Goren, D., Abra, R. M., Huang, A., New, R., and Barenholz, Y.**, Preparation and characterization of liposomal-doxorubicin for human use, *J. Liposome Res.*, 2, 93, 1992.
18. **Sai-lung, L.**, Microelectrophoresis of liposomes, in *Liposome Technology*, Vol. 1, 2nd ed., Gregoriadis, G., Ed., CRC Press, Boca Raton, FL, 1993, 331.
19. **Amselem, S., Gabizon, A., and Barenholz, Y.**, A large scale method for the preparation of sterile and nonpyrogenic liposomal formulations of defined size distributions for clinical use, in *Liposome Technology*, Vol. 1, 2nd ed., Gregoriadis, G., Ed., CRC Press, Boca Raton, FL, 1993, 501.
20. **Brand, M. M., Bachmann, D., Drechsler, M., and Bauer, K.**, Liposome preparation using high-pressure homogenizers, in *Liposome Technology*, Vol. 1, 2nd ed., Gregoriadis, G., Ed., CRC Press, Boca Raton, FL, 1993, 49.
21. **Domb, A. J., Amselem, S., Shah, J, and Maniar, M.**, Degradable polymers for site-specific drug delivery, *Polymers Adv. Technol.*, 3, 279,1992.
22. **Domb, A. J., Amselem, S., Shah, J., and Maniar, M.**, Polyanhydrides: synthesis and characterization, *Adv. Polym. Sci.*, 107, 93, 1993.
23. **Domb, A., Amselem, S., and Maniar, M.**, Biodegradable polymers as drug carrier systems, in *Polymeric Biomaterials*, Dimitriu, S., Ed., Marcel Dekker, New York, 1994, 399.
24. **Domb, A. and Amselem, S.**, Antibiotic delivery systems for the treatment of chronic bone infections, in *Polymeric Site-Specific Pharmacotherapy*, Domb, A. J., Ed., John Wiley & Sons, Chichester, 1994, 244.
25. **Domb, A., Amselem, S., Langer, R., and Maniar, M.**, Polyanhydrides as carriers of drugs, in *Biomedical Polymer: Designed to Degrade Biomedical Polymer Systems*, Shalaby, S.W., Ed., Hanser Publishers, New York, 1994, 69.
26. **Kaminski, R., Grate, S., Aboud-Pirak, E., Hooper, C., Levin, T., Weiss, I., Amselem, S., Arnon, R., and Lowell, G.**, Comparison of induction of respiratory IgA and serum Ig by intranasal and parenteral proteosome vaccines formulated with staphylococcal enterotoxin B toxoids or ricin peptides, in *Proc. Med. Defense Biosci. Rev.*, Baltimore, MD, 1993.
27. **Lowell, G. H., Kaminski, R., Mallett, C., Orr, N., Hartmann, A., Cohen, D., Amselem, S., Echeverria, P., and Hale, T. L.**, Progress in the development of intranasal and oral proteosome *Shigella* vaccines, paper presented at the Sixth Annual Joint IDF-US Army Conference on Infectious Diseases and Vaccines of Military Importance, Herzlia, Israel, 1994.
28. **Kaminski, R., Amselem, S., Levi, M., Loomis, L., Kersey, K., Vancott, T., Friedman, D., Smith, G., Wahren, B., Redfield, R., Birx, D., and Lowell, G.**, Parenteral or intranasal immunization with HIV gp160 formulated with proteosomes and/or pharmos adjuvants enhances epitope-specific IgG or IgA, in *Proc. ASM Conf.*, Las Vegas, NV, 1994.
29. **Colleton, C., Kaminski, R., Smith, A., Charney, C., Slike, B., Foster, O., Amselem, S., Grate, S., and Lowell, G.**, Development of a murine model for respiratory challenge with staphylococcal enterotoxin B to evaluate vaccine efficacy, in *Proc. ASM Conf.*, Las Vegas, NV, 1994.
30. **Breuer, M. M.**, Cosmetic emulsions, in *Encyclopedia of Emulsion Technology*, Vol. 2, Becher, P., Ed., Marcel Dekker, New York, 1985, 385.
31. **Siciliano, A. A.**, Topical liposomes, an update and review of uses and production methods, *Cosmetics Toiletries*, 100, 43, 1985.

32. **Perrier, P. and Redniziak, G.,** Liposomes in cosmetics, *Soap Perfum. Cosmetics*, 62, 29, 1989.
33. **Brooks, G. J. and McManus, R.,** Finding new uses for liposomes in cosmetics, *Inform*, 1, 891, 1990.
34. **Krog, N. J. and Riison, T. H.,** Applications in the food industry I, in *Encyclopedia of Emulsion Technology*, Vol. 2, Becher, P., Ed., Marcel Dekker, New York, 1985, 321.
35. **Jaynes, E. N.,** Applications in the food industry II, in *Encyclopedia of Emulsion Technology*, Vol. 2, Becher, P., Ed., Marcel Dekker, New York, 1985, 367.
36. **Becher, D. Z.,** Applications in agriculture, in *Encyclopedia of Emulsion Technology*, Vol. 2, Becher, P., Ed., Marcel Dekker, New York, 1985, 239.

Chapter 12

Applications of the Nano-Sized Matrix of the Bicontinuous Cubic Phase

Sudhakar Puvvada and Banahalli R. Ratna

CONTENTS

A variety of surfactants and lipids, when dissolved in water, are known to exhibit liquid-crystalline bicontinuous cubic phases[1,2] which are part of a larger class of self-assembled structures that include micelles, microemulsions, vesicles, and other liquid-crystalline phases such as lamellar and hexagonal phases. They are described as bicontinuous because a convoluted lipid bilayer divides space into two disconnected regions that are simultaneously continuous.[3,4] The structures of these phases have been described in terms of infinite periodic minimal surfaces (IPMS).[5] These phases are unusual in that the fluid bilayer with a liquid-like disorder exhibits long-range three-dimensional order with cubic symmetry. The existence of cubic symmetry in these phases and the continuity of both the lipid and aqueous regions have been well established.[6] So far, bicontinuous cubic phases exhibiting double diamond (Pn3m), primitive (Im3m), and gyroid (Ia3d) symmetries[7] have been found.

The cubic phases exhibit certain unique properties. They consist of uniform aqueous pores which are multiconnected. The sizes of the pores can be controlled by varying water concentration and sample temperature, by the addition of polar and nonpolar additives, and by mixing the lipid/surfactant with other lipids and surfactants. The cubic phases also exhibit a large interfacial area between the hydrophobic and hydrophilic regions which as mentioned earlier are simultaneously continuous. These properties make the cubic phase a desirable matrix for a number of technological applications[8] such as catalysis, bioremediation, controlled release,[9] nanoparticle synthesis,[10] and ultrafiltration. Over the last year, we have utilized the cubic phase matrix to synthesize[10] metal nanoparticles and to release[11] proteins and drugs in a controlled manner.

I. SYNTHESIS OF METAL PARTICLES

A. INTRODUCTION

The technological advances in controlling the microstructure of inorganic materials in the nanometer scale have led to tremendous success in the preparation of ceramics, synthetic layered structures, and semiconductors. Our interest lies in the development of methods by which nanometer-sized particles can be synthesized in the confined aqueous regions of the cubic phase. Materials with particles in the size range of 1 to 100 nm exhibit novel electronic, optical, magnetic, and chemical properties due to their extremely small dimensions. These nanoparticles have potential application in industrial processes such as catalysis or in technological products such as magnetic and electronic devices.

Nanoparticles have been synthesized using various approaches which include both physical and chemical methods.[12] The most common method of synthesis is by precipitation from homogeneous solutions.[13] One major problem is that the shape and composition of the synthesized particles is very sensitive to experimental conditions and cannot be well controlled. The chemical methods, currently used

0-8493-4012-8/96/$0.00+$.50
© 1996 by CRC Press, Inc.

in producing fine particles, include the use of appropriate reducing agents such as hydrogen, borohydride, or polyols. In the polyol process, a metal salt is reduced by liquid polyols at temperatures between 100 to 200°C. The polyol process has also been used[14,15] to intercalate palladium crystallites in smectite clays such as montmorillonite for use as catalysts.

The various microstructures into which lipids and surfactants self-assemble provide compartmentalized regions which can be used for the synthesis of fine particles.[16,17] The particles that are produced inside these phases have compositions and crystalline structures that are quite different from those formed in solutions. The larger structures such as vesicles, tubules, and microemulsions exhibit pores whose sizes are typically in the 0.1 to 1 μm range. Various kinds of inorganic materials in this size range have been synthesized inside microemulsions[18,19] and vesicles.[20] The liquid-crystalline phases such as the lamellar, hexagonal, and cubic phases consist of aqueous regions in the nanoscale. The hexagonal phase has been used to produce very uniform particles of copper nitrate and copper sulfate.[21] It has also been used as a template to produce mesoporous molecular sieves.[22,23] We have demonstrated[10] for the first time the possibility of utilizing the bicontinuous cubic phase of a lipid–water system to produce controlled-size metal nanoparticles. The uniformity in the size of the formed particles that we have been able to achieve underscores the critical role played by the lipid matrix of the bicontinuous cubic phase. The convoluted lipid bilayer of the cubic phase not only provides spatially localized compartments for metal reduction but also minimizes the degree of particle aggregation. Further, it imposes an upper limit to the particle size. In contrast, the particles synthesized in the aqueous space between the bilayers in a lamellar phase have a wider size distribution and appear to form plate-like particles.

B. GENERAL MECHANISM

The general approach that was used to produce the palladium particles is a modified polyol process[10] where the metal salt was reduced by the glyceride headgroup of the lipid, glycerol monooleate (GMO). The cubic or lamellar phase was first formed by mixing the lipid with aqueous solutions of the palladium salt. Reduction, which occurs at room temperature, is quite rapid, and the formation of the particles can be seen within an hour by the appearance of a black coloration to the sample. The reducing action of GMO was identified by examining the FTIR spectra of the extracted lipid before and after the redox reaction. The spectra suggest a partial oxidation of the primary $-CH_2OH$ group in GMO to an aldehyde ($-CHO$). The oxidation of the lipid is accompanied by a simultaneous reduction of the chloropalladate salt to metallic palladium.

C. FORMED PARTICLES CHARACTERISTICS

Particles produced by this process have a mean size of between 20 and 100 Å depending on the lipid concentration and the liquid-crystalline phase of the sample. Wide-angle X-ray scattering, electron diffraction, and high resolution TEM revealed that the formed particles are metallic palladium exhibiting a face centered cubic (fcc) structure with a unit cell of 3.9 Å. Figure 1a shows a bright field TEM image and an histogram of the palladium particles formed inside a cubic phase containing 60% GMO (with ~50 Å diameter aqueous channels) for a salt concentration of 50 m*M*. The images were analyzed by separating the particles into two classes based on a parameter called the roundedness. Class I consists of approximately circular particles which have a roundedness of less than 1.1. Particles with a larger aspect ratio (roundedness value greater than 1.1) are included in class II. More than 80% of the formed particles fall under class I and have an average size of 38.9 Å with a standard deviation of 7.4 Å. Figure 1b shows the dark field TEM image of the same sample (shown in Figure 1a) from which the crystallite size can be estimated. Using the same classification as before, crystallite size of particles in class I is estimated to be 39.6 ± 8.6 Å, which is almost the same as the particle size. This suggests that primarily, monocrystalline particles are formed. This was confirmed by high resolution TEM which clearly showed a single set of the (111) lattice planes. Some amorphous particles are however seen in the high resolution electron microscopy images.

Particles were also synthesized inside the lamellar phase (Figure 2) exhibited by GMO at relatively high lipid concentrations. The aqueous regions, where the reduction occurs, in the lamellar phase are not compartmentalized as short cylindrical pores but as flat sheets whose thickness is determined by the lipid/water ratio. For a sample containing 85% GMO, the thickness of the aqueous region would be between 10 to 15 Å. However, the particles that form are considerably larger. Approximately 30% of the formed particles belong to class I (roundedness <1.1) with a mean size of 47.3 Å and a standard deviation of 8 Å. The remaining 70% particles belong to class II with a size of 63.3 ± 21.9 Å. The larger size of the formed

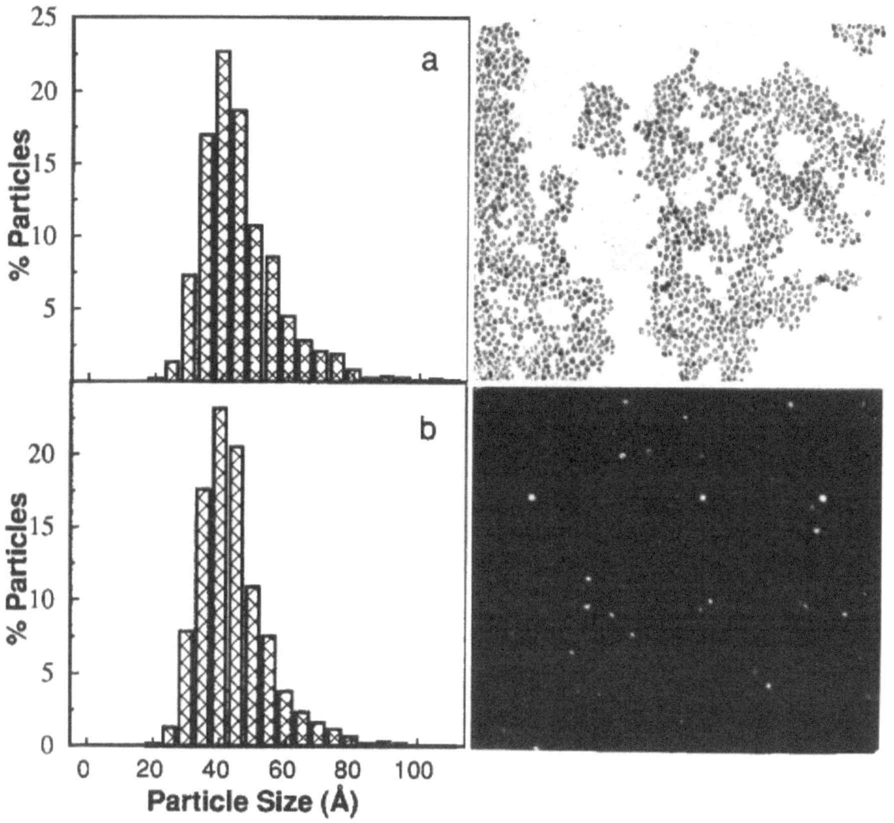

Figure 1 Palladium particles formed in the cubic phase. The sample contained 60 wt% GMO and a salt concentration of 50 m*M*. (a) Bright-field TEM image and the corresponding histogram of the formed particles and (b) dark-field TEM image and an histogram of the palladium crystallites.

particles compared to the thickness of the water layer probably indicates the formation of particles in the shape of flat plates. The actual shape cannot be inferred from the two-dimensional TEM images if the flat plates lie parallel to the EM grid. Presently, we are carrying out AFM studies to determine the third dimension of the particles. The formed particles also seem to consistently cluster in contrast to the cubic phase particles where clustering is minimal. The dramatic difference in the degree of particle aggregation can be directly related to the topological differences between the lamellar and cubic phases. In the cubic phase, the presence of nodes in the cubic structure imposes a three-dimensional restriction on the movement of the particles. On the other hand, in the lamellar phase, the restriction is limited only in the direction of the layer normal and the formed particles are free to move within the flat aqueous regions between lipid bilayers.

II. CONTROLLED RELEASE USING THE CUBIC PHASE

A. INTRODUCTION

A wide variety of systems have been designed for delivering, in a controlled manner, active substances such as hormones, enzymes, anticancer agents, opthalmics, fertilizers, and pesticides.[24] These compounds usually have a concentration range above which they are toxic and below which they are ineffective. Often the two concentrations are close together and the active compound has to be used over a very narrow concentration range. When administered conventionally, the initial concentration of the active substance is much larger than required, resulting in often toxic side effects. The concentration then decreases rapidly to values where it is ineffective. The object of any controlled release system is to deliver the active substance within the narrow concentration range for prolonged periods of time, thus minimizing the side effects associated with the compound. In addition to reducing the toxicity, release systems are also designed to protect the compound and enhance its stability. The various methods that have been

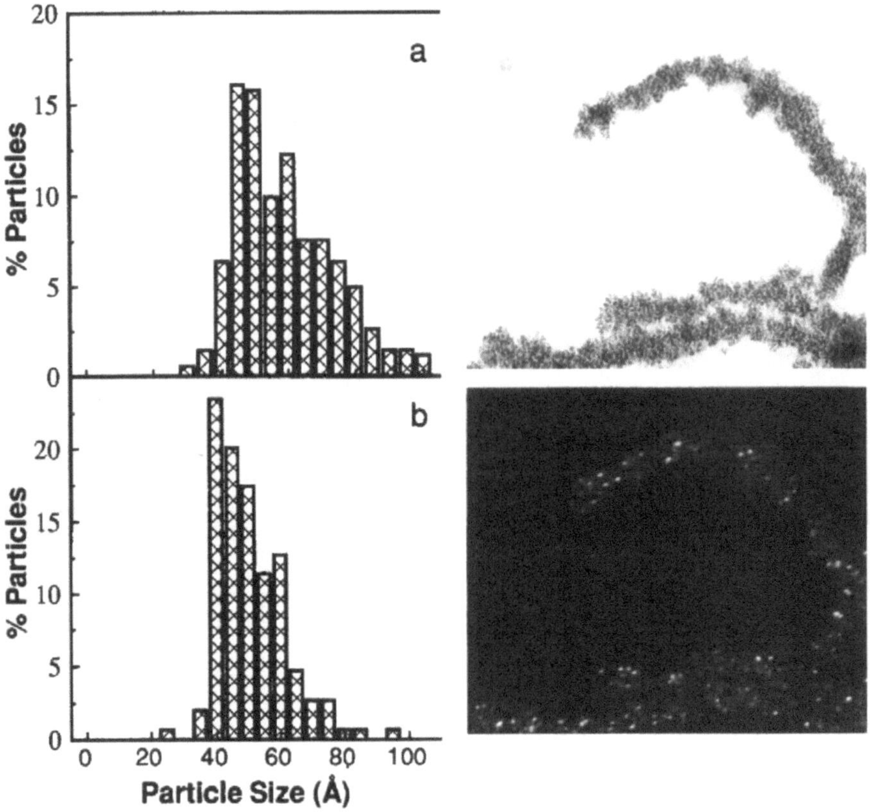

Figure 2 Palladium particles formed inside the lamellar phase. The sample contained 85 wt% GMO and a salt concentration of 50 m*M*. (a) Bright-field TEM image and the corresponding histogram of the formed particles and (b) dark-field TEM image and an histogram of the palladium crystallites.

developed include (i) chemical modification of the active compound by attaching it to macromolecules such as proteins and polysaccharides, (ii) mechanical systems where the release is induced by external forces, and (iii) incorporating the active compound into a matrix consisting of polymers, liposomes, or more recently the bicontinuous cubic phase.

Chemical modification[25] is used to selectively modify the properties of active compounds such as their size, stability, and solubility. For this purpose, drugs have been attached by degradable links to soluble macromolecules such as proteins and polysaccharides. By chemically attaching to a monoclonal antibody, drugs can be targeted to specific organs in the body. However, the presence of a larger group can modify and reduce the activity of the substance.

Mechanical systems such as pumps are used to deliver the drugs precisely when and where required. The release rate in these systems is determined by the system and not by the environment. Pumps, however, are bulky, expensive, and have to be implanted surgically. Other systems where the release is controlled externally by a magnetic field or by ultrasonic energy are also available. In these systems, the release rate is modulated by systematic application of the external force.

Delivery systems where the active substance is incorporated into matrices composed of polymers,[26] hydrogels,[27] liposomes,[28] and tubules[29] are used to slow down the release rate. In these systems, the active substance is uniformly dispersed in the matrix and released over time by diffusion out of the matrix. Various polymeric matrices have been developed where the release is initiated and controlled by degradation of the polymer, by the swelling of the polymer matrix with solvent, or by osmotic gradients through a semipermeable membrane. Liposomes can be formulated using a number of lipids which are nontoxic and biodegradable. Their stability has been enhanced by using polymerizable lipids. Hydrogel microspheres have also been used to incorporate proteins and drugs. The release from these systems is typically first-order with most of the drug released at the start. However, the release has been slowed down and the profiles modified by coating them with polymeric membranes.

More recently, the bicontinuous cubic phases have been suggested[9] as vehicles for drug delivery. These phases consist of highly branched uniform water channels whose size (20 to 200 Å) is in the same range as many proteins and drugs. This will permit a control of the release rate based on the size of the active substance. Cubic phases are self-assembled structures and can be easily prepared by mixing the lipid, water, and the active compound in the right proportions. They are very versatile and can be used to incorporate proteins of very different sizes and polarity without losing their structure. Monoolein, the monoglyceride of oleic acid, has been the lipid of choice in many of these studies because it exhibits the cubic phase at room temperature, and more importantly, it retains the cubic structure in the presence of excess water. Release studies on this lipid system have shown that the protein is released gradually over more than a day without any loss in activity.[9] The release was found to be diffusion controlled and exhibited the typical square-root dependency of a matrix system. Recently, we have been able to achieve controlled release over extended periods of time by incorporating alginate, a reversible hydrogel, into the cubic structure.[11] However, in the presence of alginate, the release is found to be more complicated, involving at least two release mechanisms.

B. CONTROLLED RELEASE BASED ON SIZE

The parameters that influence the release characteristics include (i) factors that depend on the solute such as its diffusion constant, solubility, and partition coeffecient[30] and (ii) factors independent of the solute such as porosity, tortuosity of the pores, and the geometry of the matrix. However, in these systems in which the solute size is comparable to the pore size,[31,32] the diffusion constant, which is strongly modified by the geometry of the matrix, is the predominant factor in determining the release rate. It is a function of the molecular size, shape, and flexibility of the diffusing molecule. This effect has been studied using dextran as the diffusing molecule through the cubic matrix of monoolein. The results for four different dextran molecular weights are shown in Figure 3. The initial rate for all four dextrans is quite large and the polymer gets released at a rate of about 0.1 mg/h. Subsequently, the release rate decreases exponentially to

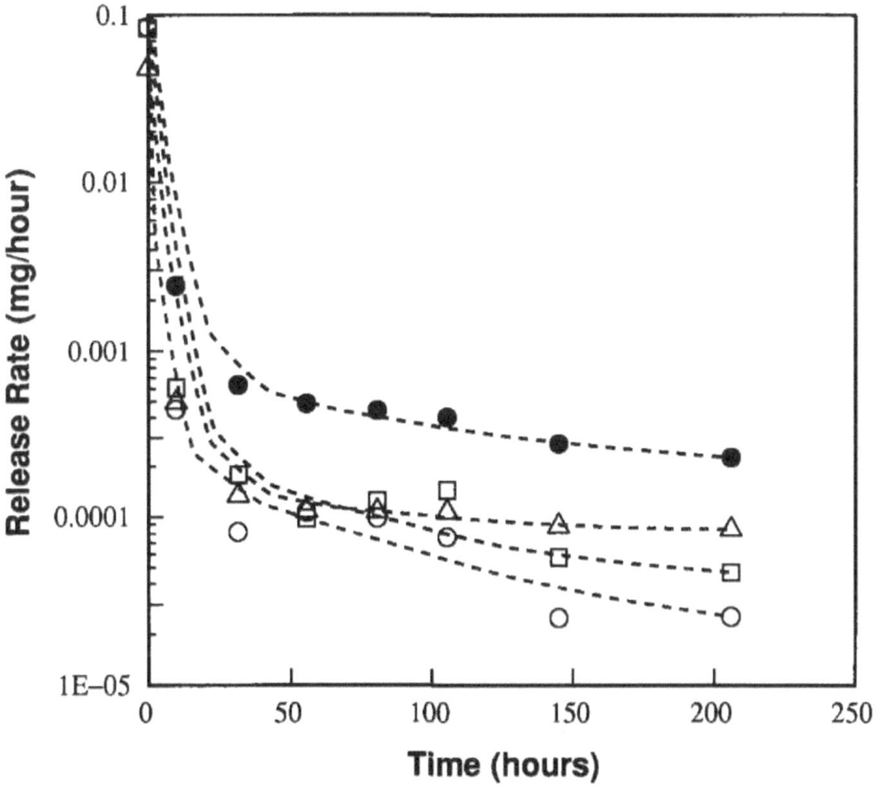

Figure 3 Release rate of dextran from the cubic matrix as a function of molecular weight: 20 kDa (●), 40 kDa (□), 71.2 kDa (△), and 150 kDa (○).

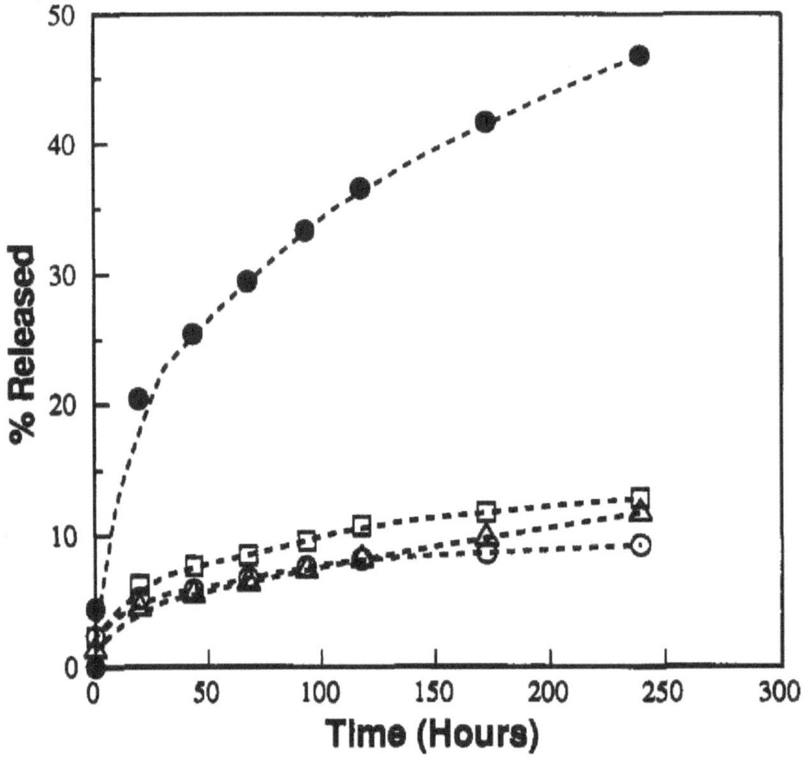

Figure 4 The cumulative release profiles for the four different dextrans. Notation is the same as in Figure 3.

values which are significantly lower and approximately constant. This constant value is almost 10 times larger in the case of dextran with a molecular weight of 20 kDa as compared to the higher molecular weight dextrans (40, 71.2, and 150 kDa). The initial high release rate can be attributed to the dextran at the surface of the sample diffusing out. Subsequently, the polymer has to diffuse from within the bulk of the cubic sample where it is hindered, thus resulting in lower release rate. Figure 4 shows the cumulative release for the four dextrans. It is evident that very little of the high molecular weight dextrans is released, whereas nearly 50% of the 20 kDa dextran comes out in the same time-scale.

The dramatic difference in release rates suggests that the bicontinuous cubic matrix behaves like a sieve. Structural analysis using X-ray scattering reveals that the radius of the aqueous pores is approximately 25 Å. This is considerably smaller than the Stokes radii of the 4 dextrans which are 32, 45, 59, and 85 Å,[32] respectively. The molecules, however, are released through the cubic pores by diffusion because dextran molecules are elongated and rod-like and the Stokes radius is not a reliable indicator of molecular size. Indeed, it has been found in other membrane systems that complete hinderance of dextran occurs only when the Stokes radius is two to three times the pore size.[32] The large difference in the release profiles between the 20 kDa dextran and the rest of the larger ones (see Figure 4) appears to conform to this earlier observation.

C. CONTROLLED RELEASE USING ALGINATE

For a given system, the relative sizes of the pore and the drug/protein determines the release rate, as shown in the previous section. This relative size is fixed, and thus the release rate cannot be altered at will. We have developed a novel method[11] to vary the effective pore size and as a result the release rate ionotropically by the addition of salts which modulate the gelation and degelation of an hydrogel, sodium alginate, which is incorporated in the cubic matrix.

Sodium alginate is a linear polysaccharide composed of mannuronate and guluronate residues[33] in varying proportions. An important property of alginate is its ability to form a reversible gel by reaction with divalent ions such as calcium. These gels, which resemble a solid in retaining their shape and

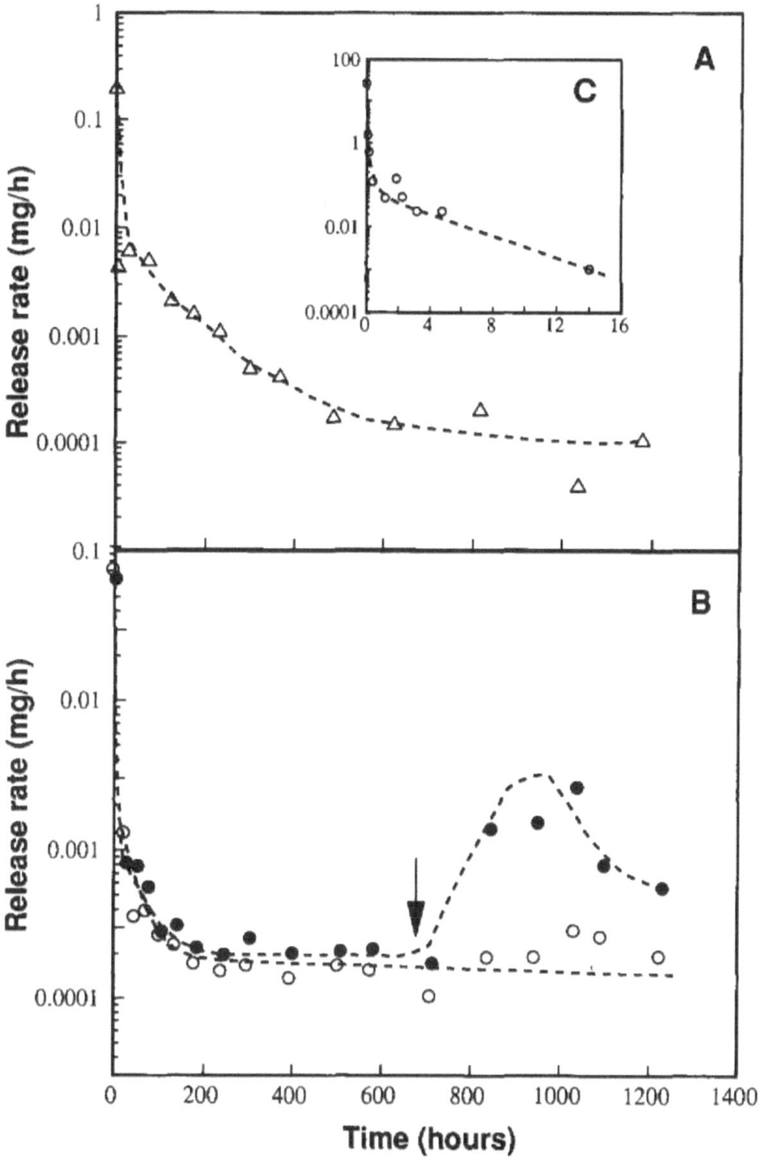

Figure 5 Release rate of BSA from cubic phase samples containing alginate. (A) Ungelled alginate samples. (B) Gelled alginate samples. The arrow represents the time at which the gelled samples (represented by ●) were degelled by adding sodium citrate. Each symbol in the figure represents an average of three samples. (C) For comparison the release from gelled alginate containing no lipid is shown as an inset in Figure 5A. All release rates have been normalized with respect to the total protein added so that the release rates from different samples can be compared.

resisting shear stress, consist mainly of water with only 0.5 to 1 wt% alginate. The bicontinuous cubic structure formed by mixing the lipid in an alginate solution was found[11] to retain its structure.

Figure 5 shows the release rate (normalized with the total protein weight) of BSA from cubic phase samples containing ungelled and gelled alginate. The cumulative release profiles are shown in Figure 6. In the case of ungelled alginate, the release rate is initially very high (0.2/h) and more than 90% of the protein is released in the first 200 h. As the protein is released, its concentration within the cubic phase decreases, and thus the rate of release decreases to values as low as 0.0001/h. This rate is considerably slower than that observed in other matrix systems such as pure gelled alginate (see inset of Figure 5). In the case of pure alginate, almost all of the protein is released in the first few hours. The initial release rates

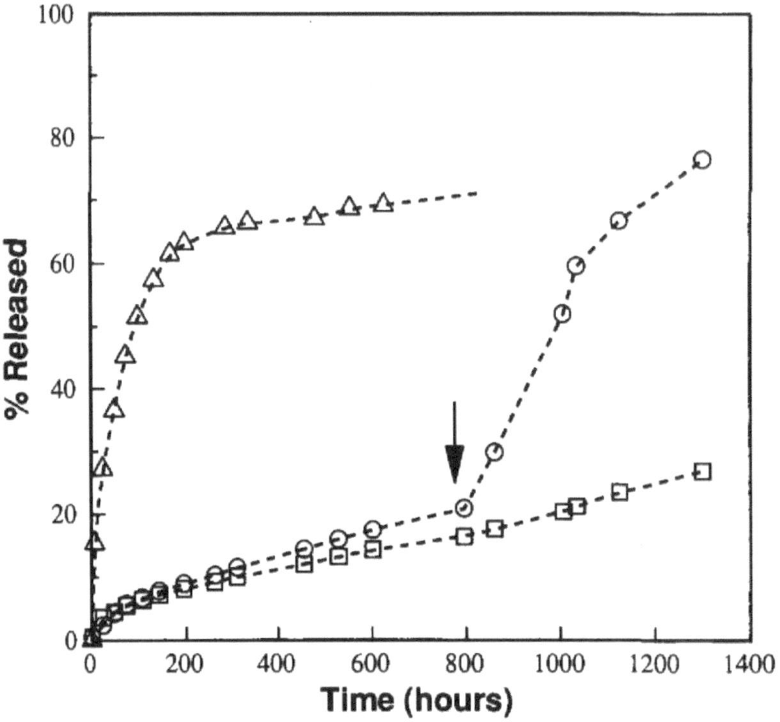

Figure 6 Cumulative release profiles of BSA from gelled (○,□) and ungelled (△) cubic phases. The arrow represents the time at which gelled sample (○) was degelled.

are also considerably higher (10 to 100/h). In both cases, the release rate is large initially and then decreases rapidly to very small values as the concentration of protein in the matrix decreases. This is typical of any matrix system and results in a parabolic cumulative release profile (see Figure 6).

However, when the sample is gelled, the initial release rate decreases by a factor of approximately 30. The rate also remains approximately constant with time (0.0002/h) as compared to the rapid decrease in the ungelled sample. Due to the slower initial release, less than 20% of the protein is released over 1000 h (Figure 5). Upon degelling the sample at around 700 h, the release rate increases to around 0.003/h. This suggests that by controlled degellation, the protein can be released as and when required.

We have modeled the release from the cubic matrices containing gelled or ungelled alginate[34] as occurring by diffusion through two distinct regions, a lipid-rich region with slow diffusion and an alginate-rich region with a fast diffusion. If the release from these two regions is independent and parallel, the cumulative release S(t) can be modeled as the sum of two Fourier series

$$S(t) = 100\left[1 - \sum_{n=0}^{\infty} \frac{2}{\lambda_n^2}\left(xe^{-\lambda_n^2 D_1 t/l^2} - (1 - x)e^{-\lambda_n^2 D_2 t/l^2}\right)\right] \quad (1)$$

where $\lambda_n = (2n + 1)\pi/2$, t is the time elapsed from the start of the release measurement, l the length of the cylindrical sample, D_1 and D_2 the diffusion constants in the lipid-rich and alginate-rich regions respectively, and x the volume fraction of the lipid-rich region (Table 1). For the ungelled samples $x \approx 0.39 \pm 0.02$, $D_1/l^2 \approx 1.4(\pm 0.25) \times 10^{-8}$/s, and $D_2/l^2 \approx 1.6(\pm 0.18) \times 10^{-6}$/s. The value of D/l^2 in the alginate-rich region is approximately 30 times smaller than that in pure gelled alginate and could be attributed to the enhanced tortuosity. However, upon gelling, the value of x increases to 1 without a significant decrease in the diffusion constant of the lipid-rich region. In other words, the volume fraction of the alginate-rich region decreases to zero upon gelling the sample and very little of the protein is released.

In the absence of any alginate, the sample consists primarily of lipid-rich cubic regions with very small diffusion constants. The addition of alginate generates defects with faster diffusion constants and thus the protein is released at a much faster rate. Upon gelling the alginate, the defects disappear and the protein

Table 1 *Values of x, D_1/l^2, and D_2/l^2 for Cubic Samples Containing no Alginate, 0.5% Ungelled Alginate, and 0.5% Gelled Alginate*

	x	$10^9 \times D_1/l^2$(/sec)	$10^6 \times D_2/l^2$(/sec)
Cubic phase with no alginate	0.906 ± 0.02	9.9 ± 2.6	3.9 ± 0.7
Cubic phase with 0.5% ungelled alginate	0.39 ± 0.02	14 ± 2.5	1.6 ± 0.2
Cubic phase with 0.5% gelled alginate	0.999 ± 0.01	9.6 ± 2.4	4.5 ± 0.1
Pure gelled alginate	1.0		46 ± 5

is once again released at a slow rate. X-ray scattering studies at various alginate and water concentrations reveal that there is a competition between the ungelled alginate and the lipid for the added water. Based on the decrease in the unit cell size of the ungelled samples, we find that approximately 21 mg of water is associated per mg alginate in the alginate-rich regions and is not available to the lipid. Similar X-ray scattering studies on gelled samples do not show any substantial dehydration of the lipid by the gelled alginate, thus suggesting a decrease in volume fraction of the alginate-rich defects. This is in agreement with the release data which also indicate a decrease in volume fraction of the alginate-rich regions and a dramatic decrease in the release rate. The gelled alginate samples could then be degelled, thus regenerating the alginate-rich defects, and the release rates increased (Figure 5). Therefore, by controlling gelling and degelling of the alginate, the protein can be released as and only when required.

III. SUMMARY

Of late, there has been a resurgence in the development of advanced organic materials with enhanced or novel properties that are derived from the self-assembling nature of the constituent molecules. The bicontinuous cubic phase of certain lipids or surfactants in water offers unique features which can be tailored to develop novel, specific technology-driven materials. In this paper, we have discussed two specific directions that we have taken in utilizing the nanostructure of the cubic phase. First, we have demonstrated that not only does the cubic phase provide small compartmentalized regions for producing uniform nanoparticles, but that it serves as a dispersing medium and prevents particle aggregation during their formation. Second, we have shown that the bicontinuous cubic phase is highly versatile and can be used in the development of innovative controlled release systems. In addition to these two applications, a number of other applications have been suggested and explored. It is expected that in the coming years other novel methods of utilizing this unique nanostructure will emerge in order to meet the growing technological needs.

REFERENCES

1. **Lindblom, G. and Rilfors, L.**, Cubic phases and isotropic structures formed by membrane lipids — possible biological relevance, *Biochim. Biophys. Acta*, 998, 221, 1989; **Larson, K.**, Cubic lipid–water phases: structures and biomembrane aspects, J. Phys. Chem., 93, 7304, 1989.
2. **Seddon, J. M. and Templer, R. H.**, Cubic phases of self-assembled amphiphilic aggregates, *Phil. Trans. R. Soc. Lond. A*, 344, 377, 1993; **Tate, M. W., Eikenberry, E. F., Turner, D. C., Shyamsunder, E., and Gruner, S. M.**, Nonbilayer phases of membrane lipids, *Chem. Phys. Lipids*, 57, 147, 1991; Fontell, K., Cubic phases in surfactant and surfactant-like lipid system, *Coll. Polym. Sci.*, 268, 264, 1990.
3. **Luzzati, V., Tardieu, A., Gulik-Kryzwicki, T., Rivas, E., and Reiss-Husson, F.**, Structure of the cubic phases of lipid-water systems, *Nature*, 220, 485, 1968; **Lindblom, G., Larsson, K., Johansson, L., Fontell, K., and Forsen, S.**, The cubic phase of monoglyceride-water systems. Arguments for a structure based upon lamellar bilayer units, *J. Am. Chem. Soc.*, 101, 5465, 1979.
4. **Longley, W. and McIntosh, T. J.**, A bicontinuous tetrahedral structure in a liquid-crystalline lipid, *Nature*, 303, 612, 1983; Charvolin, J., Crystals of interfaces: the cubic phases of amphiphile/water systems, *J. Physique Coll.*, C3, 173, 1985.
5. **Scriven, L. E.**, Equilibrium bicontinuous structure, *Nature*, 263, 123, 1976; **Larsson, K., Fontell, K., and Krog, N.**, Structural relationships between lamellar, cubic, and hexagonal phase in monoglyceride-water systems. Possibility of cubic structures in biological systems, *Chem. Phys. Lipids*, 27, 321, 1980; **Hyde, S. T., Andersson, S., Ericsson, B., and Larsson, K.**, A cubic structure consisting of a lipid bilayer forming infinite periodic minimum surface of the gyroid type in the glycerolmonooleate-water system, *Zeitschrift fur Kristallographie*, 168, 213, 1984.

6. **Mariani, P., Luzzati, V., and Delacroix, H.,** Cubic phases of lipid-containing systems. Structure analysis and biological implications, *J. Mol. Biol.*, 204, 165, 1988.
7. **Schoen, A. H.,** Infinite periodic minimal surfaces without self-intersections, NASA Tech. Rep. No. D5541, 1970.
8. **Anderson, D. M. and Strom, P.,** Polymerization of lyotropic liquid crystals, in *Polymer Association Structures: Liquid Crystals and Microemulsions*, ACS Symposium Series, El-Nokaly, M., Ed., American Chemical Society, Washington, D.C., 1988, 204.
9. **Ericsson, B., Eriksson, P. O., Lofroth, J. E., and Engstrom, S.,** Cubic phases as delivery systems for peptide drugs, *In Polymeric Drugs and Drug Delivery Systems*, ACS Symposium Series No. 469, Dunn, R. L. and Ottenbrite, R. L., Eds., American Chemical Society, Washington, D.C., 1990, 251; **Wyatt, D. M. and Dorschell, D.,** A cubic-phase system composed of glycerol monooleate and water for sustained release of water soluble drugs, *Pharm. Tech.*, 116, Oct. 1992.
10. **Puvvada, S., Baral, S., Chow, G. M., Qadri, S. B., and Ratna, B. R.,** Synthesis of palladium metal nanoparticles in the bicontinuous cubic phase of glycerol monooleate, *J. Am. Chem. Soc.*, 116, 2135, 1994.
11. **Puvvada, S., Qadri, S. B., Naciri, J., and Ratna, B. R.,** Ionotropic gelation in a bicontinuous cubic phase, *J. Phys. Chem.*, 97, 11103, 1993; **Puvvada, S., Naciri, J., and Ratna, B. R.,** Ionotropically gelled bicontinuous cubic phase as a matrix for controlled release, in *Biomaterials for Drug and Cell Delivery*, Mikos, A. G., Murphy, R. M., Bernstein, H., Pappas, N. A., Eds., Materials Research Society, Boston, 1994.
12. **Andres, R. P., Averback, R. S., Brown, W. L., Brus, L. E., Goddard, W. A., Kaldor, A., Louie, S. G., Moscovito, M., Peercy, P. S., Riley, S. J., Siegel, R. W., Spaenen, F., and Wang, Y.,** Research opportunities on clusters and cluster-assembled materials, Department of Energy Council on Materials Science Panel Report, *J. Mat. Res.*, 4(3), 704, 1989.
13. **Matijevic, E.,** Preparation of well defined finely dispersed particles, *Faraday Discuss.*, 92, 229, 1991.
14. **Malla, P. B., Ravindranathan, P., Komarneni, S., and Roy, R.,** Intercalation of copper metal layers in montmorillonite, *Nature*, 351, 355, 1991; **Malla, P. B., Ravindranathan, P., Komarneni, S., E. Breval, E., and Roy, R.,** Reduction of copper acetate hydroxide interlayers in montmorillonite by a polyol process, *J. Mater. Chem.*, 2, 559, 1992.
15. **Crocker, M., Buglass, J. G., and Herold, R. H. M.,** Synthesis and characterization of palladium crystallite intercalated in montmorillonite, *Chem. Mater.*, 5, 105, 1993.
16. **Ward, A. J. I. and Friberg, S. E.,** Preparing narrow size distribution particles from amphiphilic association structures, *MRS Bulletin*, December 1989.
17. **Friberg, S. E. and Ma, Z.,** Amphiphilic association structures: some recent applications, *Adv. Coll. Interface Sci.*, 41, 45, 1992.
18. **Kurihara, K., Kizling, J., Stenius, P., and Fendler, J. H.,** Laser and pulse radiolytically induced colloidal gold formation in water and water-in-oil microemulsions, *J. Am. Chem. Soc.*, 105, 2574, 1983.
19. **Lopez-Quintela, M. A. and Rivas, J.,** Chemical reactions in microemulsions: a powerful method to obtain ultrafine particles, *J. Colloid Interface Sci.*, 158, 446, 1993.
20. **Kurihara, K. and Fendler, J. H.,** Electron-transfer catalysis by surfactant vesicle stabilized colloidal platinum, *J. Am. Chem. Soc.*, 105, 6152, 1983; **Ferrar, W. T., O'Brien, D. F., Warshawsky, A., and Voycheck, C. L.,** Metallization of lipid vesicles via electroless plating, *J. Am. Chem. Soc.*, 110, 288, 1988.
21. **Friberg, S. E. and Wang, J. F.,** A new method to prepare uniform particles. Precipitation from lyotropic liquid crystals, *J. Disp. Sci. Tech.*, 12, 387, 1991.
22. **Beck, J. S., Vartuli, J. C., Roth, W. J., Leonowicz, M. E., Kresge, C. T., Schmitt, K. D., Chu, C. T. W., Olson, D. H., Sheppard, E. W., Mccullen, S. B., Higgins, J. B., and Schlenker, J. L.,** A new family of mesoporous molecular-sieves prepared with liquid-crystal templates, *J. Am. Chem. Soc.*, 114, 10834, 1992; Kresge, C. T., Leonowicz, M. E., Roth, W. J., Vartuli, J. C., and Beck, J. S., Ordered mesoporous molecular-sieves synthesized by a liquid-crystal template mechanism, *Nature*, 359, 710, 1992.
23. **Huo, Q., Margolese, D. I., Clesla, U., Feng, P., Gier, T. E., Sieger, P., Leon, R., Petroff, P. M., Shuth, F., and Stucky, G. D.,** Generalized synthesis of periodic surfactant/inorganic composite materials, *Nature*, 368, 317, 1994.
24. **Patawardan, S. A. and Das, K. G.,** An overview of controlled release technology, in *Controlled Release Technology, Bioengineering Aspects*, Das, K. G., Ed., CRC Press, Boca Raton, FL, 1980.
25. **Langer, R.,** New methods of drug delivery, *Science*, 249, 1527, 1990.
26. **Dunn, R. L.,** Polymeric matrices, in *Polymeric Drugs and Drug Delivery Systems*, ACS Symposium Series 469, Dunn, R. L. and Ottenbrite, R. M., Eds., American Chemical Society, Washington, D. C., 1991, chap. 2.
27. **Wheatley, M. A., Chang, M., Park, E., and Langer, R.,** Coated alginate microspheres: factors influencing the controlled delivery of macromolecules, *J. Appl. Polym. Sci.*, 43, 2123, 1991.
28. **Oku, N.,** Liposomes, *Polymeric Drugs and Drug Delivery Systems*, ACS Symposium Series 469, Dunn, R. L. and Ottenbrite, R. M., Eds., American Chemical Society, Washington, D.C., 1991, chap. 3.
29. **Schnur, J. M., Price, R., and Rudolph, A. S.,** Biologically engineered microstructures: controlled release applications, *J. Controlled Release*, 28, 3, 1994.
30. **Zeoli, L. T. and Kydonieus, A. F.,** Physical methods of controlled release, *in Controlled Release Technology, Bioengineering Aspects*, Das, K. G., Ed., CRC Press, Boca Raton, FL, 1980, chap. 4.

31. **Nicholson, C. and Tao, L.,** Hindered diffusion of molecular weight compounds in brain extracellular microenvironment measured with integrative optical imaging, *Biophys. J.*, 65, 2277, 1993.
32. **Bohrer, M. P., Patterson, G. D., and Carroll, P. J.,** Hindered diffusion of dextran and ficoll in microporous membranes, *Macromolecules*, 17, 1170, 1984.
33. **Huag, A., Larsen, B., and Smidsrod, O.,** A study of the constitution of alginic acid by partial acid hydrolysis, *Acta Chem. Scand.*, 20, 183, 1966.
34. **S. Puvvada and Ratna, B. R.,** in preparation.

LIPOSOMES IN CATALYSIS

Heterogeneous catalysis is essential in many industries as well as in life processes. Many of these reactions also occur at hydrophilic/hydrophobic boundaries where molecules may achieve certain orientation which can improve the action of the catalyst, such as proton, hydroxyl, metal ions, or an enzyme. Liposomes were thought of as a very promising catalytic substrate because they have a large surface area (for instance, for small unilamellar vesicles one gets >200 m^2/g of surface area), hydrophilic/hydrophobic boundary, and can compartmentalize the system. In addition, surface charge (potential) can be easily tailored and co-enzymes can be either entrapped or embedded into the membrane. Furthermore, membrane permeability can be controlled and specific ionophores can be inserted into bilayers. Also, the chirality of the substrate can be controlled by using optically pure lipid enantiomers.

These advantages, however, have to be weighed with some disadvantages which encompass mainly relatively low stability (especially at extreme values of pH), irreproducible size distribution, and possible flocculation/coagulation resulting in precipitation. Nevertheless, liposomes were used in the preparation of some colloidal particles, in the studies of bone growth, and calcification as described in these volumes. Colloidal platinum, as described by Fendler, was encapsulatd into liposomes for hydrogenization of ethylene and cyclohexane. The catalyst was stabilized at the expense of the free diffusion of reacting species. With proper ionophores, however, these rates can be controlled. Similarly, the thiolysis of p-nitrophenyl octanoate was studied and it was shown that small differences in the lipid packing, reflecting the accessibility to the hydrophilic/hydrophobic boundary had relatively large effect on the reaction rates. Similarly, colloidal gold was prepared in liposomes for imaging studies by Papahadjopoulos an collaborators. A different approach was tried in the prevention of the lipolysis of triglycerides in the human intestine. Stable liposomes were suggested by Hauton to be ingested in order to trap digestive lipases and thus reduce the rate of fatty acid generation and reduce the growth of atheriosclerotic plaques on blood vessels walls. Other approaches involve vesicle coating with metal ions, either directly or via membrane residing chelating agents.

Most of the progress, however, we anticipate in the studies of biological reactions occuring on the cellular level, which are responsible for vesicle formation and fusion in cell signaling and inter- and intracellular traffic.

D. D. Lasic
Y. Barenholz

Chapter 13

Liposomes in Catalysis

Iolanda M. Cuccovia and Hernan Chaimovich

CONTENTS

I. INTRODUCTION

Amphiphiles form a variety of supramolecular structures such as aqueous and reverse micelles, monolayers, multibilayers, liposomes and vesicles (Scheme 1). These aggregates are adequate reaction media for the study of a part of real world chemistry, since many biological and industrial reactions occur in interfaces and not in dilute solution. The increasing understanding of interface and aggregate effects on rates and product composition has expanded the applications in chemistry and contributed to several other fields of knowledge.[1,2]

Natural phospholipids are amphiphiles that spontaneously form bilayers in excess water.[1-4] The first closed systems containing an aqueous compartment encompassed by a lipid bilayer was described by Bangham and co-workers in the early 1960s.[5] The interest and applications of these supramolecular structures have evolved from the purely academic use to liposome-based industries.

The property of forming bilayers and, as a consequence, bilayer-encompassed assemblies is not, however, a unique feature of natural phospholipids but a general property of amphiphiles of appropriate structures.[3] Among others,[6] amphiphiles containing two long hydrophobic chains and a relatively small hydrophilic head group can form, in excess water, spheroidal aggregates containing one or more bilayers (Schemes 1, 2).[3,4,7] These aggregates, with diameters varying from 25 nm to microns are generally referred as vesicles if the internal aqueous compartment is surrounded by a single bilayer. Aggregates containing more than one bilayer are known as multibilayer vesicles or liposomes. Vesicles and liposomes vary in size and homogeneity according to the preparation method.

The first vesicles formed by synthetic amphiphiles, described in the late 1970s, were prepared by sonic dispersion of dialkyldimethylammonium or dialkyl phosphate salts in aqueous solutions[8,9] (Scheme 2). The formation of vesicle-like structures in organic solvents was described recently.[10] Sucrose monoalkanoates, dispersed by agitation or sonication in hexanol-decane-water, form "reverse vesicles," closed structures where an inner pool of organic solvent is surrounded by an amphiphile bilayer exhibiting a structure reminiscent of reverse micelles.[10] The bilayer of reverse vesicles, stabilized by headgroup-water interactions, projects the hydrophobic moieties into the organic solvent (Scheme 1).

Bilayer-forming amphiphiles, with a variety of chemical structures, form vesicles or liposomes of a wide range of sizes and complexity. Many of the known properties of aqueous and reverse micelles, a well-understood supramolecular aggregate, are found at the surface(s) of vesicles and liposomes. Evidently, micellar properties can be extended in a more stable and compartmentalized system. The interest in vesicles and liposomes as microreactors for controlling rates and compartmentalizing reagents increases steadily. In this review we will focus on the effect of vesicles on rates and product composition of ground state reactions.

0-8493-4012-8/96/$0.00+$.50
© 1996 by CRC Press, Inc.

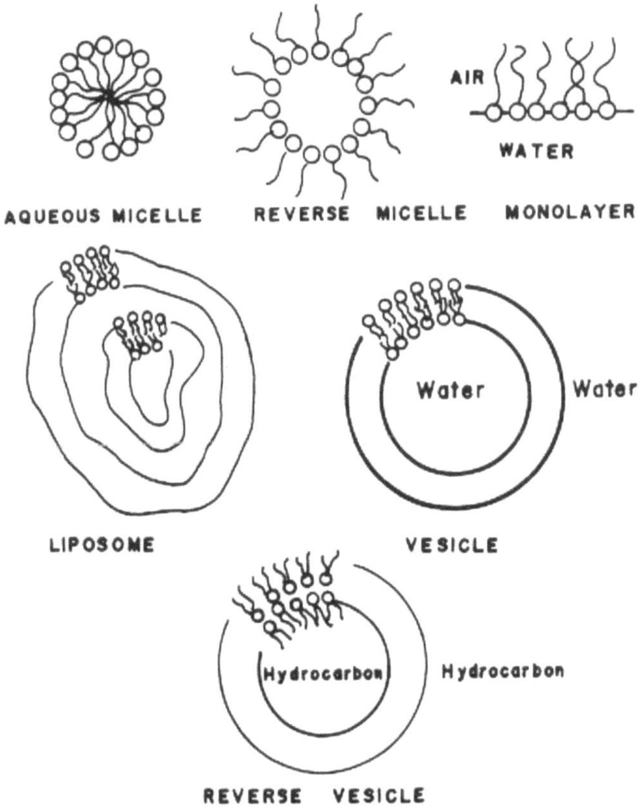

Scheme 1 Some supramolecular structures formed by amphiphiles.

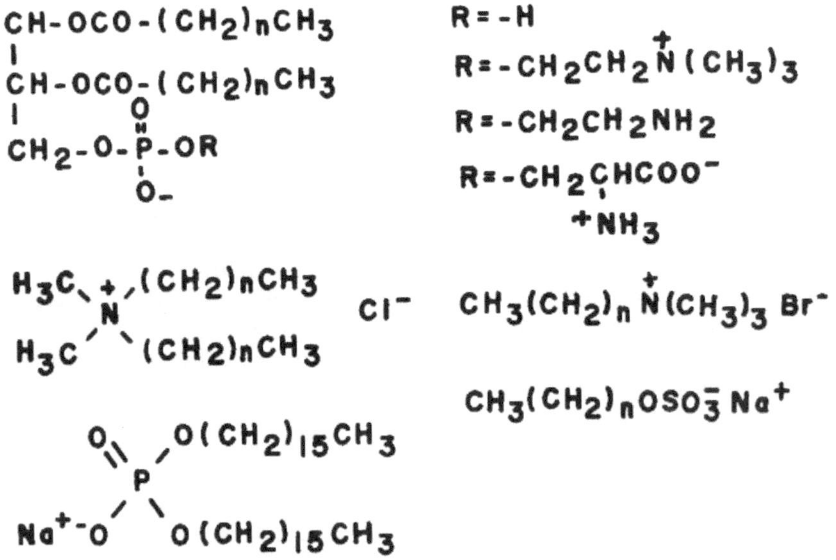

Scheme 2 Commonly used amphiphiles.

II. VESICLE PREPARATION AND STABILITY

Vesicles and liposomes can be prepared by several methods and the average size, and population homogeneity, depends on the particular preparation method.[11] We will briefly describe some of the most common methods for obtaining vesicles and liposomes prepared with synthetic amphiphiles.

The simplest method of liposome preparation consists in the deposition of a film, obtained by (solvent) evaporation of a solution of an amphiphile in an organic solvent, hydration with excess aqueous solution, and gentle or vigorous shaking. The liposomal size distribution, as could be expected from this procedure, is wide, the particle's diameters varying from 0.1 to several micrometers.

Techniques are available for the preparation of unilamellar vesicles (SUV) measuring from 20 to thousands of nm in diameter. Sonication is widely used for obtaining SUVs.[11] SUVs can also be obtained by extrusion through controlled pore membranes and by high pressure passage through small orifices.[11]

Large unilamellar vesicles (LUV) (hydrodynamic diameters [Dh] >100 nm) can be prepared, among other methods, by reverse phase evaporation, solvent injection, or extrusion.[11]

Spontaneous vesiculation, suggesting the formation of thermodynamically stable vesicles, has been reported using mixtures of phospholipids and deposited films of charged synthetic amphiphiles.[11-13]

Vesicles prepared with synthetic amphiphiles are generally metastable and revert to multilamellar structures with time.[11] Salts and organic additives destabilize synthetic amphiphile vesicles causing an increase in the mean Dh.[14,15] The mechanisms of vesicle destabilization are complex and can lead to aggregation and fusion.[14,15] The stability of dioctadecyldimethylammonium chloride (DODAC) and dihexadecylphosphate (DCP) vesicles has been analyzed in detail.[14b] The physical changes resulting from vesicle aggregation or fusion can, at times, complicate the analysis of the effect of these aggregates on rates of chemical reactions.

Small unilamellar vesicles have been obtained by slow injection of an ethanolic solution of DODAC into aqueous solutions maintained above the critical phase transition temperature (Tc) of vesicles of the amphiphile.[16] It is not clear if the aggregates obtained by this method are closed vesicles. Solutions of ethanolic "vesicles" are optically clear in water or at low ionic strength (<10 m*M*) and slowly flocculate. The aggregates do not show osmometric behavior and fail to incorporate water soluble substrates in internal compartments. Early results showing high sucrose incorporation in the internal aqueous compartment[16] were in error due to negatively charged contaminants in the radioactive sucrose used to estimate incorporation. However, definite demonstrations of the vesicular nature of the preparations obtained by ethanol injection are lacking. The small diameter of the aggregate and, consequently, the high radius of curvature may render the bilayer very permeable. Hence, measuring the incorporation of water-soluble substrates in the vesicle may be difficult. The ethanol injection method is used for the preparation of sulfhydryl-containing amphiphiles or substrates, since vesicles are obtained rapidly, with no metal contamination (as with tip-sonicated preparations), under very mild conditions and, therefore, the oxidation of SH groups is minimal.

Sonicated vesicles of DODAC and DCP are larger than ethanol-injected and more compact.[17] The incorporation of water-soluble substrates in the internal aqueous compartment is also very small.[18] As in the case of ethanol-injected vesicles, it is difficult to affirm whether these aggregates are very permeable or are not closed structures. The formation of open structures that slowly, and spontaneously, form larger closed vesicles has been suggested.[19]

Large unilamellar vesicles of DODAC and DCP (LUV, Dh >50 nm) can be obtained by (slow) injection of chloroform solution into an aqueous medium thermostated above the boiling point of the solvent.[14b] The properties of LUVs of DODAC and DCP have been analyzed in detail and there is wide agreement that these vesicles are closed, retain water soluble substrates, are very impermeable to ions and exhibit ideal osmometric behavior.[14b] The size and properties of ethanolic, sonicated and chloroform injected vesicles of DODAC, dioctadecyldimethylammonium bromide (DODAB) and DCP have been analyzed.[17] Physical parameters such as Tc of the vesicle bilayer, degree of ion dissociation (α) and selectivity constants for ion exchange at the external surface ($K_x/_y$) have been determined.[17,20,21] These parameters, as will be discussed below, are essential for quantitative analysis of the effect of vesicles on reaction rates. Some of the physical properties of DODAC, DODAB, and DCP vesicles are summarized in Table 1.

Broader description of vesicle solutions in electrolyte-containing water requires a knowledge of ion distribution in the intervesicular solution and in the vesicle interior. Ion distribution in micellar solutions has not only broadened our understanding of micellar effects on reaction rates,[22] but helped to describe interparticle interactions.[23] Recently, we reported a mean field description of ion distribution in electrolyte solutions containing charged vesicles.[24] Applications of this and other more general models for the analysis of vesicular effects, vesicle–vesicle interactions as well as experimental approaches to the determination of free ion contents of charged vesicle's interior are in progress (Cuccovia, I. M., et al., unpublished).

It is important to emphasize that the differences in physical properties of vesicles (obtained by different methods) influence the effect of vesicles on rates and/or product composition of chemical reactions. This sometimes overlooked property has complicated comparisons of vesicle effects on

Table 1 Selected Physical Parameters of Synthetic Amphiphile Vesicles

Method and Conditions[a]	Dh (nm)	α^b
	DODAC	
EtOH injection[c]		
No salt (0.92)	23	0.29
No salt (8.0)	22	0.35
5 m*M* NaCl (9.0)	25	0.23
Sonication		
No salt (5.0)	53	0.21
No salt (1.0)	53	0.22
0.93 m*M* NaCl (5.0)		0.20
4.76 m*M* NaCl (5.0)	38	0.13
9.27 m*M* NaCl (5.0)	40	0.11
$CHCl_3$ injection		
No salt (0.3)	285	0.044
9.0 m*M* NaCl (0.3)	350	
	DODAB	
Sonication		
No salt (5.0)	65	0.13
$CHCl_3$ injection		
No salt (0.3)	275	0.043
	DCP Na	
Sonication		
No salt (5.0)	67	0.22
	DCP	
No salt (4.0)	116	0.068

[a] Amphiphile concentrations (m*M*) in parentheses.
[b] α of the external vesicular surface.
[c] 5% (v/v) EtOH.

Data from Reference 17.

reactions and hampered reproducibility of experimental results. Therefore it is essential that, in this system, experimental results be compared with preparations obtained with identical methods and, especially in the case of the more unstable preparations, adequate attention be given to the fact that vesicle instability may obscure and complicate kinetic analysis.

III. REACTIONS IN VESICLES

Vesicles are efficient and complex microreactors. It is not trivial to identify reaction sites and/or substrate solubilization sites. Therefore, in a large number of cases, the effect of vesicles on a particular reaction is clear, but the reaction site and molecular details of the vesicle's effect are unknown.

Vesicles, when compared with simpler interfacial models such as micelles, contain a much larger number of potential reaction sites. Furhop and Mathieu[2] proposed a classification where nine potential reaction sites can be distinguished in a solution containing vesicles (Scheme 3): (1) intervesicle or bulk; (2) aqueous volume in the vicinity of the head groups and/or Gouy-Chapman layer if the vesicle is charged; (3) outer head groups; (4) the hydrophobic membrane close to the outer head groups; (5) the interior of the hydrophobic membrane; and (6–9) the inner regions corresponding to the outer regions 4-1 (Scheme 3).

In vesicles the reaction site has been determined with confidence only in a limited number of cases (see below). The distinction between regions 1 and 2 (or 8 and 9) (Scheme 3) may be arbitrary for solutions of vesicles of charged amphiphiles, especially at low ionic strength.[24] The nature of the monomer, even if formally neutral, may dominate electrostatic properties at the interface,[25] making it impossible to distinguish between regions 3 and 4 (or 6 and 7) (Scheme 3). In spite of these (among other) questions, the classification proposed by Furhop and Mathieu[2] raises some of the major difficulties in

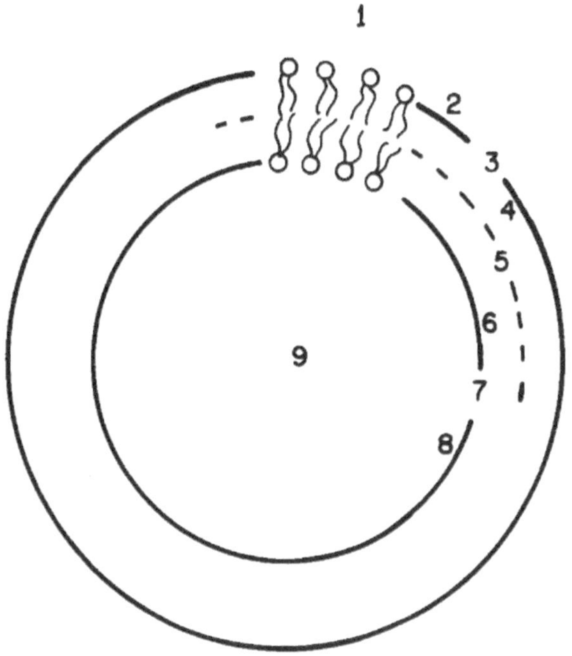

Scheme 3 The nine regions of vesicle solutions.[2]

pinpointing reaction sites in vesicles and shows, at the same time, the potential of these microreactors for reaction modification and control.

More in line with the type of experimental data available for reaction modifications by vesicles we have divided this review in two sections:

1. Noncompartmentalized reactions, where all vesicular sites are potentially accessible to substrates and
2. Compartmentalized reactions, where specific sites can be defined and rates may be limited by substrate diffusion into (or through) the vesicle.

The kinetic effects produced by vesicles can be analyzed quantitatively and the understanding of these effects depends on the determination of rate constants at the reaction site(s) from the observed data. Here we will present a concise synopsis of the model most commonly used to analyze quantitatively rate data in vesicles.

IV. MODELS FOR QUANTITATIVE ANALYSIS OF REACTIONS IN VESICLES

Quantitative analysis of rate changes produced by the introduction of supramolecular aggregates to a solution has been extensively investigated with micelles.[26] One of the aims of the analysis is the determination of rate constants of the reaction of interest at the micellar site from rate data obtained in a system that contains aggregates and, in general, a large excess of the inter-aggregate bulk phase.

In a micellar-accelerated chemical reactions the observed rate constant (k_ψ) increases with detergent concentration above the critical micelle concentration (cmc). In a monomolecular reaction, k_ψ increases to a plateau, and the value of the (micelle-modified) rate constant is defined as the maximum rate constant in the presence of micelles ($k_{\psi max}$). Note that $k_{\psi max}$ only represents the value of the rate constant in the micellar phase for neutral substrates in ionic, zwitterionic, or neutral micelles, where, as a limit, all substrate is micelle-bound at infinite detergent concentration. This limit does not necessarily hold for charged substrates in ionic or zwitterionic micelles due to ion exchange at the interface.[26c] For bimolecular reactions micellar acceleration produces an increase in k_ψ, reaching a maximum value ($k_{\psi max}$) and then decreasing at higher detergent concentration due to substrate dilution and/or ion exchange.

The maximum micellar effect may be expressed as the ratio of $k_{\psi max}$ and the rate constant obtained under the same conditions without micelles (k_o) (i.e., $k_{\psi max}/k_o$).[1a]

The two most extensively used models for determining micellar rate constants are pseudophase models that consider the micellar ensemble as a separate pseudophase, with explicit consideration of ion exchange (PPIE) and coulombic models. Coulombic models use cells containing a single micelle to describe the micellar ensemble, solve the nonlinear Poisson–Boltzmann equation and use Volmer or

Langmuir isotherms for the analysis of ion distribution (PBE). Detailed descriptions of the models, as well as extensive critical analysis of their limitations have been published.[26,27] Here we will present the general basis of the PPIE model, since this analysis has been used to examine quantitatively the effect of vesicles on chemical reactions.

The most general approach for the quantitative analysis of reaction rates in micellar systems begins by considering that the system can be treated as composed of two separate pseudophases, micelles and intermicellar aqueous phase, and that reaction rates can be described as a sum of separate reactions occurring in the two compartments. This model further assumes that the rate of equilibration of the substrates between the pseudophases is much higher than the reaction rate, i.e., equilibrium partitioning. The velocity (v) of a spontaneous monomolecular decomposition of a substrate S is appropriately described by:[26]

$$v = k_\psi \times S_t = k_m \times S_b + k_w \times S_f \tag{1}$$

where k is the rate constant, S is the substrate concentration, and the subscripts m, w, t, f and b indicate micelle, water, total, free and bound, respectively. In this particular case the observed rate constant (k_ψ) is:

$$k_\psi = k_m \times (S_b/S_t) + k_w \times (S_f/S_t) \tag{2}$$

The unequivocal calculation of the rate constant in the micellar pseudophase (k_m) depends on the (independent) determination of the substrate micelle–water partition coefficient or incorporation constant (K_s) defined as:

$$K_s = \frac{S_b}{S_f \times C_d} \tag{3}$$

where C_d is the concentration of micellized detergent. C_d is the total concentration of added detergent (C_t) minus the cmc (i.e., $C_d = C_t - \text{cmc}$).

For higher order reactions, the models generally assume independent substrate partitioning and apply a similar analysis to the description of the reaction rate. For neutral substrates, at low micellar occupancy, independent partitioning has been demonstrated in a large number of cases.[28] Note that independent partitioning needs not to apply even in micelles, especially if the mole fraction of one of the partitioning substrates in the aggregate is significant.[29] With vesicles the problem may be more complicated since even small mol fractions of bulky substrates (i.e., cholesterol-derived) may drastically affect bilayer properties and, therefore, partitioning of other substrates.[30]

The simple pseudophase distribution model describes quantitatively a number of reactions between neutral species in micellar solutions but fails to fit kinetic data where one of the reactive species is ionic, especially in the case of small hydrated ions such as OH^-, halides, or H^+.[26]

The failure of simple pseudophase models for the description of bimolecular reactions with hydrated ions, or unimolecular reactions of water soluble ions led to the recognition that ion binding to the micelle surface cannot be treated as a simple partitioning, especially in the case of ionic micelles. Since the ion dissociation from ionic micelles varies little with salt or detergent concentration the surface can be considered counterion-saturated and ion binding treated as an ion-exchange phenomena.[26d,e] Furthermore, since the effect of ions on micellar properties depends on the nature of the ion both binding and specificity can be analyzed quantitatively using an ion selectivity constant ($K_{x/y}$) for the exchange between ions X and Y.

$$K_{x/y} = \frac{[X_b] \times [Y_f]}{[X_f] \times [Y_b]} \tag{4}$$

In the analysis it may be further assumed that the degree of ion dissociation from the micelle (α) is constant. α Equals $1 - \beta$, where β is the degree of ion coverage of the micelle (i.e., β = [concentration of bound ions]/C_d).

By definition, calculation of second order rate constants (k_2) requires the use of a reaction volume element to estimate the local concentration of the reagents. This requirement complicates the calculation of a k_2 in the micelle (k_{2m}), since the reaction volume has to be defined in a supramolecular aggregate.[26f]

In the case of vesicles, the problem of quantitative description of reaction rates is further complicated by the number of potentially different reaction sites (Scheme 3). Furthermore, some of the assumptions of the above described model may not apply to vesicles. For example, it is not evident that α is constant and that all the potential reaction sites for a particular reaction are accessible to all substrates. This requires that, for reactions that are not first order, equilibrium distribution has to be proven rather than assumed. In spite of all these problems quantitative analysis has been attempted in a reduced number of cases.

V. NONCOMPARTIMENTALIZED REACTIONS

A. PHOSPHOLIPID SYSTEMS

Vesicles prepared with natural lipids, specially phosphatidylcholines (PC) are used as microreactors, with special emphasis in using PC vesicles as drug carriers. From a pharmaceutical standpoint it is essential that drug carriers inhibit drug decomposition. Recently the effect of detergents and other amphiphiles on the stability of pharmaceutical drugs has been reviewed.[31]

Kinetic effects of PC vesicles and/or liposomes are usually small and the magnitude is comparable to those observed with micelles of zwitterionic detergents.

PC vesicles can affect the rate of hydrolysis of procaine, a local anesthetic.[32] At neutral pH the rate of the water-catalyzed spontaneous hydrolysis of procaine is not modified by PC vesicles, probably because the protonated form of the drug does not significantly dissolve in the bilayer. Above pH 8 vesicles inhibit twofold the OH^- attack and this inhibition, quantitatively analyzed using a pseudophase separation model, was attributed to the solubilization of dissociated procaine in the vesicles. At the vesicular reaction site the calculated second order rate constant is three to seven times lower than in water depending on the external pH.

The rate of alkaline hydrolysis of tempobenzoate, a spin label analog of a local anesthetic, decreases with concentration of PC liposomes.[33] A phase separation model[26b] was used to suggest that tempobenzoate does not react with OH^- in the bilayer. Hydroxide ion is excluded from the zwitterionic aggregate only at high ionic strength, since anions displace OH^- from the zwitterionic interface.[34]

B. SYNTHETIC AMPHIPHILES

One of the first observations exploring catalytic effects of vesicles prepared with synthetic amphiphiles was the study of the hydrolysis of p-nitrophenyl acetate (NPA) and p-nitrophenyl palmitate (NPP) catalyzed by the cholesteryl ester of imidazolecarboxylate (Ch-Im) in didodecyldimethylammonium bromide (DDAB) sonicated vesicles and hexadecyltrimethylammonium bromide (CTAB) micelles.[35] Vesicles were prepared by sonication of solutions containing either NPA, Ch-Im or NPP and DDAB in mildly acidic water and the reaction initiated by mixing the ester, the nucleophile and alkalinization. The reaction of NPA and Ch-Im in DDAB vesicles is accelerated tenfold over the rate in CTAB. The reaction of NPP and Ch-Im is only catalyzed when the vesicles are cosonicated after mixing and the rate is 40-fold higher than that in CTAB micelles. The reaction rate of Ch-Im with NPP solubilized in separately prepared DDAB vesicles is 200-fold lower than the cosonicated preparation. Clearly, NPA can rapidly equilibrate in the vesicle population while NPP can only react with Ch-Im when both are solubilized in the same vesicle.

Synthetic amphiphile vesicles can accelerate reactions more efficiently than micelles. The thiolysis of NPA by alkyl mercaptans (RSH, R = C_2H_5 to $C_{12}H_{25}$) are accelerated ($k_{\psi max}/k_o = 10^4$) by steryltrimethylammonium bromide micelles.[36]

For ethanol-injected DODAC vesicles $k_{\psi max}/k_o$ is 7×10^4 in the thiolysis of NPA by heptyl mercaptan (HM) (Scheme 4) while, under the same conditions, $k_{\psi max}/k_o$ is 4×10^3 with CTAB micelles.[16] The maximum increase in reaction rate produced by DODAC vesicles in the thiolysis of *p*-nitrophenyloctanoate (NPO) by HM reaches 7.7×10^6-fold, while $k_{\psi max}/k_o$ for CTAB micelles is 1.5×10^4.[37,38] DODAC vesicles also produce significant changes in the apparent pKa of the mercaptan. The apparent acid dissociation constant (Kap) of the mercaptan was defined as:[26c]

$$\text{Kap} = [H_f] \times \frac{[A_b] \times [A_f]}{[HA_b] \times [HA_f]} \qquad (5)$$

The undissociated weak acid (HA) and corresponding anion (A) can be either free (f) or bound (b) to the micelle at an H^+ concentration, (H_f) determined from the measured pH. Note that the distribution of the

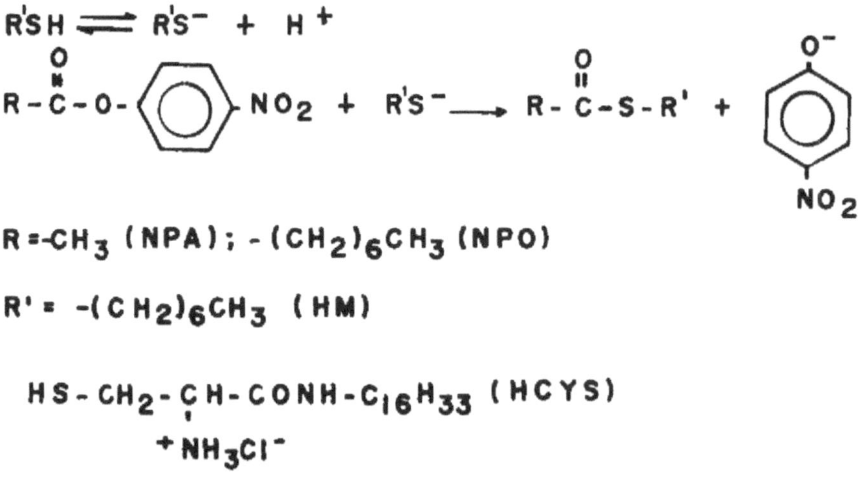

Scheme 4 Ester thiolysis and mercaptans.

dissociated species, if charged, has to be analyzed with an ion exchange formalism while HA, if uncharged, will partition as a neutral species.

The similarity between micellar and vesicular effects on rates and acid dissociation led to the application of the PPIE model to the data obtained with vesicles. The determination of the value of the second-order rate constants for thiolysis of NPO in the vesicular pseudophase (k_{2v}) required assumptions besides those made by the PPIE model. The major additional assumptions were (i) substrates equilibrate between the reaction sites faster than reaction rate; (ii) both internal and external interfaces are available for reaction; and (iii) the values of α and selectivity constants for ion exchange on the DODAC interfaces are constant and comparable to those obtained in CTAB micelles. On the basis of these assumptions, the calculated ratio between the second-order rate constants in water (k_{2w}) and k_{2v} (k_{2v}/k_{2w}) was 15. For the same reaction in CTAB micelles, the second-order rate constants in the aqueous phase and in the micelle are equal and $k_{2m}/k_{2w} = 1.0$. The spectacular acceleration ($k_{\psi max}/k_o > 10^6$) produced by vesicles in the thiolysis reaction, therefore, results from several contributions: (i) local concentration of the hydrophobic reagents in the vesicle bilayer; (ii) increase of the dissociation of the mercaptan under the influence of the positively charged interface; and (iii) a moderate decrease in the activation energy for the rate limiting attack of the mercaptide ion on the ester.

The maximum rate acceleration produced by DODAC vesicles on thiolysis is even larger using N-hexadecyl cysteinamide (HCys) as a nucleophile (Scheme 4).[38] The pKa of the SH group in HCys (pKa = 8.5) is lower than that of HM (pKa = 10.7); therefore, the thiolysis reaction is fast even at pH 2.3. At such a low pH the calculated extent of mercaptan dissociation in water is extremely low and $k_{\psi max}/k_o$ is about 3×10^6. The maximum acceleration[38] reaches 10^7 in the presence of a weakly interacting buffer that increases the local concentration of OH^- ion[39] at the positively charged interface of DODAC vesicles. For the thiolysis of NPO by HCys the value of k_{2v}/k_{2w} is 50, indicating a moderate catalysis in the rate limiting step. Again the major factors involved in the rate enhancement are local concentration and pKa shifts.

The analysis of the thiolysis of esters by mercaptans provided one of the first quantitative analysis of vesicle-modified reaction rates and indicated that the factors contributing to rate acceleration are comparable to those that had been described for micelles. In addition, the magnitudes of the maximum rate accelerations show that enzyme-like rate effects can be reached in very simple biomimetic systems (Table 2).

From the data in Table 2 it is attractive to speculate that charge and fluidity controls (reproducing the interfaces ranging from sodium dodecyl sulfate micelles to DODAC vesicles) could control, in a single system, the reaction rate a billionfold. That type of control is known in living cells, however, even though "mother nature needs not to be slavishly reproduced,"[1a] it is possible to envisage that such a control could be feasible in a biomimetic system where interfacial charge and head group separation can be varied.

The alkaline decomposition of 5,5′-dithiobis(2-nitrobenzoic acid) (DTNB) (Scheme 5) has been extensively used to analyze vesicular effects in reaction rates. With sonicated DODAC vesicles $k_{\psi max}/k_o = 1500$ and vesicles are 100 times more efficient than CTAB micelles.[40] This reaction was quantitatively analyzed

Table 2 Maximum Rate Enhancements and Calculated Second-Order Rate Constants for the Thiolysis of NPO in the Presence of Amphiphile Aggregate

Thiol	Detergent (pH)	$k_{\psi max}/k_{\psi}$	k_{2m}/k_{2w}
HM	SDS[a] (11.3)	0.09	0.27
HM	HPS[a] (10.5)	266	1.0
HM	CTAB[a] (8.9)	1.6×10^4	1.2
HM	DODAC[b] (4.0)	7.7×10^6	15
HCys	CTAB[a] (4.05)	1.0×10^6	15
HCys	DODAC[b] (3.75)	9.0×10^6	50

[a] Micelles.

[b] Vesicles prepared by ethanolic injection.

Data from References 37 and 38.

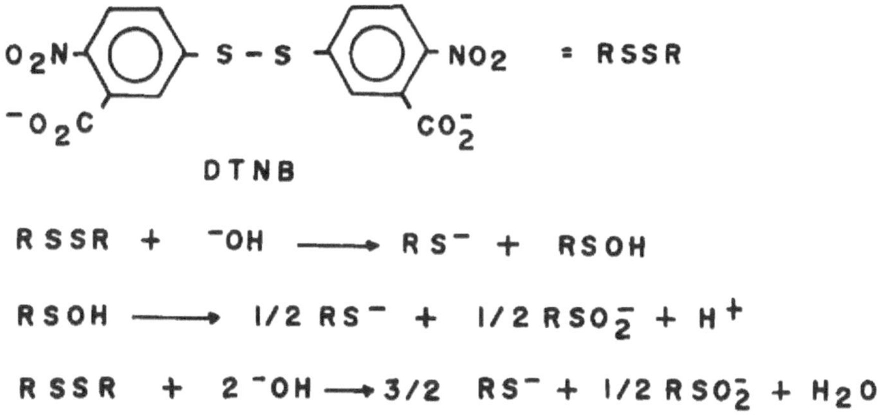

Scheme 5 Alkaline hydrolysis of DTNB.

using both pseudophase and PPIE models.[26a,d] Rate acceleration of OH^- attack on DTNB by DODAC vesicles is exclusively due to reagent concentration at the vesicular surface.

C. EFFECT OF THE AGGREGATE STRUCTURE ON RATE AND PRODUCT COMPOSITION

The use of synthetic amphiphiles for vesicle preparation allows the analysis of the effect of changes of the monomer structure on the architecture of the aggregates and, as a consequence, effects of aggregates of varying structure on chemical reactions.

The effects of structural changes on rates have been examined in several reactions using vesicles of didodecyldimethylammonium bromide (DDAB) and DODAB, micelles of CTAB, and aggregates of trioctadecylmethylammonium chloride (TMAC).[41] Nucleophilic attack by *N*-dodecylbenzohydroxamic acid (C_{12}-BHA), *N*-dodecyl-4-imidazolecarboxamide (C_{12}-ImAm), *N*-(methylcholohydroxamic acid (Chol-HA), *N*-(4-imidazolylmethyl)-cholohydroxamic acid (Chol-ImHA), *N*-(2-(4-imidazolyl)ethyl)cholamide (Chol-His), and cholesteryl-4-imidazolecarboxylate (Chol-Im) on NPA and *p*-nitrophenyl nonanoate (NPN) have been examined. The alkaline hydrolysis of NPA is slower than with NPN, reflecting the higher partition coefficient of (the more hydrophobic) NPN. The $k_{\psi max}/k_0$ obtained with NPN is similar in all the aggregates. Reactions of hydroxamates and imidazol derivatives with NPA yield values of $k_{\psi max}/k_0$ varying from 10^2 to 10^3 depending on the nucleophile and amphiphile. The reactions with NPA are accelerated in an order reflecting the hydrophobicity of the aggregates, with the exception of Chol-HA and Chol-Im. The aggregates also decrease the apparent pKa of the nucleophiles from 0.3 to 3 pH units.[41]

The acceleration by CTAB and TMAC are determined predominantly by the hydrophobicity of the nucleophiles and the aggregates. Vesicle effects were explained by specific substrate binding, especially with the cholesterol-derived nucleophiles.[41]

Rate control by phase separation can be verified in the reaction of alkaline hydrolysis of $C_{12}AzoOCOC_{10}N$ (*2*) and the hydrolysis of *p*-nitrophenyl *N*-(benzoxycarbonyl)-phenylalaninate (*4*) catalyzed by the nu-

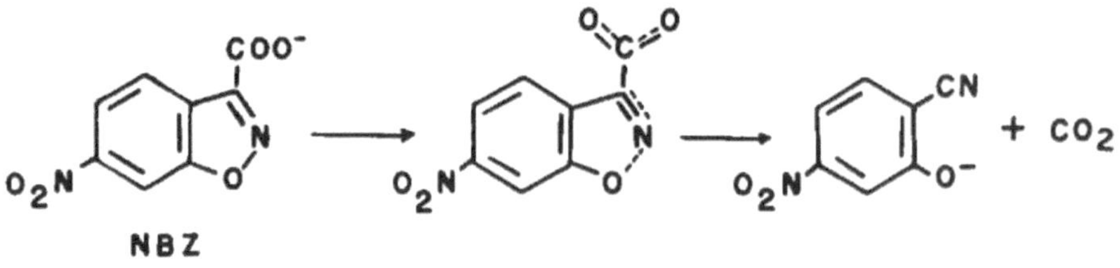

Scheme 6 Decomposition of NBZ.

cleophile $C_{12}AzoC_{11}His$ (*3*) in sonicated vesicles and micelles.[42] Above the Tc of DODAB, dihexadecyldimethylammonium bromide (DHDAB) and ditetradecyldimethylammonium bromide (DTDAB) vesicles, compound (2) is distributed as a monomer, while below the Tc (2) forms clusters. The rate of hydrolysis of clustered substrate is about 1/20 of that of the monomers. In these aggregates the temperature dependence of the rate of alkaline decomposition of (2) shows breaks reflecting changes in the free monomer/cluster ratios. Arrhenius plots for the alkaline decomposition of (*4*) in DODAB vesicles are linear while the catalysis by (*3*) in the same vesicular systems shows a discontinuity at the phase transition of clusters of (*4*). These results clearly demonstrate that physical changes at the bilayer, including lateral phase separation, can produce reactivity changes and rate control.

The effect of modifications in the interface structure on chemical reactivity has been analyzed probing aggregate's effects on the rate of spontaneous decomposition of 6-nitrobenzisoxazole-3-carboxylate (NBZ) (Scheme 6).

The reaction rate for the (monomolecular) decarboxylation of NBZ depends critically on the relative stability of the initial charge-localized carboxylate anion and the charge-delocalized transition state. This is a specially useful reaction since the rate varies by more than 10^6 fold in different solvents.[43] Not only the medium polarity affects the reaction but the hydrogen bonding capacity severely influences the solvent effect. Micelles catalyze the decomposition of NBZ and the effect is attributed to a decrease in local polarity at the reaction site as well as a relative stabilization of the charge delocalized transition state.[44]

The decomposition of NBZ in sonicated vesicles of DDDAB, DTDAB, DHDAB, and DODAB, aggregates of TMAC and CTAB micelles have been compared.[45] The $k_{\psi max}/k_0$ ratios produced by the aggregates varies from 50 to 1300. The catalytic effect increases in the order micelles < vesicles < MTA aggregates. With vesicles a clear phase transition is observed with increasing temperature. The data were not analyzed quantitatively; therefore, catalytic effects were not separated from possible differences in partition coefficients (or ion exchange) of NBZ between the aqueous pseudophase and the aggregates. Clearly, however, both factors contribute to the differences observed with changing aggregate structure and fluidity. Vesicles, therefore, furnish a more versatile microreactor where rate can be controlled by structure as well as temperature.

Reactions of thermal isomerization provide an interesting system to analyze the effect of physical constraints on reaction rates. The thermal isomerization of substituted spiropyrans (Scheme 7) in sonicated DODAB vesicles as a function of temperature has been analyzed.[46]

The rate constant for isomerization shows an abrupt change at the Tc of the (vesicle) bilayer and the reaction rate increases sharply above the Tc. For the more hydrophobic spiropyrans, a biphasic kinetic behavior suggests that the substrate occupies two solubilization sites that are not in equilibrium during

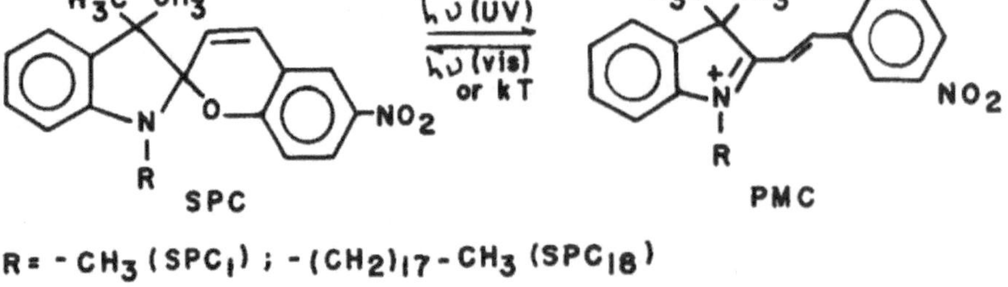

R = $-CH_3$ (SPC_1) ; $-(CH_2)_{17}-CH_3$ (SPC_{18})

Scheme 7 Isomerization of spiropyrans.

the time of the experiment. Clearly, this isomerization depends not only on the microscopic polarity at the reaction site, but also on the physical state of the bilayer.

Nonequilibriating solubilization sites in vesicle bilayers have also been demonstrated in the bromination of substituted *trans*-stilbenes.[47] In micellar sodium dodecyl sulfate (SDS) the bromination reaction exhibits a fast single kinetic transient while in DCP and dipalmitoyl phosphatidyl choline (DPPC) vesicles the reaction is slower, very dependent on substrate structure, and the kinetics are biphasic.

The effect of headgroup density on reaction rates was described decades ago in reactive monolayers. The rate constant for acid catalyzed lactonization of γ-hydroxystearic acid in monolayers formed by the same substrate is lateral pressure-independent up to 8 dynes/cm.[48] The twofold decrease of km upon pressure increase was attributed to changes in accessibility of the reaction center to H^+.

Several descriptions of fluidity modifications upon temperature changes or by the addition of a bilayer disrupting agent to the system have been described. At times, however, it is difficult to separate the effect of the additive, or temperature, on the basic structure from the effect on the reaction itself. The bilayer of vesicles prepared with the same amphiphile shows distinct physical properties depending on vesicle size (see above). This feature was utilized to probe the effect of vesicle changes on reaction rate, using the same amphiphile, maintaining temperature and constant medium composition.

Ester thiolysis and hydrolysis have been extensively analyzed in both micelles and vesicles. The rate determining step for OH^- [49] and alkyl mercaptide[50] attack on activated esters, such as *p*-nitrophenyl alkanoates, is nucleophilic attack both in water and in CTAB micelles. Therefore, it is likely that there is no change in the rate limiting step at the interface of DODAC vesicles. These reactions were chosen to investigate the effect of vesicle size on reaction rate.[51] Selected data showing the effects of DODAC vesicles on the pK of the nucleophile and on reaction rate are presented in Figures 1 and 2.

Quantitative analysis of the effect of vesicles on the apparent pKa of the mercaptan and reaction rate was done using the PPIE model. The appropriate equations to describe the effect of DODAC on the pKa and observed rate constants[26d] are Equations 6 and 7, respectively.

$$K_{ap} = K_a \times \frac{1 + K_{M/Y}(Y_b/Y_f)}{1 + K_{HM}C_d} \tag{6}$$

$$k\psi = [HM_T] \times Ka \times \frac{(k_{2v}/V) \times (K_s K_{M/Y})(Y_b/Y_f) + k_{2o}}{(1 + K_{HM}C_d) \times ([H_f] + K_{ap}) \times (1 + K_s C_d)} \tag{7}$$

The mercaptan-DODAC association constant (K_{HM}) was determined by fluorescent quenching and the mercaptide halide selectivity coefficient for ion exchange ($K_{M/Y}$) was calculated by fitting data relating the variation of K_{ap} with [DODAC] to Equation 6[51] (Figure 1). The satisfactory fit of the data to Equation 6 ensures that the PPIE model can be used for vesicles. Note that the effect of DODAC concentration on Kap varies with vesicle size, the largest effect being produced by sonicated and alcohol-injected vesicles. The effect of vesicle concentration on the rate of thiolysis, analyzed with Equation 6, employed k_{2v} as the only adjustable parameter. All the other terms in the equation including H_f, the NPO-DODAC association constant (K_s), variation of bound and free chloride (Y_b/Y_f), k_{2w}, partial reaction volume (V) were calculated independently, taken as constants or had been calculated in separate experiments. The excellent fit of the data to PPIE with a single adjustable parameter permitted the determination of k_{2v} (Figure 2). The effect of one substrate (HM) on the partition coefficient of the second (NPO), as well as fast substrate equilibration were measured. The overall rate enhancement increases with decreasing vesicle diameter. This effect, however, is a consequence of a decrease in the association constants with bilayer compactness (Table 3). On the other hand, the calculated second-order rate constants, increase with vesicle size only in DODAC vesicles (Table 3).

These results emphasize the importance of quantitative analysis in the interpretation of the effect of structure of the aggregate on rates of chemical reactions.

Micelles and vesicles can dramatically change the product composition of a chemical reaction. For example, OH^- reacts with *N*-alkyl-4-cyanopyridinium ions at the CN group and at C_4 of the pyridinium ring, producing, at pH 10, similar yields of the corresponding pyridone and carboxyamide.[52a] Incorporation of *N*-dodecyl-4-cyanopyridinium ions in CTAB micelles produce almost exclusively the corresponding pyridone upon reaction with OH^-.[52b]

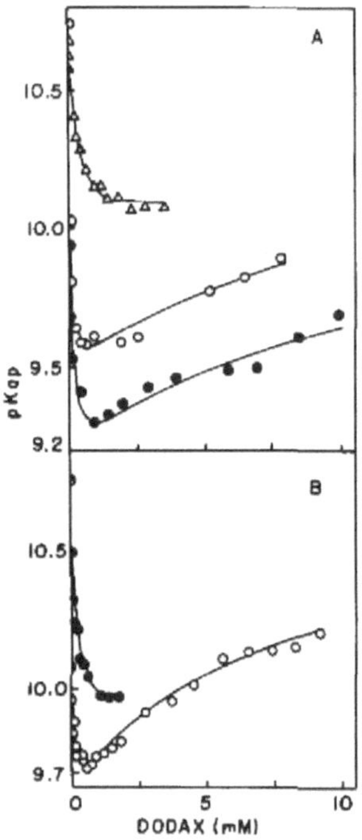

Figure 1 Effect of vesicles on the apparent pKa of heptyl mercaptan. [HM] = (5 − 8) × 10^{-5} *M*. (A) DODAC: (○) ethanolic (Tris-acetate buffer, 0.05 *M*, pH = 9.02); (●) sonicated (Tris-acetate buffer, 0.05 *M*, pH = 9.2); (△) large vesicles (borate buffer, 0.01 *M*, pH = 10.14). (B) DODAB: (○) sonicated (borate buffer, 0.01 *M*, pH = 9.18); (●) large vesicles (borate buffer, 0.06 *M*, pH = 9.86). (Lines were calculated from Reference 51.)

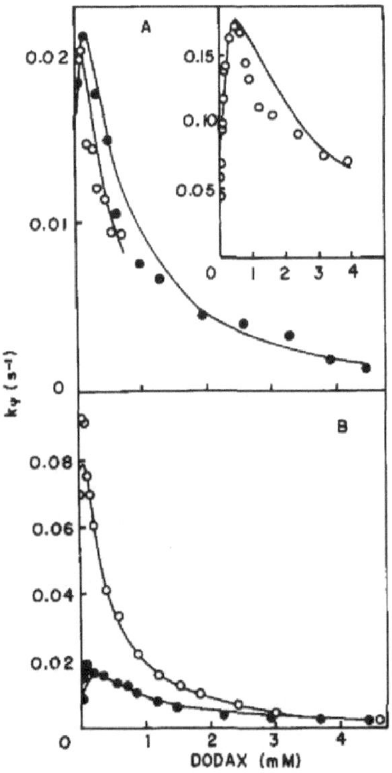

Figure 2 Effect of vesicles on the thiolysis of *p*-nitrophenyl octanoate by heptyl mercaptan. [NPO] 5 × 10^{-6} *M*. (A) DODAC: (○) ethanolic (Tris-acetate buffer, 0.05 *M*, pH = 5.35, [HM] = 5 × 10^{-5} *M*); (●) sonicated (Na^+ acetate buffer, 0.05 *M*, pH = 5.35, [HM] = 5 × 10^{-5} *M*). Inset shows large vesicles (Tris-acetate buffer, 0.006 *M*, pH = 7.17, [HM] = 8.2 × 10^{-5} *M*). (B) DODAB: (○) sonicated (piperazine-acetate buffer, 0.006 *M*, pH = 6.46, [HM] = 8.2 × 10^{-5} *M*); (●) large vesicles (piperazine-acetate buffer, 0.006 *M*, pH = 6.57, [HM] = 8.4 × 10^{-5} *M*). (Lines were calculated from Reference 51.)

Chlorination or bromination of substituted phenyl ethers (Phe-O-R, R = C_5H_{11}, C_9H_{19}, $C_{12}H_{25}$) produces the *o*- and *p*-monosubstituted products.[53] The *para/ortho* (p/o) ratio for monochlorination of Phe-O-C_5H_{11} is 1.3 in water, 4.6 in SDS micelles, and 4.0 in vesicles of a synthetic acid unstable amphiphile (sodium 3-(2,2-heptadecyl-1,3-dioxalan-4-yl)methoxy-1-propane sulfonate (DHPS) and DPPC. With a more hydrophobic phenyl ether (Phe-O-$C_{12}H_{25}$) the p/o ratio is 7.0 in SDS micelles, 6 in DHPS vesicles, and reaches 9.0 in DPPC vesicles. The small but significant differences in the p/o ratios in the

Table 3 Thiolysis of NPO by HM in Vesicles

Vesicle (method)	Dh (nm)	K_{HM} (M^{-1})	$K_S * 10^3$ (M^{-1})	$k_{\psi max}/k_o$	k_{2m}/k_{2w}
DODAC (EtOH)	22	3,500	12.5	3.4×10^6	28
DODAC (sonic)	52	2,000	7.0	3.6×10^6	38
DODAC ($HCCl_3$)	285	500	2.5	2.7×10^5	68
DODAB (sonic)	65	2,500	30	7.4×10^5	20
DODAB ($HCCl_3$)	275	600	10	1.3×10^5	11

Data from Reference 38.

vesicles can be attributed to differences in solubilization sites or to the detailed structure of the interface at the reaction site.

Considering the wide structural range of vesicle-forming monomers and the number of potential reaction sites, the development of regioselective vesicle systems where reaction products can be controlled (for example by changing the temperature) is an attractive research area.

D. FUNCTIONALIZED VESICLES

Synthetic vesicle-forming monomers can contain a variety of functional groups providing a vast research (and application) field in the study of vesicle properties and vesicular effects on chemical reactions.

Vesicles of N,N-ditetradecyl-N$^{\alpha}$ (6-trimethylammoniohexanoyl)-L-histidinamide bromide (THB) or N,N-ditetradecyl-N$^{\alpha}$-(6-trimethylammoniohexanoyl)-L-alaninamide bromide (TAlaB) and CTAB micelles increase the pyridoxal-5′-phosphate catalyzed transamination of N-dodecyl-L-alaninamide (Ala C12).[54] The $k_{\psi max}/k_o$ ratio for transamination in THB vesicles is 200, 85 for TAlaB vesicles, and 150 in CTAB micelles. Part of the rate enhancement was attributed to OH^- concentration at the aggregate's surface. The increased efficiency of THB was attributed to general base catalysis by the interfacial imidazol moiety.

Vesicles of (N,N-didodecyl-N$^{\alpha}$-6-(trimethylammonio) hexanoyl) histidinamide bromide (*1*) and the micelle forming analog with only one dodecyl chain (*2*) estereoselectively catalyze the hydrolysis of L- and D- *p*-nitrophenyl benzyloxycarbonyl phenylalaninates.[55] Micelles of (*2*) preferentially increase the rate of hydrolysis of the L substrate (L/D = 2.5) and the stereoselective ratio is identical at 4 and 20°C. The rate is higher in vesicles of (*1*) and the reaction is more stereoselective (L/D = 4.4 at 4°C and 3.3 at 20°C) both below and above the Tc (Tc <15°C). Clearly, enantiomer selection is more efficient with vesicles, moreover, stereoselectivity in vesicles can be temperature-modulated.

Vesicles prepared with N,N-dihexadecyl-(β-mercaptoethyl)-N-methylammonium chloride (DHASH) and the micelle-forming monohexadecyl analog (*1*) react with NPA by thiolysis.[56] Alcohol-injected vesicles of DHASH increase the thiolysis rate two times more than micelles of (*1*). Vesicles of DHASH are four times more effective than micelles of (*1*). The higher efficiency of vesicles (compared to micelles) can be due to changes in the partition coefficient of the substrate, decrease in the apparent pKa of the nucleophile or catalysis, i.e., decrease in the activation energy for nucleophilic attack.

VI. COMPARTMENTALIZED REACTIONS

The separation of the effects of the external and internal interfaces of synthetic amphiphile vesicles on kinetics and mechanism has been a subject of interest. In a related system, exhibiting an internal aqueous compartment, i.e., reverse micelles, the properties of water change very sharply only before completion of the hydration of the head groups.[57] Since vesicles are obtained in excess water, the head groups in the internal aqueous compartment must be fully hydrated. The interest in differentiating reaction features in the internal vesicular compartment stems partly from the expected differences in head group packing at the internal interface, especially for small vesicles, and the possible differences in electrostatic features in a closed compartment.[17,24] Moreover, substrate diffusion through the bilayer may determine access to internal reaction sites and the study of passive transport becomes important.

Thiolysis of NPA, *p*-nitrophenylhexanoate (NPH), *p*-nitrophenyl palmitate (NPP), and 4-acetoxy-3-nitrobenzenesulfonate (ANBS) were studied in ethanol-injected DHASH vesicles.[58] The $k_{\psi max}$/[DHASH] ratios were 1300, 9000, and 10^4 for NPA, NPH, and ANBS, respectively. Quantitative analysis could clarify if these variations reflect partition or solubilization site differences.

The difficulties of employing dialkyldimethylammonium-based vesicles for the analysis of endovesicular reactions were clearly shown by Moss and co-workers[58,59] demonstrating the artifactual nature of some slow kinetic transients arising from salt and/or osmotic derived vesicle aggregation. The work of Moss and co-workers[58,59] in these systems was also essential to demonstrate that the separation of reactions in the vesicle's external interface (exovesicular reaction) and in the internal interface (endovesicular reaction) must take into account the effect of ionic strength and osmolarity in the stability of the vesicles, especially with sonicated and alcohol-injected preparations.

Reactions of a series of nucleophiles with DTNB have been used to monitor the effect of external and internal vesicular interfaces on rates.[60] DTNB reacts with dithionite (excess) to yield clean pseudo first-order kinetics in CTAB micelles.[60] Using DHDB sonicated vesicles, the rate is threefold faster than in micelles and reaction of DTNB with sulphite and thiophenoxide produces biphasic kinetics due to fast exovesicular rates and slow redistribution of either DTNB or nucleophile. The proportion of the (slow) endovesicular reaction depends on vesicle-DTNB preincubation time and on the presence of cholesterol in the bilayer. With OH^- the reaction is monophasic, suggesting that OH^- permeated DHDB vesicles faster than reaction rate.[60]

The reaction of o-iodosobenzene with 2-nitro-5-thiolatobenzoate (TNB) was studied in CTAB micelles and small (sonicated) and large (chloroform-injected) DODAC vesicles.[61] Vesicles containing TNB in the external and internal aqueous compartments show fast exo- and slow endovesicular rates, attributed to slow diffusion of iodosobenzoate across the bilayer. Elimination of external TNB, by Sephadex filtration of vesicles prepared with the substrate, leads to a single (slow) kinetic transient. TNB does not leak from the vesicle upon incubation for 3 days at room temperature and addition of iodosobenzene to these vesicles reproduces the slow kinetic transient. Therefore, in this case the permeating species is the nucleophile.

Clear demonstration of modulation of exo- and endovesicular rates by reagent permeation, vesicle preparation method, and temperature in several vesicular systems has been extensively described by Moss and co-workers[60–62] in a variety of reaction systems.

Reagent compartmentalization in vesicular microreactors permitted the rational design of systems where rates are modulated by bilayer permeation. Rates of permeation are also important data for the quantitative description of reactions occurring in the internal vesicle compartment.

The rate of OH^- permeation through DODAC bilayers was determined using a reaction-type probe (*N*-methyl-4-cyanopyridinium bromide, MCP) and a positively charged pH indicator (2-hydroxy-5-(2-trimethylammonium)acetyl benzoate (PAB).[63] Independent data, obtained in CTAB micelles, demonstrated that (the positively charged) MCP is coulombically excluded from the positively charged interface.[64] This control is necessary since interfacial binding of a OH^- (or H^+) sensitive probe can lead to very complex results when measuring an attribute of the internal water compartment, especially when analyzing any property depending on the concentration of a counterion. The local concentration of counterions at the interface of a charged bilayer can reach 2 to 4 *M*, in a solution containing m*M* bulk ion concentration.[65] Therefore, specific binding of even a small proportion of a probe at the interface will report a local concentration that can exceed, in orders of magnitude, that in the bulk phase. Hence, in the case of a dimensionally restricted system, such as the internal compartment of a vesicle, it is essential to demonstrate that the probe does not bind to the interface. Reaction of OH^- with MCP shows the added advantage that interfacial binding of cyanopyridinium derivatives results in a change in product distribution[66] that depends on the relative distance of the reacting center to the interface.[67] The similarity of PAB and MCP structure, and the results of rates of permeation measured with both probes indicate that PAB is also excluded from the internal interface. PAB and MCP were incorporated in the internal aqueous compartment of large (chloroform-injected) vesicles, and the excess reagent eliminated by Sephadex filtration. Slow OH^- permeation was measured both by PAB deprotonation and reaction of MCP with OH^-. In DODAC vesicles the half-life for OH^- equilibration is about 7 min at 30°C. Buffers can accelerate OH^- permeation and, as a limit, the rate of reaction of MCP with OH^- in the internal aqueous compartment reaches that in bulk aqueous solution.

The reaction of DTNB with OH^- ion was also used to determine the exo- and endovesicular reactivity of OH^- at the interfaces.[68] Vesicles containing DTNB exclusively in the internal interface can be prepared

Table 4 Effect of Buffer Composition on the Endo- and Exovesicular Rate Constant Ratio (k_{endo}/k_{exo}) for the Reaction of OH^- with DTNB in DODAC Vesicles

Buffer Composition	pH	k_{endo}/k_{exo}
Borate 5 m*M*, NaCl 2.5 m*M*	9.2	2.8
Borate 5 m*M*, NaBr 2.5 m*M*	9.2	2.1
NaOH 1 m*M*, NaCl 4 m*M*	10.9	90
NaOH 1 m*M*, NaBr 4 m*M*	10.8	13.3
Triethylamine.HCl 5 m*M*	9.7	1.7
Triethylamine.HBr 5 m*M*	9.7	0.6
Triethylamine.HCl 10 m*M*	10.8	2.2
Triethylamine.HBr 10 m*M*	10.9	0.6

Note: Vesicles prepared by $HCCl_3$ injection.

Data from Reference 68.

by (external) ion exchange. The rate of OH^- reaction with internally adsorbed DTNB is slower than that obtained with externally added DTNB by orders of magnitude. Definite demonstration that this rate difference is due to slow OH^- permeation was obtained by addition of buffers that increase the rate of OH^- permeation. By using ions that displace OH^- from the external surface, and ions that do not compete as efficiently with OH^- in the internal compartment, the endovesicular reaction of OH^- with DTNB can be faster than that in the external surface (Table 4).

Differentiation of endo- and exovesicular interfaces has also been elegantly explored by Moss and co-workers using an approach that relies in the synthesis of reactive amphiphile monomers. Vesicles of DHDB containing 10 mol% of N,N-dihexadecyl-N-methyl-N-methane-4-nitrophenyl benzoate (*1S*) or N,N-dihexadecyl-N-ethyl *p*-nitrophenyl carbonate (*2S*) (Scheme 8) were prepared by ethanol injection at pH 3.0.[69] The reactive groups of the amphiphiles are stable at this pH. Alkaline hydrolysis or thiolysis by thiophenoxide of the reactive amphiphiles produced single kinetic transients, again indicating that dihexadecyl ammonium derived vesicles are permeable to both OH^- and thiophenol.

Increasing the chain length of the two alkyl chains of (*1S*) from diC_{16} to diC_{18} and diC_{20} and incorporation of these reactive amphiphiles in sonicated vesicles of DHDAB, DODAB, and dieicosyldimethylammonium bromide (DEDAB) led to clear endo- and exovesicular differentiation in the cases of DODAB and DEDAB. In contrast, endo- and exovesicular sites are not differentiated in DHDAB vesicles.[70]

Especially interesting are the properties of 1,2-bis (palmitoyloxy)-3-(propyl)dimethyl (β-(*p*-nitrophenyl carbonate) ethyl)ammonium bromide (*1*) (Scheme 8).[71] When sonicated vesicles of (*1*) prepared in dilute acid are mixed with an alkaline buffer, two clear transients are observed. In the first process (half-life 5 s) 50% of product is formed. In a second kinetic process (half-life 20 min) the remaining substrate is hydrolyzed. Diffusion-limited OH^- attack on the internally facing monomers clearly explain the observed results (Scheme 8). Note that the charge of the vesicle remains positive upon reaction.

The notion of reactive functional amphiphiles was used further by Moss and co-workers[72–74] not only to extend the studies of permeation-limited reactions but to demonstrate and determine the rates of monomer flip-flop in vesicles of synthetic amphiphiles. The rate of flip-flop of reactive substrates was measured allowing all the external substrate to react with OH^- or gluthathione under conditions where the vesicle allows only very slow OH^- permeation. Thereby the pH is lowered (making the reactive detergent stable) and the vesicles preincubated at low pH for different times. Realcalinization results in further reaction only with those monomers that have flip-flopped during the preincubation time. A similar rationale was used to demonstrate that chain ramification enhances the flip-flop rate[75] and to demonstrate that the half-life of flip-flop of synthetic amphiphiles increases in the order saturated (30 min) << trans (4 min) < cis (1 min).[74] It is important to note that the flip-flop rate of monomers in these synthetic amphiphile vesicles is orders of magnitude faster than those of dioleylphosphatidylcholine (days above the Tc).[76] The difference in flip-flop rates between natural lipids and synthetic amphiphiles containing tetramethylammonium groups may be related also to the differences in vesicle stability and is poorly understood.

Chain interdigitation in synthetic amphiphile vesicles increases vesicle stability.[77] The rate of flip flop of fluorocarbon-derived amphiphiles was studied using a strategy similar to that employed by Moss and co-workers.[78]

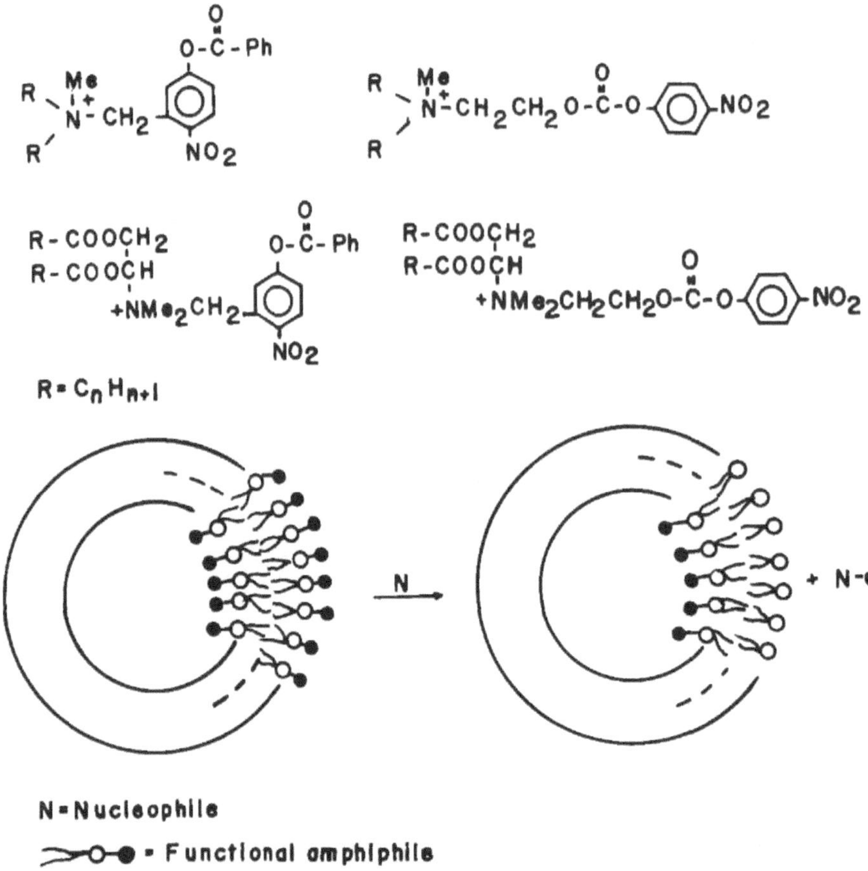

Scheme 8 Representation of reactions of functional amphiphiles in the outer vesicular interface.

In conclusion, vesicles of synthetic amphiphiles can accelerate reactions more efficiently than micelles. The increase in the reaction rate is due, in large part, to reagent concentration in the bilayer, changes in the pKa of weak acid-derived nucleophiles, and in some cases, decrease of the free energy of activation of the reaction. The reaction rate can be controlled by vesicle size, physical state of the bilayer, and medium composition. Furthermore, vesicles allow substrate compartmentalization and hence, by selectively modulating permeation, added reaction control. Synthesis of functional amphiphiles has extended the use of vesicles and included systems where the bilayer is totally asymmetric. The complexity of vesicles, and corresponding reaction sites, make this system a model where much more quantitative work is needed to fully exploit its capabilities and, at the same time, an assembly where biological paradigms can be exploited to both understand and apply the principles of control by complexity.

ACKNOWLEDGMENTS

The authors thank the collaborators and students cited in our publications. Discussions with Professors M. J. Politi, C. A. Bunton, L. S. Romsted, E. Lissi, E. Abuin, F. Nome, and R. A. Moss have helped us. Our work is partially supported by the following Brazilian agencies: Fundação de Amparo à Pesquisa do Estado de São Paulo (FAPESP), Conselho Nacional de Desenvolvimento Científico e Tecnológico (CNPq), Financiadora de Estudos e Projetos (FINEP) and PADCT.

REFERENCES

1. **Fendler, J. H.,** *Acc.Chem.Res.* 13, 7, 1980; **Fendler, J. H.,** *Membrane Mimetic Chemistry,* Wiley-Interscience, New York, 1982; **Fendler, J. H.,** *Chem. Eng. News,* 62, 25, 1984.
2. **Fuhrhop, J. and Mathieu, J.,** *Angew. Chem. Int. Ed. Engl.,* 23, 100, 1984.

3. **Israelachvili, J. N.,** *Intermolecular and Surface Forces*, 2nd ed., Academic Press, San Diego, 1991.
4. **Tanford, C.,** *The Hydrophobic Effect: Formation of Micelles and Biological Membranes*, 2nd ed., John Wiley & Sons, New York, 1980.
5. **Bangham, A. D. and Horne, R. W.,** *J. Mol. Biol.*, 8, 660, 1964.
6. **Hupfer, B., Ringsdorf, H., and Schupp, H.,** *Chem. Phys. Lipids*, 33, 355, 1983.
7. **Ringsdorf, H., Schlarb, B., and Venzmer, J.,** *Angew. Chem.*, 27, 113, 1988.
8. **Kunitake, T., Okahata, Y., Tamaki, K., Kumamuru, F., and Yakayanagi, M.,** *Chem. Lett.*, 387, 1977.
9. **Mortara, R. A., Quina, F. H., and Chaimovich, H.,** *Biochem. Biophys. Res. Commun.*, 81, 1080, 1978.
10. **Kunieda, H., Akimaru, M., Ushio, N., and Nakamura, K.,** *J. Coll. Interface Sci.*, 156, 446, 1993.
11. **Lasic, D. D.,** *Biochem. J.*, 256, 1, 1988.
12. **Lacouturière, S. and Piekarski, S.,** *J. Chimie. Phys.*, 84, 941, 1987.
13. **Brady, J. E., Evans, D. F., Kachar, B., and Ninham, B. W.,** *J. Am. Chem. Soc.*, 106, 4279, 1984.
14. **Carmona-Ribeiro, A. M. and Chaimovich, H.,** *Biophys. J.*, 50, 621, 1986; **Carmona-Ribeiro, A. M.,** *Chem. Soc. Revs.*, 209, 1992.
15. **Madani, H. and Kaler, E. W.,** *Langmuir*, 6, 125, 1990.
16. **Cuccovia, I. M., Aleixo, R. M. V., Mortara, R. A., Berci Filho, P., Bonilha, J. B. S., Quina, F. H., and Chaimovich, H.,** *Tetrahedron Lett.*, 3065, 1979.
17. **Cuccovia, I. M., Feitosa, E., Chaimovich, H., Sepulveda, L., Reed, W. F.,** *J. Phys. Chem.*, 94, 3722, 1990.
18. **Carmona-Ribeiro, A. M., Yoshida, L. S., and Chaimovich, H.,** *J. Phys. Chem.*, 89, 2928, 1985.
19. **Pansu, R. B., Arrio, B., Roncin, J., and Faure, J.,** *J. Phys. Chem.*, 94, 796, 1990.
20. **Lissi, E. A., Abuin, E. B., Zanocco, A., Backer, C. A., Whitten, D. G.,** *J. Phys. Chem.*, 93, 4886, 1989.
21. **Cuccovia, I. M., Chaimovich, H., Lissi, E. A., and Abuin, E.,** *Langmuir*, 6, 1601, 1990.
22. **Bunton, C. A.,** in *Surfactants in Solution*, Vol. 1, Mittal, K. L. and Shah, D. O., Eds., Plenum Press, New York, 1991, 17–40.
23. **Egorova, E. M., Dukhin, A. S., and Svetlova, I. E.,** *Biochim. Biophys. Acta*, 1104, 102, 1992.
24. **Feitosa, E., Agostinho-Neto, A., and Chaimovich, H.,** *Langmuir*, 9, 702, 1993.
25. **Baptista, M. S., Cuccovia, I. M., Chaimovich, H., Politi, M. J., and Reed, W. F.,** *J. Phys. Chem.*, 96, 6442, 1992.
26. **Berezin, I. V., Martinek, K., and Yatsimirski, A. K.,** *Russ. Chem. Rev. (Engl. Transl.)*, 42, 787, 1973; **Menger, F. M. and Portnoy, C. E.,** *J. Am. Chem. Soc.*, 89, 4698, 1985; **Bunton, C. A. and Moffatt, J. R.,** *J. Phys. Chem.*, 89, 4166, 1985; **Quina F. H. and Chaimovich, H.,** *J. Phys. Chem.*, 83, 1844, 1979; **Romsted, L. S.,** *Micellization, Solubilization and Microemulsions*, Vol. 1, Mittal, K. L., Ed., Plenum Press, New York, 1977, 509; **Bunton, C. A., Wright, S., Holland, P. M., and Nome, F.,** *Langmuir*, 9, 117, 1993.
27. **Bunton, C. A., Romsted, L. S., and Savelli, G.,** *J. Am. Chem. Soc.*, 101, 1253, 1979; **Ferreira, L. C. M., Zucco, C., Zanette, D., and Nome, F.,** *J. Phys. Chem.*, 96, 9058, 1992.
28. **Sepulveda, L., Lissi, E., and Quina, F. H.,** *Adv. Colloid Interface Sci.*, 25, 1, 1986.
29. **Abuin, E. B. and Lissi, E. A.,** *J. Coll. Interface Sci.*, 95, 198, 1983.
30. **Moss, R. A. and Hui, Y.,** *Tetrahedron Lett.*, 24, 3961, 1983.
31. **Oliveira, A. G. and Chaimovich, H.,** *J. Pharm. Pharmacol.*, 45, 850, 1993.
32. **Yotsuyanagi, T., Hamada, T., Tomida, H., and Ikeda, K.,** *Acta Pharm. Suec.*, 16, 271, 1979.
33. **Bianconi, M. L. and Schreier, S.,** *J. Phys. Chem.*, 95, 2483, 1991.
34. **Brochztein, S., Berci-Filho, P., Toscano, V. G., Chaimovich, H., and Politi, M. J.,** *J. Phys. Chem.*, 94, 6781, 1990.
35. **Kunitake, T. and Sakamoto, T.,** *J. Am. Chem. Soc.*, 100, 4615, 1978.
36. **Tagaki, W., Amada, T., Yamashita, Y., and Yano, Y.,** *J. Chem. Soc. Chem. Comm.*, 1131, 1972.
37. **Cuccovia, I. M., Quina, F. H., and Chaimovich, H.,** *Tetrahedron*, 38, 917, 1982.
38. **Chaimovich, H., Bonilha, J. B. S., Zanette, D., and Cuccovia, I. M.,** in *Surfactants in Solution*, Vol. 2, Mittal, K. L. and Lindman, B., Eds., Plenum Press, New York, 1984, 1121–1138.
39. **Quina, F. H., Politi, M. J., Cuccovia, I. M., Baumgarten, E., Martins-Franchetti, S. M., and Chaimovich, H.,** *J. Phys. Chem.*, 84, 361, 1980.
40. **Fendler, J. H. and Hinze, W. L.,** *J. Am. Chem. Soc.*, 103, 5439, 1981.
41. **Okahata, Y., Ando, R., and Kunitake, T.,** *Bull. Chem. Soc. Jpn.*, 52, 3647, 1979.
42. **Kunitake, T., Ihara, H., and Okahata, Y.,** *J. Am. Chem. Soc.*, 105, 6070, 1983.
43. **Kemp, D. S., Cox, D. D., and Paul, K. G.,** *J. Am. Chem. Soc.*, 97, 7305, 1975.
44. **Bunton, C. A., Minch, M. J., Hidalgo, J., and Sepúlveda, L.,** *J. Am. Chem. Soc.*, 95, 3262, 1973.
45. **Kunitake, T., Okahata, Y., Ando, R., Shinkai, S., and Hirakawa, S.,** *J. Am. Chem. Soc.*, 102, 7877, 1980.
46. **Seki, T. and Ichimura, K.,** *J. Coll. Interface Sci.*, 129, 353, 1989.
47. **Mizutani, T. and Whitten, D. G.,** *J. Am. Chem. Soc.*, 107, 3621, 1985.
48. **Davies, J. T.,** *Trans. Faraday Soc.*, 45, 448, 1949.
49. **Correia, V. R., Cuccovia, I. M., and Chaimovich, H.,** *J. Phys. Org. Chem.*, 4, 13, 1991.
50. **Correia, V. R., Cuccovia, I. M., and Chaimovich, H.,** *J. Phys. Org. Chem.*, 6, 7, 1993.
51. **Kawamuro, M. K., Cuccovia, I. M., Abuin, E., Lissi, E., and Chaimovich, H.,** *J. Phys. Chem.*, 95, 1458, 1991.

52. Kosower, E. M. and Patton, J. W., *Tetrahedron*, 22, 2081, 1966; Politi, M. J., Cuccovia, I. M., Chaimovich, H., Almeida, M. L. C., Bonilha, J. B. S., and Quina, F. H., *Tetrahedron Lett.*, 115, 1978.
53. Jaeger, D. A., Clennan, M. W., and Jamrozik, J., *J. Am. Chem. Soc.*, 112, 1171, 1990.
54. Murakami, Y., Nakano, A., and Akiyoshi, K., *Bull. Chem. Soc. Jpn.*, 55, 3004, 1982.
55. Murakami, Y., Nakano, A., Yoshimatsu, A., and Fukuya, K., *J. Am. Chem. Soc.*, 103, 728, 1981.
56. Moss, R. A. and Bizzigotti, G. O., *J. Am. Chem. Soc.*, 103, 6512, 1981.
57. El Seoud, O. A., El Seoud, M. I., and Mickiewicz, J. A., *J. Coll. Interface Sci.*, 163, 87, 1994.
58. Moss, R. A., Hendrickson, T. F., and Bizzigotti, G. O., *J. Am. Chem. Soc.*, 108, 5520, 1986.
59. Moss, R. A., Swarup, S., Hendrickson, T. F., and Hui, Y., *Tetrahedron Lett.*, 25, 4079, 1984.
60. Moss, R. A. and Schreck, R. P., *J. Am. Chem. Soc.*, 107, 6634, 1985.
61. Moss, R. A., Swarup, S., and Zhang, H., *J. Am. Chem. Soc.*, 110, 2914, 1988.
62. Moss, R. A. and Swarup, S., *J. Org. Chem.*, 53, 5860, 1988.
63. Cuccovia, I. M. and Chaimovich, H., *Bol. Soc. Chil. Quim.*, 35, 39, 1990.
64. Chaimovich, H., Bonilha, J. B. S., Politi, M. J., and Quina, F. H., *J. Phys. Chem.*, 83, 1951, 1979.
65. Bunton, C. A., Nome, F., Quina, F. H., and Romsted, L. S., *Accs. Chem. Res.*, 24, 357, 1991.
66. Politi, M. J. and Chaimovich, H., *J. Phys. Org. Chem.*, 4, 207, 1991.
67. Zanette, D. and Chaimovich, H., *J. Phys. Org. Chem.*, 5, 341, 1992.
68. Cuccovia, I. M., Kawamuro, M. K., Krutman, M. A. K., and Chaimovich, H., *J. Am. Chem. Soc.*, 111, 365, 1988.
69. Moss, R. A. and Schreck, R. P., *Tetrahedron Lett.*, 26, 6305, 1985.
70. Moss, R. A., Fujita, T., and Ganguli, S., *Langmuir*, 6, 1197, 1990.
71. Moss, R. A., Bhattacharya, S., Scrimin, P., and Swarup, S., *J. Am. Chem. Soc.*, 109, 5740, 1987.
72. Moss, R. A., Bhattacharya, S., and Chatterjee, S., *J. Am. Chem. Soc.*, 111, 3680, 1989.
73. Moss, R. A., Gangulli, S., Okumura, Y., and Fujita, T., *J. Am. Chem. Soc.*, 112, 6391, 1990.
74. Moss, R. A., Fujita, T., and Okumura, Y., *Langmuir*, 7, 440, 1991.
75. Moss, R. A. and Fujita, T., *Tetrahedron Lett.*, 31, 7559, 1990.
76. de Kriujff, B. and Wirtz, K. W. A., *Biochim. Biophys. Acta*, 468, 318, 1977.
77. Moss, R. A. and Bhattacharya, S., *J. Phys. Org. Chem.*, 5, 467, 1992.
78. Liang, K. and Hui, Y., *J. Phys. Org. Chem.*, 5, 715, 1992.

Chapter 14

Homogeneous and Heterogeneous Hydrogenation of Phospholipid Acyl Chains in Liposomes: A Comparison with Biological Membranes

Ferenc Joó and László Vígh

CONTENTS

I. INTRODUCTION: COMPOSITION OF MEMBRANES AND THE ROLE OF UNSATURATED POLAR LIPIDS

Fluidity of biomembranes is determined by a delicate balance of the amount and of the properties of its constituents: polar and nonpolar lipids, proteins, and carbohydrates. Depending on their head-group, the various lipid molecular species can be classified into lipid classes, and most of them have unsaturated fatty acyl residues. This balance may be modified by several methods,[1] however, in living organisms there seems to be a mechanism by which the average fluidity — characteristic to certain environmental conditions — is maintained (homeoviscous adaptation). This indirectly means, that if for instance the average unsaturation is changed by dietary manipulations, this should be accompanied by changes in the ratio of different lipid classes, the average fatty acid chain length, isomer distribution, etc. In principle, hydrogenation allows the fast and selective modification of one important element in this balance: unsaturation. Obviously, when liposomes are hydrogenated, the product maintains the new fatty acid distribution after the treatment. However, if otherwise the conditions are kept unchanged, the regulatory mechanism of living cells will restore the original (for some reason "optimal") distribution, and the result of hydrogenation will be eliminated in rather fast compensatory processes.[2] It is the "time window" allowed by such compensatory processes which can be used for investigation of the properties of cells with selectively changed fatty acid distribution (and membrane fluidity).

II. HYDROGENATION AS A METHOD FOR FLUIDITY ALTERATIONS

Hydrogenation is a well-known method in the chemistry and technology of lipids, fats, and oils; however, it has been mostly restricted to treatment of nonaqueous solutions or solvent-free liquid substrates (melts). In general, heterogeneous catalysts give low rates when applied to aqueous lipid dispersions, and the classical homogeneous catalysts are usually insufficiently soluble in water to be useful for hydrogenation of liposomes and biomembranes. Recently, research into aqueous organometallic chemistry led to the discovery of active and selective water-soluble hydrogenation catalysts[3] and this changed the scene for liposome hydrogenation.

0-8493-4012-8/96/$0.00+$.50
© 1996 by CRC Press, Inc.

There is a vast number of soluble transition metal catalysts which are capable for activating molecular hydrogen and catalyze the reduction of unsaturated bonds. The latter include not only olefinic but C=O and C=N functions, as well. However, there are strict requirements which a catalyst for biomembrane hydrogenation should meet.

- Must be sufficiently stable under physiological condition (stability in aqueous systems, preferably air stability).
- Must not influence the properties of the membranes on ways other than hydrogenation.
- Must not be reactive on ways other than hydrogenation (chemoselectivity). Possible side reactions include isomerization, peroxidation, and oxidative fission of unsaturated fatty acyl chains in lipids, but catalytic hydrogenolysis and hydration may also take place (though this is rather improbable with homogeneous catalysts). Ligand exchange with biomolecules or with components of the buffers may lead to the formation of new metal complexes with unwanted properties.
- Should be removable subsequent to hydrogenation.

Obviously, liposome technology includes several areas where toxicity problems do not arise. However, in pharmaceutical, food, cosmetic, etc., industries, and in manipulation of biomembranes or whole cells an *a priori* requirement is that neither the catalysts or its constituents, nor the products (and eventual byproducts) of hydrogenation be toxic.

A central question is the activity and selectivity of the catalyst (the latter has been already mentioned in the context of chemoselectivity). With a stable liposome, the rate of hydrogenation may not be a crucial question and belongs in the considerations of productivity. In many (though not in all) cases, the intrinsic rate of H_2-activation increases considerably with increasing temperature and the overall rate of liposome hydrogenation is further raised by the higher fluidity of the lipid membranes at higher temperatures. However, the study of biomembranes often requires low temperatures, either for keeping up the physical integrity and function of isolated membranous structures (e.g., thylakoid) or for study of whole cell functions in cold (e.g., temperature acclimation studies). In such cases the success of hydrogenation experiments critically depends on the activity of the catalyst.

III. HOMOGENEOUS HYDROGENATION

Homogeneous hydrogenation is one of the most investigated and most deeply understood fields of catalysis, and the number of soluble catalysts capable for activating molecular hydrogen is literally countless.[3,4] However, in the light of the above requirements the vast majority of them is unsuitable for liposome and biomembrane hydrogenation. The much studied and rather active $K_3[Co(CN)_5]$, for example, is stable only in the presence of an excess of cyanide. Others, like $RuCl_3$·aq, require elevated temperatures and high chloride concentration, and even then, the hydrogenation of olefins is slow. The classical organometallic compounds, containing low-valent transition metal ions in metal-carbon or metal-hydrogen bonds, are usually unstable in aqueous solutions.

A. CATALYSTS WITH PHOSPHINE LIGANDS. THE MECHANISM OF HYDROGENATIONS WITH SUCH CATALYSTS

The family of complexes containing triphenylphosphine (PPh_3) ligands shows extraordinary features with respect to stability and selectivity. The most prominent members of this group are $[RhCl(PPh_3)_3]$ (Wilkinson's catalyst) and $[RuCl_2(PPh_3)_3]$ which activate H_2 by differing mechanisms (see later). A major drawback, however, is in that tertiary phosphines like triphenylphosphine and their complexes are insoluble in water which makes use of co-solvents (e.g., tetrahydrofuran, THF) or solvent vectors (e.g., dimethylsulfoxide, DMSO) unavoidable.

Soya lecithin was hydrogenated successfully by this catalyst at 35°C and atmospheric pressure in water-THF mixtures.[5,6] In all cases, complete hydrogenation of unsaturated fatty acyl chains could be achieved during sufficiently long times or under increased pressure of H_2 (e.g., 9 bar). Increasing the proportion of water in the solvent mixture considerably decreased the initial rate of hydrogenation and in pure aqueous dispersions this rate was only about 10% of that in anhydrous THF. The catalyst showed the expected selectivity, i.e., 18:3 and 18:2 was preferentially reduced over 18:1.

In a nonaqueous environment, and probably in the lipophilic inner core of lipid bilayers, $[RhCl(PPh_3)_3]$ acts[3,4] by the following mechanism:

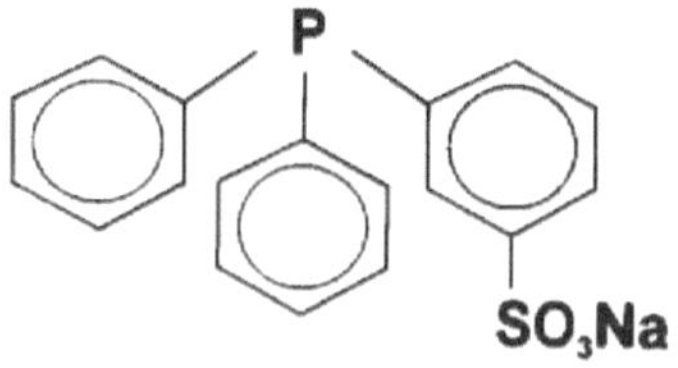

Figure 1 *m*-Sulfonatophenyldiphenylphosphine, TPPMS.

$$[RhCl(PPh_3)_3] + H_2 \rightleftharpoons [H_2RhCl(PPh_3)_2] + PPh_3 \quad (1)$$

$$[H_2RhCl(PPh_3)_2] + -CH{=}CH- \rightleftharpoons [H_2Rh(-CH{=}CH-)Cl(PPh_3)_2] \quad (2)$$

$$[H_2Rh(-CH{=}CH-)Cl(PPh_3)_2] \rightleftharpoons [HRh(-CH-CH_2-)Cl(PPh_3)_2] \quad (3)$$

$$[HRh(-CH-CH_2-)Cl(PPh_3)_2] + H_2 \rightarrow\rightarrow\rightarrow [H_2RhCl(PPh_3)_2] + -CH_2-CH_2- \quad (4)$$

There are several important features of this mechanism. First, it is seen that Equations 1 to 3 are written as reversible, and indeed, model studies with simple olefins showed this being the case. Under ordinary circumstances reaction 4 is irreversible. Second, the major catalytic species is a rhodium dihydride, formed in reaction 2 (therefore $[RhCl(PPh_3)_3]$ is said to act through a *dihydridic mechanism*). This species transfers its hydride ligands to the substrate olefin in a stepwise manner: first a rhodium-alkyl complex is formed (reaction 3) the further reactions of which supplies the product ($-CH_2-CH_2-$) and gives back the intermediate dihydride (reaction 4 represents at least three elementary steps). This is a very rudimentary description of what really happens in such systems (further details can be found in References 3 and 4) but it is sufficient to explain two important features: (a) In reaction 3 a dihydridorhodium-olefin species is reversibly rearranged to a monohydridorhodium-alkyl complex. If the starting olefin is one of a pair of geometric isomers, there is no obstacle to get the other isomer through the reverse reaction. Indeed, this is often the case, especially when the starting olefin has the thermodynamically less stable *cis*-geometry, as the unsaturated fatty acids in natural lipids do. (b) Proton is not involved in the reactions shown and if there is no fast exchange of hydrogens between the rhodium hydrides and the solvent (e.g., H_2O or alcohols) all hydrogens appearing in the saturated product come from the H_2 of the gas phase. By the same virtue, when D_2 is used as the hydrogen source, all products should be completely dideuterated at each former C=C bond.

There are several ways to prepare water-soluble tertiary phosphine complexes by appropriate substitution of the parent phosphine (for a comprehensive review see Reference 3). However, only the complexes of sulfonated triphenylphosphine were used in liposome and biomembrane hydrogenation; therefore we restrict our attention to this class of catalysts.

Sulfonation of PPh_3 can be directed to yield mainly the monosulfonated product, *m*-sulfonatophenyldiphenylphosphine, or triphenylphosphine-monosulfonate (TPPMS) (Figure 1). This ligand can replace PPh_3 in most of its derivatives, giving water-soluble hydrogenation catalysts, *inter alia* $[RhCl(TPPMS)_3]$ and $[RuCl_2(TPPMS)_2]$.

According to detailed investigations $[RuCl_2(PPh_3)_3]$ and $[RuCl_2(TPPMS)_2]$ catalyze the hydrogenation of olefins by the same mechanism,[4,7] in which the active species has the $[RuCl_2(P)_2]$ composition (P=PPh_3 or TPPMS):

$$[RuCl_2(P)_2] + H_2 \rightleftharpoons [HRuCl(P)_2] + H^+ + Cl^- \quad (5)$$

$$[HRuCl(P)_2] + -CH{=}CH- \rightleftharpoons [HRu(-CH{=}CH-)Cl(P)_2] \quad (6)$$

$$[HRu(-CH{=}CH-)Cl(P)_2] \rightleftharpoons [Ru(-CH-CH_2)Cl(P)_2] \quad (7)$$

$$[Ru(-CH-CH_2)Cl(P)_2] + H^+ + Cl^- \rightarrow\rightarrow\rightarrow [HRuCl(P)_2] + -CH_2-CH_2- \quad (8)$$

This is the so-called *monohydridic mechanism* since only monohydridoruthenium species appear in the equations (again, reaction 8 represents several elementary steps and is regarded as irreversible). As above, there is a possibility of substrate isomerization (reaction 7). However, the product is liberated from the intermediate alkylruthenium species by protonation (reaction 8) and this has an important consequence: half of the hydrogens built into the product of hydrogenation now comes from the solvent as H^+ (reaction 8). In organic solutions this proton originates from the gas phase (reaction 5). However, when the hydrogenation is carried out in aqueous solutions there is a much higher chance for a proton from the bulk aqueous phase to be built into the product and therefore high proportion of saturated fatty acids monodeuterated at each former –CH=CH– bond can be obtained. The same is true for aqueous lipid dispersions, i.e., liposomes.

Very recently, it has been shown that in anhydrous solutions, $[RhCl(TPPMS)_3]$ catalyzes the hydrogenation of simple olefins by exactly the same mechanism as $[RhCl(PPh_3)_3]$.[8] However, when using more and more aqueous solvent systems (e.g., mixtures of water and ethylenglycol dimethyl ether, diglyme) there is a switch in the mechanism from the dihydridic to the monohydridic pathway:

$$[RhCl(PPh_3)_3] + H_2 \rightleftharpoons [HRh(PPh_3)_3] + H^+ + Cl^- \tag{9}$$

$$[HRhCl(PPh_3)_3] + -CH{=}CH- \rightleftharpoons [HRh(-CH{=}CH-)(PPh_3)_3] \tag{10}$$

$$[HRh(-CH{=}CH-)(PPh_3)_3] \rightleftharpoons [Rh(-CH{-}CH_2-)(PPh_3)_3] \tag{11}$$

$$[Rh(-CH{-}CH_2-)(PPh_3)_3] + H^+ + Cl^- \rightarrow\rightarrow\rightarrow [RhCl(PPh_3)_3] + -CH_2{-}CH_2- \tag{12}$$

It is seen from reaction 12 that there is a chance for deuteration if there is a supply of D^+. Detailed mechanistic studies proved that in truly aqueous systems, using $[RhCl(TPPMS)_3]$ as catalyst, hydrogenations take place by both monohydridic and dihydridic mechanisms and the ratio of the products formed on the two pathways depends on the pH of the solutions and on the polarity of the solvent. It should be borne in mind that in hydrogenations of liposomes or biomembranes the immediate local environment around the catalyst may be much less polar than the bulk of the aqueous phase.

Concerning the selectivity, it is a well-established general feature of these catalytic hydrogenations[3,4] that terminal olefins react faster than internal ones, of which *cis*-olefinic bonds are reduced faster than those with *trans*-geometry, and that hydrogenation of other functional groups (e.g., C=O and C=N) is very slow compared to the reaction of C=C bonds, if it happens at all. Under ambient conditions hydrogenolysis of ethers, esters, alcohols, or amines does not take place.

Tertiary phosphines and their complexes with low valent transition metal ions are easily oxidized by molecular oxygen, especially in aqueous solutions. Therefore, care should be taken to exclude traces of O_2 from reaction mixtures to avoid catalyst losses. This is best attained when the catalyst is dissolved under an inert atmosphere (N_2, Ar, or H_2) in oxygen-free water and added to the liposome or membrane (cell) suspension previously deoxygenated by several evacuation/hydrogen refill cycles.

Both $[RhCl(TPPMS)_3]$ and $[RuCl_2(TPPMS)_3]$ were used for hydrogenation of soy-bean phosphatidyl choline and wheat chloroplasts.[9,10] Conversions of C=C double bonds of PC liposomes up to 60% could be achieved under 1 bar H_2 in 1 h. A temperature variation study[10] showed that $[RuCl_2(TPPMS)_2]$ retained considerable hydrogenation activity even below 20°C, whereas the activity of $[RhCl(TPPMS)_3]$ was close to zero at this temperature (Figure 2).

The rate of hydrogenation continuously decreased with increasing sterol to phospholipid ratio[10] both with stigmasterol and cholesterol (Figure 3). This reflects the diminished ability of the catalyst to have an access to the unsaturated bonds in the more rigid membranes of increasing sterol content.

The hydrogenation pattern[10] of wheat thylakoid and that of the liposomes of extracted thylakoid total lipids was very much the same; however there is a substantial variation in the hydrogenation pattern of individual lipid classes (Table 1). The highest level of hydrogenation was achieved on the galactolipids, which comprise the most unsaturated representatives of thylakoid lipids; however, digalactosyl diacylglycerol (DGDG) was less hydrogenated than monogalactosyl diacylglycerol (MGDG) despite being considerably more unsaturated. These results certainly reflects the reduced accessibility of DGDG in thylakoid with respect to MGDG.

Discrimination by the catalyst between the variously accessible lipid pools makes the technique of hydrogenation suitable for a structural probe. Indeed, it was demonstrated using $[RhCl(TPPMS)_3]$ that distinction could be made between the lipid clusters around very low density, low density and high density

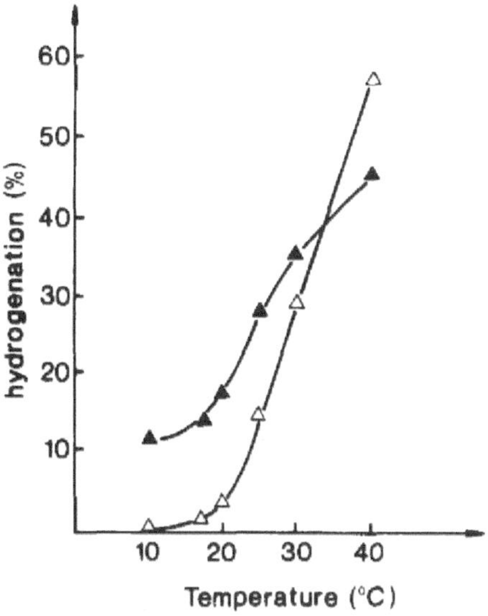

Figure 2 Percentage hydrogenation of soybean phosphatidyl choline, catalyzed by $[RuCl_2(TPPMS)_2]$ (▲) and $[RhCl(TPPMS)_3]$ (△), as a function of temperature. Total pressure = 0.1 MPa, t = 1 h, [Ru or Rh]/lipid = 0.06. (Reproduced from Vígh, L., et al., *J. Mol. Catal.*, 22, 15, 1983. With permission.)

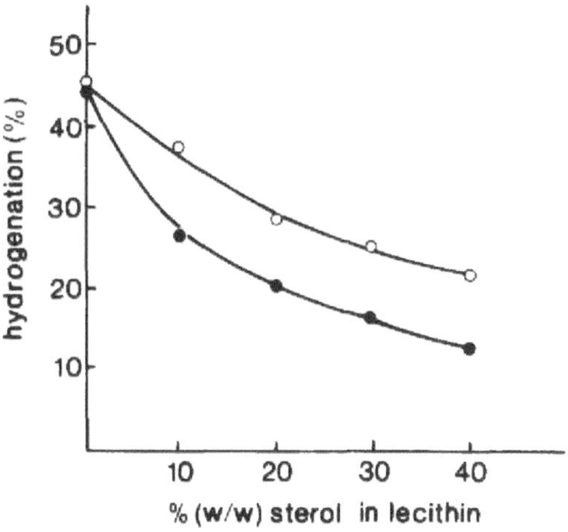

Figure 3 The effect of increasing proportions of cholesterol (●) and stigmasterol (○) on the hydrogenation of soybean lecithin, catalyzed by $[RuCl_2(TPPMS)_2]$. Total pressure = 0.1 MPa, T = 20°C, t = 3 h, [Ru]/lipid = 0.06. (Reproduced from Vígh, L., et al., *J. Mol. Catal.*, 22, 15, 1983. With permission.)

plasma lipoproteins following the pattern of hydrogenation.[11] Also, hydrogenation experiments with $[Pd(QS)_2]$ catalyst revealed the tight specific association of cardiolipin with membrane proteins in rat liver mitochondria which became looser on double bond reduction.[12] Similar experiments on liposomes and thylakoids originating from triazine-resistant and -susceptible lines of *Solanum nigrum* point to the possible existence of some intramembraneous domains of acyl lipids, relatively protected against hydrogenation and present only in the atrazine-susceptible biotype.[13]

A most interesting conclusion based on results of hydrogenation of protoplasts from *Nicotiana plumbaginifolia* with $[RuCl_2(TPPMS)_2]$ as catalyst is in that living systems strive to maintain their fatty acid composition even under conditions of hydrogenation; desaturation of existing long-chain fatty acids and *de novo* synthesis of unsaturated ones may even temporarily overcompensate the effect of hydrogenation.[2]

The rate of hydrogenation may be severely limited by diffusion, primarily by dissolution of H_2 into the bulk of the aqueous phase. Therefore, an efficient stirring is required which — in case of mechanically sensitive systems — may be achieved by gentle rotation of the hydrogenation vessels somewhat tilted relative to the vertex.[14] Molecular hydrogen fast diffuses and dissolves into the aqueous film on the wall of the reactor without strong mechanical stirring of the liposome. Limited availability of dissolved hydrogen may even lead to a change in the reaction pattern. It was shown that in unstirred solutions the *cis*-fatty acids on DOPC liposomes were mainly isomerized to the respective *trans*-isomers before being hydrogenated.[15] The catalyst in these studies was a Rh(I) complex of an amphiphilic tertiary phosphine: diphenylphosphinoundecylphosphine (DPUP). The preferential isomerization is in accord with the pos-

Table 1 Fatty Acid Composition of Wheat Chloroplast Thylakoid and the Extracted Thylakoid Total Lipids Before and After Partial Hydrogenation[a]

Sample		Fatty Acid (wt.%)							Hydrogenation (%)
		16:0	Δ^3-trans 16:1	16:3	18:0	18:1	18:2	18:3	
Thylkoid	I	8.9	3.4	2.4	1.0	0.9	3.7	79.8	
	H	11.0	2.5	—	5.7	8.3	12.7	59.8	15.7
Total lipids	H	12.7	0.8	—	6.3	9.2	11.8	59.2	17.5
MGDG[b]	I	—	—	20.2	—	—	1.3	78.4	
	H	12.5	—	10.8	2.1	3.4	10.7	60.3	20.3
DGDG[b]	I	3.9	—	3.1	0.6	1.0	1.7	89.7	
	H	6.3	—	—	4.7	18.8	11.5	68.7	15.9
PG[b]	I	16.1	33.0	—	0.8	1.4	3.5	45.2	
	H	43.3	3.0	—	3.2	3.8	13.9	38.8	14.7
PC[b]	I	19.8	—	—	1.4	13.1	14.8	50.9	
	H	21.2	—	—	1.7	14.8	25.3	37.0	10.2
SL[b]	I	38.7	—	—	1.3	2.4	8.9	49.3	
	H	39.5	—	—	1.4	4.3	15.4	39.4	8.8

[a] Reactions were performed for 3 h at 20°C, under 0.1 MPa total pressure. [Ru]/[lipid] = 0.6. Different lipid classes were separated from the initial (I) and hydrogenated (H) thylakoids.

[b] MGDG = monogalactosyl diacylglycerol; DGDG = digalactosyl diacylglycerol; PG = phosphatidyl glycerol; PC = phosphatidylcholine; SL = sulpholipids.

From Vígh, L., et al., *J. Mol. Catal.*, 22, 15, 1983. With permission.

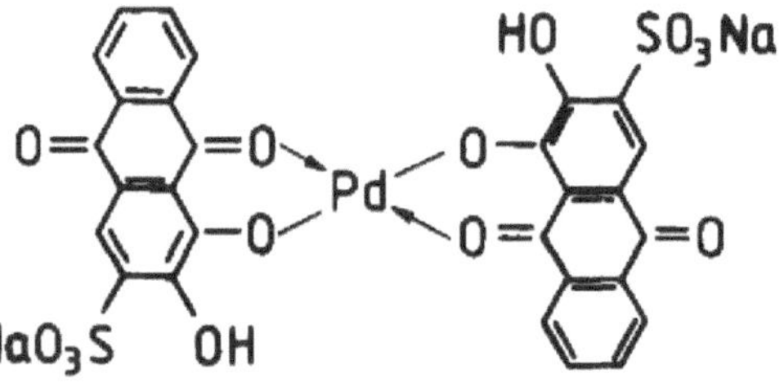

Figure 4 Palladium(II)-*di*(sodium alizarinmonosulfonate), $[Pd(QS)_2]$.

sible mechanisms discussed above, i.e., the reactions shown in Equations 2 or 10 may become dominating in solutions having low concentrations of dissolved H_2.

B. $[Pd(QS)_2]$ AND ITS MECHANISM OF HYDROGENATION

A major step towards wider utilization of hydrogenation as a technique of biochemistry was the introduction of the palladium complex of sulfonated alizarin, $[Pd(QS)_2]$ (Figure 4).[16] In addition to being commercially available (from Molecular Probes, Eugene, OR) this compound is fairly easy to prepare and has remarkable properties. The shiny black, crystalline material is stable to air, with virtually unlimited shelf-life. It dissolves well in water, dimethyl sulfoxide, and dimethyl formamide; however, it is insoluble in apolar organic solvents. Being a dye known since ancient times, it is not surprising that the color of alizarin, its derivatives, and their metal complexes is pH-dependent; the colors quoted below usually refer to pH 6 to 8. The following is a brief description of the general features of hydrogenations with $[Pd(QS)_2]$ (more detailed information can be obtained from References 16 to 18).

When *properly deoxygenated* aqueous solutions of $[Pd(QS)_2]$ are brought into contact with H_2 a fast color change from blue/violet to yellow signals the transformation of the catalyst from its initial form (**A**) to an active form (**C**). This reaction is autocatalytic in **C**. When the yellow solution is kept further under H_2, **C** is gradually replaced by a green compound, **D**, which slowly reacts further with H_2 giving colorless, unidentified compounds. Reactions of **C** and **D** with O_2 give rise to the formation of different products, the strong pink-colored **B** and the deep-red **E**, respectively. Unfortunately, the precise chemical composition of these intermediates are still not known; however, they can be unambiguously characterized by their UV-vis spectra. Also, it has been established that while **A**, **B**, and **D** are not paramagnetic, **C** and **E** are free radicals (ESR spectra shown on Figure 5). The catalytically active species is most probably **C**.

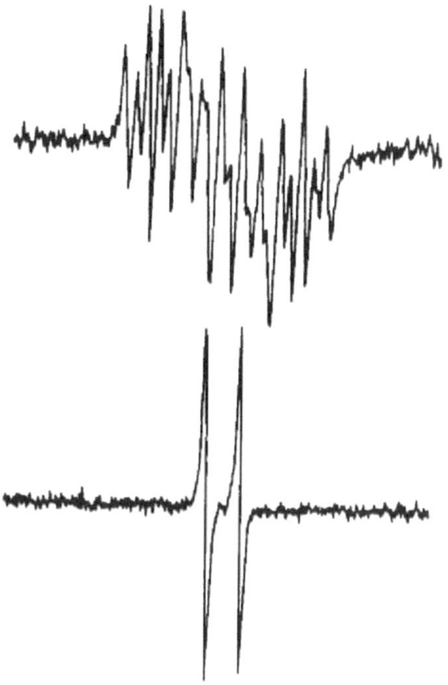

Figure 5 ESR spectra of solutions containing the catalytically active **C** form of $[Pd(QS)_2]$ (above) and its oxidized product, **E** (below). 5 mg/ml $[Pd(QS)_2]$, pH 7.0, sodium phosphate buffer (10 m*M*). Scan width: 1 mT (**C**); 2 mT (**E**). (Reproduced from Joó, F., et al., *Anal. Biochem.*, 194, 34, 1991. With permission.)

There are several important consequences of the above features:

1. since the formation of **C** is autocatalytic, and since **C** is a free radical, all radical scavengers, including O_2, inhibit catalytic hydrogenations by $[Pd(QS)_2]$. Since O_2 can be reduced by H_2, traces of oxygen cause only induction periods of varying length (from minutes to hours).
2. Neither **C** nor **D** are practical to use as starting form of the catalyst, since they are further degraded by H_2 and also react with O_2
3. **B**, which can be obtained by oxygenation of **C**, i.e., by short, intensive shaking of the yellow solutions on air, is indefinitely stable in aqueous solutions and can be regarded as a resting state of the catalyst. It reacts with H_2 in an extremely fast reaction providing **C**, which in turn hydrogenates olefinic substrates by a radical mechanism.

Such transformations of the catalyst are manifested in the rate of catalyzed hydrogenations as a function of catalyst prehydrogenation. Figure 6 demonstrates, under the conditions stated there, that multilamellar vesicles of DOPC are almost uneffected by $[Pd(QS)_2]$ with no prehydrogenation (**A**). When the catalyst was added in form **B** or **C** the overall rate is practically the same; this is explained by the fast transformation of **B** into **C** under H_2. However, because of the slow degradation processes mentioned before, both **D** and **E** gave lower hydrogenation rates.

As a final conclusion, it can be stated that prehydrogenation of $[Pd(QS)_2]$ to form **C** followed by air oxidation to form **B** gives an air-stable solution of an active hydrogenation catalyst with acts with no induction period.

The most possible mechanism of hydrogenations catalyzed by $[Pd(QS)_2]$ involves the formation of a palladium(II) monohydride complex:

$$[Pd(QS)_2] + H_2 \rightleftharpoons [HPd(QS)_2]^- + H^+ \tag{13}$$

which rearranges to give the Pd(II) complex of the corresponding anthrasemiquinone radical anion. Hydrogenation of C=C bonds takes place by this latter compound and therefore is characterized by several important features:

- Olefin hydrogenations are fast and their rate is practically independent of temperature (characteristic to free radical reactions)
- Radical scavengers inhibit hydrogenation
- Highly selective monodeuteration can be achieved:

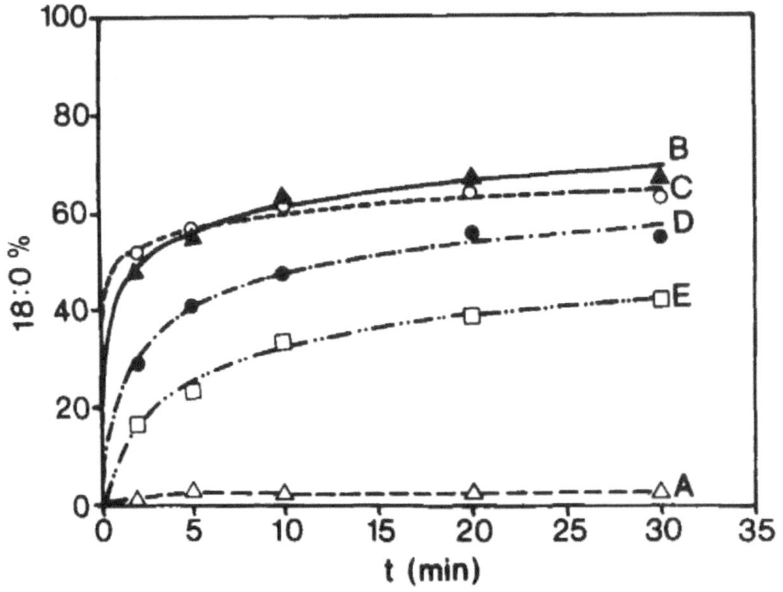

Figure 6 Hydrogenation of DOPC catalyzed by different forms of $[Pd(QS)_2]$ catalyst. [Pd] = 0.12 m*M*, 0.1 mg/ml DOPC, pH 6.5, sodium phosphate buffer (10 m*M*), 25°C, total pressure = 0.1 MPa. (Reproduced from Joó, F., et al., *Anal. Biochem.*, 194, 34, 1991. With permission.)

$$[HPd(QS)_2]^- + C{=}C + H^+ (D^+) \rightarrow CH{-}CH\ (CH{-}CD) + [Pd(QS)_2] \quad (14)$$

- As can be assumed from Equation 14, the rate of hydrogenation is strongly influenced by the pH of the medium; the catalyst can be used conveniently between pH 5 to 9.

The physical state of the membrane has a profound effect on the rate of hydrogenation.[18] This is demonstrated by the data of Figure 7. When multilamellar vesicles of DOPC, DLPC, and DLnPc were hydrogenated by $[Pd(QS)_2]$ under the same conditions, the originally rapid hydrogenation slowed down around 60% conversion of all double bonds. However, when such hydrogenated liposomes were extracted by chloroform, dried, and resonicated in the same buffer, additional loss of the double bonds led to an almost complete hydrogenation of the fatty acyl residues. It seems reasonable, that in the first run of hydrogenation approximately 30 to 40% of the double bonds was protected against hydrogenation, presumably because of their inaccessibility inside the lipid bilayer of the vesicles. During extraction followed by reconstruction of the vesicles, most of these double bonds became exposed to the catalyst and were reduced accordingly.

As can be expected from the reaction mechanism, hydrogenations by $[Pd(QS)_2]$ are usually accompanied by isomerization, exemplified here by the hydrogenation of 18:2/18:2 PC liposomes (Figure 8).[18] Under the conditions used, a transient, but well detectable amount of *cis*-Δ^{12}-18:1 appeared. However, most of the product 18:1 was comprised of the natural *cis*-Δ^9 accompanied by the *trans*-Δ^9-18:1 isomer. According to the general rule of selectivity, *cis*-D^9-18:0 is hydrogenated faster to the saturated product than its *trans*-isomer, and this leads to the relative accumulation of *trans*-18:1 PC during hydrogenation of 18:2.

It has been observed in several cases that hydrogenation of PE liposomes followed a slower time-course than the hydrogenation of the analogous PC systems. The lower rate might be attributed to the different affinities of the different polar head-groups for the catalyst. However, it is more probable that the different physical state of the membranes in question manifests itself in the different accessibility of their constituent lipids by $[Pd(QS)_2]$. At 20°C 18:1/18:1 PE exists predominantly in the hexagonal$_{II}$ phase while PC forms bilayers at this temperature. Hydrogenation of mixtures of PC and PE gave rise to the same conclusion.[18]

C. REDUCTION WITH HYDROGEN TRANSFER

There are several systems known where catalytic reduction of C=C bonds is achieved by using a so-called hydrogen-donor molecule, H_2A,[3,4] instead of hydrogen gas:

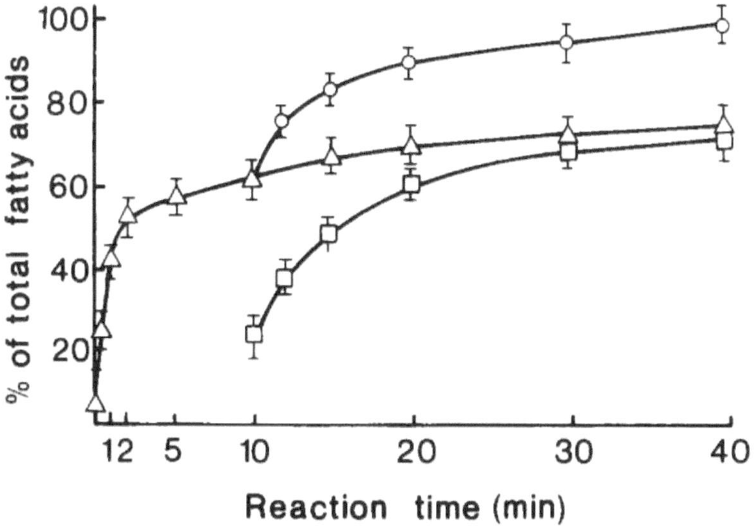

Figure 7 The effect of reaction conditions on the hydrogenation characteristics of DOPC liposomes. After 10 min hydrogenation of a freshly prepared DOPC liposome (△) an aliquot was removed and the lipids were extracted into organic solvents. Fresh liposomes were then prepared from the partially saturated lipids by sonication and the preparation was further hydrogenated (○). In separate experiments, hydrogenations were carried out for 10 min (△) at which time a second aliquot of freshly prepared DOPC liposomes was added (raising the lipid concentration to 0.2 mg/ml), and the hydrogenation was continued for an additional 30 min (□). All experiments were performed at 20°C, with a catalyst concentration of 0.1 mg/ml. (Reproduced from Vígh, L., et al., *Biochim. Biophys. Acta*, 921, 167, 1987. With permission.)

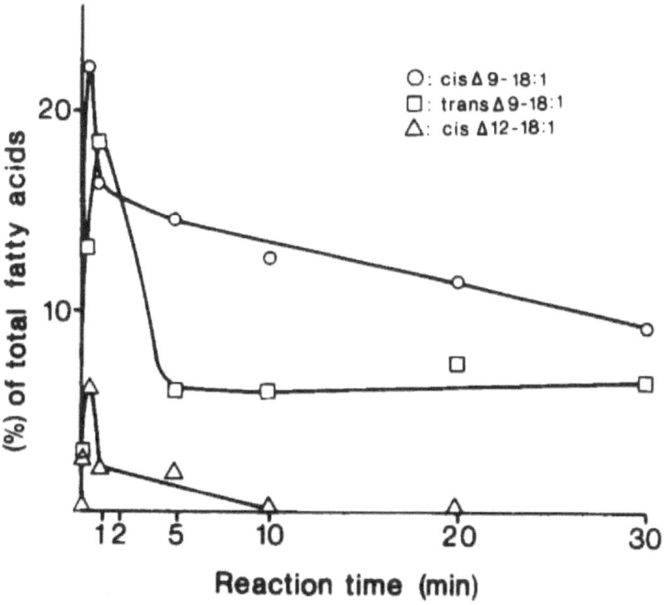

Figure 8 The formation of individual 18:1 isomers, *cis*-Δ^9-18:1 (○), *trans*-Δ^9-18:1 (□), and *cis*-Δ^{12}-18:1 (△) during hydrogenation of 18:2/18:2 PC liposomes at 20°C. (Reproduced from Vígh, L., et al., *Biochem. Biophys. Acta*, 921, 167, 1987. With permission.)

$$-CH{=}CH- + H_2A \rightarrow\rightarrow\rightarrow -CH_2{-}CH_2- + A \qquad (15)$$

Though several classes of compounds may serve as H-donors (alcohols, sugars, silicon, or tin hydrides) most of them are not biocompatible. Since the hydrogen transfer happens with breaking the H–A bond, transfer hydrogenations usually require relatively high activation energy, i.e., high temperature or photochemical activation. The relatively low rates of hydrogen transfer reactions may be compensated for by two important features: (a) a biocompatible H-donor can be used in much higher concentration than dissolved H_2 (the solubility of H_2 in water is 8×10^{-4} *M* at 20°C) or (b) hydrogen gas is unsuitable for lipid modifications in higher organisms; on the other hand, a solution containing the catalyst and an H-donor may be applied locally.

There is only one example in the literature for liposome hydrogenation with hydrogen transfer.[19] DOPC liposomes were reduced by ascorbate as H-donor and [RhCl(TPPMS)$_3$] as catalyst. The reaction requires photochemical activation: with visible light [RuCl$_2$(bipy)$_3$] served as photosensitizer (bipy=2,2′-bipyridyl) and the rate of the reaction was found dependent on the intensity of irradiation (Figure 9). Again, the extent of reduction was influenced by the physical state of the membrane. While liposomes of egg-yolk lecithin were reduced very fast, cod liver oil, co-dispersed with dipalmitoylphosphatidylcholine reacted sluggishly (Figure 10). Very interestingly, when *Tetrahymena pyriformis* was hydrogenated with this method some reduction of membrane lipids was observed even in the *absence* of ascorbate (in a typical experiment 12% conversion of C=C bonds was achieved, as opposed to 20% with ascorbate). This result implies that either the ascorbate sources of the cells or their other H-donor constituents were utilized for the reaction. Care should be taken therefore to avoid unwanted side reactions by such H-transfer.

D. DEUTERATION OF UNSATURATED LIPIDS

As was already mentioned, hydrogenation of biomembranes may be used as a structural probe since the accessibility of lipids critically depends on their microenvironment. The sensitivity of such a probe is further increased by labeling the lipids with deuteration and following their properties by infrared spectroscopy.

Fourier transform infrared spectroscopy (FTIR) became a powerful tool for studying thermotropic phase transitions in liposomes or native biomembranes.[20] This technique is based on the sensitivity of the stretching frequencies of CH_2 units to the packing of the alkyl chains in the lipid bilayer. Gel to liquid-crystalline phase transitions reveal themselves in batochromic shifts of νCH_2 in the 2800 to 3000 cm^{-1} spectral region.

A straightforward method of deuteration is the use of D_2 instead of H_2 in catalytic hydrogenations. Indeed, with [Pd(QS)$_2$] as catalyst, DOPC could be easily deuterated[21] up to 80% conversion of C=C bonds (Figure 11).

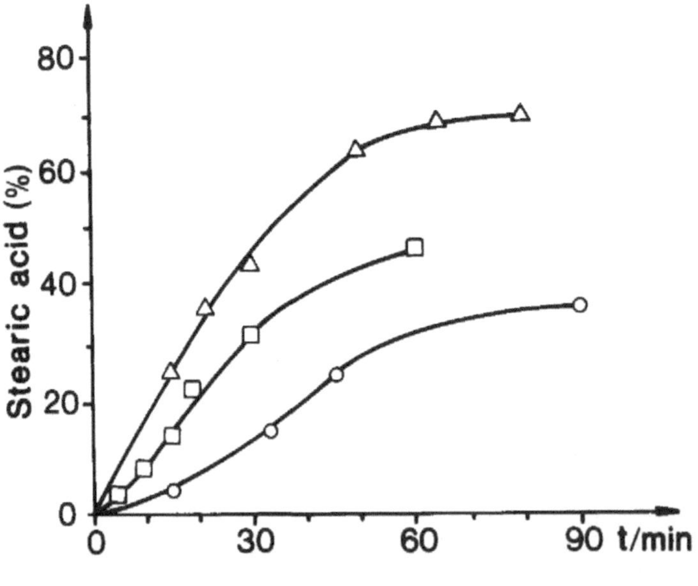

Figure 9 Photochemical transfer hydrogenation of DOPC liposomes as a function of time and irradiation intensity. Intensities are characterized by the nominal power of projector lamps: (○) 75 W; (□) 120 W; (▲) 150 W. [DOPC] = 0.2 mg/ml, [RuCl$_2$(bipy)$_3$] = [RhCl(TPPMS)$_3$] = 0.2 m*M*, 2 ml ascorbate buffer (pH 5), T = 35°C. (Reproduced from Joó, F., et al., *J. Mol. Catal.*, 49, L1, 1988. With permission.)

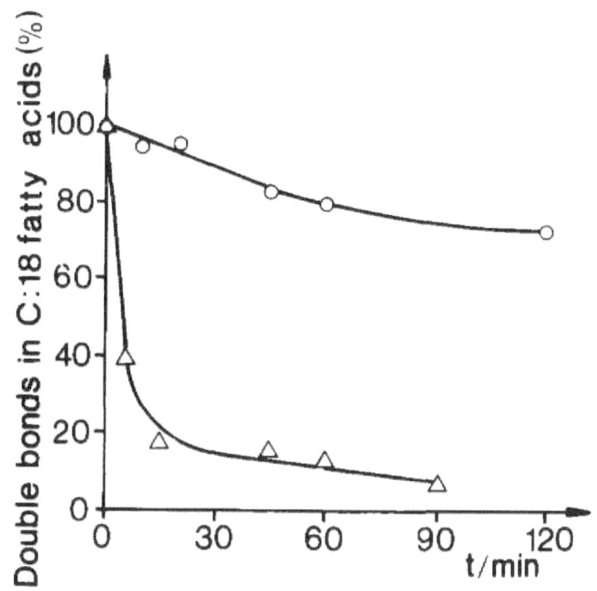

Figure 10 Photochemical hydrogenation of egg-yolk lecithin (△) and cod liver oil (○) liposomes by hydrogen transfer from ascorbate. [Substrate] = 0.2 mg/ml, [$RuCl_2(bipy)_3$] = [$RhCl(TPPMS)_3$] = 0.2 m*M*, T = 25°C, 2 ml of ascorbate buffer (pH = 5), 150 W. (Reproduced from Joó, F., et al., *J. Mol. Catal.*, 49, L1, 1988. With permission.)

The 2050 to 2250 cm^{-1} region of the infrared spectrum shows the presence of both CD_2 and CHD units (a definite advantage of recording νCD is in that this region is free from interference from other vibrational modes, except for the water association band). Upon increasing the temperature of the partially hydrogenated DOPC multilamellar system, both νCH and νCD shift towards higher wave numbers; however, the sensitivity of the νCD to the temperature change is about twice of νCH.

This latter phenomenon deserves special comment. It is known that the molecular mobility of the alkyl chain segments in lipid bilayers increases with the distance from the head-group, and that this increase is especially fast after the first nine CH_2 units. First, this means, that in case of Δ^9-18:1, deuteration labels the molecule at a position which is most sensitive to changes in the chain conformational order. Second, the primary signal of such conformational change is observed unbuffered; while νCH appears as an *average signal* of all C–H groups along the lipid alkyl chains, C–D groups are all located at Δ^9 and/or

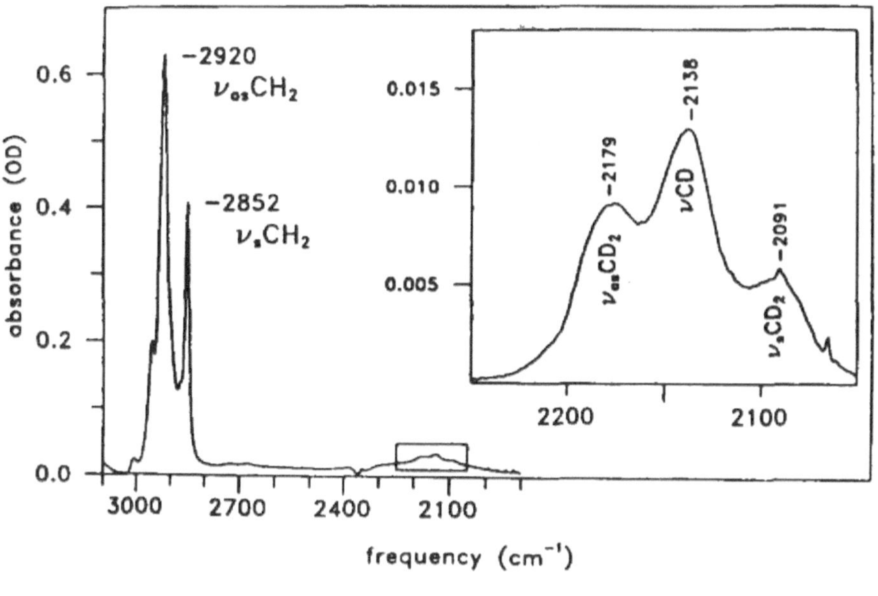

Figure 11 Fourier transform infrared (FTIR) spectrum of 80% deuterated DOPC. Inset shows enlarged the νCD region of the spectrum. (Reproduced from Török, Zs., et al., *Biochem. Biophys. Res. Commun.*, 192, 518, 1993. With permission.)

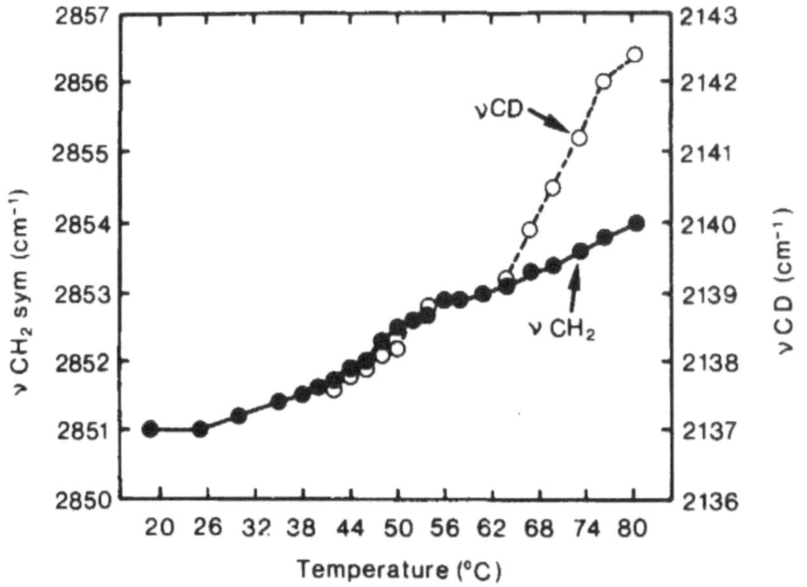

Figure 12 Frequency shifts of the νCD and the $\nu_s CH_2$ modes of partially, 55%, deuterated DOPE multilamellar system upon increasing the temperature from 20 to 80°C. (Reproduced from Török, Zs., et al., *Biochem. Biophys. Res. Commun.*, 192, 518, 1993. With permission.)

Δ^{10} positions (or only in a few positions when hydrogenating more unsaturated lipids) and the shift in the stretching frequency can be directly related to the conformational mobility of these segments.

In a mixture of phase separated lipids the individual domains undergo phase transitions (melt) separately. In many cases the enthalpy changes associated with these phase transitions of the domains are too low to be detected by differential scanning calorimetry (DSC); in other cases the average νCH change is too small to reveal the process of melting. The specific position of deuterium labels in the alkyl chain and its higher sensitivity to changes in conformational mobility makes possible the observation of such phase transitions by FTIR. Figure 12 shows the frequency shifts of a partially hydrogenated DOPE sample. Both νCH and νCD refers to a phase transition with a midpoint of 45°C. However, while there was no characteristic further change in νCH, νCD showed a large shift centered around 73°C. The occurrence of an endothermic phase transition at 73°C was, indeed, confirmed by DSC.[21]

IV. HETEROGENEOUS HYDROGENATION

There are two equally important, main reasons for using heterogeneous catalysts for liposome and biomembrane modification. One is the circumvention of the eternal problem of homogeneous catalysis: removal (recovery) of the soluble catalyst from the product mixture. The other is the topological selectivity which can be observed when using nonpermeable solid catalysts. Though heterogeneous catalytic reactions are inherently slower than the comparable homogeneous ones because of diffusion limitations, in many cases these two features counterbalance the loss in reaction rate.

A. HYDROGENATION WITH HETEROGENIZED METAL COMPLEX CATALYSTS

In general, soluble catalysts like those discussed in the preceding chapters show a much higher degree of chemo- and stereoselectivity than metal or metal oxide heterogeneous catalysts. It became, therefore, a special field of catalysis to devise ways of anchoring soluble compounds on surfaces of various supports[3,22] (both organic and inorganic solids) and study the effect of such heterogenization on the activity and selectivity of the catalyst.

Complexes with sulfonated phosphine (like TPPMS) and alizarin (e.g., QS) ligands bind strongly via the sulfonate group on commercially available anion exchangers (Dowex, Sephadex, Molselect, etc.).[23] These anionites usually carry protonated amino (e.g., diethylammonium-ethyl, DEAE) or quaternary ammonium functional groups, which may not be innocent in a liposome or toward the constituents of biomembranes. Therefore, covering the residual exchange groups by strongly binding ions (e.g., $[SO_4]^{2-}$)

Table 2 Hydrogenation of Soybean Lecithin with $[Pd(QS)_2]$ Heterogenized on QAE-Sephadex Anion Exchanger[24]

Time (min)	Fatty Acids (%)		
	18:0	18:1	18:2
0	10.2	27.2	62.6
15	37.3	44.8	17.9
45	40.3	54.5	5.2
60	49.3	50.7	—

Note: Conditions: 1 mg lipid in 5 ml Na-phosphate buffer (pH = 7), 3 mg $[Pd(QS)_2]$ on 100 mg QAE-Sephadex, T = 37°C, total pressure = 0.1 MPa.

is recommended. Detailed kinetic investigations[23] have established that $[RuCl_2(TPPMS)_2]$ exchanged onto Molselect-DEAE catalyzed hydrogenation of olefins with the same reaction mechanism and selectivity as in soluble form.

$[Pd(QS)_2]$ on Sephadex-QAE (QAE = quaternary ammonium-ethyl) was applied for the hydrogenation of soybean lecithin in aqueous dispersions.[24] At 37°C it actively removed 18:3 and most of 18:2 in 45 min under 1 bar hydrogen pressure with the concomitant increase in the proportion of 18:1 and 18:0 (Table 2).

The heterogenized $[Pd(QS)_2]$ catalyst undergoes the same color changes on the surface of the ion exchanger than in homogeneous solutions, therefore the same reaction mechanism and selectivities can be expected as from the dissolved palladium complex.

B. HYDROGENATION WITH CONVENTIONAL HETEROGENEOUS CATALYSTS

During the early studies of membrane hydrogenation, efforts were made to use Adams catalyst, PtO_2, for such purpose.[5] However, despite being active for the hydrogenation of soya lecithin in tetrahydrofuran solution, the catalyst failed to reduce the same substrate in sonicated (presumably monolamellar) aqueous dispersions.

In aqueous solutions poly(vinylpirrolidone) (PVP) forms complexes with several transition metal ions [Ni(II), Pd(II), Pt(II), etc.] and when these solutions are reacted with H_2 PVP-protected metal colloids are formed, capable of catalyzing the hydrogenation of a wide range of substrates.[25] By the same principle, when an aqueous solution of $(NH_4)_2[PdCl_4]$ was hydrogenated in the presence of water-*insoluble* polyvinyl-polypirrolidone (PVPP; e.g., Plasdone XL from Fluka) a heterogeneous hydrogenation catalyst was obtained.

The Pd-PVPP catalyst was successfully applied for the hydrogenation of DOPC liposomes and for the modification of thylakoid from *Synechocystis PCC 6803* (Table 3).[26] In both cases the catalyst could be separated easily and completely from the rest of the reaction system, either by filtration (liposome) or by gentle centrifugation.

The surface selectivity of hydrogenation, provided by the heterogeneous Pd-PVPP catalyst was utilized in studying the asymmetric spatial distribution of polyunsaturated fatty acid enriched phospholipids among the external and internal leaflet of the plasma membrane of rat platelets.[27] Platelets were hydrogenated either before or after their activation by calcium ionophore. Following the pattern of hydrogenation catalyzed by the nonpermeating catalyst using desorption-chemical soft ionization-mass spectrometry (DCI-MS), the surface exposure upon activation of certain polyunsaturated lipid species could be unambiguously detected.

V. CONCLUSIONS

Hydrogenation of lipids either in the form of aqueous dispersions (liposomes) or as constituents of biomembranes (including those of living cells) can be achieved under biologically acceptable conditions. There are several well characterized catalysts available, both water-soluble and heterogeneous. Heterogeneous catalysis allows selective modification of different lipid pools found in asymmetric spatial distribution inside the membranes (cells). The newly emerging technique of deuteration coupled with infrared spectroscopy detects the conformational order around specifically labeled segments of the acyl chains inside the lipid bilayer.

Table 3 Fatty Acid Distribution of Lipids in Thylakoid from *Synechocystis PCC 6803* During Hydrogenation with Pd-PVPP

	Reaction Time (min)				
Fatty acids (%)	**0**	**5**	**10**	**20**	**30**
16:0	43.1	44.7	45.1	45.0	46.3
16:1 (Δ9)	3.3	2.8	3.0	2.3	2.0
18:0	5.3	13.2	13.5	14.7	20.9
18:1 (Δ9)	3.6	6.5	6.7	9.5	12.9
18:2 (Δ9,12)	5.6	5.0	5.7	5.5	4.1
α18:3 (Δ9,12,15)	6.5	4.6	4.0	3.6	1.9
γ18:3 (Δ6,9,12)	27.0	20.8	19.6	17.3	10.8
18:4 (Δ6,9,12,15)	5.6	2.4	2.4	2.1	1.1

Note: Conditions: 20 mg $[(NH_4)_2PdCl_4]$ on 1 g PVPP, $C_{chlorophyll}$ = 10 μg/ml, 22°C, V = 25 ml.

ACKNOWLEDGMENTS

The authors acknowledge the contribution to this project of their students and co-workers whose names appear in the references. Financial support was provided by the Hungarian National Scientific and Research Foundation (OTKA) through grant Nos. 897 and 1699, and by the Ministry of Culture and Education through grant No. 209/92, and is gratefully acknowledged. We are grateful to Johnson Matthey Plc for loans of $RuCl_3$ and $(NH_4)_2[PdCl_4]$. We also thank Elsevier Science Publishers B.V. and Academic Press Inc. for the permission to reproduce previously published material.

REFERENCES

1. **Quinn, P. J., Joó, F., and Vígh, L.,** The role of unsaturated lipids in membrane structure and stability, *Prog. Biophys. Mol. Biol.*, 53, 71, 1989.
2. **Vígh, L., Joó, F., and Csépl̋ő, Á.,** Modulation of membrane fluidity in living protoplasts of *Nicotiana plumbaginifolia* by catalytic hydrogenation, *Eur. J. Biochem.*, 146, 241, 1985.
3. **Chaloner, P. A., Esteruelas, M. A., Joó, F., and Oro, L. A.,** *Homogeneous Hydrogenation (Catalysis by Metal Complexes)*, Kluwer Academic Publishers, Dordrecht, The Netherlands, 1994.
4. **James, B. R.,** *Homogeneous Hydrogenation*, John Wiley & Sons, New York, 1973.
5. **Chapman, D. and Quinn, P. J.,** A method for the modulation of membrane fluidity: homogeneous catalytic hydrogenation of phospholipids and phospholipid-water model biomembranes, *Proc. Natl. Acad. Sci. U.S.A.*, 73, 3971, 1976.
6. **Vigo, C., Goni, F. M., Quinn, P. J., and Chapman, D.,** The modulation of membrane fluidity by hydrogenation processes. II. Homogeneous catalysis and model biomembranes, *Biochim. Biophys. Acta*, 508, 1, 1978.
7. **Tóth, Z., Joó, F., and Beck, M. T.,** Homogeneous hydrogenations in aqueous solutions catalyzed by ruthenium-phosphine complexes, *Inorg. Chim. Acta*, 42, 153, 1980.
8. **Joó, F., Csiba, P., and Bényei, A.,** Effect of water on the mechanism of hydrogenations catalyzed by rhodium phosphine complexes, *J.C.S. Chem. Commun.*, 1602, 1993.
9. **Madden, T. D., Peel, W. E., Quinn, P. J., and Chapman, D.,** The modulation of membrane fluidity by hydrogenation processes. IV. Homogeneous catalysis of liposomes using a water-soluble catalyst, *J. Biochem. Biophys. Methods*, 2, 19, 1980.
10. **Vígh, L., Joó, F., van Hasselt, P. R., and Kuiper, P. J. C.,** Hydrogenation of model and biomembranes using a water-soluble ruthenium phosphine catalyst, *J. Mol. Catal.*, 22, 15, 1983.
11. **Katagiri, C., Owen, J. S., Quinn, P. J., and Chapman, D.,** Hydrogenation of plasma lipoproteins by a water-soluble catalyst; its use as a structural probe, *Eur. J. Biochem.*, 118, 335, 1981.
12. **Schlame, M., Horváth, L., and Vígh, L.,** Relationship between lipid saturation and lipid–protein interaction in liver mitochondria modified by catalytic hydrogenation with reference to cardiolipin molecular species, *Biochem. J.*, 265, 79, 1990.
13. **Mayor, J.-P., Török, Zs., Siegenthaler, P.-A., and Vígh, L.,** Differential accessibility of hydrogenation catalyst to acyl lipids in thylakoid membranes from atrazine-resistant and -susceptible *Solanum nigrum* biotypes, *Plant Cell Physiol.*, 33, 209, 1992.
14. **Vígh, L. and Joó, F.,** Modulation of membrane fluidity by catalytic hydrogenation affects the chilling susceptibility of the blue-green alga, *Anacystis nidulans*, *FEBS Lett.*, 162, 423, 1983.
15. **Farin, F., van Gaal, H. L. M., Bonting, S. L., and Daemen, F. J. M.,** Studies on the catalytic hydrogenation of biomembranes. Hydrogenation of phosphatidylcholine liposomes by two water-soluble rhodium-phosphine catalysts, *Biochim. Biophys. Acta*, 711, 336, 1982.

16. **Joó, F., Balogh, N., Horváth, L. I., Filep, Gy., Horváth, I., and Vígh, L.,** Complex hydrogenation/oxidation reactions of the water-soluble hydrogenation catalyst palladium di(sodium alizarinmonosulfonate) and details of homogeneous hydrogenation of lipids in isolated biomembranes and living cells, *Anal. Biochem.*, 194, 34, 1991.
17. **Bulatov, A. V., Izakovich, E. N., Karklin, L. N., and Khidekel, M. L.,** Palladium(II) complexes of alizarin and alizarin red (in Russian), *Izv. Akad. Nauk. SSSR Ser. Khim.*, 2032, 1981.
18. **Vígh, L., Horváth, I., Joó, F. and Thompson, Jr., G. A.,** The hydrogenation of phospholipid-bound fatty acids by a homogeneous, water-soluble palladium catalyst, *Biochim. Biophys. Acta*, 921, 167, 1987.
19. **Joó, F., Csuhai, É., Quinn, P. J., and Vígh, L.,** Hydrogenation of membrane lipids by catalyzed hydrogen transfer from ascorbate, *J. Mol. Catal.*, 49, L1, 1988.
20. **Crowe, H. J., Hoekstra, F. A., Crowe, L. M., Anchordoguy, T. J., and Drobnis, E.,** Liquid phase transitions measured in intact cells with Fourier transform infrared spectroscopy, *Cryobiology*, 26, 76, 1989.
21. **Török, Zs., Szalontai, B., Joó, F., Wistrom, C. A., and Vígh, L.,** Homogeneous catalytic deuteration of fatty acyl chains as a tool to detect lipid phase transitions in specific membrane domains: a Fourier transform infrared spectroscopic study, *Biochem. Biophys. Res. Commun.*, 192, 518, 1993.
22. **Hartley, F. R.,** *Supported Metal Complexes (Catalysis by Metal Complexes)*, Reidel, Dordrecht, The Netherlands, 1985.
23. **Joó, F. and Beck, M. T.,** Rhodium(I) and ruthenium(II) complexes of sulphonated triphenylphosphine immobilized on ion-exchange materials, *J. Mol. Catal.*, 24, 135, 1984.
24. **Joó, F., Papp, É., Benkő, S., Horváth, I., and Vígh, L.,** unpublished data, 1994.
25. **Bayer, E. and Schumann, W.,** Liquid phase polymer-based catalysis for stereo- and regio-selective hydrogenation, *J.C.S. Chem. Commun.*, 949, 1986.
26. **Joó, F., Benkő, S., Horváth, I., Török, Zs., Nádasdi, L., and Vígh, L.,** Hydrogenation of biological membranes using a polymer anchored colloidal catalyst, *React. Kin. Catal. Lett.*, 48, 619, 1992.
27. **Joó, F., Chevy, F., Colard, O., and Wolf, C.,** The activation of rat platelets increases the exposure of polyunsaturated fatty acid enriched phospholipids on the external leaflet of the plasma membrane. Study of the sidedness of the molecular species of different phospholipid classes in resting or in ionophore stimulated platelets using a non permeating palladium hydrogenation catalyst, *Biochim. Biophys. Acta*, 1149, 231, 1993.

Chapter 15

Liposomes as Biomimetic Microreactors

Giuseppe Mossa, M. Cristina Annesini, Luisa Di Marzio

CONTENTS

I. INTRODUCTION

Liposomes are the closest approach to biomembranes[1,2] and can be used for several biophysical and biochemical purposes, e.g., to measure membrane permeability, to reconstruct active membrane proteins, and to study surface recognition reactions or dynamic membrane processes. The use of the liposome as a structural and/or functional biomimetic model of the cell represents in fact a consolidated trend. In this context, we would mention the papers on supramolecular chemistry which have reported on the use of polymerized liposomes to probe and exploit ligand–receptor recognition[3] or those describing the incorporation of "bouquet molecules" into the phospholipid membrane of liposomes[4] to mimic the transport of ions across membranes by a channel mechanism. Recent papers have reported the use of vesicles as models of the prebiotic world to study encapsulated macromolecules in the protocellular systems[5] and autopoietic phenomena.[6]

Eukaryotic cells have a great of internal membranes that define different segregated compartments, each with its own distinct protein composition and well-defined function. Specific biochemical reactions occur in each organelle, depending on its enzymatic content. Furthermore, the processes must be highly regulated to carry out a specific cellular activity. The generally accepted models of transport between the different intracellular compartments maintain that transport is mediated by small carrier vesicles, even if recent data seem to suggest the participation of tubulovesicular structures in membrane transport.[7]

One of the major challenges of cell biology is to understand the physicochemical mechanisms of different cellular processes that take place in each organelle. Different research strategies are available for this purpose. Many experiments are designed to reconstitute such biological processes *in vitro*, using subcellular organelles isolated from the whole cells. While such *in vitro* systems are not too far removed from real biological systems, in some cases they are still too complex to isolate single biochemical events. The use of liposomes as functional biomimetic systems is very promising as a means of identifying the rules that govern organelle functions. While in their simplest form liposomes do not necessarily constitute a close and proper model of biological systems, liposomes of increasing complexity can be designed incorporating those components that are supposed to play a key role in biological events. In this way it is possible to isolate individual phenomena and to mirror biological reality as close as possible. Of course, the biological relevance of such studies depends on the extent to which the molecular details of the processes reconstituted correspond to those that occur inside the cells, which in many cases remains to be determined.

No matter what the strategy of the research project, using cell-free or liposome systems, the most rewarding results can be obtained combining the biological experiments with mathematical models that allow one to fully exploit the values of the experimental data.

0-8493-4012-8/96/$0.00+$.50
© 1996 by CRC Press, Inc.

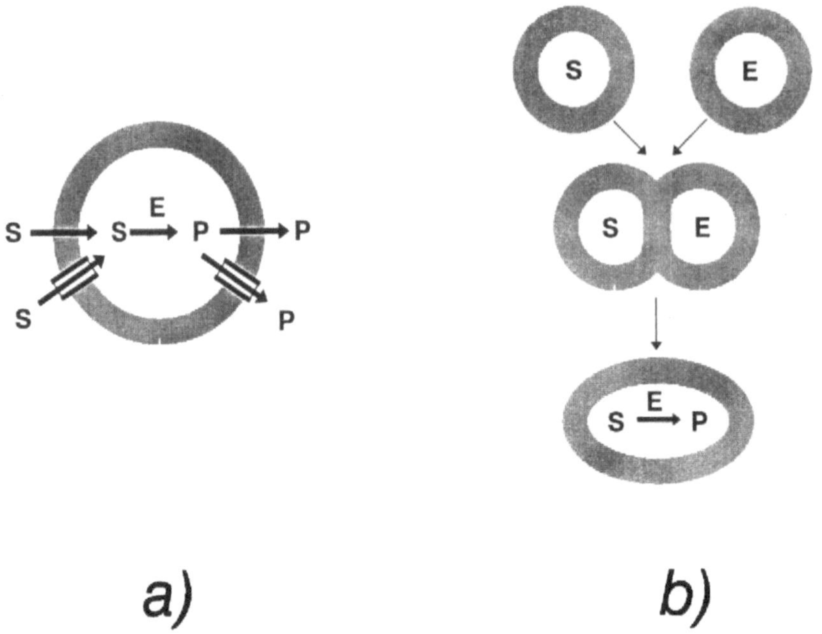

Figure 1 Schemata of two different types of bioprocess. (A) Diffusion-mediated process. The biochemical reaction occurs inside the membrane-bounded compartments where the enzyme (E) is located. The substrate (S) and the products (P) cross the boundary membrane by passive diffusion through the lipid bilayer or channels or they are pumped by transporters. (B) Fusion-mediated process. Enzyme- and substrate-loaded vesicles aggregate and subsequently fuse with membrane merging and mixing of the aqueous contents. Fusion allows the interaction of substrate and enzyme and, therefore, the enzymatic reaction.

This chapter will focus on the use of liposomes as microreactors to mimic some biochemical processes in cells or membrane-bounded subcellular compartments. These processes may be divided in two distinct types, as depicted in Figure 1. The first concerns diffusion-mediated processes (Figure 1a) in which the identity of the compartment is conserved during the processes. In this case, the substrate diffuses through the lipid membrane to react with the enzyme inside the vesicle. The products of the reaction diffuse back into the external solution. The second concerns fusion-mediated processes (Figure 1b), in which the identity of the compartments is destroyed during the fusion process. Two different populations of vesicles — one entrapping the substrate and the other entrapping the enzyme — fuse, allowing the interaction of substrates and enzymes.

This chapter is organized as follows. First, we discuss some examples of cellular biochemical processes occurring in membrane-bounded systems and some *in vitro* studies aimed at identifying the fundamental mechanisms of such processes. Second, we discuss two models which refer to the two different types of processes discussed above. Finally, we report on our experimental and theoretical research on liposomes entrapping carbonic anhydrase as a very simplified model of the process of carbon dioxide hydration in erythrocytes or other subscellular compartments.

II. DIFFUSION-MEDIATED AND FUSION-MEDIATED PROCESSES IN CELLS AND SUBCELLULAR COMPARTMENTS

In this section we highlight some biochemical processes in cells or membrane-bounded subcellular compartments in the light of the two schemata of processes described above. We also report on some *in vitro* experiments carried out with cell-free systems and liposomes in order to elucidate some specific aspects of the molecular machinery of organelle processes.

The process of carbon dioxide hydration by intracellular carbonic anhydrase is a typical diffusion-mediated process. In fact, most forms of the carbonic anhydrase (CA) enzyme are intracellular and enzymatic reaction requires CO_2 diffusion across the cell membrane. Figure 2 shows the schema of CO_2 hydration in erythrocytes.[7a] In particular, carbon dioxide diffuses through the lipidic membrane and is converted into carbonic acid by CA. The carbonic acid then dissociates and the released proton is taken

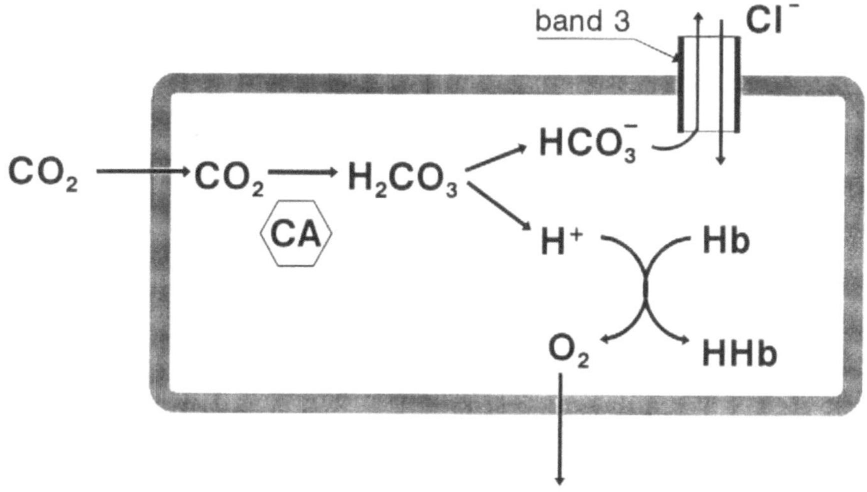

Figure 2 Schema of CO_2 hydration in erythrocytes in peripheral tissues. In pulmonary capillary, reactions proceed in the opposite direction. CA, carbonic anhydrase; HHb, hemoglobin; Hb, deoxyhemoglobin. (Adapted from Stryer, L., *Biochemistry*, 3rd ed., 305, Copyright © 1988 by Lubert Stryer. Used with permission of W. H. Freeman and Company.)

up by the deoxyhemoglobin, whereas HCO_3^- leaves the red cell through the anion channel (band 3).[8] In terms of the schemata shown in Figure 1a, we thus have a diffusion-mediated process, with the substrate diffusion across the boundary membrane and the enzymatic reaction inside the cell. One of the products (proton) is taken up by deoxyhemoglobin while the others (oxygen and bicarbonate ion) move out respectively through the lipid bilayer and through a channel. A similar process seems to occur in mitochondria, where the carbonic anhydrase in the matrix lumen readily converts the CO_2 taken up from the external medium into bicarbonate ion or converts bicarbonate into CO_2 to be released into the external medium.[9]

Peroxisomes represent another example of diffusion-mediated bioreactors. In these cellular organelles, different enzymes degrade substrate molecules, producing hydrogen peroxide. This component, in turn, is used by catalase to oxidize compounds of low molecular weight, like alcohols, or is degraded to water and molecular oxygen by the same enzyme to avoid any damage to the cell. In this way, peroxisomes in liver and kidney cells exploit their detoxification function and oxidize toxic molecules (like ethanol) that diffuse across the peroxisomal membrane.[10]

The behavior of other organelles follows the fusion-mediated process schema (Figure 1b). Let us consider, for example, the behavior of lysosomes for the degradation of extracellular material (Figure 3). Lysosmes are small cellular organelles which contain acid hydrolases. During endocytosis, a region of the plasma membrane invaginates and forms a vesicle containing extracellular material. These vesicles may fuse with each other and, finally, fuse with primary lysosomes containing digestive enzymes, thus forming secondary lysosomes where the substrate is degraded.[11] Products of low molecular weight are released into the cytoplasm by passive diffusion across the lipid bilayer or by transport mediated by specific proteins. While the main substrate uptake occurs through a fusion process, the transport system across the membrane has a fundamental function for the lysosomal activity.[12] In comparison with the great amount known about lysosomal hydrolases, little is known about the mechanism of fusion of primary lysosomes with endocytic vesicles or about transport mechanisms across the lysosomal membrane. A deeper insight into such processes can be gained by *in vitro* experiments of the fusion of lysosomes with vesicles of endocytic origin or by biomimetic experiments with properly designed liposomes. There are few reports on *in vitro* lysosome fusion. Altstiel and Branton[13] used a fluorescence assay to investigate the function of the clathrin coat of vesicles in the fusion process between lysosomes and coated or stripped vesicles. Their interesting experiments clearly show that vesicles stripped of their clathrin coats fuse with lysosomes faster than coated vesicles.

The late stage of the endocytotic pathway, i.e., the interaction between endosomes and lysosomes, has been studied *in vitro* by Mullock et al.[14] in a cell-free system containing endosomes loaded with a radiolabeled ligand and lysosomes. The results obtained suggest that transfer between endosomes and lysosomes can be reconstituted *in vitro* and provide some evidence that the process is energy-, cytosol-, and temperature-dependent.

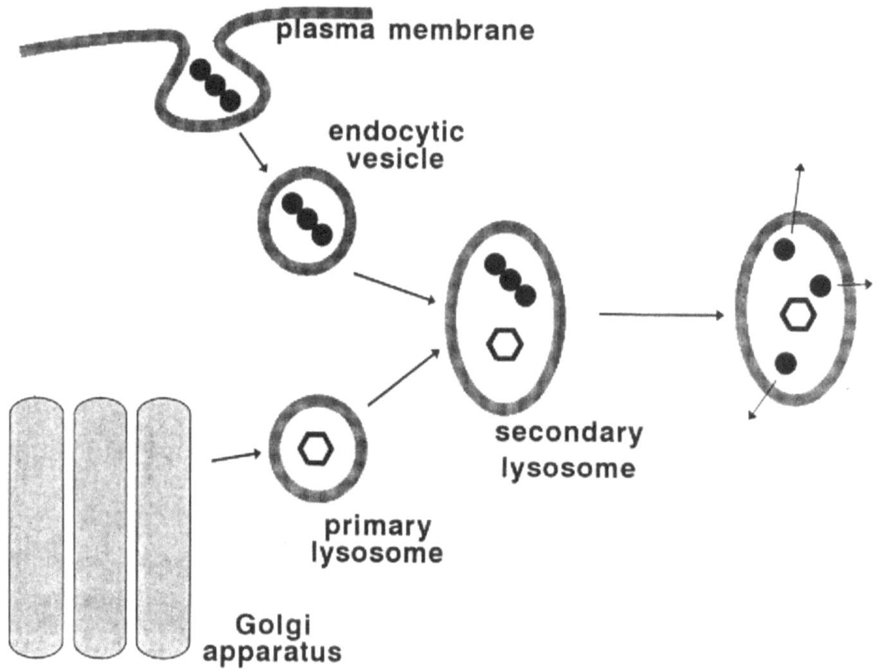

Figure 3 Schema of interaction between the endocytic vesicles, containing the substrate, and the lysosomes, containing acid hydrolases. The endocytic vesicle fuse with the primary lysosomes, forming secondary lysosomes where the substrate is degraded. The reaction products are released from the secondary lysosomes.

While it is clear that endosomes fuse with lysosomes *in vivo*, little is known about the mechanisms of fusion. A significant role may well be played by the pH jump between the internal lysosome medium and the cytosol, which has been proved to occur in liposomes.[15] Furthermore, as in other membrane fusion processes,[16–18] some fusogenic membrane proteins are likely to be involved in lysosome fusion. An interesting study in this direction has been recently reported by Vaccaro et al.[19] These authors studied the interaction of Saposin C, a protein associated with the lysosomal membrane, with phosphatidylserine liposomes. Their results suggest that this protein has a destabilizing effect on the membrane and that it induces membrane fusion in an acidic environment.

Fusion-mediated processes also take place in the endoplasmic reticulum and the Golgi apparatus. In fact, newly synthesized proteins from the endoplasmic reticulum are packaged into transport vesicles. A complex machinery of docking and fusion processes transports proteins to the Golgi apparatus, where further maturation of proteins occurs. From a schematic point of view, the Golgi apparatus may be described as a series of bioreactors in each of which substrate influx and product efflux occur respectively through processes of membrane fusion and membrane budding (see Figure 4). In a recent review, Mellman and Simons[20] summarize how the Golgi apparatus is commonly believed to work and the evidence that underlies these views. Many considerations about the Golgi working come from *in vitro* experiments,[21] some of which use liposomes as biomimetic system. Kobayashi and Pagano[22] used liposomes and permeabilized cell preparation as a model system to study the molecular mechanism of biochemical processes taking place in the Golgi apparatus. These authors studied the influence of physicochemical vesicle characteristics (size and composition) on the process of vesicle fusion with the Golgi apparatus. The authors stress the biological significance of their results: in fact liposome fusion with the Golgi apparatus occurs only when the vesicles are approximately the same size as the transport vesicles in cells. Furthermore, while liposomes containing phosphatidylcholine — which is the major constituent of the endoplasmic reticulum and thus probably of transport vesicles — readily fuse with the Golgi apparatus, addition of phosphatidylserine or sphingomyelin inhibits the fusion. In a more recent paper, Kagiwada and co-workers[23] again used an *in vitro* model system consisting of liposomes and rabbit liver Golgi membranes. Their experimental results show that the treatment of the Golgi membrane with protease inhibits both the binding and the fusion of liposomes, whereas these are not affected by extraction of peripheral proteins from the Golgi membrane. It can therefore be argued that fusion is mediated by integral proteins in the Golgi membrane.

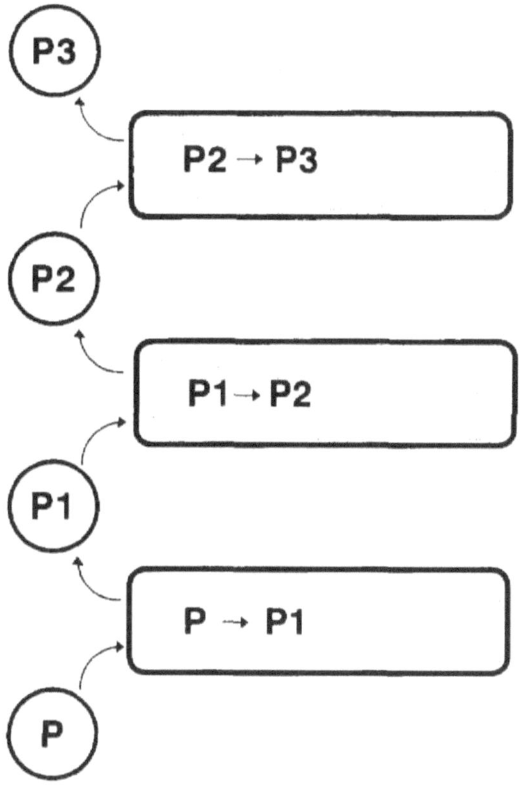

Figure 4 Schema of protein maturation process occuring in the Golgi apparatus. It is commonly thought that proteins produced in the endoplasmic reticulum undergo a series of sequential reactions in the Golgi compartments, as if on an assembly line. Transport between different Golgi stacks seems to occur via vesicular carriers.

III. THEORETICAL MODELS

The examples discussed in the previous sections clearly show that most of the enzyme-containing compartments may be viewed as small chemical bioreactors where diffusion- or fusion-mediated processes occur.

It is clear that biological systems are very complex and strictly controlled and the schemata shown in Figure 1 should be regarded only as very simplified approaches. In fact, many different reactions may occur. Transport across the membrane can be regarded not only as a passive process through the bilayer or channels, but also as a facilitated or active transport process mediated by transports. Fusion may be mediated by specific membrane protein and so on.

In any case, on the basis of the schemata described, it is possible to develop some mathematical models that can help us to understand the results obtained by *in vitro* experiments with cell-free systems, to quantify the effects of elemental phenomena on overall behavior, and to guide further experiments to obtain more significant results, also from a quantitative point of view. Theoretical models may also be useful in studying the behavior of real biological systems, even if, in this case, the greater complexity of the system under examination demands a greater complexity of the theoretical model used to simulate the behavior of the system.

Here we refer to mathematical models developed in order to analyze the experiments carried out with liposomes as biomimetic reactors. In terms of the previous classification in diffusion- and fusion-mediated bioprocesses, we also consider two different types of theoretical models. Within this framework, specific mathematical models can be developed to simulate the behavior of different systems.

A. DIFFUSION-MEDIATED BIOREACTORS

The theoretical model for these systems, where the enzyme is entrapped inside the liposomes, is similar to the classical model for heterogeneous chemical reactors. The reaction process involves three steps: (1) substrate diffusion across the vesicle membrane; (2) substrate diffusion in the internal liquid phase; and (3) chemical reaction of the substrate catalyzed by the enzyme. The first step is in series with respect to the other two, which are parallel to each other. The kinetic behavior of such a system can be described by the relationships listed in Table 1. We refer to the passive permeation of the substrate through the lipid bilayer with a permeation coefficient, *P*, described in terms of a solution–diffusion model. Different

Table 1 Relationships for Diffusion-Mediated Bioreactors

Substrate permeation through the boundary membrane:

$$N = P(S_e - S_i) \qquad P = KD_m/\delta \qquad \tau_m = R/3P \tag{I-1}$$

Substrate diffusion in the inner aqueous solution:

$$N = -D\left(\frac{dS}{dr}\right) \qquad \tau_i = R^2/D \tag{I-2}$$

Enzymatic reaction inside the aqueous core:

$$r_e = k_3 E \frac{S}{K_M + S} \tag{I-3}$$

Symbols: D, diffusivity in the aqueous phase; D_m, diffusivity in the membrane; E, enzyme concentration inside the vesicle; K, partition coefficient between the membrane and the liquid phase; k_3, kinetic constant; K_M, Michaelis constant, N, substrate flux; P, permeation coefficient; r, radial coordinate; R, vesicle radius; r_e, enzymatic reaction rate; S, substrate concentration in the internal liquid phase, S_e substrate concentration in the external liquid phase; S_i, substrate concentration in the internal membrane surface; δ, membrane thickness; τ_m, characteristic time for substrate diffusion through the membrane; τ_i, characteristic time for substrate diffusion in the aqueous phase.

expressions may be used for substrate permeation through the membrane if facilitated or active transport phenomena occur. Fickian diffusion is assumed in the inner solution and the reaction kinetics is given by the Michaelis–Menten equation. Details of this model and similar models developed for different systems (e.g., for the enzyme entrapped in the bilayer instead of in the aqueous core) are provided by Cioci and Lavecchia in Chapter 16 of this book. Here we are concerned solely with the simplified model obtained in the case of negligible diffusional limitation in the internal phase. This is the case that occurs when the ratio between the characteristic time of internal diffusion, τ_i, and that of permeation across the membrane, τ_m, is much smaller than 1. Since $\tau_i/\tau_m = RP/D$ (see Table 1), the internal diffusion limitations are negligible when $R \ll D/P$. Assuming typical values of low molecular weight substrate diffusivity inside the cells ($\sim 10^{-6}$ cm^2/s)[24] and their permeability across the lipid membrane ($<10^{-6}$ cm/s),[25] it can be estimated that this assumption holds true if $R \ll 1$ cm, i.e., much more than the radius of liposomes as well as many subcellular organelles or cells. On this assumption, we can assume that substrate concentration can be taken as uniform in the internal solution (S_i) and depends on the external substrate concentration (S_e) and on the membrane permeability:

$$\beta\left(S_e - S_i\right) = \frac{S_i}{K_M + S_i} \tag{1}$$

where $\beta = 3P/Rk_M E$ (see Table 1 for symbols). High β values correspond to kinetic control of the overall process, whereas low β values correspond to strong diffusional limitations, due to low membrane permeability or to a high intrinsic reaction rate. A high catalytic efficiency of membrane-bounded biological or artificial vesicles entrapping enzymes requires both a high inner enzyme concentration and a high β value, which in turn requires a high membrane permeability to the substrate. This is the case of CO_2 hydration in erythrocytes due to the high lipid bilayer permeability to CO_2 (P = 0.3 to 0.4 cm/s).[26]

B. FUSION-MEDIATED BIOREACTORS

Let us consider a system composed of both enzyme-loaded and substrate-loaded vesicle populations. According to the schema shown in Figure 1b, the overall reaction process occurs in three steps: (1) reversible vesicle aggregation resulting in a close apposition of the vesicle surfaces; (2) irreversible fusion of the aggregated vesicles resulting in the mixing of encapsulated aqueous contents and membrane lipids; and (3) enzymatic reaction of the substrate inside the fused vesicles. The overall reaction rate derives from the interaction of the fusion kinetics and the enzymatic reaction rate. Given the high enzyme and substrate concentrations in the vesicles, it is probably that aggregation and fusion are rate controlling steps and that the enzymatic reaction may be regarded as an instantaneous reaction. Many theoretical models on the aggregation and fusion kinetics are reported in the literature.[27,28] These are mainly mass-action kinetic models that result in a large set of differential equations to be solved numerically. In a previous work[29] we suggest a simplified model to describe the aggregation and fusion of vesicles. This model is based on the idea of the formation of clusters composed of vesicles joined at aggregation or fusion sites. Collisions of clusters result in the formation of aggregation sites, which in turn can break down (reversible aggregation) or be transformed into fusion sites. In the latter case, a larger compartment (fusion product) is formed; mixing of the aqueous contents and enzymatic reaction occur instantaneously.

The mathematical model to describe these systems is summarized in Table 2. This model is based on three points: (1) evaluation of the number of cluster and fusion sites as a function of time; (2) probabilistic analysis of the size distribution of the fusion product; and (3) evaluation of the amount of substrate converted into the reaction product.

The kinetics of the formation of aggregation and fusion sites can be described by the set of differential Equations II-1 (Table 2). These equations account for the kinetics of aggregation and deaggregation, with kinetic constants C and D, respectively, as well as for the kinetics of the transformation of an aggregation site into a fusion site, with the kinetic constant f. Aggregation is a second-order process with respect to the number of clusters, n. From the vesicle mass balance and assuming that inside each cluster two vesicles are connected by just one site (of aggregation or fusion), the number of clusters, n, and the overall number of aggregation, σ, and fusion sites, π, are related by Equation II-2 (Table 2). The set of differential Equations II-1 (Table 2) can therefore be solved numerically with the initial conditions $t = 0$, $\sigma = \pi = 0$ in order to obtain the number of clusters and the overall number of aggregation and fusion sites vs. time.

From these data, probabilistic considerations allow us to evaluate the size distribution of clusters and of fusion products. In fact, a system made up of X_0 vesicles grouped into n clusters results in a distribution of cluster sizes. Furthermore, as depicted in Figure 5, a cluster made up of x vesicles may have different structures with different numbers of vesicles collapsed into fusion products.

In order to study the reaction kinetics, only the fusion products must be considered. In fact, the integrity of the vesicles is conserved during aggregation and no mixing of aqueous contents occurs. The number of fusion products made up of r vesicles, $F(r)$, is given by Equation II-6 (Table 2), where $CL(x)$ is the number of clusters made up of x vesicles, $p(k_a,x)$ the probability of finding k_a aggregation sites in a cluster of x vesicles and $p(r,k_a,x)$ the probability of finding a fusion product made up of r vesicles in a cluster of x vesicles with k_a aggregation sites. Expressions for $CL(x)$, $p(k_a,x)$ and $p(r,k_a,x)$ are derived elsewhere.[29] Finally, we must consider the fact that a fusion product made up of r vesicles can be obtained by different numbers of substrate- and enzyme-loaded vesicles. The probability $p(s,r)$ of finding a fusion product made up of s substrate-loaded vesicles (s = 1,2...r) and r-s enzyme-loaded vesicles is given by Equation II-7 (Table 2) as a function of the fraction, y_s, of substrate-loaded vesicles in the overall vesicle population.

We assume that all the substrate is converted instantaneously if at least one enzyme-loaded vesicle is present in the fusion product. The overall amount of converted substrate is therefore given by:

$$R_s = s_0 \sum_{r=2}^{m} F(r) \sum_{s=1}^{r-1} s \cdot p(s,r) \tag{2}$$

Figure 6 shows some typical trends of the amount of converted substrate vs. time. These curves were obtained using the values of the kinetic constants of aggregation, deaggregation, and fusion rate obtained from fusion experiments with phosphatidylserine-cholesterol liposomes.[29] The plot shows the effect of the ratio between substrate- and enzyme-loaded vesicles, with a given number of vesicles, X_0. Initially, higher substrate consumption rates are predicted for 1:1 population ($y_s = 0.5$), whereas, of course, the ultimate substrate consumption increases as y_s increases. An increase in the CX_0 or an increase in the fusion constant, f, leads to an increase in the substrate consumption rate as shown in Figure 7.

IV. LIPOSOMES ENTRAPPING CARBONIC ANHYDRASE AS BIOMIMETIC REACTORS

Carbonic anhydrase (CA) is one of the most abundant enzymes and many of its forms are present inside cells (especially in the erythrocytes) or inside cellular organelles (e.g., in mitochondria). We therefore decided to study the behavior of this enzyme in membrane-bounded compartments using liposomes as

Table 2 Relationships for Fusion-Mediated Bioreactors

Number of cluster, aggregation and fusion sites:

$$\frac{d\sigma}{dt} = Cn^2 - D\sigma - f\sigma$$

$$\frac{d\pi}{dt} = f\sigma \qquad t = 0 \quad n = X_0 \quad \sigma = \pi = 0 \tag{II-1}$$

$$X_0 = n + \sigma + \pi \tag{II-2}$$

Number of cluster made up of x vesicles, $CL(x)$:

$$CL(x) = n\frac{\exp\left(-\frac{X_0 - n}{n}\right)\left(\frac{X_0 - n}{n}\right)^{x-1}}{(x-1)!} \tag{II-3}$$

Probability of having k_a aggregation sites in a cluster made up of x vesicles:

$$p(k_a, x) = \binom{x-1}{k_a}\left(\frac{\sigma}{\sigma + \pi}\right)^{k_a}\left(\frac{\pi}{\sigma + \pi}\right)^{x-1-k_a} \tag{II-4}$$

Probability of having a fusion product made up of r vesicles within a cluster made up of x vesicles with k_a aggregation sites:

$$p(r, k_a, x) = \frac{\binom{x-1-k_a}{r-1}}{\binom{x-1}{r-1}} - \frac{\binom{x-1-k_a}{r}}{\binom{x-1}{r}} \tag{II-5}$$

Number of fusion product formed of r vesicles, $F(r)$:

$$F(r) = \sum_{x=1}^{\infty}\left[xCL(x)\left(\sum_{k_a=0}^{x-1} p(k_a, x)p(r, k_a, x)\right)\right] \tag{II-6}$$

Probability to find s substrate-loaded vesicles within a fusion product formed by r vesicles:

$$p(s, r) = \binom{r}{s} y_s^s (1 - y_s)^{r-s} \tag{II-7}$$

Consumption of substrate, R_s:

$$R_s = s_0 \sum_{r=2}^{m} F(r) \sum_{s=1}^{r-1} s \cdot p(s, r) \tag{II-8}$$

Symbols: C, kinetic constant of aggregation; X_0, number of vesicles; D, kinetic constant of deaggregation; y_s, fraction of substrate-loaded vesicles; f, kinetic constant of fusion; p number of fusion sites; n, number of cluster; s, number of aggregation sites; s_0, substrate amount in each substrate-loaded vesicle.

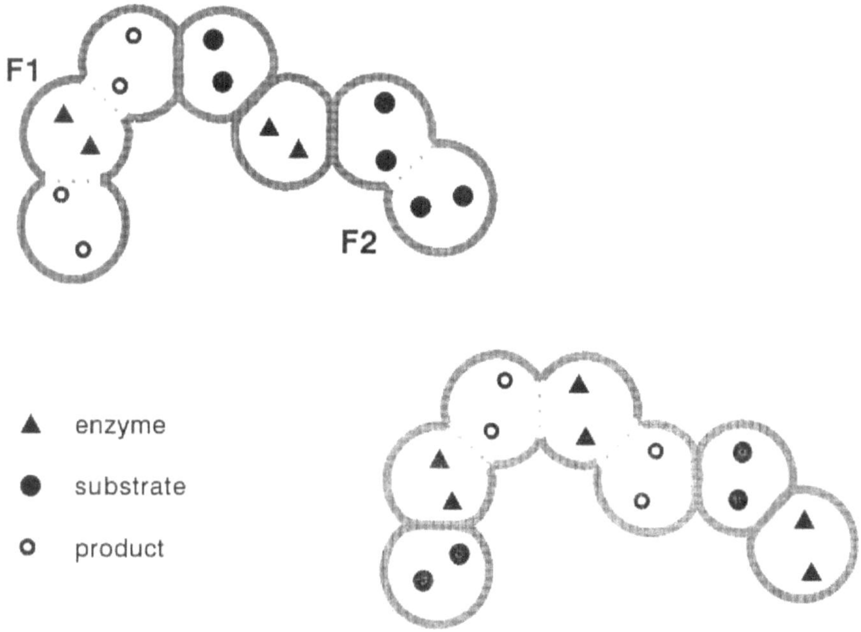

Figure 5 Schemata of clusters made up of $x = 7$ vesicles joined at $k_a = 3$ aggregation sites and 3 fusion sites. Within the cluster on the left, we have a fusion product (F_1) made up of $r = 3$ vesicles and a fusion product (F_2) made up of $r = 2$ vesicles. F_1 is made up of one enzyme-loaded and two substrate-loaded vesicles; the substrate (full circles) is therefore converted to the product (empty circles) inside F_1. F_2 is made up of two substrate-loaded vesicles and does not contain the enzyme; therefore, the substrate is not converted inside F_2. A different cluster made up of $x = 7$ vesicles joined at $k_a = 3$ aggregation sites and 3 fusion sites is reported on the right. For sake of clarity, the aggregate shape is also retained in the fusion product, which is almost spherical in shape.

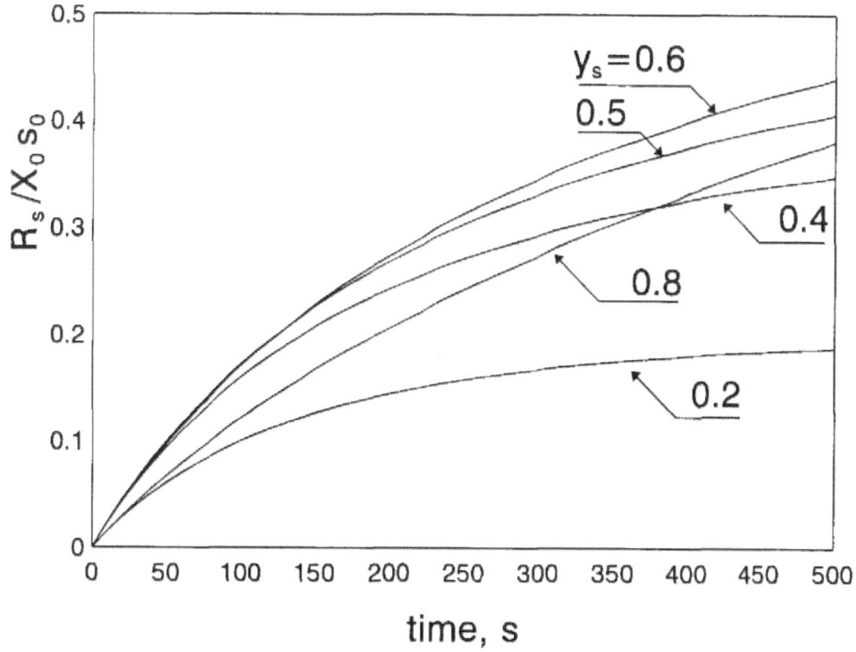

Figure 6 Kinetics of substrate consumption in a fusion-mediated process. Kinetics parameters ($CX_0 = 0.012$, $D = 0.56$, $f = 0.79$), are obtained from fusion experiments with phosphatidylserine-cholesterol liposomes.[29]

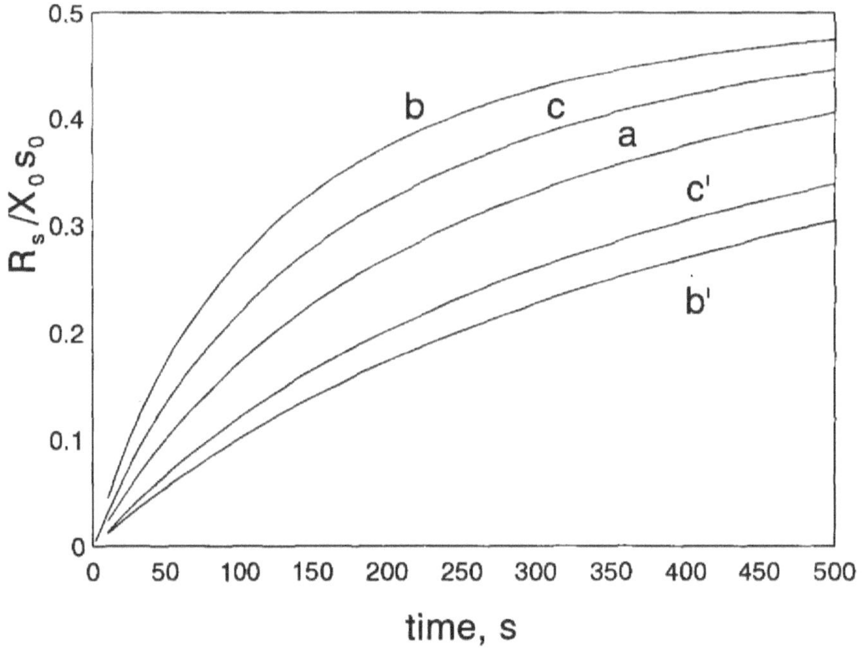

Figure 7 Effects of the aggregation and fusion constant on the substrate consumption in a fusion-mediated process ($y_s = 0.5$). (a) Kinetic constants as in Figure 6; (b) $CX_0 = 0.024$, $D = 0.56$, $f = 0.79$; (b′) $CX_0 = 0.006$, $D = 0.56$, $f = 0.79$; (c) $CX_0 = 0.012$, $D = 1.12$, $f = 0.79$; (c′) $CX_0 = 0.012$, $D = 0.28$, $f = 0.79$.

biomimetic systems. Simple phospholipid vesicles with CA enzyme have been used until now and, of course, such a system is far removed from the real biological systems to be simulated since both erythrocytes and mitochondria contain a lot of soluble or membrane-bound proteins. The biomimetic model worked out is therefore still very rudimentary. Work is in progress to approach the structure of a real biological system more closely and therefore to shed light on some aspects of their behavior. It is worth noting that this research may also have some biotechnological fall out. In fact, liposome-encapsulated hemoglobin has been shown to provide a red blood cell substitute in medical emergencies[30] and the inclusion of carbonic anhydrase in liposome formulation should enhance the oxygen release by Bohr effect.

The composition of the lipid cocktail used in liposome preparation is a paramount importance as regards CA-loading and CA-location in liposomes.[31] While neutral liposomes (dipalmitoylphosphatidylcholine (DPPC):cholesterol (CHO) = 7:3 molar ratio) do not trap CA at pH = 7.55, cationic liposomes (DPPC:CHO:stearylamine (SA) = 6:3:1 or 5:3:2) do. Measurements of the esterase activity of CA in DPPC:CHO:SA = 6:3:1 liposomes or of CA released from liposomes show that about 38 μg CA/μmol of lipids remain on the external surface of liposomes while about 30 μg CA/μmol of lipids are trapped inside the vesicles. On the other hand, the same experiments carried out with DPPC:CHO:SA = 5:3:2 showed that about 28 μg CA/μmol of lipids are trapped inside the vesicles, but no enzyme can be detected on the external surface. The presence of CA seems to have two different effects on liposome size: an increase in the mean diameter from 47 nm (without enzyme) to about 61 nm (trapping CA) was observed with DPPC:CHO:SA = 6:3:1 liposomes, while a reduction from 40 nm (without enzyme) to about 29 nm (trapping CA) was observed with DPPC:CHO:SA = 5:3:2 liposomes. Furthermore, acetazolamide, which is a strong inhibitor of CA, does not affect the empty liposome volume but inhibits the volume increase of DPPC:CHO:SA = 6:3:1 enzyme-loaded liposomes. These results suggest that the increase in the volume of DPPC:CHO:SA = 6:3:1 liposome loaded with CA may be due to the activity of the enzyme protruding toward the external solution. It is interesting to note that these results may have some physiological significance, since CA seems to give the matrix lumen of liver mitochondria a capacity for very rapid volume regulation.[9]

The kinetics of CO_2 hydration with CA entrapped in liposomes has also been studied.[32] DPPC:CHO:SA = 5:3:2 enzyme-loaded liposomes were used in these experiments in order to avoid the presence of CA on the external surface of liposomes, which can conceal the behavior of the inner enzyme.

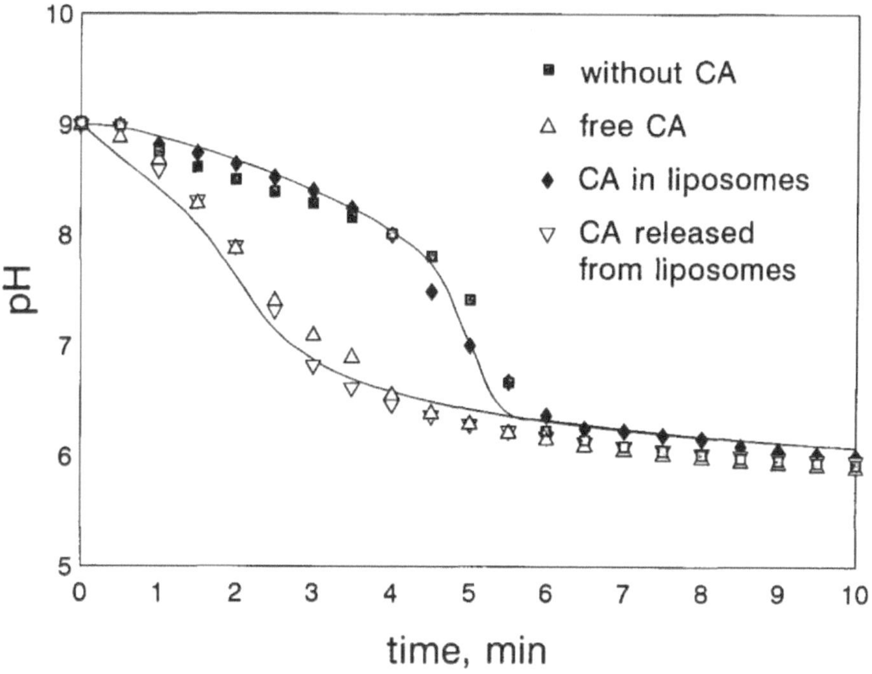

Figure 8 Different trends of pH vs. time with or without carbonic anhydrase. (Adapted from Annesini, M. C., et al., *J. Liposome Res.*, 3(3), 645, 1993. Reprinted by courtesy of Marcel Dekker, Inc.)

The pH variation vs. time was monitored in a slurry reactor with continuous CO_2 flux. The curves obtained without CA, with free CA, and with CA in liposomes or released from liposomes are compared in Figure 8.[33] A substantial increase in the hydration rate is observed with free CA or CA released from liposomes, but no significant change of pH vs. time curve was detected with CA entrapped in liposomes. In order to understand this behavior, we developed a theoretical model for the CO_2 hydration catalysed by CA entrapped in liposomes. This theoretical model accounts for the following elemental phenomena (see Table 3):

a. CO_2 transfer from gas to liquid
b. chemical reactions in the bulk liquid phase that do not contain CA; in particular, CO_2 reaction to yield carbonic acid, CO_2 reaction to yield the bicarbonate ion, and equilibrium reactions between carbonic acid, bicarbonate, and carbonate ions
c. CO_2 transfer from the external liquid phase into the aqueous core of liposomes
d. all the reactions reported in (b) in the aqueous core of liposomes together with the enzymatic hydrolysis of CO_2 to carbonic acid

Details of the kinetic models are given elsewhere.[32] Calibration experiments carried out without the enzyme permit preliminary evaluation of the unknown model parameters (all the equilibrium constants and the rate of uncatalyzed reaction are given in the literature[34]), i.e., the CO_2 solubility in the buffer solution and the CO_2 transfer coefficient from the gas to liquid phase. In the presence of free CA, the experimental data show that, at least for an enzyme concentration greater than 30 μg/l, it is possible to assume that the overall kinetics is controlled by the CO_2 transfer rate and that equilibrium is achieved between CO_2 and carbonic acid. We overlook the increase in the mass transfer coefficient itself due to the increase in the CO_2 hydration rate, i.e., we use the same value of k_La determined from the experiments with buffer solution only. The theoretical model thus makes it possible to predict the pH curve vs. time completely. The model prediction obtained on this assumption agrees closely with the experimental data as reported in Figure 9.[35]

As regards CA entrapped in liposomes, we assume that the enzyme is located only inside the liposomes and that the enzymatic reaction inside the liposome is fast enough to ensure that equilibrium is achieved between carbon dioxide and carbonic acid. The high lipid bilayer permeability for CO_2 (about 0.3 to

Table 3 Relationships for CO_2 Hydration by Carbonic Anhydrase Entrapped in Liposomes

Chemical reactions:

$$(i)\quad CO_2 + H_2O \rightleftharpoons H_2CO_3$$
$$(ii)\quad CO_2 + OH^- \rightleftharpoons HCO_3^-$$
$$(ii)\quad H_2CO_3 \rightleftharpoons H^+ + HCO_3^-$$
$$(iv)\quad HCO_3^- \rightleftharpoons H^+ + CO_3^=$$

CO_2 flux, J_{CO2}, from gas to liquid:

$$J_{CO_2} = k_L a\left(C^*_{CO_2} - C_{CO_2}\right) \qquad \text{(III-1)}$$

Reaction rate of carbonic acid production (reaction *i*):

$$r_1 = k_1\left[C_{CO_2} - \frac{C_{H_2CO_3}}{K_1}\right] + E_0 k_e\left[\frac{C_{CO_2}}{K_M + C_{CO_2}} - \frac{C_{H_2CO_3}}{K_M K_1 + C_{H_2CO_3}}\right] \qquad \text{(III-2)}$$

Reaction rate of bicarbonate production (reaction *ii*):

$$r_2 = k_2\left[C_{CO_2} C_{OH^-} - \frac{C_{HCO_3^-}}{K_2}\right] \qquad \text{(III-3)}$$

Equilibrium between carbonic acid, bicarbonate, and carbonate ions:

$$K_3 = \frac{C_{HCO_3^-} C_{H^+}}{C_{H_2CO_3}} \qquad \text{(III-4)}$$

$$K_4 = \frac{C_{H^+} C_{CO_3^=}}{C_{HCO_3^-}} \qquad \text{(III-5)}$$

CO_2 flux, J'_{CO2}, from the external liquid phase into the aqueous core of liposomes:

$$J'_{CO_2} = Pa\left(C_{CO_2} - C'_{CO_2}\right) \qquad \text{(III-6)}$$

Symbols: *C*, concentrations in the liquid phase; C_{CO2}, CO_2 concentration at equilibrium with the gas phase; C'_{CO2}, CO_2 concentration inside the liposomes; E_0, enzyme concentration; *k*, kinetic constants; *K*, equilibrium constants; k_La mass transfer coefficient from gas to liquid; K_M, Michaelis constant; *P*, liposome membrane permeability.

0.4 cm/s)[26] and the small vesicle diameter (about 50 nm) yield a very small characteristic time for CO_2 transfer inside the liposomes (order of magnitude $\sim 10^{-6}$ min). It is therefore possible to overlook the difference between the CO_2 concentration in the external solution and in the aqueous core of liposomes. On these assumptions, the model makes it possible to predict the pH drop inside and outside the liposomes. In agreement with the experimental data, the model predicts that, in spite of the fast decrease of the internal pH, the trend of the external pH is quite similar to that obtained without the enzyme, since CA-loaded liposomes behave as a sink for CO_2 without releasing the reaction product. On the other hand, controls carried out by disrupting the liposome structure and releasing the enzyme showed a significant increase in the CO_2 absorption rate. This behavior confirms that a significant amount of the enzyme is trapped in DPPC:CHO:SA = 5:3:2 liposomes and that there is no enzyme on the liposome external surface.

Carbonic anhydrase entrapped inside the vesicle seems to be unable to exploit its influence on the extravesicle pH unless a mechanism of ion flux is present. Work is in progress to verify this result by detecting also the intravesicle pH and by modifying the vesicle structure to allow the ionic fluxes. For this purpose, kinetic experiments with liposomes with reconstituted membrane proteins, such as band 3 should be very interesting for their biological significance.

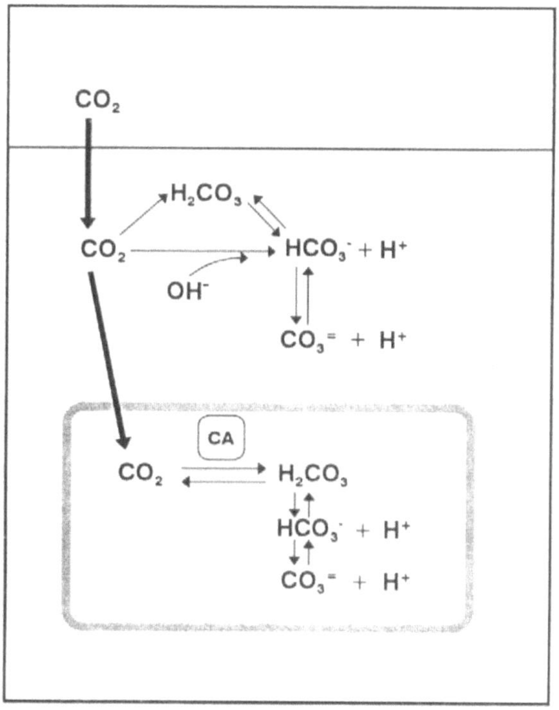

Figure 9 Schema of CO_2 hydration in the presence of carbonic anhydrase entrapped in liposomes. (Adapted from Annesini, M. C., et al., *J. Liposome Res.*, 3(3), 642, 1993. Reprinted by courtesy of Marcel Dekker, Inc.)

V. CONCLUSIONS

Even though considerable information already exists concerning the mechanisms of cellular processes that take place in subcellular organelles, there is still much to be understood. For this purpose, much can be known from *in vitro* experiments using liposomes as biomimetic systems and from proper mathematical models that allow us to fully exploit the meaning of the experimental data.

Conclusions derived from the investigation of liposomes cannot always be directly extrapolated to real biological systems, because liposomes are simpler than their biological counterpart. However, an increased refinement of lipid vesicle systems, incorporating components that play a key role in biological events, can lead to a closer biomimetic system and allow the mechanisms of biological processes to be understood in detail.

ACKNOWLEDGMENT

The work of Maria C. Annesini was financially supported by Ministero dell'Università e della Ricerca Scientifica e Tecnologica. The authors are indebted to Professor Alessandro Finazzi-Agró, University of Rome "Tor Vergata", for helpful discussions of the ideas that are the basics of our research.

REFERENCES

1. **Ringsdorf, H., Schlarb, B., and Venzmer, J.,** Molecular architecture and function of polymeric oriented systems: models for the study of organization, surface recognition, and dynamics of biomembranes, *Angew. Chem. Int. Ed. Engl.*, 27, 113, 1988.
2. **Bader, H., Dorn, K., and Ringsdorf, H.,** Polymeric monolayers and liposomes as model for biomembranes, *Adv. Polym. Sci.*, 64, 1, 1985.
3. **Dodrer, N. and Regen, S. L.,** Polymerized liposomes designed to probe and exploit ligand-receptor recognition at the supramolecular level, *J. Am. Chem. Soc.*, 112, 2829, 1990.
4. **Pregel, M. J., Jullien, L., and Lehn, J. M.,** Towards artificial ion channels: transport of alkali metal ions across liposomal membrane by "bouquet" molecules, *Angew. Chem. Int. Ed. Engl.*, 31, 1637, 1992.
5. **Deamer, D. W., Mahon, E. H., and Bosio, G.,** Self-assembly and function of primitive membrane structures, in *Early Life on Earth*, Bengtson, S., Ed., Columbia U.P., New York, 1993.

6. **Luisi, P. L.,** Defining the transition to life: self-replicating bounded structures and chemical autopoiesis, in *Thinking About Biology,* Stein, W. and Varela, F. J., Eds., Addison-Wesley, Reading, MA, 1993.
7. **Ayala, J. S.,** Transport and internal organization of membranes: vesicles, membrand networks and GTP-binding proteins, *J. Cell Sci.,* 107, 735, 1994.
7a. **Stryer, L.,** *Biochemistry,* 3rd ed., Freeman and Co., New York, 1988, 305.
8. **Klocke, A.,** Velocity of CO_2 exchange in blood, *Ann. Rev. Pysiol.,* 50, 625, 1988.
9. **Storey, B. A. and Forster II, R. E.,** Mediation of mitochondrial volume changes by carbonic anhydrase as followed by light scattering changes, in *Carbonic Anhydrase. From Biochemistry and Genetics to Physiology and Clinical Medicine,* Botrè, F., Gros, G., and Storey, B. T., Eds., VCH, Weinheim, 1991, 299.
10. **Alberts, B., Bray, D., Lewis, J., Raff, M., Roberts, K., and Watson, J. D.,** *Molecular Biology of the Cell,* Garland, New York, 1989.
11. **Darnell, J., Lodish, H., and Baltimore, D.,** *Molecular Cell Biology,* Scientific American Books, New York, 1986.
12. **Pisoni, R. L. and Thoene, J. L.,** The transport systems of mammalian lysosomes, *Biochim. Biophys. Acta,* 1071, 351, 1991.
13. **Altstiel, L. and Branton, D.,** Fusion of coated vesicles with lysosomes: measurements with a fluorescence assay, *Cell,* 32, 921, 1983.
14. **Mullock, B. M., Branch, W. J., van Schaik, M., Gilbert, L. K., and Luzio, J. P.,** Reconstitution of an endosome-lysosome interaction in a cell-free system, *J. Cell Biol.,* 108, 2093, 1989.
15. **Wilschut, J., Scholman, J., Eastman, S. J., Hope, M. J., and Cullis, P. R.,** Ca^{2+} induce fusion of phospholipid vesicles containing free fatty acids: modulation by transmembrane pH gradients, *Biochemistry,* 31, 2629, 1992.
16. **Bark, J. C. and Wilson, M. C.,** Regulated vesicular fusion in neurons: snapping together the details, *Proc. Natl. Acad. Sci. U.S.A.,* 91, 4621, 1994.
17. **Hurtley, S. M.,** Membrane proteins involved in targetted membrane fusion, *TIBS,* 18, 453, 1993.
18. **Stegmann, T., Doms, R. W., and Helenius, A.,** Protein-mediated membrane fusion, *Annu. Rev. Biophys. Biophys. Chem.,* 18, 187, 1989.
19. **Vaccaro, A. M., Tatti, M., Ciaffoni, F., Salvioli, R., Serafino, L., and Barca, A.,** Saposin C induces pH dependent destabilization and fusion of phosphatidylserine-containing vesicles, *FEBS Lett.,* 349, 181, 1994.
20. **Mellman, I. and Simons, K.,** The Golgi complex: in vitro veritas?, *Cell,* 68, 829, 1992.
21. **Rothman, J. E.,** Enzymology of intracellular membrane fusion, *Klin. Wochenschr,* 69, 98, 1991.
22. **Kobayashi, T. and Pagano, R. E.,** ATP-dependent fusion of liposomes with the Golgi apparatus of perforated cells, *Cell,* 55, 797, 1988.
23. **Kagiwada, S., Murata, M., Hishida, R., Tagaya, M., Yamashina, S., and Ohnishi, S.,** In vitro fusion of rabbit liver Golgi membranes with liposomes, *J. Biol. Chem.,* 268, 1430, 1993.
24. **Okamoto, G. H. and Lightfoot, E. N.,** Energy cost of intracellular organization, *Ind. Eng. Chem. Res.,* 31, 732, 1992.
25. **Gennis, R. B.,** *Bomembranes. Molecular Structure and Function,* Springer-Verlag, New York, 1989.
26. **Gutknecht, J., Bisson, M. A., and Tosteson, F. C.,** Diffusion of carbon dioxide through lipid bilayer membranes, *J. Gen. Physiol.,* 69, 779, 1977.
27. **Bentz, J. and Nir, S.,** Mass action kinetics and equilibria of reversible aggregation, *J. Chem. Soc. Faraday Trans.,* 77, 1249, 1981.
28. **Bentz, J., Nir, S., and Wilschut, J.,** Mass action kinetics of vesicle aggregation and fusion, *Colloids Surfaces,* 6, 333, 1983.
29. **Annesini, M. C., Di Marzio, L., Trazza, C., and Mossa, G.,** A simplified model for liposome aggregation and fusion, *J. Liposome Res.,* 3, 687, 1993.
30. **Zheng, S., Beissinger, R., Sherwood, R. L., McCormick, D. L., Lasic, D. D., and Martin, F. J.,** Liposome-encapsulated hemoglobin: a red blood cell substitute, *J. Liposome Res.,* 3, 575, 1993.
31. **Annesini, M. C., Di Marzio, L., Finazzi-Agrò, A., Serafino, A. L., and Mossa, G.,** Interaction of cationic phospholipid vesicles with carbonic anhydrase, *Biochem. Mol. Biol. Intern.,* 32, 87, 1994.
32. **Annesini, M. C., Di Giorgio, L., Di Marzio, L., Finazzi-Agrò, A., Serafino, A. L., and Mossa, G.,** Carbonic dioxide hydration with liposomes entrapping carbonic anhydrase, *J. Liposome Res.,* 3, 639, 1993.
33. **Annesini, M. C., et al.,** Carbon dioxide hydration with liposomes entrapping carbonic anhydrase, *J. Liposome Res.,* 3(3), 645, 1993.
34. **Dean, D. N., Fuchs, M. J., Schaffer, J. M., and Carbonell, R. G.,** Batch adsorption of CO2 by free and microencapsulated carbonic anhydrase, *Ind. Eng. Chem. Fund.,* 16, 452, 1977.
35. **Annesini, M. C., et al.,** Carbon dioxide hydration with liposomes entrapping carbonic anhydrase, *J. Liposome Res.,* 3(3), 642, 1993.

Chapter 16

Enzyme-Loaded Liposomes as Microreactors

Federico Cioci and Roberto Lavecchia

CONTENTS

I. ABSTRACT

A theoretical description of the kinetic behavior of enzyme-loaded liposomes is presented. Liposomes have been regarded as microscopic bioreactors in which the enzymatic conversion of substrates into products is accomplished. The classical theory of chemical reactors, based on the mass-balance equations and the rate laws, has been used to describe the kinetics of the microencapsulated enzyme. Particular attention has been devoted to the influence of transport phenomena on the macroscopic behavior of the vesicle, aimed at evidencing the risks of misinterpreting the experimental results when diffusional limitations are present.

Notwithstanding the complexity of the phenomena involved in the overall process, simplified models have been developed which account for the main physicochemical mechanisms by which substrate conversion occurs. In this context it is shown how a qualitative analysis of an enzyme-loaded liposome allows you to identify the elementary steps to be considered and hence to simplify the formulation of the mathematical model.

The most common situations, relative to the confinement of the enzyme in a well-defined region of the vesicle, namely the external surface, the liposomal membrane, and the internal core, have been investigated. Furthermore, a mathematical analysis of carrier-mediated transport phenomena has been carried out together with the modeling of the transient behaviour of a liposome entrapping a homogeneous enzyme solution.

The results obtained by numerical simulations have been used to draw some conclusions on the catalytic properties of the vesicles which could be helpful for planning the experimental work or optimizing the kinetic behavior of enzyme-loaded liposomes.

0-8493-4012-8/96/$0.00+$.50
© 1996 by CRC Press, Inc.

II. INTRODUCTION

In the course of evolution living organisms have developed highly organized biological structures in which chemical transformations and energy transduction are efficiently carried out. A large number of these specialized functions have been made possible by the segregation of enzymes into membrane-enclosed domains. Some well-known examples are represented by lysosomes, chloroplasts, and mitochondria, in which optimal microenvironmental conditions are realized for the reactions taking place in their interior.[1,2] This compartmentation contributes significantly to the coordination of the network of enzyme-catalyzed reactions and hence to the optimization of the cell behavior.[3] We therefore can infer that chemical reactions coupled with membrane-mediated processes should represent a powerful means for enhancing the performance of enzymes, in natural as well as artificial systems.

Liposomes are the simplest artificial structures, resembling natural membrane-delimited domains. When enzymes are entrapped in liposomes the resulting supramolecular structures exhibit very interesting properties from both a fundamental and a practical point of view.[4–6] Enzyme-loaded liposomes can be regarded as simplified models of biomembranes and hence they can be utilized for investigating the behavior of cells or cellular organelles.[7] On the other hand, they can be used as highly specific microreactors for exploiting the potential of enzymes in a variety of applications.[8,9]

The entrapment of enzymes in liposomes is a relatively simple procedure and a widely general method, provided that the enzyme is not adversely affected by the immobilization process. To date, numerous enzymes have been encapsulated in liposomes, including lysozyme,[10] glucose oxidase,[11] peroxidase,[12] neuraminidase,[13] β-galactosidase,[14] asparaginase,[15] and many others. Although most of the initial research work on enzyme-loaded liposomes was aimed at medical applications, particularly enzyme replacement therapies for the treatment of genetically inherited diseases,[16,17] novel potential uses are now appearing in far different biotechnological fields.[18–23]

The modalities of entrapment are dependent on the size of the protein and the morphology of the vesicle as well as on electrostatic and hydrophobic interactions. Hydrophilic enzymes are preferentially enclosed within the internal aqueous core, whereas hydrophobic or amphiphilic macromolecules are incorporated, at least partially, into bilayers. In some instances enzymes can interact with liposomes in a more complex way, through binding or adsorption on the surfaces of the bilayer,[24,25] where they may be immobilized or move laterally.

These interactions determine also the trapping efficiency of the loaded liposome. It has been shown that proteins having molecular weight up to 100,000 can be entrapped in LUV and MLV with very high efficiencies,[26,27] whereas SUV are practically unsuitable for macromolecules with a molecular weight above 40,000.

For many practical applications the entrapment of enzymes in liposomes could provide remarkable advantages over other immobilization modalities. One of the most attractive peculiarity of this method is the possibility to control the characteristics of the internal microenvironment through the selective barrier determined by the liposomal membrane. Experimental studies on lipid vesicles have shown that the range of permeability coefficients is very wide. Permeabilities of small nonpolar molecules are high when compared to those of hydrophilic charged compounds.[28,29] Ions have extremely low permeabilities[30] and this can be exploited for protecting enzymes sensitive to inhibitory ions.[31] On the other hand, substrates of very low permeability could be transported through the liposomal membrane by virtue of facilitated diffusion mechanisms,[32] which are very common in living systems. Recently it has been shown that engineered liposomes containing a selective ionophore in the lipid bilayer can be utilized to remove metal ions from dilute solutions.[33,34] Therefore, it should be expected that the insertion of a specific substrate transporter in the liposomal membrane could assure a sufficiently high diffusive flux for low-permeability substrates across the lipid bilayer.

Further interest in enzyme-loaded liposomes results from the possibility to reconstitute functionally active membrane-bound enzymes in the liposomal membrane. Although the catalytic properties of the enzyme may be negatively affected during solubilization from biological membranes, after reconstitution the protein exhibits a far enhanced activity and stability in comparison with the immobilized enzyme in its free state.[35,36]

Finally, liposomes could be used as a reservoir for accommodating multi-enzyme systems or retaining quite expensive coenzymes, like NAD^+ or $NADP^+$, in the reaction medium. Co-immobilization of enzymes catalyzing multi-step reactions appears particularly promising for the improved efficiency which would result from the low entrapped volume.

From the chemical engineering point of view enzyme-loaded liposomes can be regarded as microscopic chemical reactors.[37,38] In these miniaturized devices enzymes are likely to operate under conditions similar to those prevailing in the cellular environment, so that conformational changes in the protein structure or steric hindrances in the vicinity of the enzyme are generally limited. However, it should be stressed that the behavior of enzymes in liposomes is quite complex and difficult to predict owing to the heterogeneous nature of these systems which leads to mass-transfer limitations interfering with chemical kinetics.[39]

This chapter is devoted to the analysis of the kinetic behavior of enzyme-loaded liposomes as can be observed from macroscopic measurements. There are several reasons for modeling such behavior. As will be pointed out in the successive paragraphs, the design and the use of liposomes as enzymatic microreactors require an understanding of the overall kinetic properties of these systems. When enzymes are entrapped in liposomes the appearance of mass-transfer limitations and/or partition effects may affect their behavior and give rise to catalytic properties significantly different from those they exhibit in free solution. As a result, if these perturbations are not taken into account, the kinetic behavior of the immobilized enzyme can be seriously misinterpreted.

Modeling enzyme-loaded liposomes may be helpful not only to estimate correctly the kinetic parameters from the experimental data or to compare experiments performed under different conditions, but also to design liposomes with the desired properties. It is quite obvious that the prediction of the catalytic properties of a microencapsulated enzyme from those of the same enzyme in free solution is still distant from the actual possibilities. Nevertheless, it should be considered that the elucidation of the combined effect of transport phenomena and chemical kinetics on the macroscopic behavior of enzyme-loaded liposomes is perhaps the first requisite for designing these vesicles at the molecular level.

In spite of the fundamental and practical implications of such questions, only marginal attention has been so far devoted to the modeling of enzyme-loaded liposomes. In the following sections some guidelines will be presented for the analysis of liposomes as bioreactors. Of course, the development of a reliable model requires a full understanding of the physicochemical behavior of these systems. Consequently, the effects of microencapsulation on the enzyme behavior will be discussed first. Then the bases of modeling will be presented and applied to some simple but representative situations which may be encountered when dealing with enzyme-loaded liposomes.

III. LITERATURE SURVEY ON LIPOSOME MODELING

Very few literature works dealing with liposome modeling have been thus far published. Some papers appeared in the early 1970s describing the catalytic properties of enzyme-loaded semipermeable microcapsules.[40–42] Although such structures, consisting of an enzymatic solution enclosed inside a membrane of nylon or cellulose, are quite different from lipidic vesicles, their catalytic behavior is very similar to that of liposomes.

Mogensen and Vieth[43] analyzed the transient behavior of microcapsules containing catalase and estimated the permeability of the collodion membrane to hydrogen peroxide by assuming a first-order approximation for the reaction kinetics. The authors investigated also the steady-state operation of a column packed with urease microcapsules to provide the membrane permeability for urea.

Sundaran[44] studied the kinetics of nylon microcapsules enclosing an aqueous solution of urease and prepared by interfacial polimerization. A theoretical treatment of the experimental data was carried out aimed at evidencing the effects of mass-transfer on the overall kinetics. The urea permeability was determined by transient experiments on the substrate uptake in the absence of reaction. Then the Michaelis–Menten parameters of the encapsulated enzyme were estimated from initial rate measurements by assuming membrane diffusion and chemical reaction as elementary steps.

Wadiak and Carbonell[45] analyzed the kinetic behavior of a continuous stirred and a packed bed reactor containing cellulose nitrate microcapsules loaded with β-galactosidase. Making use of the concept of effectiveness factor the authors investigated the effects of mass-transfer on the observed kinetics and estimated the membrane permeability of the microcapsules from the reactor data. The influence of mass-transfer resistances on the effectiveness factor for an enzyme solution constrained within a spherical membrane and exhibiting Michaelis–Menten kinetics was discussed by Fink et al.[46] This approach was then extended by Wadiak and Carbonell[47] to include the cases of substrate and product inhibition. The authors investigated also the existence of multiple steady states and provided a criterion for the uniqueness of steady states.

Dean et al.[48] studied the batch absorpion of carbon dioxide by free and microencapsulated carbonic anhydrase. The process was described by a pseudo-steady-state model which enabled to determine the mass-transfer coefficients and the effectiveness factor for the entrapped enzyme.

Mossa et al.[49] developed a simplified theoretical model of liposomes describing the interactions between transport phenomena and chemical kinetics. The model assumes that a homogeneous enzyme solution is entrapped inside the vesicle and that both the external and the internal mass-transfer resistances are negligible. A test on the proposed model was carried out on the activity data reported by Madeira[50] and relative to urease-loaded liposomes. By assuming a first-order approximation for the Michaelis–Menten equation and using the kinetic parameters determined for the free enzyme the authors estimated the urea permeability, which was found to be very close to the literature data.

Annesini et al.[51] analyzed the kinetic behavior of unilamellar liposomes containing the plant enzyme ascorbate oxidase. The authors developed a model which assumes, according to the experimental results of a preliminary investigation,[52] that a fraction of the enzyme is embedded in the lipid bilayer and in contact with the external solution and that the remaining protein molecules are located in the aqueous core. The maximum specific reaction rate and the Michaelis constant for the free enzyme were derived from experiments in homogeneous solution. These parameters were then used in the fitting of the kinetic data obtained with the microencapsulated enzyme: this procedure enabled estimation of the membrane permeability of ascorbate.

More recently the same authors studied the kinetics of cationic liposomes entrapping erythrocyte carbonic anhydrase.[53,54] This enzyme catalyzes the reversible hydration of carbon dioxide and plays a key role in the transport of hydrogen and bicarbonate ions across the red cell membrane. The model proposed by the authors accounts for the transfer of carbon dioxide from the gas to the internal liquid solution and for both the uncatalyzed and enzymatic reaction. The solution of the unsteady mass-balance equations allows prediction of the pH drop inside and outside the vesicle. Furthermore, from a comparison between the kinetic behavior of free and encapsulated carbonic anhydrase, conclusions were drawn on the location of the enzyme in the liposome.

Finally, in a recent paper, we developed a simplified model which describes the unsteady behavior of liposomes entrapping the enzyme urease.[55] The work, aimed at providing preliminary information on the potential of these vesicles for the removal of blood urea, allowed us to predict the variations with time of the reactant concentrations, the internal pH, and the catalytically active enzyme fraction. Moreover, it was shown how numerical simulations carried out by the kinetic model could be used for determining the optimal operating conditions of these enzymatic bioreactors.

IV. EFFECT OF MICROENCAPSULATION ON THE ENZYME BEHAVIOR

The kinetic behavior of an enzyme inside a constraining structure, like a liposome, may be affected by the interactions between such structure and the enzyme and/or the various species which are present in the reaction medium. These phenomena result in more or less pronounced perturbations of the kinetics of the entrapped enzyme with respect to that of the free enzyme.

Enzyme–liposome interactions can determine conformational changes in the protein structure and steric hindrances, whereas substrate–liposome interactions can give rise to partition effects and mass-transfer limitations. This latter kind of interactions is always present, being a consequence of the heterogeneous nature of the system. In fact when enzymes are contained in a well-stirred homogeneous solution all intensive variables, like temperature and concentration, are uniform throughout the system. On the contrary, when enzymes are entrapped in liposomes, concentration nonuniformities may appear in the reaction system leading to the formation of a zone in the immediate vicinity of the enzyme, the *microenvironment*, which may differ significantly from the bulk solution, the *macroenvironment*.

It is usually assumed that the kinetic behavior of enzymes in liposomes reflects only the effects of diffusional limitations rather than true changes in the enzyme activity. For a quantitative analysis of these phenomena, the microenvironment and mass-transfer resistance concepts represent the most adequate bases actually available. In the following paragraphs the basic methodologic tools will be presented for modeling the interactions between transport phenomena and chemical kinetics in enzyme-loaded liposomes.

A. CONFORMATIONAL CHANGES AND STERIC HINDRANCES

The effects of conformational changes and steric hindrances on the enzyme molecule are schematically illustrated in Figure 1.

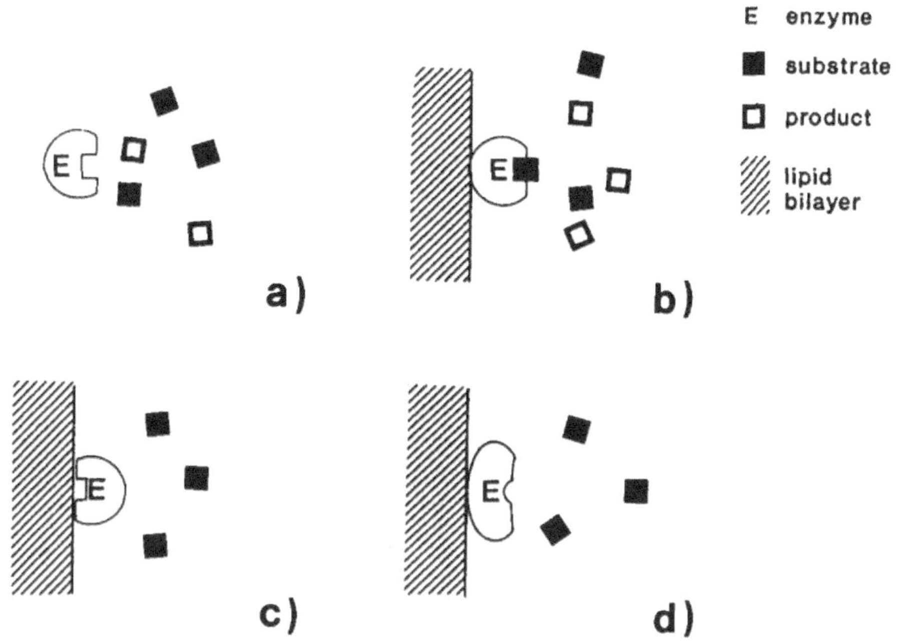

Figure 1 Effects of immobilization: (a) enzyme in free solution; (b) immobilized enzyme; (c) steric hindrance; (d) conformational change.

Conformational changes can range from alterations of local parts of macromolecules, one or a few aminoacidic residues, to protein unfolding, which causes a distortion of the whole spatial structure. These modifications can influence the properties of the active and allosteric sites of the enzyme and hence its activity and selectivity. Consequently, the intrinsic catalytic behavior of the enzyme, as revealed for instance by the value of the Michaelis–Menten parameters, may be considerably different from that of the free enzyme.

Steric hindrances are caused by the shielding effect of the liposomal membrane, which may reduce the accessibility to the substrate or effectors when the enzyme molecule interacts with the double layer. Although the intrinsic kinetic parameters are not modified, a reduction of the reaction rate can be equally observed owing to the fact that only a fraction of the enzyme molecules is available for substrate conversion.

In spite of the marginal attention which has been thus far devoted to the modeling of enzyme–liposome interactions it is expected that the growing interest in enzyme-loaded vesicles will result in a deeper investigation of these phenomena in the near future.

B. PARTITION EFFECTS

Partition effects are the result of hydrophobic and/or electrostatic interactions between the liposomal membrane and the components of the reaction medium. These interactions may cause sudden discontinuities at the liquid solution–membrane interface (Figure 2). In particular, if a solute is less soluble in the membrane than in the adjacent solution its concentration drops (Figure 2a), whereas if the solute is more soluble in the membrane its concentration increases (Figure 2b).

When fixed charges are present on the liposomal membrane, the electrostatic interactions between them and the mobile charged species in the aqueous solution may give rise to nonuniform distributions of such substances. An interesting case is represented by the partitioning of the hydrogen ions, which produces pH discontinuities at the membrane interfaces.

Partition phenomena are usually described by considering an equilibrium quantity, the partition or distribution coefficient (K_P), defined as

$$K_p = \frac{c}{c_0} \tag{1}$$

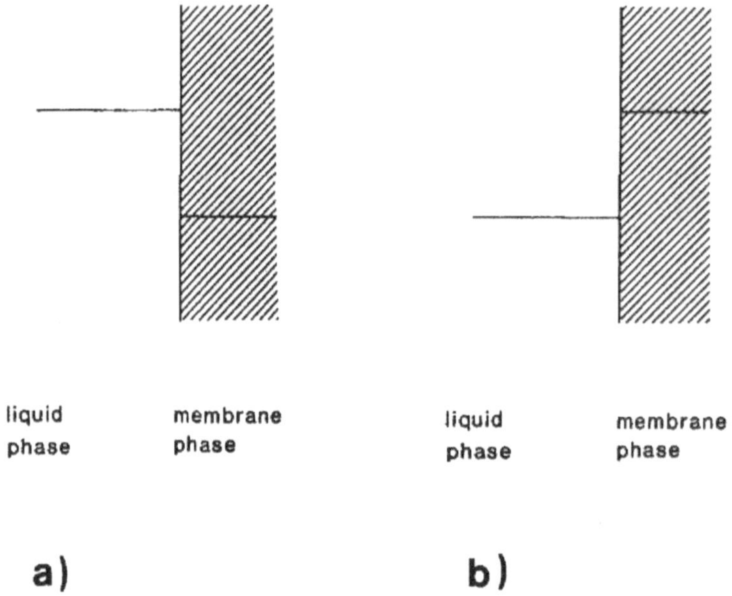

Figure 2 Partition effects: (a) the solute is more soluble in the liquid phase (K_P <1); (b) the solute is more soluble in the membrane phase (K_P >1).

where c and c_0 are the equilibrium concentrations of the species in the micro- and macroenvironment, respectively. The partition coefficient depends on temperature and on the medium composition; for dilute systems it is generally assumed to be constant.

C. MASS-TRANSFER LIMITATIONS

Mass-transfer limitations are a consequence of the finite transport rate of a given species from the bulk solution to the environment surrounding the enzyme. Mass-transfer can be concerned with the movement of individual molecules by virtue of their thermal energy (molecular diffusion) or with the motion of relatively large molecular aggregates in the form of eddies (turbulent diffusion). Molecular diffusion is a very slow process occurring in stagnant fluids or in laminar flow, although it is nevertheless always present, even in highly developed turbulent flow. Diffusion in the internal core of liposomes and across the lipid bilayer can be assumed to proceed by a molecular mechanism, whereas in the external liquid solution mass-transfer is likely to be turbulent. Both transport mechanisms lead to concentration gradients in the reaction system. Figure 3 shows the concentration profiles in the liquid phase and in the membrane, together with the effects of simultaneous diffusion and partitioning.

The rate at which a solute moves from one point to another in a given direction is generally described in terms of molar flux. The rigorous analytical description of mass-transfer is not possible, since the integration of the mass-balance equations for the diffusing solute would require a detailed knowledge of the local hydrodynamic field. However, several simplifications can be made which allow a mathematical treatment of both molecular and turbulent diffusion.

For dilute solutions the diffusive flux in the z direction can be evaluated from Fick's first law

$$J = -D\frac{\partial c}{\partial z} \tag{2}$$

where the negative sign indicates that diffusion occurs in the direction where the concentration decreases. D is the mass diffusivity, or diffusion coefficient. This quantity is a characteristic of the diffusing substance and its environment (temperature, pressure, concentration and chemical nature of all constituents).

Under turbulent flow conditions the molar flux from the bulk liquid phase to the liposome surface can be expressed, according to the film theory, as

$$J = k_c(c_0 - c) \tag{3}$$

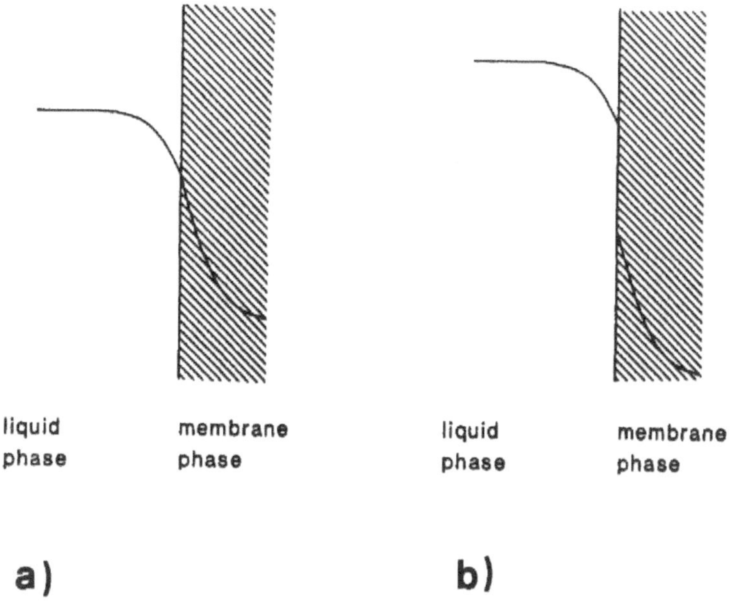

Figure 3 Concentration profiles in the absence (a) and in the presence (b) of partition effects.

where k_c is the mass-transfer coefficient for the diffusing solute. This quantity depends on the physico-chemical properties of the system as well as on the hydrodynamic conditions of the flow.

Some widely used correlations for the evaluation of the diffusion and mass-transfer coefficients for a spherical unilamellar liposome are reported in the Appendix.

The flux across the liposomal membrane can be calculated by the "solubility-diffusion" model,[56] which assumes that the rate-limiting step of the overall process is the molecular diffusion of the substrate through the lipid bilayer and that the interfacial resistances are negligible. For a spherical membrane the following expression is obtained

$$J = \frac{D_m K_P (R - \delta)}{R\delta}(c_2 - c_1) \tag{4}$$

where D_m and K_P are the diffusion and partition coefficients for the solute in the membrane interior, δ is the membrane thickness, R is the liposome radius, c_1 and c_2 are the solute concentrations on the two sides of the membrane.

The first term at the second member is often referred to as membrane permeability

$$P = \frac{D_m K_P (R - \delta)}{R\delta} \tag{5}$$

This quantity reflects both the solubility and the diffusive behavior of a solute in the membrane interior: the former is determined mainly by the hydrophobic and electrostatic interactions, whereas the latter depends significantly on the solute molecular size.

If the membrane thickness is much smaller than the liposome radius ($\delta \ll R$) curvature effects can be neglected. As a result, the permeability simplifies into the following expression

$$P = \frac{D_m K_P}{\delta} \tag{6}$$

For a large number of substances the permeability coefficients through biomembranes or model membranes are available. It has been shown that for nonelectrolytes with molecular weight between 50 and 300 the permeability across a phoshpolipid bilayer can be correlated with the partition coefficient of the solute between water and n-hexadecane.[29] Deviations observed for very small molecules can be explained in terms of size dependence of the membrane diffusion coefficient.[57] These experimental evidences are consistent with the "solubility-diffusion" model, which predicts correctly the dependence of the permeability coefficient on the hydrophobicity and the molar volume of the solute.

V. PRELIMINARY ANALYSIS OF AN ENZYME-LOADED LIPOSOME

Owing to the complex kinetic behavior of enzyme-loaded liposomes, some simplifying assumptions must necessarily be made. In the successive developments we shall focus our attention on a spherical unilamellar liposome of constant radius. Furthermore we shall assume that the vesicle is not permeable to the enzyme, i.e., the enzyme molecules are not released into the external environment, and that the thickness of the liposomal membrane is uniform and much smaller than the curvature radius.

The formulation of a kinetic model for a similar membrane-enclosed system can be further simplified if one recognizes that the overall process giving rise to the conversion of substrates into products can be broken down into a limited number of elementary steps, a combination of which may be used to model the different situations. The typical steps involved include mass-transfer in the external and internal liquid solution, diffusion through the lipid bilayer, and chemical reaction. Depending on the enzyme location these four steps may occur in succession or they may be arranged into a series–parallel scheme. In general the following cases can be contemplated (Figure 4):

a. the enzyme is adsorbed on the outer surface of the vesicle;
b. the enzyme is incorporated in the liposomal membrane;
c. the enzyme is enclosed in the aqueous core.

Although this outline does not account for the simultaneous presence of the enzyme in more than one region, it can be considered fairly general, since in most cases enzymes are preferentially located in one of the three zones.

If the enzyme is adsorbed in on the external surface of the vesicle, mass-transfer of substrate takes place from the bulk solution to the outer surface of the liposome, where it is converted by the adsorbed enzyme (Figure 5a). If the enzyme is segregated in the liposomal membrane, the external mass-transfer of substrate is followed by the simultaneous chemical reaction and membrane diffusion (Figure 5b). Finally, if the enzyme is confined in the aqueous core, the substrate diffuses from the bulk solution to the outer surface and through the lipid bilayer. Then internal diffusion and chemical reaction take place, in

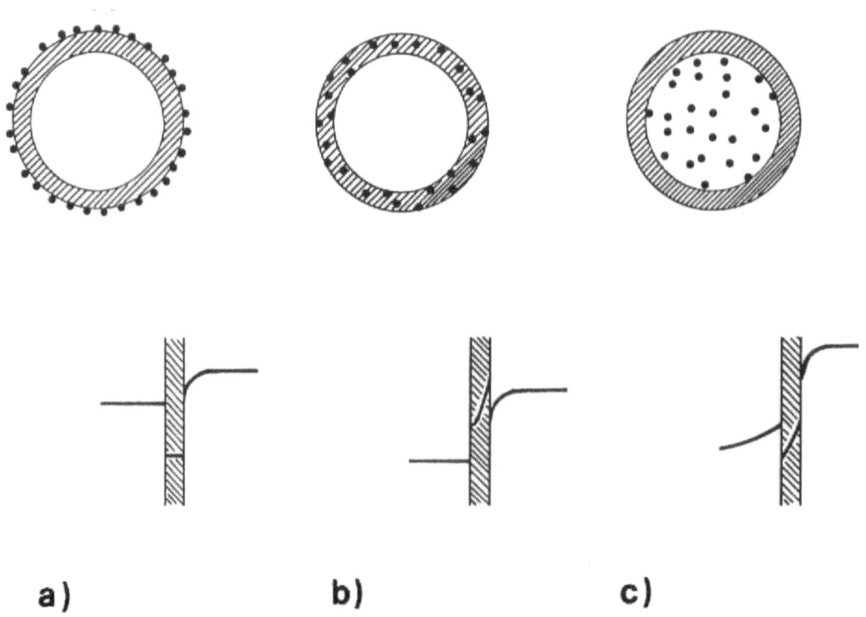

Figure 4 Schematic illustration of different immobilization topologies with the corresponding substrate profiles: (a) enzyme on the outer surface; (b) enzyme in the membrane; (c) enzyme in the aqueous core.

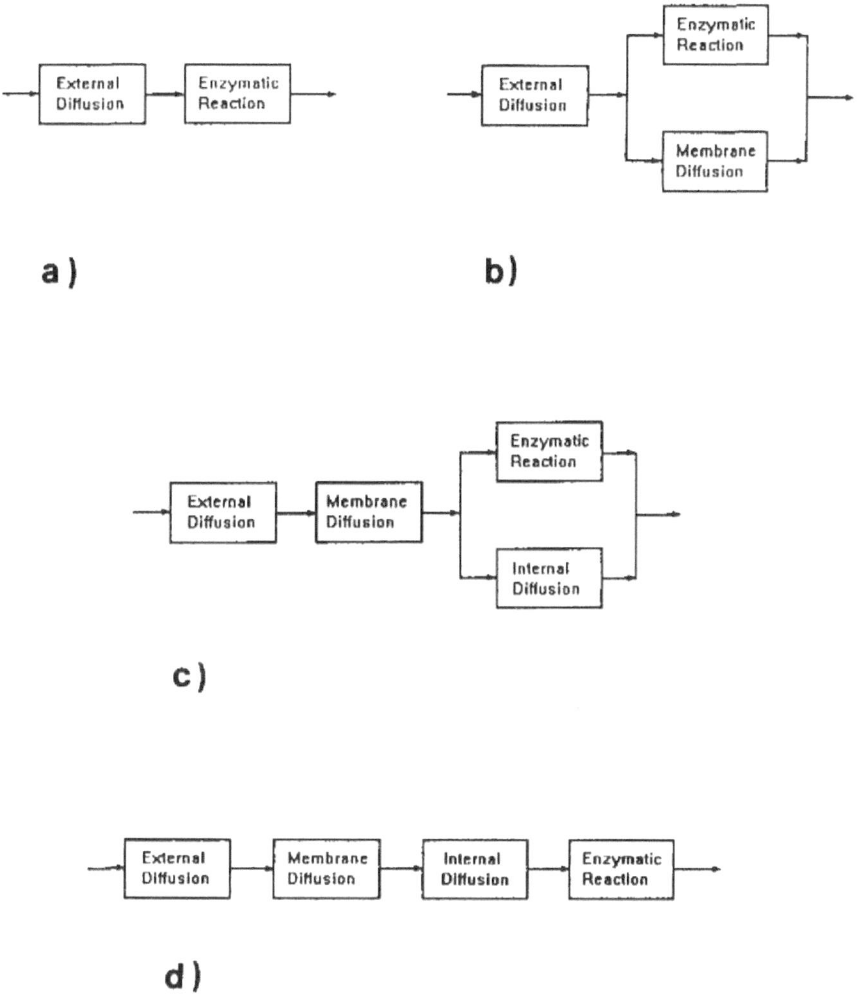

Figure 5 Elementary steps involved in the overall process: (a) enzyme on the outer surface; (b) enzyme in the membrane; (c) homogeneous enzyme solution in the aqueous core; (d) single enzyme molecule in the aqueous core.

parallel or in series, depending on the sizes of the enzyme molecule and the vesicle. In particular two limiting situations can be considered. The first one arises when the enzyme molecule is much smaller than the liposome. In this case we can assume that a homogeneous enzyme solution is enclosed in the vesicle. In such conditions internal diffusion and chemical reaction proceed simultaneously (Figure 5c). Conversely, if the molecular size of the enzyme is comparable with the liposome radius, we can suppose that the vesicle contains one molecule of enzyme at the most (some vesicles may be empty) and hence internal diffusion and chemical reaction are in series (Figure 5d).

An important point to be stressed is that each basic step occurs over a particular time length, the *characteristic* or *relaxation time*. Since only the steps with higher characteristic times are likely to affect the rate of the overall process, a preliminary estimate of these quantities can provide useful indications for the development of the kinetic model. In particular, if all the characteristic times are comparable, i.e., of the same order of magnitude, every elementary step should be taken into consideration. On the contrary, if one or more steps have characteristic times much smaller than those of the others, they could be excluded from the model, due to the limited influence exerted on the overall process.

The characteristic time relative to each step can be easily evaluated by considering the unsteady mass-balance equation for the substrate in a process consisting solely of that step. The expressions so obtained are summarized in Table 1.

Table 1 Characteristic Times for the Elementary Steps Involved in the Overall Process

Elementary Step	Characteristic Time
External diffusion	$\tau_e = R/(3\ k_c)$
Internal diffusion	$\tau_i = R^2/D$
Membrane diffusion	$\tau_m = R/(3\ P)$
Chemical reaction	$\tau_r = S_0/-r(S_0)$

In these relationships R is the liposome radius, k_c is the external mass-transfer coefficient, D is the diffusion coefficient, P is the substrate permeability across the liposomal membrane, S_0 is the maximum substrate concentration, and $-r(S_0)$ is the reaction rate corresponding to that concentration. The negative sign is determined by the convention that the reaction rate is negative for a substance disappearing from the system.

The final expression of the characteristic reaction time depends on the rate law adopted to model the enzyme kinetics. If the reaction rate is expressed by the Michaelis–Menten equation τ_r takes the form

$$\tau_r = \frac{K_M + S_0}{v_{max}} \tag{7}$$

where K_M is the Michaelis constant and v_{max} is the maximum reaction velocity, which is given by

$$v_{max} = ke_0 \tag{8}$$

where k is a rate constant and e_0 is the total enzyme concentration.

Depending on the values of S_0 and K_M the following simplifications can be made

$$\frac{S_0}{K_M} \ll 1 \quad : \quad \tau_r = \frac{K_M}{v_{max}} \tag{9}$$

$$\frac{S_0}{K_M} \gg 1 \quad : \quad \tau_r = \frac{S_0}{v_{max}} \tag{10}$$

We can therefore observe that at low substrate concentration (with respect to K_M) the first-order approximation for the Michaelis–Menten kinetics leads to a characteristic time which is independent of the substrate concentration. Conversely, for the zero-order approximation the characteristic time does not depend on the value of K_M, being solely a function of S_0 and v_{max}.

The characteristic diffusion times (τ_e, τ_i, and τ_m) are dependent on the liposome radius and a transport property (k_c, D, and P, respectively). Some values of the characteristic times relative to the enzyme urease entrapped in liposomes of different sizes have been calculated and reported in Table 2. As can be seen, the characteristic times for the external and internal diffusion are much smaller than those relative to the permeation across the liposomal membrane and to chemical reaction. These results, which can be

Table 2 Characteristic Times (s) for a Urease-Loaded Liposome

		Liposome Radius (nm)		
Elementary Step		200	500	1000
External diffusion		$9.73\ 10^{-6}$	$6.08\ 10^{-4}$	$2.43\ 10^{-4}$
Internal diffusion		$2.92\ 10^{-5}$	$1.82\ 10^{-4}$	$7.30\ 10^{-4}$
Membrane diffusion		1.67	4.17	8.33
Chemical reaction:	$S_0 = 10^{-6}\ M$	1.31	1.31	1.31
	$S_0 = 10^{-4}\ M$	1.37	1.37	1.37
	$S_0 = 10^{-2}\ M$	6.77	6.77	6.77

Note: $P = 4\ 10^{-6}$ cm/s (29), $D = 1.37\ 10^{-5}$ cm²/s (58), $k/K_M = 7.64\ 10^6\ M^{-1}\ s^{-1}$ (44), $e_0 = 10^{-7}\ M$.

considered quite general, are a consequence of the small liposome sizes. However they should be verified for each particular situation, because of the wide range of variability of both substrate permeability and Michaelis–Menten parameters. In any case, when external and internal mass-transfer limitations are negligible with respect to membrane diffusion and chemical reaction, the kinetic analysis can be restricted to these latter steps. Turning attention to this situation, we can also say that if the characteristic reaction time is much smaller than the characteristic permeation time ($\tau_r/\tau_m \ll 1$) membrane diffusion is the rate-limiting step and the global process is said to be *diffusion-controlled*. As will be pointed out in the next section, under these conditions the observed kinetics no longer exhibits Michaelis–Menten dependence on S_0 and should be characterized by apparent parameters which are different from the true or intrinsic ones. Conversely, if the characteristic permeation time is much smaller than the characteristic reaction time ($\tau_m/\tau_r \ll 1$) the enzymatic reaction is the rate-limiting step. Under such conditions the process is said to be *reaction-controlled*: the substrate concentration inside the liposome is close to that in the bulk solution and therefore the observed and intrinsic kinetics tend to coincide. In the intermediate cases, where τ_m and τ_r are of the same order of magnitude, the liposome behavior lies within the confines of these two limiting situations.

It is worth noting that for a given system, i.e., for assigned values of R, P, k, and K_M, the ratio between τ_m and τ_r, which is defined by

$$\frac{\tau_m}{\tau_r} = \frac{Rke_0}{3P(K_M + S_0)} \tag{11}$$

depends solely on S_0 and e_0. Since the substrate concentration in the external solution diminishes as the reaction proceeds, situations may appear in which the controlling mechanism changes with time. Analogously, if the reaction system contains liposomes with different amounts of enzyme, the controlling regime might change from one liposome to another. These considerations are illustrated in Figure 6, where τ_m/τ_r has been plotted against e_0 at different S_0 values.

An important point to be stressed is that the analysis developed thus far has been addressed to a single enzyme-loaded liposome. Clearly, significant complexities may arise when a non-homogeneous distribution of vesicles is involved. The kinetic behavior of each liposome, in fact, is influenced by its own microenvironment and the net rate of substrate conversion at a given instant of time can differ from that of the other liposomes. Of course, it is always possible to consider liposomes with average structural and kinetic properties. This leads to mathematically tractable models which are generally adequate for most

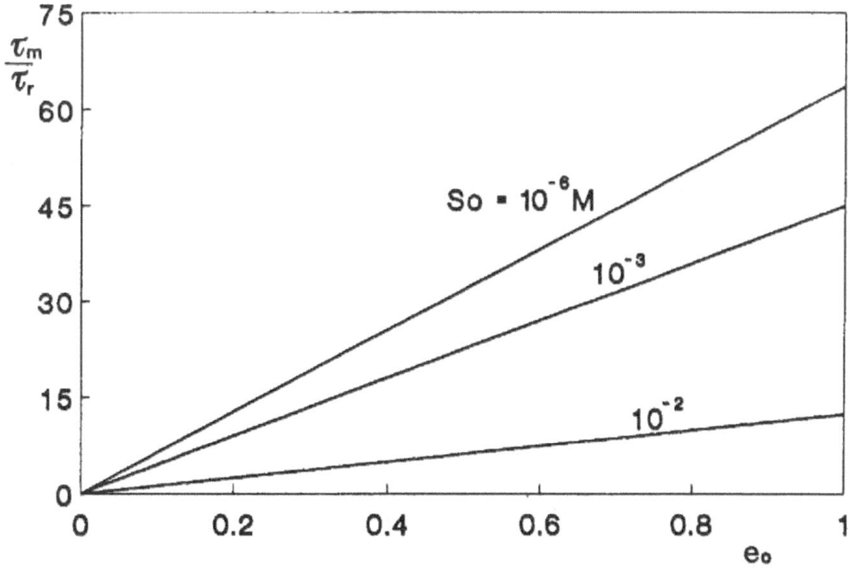

Figure 6 Effect of the substrate (S_0) and the enzyme (e_0) concentrations on the ratio between the characteristic times τ_m and τ_r.

microencapsulated enzymes. More rigorous models should account for the distribution of liposome sizes and properties as well as for the fact that the initial size distribution could evolve during time due to liposome coalescence or fusion. These models are obviously much more complex and will not be discussed in the successive sections.

VI. KINETIC MODELS OF ENZYME-LOADED LIPOSOMES

The previous section has been concerned with a qualitative investigation of the kinetic behavior of enzyme-loaded liposomes, aimed at evaluating the influence of the different elementary steps on the overall rate of substrate conversion. Here we shall develop some simplified mathematical models which allow us to analyze the kinetics of these vesicles under steady-state conditions. In general, when liposomes are confined in a well-stirred vessel containing the substrate solution, steady state can be regarded as a good approximation of the real situation if the volume of the liquid solution is sufficiently large, so that the variations of the substrate concentration in the bulk solution can be neglected. However, in some instances this approximation can no longer be maintained, even if the concentration of all substances in the external solution do not change with time. The modeling of the transient behavior of an enzyme-loaded liposome will be presented in the last part of this section.

A. ENZYME IMMOBILIZED ON THE OUTER SURFACE OF THE MEMBRANE

We have already seen that when the enzyme is immobilized on the outer surface of the liposome the elementary steps to be considered are substrate diffusion in the external solution and chemical reaction at the surface. At steady state the molar flux of substrate from the bulk fluid to the liposome must be equal to the rate of substrate consumption per unit area (–r)

$$-r = k_c\left(S_0 - S\right) \tag{12}$$

where S_0 and S are the substrate concentration in the macro- and microenvironment, respectively. The reaction rate can be assumed to depend on the substrate concentration according to the Michaelis–Menten equation

$$-r = \frac{v'_{max} S}{K_M + S} \tag{13}$$

where v'_{max} represents the saturation rate per unit area of the membrane.

Equations 12 and 13 allow us to determine S and –r once S_0, k_c, K_M, and v'_{max} are known. It may be convenient to make them dimensionless by defining a dimensionless substrate concentration (σ) and a dimensionless reaction rate (ρ)

$$\sigma = \frac{S}{K_M} \quad ; \quad \rho = \frac{-r}{v'_{max}} \tag{14}$$

Using these new variables Equations 12 and 13 assume the following form

$$\rho = \psi\left(\sigma_0 - \sigma\right) \tag{15}$$

$$\rho = \frac{\sigma}{1+\sigma} \tag{16}$$

where ψ is a dimensionless parameter given by

$$\psi = \frac{k_c K_M}{v'_{max}} \tag{17}$$

It should be noted that this parameter is equal to the ratio between the characteristic reaction time for $S_0/K_M \ll 1$ and the characteristic diffusion time. Consequently, its value can provide an estimate of the

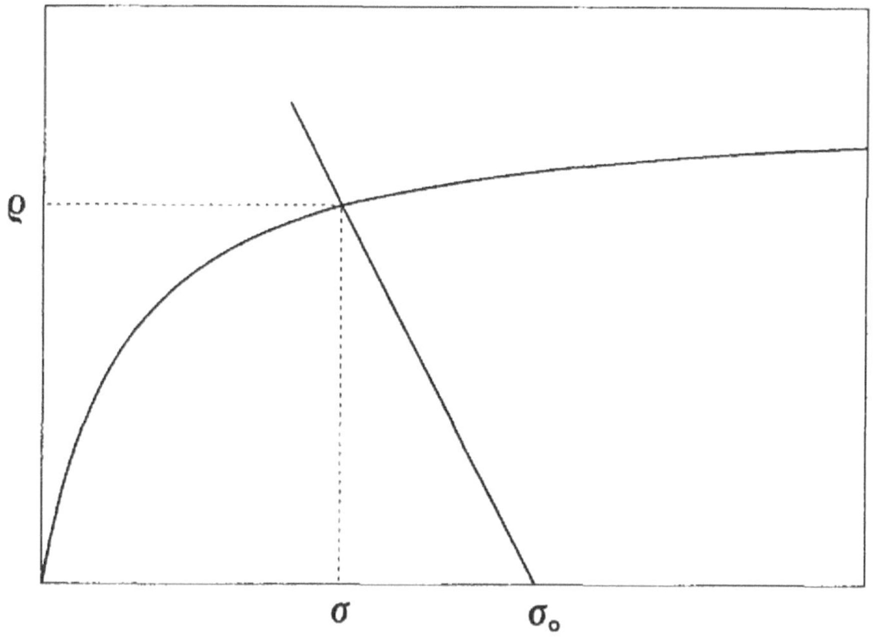

Figure 7 Graphical determination of the microenvironmental substrate concentration (σ) and the effective reaction rate (ρ) corresponding to an assigned substrate concentration in the bulk solution (σ_0).

influence of diffusional limitations on the enzyme kinetics. In particular, when $\psi \rightarrow 0$ the process is controlled by diffusion, whereas when $\psi \rightarrow \infty$ the process is controlled by the enzymatic reaction.

Once assigned σ_0 and ψ, the solution of the system, Equations 15 and 16, yields the substrate concentration at the liposome surface (σ) and the relative reaction rate (ρ). A graphical method for solving this system is shown in Figure 7, where the curves corresponding to the two equations have been plotted. Equation 15 represents a straight line with slope $-\psi$, whereas Equation 16 gives rise to a saturation curve with a horizontal asymptote at $\rho = 1$. The coordinates of the intersection point yield the actual surface concentration (σ) and the reaction rate at this concentration (ρ). By repeating such procedure for a number of σ_0 values, the curves plotted in Figure 8 are obtained.

If one refers to the range of variability of ψ, two limiting behaviors are easily evidenced. The first one corresponds to a sufficiently large ψ, i.e., to a substrate diffusion much faster than chemical reaction. Under these conditions the substrate concentration at the liposome surface is practically equal to that in the bulk solution (the straight line in Figure 7 is vertical) and the kinetics of the immobilized enzyme is the same of the free enzyme.

For very small values of ψ, i.e., when the reaction is much faster than diffusion, all substrate molecules arriving at the liposome surface are instantaneously converted, so that the surface concentration drops to zero (the straight line in Figure 7 is horizontal). It is quite obvious that under these conditions the activity of the bound enzyme is practically independent of the kinetic parameters of the reaction, being the substrate conversion controlled solely by the transport properties of the system.

At intermediate values of ψ the process is influenced by both the transport and reaction properties of the system. These considerations are illustrated in Figure 8, where the dependence of ρ on σ_0 is shown. We note that when $\psi \rightarrow \infty$ the conversion rates attainable are maximum, whereas in all other cases the reaction rates are reduced, owing to the limited velocity of substrate diffusion. This phenomenon is often referred to as *diffusional inhibition*.

An important point to be stressed is that, when diffusional limitations are present, the relationship between the apparent reaction rate and the substrate concentration in the macroenvironment no longer obeys the Michaelis–Menten equation. This occurrence is determined by the fact that when the process is diffusional inhibited, the substrate concentration at the liposome surface is quite different from that in the solution. The departure from Michaelis–Menten kinetics is shown in Figure 9, where ρ/σ_0 has been plotted against ρ. This plot yields a straight line only when the Michaelis–Menten equation is obeyed, i.e., for $\psi \rightarrow \infty$. As ψ

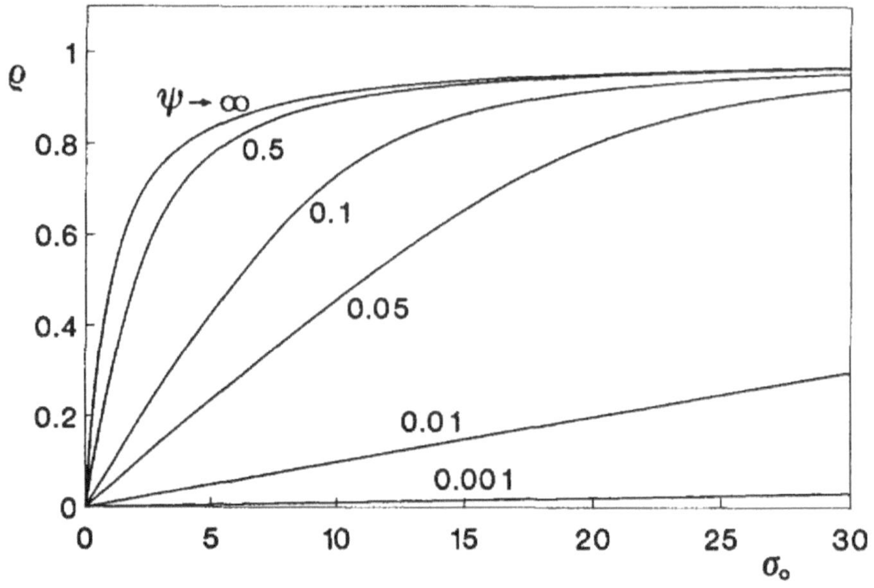

Figure 8 Dimensionless overall reaction rate (ρ) as a function of the dimensionless substrate concentration in the macroenvironment (σ_0) at different values of ψ.

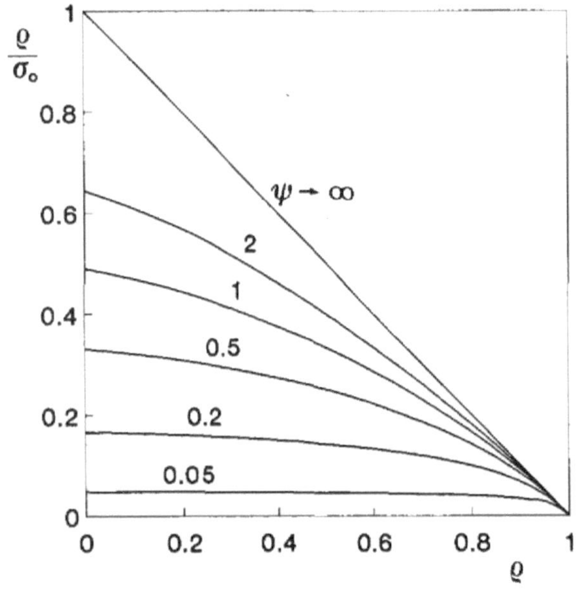

Figure 9 Departure of the effective rate from Michaelis–Menten kinetics at different values of ψ.

decreases, the resulting curves deviate from straight lines, particularly when wide concentration ranges are involved.

These considerations clearly indicate that the observed rates do not always reflect the intrinsic kinetic behavior of the enzyme. Consequently, when the kinetics of the enzyme-loaded liposome is investigated experimentally, particular care must be devoted to estimating the kinetic parameters. It should be remembered that the most adequate experimental procedure is to carry out measurements in the kinetically controlled regime. If this is not possible, owing for instance to large mass-transfer limitations, diffusional effects need to be separated from the observed kinetics. This can be made by determining the external transport parameters from appropriate experiments or empirical correlations. Once the transport coefficient of the substrate is known, the analysis of the experimental data by the equations previously developed enables us to estimate the kinetic parameters K_M and v'_{max}. If these latter differ from those of the free enzyme, conformational changes or steric hindrances are likely to occur upon immobilization.

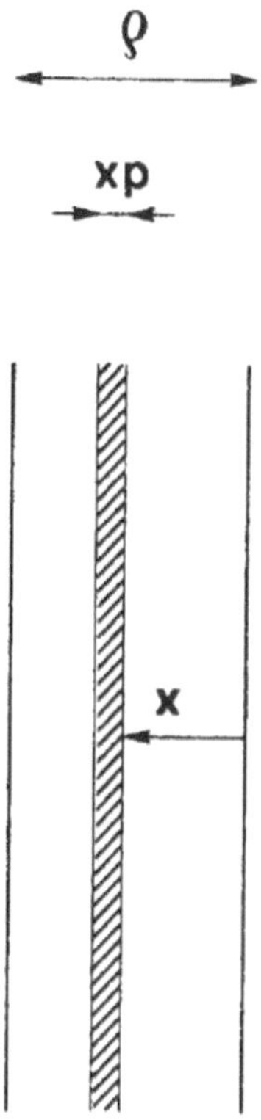

Figure 10 Elemental volume in the liposomal membrane for deriving the differential mass-balance equation.

B. ENZYME INCORPORATED IN THE LIPOSOMAL MEMBRANE

When the enzyme is incorporated in the lipid bilayer, the elementary steps to be taken into account are substrate diffusion in the membrane and chemical reaction. The third step, substrate diffusion in the external solution, can be usually neglected due to its small characteristic time. At steady state the mass-balance for the substrate in a flat membrane gives rise to the following differential equation

$$D_m \frac{d^2S}{dx^2} = \frac{v''_{max} S}{K_M + S} \tag{18}$$

where D_m is the effective diffusivity of the substrate through the lipid bilayer, x is the distance from the external surface (Figure 10), and v''_{max} is the saturation rate per unit volume of the membrane. In most cases the assumption of flat membrane is not too stringent, since the membrane thickness is generally much smaller than the liposome radius.

Equation 18 can be integrated after assigning the appropriate set of boundary conditions

$$\begin{aligned} S &= S_m \quad && at \quad x = 0 \\ \frac{dS}{dx} &= 0 \quad && at \quad x = \delta \end{aligned} \tag{19}$$

where S_m is the substrate concentration on the outer side of the membrane and δ is the membrane thickness. It may be convenient to put the mass-balance equation in dimensionless form by defining a dimensionless distance (ξ) and a dimensionless substrate concentration (σ)

$$\xi = \frac{x}{\delta} \quad ; \quad \sigma = \frac{S}{K_M} \tag{20}$$

The rearrangement of Equation 18 yields

$$\frac{d^2\sigma}{d\xi^2} = \Phi^2 \frac{\sigma}{1+\sigma} \tag{21}$$

with the following dimensionless boundary conditions

$$\begin{aligned} \sigma &= \sigma_m \quad at \quad \xi = 0 \\ \frac{d\sigma}{d\xi} &= 0 \quad at \quad \xi = 1 \end{aligned} \tag{22}$$

Φ is a dimensionless parameter given by

$$\Phi = \delta\left(\frac{v''_{\max}}{K_M D_m}\right)^{0.5} \tag{23}$$

We note that even in this case, the dimensionless parameter obtained has a clear physical meaning. In fact it can be easily verified that Φ^2 is equal to the ratio between the characteristic permeation time and the characteristic reaction time for $S_0/K_M \ll 1$. Consequently, its value reflects the influence of diffusional limitations through the lipid bilayer on the enzyme kinetics. In particular, when $\Phi \to 0$ the process is controlled by the enzymatic reaction, whereas when $\Phi \to \infty$ the process is controlled by membrane diffusion.

The integration of Equation 21 yields the dimensionless substrate concentration profile in the membrane. The results of some simulations are shown in Figure 11: we observe that the concentration gradients and hence the extent of substrate depletion increase with Φ. In particular, for $\Phi > 100$ the substrate concentration drops to zero at the membrane entrance, whereas for $\Phi < 0.1$, the substrate molecules cross the membrane without being converted.

Once the concentration profile in the membrane is known, the overall conversion rate can be determined by integrating the local reaction rate over the membrane thickness. It should be pointed out that only at low Φ (less than 0.1 in the case shown in Figure 11) the process is kinetically controlled and the experimental data can be used to obtain the intrinsic kinetic parameters. At high Φ the observed kinetics are influenced also by the transport properties of the substrate in the bilayer. In these cases diffusional effects should be separated from the kinetic ones prior to estimating the intrinsic kinetic

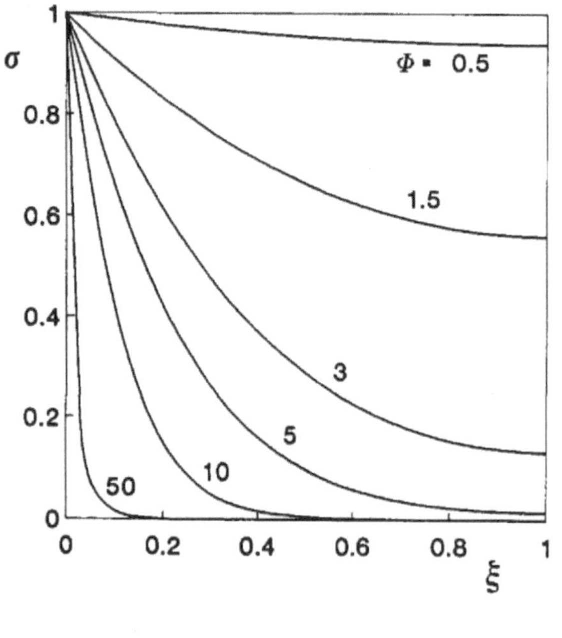

Figure 11 Dimensionless concentration profiles in the lipid bilayer at different values of Φ.

parameters; this can be accomplished by measuring the effective diffusion coefficient of the substrate in the membrane in the absence of reaction.

The effects of diffusional limitations in the lipid bilayer can be described by a quantity named *effectiveness factor* and defined as the ratio of the effective reaction rate to that which would be attained if all the immobilized enzyme molecules were exposed to the macroenvironmental concentration.[37] This parameter is a function of Φ and σ_0. It can be easily shown that in the absence of mass-transfer limitations, i.e., when the effectiveness factor approaches one, the dependence of the effective rate on the macroenvironmental concentration obeys Michaelis–Menten law. When the membrane resistance is not negligible a departure is observed from Michaelis–Menten kinetics which increases with increasing Φ.

C. ENZYME CONFINED IN THE AQUEOUS CORE

We have previously noted that when the enzyme is confined in the aqueous core the resulting kinetic behavior may be more complex than in other cases. In particular, two limiting situations should be considered, depending on the ratio between the sizes of the enzyme and the liposome. If this ratio is close to one, i.e., the enzyme and the liposome have comparable sizes, the internal diffusion of the substrate is in series with the chemical reaction. Conversely, if this ratio is much lesser than one, i.e., the enzyme molecule is much smaller than the vesicle, substrate diffusion and chemical reaction occur in parallel. However, it should be remembered that if the characteristic diffusion time is much smaller than the characteristic permeation time the internal diffusional resistance can be neglected and the kinetic scheme to be considered is no longer dependent on the relative molecular size of the enzyme.

Turning attention to this situation, we shall suppose that a homogeneous enzyme solution is entrapped in the liposome. The elementary steps to be taken into account are substrate diffusion across the lipid bilayer and chemical reaction in the aqueous core. At steady state, the flow-rate of the substrate through the membrane must be equal to the conversion rate

$$aP(S_0 - S) = \frac{v_{max} S}{K_M + S} \tag{24}$$

where P is the substrate permeability, a is the surface area of the vesicle (equal to 3/R for a spherical liposome) and v_{max} is the saturation rate per unit volume of the internal solution. Equation 24 can be made dimensionless by defining the following dimensionless variable

$$\sigma = \frac{S}{K_M} \tag{25}$$

We obtain

$$\zeta(\sigma_0 - \sigma) = \frac{\sigma}{1 + \sigma} \tag{26}$$

where the dimensionless parameter ζ is given by

$$\zeta = \frac{aPK_M}{v_{max}} \tag{27}$$

It is worth noting that this parameter is equal to the ratio between the characteristic reaction time for $S_0/K_M \ll 1$ and the characteristic permeation time. High values of ζ correspond therefore to low diffusional resistances, whereas low values of ζ are indicative of significant diffusional inhibitions. When $\zeta \to \infty$ no difference exists between the substrate concentration in the external and in the internal solution: under these conditions the process is kinetically controlled. When $\zeta \to 0$ the internal substrate concentration approaches zero and the process is diffusion controlled. Of course, at intermediate values of ζ the observed kinetics is affected by both the transport properties of the substrate in the liposomal membrane and the catalytic properties of the enzyme. Some curves of ρ against σ_0 at different ζ are reported in Figure 12: we note that diffusional limitations through the lipid bilayer cause a reduction of the overall reaction rate even though at high substrate concentrations the reaction rate tends to v_{max} for any value of ζ.

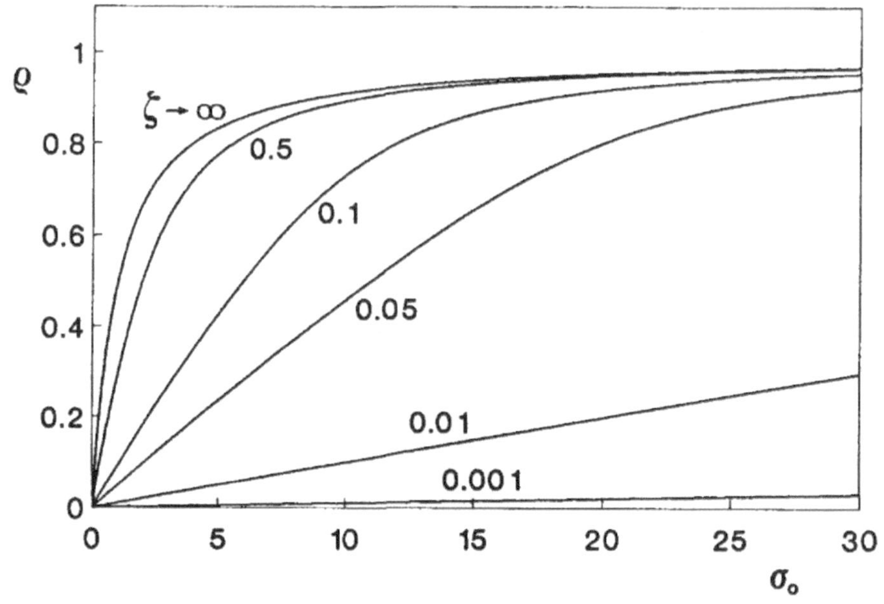

Figure 12 Dimensionless overall reaction rate (ρ) as a function of the dimensionless substrate concentration in the macroenvironment (σ_0) at different values of ζ.

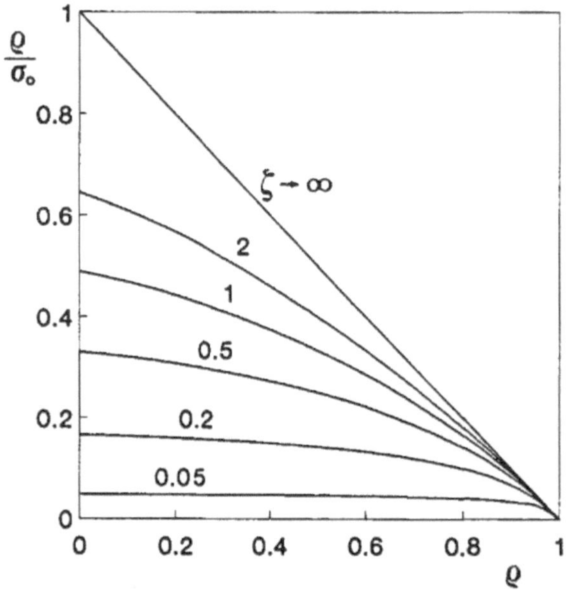

Figure 13 Departure of the effective rate from Michaelis–Menten kinetics at different values of ζ.

It is therefore apparent that the measured rates reflect the intrinsic enzyme kinetics only when the process is not affected by diffusional limitations. In such situations the observed reaction rates depend on the macroenvironmental substrate concentration according to the Michaelis–Menten equation and this yields a straight line in the plot shown in Figure 13. In all other cases diffusional effects should be separated from the observed enzyme behavior to obtain the true kinetic parameters. This can be performed, as already pointed out, by determining the substrate permeability across the membrane from separated experiments and substituting its value in ζ to estimate K_M and v_{max}. On the other hand, if the kinetic parameters for the free enzyme are known and conformational changes can be excluded to occur, the experimental data can be correlated by Equation 24 to provide an estimate of the substrate permeability through the liposomal membrane.

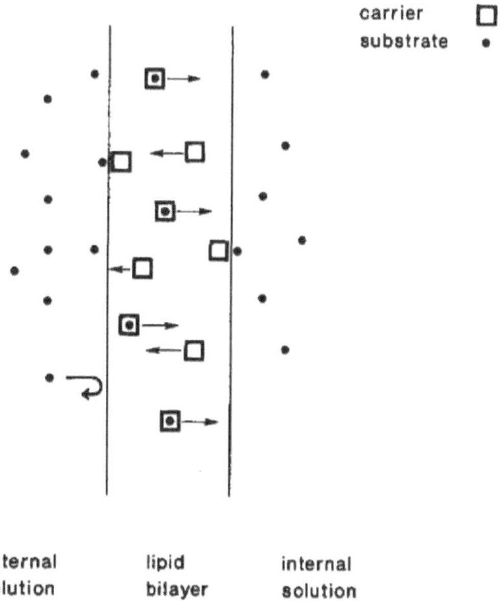

Figure 14 Schematic illustration of the facilitated diffusion mechanism.

D. CARRIER-MEDIATED TRANSPORT PHENOMENA IN ENZYME-LOADED LIPOSOMES

We have seen that the permeabilities of most polar molecules and high molecular weight substances through the liposomal as well as the cell membranes are generally very low. In living organisms such a diffusive barrier is often passed by superimposing a facilitated diffusion mechanism to the ordinary one.[59,60] Some well-known examples are represented by the transport of amino acids, nucleotides, and sugars across the cell membrane. The same principle should be exploited, at least from a theoretical point of view, for extending the use of enzyme-loaded liposomes to low-permeability substrates. For a similar application, vesicles should be prepared containing a mobile carrier in the bilayer which is capable to bind the substrate and to allow its diffusion across the lipid bilayer.

In most cases the facilitated solute diffusion involves a reversible chemical reaction between the solute and the carrier. A simplified schematization of the mechanism by which the substrate is transported is illustrated in Figure 14. The first step of the process is represented by the reaction leading to the formation of the solute-carrier complex

$$\text{solute} + \text{carrier} \rightleftharpoons \text{complex}$$

The complex diffuses through the lipid bilayer and on the opposite side of the membrane the reaction is reversed and the solute is released. Then the carrier diffuses back to the membrane and reacts with a new solute molecule.

In the modeling of carrier-mediated transport phenomena, the following limiting situations can be considered:

a. the forward and reverse reactions are much faster than diffusion;
b. the complex diffusion is much faster than reaction;
c. the membrane contains an excess carrier, i.e., both solute and carrier concentrations are much less than that of the uncomplexed carrier.

In this section we shall analyze only the first case, which has been found to describe satisfactorily the behavior of a large number of biological systems.[32] Since the chemical reaction is much faster than diffusion, we can assume that at each point of the membrane thermodynamic equilibrium is established, so that

$$c_{AB} = K_1 c_A c_B \tag{28}$$

where K_1 is the equilibrium constant for the complex formation, c_{AB} is the concentration of the complex, c_A is the concentration of the solute, and c_B is the concentration of the carrier.

If we assume for simplicity that the diffusion coefficient of the solute–carrier complex is equal to that of the solute, the facilitated diffusive flux (J_f) takes the form

$$J_f = \frac{D_m K_P}{\delta} \frac{K_1 c_t (S_0 - S)}{(1 + K_1 K_P S_0)(1 + K_1 K_P S)} \tag{29}$$

where c_t is the total carrier concentration in the membrane.

The total flux of solute (J_t) is the sum of this term and that relative to the uncomplexed solute

$$J_t = \frac{D_m K_P}{\delta} \left[1 + \frac{K_1 c_t}{(1 + K_1 K_P S_0)(1 + K_1 K_P S)} \right] (S_0 - S) \tag{30}$$

or equivalently

$$J_t = P(1 + \alpha)(S_0 - S) \tag{31}$$

where

$$\alpha = \frac{K_1 c_t}{(1 + K_1 K_P S_0)(1 + K_1 K_P S)} \tag{32}$$

can be regarded as an enhancement factor by which the membrane permeability of the substrate is raised. In general the terms $K_1K_PS_0$ and K_1K_PS are much smaller than one. Therefore, Equation 32 becomes

$$\alpha = K_1 c_t \tag{33}$$

Under these conditions the enhancement factor is given by the product K_1c_t and this indicates that the solute permeability can be increased by acting either on the chemical nature of the carrier (K_1) or on its concentration (c_t).

At this point, if we wish to model the kinetic behavior of an engineered enzyme-loaded liposome in which substrate diffusion occurs via a facilitated diffusion mechanism, we need to make an assumption on the enzyme location. If we hypothesize that all the enzyme molecules are confined in the aqueous core, the elementary steps to be considered are solute diffusion in the membrane and chemical reaction in the solution containing the enzyme.

At steady state the flow-rate of the substrate through the lipid bilayer must be equal to the conversion rate

$$aP(1 + \alpha)(S_0 - S) = \frac{V_{max} S}{K_M + S} \tag{34}$$

This equation can be made dimensionless by defining the following dimensionless variable

$$\sigma = \frac{S}{K_M} \tag{35}$$

We obtain

$$\zeta(\sigma_0 - \sigma)\left[1 + \frac{\omega}{(1 + \chi\sigma_0)(1 + \chi\sigma)} \right] = \frac{\sigma}{1 + \sigma} \tag{36}$$

where ζ, ω, and χ are dimensionless parameters given by

$$\zeta = \frac{aPK_M}{V_{max}} \quad ; \qquad \omega = K_1 c_t \quad ; \qquad \chi = K_1 K_P K_M \tag{37}$$

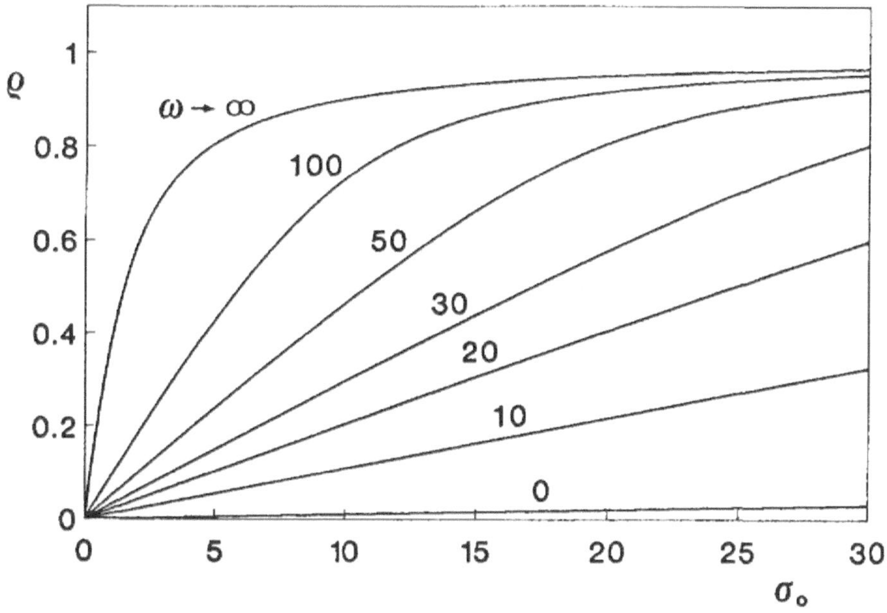

Figure 15 Dimensionless overall reaction rate (ρ) as a function of the dimensionless substrate concentration in the macroenvironment (σ_0) at different values of ω.

It should be noted that ζ is equal to the ratio between the characteristic reaction time for $S_0/K_M \ll 1$ and the characteristic permeation time for the uncomplexed solute, whereas ω and χ account for the concentration and the chemical nature of the carrier. It may be interesting to evaluate how these quantities affect the macroscopically observed reaction rates. Some plots of ρ against σ_0 at different values of ω are shown in Figure 15. As indicated by the figure, two limiting situations can be evidenced. The first one, relative to $\omega = 0$, corresponds to the absence of the carrier. In this case the diffusive flux of the substrate is very low and hence the reaction rates are extremely reduced. In the second case ($\omega \to \infty$) the facilitated flux is infinitely large so that the substrate concentration in the liposome interior is equal to that in the external solution. Under these conditions the measured rates are equal to those attainable with the free enzyme and therefore the observed kinetics reflect the intrinsic behavior of the enzyme. In general, the behavior of the entrapped enzyme lies within the limits determined by the two curves, even though above a given ω (close to 200 for the situation illustrated in Figure 15) the observed kinetics is practically coincident with that corresponding to $\omega \to \infty$.

The effects of the carrier on the diffusive flux have been simulated by considering a hypothetical liposome enclosing a homogeneous enzyme solution. Some results obtained at different carrier concentrations in the membrane are summarized in Table 3. It can be seen that the presence of the transporter yields a significant enhancement of the substrate flux across the liposomal membrane: using a carrier concentration 10^{-2} *M* a facilitated flux (J_f) 100 times greater than the ordinary one (J_o) is obtained. Another important point to be stressed is that the ratio J_f/J_o is independent of the substrate permeability, being equal to $K_1\ c_t$.

Table 3 Facilitated Diffusion in a Liposome Containing a Mobile Carrier

c_t (*M*)	J_t (μmol/cm² s)	J_o (μmol/cm² s)	J_f/J_o
0	1.00 10^{-8}	0	0
10^{-4}	2.00 10^{-8}	10^{-8}	1
10^{-3}	1.10 10^{-7}	10^{-7}	10
10^{-2}	1.01 10^{-6}	10^{-6}	100

Note: $S_0 = 1$ m*M*, $P = 10^{-8}$ cm/s, R = 1000 nm, $K_1 = 10^4\ M^{-1}$, $K_M/v_{max} = 2$ s.

E. TRANSIENT BEHAVIOR OF AN ENZYME-LOADED LIPOSOME

So far we have analyzed the kinetics of microencapsulated enzymes under steady-state conditions, i.e., when the behavior of the system is time independent. This assumption implies that the concentrations of the various species both outside and inside the liposome do not vary with time. Although such conditions are generally fulfilled, in some instances, like the starting of the reaction or when one or more products accumulate in the vesicle, the system behavior is typically nonstationary.

In this section we shall focus attention on the transient behavior of an enzyme-loaded liposome which is suddenly contacted with the substrate solution, with the aim to evaluate the time needed to reach the steady state. We shall assume that the substrate concentration in the external solution is constant (i.e., the volume of the solution surrounding the vesicle is sufficiently large) and that the liposome contains a homogeneous enzyme solution in its aqueous core.

The unsteady mass-balance equation for the substrate in the liposome yields

$$\frac{dS}{dt} = aP(S_0 - S) - \frac{V_{max}S}{K_M + S} \tag{38}$$

where S_0 is the external substrate concentration and S is the substrate concentration inside the vesicle at time t. The term dS/dt represents the substrate accumulation in the vesicle and goes to zero only when the system reaches a steady state. In such conditions Equations 38 and 24 are obviously coincident.

The integration of Equation 38 with the condition that at time t = 0 no substrate molecules are present

$$S = 0 \qquad at \qquad t = 0 \tag{39}$$

allows us to determine the trend of the internal substrate concentration with time.

If we define the following dimensionless variables

$$\sigma = \frac{S}{K_M} \quad ; \qquad \vartheta = \frac{V_{max}t}{K_M} \tag{40}$$

we obtain

$$\frac{d\sigma}{d\vartheta} = \zeta(\sigma_0 - \sigma) - \frac{\sigma}{1 + \sigma} \tag{41}$$

where ζ is the same dimensionless parameter defined by Equation 27.

The dimensionless differential equation can be solved once assigned the dimensionless initial condition

$$\sigma = 0 \qquad at \qquad \vartheta = 0 \tag{42}$$

A plot of σ/σ_∞ against ϑ at different ζ is shown in Figure 16; σ_∞ is the steady-state substrate concentration inside the vesicle, i.e., the concentration reached after an infinite time. As can be seen, all the curves exhibit a saturation trend, being $\sigma = \sigma_\infty$ a horizontal asymptote, but the rapidity with which they tend to this value is dependent on the parameter ζ. In particular, as ζ increases, the system approaches the steady-state concentration in less time. Two limiting situations can be evidenced. The first one, for $\zeta \to \infty$, corresponds to an internal substrate concentration instantaneously reached. The second one, for $\zeta \to 0$, gives rise to the lower curve plotted in Figure 16, on which the point representing the actual concentration moves. For $0 < \zeta < \infty$ the curves exhibit an intermediate behavior, with an initial slope increasing with ζ.

If the values of ϑ corresponding to $\sigma/\sigma_\infty = 0.99$ are plotted against ζ, the curve shown in Figure 17 is obtained. As can be seen, this curve displays a sigmoidal shape; moreover, for $\zeta < 0.1$ and $\zeta > 100$ the system behavior is no longer dependent on ζ.

It seems interesting to examine the physical meaning of a high ζ. If we consider the quantities appearing in this parameter we can observe that high values of ζ correspond to

- high substrate permeability
- low enzyme activity
- small vesicle size

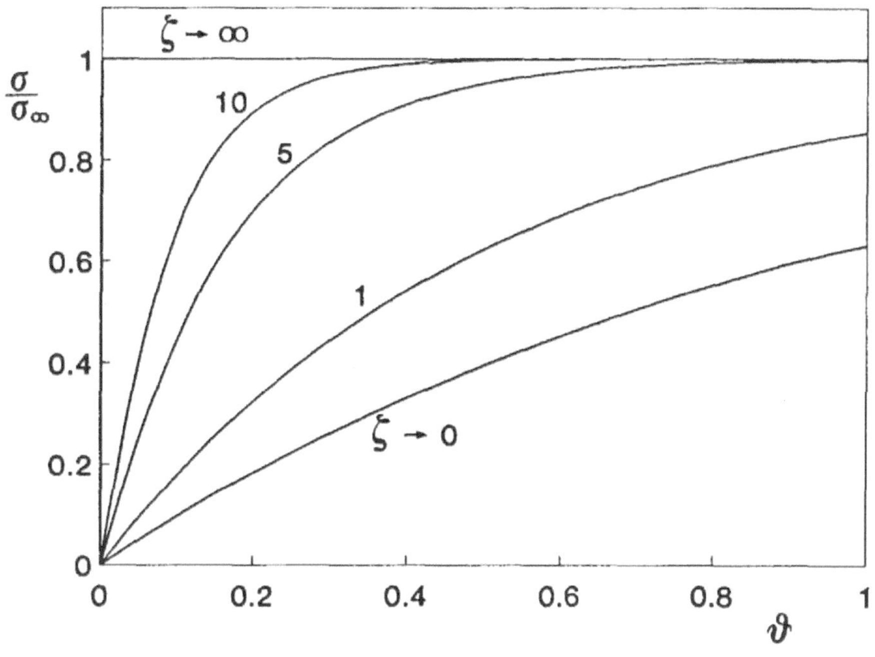

Figure 16 Transient behavior of an enzyme-loaded liposome: normalized substrate concentration (σ/σ_∞) as a function of the dimensionless time (ϑ) at different values of ζ.

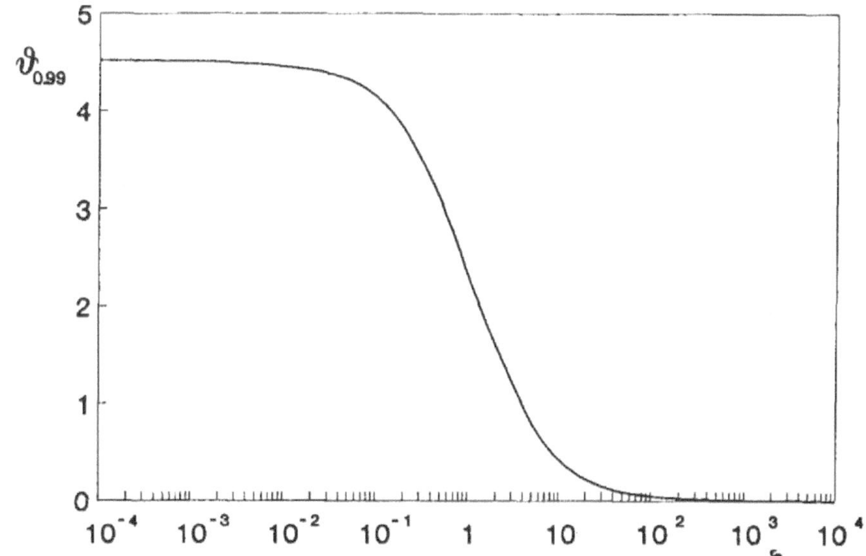

Figure 17 Effect of ζ on the dimensionless time ($\vartheta_{0.99}$) corresponding to $\sigma/\sigma_\infty = 0.99$.

Therefore, for a given enzymatic system, i.e., once assigned the Michaelis–Menten parameters and the substrate permeability, the value of ζ is determined solely by the liposome radius and hence the system approaches the steady state with a rapidity which is inversely related to the vesicle size. The effects of v_{max}/K_M, P, and R on the time needed to reach the 99% of the steady-state concentration are illustrated in Tables 4 and 5. It should be emphasized that, at a given liposome radius, this time increases with the ratio v_{max}/K_M and decreases as P increases. Furthermore, above a critical value of v_{max}/K_M or below a

Table 4 Time (s) Needed to Reach the 99% of the Steady-State Concentration: Effect of Enzyme Activity

	Liposome Radius (nm)		
v_{max}/K_M (s^{-1})	200	500	1000
10^{-4}	$9.73\ 10^{-6}$	$6.08\ 10^{-4}$	$2.43\ 10^{-4}$
10^{-3}	$2.92\ 10^{-5}$	$1.82\ 10^{-4}$	$7.30\ 10^{-4}$
10^{-2}	1.67	4.17	8.33
10^{-1}	1.31	1.31	1.31
1	1.37	1.37	1.37
10	6.77	6.77	6.77

Note: $P = 10^{-6}$ cm/s, $S_0 = 0.1\ K_M$.

Table 5 Time (s) Needed to Reach the 99% of the Steady-State Concentration: Effect of Substrate Permeability

	Liposome Radius (nm)		
P (cm/s)	200	500	1000
10^{-10}	458	458	459
10^{-8}	406	437	448
10^{-7}	190	298	364
10^{-6}	28.9	66.9	117
10^{-5}	3.0	7.5	14.9
10^{-4}	0.3	0.76	1.52
10^{-3}	0.03	0.08	0.15

Note: $v_{max}/K_M = 10^{-2}\ s^{-1}$, $S_0 = 0.1\ K_M$.

Table 6 Transient Behavior of Liposomes Containing Different Enzymatic Systems

Enzymatic System	Ref.	S_∞ (*M*)	$t_{0.5}$ (s)	$t_{0.9}$ (s)	$t_{0.99}$ (s)
Carbonic anhydrase — CO_2		$9.99\ 10^{-4}$	$5.75\ 10^{-5}$	$1.91\ 10^{-4}$	$3.82\ 10^{-4}$
$P = 2\ 10^{-1}$ cm/s	61				
$k/K_M = 1.18\ 10^7\ M^{-1}\ s^{-1}$	62				
Urease — urea		$2.58\ 10^{-4}$	0.77	2.59	5.24
$P = 4\ 10^{-6}$ cm/s	29				
$k/K_M = 7.64\ 10^6\ M^{-1}\ s^{-1}$	44				
Glucose oxidase — glucose		$1.29\ 10^{-4}$	$1.49\ 10^5$	$4.93\ 10^5$	$9.86\ 10^5$
$P = 1\ 10^{-11}$ cm/s	30				
$k/K_M = 40.6\ M^{-1}\ s^{-1}$	63				

Note: $S_0 = 1$ m*M*, R = 500 nm, $e_0 = 10^{-7}$ *M*.

critical value of P (0.1 s^{-1} and 10^{-10} cm/s, respectively, for the examples reported in Tables 4 and 5) the transient behavior appears to be independent of the vesicle size.

Table 6 shows the results of some simulations carried out on three enzymatic systems characterized by different substrate permeabilities. It is interesting to note that, although the values of the steady-state concentration in the internal solution are all comparable, the transient behavior is very different. In particular, the times needed to achieve a given substrate concentration (0.5, 0.9, and 0.99 S_∞ in the examples reported in Table 6) for liposomes containing carbonic anhydrase and glucose oxidase differ by about ten orders of magnitude. Urease-loaded vesicles exhibit an intermediate behavior. As can be seen, such differences reflect the high variability of permeabilities, which range from 10^{-11} cm/s for glucose to 2×10^{-1} cm/s for carbon dioxide.

These results appear very important and should be taken into account when the experimental work is planned. It is quite obvious, in fact, that when the system approaches the steady state at an extremely reduced rate, significant errors can be made if steady-state conditions are tacitly assumed.

VII. CONCLUSIONS

Although it is now well established that enzymes can be effectively enclosed into liposomes and that the resulting supramolecular structures exhibit very interesting catalytic properties, only limited attention has been thus far devoted to the modeling of their kinetic behavior.

From a fundamental point of view, modeling enzyme-loaded liposomes may be helpful for investigating some isolated aspects of a large number of biological phenomena mediated by biomembranes. On the other hand, the development of any biotechnological application based on enzyme-loaded liposomes needs a preliminary quantitative analysis of their catalytic behavior.

The recognition that a vesicle encapsulating one or more enzymes behaves as a microscopic bioreactor allows us to extend the classical theory of chemical reactors to such devices and to analyze their behavior by the usual mathematical relationships expressing the mass-balance equations and the kinetic laws. Notwithstanding the relevant complexity of the phenomena involved in the overall process, simplified models can be formulated accounting for the main physicochemical mechanisms by which the conversion of substrates into products is accomplished.

Theoretical models can be used as a predictive tool to optimize the molecular behavior of such miniaturized biodevices or to provide an efficient planning of the experimental work. Furthermore, extrapolations can be made for systems not yet investigated which enable us to estimate the overall reaction rate, the degree of substrate conversion, or other performance criteria of practical importance.

The analysis developed in the previous sections has evidenced the key role played by the interaction between transport phenomena and chemical kinetics as well as the risks of misinterpreting the experimental results when diffusional limitations are not taken into account. In this context it seems worth summarizing the most important aspects which have emerged:

- the behavior of an enzyme-loaded liposome as can be observed from macroscopic measurements reflects the intrinsic enzyme kinetics only when the process is not affected by mass-transfer limitations;
- in all other cases, i.e., when mass-transfer resistances are not negligible, the observed reaction rates are influenced by both the transport and catalytic properties of the system;
- the most adequate procedure to derive the true kinetic parameters from experimental measurements is to carry out experiments in the kinetically controlled regime;
- when this is not possible, diffusional effects should be separated from the observed kinetics by measuring the transport parameters in the absence of reaction or by estimating their values by semiempirical correlations.

If the experiments are performed under these conditions the analysis of the kinetic data allows us to estimate the intrinsic kinetic parameters of the microencapsulated enzyme, once the membrane permeability is known. These parameters can be compared with those relative to the free enzyme in order to evidence possible conformational changes resulting from immobilization. On the other hand, if the intrinsic kinetic parameters are known the experimental data can be used to derive the substrate permeability through the liposomal membrane. Moreover, some indications can be given of the enzyme location.

It is quite obvious that the models presented here could be significantly complicated in order to account for a number of phenomena actually neglected. For instance, they should include the effects of structural changes in the membrane or dimensional variations resulting from liposome aggregation or fusion. This upsurge of complexity leads to an increase of the number of parameters to be estimated from the experiments so that much more care should be devoted to the interpretation of the experimental results.

In any case it can be concluded that modeling the phenomena involved in enzyme-loaded liposomes may contribute to shed light on the behavior of membrane-enclosed living systems as well as to exploit the peculiar features of these vesicles as highly specific microreactors for a variety of technological applications. To date, industrial applications of enzyme-loaded liposomes have been poorly investigated, but it should be expected that this scenery will change in the near future with the appearance of liposome-based analytical devices and the development of novel processes in the drug and chemical industries.

APPENDIX: EVALUATION OF TRANSPORT COEFFICIENTS FOR A SPHERICAL LIPOSOME

In the rate equations describing the transport of a solute from the external solution to the aqueous core of a liposome three parameters are present, the diffusion coefficient (D), the membrane permeability (P), and the external mass-transfer coefficient (k_c), whose values depend on the physicochemical properties of the system. These coefficients must be determined, either from experimental measurements or by using suitable estimation methods, in the form of simplified theoretical models or semiempirical correlations. In this section some methods are presented for the evaluation of the diffusion and the mass-transfer coefficients.

Estimation of Diffusion Coefficient[64,65]

Diffusion coefficients in liquid systems cannot be determined with the same accuracy attainable for gases because of the far more complex behavior of liquids. Moreover, the diffusion coefficient varies significantly with solute concentration and sometimes it exhibits a maximum or a minimum. The diffusivity in concentrated solutions differs from that in dilute solutions because of changes in viscosity and in the degree of nonideality. The models currently used for predicting diffusion coefficients are relative to infinitely dilute systems, as are nearly all biological systems.

The common basis for estimating the diffusion coefficient is the Stokes–Einstein equation, derived by assuming a rigid solute sphere diffusing in a continuum of solvent

$$D = \frac{k_B T}{6\pi\mu R_0} \tag{A1}$$

where k_B is the Boltzmann's constant (equal to 1.38 10^{-23} J/K), T is the absolute temperature, μ is the solvent viscosity, and R_0 is the solute radius. This latter value can be estimated from other transport properties or, if available, from the collision diameter. Diffusion coefficients calculated with this equation are accurate to only about 20%. When the solute size is less than five times that of the solvent the equation fails, particularly in high-viscosity solvents.

Alternatives have been proposed to the Stokes–Einstein equation. One of the most general correlations available to date is the Wilke–Chang equation, which usually predicts diffusivities to better than 10 to 15%

$$D = 7.4\ 10^{-8}\ \frac{T(\varphi M)^{0.5}}{\mu v^{0.6}} \tag{A2}$$

where D is the diffusion coefficient in $cm^2\ s^{-1}$, M is the solvent molecular weight, v is the molar volume of the solute at its normal boiling point, and φ is an association factor for the solvent (ranging from 1 for unassociated solvents to 2.26 for water). The molar volume of the solute can be estimated, if necessary, from the contributions of the constituent atoms, for instance by the Le Bas method. For solvents different from water, if the association factor is in doubt, the Scheibel equation can be used.

Estimation of Mass-Transfer Coefficient[66,67]

The external mass-transfer coefficient is normally evaluated from empirical correlations among dimensionless groups. As indicated by dimensional analysis the characteristic numbers involved in mass-transfer are

Sherwood number : $$Sh = \frac{k_c L}{D} \tag{A3}$$

Reynolds number : $$Re = \frac{duL}{\mu} \tag{A4}$$

Schmidt number : $$Sc = \frac{\mu}{Dd} \tag{A5}$$

where L and u are the characteristic length and velocity, respectively, d is the liquid density, and μ is the viscosity of the solution.

For a fluid flowing around a single sphere the correlation takes the form

$$Sh = 2 + f(Re,Sc) \tag{A6}$$

In unstirred systems a limiting (minimum) value for Sh is given by

$$Sh = 2 \tag{A7}$$

The mass-transfer coefficient can be determined from this expression after assuming the liposome diameter as the characteristic length. We obtain

$$k_c = \frac{D}{R} \tag{A8}$$

where D is the diffusion coefficient for the solute in the external solution and R is the liposome radius. Of course, also the estimate so obtained for k_c represents a limiting value, being relative to laminar flow conditions.

If liposomes are contained in an agitated vessel the following correlation, valid for small gas bubbles or liquid droplets in stirred solutions, can be used

$$Sh = 0.13\, Re^{3/4}\, Sc^{1/3} \tag{A9}$$

When the power delivered by the impeller per unit volume of the vessel (W/V) is known, the above equation can be transformed as

$$\frac{k_c b}{D} = 0.13\left(\frac{b^4 d^2 (W/V)}{\mu^3}\right)^{1/4}\left(\frac{\mu}{Dd}\right)^{1/3} \tag{A10}$$

In this case the characteristic length in the Sherwood number is given by the impeller diameter (b). The mass-transfer coefficient resulting from the equation is

$$k_c = 0.13\left(\frac{W}{V}\right)^{1/4} d^{1/6} \mu^{-5/12} D^{2/3} \tag{A11}$$

NOMENCLATURE AND UNITS

a	Surface area	m^2/m^3
b	Impeller diameter	m
c	Solute concentration	mol/m^3
D	Diffusion coefficient	m^2/s
d	Liquid density	kg/m_3
e_0	Total enzyme concentration	mol/m^3
J	Diffusive flux	$mol/m^2\ s$
k	Rate constant	s^{-1}
k_B	Boltzmann's constant	J/K
k_c	Mass-transfer coefficient	m/s
K_I	Equilibrium constant for the complex formation	m^3/mol
K_M	Michaelis-Menten constant	mol/m^3
K_P	Partition coefficient	—
L	Characteristic length	m
M	Molecular weight	—
P	Membrane permeability	m/s
r	Reaction rate	$mol/m^3\ s$
R	Liposome radius	m
Re	Reynolds number (Equation A4)	—
S	Substrate concentration	mol/m^3
Sc	Schmidt number (Equation A5)	—
Sh	Sherwood number (Equation A3)	—
t	Time	s
T	Absolute temperature	K

u	Characteristic velocity	m/s
v	Solute molal volume	m^3/mol
v_{max}	Maximum reaction rate	mol/m^3 s
W	Impeller power	W
x	spatial coordinate	m
z	spatial coordinate	m

Greek Symbols

α	Enhancement factor (Equation 32)	—
δ	Membrane thickness	m
ζ	Dimensionless parameter (Equation 27)	—
ϑ	Dimensionless time (Equation 40)	—
μ	Liquid viscosity	Kg/m s
ξ	Dimensionless distance (Equation 20)	—
ρ	Dimensionless reaction rate (Equation 14)	—
σ	Dimensionless concentration (Equation 14)	—
τ	Characteristic time	s
φ	Association factor	—
Φ	Dimensionless parameter (Equation 23)	—
χ	Dimensionless parameter (Equation 37)	—
ψ	Dimensionless parameter (Equation 17)	—
ω	Dimensionless parameter (Equation 37)	—

Subscripts

A	Solute
B	Carrier
AB	Solute-carrier complex
e	External
f	Facilitated
i	Internal
m	Membrane
o	Ordinary
r	Reaction
t	Total
0	Macroenvironmental

REFERENCES

1. **Gennis, R. B.,** *Biomembranes: Molecular Structure and Function*, Springer-Verlag, New York, 1989, 1.
2. **Alberts, B., Bray, D., Lewis, J., Raff, M., Roberts, K., and Watson, J. D.,** *Molecular Biology of the Cell*, Garland Publishing, New York, 1983, chap. 7.
3. **Stryer, L.,** *Biochemistry*, W. H. Freeman and Company, New York, 1985, 315.
4. **Fendler, J. and Fendler, E. J.,** *Catalysis in Micellar and Macromolecular Systems*, Academic Press, New York, 1975, 1.
5. **Fendler, J.,** *Membrane Mimetic Chemistry*, John Wiley & Sons, New York, 1982, 1.
6. **Lasic, D. D.,** *Liposomes: from Physics to Applications*, Elsevier, Amsterdam, 1993, 173.
7. **Bangham, A. D., Hills, M. W., and Miller, N. G. A.,** Preparation and use of liposomes as models of biological membranes, in *Methods in Membrane Biology*, Korn, E. D., Plenum Press, New York, 1974.
8. **Sada, E., Katoh, S., Tarashima, M., and Kikuchi, Y.,** Reaction properties of hydrophobic enzymes and their immobilization on phospholipid composite membranes, *Biotechnol. Bioeng.*, 30, 117, 1987.
9. **Belfort, G.,** Membranes and bioreactors: a technical challenge in biotechnology, *Biotechnol. Bioeng.*, 33, 1047, 1989.
10. **Sessa, G. and Weissmann, G.,** Incorporation of lysozyme into liposomes, *J. Biol. Chem.*, 245, 3295, 1988.
11. **Dapergolas, G., Neerunjun, E. D., and Gregoriadis, G.,** Penetration of target areas in the rat by liposome-associated bleomycin, glucose oxidase and insulin, *FEBS Lett.*, 63, 235, 1976.
12. **Magee, W. E., Goff, C. W., Schoknecht, J., and Smith, M. D.,** The interaction of cationic liposomes containing entrapped horseradish peroxidase with cells in culture, *J. Cell. Biol.*, 63, 492, 1974.
13. **Gregoriadis, G., Putman, D., Louis, L., and Neerunjun, E. D.,** Comparative effects and fate of non-entrapped and liposome-entrapped neuraminidase injected into rats, *Biochem. J.*, 140, 323, 1974.

14. **Umezawa, F., Eto, Y., Tokoro, T., Ito, F., and Mackawa, K.,** Enzyme replacement with liposomes containing β-galactosidase from charonia lumpas in murine globoid cell leucodystrophy (Twitcher), *Biochem. Biophys. Res. Commun.*, 127, 663, 1985.
15. **Fishman, Y. and Citri, N.,** L-asparaginase entrapped in liposomes: preparation and properties, *FEBS Lett.*, 60, 17, 1975.
16. **Patel, H. M. and Ryman, B. E.,** Systemic and oral administration of liposomes, in *Liposomes: From Physical Structure to Therapeutic Applications*, Knight, C. G., Ed., Elsevier, Amsterdam, 1981, 417.
17. **Torchilin, V. P.,** Immobilised enzymes as drugs, *Adv. Drug Delivery Rev.*, 1, 41, 1987.
18. **Koide, K. and Karel, M.,** Encapsulation and stimulated release of enzymes using lecithin vesicles, *Int. J. Food Sci. Technol.*, 22, 707, 1987.
19. **El Soda, M., Pannell, L., and Olson, N.,** Microencapsulated enzyme systems for the acceleration of cheese ripening, *J. Microencaps.*, 6, 319, 1989.
20. **Kirby, C.,** Delivery systems for enzymes, *Chem. Britain*, 26, 847, 1990.
21. **Kirby, C.,** Microencapsulation and controlled delivery of food ingredients, *Food Sci. Tech.*, 5, 74, 1991.
22. **Rosenberg, M. F., Jones, M. N., and Vadgama, P. M.,** A liposomal enzyme electrode for measuring glucose, *Biochim. Biophys. Acta*, 1115, 157, 1991.
23. **Jones, M. A., Kilpatrick, P. K., and Carbonell, R. G.,** Preparation and characterization of bifunctional unilamellar vesicles for enhanced immunosorbent assays, *Biotechnol. Prog.*, 9, 242, 1993.
24. **Kimelberg, H. K. and Papahadjopoulos, D.,** Interactions of basic proteins with phospholipid membranes, *J. Biol. Chem.*, 46, 1142, 1971.
25. **Dombrose, F. A., Gitel, S. N., Zawalich, K., and Jackson, G. M.,** The association of bovine prothrombin fragment I with phospholipids, *J. Biol. Chem.*, 254, 5027, 1979.
26. **Adrian, G. and Huang, L.,** Entrapment of proteins in phosphatidylcholine vesicles, *Biochemistry*, 18, 5610, 1979.
27. **Philippot, J., Mutaftschiev, S., and Liautard, J. P.,** A very mild method allowing the encapsulation of very high amounts of macromolecules into very large (1000 nm) unilamellar liposomes, *Biochim. Biophys. Acta*, 734, 137, 1983.
28. **Orbach, E. and Finkelstein, A.,** The nonelectrolyte permeability of planar lipid bilayer membranes, *J. Gen. Physiol.*, 75, 427, 1980.
29. **Walter, A. and Gutknecht, J.,** Permeability of small nonelectrolytes through lipid bilayer membranes, *J. Membrane Biol.*, 90, 207, 1986.
30. **Cevc, G. and Marsh, D.,** *Phospholipid Bilayers: Physical Principles and Models*, John Wiley & Sons, New York, 1987, 190.
31. **Sada, E., Katoh, S., Tarashima, M., and Tsukiyama, K.-I.,** Entrapment of an ion-dependent enzyme into reverse-phase evaporation vesicles, *Biotechnol. Bioeng.*, 32, 826, 1988.
32. **Cussler, E. L.,** *Diffusion: Mass-transfer in Fluid Systems*, Cambridge University Press, Cambridge, 1984, 378.
33. **van Zanten, J. H. and Monbouquette, H. G.,** Biomimetic metal-sorbing vesicles: Cd^{2+} uptake by phosphatidylcholine vesicles doped with ionophore A23187, *Biotechnol. Prog.*, 8, 546, 1992.
34. **Walsh, A. J. and Monbouquette, H. G.,** Extraction of Cd^{2+} and Pb^{2+} from dilute aqueous solution using metal-sorbing vesicles in a hollow-fiber cartridge, *J. Membr. Sci.*, 84, 107, 1993.
35. **Sada, E., Katoh, S., Terashima, M., Kheirolomoom, A., and Sawai, H.,** The effect of liposomization on reaction characteristics of prostaglandin synthetase, *J. Chem. Eng. Jpn.*, 20, 531, 1987.
36. **Kheirolomoom, A., Katoh, S., Sada, E., and Yoshida, K.-I.,** Reaction characteristics and stability of a membrane-bound enzyme reconstituted in bilayers of liposomes, *Biotechnol. Bioeng.*, 37, 809, 1991.
37. **Engasser, J.-M. and Horvath, C.,** Diffusion and kinetics with immobilized enzymes, in *Applied Biochemistry and Bioengineering*, Wingard L. B., Katchalski-Katzir, E., and Goldstein, L., Eds., Academic Press, New York, 1976, 127.
38. **Bailey, J. E. and Ollis, D. F.,** *Biochemical Engineering Fundamentals*, McGraw-Hill, New York, 1986, 157.
39. **Atkinson, B.,** *Biochemical Reactors*, Pion Limited, London, 1974, 191.
40. **Chang, T. M. S.,** Stabilization of enzymes by microencapsulation with a concentrated protein solution or by microencapsulation followed by cross-linking with glutaraldehyde, *Biochem. Biophys. Res. Commun.*, 44, 1531, 1971.
41. **Chang, T. M. S.,** Microencapsulation of enzymes and biologicals, *Methods Enzymol.*, 44, 201, 1976.
42. **Chang, T. M. S.,** Artificial cells: biotechnology and medical applications, *Appl. Biochem. Biotechnol.*, 10, 5, 1984.
43. **Mogensen, A. O. and Vieth, W. R.,** Mass-transfer and biochemical reaction with semipermeable microcapsules, *Biotechnol. Bioeng.*, 15, 467, 1973.
44. **Sundaram, P. V.,** The kinetic properties of microencapsulated urease, *Biochim. Biophys. Acta*, 321, 319, 1973.
45. **Wadiak, D. T. and Carbonell, R. G.,** Kinetic behaviour of microencapsulated β-galactosidase, *Biotechnol. Bioeng.*, 17, 1157, 1975.
46. **Fink, D. J., Na, T.-Y., and Schultz, J. S.,** Effectiveness factor calculations for immobilized enzyme catalysts, *Biotechnol. Bioeng.*, 15, 879, 1973.
47. **Wadiak, D. T. and Carbonell, R. G.,** Effectiveness factor for substrate and product inhibition, *Biotechnol. Bioeng.*, 17, 1761, 1975.
48. **Dean, D. N., Fuchs, M. J., Schaffer, J. M., and Carbonell, R. G.,** Batch absorption of CO_2 by free and microencapsulated carbonic anhydrase, *Ind. Eng. Chem. Fundam.*, 16, 452, 1977.

49. **Mossa, G., Annesini, M. C., Di Giulio, A., Dini, L., and Finazzi-Agrò, A.,** Liposomes as bioreactors: transport phenomena in proteoliposomes, in *Biological and Synthetic Membranes*, Butterfield, D. A., Ed., Alan Liss, New York, 1989, 292.
50. **Madeira, V. M. C.,** Incorporation of urease into liposomes, *Biochim. Biophys. Acta*, 499, 202, 1977.
51. **Annesini, M. C., Di Giulio, A., Di Marzio, L., Finazzi-Agrò, A., and Mossa, G.,** Diffusion and catalysis processes in unilamellar liposomes, *J. Liposome Res.*, 2, 455, 1992.
52. **Mossa, G., Di Giulio, A., Dini, L., and Finazzi-Agrò, A.,** Interaction of DPPC-cholesterol vesicles with ascorbate oxidase, *Biochim. Biophys. Acta*, 986, 310, 1989.
53. **Annesini, M. C., Di Giorgio, L., Di Marzio, L., Finazzi-Agrò, A., Serafino, A. L., and Mossa, G.,** Carbon dioxide hydration with liposomes entrapping carbonic anhydrase, *J. Liposome Res.*, 3, 639, 1993.
54. **Annesini, M. C., Di Marzio, L., Finazzi-Agrò, A., Serafino, A. L., and Mossa, G.,** Interaction of cationic phospholipid vesicles with carbonic anhydrase, *Biochem. Mol. Biol. Int.*, 32, 87, 1994.
55. **Cioci, F. and Lavecchia, R.,** Urease-loaded liposomes as detoxifying microreactors: effect of ammonia accumulation on enzyme kinetics, *J. Liposome Res.*, 3, 725, 1993.
56. **Stein, W. D.,** *Transport and Diffusion Across Cell Membranes*, Academic Press, New York, 1986, 69.
57. **Lieb, W. R. and Stein, W. D.,** Biological membranes behave as non-porous polymeric sheets with respect to the diffusion of non-electrolytes, *Nature*, 224, 240, 1969.
58. **Cussler, E. L.,** *Diffusion: Mass-Transfer in Fluid Systems*, Cambridge University Press, Cambridge, 1984, 116.
59. **Finean, J. B., Coleman, R., and Michell, R. H.,** *Membranes and Their Cellular Functions*, Blackwell, London, 1978, 42.
60. **Wilson, D. B.,** Cellular transport mechanisms, *Ann. Rev. Biochem.*, 47, 933, 1978.
61. **Gutknecht, J., Bisson, M. A., and Tosteson, F. C.,** Diffusion of carbon dioxide through lipid bilayer membranes, *J. Gen. Physiol.*, 69, 779, 1977.
62. **Sanyal, G.,** Comparative carbon dioxide hydration kinetics and inhibition of carbonic anhydrase isoenzymes in vertebrates, in *Biology and Chemistry of the Carbonic Anhydrases*, Tashian, R. E. and Hewett-Emmett, D., Eds., The New York Academy of Sciences, New York, 1984, 165.
63. **Gibson, Q. H., Swoboda, B. E. P., and Massey, V.,** Kinetics and mechanism of action of glucose oxidase, *J. Biol. Chem.*, 11, 239, 1964.
64. **Reid, R. C., Prausnitz, J. M., and Sherwood, T. K.,** *The Properties of Gases and Liquids*, McGraw-Hill, New York, 1977, 566.
65. **Brodkey, R. S. and Hershey, H. C.,** *Transport Phenomena: A Unified Approach*, McGraw-Hill, New York, 1988, 738.
66. **Treybal, R. E.,** *Mass-Transfer Operations*, McGraw-Hill, New York, 1980, 72.
67. **Fair, J. R., Steinmeyer, D. E., Penney, W. R., and Crocker, B. B.,** Liquid-gas systems, in *Perry's Chemical Engineer's Handbook*, Green, D. W. and Maloney, J. O., Eds., McGraw-Hill, New York, 1984, chap. 18.

Chapter 17

Biotransformations in a Liposomal Medium

Raphael Bar

CONTENTS

I. INTRODUCTION

Biotransformations refer primarily to chemical modifications of organic substrates, catalyzed either by growing or resting cells or by an enzyme. Common substrates of these bioconversions are steroids and sterols which serve as raw material for pharmaceutically important drugs.[1] In bioconversion technology one commonly strives at attaining the goals of faster kinetics, higher product concentration, thus allowing for low-volume bioreactors and higher substrate conversion, and thus lowering cost of product separation and purification. However, the efficiency of microbe- or enzyme-mediated transformations is often seriously impeded by two major obstacles:

a. *Poor accessibility of lipophilic substrates (liquids or solids) to the biocatalyst.* This is expressed not only by a poor aqueous solubility of the hydrophobic compound such as that of a steroid, but also by a slow dissolution rate of organic liquid or solid into the aqueous environment of the biocatalyst.
b. *Toxic or inhibitory effects of substrates exerted upon the microbe or enzyme.* This second obstacle stems from either the molecular toxicity exerted by dissolved substrate molecules or the so-called[2] "phase toxicity" exerted by a separate organic liquid substrate phase in contact with cells. The latter is a result of extraction of cell wall and membrane components by the organic liquid, thus causing cell disruption and inactivation of the biocatalytic activity.

Attempts to alleviate or overcome these two obstacles include the use of surfactants such as Tween 80, water-miscible solvents such as dimethylsulfoxide to form a one-liquid phase system, or water-immiscible solvents such as alkanes to form a water-organic solvent two-liquid phase system.[3] However, toxic or inhibitory effects introduced by these organic solvents and surfactants may again partly or fully inactivate the biocatalyst. Another attempt consists of employing water-soluble, clathrate-forming cyclodextrins in the bioconversions of toxic- or water-insoluble substrates.[4] This chapter discusses the use of liposome-forming diacyl phospholipids as carriers of sparingly soluble hydrophobic compounds to be transformed by either an enzyme or a whole cell. It is not, however, meant to be an exhaustive review of all enzyme- or cell-mediated reactions involving liposomes, most of which may be viewed as medically related. Rather, it is aimed at highlighting, while referring to relevant studies, the features of liposomes as a biotransformation medium and the major factors governing the performance of biotransformations in such a medium.

0-8493-4012-8/96/$0.00+$.50
© 1996 by CRC Press, Inc.

II. ADVANTAGES OF LIPOSOMES IN A BOTRANSFORMATION MEDIUM

The liposomes confer onto the bioconversion medium the following advantages:

a. *Solubility enhancement.* Incorporation of organic compounds in the hydrophobic bilayers of liposomes increases their effective solubility. For instance, cholesterol, which has a maximum aqueous solubility of 1.8 mg/l may be solubilized at a concentration of 20 g/l in liposome preparations.[5] The liposomes serve as a well-dispersed pool of organic substrates in a dynamic exchange equilibrium with the aqueous environment where the biocatalyst resides.
b. *Biocompatibility.* Phospholipids which are also the main constituents of natural cellular membranes are generally held to be nontoxic. Thus, the liposomal medium appears to provide the cell or enzyme a pseudo-biological or a biomimetic environment.

Interestingly, the liposomal medium appears to combine features characteristic of the three aforementioned hydro-organic systems employed for biotransformation. The liposome is made of amphiphiles like the surfactant. Liposomal media are visibly homogenous at the macroscopic level like water-cosolvent systems, yet heterogeneous at the microscopic level like two-liquid phase systems. However, it is the combination of biocompatibilty and high solubilization capacity which makes liposomal media worthwhile for biotransformations.

III. INTEGRITY OF LIPOSOMES IN A BIOTRANSFORMATION MEDIUM

For the previously described advantages to be exploited, it is imperative to maintain the chemical and structural integrity of the liposomes throughout the biotransformation process, particularly when the biocatalyst is microbial cells. Phospholipids may be degraded by lipases or phospholipases borne or secreted by many bacteria,[6] thus generating intermediates such as lysolecithin and fatty acids. The latter hydrolysis products could transform the liposomal medium to an undefined mixture of liposomes and mixed micelles. Furthermore, the degradation products may be utilized by these bacteria as sources of carbon, nitrogen, and phosphorus.[6] Thus, in every application of liposomes involving whole cells, one should:

a. Test for the inherent ability of the cells to hydrolyze phospholipids and/or egg yolk and to grow on them as a sole carbon and energy source. Such a test was performed for instance, by Goetschel and Bar[7] who demonstrated that *Arthrobacter simplex* did not hydrolyze an egg yolk–agar medium prior to using it as a biocatalyst in a liposomal medium. If, however, the microbe turns out to be lipolytic, its cultivation in a lipid-free medium may repress the formation of inducible lipases.
b. Monitor closely the concentration of phospholipids throughout the process. For instance, Goetschel et al.,[8] who investigated the microbial oxidation of liposome-borne cholesterol, measured quantitatively the amount of PC present during the bioprocess. Figure 1 shows that after the biotransformation was almost complete, the PC content gradually dropped to 50% of the initial value. As these authors could not detect intermediate degradation products such as lysolecithin, phosphatidic acid, or diglycerides, they concluded that *Rhodococcus* cells utilized PC upon prolonged incubation.

Thus, with the chemical and structural integrity of the liposomes preserved, one can perform enzyme or mediated conversions.

IV. ENZYMIC TRANSFORMATIONS

These transformations were performed with liposomes containing either the organic substrate or the enzyme. In the former case, a sparingly soluble compound is incorporated in the bilayers while the enzyme — usually a hydrophilic one — is freely dissolved in the aqueous milieu, i.e., the trapped volume and/or extravesicular aqueous medium. In the latter case, the reverse holds true for a water-soluble compound and a hydrophobic, often reconstituted, enzyme.

A. BIOTRANSFORMATIONS OF LIPOSOME-BORNE SUBSTRATES

A number of these studies[9–11] were concerned with the oxidation of liposome-borne cholesterol (CL) to cholest-4-en-3-one by cholesterol oxidase dissolved in the aqueous medium. The main objective of these studies was to elucidate the mechanism and the kinetics of the transbilayer exchange of the sterol as well as its exchange between liposomes and to this purpose, the biotransformation of cholesterol served in fact

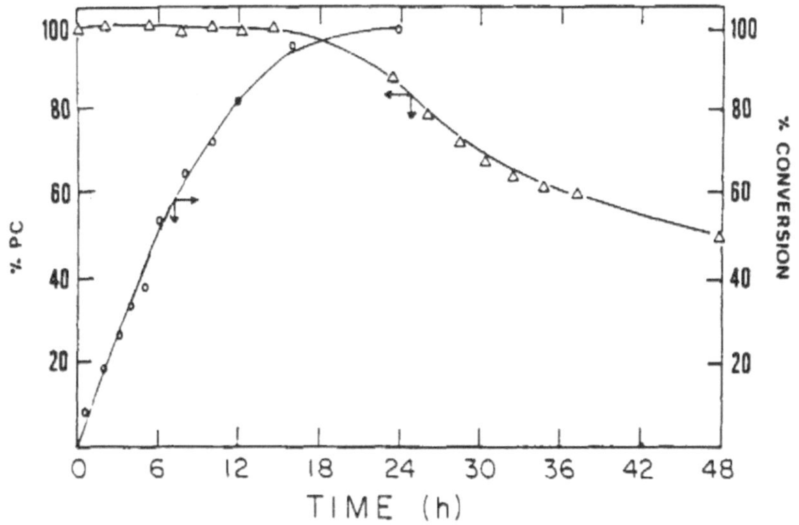

Figure 1 Remaining phosphatidylcholine (Δ) during the oxidation of 5 g/l cholesterol (O) incorporated in SUVs (PC:CL::1:1) by cells of *Rhodococcus erythropolis* (10 g dry weight/l).[8]

as an enzymatic probe. Nevertheless, these studies are helpful for understanding the performance of liposomes as a site of biotransformation.

Moore et al.[9] reported that cholesterol oxidase could not oxidize more than 5% of cholesterol incorporated in small unilamellar vesicles (SUVs) made up of egg phosphatidylcholine (molar ratio Cl:PC::1:3). However, a prior or simultaneous treatment of the vesicles with phospholipase C resulted in more than 90% sterol oxidation. This enzyme hydrolyzed away the phospholipid polar head groups and consequently rendered the sterol β–OH exposed and susceptible to attack by the oxidase. In contrast, Backer and Dawidowicz[10] found cholesterol oxidase capable, despite the proven absence of phospholipase C activity, of oxidizing more than 90% of cholesterol incorporated in egg lecithin SUVs at a molar PC:CL ratio of 1:0.7. Since CL in the SUVs is present in both sides of the bilayer and since only CL in the outer vesicle surface is oxidized, the authors utilized the kinetics of the biotransformation to set up an upper limit of 1 min on the half-life time of the transbilayer CL exchange. The discrepancy with regards to the role of phospholipase C was ultimately settled by Pal et al.,[11] who reported that CL in PC vesicles becomes accessible to CL oxidase when its concentration exceeds 42 to 46 mol%. Below these concentrations, the β–OH of CL is not exposed to the oxidase unless the vesicles are subjected to phospholipase C. In any event, all these studies definitely point out the "local substrate concentration" within the liposome, i.e., the phospholipid:substrate ratio, as an important factor to be well considered in the design of biotransformations in a liposomal medium.

B. BIOTRANSFORMATIONS BY A LIPOSOME-BORNE ENZYME

Reactions catalyzed by a hydrophobic enzyme incorporated in a vesicle bilayer are certainly reminiscent of biological membranes where membrane-bound (hydrophobic) enzymes catalyze single or well-organized chain reactions. The studies by Sada et al.[12] and Kheirolomoon et al.[13] present illustrative examples of the behavior of hydrophobic enzymes borne in small unilamellar and multilamellar vesicles.

Sada et al.[12] measured the reaction rates at various temperatures of the hydrophobic enzymes, aminopeptidase, and alkaline phosphatase, in microsomes as well as in SUVs and MLVs in view of investigating the effect of the bilayer fluidity on the enzymatic activities. The substrates of the enzymes were correspondingly L-leucine-*p*-nitroanilide and *p*-nitrophenylphosphate. The activities of the two enzymes in either microsomes or vesicles were found to be affected by the phase transition of the phospholipid. Figure 2 shows an example of such effect expressed in an Arrhenius plot of alkaline phosphatase in SUVs made up of DMPC, and it is seen that the rate curve undergoes a change of slope at the phase transition temperature (T_c) of the phospholipid. The authors concluded that the fluidity change of the phospholipid affects the substrate diffusion in the system of aminopeptidase-borne MLVs, while it affects both the enzyme activity and the substrate diffusion in the system of the phosphatase-borne MLVs.

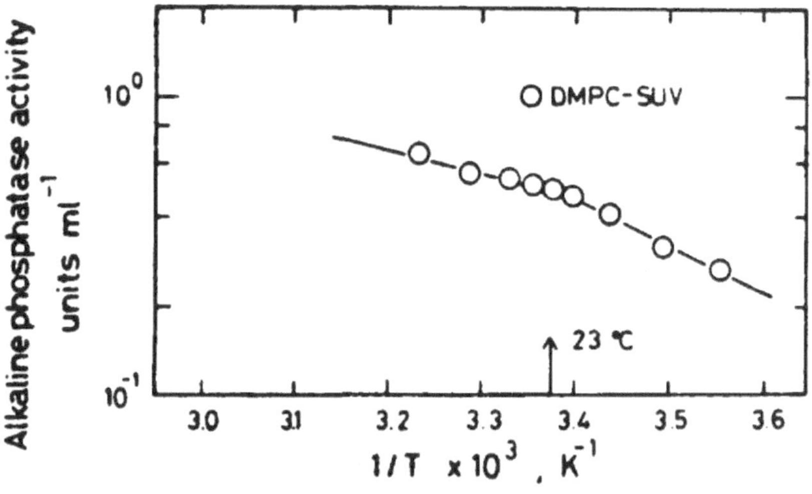

Figure 2 Arrhenius plot of alkaline phosphatase incorporated in DMPC SUVs.[12] The substrate was *p*-nitrophenylphosphate.

The enzyme sarcosine dehydrogenase has also been a subject of an extensive investigation by Kheirolomoon et al.[13] When comparing the free enzyme with 0.006% Triton X-100 and reconstituted enzyme in bilayers of SUVs made of DMPC, a 25-fold enhanced activity was reported, which the authors attributed to a favorable conformational change of the enzyme. Inclusion of the negatively charged DMPA and DMPG in those SUVs resulted in an enhanced stability of the enzyme. Similarly to the previous study, the enzymatic activity appeared to be sensitive to the physical state of the vesicle phospholipids as suggested by the break point in the Arrhenius plot of the reconstituted enzyme. Both studies indicate that the system temperature with respect to T_c is another factor which influences the kinetics of the biotransformation in the liposomal medium.

V. MICROBIAL TRANSFORMATIONS OF LIPOSOME-BORNE SUBSTRATES

Unlike the enzymic systems, microbial cells are obviously too large to be embedded in the bilayers of SUVs and therefore, these microbial systems are concerned primarily with transformation of liposome-borne hydrophobic substrates, catalyzed by cells suspended in the extravesicular aqueous medium. Two major studies involving the biotransformation of a steroid[7] and a sterol[8] were performed in these systems.

Goetschel and Bar[7] investigated the dehydrogenation of hydrocortisone to prednisolone by resting cells of *Arthrobacter simplex* in a liposomal medium made up of MLVs of either pure or crude egg phospholipids. It is important to note that both steroids have finite, albeit small, aqueous solubilities of 0.45 and 0.29, respectively,[7] and consequently, the observed biotransformation includes also a contribution from the aqueous dissolved substrate. Yet, the liposomal medium was found to promote a faster biotransformation and to a higher conversion than those obtained in an aqueous steroid dispersion. The performance of the liposomal medium was also superior to those of water-surfactant and water-cosolvent systems. Figure 3 compares bioconversion profiles in the liposomal medium and in aqueous slurries with and without two surfactants, Tween 80, and polyethylene glycol. Both surfactants had a marginal effect on the bioconversion of 2 g/l hydrocortisone while a marked enhancement took place in the liposomal medium, and this enhancement was progressively accentuated as the phospholipid loading increased from 2.5 to 20 g/l.

A more extensive study was subsequently performed by Goetschel et al.[8] utilizing cholesterol borne in MLVs or SUVs. Cholesterol, unlike the previous steroid, is extremely insoluble in water and more easily incorporated in phospholipid vesicles at substantially higher loadings.[8] Consequently, it constitutes a better example of a hydrophobic substrate for biotransformation in a liposomal medium. The authors investigated the oxidation of liposome-borne cholesterol to cholest-4-en-3-one by resting cells of *Rhodococcus erythropolis*. While the microbial oxidation of the sterol in MLVs did not take place to any significant extent, that of cholesterol borne in SUVs proceeded to completion. The virtual failure of the

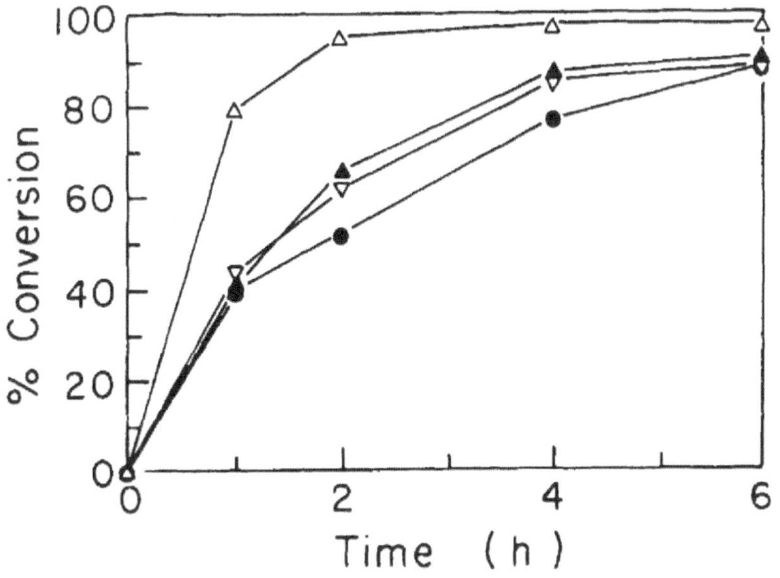

Figure 3 Biotransformation profiles[7] of hydrocortisone (2 g/l) in a liposomal medium (△, 20 g/l pure egg-PC per liter) and aqueous media without (▽) and with Tween 80 (▲, 0.1 g/l) and polyethylene glycol (●, 1 g/l). The concentration of *A. simplex* cells was 4g of DW per liter with a specific activity of 66.7 U/g DW.

bioconversion in MLVs may be related to the fact that the majority of cholesterol is "buried" within the vesicles, thus significantly hindering its accessibility to the biocatalyst. In SUVs, however, the sterol molecules in the external lamella are easily accessible and since they are in a quite rapid exchange equilibrium with those in the internal lamella,[10] all cholesterol is available for oxidation. Figure 4a shows that a liposomal medium composed of 9.82 g PC per liter and 5.0 g cholesterol per liter in the form of SUVs with a mean diameter of 40 nm clearly promoted the biotransformation at a faster rate to a considerably higher conversion than those achieved in a fine aqueous dispersion of the substrate. In the latter system, the conversion was only 33% after 24 h while that in the liposomal medium was 98% after only 16 h. Figure 4b further shows that this enhancement decreased as the PC:CL molar ratio increased for a constant CL concentration of 5.0 g/l. Interestingly, rendering SUVs negatively charged by incorporating POPG lowered the bioconversion rate.

The studies on steroid transformations point out at the overall phospholipid content in the system, the phospholipid:substrate ratio, and the electrical vesicle charge as important factors affecting the efficiency of the liposomal medium.

VI. MECHANISMS OF BIOTRANSFORMATION IN A LIPOSOMAL MEDIUM

The oxidation of liposome-borne cholesterol has equally drawn attention to the mechanism of the biotransformation. In fact, two mechanisms can be envisaged for participation of the substrate-bearing liposomes in the biotransformation process.

In the first mechanism, the liposomes could serve as a mere reservoir of substrate molecules which dissolve into the aqueous medium. These molecules then become available to the biocatalyst in the aqueous medium and are converted to the product molecules which get incorporated back into the liposomes. The movement of cholesterol between SUVs is known to take place with a half-life time of about 1.5 to 3.0 h[14] while the very interaction between the dissolved substrate and the biocatalyst is identical with that prevailing in the liposome-free medium. In this mechanism, an enhancement effect in the biotransformation is attributed to a better substrate accessibility rate of the hydrophobic substrate (*vide supra*).

In the second mechanism, the liposomes interact directly with the biocatalyst. In the studies on CL oxidation by soluble enzyme,[9,10] it was indeed concluded that the enzyme acts directly on the vesicle. The study on CL oxidation by microbial cells[8] is suggestive of an adhesion of *Rhodococcus* cells to SUVs

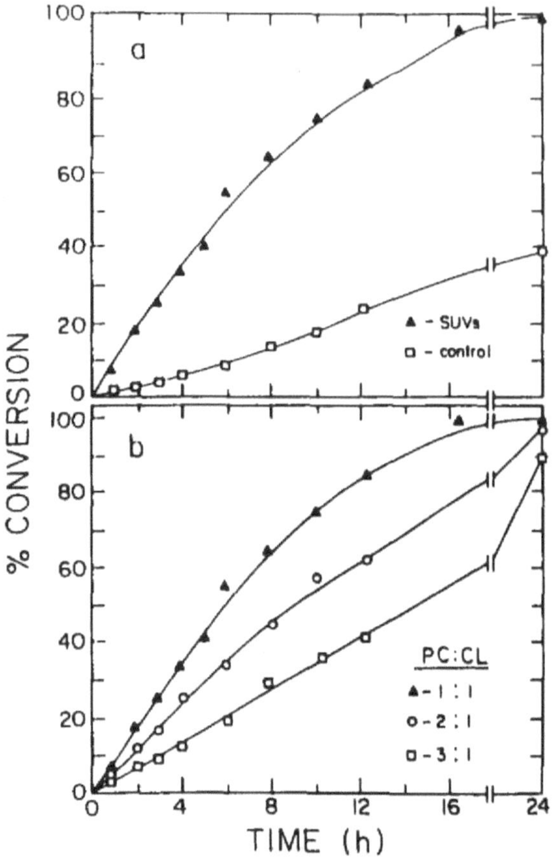

Figure 4 Oxidation of 5 g/l cholesterol (CL) incorporated in PC-SUVs by cells of *R. erythropolis*[8] (10 g of DW per liter). (A) Comparison with microbial oxidation of 5 g/l cholesterol dispersed in an aqueous medium (control). (B) Effect of the molar PC:CL ratio on the microbial oxidation.

following which the sterol would undergo a partitioning equilibrium between the vesicle and the microbial membrane. The sterol would then be converted by the membrane-bound CL oxidase and finally, the product would also undergo an exchange equilibrium with the vesicle. The interaction between the negatively charged bacterial cells and the electrically charged POPG-containing vesicles is likely to be diminished, thus leading to a diminished enhancement of the bioconversion as was indeed observed by the investigators.[8]

In the biotransformations of liposome-borne substrates, both mechanisms most likely take place simultaneously and the observed enhancement is a result of a better solubilization as well as of a favorable interaction between the biocatalyst and the substrate.

VII. CONCLUDING REMARKS

The studies discussed appear to highlight the following factors as determining the performance of the liposomal medium with respect to biotransformations:

- biodegradability of phospholipids
- phospholipid content
- type and size of vesicles
- temperature above or below T_c
- phospholipid:substrate molar ratio
- electrical charge of the vesicles

However, if liposomes are to be employed in bioconversion technology, one should also consider the technology of liposome preparation as well as phospholipid recycling. Among the three dominant forms of liposomes, MLVs, SUVs, and LUVs, the latter require a tedious and rather delicate method of preparation. In contrast, MLVs are the easiest to prepare and a large-scale preparation of SUVs in commercially available high-pressure homogenizers is already practiced.[15] A reasonable methodology would be one in which a concentrated liposome preparation of the substrate is first made in a commercial apparatus and subsequently added to a suitable bioconversion medium. The cost of phospholipds, pure or crude, certainly constitutes a decisive economic factor in biotransformations carried out in a liposomal medium. Therefore, the possibility of product recovery with a simultaneous recovery of the phospholipids, will determine the cost effectiveness of the liposomal medium with regard to the other surfactant-water or water-cosolvent systems for biotransformations.

NOTATIONS

DMPC — L-α-dimyristoylphosphatidylcholine
DMPG — DL-α-dimyristoylphosphatidylglycerol
DMPA — L-α-dimyristoylphosphatidic acid
POPG — palmitoyloleylphosphatidylglycerol
PC — phosphatidylcholine
CL — cholesterol
DW — dry weight

REFERENCES

1. **Sebek, O. K. and Perlman, D.,** Microbial transformations of steroids and sterols in microbial technology, in *Microbial Processes,* Vol. 1, Peppler, H. J. and Perlman, D., Eds., Academic Press, New York, 1979, 483.
2. **Bar R.,** Phase toxicity in a water-organic solvent two-liquid phase microbial system, in *Biocatalysis in Organic Media,* Laane, C., Tramper, J., and Lilly, M. D., Eds., Elsevier, Amsterdam, 1987, 147.
3. **Bar R.,** Liposomal solution for microbial catalysis in organic solvents?, *Trends Biotechnol.,* 6, 1, 1988.
4. **Bar R.,** Cyclodextrin-aided bioconversions and fermentations, *Trends Biotechnol.,* 7, 2, 1989.
5. **Collins, J. J. and Phillips, M. C.,** The stability and structure of cholesterol-rich codispersions of cholesterol and phosphatidylcholine, *J. Lipid Res.,* 23, 291, 1982.
6. **Mackie, R. I., White, B. A., and Bryant, M. P.,** Lipid metabolism in anarobic ecosystems, *Crit. Rev. Microbiol.,* 17, 449, 1991.
7. **Goetschel, R. and Bar, R.,** Dehydrogenation of hydrocortisone by *Arthrobacter simplex* in a liposomal medium, *Enz. Microb. Technol.,* 13, 245, 1991.
8. **Goetschel, R., Barenholz, Y., and Bar, R.,** Microbial transformations in a liposomal medium. II. Cholesterol oxidation by *Rhodococcus erythropolis, Enz. Microb. Technol.,* 14, 390, 1992.
9. **Moore, N. M., Patzer, E. J., Barenholz, Y., and Wagner, R. R.,** Effect of phospholipase C and cholesterol oxidase on membrane integrity, microviscosity and infectivity of vesicular stomatitis, *Biochemistry,* 16, 4708, 1977.
10. **Backer, J. M. and Dawidowicz, E. A.,** Transmembrane movement of cholesterol in small unilamellar vesicles detected by cholesterol oxidase, *J. Biol. Chem.,* 256, 586, 1981.
11. **Pal, R., Barenholz, Y., and Wagner, R. R.,** Effect of cholesterol concentration on organization of viral and vesicle membranes, *J. Biol. Chem.,* 255, 5802, 1980.
12. **Sada, E., Katoh, S., Terashima, M., and Kikuchi, Y.,** Reaction properties of hydrophobic enzymes and their immobilization on phospholipid composite membranes, *Bioeng. Biotechnol.,* 30, 117, 1987.
13. **Kheirolomoon, A., Katoh, S., Sada, E., and Yoshida, K. I.,** Reaction characteristics and stability of a membrane-bound enzyme reconstituted in bilayers of liposomes, *Bioeng. Biotechnol.,* 37, 809, 1991.
14. **Bar, L. K., Barenholz, Y., and Thompson, T. E.,** The fraction of cholesterol undergoing spontaneous exchange between small unilamellar phosphatidylcholine vesicles, *Biochemistry,* 25, 6701, 1986.
15. **Brandl, M., Bachmann, D., Drechsler, M., and Bauer, K. H.,** Liposome preparation by a new high pressure homogenizer Gaulin Micron lab, *Drug Dev. Ind. Pharm.,* 16, 2167, 1990.

Chapter 18

Applications of Magnetoproteoliposomes in Bioreactors Operating in High-Gradient Magnetic Fields

Marcel De Cuyper

CONTENTS

I. INTRODUCTION

A computer search[1] of the literature from 1967 to 1976 using the keywords "enzyme(s)" and "immobilization" recorded 658 hits, whereas the same search just 10 years later (1977 to 1986) scored 2538 hits. This publication bonanza has of course inspired many review article and handbook writers. Concomittantly, the unique catalytic properties exhibited by enzymes have been harnessed in many industrial applications, particularly in the design of enzyme bioreactors.[2] A close look at the type of enzyme used, however, reveals that most applications deal with so-called water-soluble enzymes, although numerous membrane-associated enzymes can also be very useful in many biotechnological processes.[3-9]

Most purification and immobilization protocols used for water-soluble biocatalysts, however, are, in general, not applicable to enzymes which *in vivo* are intimately associated with membranes. Membrane-bound enzymes indeed expose highly hydrophobic regions on their surface so that they tend to aggregate continuously during the consecutive purification and immobilization steps, often resulting in irreversible loss of enzyme activity. In addition, for an acceptable degree of activity, membrane-bound enzymes must be surrounded by a lipidic environment.[10-12]

Until now, industrial processes aimed at exploiting the catalytic potential of intrinsic membrane enzymes could only circumvent these problems by using whole cells or cell fragments. A main drawback, often encountered in using whole cells, is product contamination caused by unwanted side-reactions (e.g., by further metabolism of the desired product), due to the presence of cell fragments, or of nutrients, required to sustain the viability of the cells. In addition, the performance of immobilized cell preparations may be adversely affected by the action of intracellular proteases, by diffusional limitations, etc.

To bypass these difficulties, it is possible to envisage the immobilization of purified membrane enzymes in artificial phospholipid bilayers, such as vesicles or liposomes. However, for use in a bioreactor, the resulting proteovesicles or proteoliposomes have to meet certain special requirements. In a batch type bioreactor, for instance, the structures have to be removed repeatedly from the reaction

0-8493-4012-8/96/$0.00+$.50
© 1996 by CRC Press, Inc.

mixture in a quick and quantitative way, whereas in a continuous-flow, fixed-bed bioreactor, the enzyme–lipid complexes must be immobilized efficiently and, moreover, must withstand the high shear forces which can be involved. Two original and elegant ways to immobilize phospholipid vesicles have been recently developed by the Uppsala group of P. Lundahl.[13-15] In the first approach (the so-called "immobilization by hydrophobic interaction"), preformed vesicles were attached to the pore-walls of gel beads, which were modified by hydrophobic ligands. In the second approach (the so-called "steric immobilization" method), the authors gradually created liposomal structures in gel beads, whereupon those liposomes which attained a sufficiently large size became sterically trapped in the network of the gel matrix.

Due to the fragility of the immobilized vesicles (e.g., as a result of osmotic forces), it will often be advantageous to attach the lipids directly to a solid support, preferably by strong chemisorption or covalent forces. For instance, in the work of Rothe and Aurich,[7] a true bilayer was built up in two distinct steps: first, the headgroup of phosphatidylethanolamine molecules was covalently coupled to reactive polymer beads (resulting in monolayer-coated particles), on which, after immersion in an aqueous lipid suspension, a second layer of phospholipids was slowly deposited. The resulting structures were found to be excellent hosts for membrane enzymes.

In our approach, we have tackled the various aspects of this complex, but highly relevant, area in a different way. We start with a magnetizable solid support, consisting of magnetite (Fe_3O_4). The size of the iron oxide particles is in the lower nanometer range, so that attraction is only achieved in a high-gradient magnetic field, as outlined in Section II. In the following section, we describe the preparation of magnetoliposomes, i.e., colloids in which the iron oxide core is coated by phospholipids. Based on both theoretical considerations and experimental evidence, arguments will be presented which definitely prove that the phospholipids adopt a bilayer configuration. In Section IV, the value of magnetoliposomes as hosts for membrane enzymes is highlighted. Cytochrome *c*-oxidase, which *in vivo* is "plugged through" the inner mitochondrial membrane,[16] was chosen as a representative model for a membrane enzyme. Some crucial questions addressed in this section include

1. How is the enzymatic activity changed after incorporation of the enzyme into the magnetoliposomal coat?
2. How stable is the enzyme after insertion?
3. Is the enzyme firmly fixed in the phospholipid coat and, moreover, is the whole oxidase–magnetoliposome complex strongly attached to the bioreactor's magnetic filter device, e.g., when the reactor is operating in a continuous-flow regime?

Finally, the summary points out some advantages of magneto(proteo)liposomes and looks forward to areas of work that we believe are particularly promising.

II. MAGNETIC RESPONSITIVITY OF MATERIALS

Before describing the formation and structural characteristics of magneto(proteo)liposomes, it may be fruitful to first discuss very briefly the basics of the interaction of substances, particularly the small magnetizable Fe_3O_4 cores of magnetoliposomes, with a magnetic field.

A. MAGNETIC OR NOT?

In an externally applied magnetic field, H, the magnetization, M, of a material is given by the following relationship:

$$M = \chi_V \cdot H \tag{1}$$

in which χ_V is the difference in volume magnetic susceptibility between the particle and the surrounding fluid in which it is immersed. M refects the alignment of the individual electron spins of the material induced by the applied magnetic field, and χ_V can be envisaged as a measure of the ease by which the electron spins will be oriented.

Depending on the strength of the interaction of substances with magnetic fields, they can be classified into different categories:

1. In *ferro/ferrimagnetic* materials, the atoms have several unpaired electrons, which strongly interact under the influence of the external magnetic field. As a consequence — even at relatively weak field

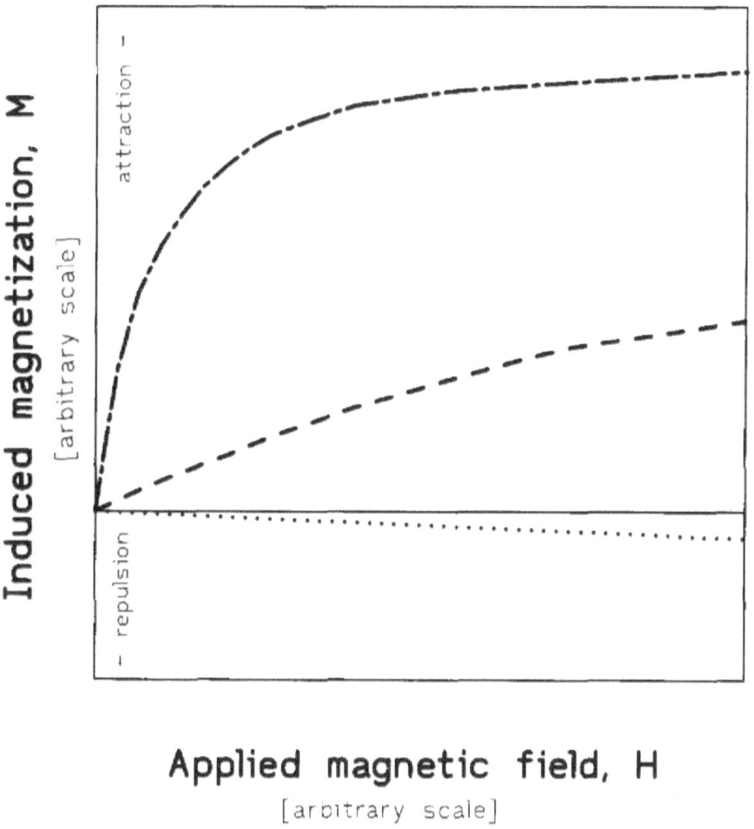

Figure 1 Qualitative representation of the responsitivity of ferro/ferrimagnetic (-·-·-·- line), paramagnetic (- - - - line), and diamagnetic substances (······· line) with respect to an applied magnetic field.

strengths — the unpaired electron spins are oriented parallel to the external field. (In ferromagnets, a small fraction adopts an antiparallel orientation.) The mutual interaction between the individual electron spin moments is so strong that a magnetic remnant remains after removal of the extand field. Examples of ferromagnetic structures are metallic iron, nickel, and cobalt. Magnetite* belongs to the ferrimagnets.

2. In contrast, the interaction of the electronic spins of neighboring *paramagnetic* atoms/molecules is much weaker or even absent in ordinary magnetic fields. Only at very high field strengths will magnetic saturation be reached. At this point, the induced magnetic moment sometimes approaches that obtained with ferromagnets. Examples of paramagnets include $NiSO_4$, many transition metal complexes, and organic free radicals.
3. *Diamagnetic* materials have no unpaired electron spins and are even slightly repelled in external magnetic fields. (As compared to ferromagnetic attraction forces, the repulsion is 10 to 100,000 times weaker.) Water, carbohydrates, proteins, and (phospho)lipids[17–19] belong to this group.

Figure 1 shows the course of the induced magnetism as a function of the strength of the applied magnetic field.

B. SUPERPARAMAGNETIC PARTICLES

As mentioned above, phospholipid molecules demonstrate diamagnetic properties. As a result, even in phospholipid clusters, such as vesicles or liposomes, they cannot be recovered from an aqueous solution by means of magnetic forces. After their association with a magnetic substance, however, this should be feasible. In this respect, magnetite is a popular material. Fe_3O_4 particles, with a diameter of 50 nm (a so-called mono-domain particle) or more, however, will express remnent magnetism after

* Many authors use the words "ferro- and ferrimagnetic" in a loose, practical sense. A solid state physicist however classifies magnetite as ferrimagnetic.

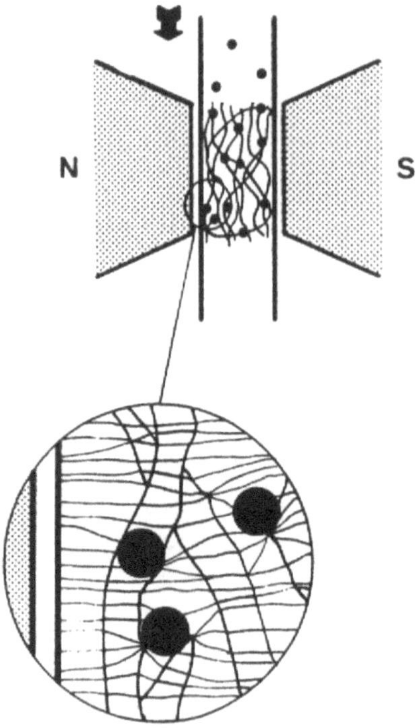

Figure 2 Schematic design of a high-gradient magnetic fractionation setup. Magnetic particles pass through over steel wool fibers, which are magnetic due to the presence of an external magnetic field. The particles stick on the grid. When a buffer stream is passed over the filter in the absence of a magnetic field, the particles are released. Inset: Detail of a high-gradient magnetic separation unit. The field lines (thin lines) between the poles of an electromagnet converge around a magnetizable filter unit (in particular around ridges or surface irregularities), thereby creating localized regions of high-gradient magnetic fields that attract and capture superparamagnetic entities.

magnetophoresis, resulting in unwanted particle clustering. To avoid this magnetic hysteresis-phenomenon, smaller particles must be used.

The problem that then arises is that the resultant magnetic dipole will be too small, so that the colloidal particles cannot be captured efficiently in an ordinary magnetic field. In this case, we have to deal with so-called superparamagnetic properties. The term "paramagnetic" means that after removal of the magnetic field the small particles no longer demonstrate any remnant magnetism (as do paramagnetic materials), while "super" means that the electronic spins remain able to orient very easily along external magnetic field lines (as do ferri/ferromagnets). This superparamagnetic behavior is also discussed in the literature under several other names, including "apparent paramagnetism," "collective paramagnetism," "quasi-paramagnetism," and "subdomain behavior."[20]

III. HIGH-GRADIENT MAGNETOPHORESIS

Superparamagnetic particles can be removed from solution in a high-gradient magnetic field.[21,22] The force a particle experiences in the direction x (F_X) equals

$$F_X = V \cdot \chi_V \cdot H \cdot \delta H/\delta x \qquad (2)$$

in which χ_V and H have the same meaning as above, and V and $\delta H/\delta x$ are the volume of the particle and the magnetic field gradient, respectively.

Since the absolute strength of the magnetic field has — for physical reasons — an upper limit, and since V and χ_V are fixed values at a given temperature, the main trick in magnetic separations is to find practical methods for creating high magnetic field gradients. In practice, this can be easily done by putting magnetizable, stainless steel fibers in the conduit tubing, placed in the throat of an (electro)magnet. In the direct vicinity of the fibers, the magnetic fieldlines are severely distorted, thereby creating strong local magnetic field gradients (Figure 2). In practical setups, the generation of these gradients by the magnetic filter is a most critical parameter for attracting small particles. After magnetophoresis, the magnet is switched off, or the filter matrix is removed from the field vicinity, and the accumulated particles are readily released from the steel wool by flushing it with a stream of liquid. A most interesting observation in this respect is that, in solution, both gravity and interparticle magnetic attraction forces are overshadowed by Brownian motion, so that the eluted particles spontaneously remain in suspension if they are sterically and/or electrostatically stabilized (see below).[23,24]

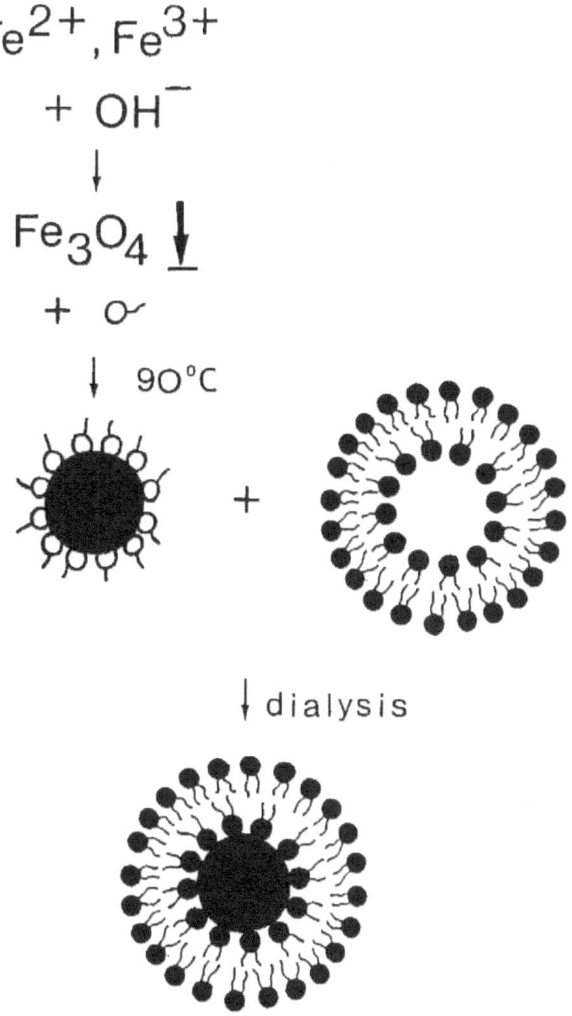

Figure 3 Outline of the preparation protocol of magnetoliposomes. (From De Cuyper, M. and Joniau, M., *Prog. Coll. Polym. Sci.*, 82, 353, 1990. With permission.)

IV. MAGNETOLIPOSOMES

A. PREPARATION RECIPE

The overall scheme for the preparation of magnetoliposomes is presented in Figure 3. The first step deals with the preparation of a so-called water-based magnetic fluid. This consists of a concentrated solution of superparamagnetic Fe_3O_4 colloids, which are prepared by the coprecipitation technique, essentially as described in patented work by Reimers and Khalafalla.[25] A solution, containing 60 g $FeCl_2 \cdot 4aq$ (dissolved in 250 ml water) is mixed with 120 g $FeCl_3 \cdot 6aq$ (also dissolved in 250 ml water). Although the stoichiometric molar ratio of Fe(II) to Fe(III) in Fe_3O_4 is 1:2, as a precaution an initial ratio of 2:3 is used, as some oxidation of Fe(II) occurs during the reaction.[26] 250 ml of a concentrated NH_4OH solution (28 to 30% on a NH_3 base) is then slowly added to the mixed Fe(II)/Fe(III) solution, while stirring with a glass rod. The Fe_3O_4 cores produced in this way are not stabilized and sink as a heavy precipitate. After magnetic decantation, the gel-like slurry is washed three times with 500 ml of an ammonia/water (5/95) solution. The resulting paste is subsequently peptizated by adding solid lauric acid (30 g) and heated in a water bath at 90°C (the heating step is included to improve the solubility of the laurate molecules). Heating is continued until foam formation ceases (approximately 7 min). Finally, the liquified slurry is dispersed in 600 ml water (pH ~9.2). Centrifuging the fluid improves its colloidal stability by removing any large aggregates that may be present. Actually, the different batches of this water-adapted magnetic fluid, which we have prepared in the past, showed no tendency to precipitate over a time scale of several years. The modal diameter of the particles indeed remained constant at a value of about 15 nm (Figure 4).

The Fe_3O_4-laurate complexes are then incubated with preformed, small unilamellar vesicles (SUVs), generated by ultrasonication in a 5 m*M* 2-((*Tris*(hydroxymethyl)methyl)amino) ethanesulfonic acid

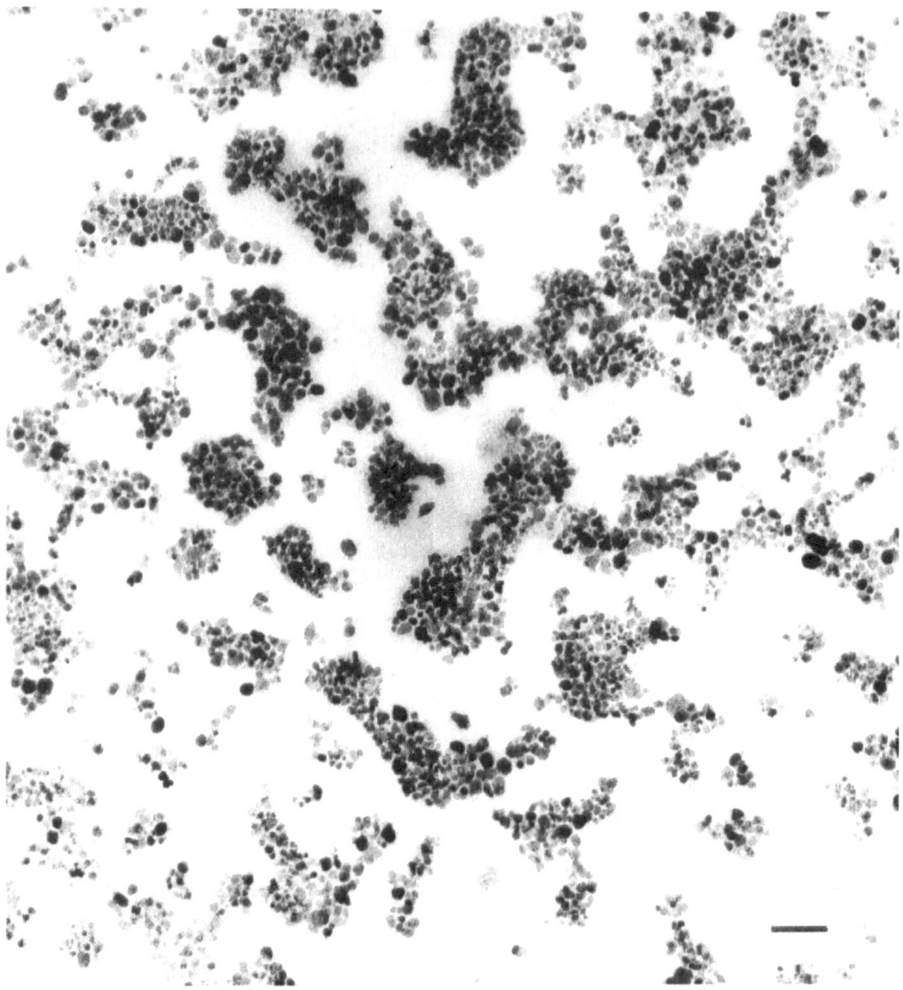

Figure 4 Electron microscopic evaluation of an unstained sample of the magnetic fluid, consisting of Fe_3O_4 cores which are enveloped by a cushion of lauric acid (Scale bar = 60 nm).

buffer, pH 7.0. During a subsequent 2- to 3-day dialysis step (molecular weight cut-off: 10,000), with regular changes of the external buffer, the laurate molecules are slowly removed from the oxide particles and concomittantly replaced by phospholipid molecules (see below).* Nonattached phospholipids are separated from the Fe_3O_4-phospholipid colloids by high-gradient magnetophoresis. Portions of 1 ml are pumped at a uniform rate of 6 ml/h through tubing (inner diameter: 0.078 in.), plugged with loosely-packed magnetic steel wool (~50 mg; Type 430; Bekaert Steel Corporation, Zwevegem, Belgium), placed in a 3-mm gap between the two pole faces of a water-cooled electromagnet (Bruker, Type BE-15, Karlsruhe, Germany). The diameter of the flat pole caps is 15 cm, so more than 25 samples can be fractionated simultaneously. (Usually the electromagnet operates at 30 A and 80 V. Under these conditions, but in the absence of the magnetic filter matrix, the magnetic field between the two poles is about 2.3 Tesla.) After separation, the retentate is thoroughly washed with buffer (0.5 ml) to remove any nonadsorbed phospholipids remaining in the capillary liquid surrounding the iron fibers. The magnetic field is then turned off and the retentate is released from the filter matrix by a high-speed (500 ml/h) buffer stream and collected.

The amount of phospholipid bound per gram of magnetite strongly depends on the initial phospholipid/Fe_3O_4 ratio. For instance, if negatively charged dimyristoylphosphatidylglycerol (DMPG) is used, saturation of the Fe_3O_4 particles (0.8 to 0.9 mmol/g Fe_3O_4) is observed if the starting DMPG/Fe_3O_4 ratio

* Another publication from our laboratory (De Cuyper, M., and Joniau, M., Mechanistic aspects of the adsorption of phospholipids onto lauric acid-stabilized Fe_3O_4 nanocolloids, *Langmuir*, 7, 647, 1991) has addressed the mechanism of magnetoliposome formation in considerable detail.

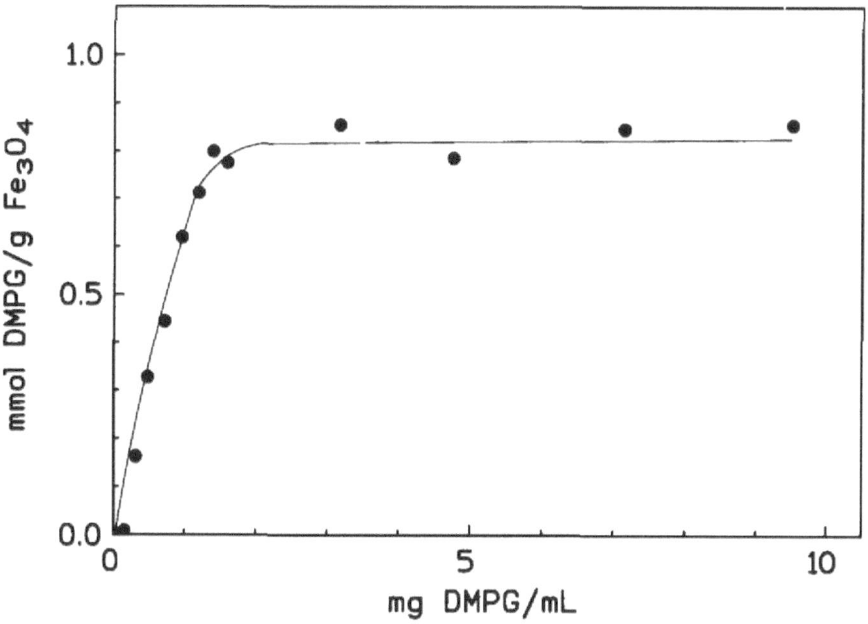

Figure 5 *Adsorption of DMPG on Fe_3O_4 nanocolloids in 5 mM Tes buffer, pH 7.0, measured after 2 days of incubation and dialysis at 37°C. The final magnetite concentration in each incubation mixture was 3.24 mg/ml. The x-axis represents the total phospholipid concentration in the incubation mixtures. A detailed analysis of the adsorption isotherm (amount adsorbed vs. free DMPG concentration) is given in References 34 and 35.*

equals at least 2.73 (weight/weight) (Figure 5). At saturation, the magnetite cores are surrounded by a phospholipid bilayer coat. As discussed below, both theoretical calculations and experimental results confirm this statement.

B. EVIDENCE FOR THE EXISTENCE OF A BILAYER

1. Theoretical Indications

For highly curved, small unilamellar vesicles, the packing of the phospholipid polar headgroups in the inner monolayer determines the lower limit of the vesicle size.[27] Given the equality in the sizes (~14 nm) of the Fe_3O_4 particles and of the SUV's internal volume, it can be reasonably assumed that the amount of phospholipid directly adjacent to the iron oxide surface is also restricted by the headgroup size. If one takes into account a surface of 615 nm^2 for the 14 nm diameter Fe_3O_4 cores and a cross sectional area of 0.62 nm^2 for a PG headgroup,[28] the calculated value for monolayer coverage is 992 molecules. As the bilayer thickness of dimyristoylphospholipids is approximately 3.5 nm,[29,30] it can be calculated that an additional 2,233 molecules are required to fill the outer layer. Based on a density of 5.1 g/cm^3 [31] for magnetite, the total phospholipid corresponds to a mmol DMPG/g Fe_3O_4 ratio of 0.73 which agrees well with the value of 0.85 found at saturation (Figure 5).

2. Experimental Evidence

The theoretical support for the existence of a bilayer configuration, as described in the preceding paragraph, is also supported experimentally.

The first piece of evidence comes from transmission electronmicrographs of DMPG magnetoliposomes, negatively stained with 2% sodium phosphotungstate.[32] As shown in Figure 6, the particles have an electron-dense nucleus, surrounded by a sheet of electron-transparent material and, at the outside, a thin electron-dense zone. Consideration of the chemical moieties involved in the complexes suggests that the nuclei contain Fe_3O_4, and that the electron light and dark sheets are the hydrophobic interior of the lipid bilayer and the polar headgroups of the outer leaflet phospholipids, respectively.

Another argument in favor of a bilayered architecture of the lipidic coat comes from an experiment in which magnetoliposomes were treated with Tween 20.* de Foresta et al.[33] had found that this nonionic detergent is able to disrupt liposomal membrane structures at a concentration of about 0.1 weight %.

* Registered Trademark of ICI Americas Inc., Wilmington, U.S.A.

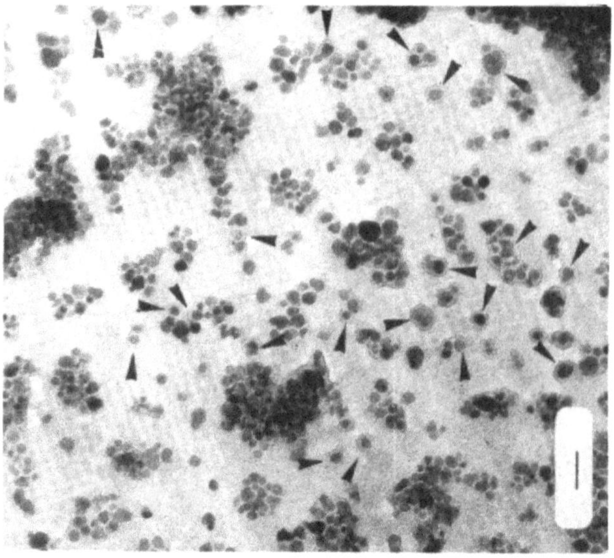

Figure 6 Transmission electron micrograph of magnetoliposomes, negatively stained with 2% phosphotungstate. Note the covering around the magnetite particles (indicated by the arrows) (Scale bar = 40 nm).

The DMPG magnetoliposomes which we used were characterized by a DMPG/Fe_3O_4 ratio of 0.90 mmol/g. 1 ml aliquots of these magnetoliposomes were incubated for 20 h at 37°C with 1 ml of Tween 20 solutions at concentrations ranging from 0 to 4%. After high-gradient magnetophoresis and analysis of the phosphate and iron contents of the retentates, it was found that the DMPG content of the colloidal particles dropped as the detergent concentration increased. The data points in Figure 7 can be fitted by two straight lines covering Tween concentration ranges of 0 to 0.1% ($r^2 = 0.99_1$) and 0.8 to 2.0% ($r^2 = 0.88_6$), respectively. The two lines intersect at a point, corresponding to a concentration of 0.15% Tween 20 (x-axis) and a DMPG/Fe_3O_4 ratio of 0.27 mmol/g (y-axis). The residual DMPG molecules, which comprised approximately one third of the amount in the original magnetoliposomes, was difficult to desorb; this points to strong chemisorption forces between the phospholipid phosphate and the iron in the magnetite surface.[34,35]

The calculations and experiments mentioned above strongly suggest that the Fe_3O_4 cores are enwrapped by a phospholipid bilayer. Consequently, it can be anticipated that the magnetoliposome coat emulates — at least to a certain extent — the lipid arrangement found in other types of membranes, such as artificial phospholipid vesicles or even natural membranes.

V. MAGNETOPROTEOLIPOSOMES

Over the last few years, reconstitution of membrane enzymes into lipid bilayers has become an important task for model membranologists, e.g., in view of potential applications in the field of biosensors and bioelectronics.[36] It is in this general context of growing interest in applied aspects of membrane biochemistry that we have started a study of the immobilization of purified membrane enzymes in the lipid coating of magnetoliposomes. The ultimate goal is to construct a bioreactor in which the resulting magnetoproteoliposomes are first captured in a high-gradient magnetic field and then a solution of substrate is passed over them.

A. LIPID-DEPLETED CYTOCHROME *C*-OXIDASE

As a model membrane enzyme, we decided to work with beef heart cytochrome *c*-oxidase, because it can be isolated in a purified and almost lipid-free form, so that, upon reconstitution, its activity can be modulated by the type of exogenous phospholipid used.[11] The lipid-depleted enzyme, used in the present report, was prepared according to the method of Yu et al.[37] The fractionation protocol for the inner mitochondrial membrane was essentially based on alternating ammonium sulfate precipitation and cholate solubilization steps. In our hands, on average 0.03 mg phospholipid per mg protein was left after the treatment. This amount of residual lipid corresponds to an 85% depletion of the lipid annulus, which is believed to consist of about 50 molecules.[38–41] The stock solution of oxidase was stored at –70°C at a

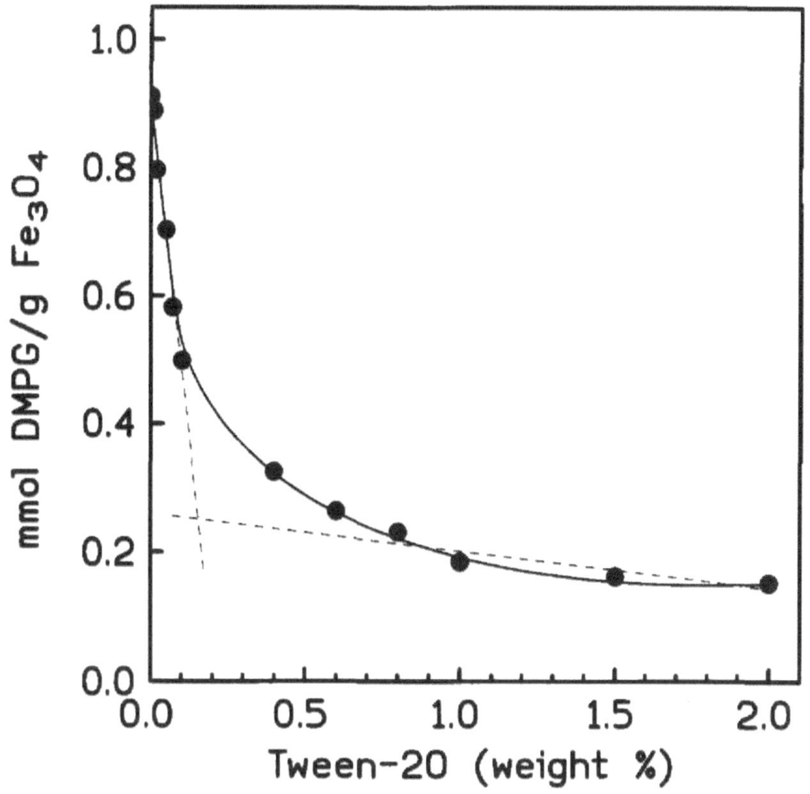

Figure 7 Extraction of DMPG-saturated Fe_3O_4 magnetoliposomes by Tween 20. The dashed lines are linear regression approximations of the extraction profile of the two subpopulations of phospholipids involved: the outer shell phospholipids at high DMPG/Fe_3O_4 ratios and the inner leaflet phospholipids at lower DMPG/Fe_3O_4 ratios. (From De Cuyper, M. and Joniau, M., *Eur. Biophys. J.*, 15, 311, 1988. With permission.)

concentration of 11.4 mg/ml in a 67 m*M* phosphate buffer, pH 7.4, containing 0.25 *M* sucrose and 1.5% cholate.

As described below, different cytochrome *c*-oxidase-phospholipid complexes were generated. In all cases, the phospholipid/protein weight ratio in the reconstituted complexes equalled 10. Assuming a phospholipid molecular weight of about 670, the presence of 3000 lipid molecules in the double-layered structures[42] and a molecular weight of 200,000 for cytochrome *c*-oxidase,[43,44] this corresponds statistically to a mean of 1 protein molecule per vesicle or magnetoliposome. In this way, unwanted intraparticle enzyme–enzyme interactions can be minimized.[45,46]

B. RESTORATION OF ENZYMATIC ACTIVITY

The most common method for measuring the activity of cytochrome *c*-oxidase consists in following the decrease in the absorption of reduced cytochrome *c* at 550 nm (at the top of the α-band in the absorption spectrum), as it is oxidized. To express the catalytic activity of the enzyme in a quantitative way in terms of pseudo-first-order rate constants, k_1, we used the conditions proposed by Smith et al.,[47,48] i.e., 60 μl (3 μg) of cytochrome *c*-oxidase (free or immobilized) was added to 3 ml of cytochrome *c* solution (42 μ*M*; 75% reduced form).

Using the above-mentioned anionic DMPG magnetoliposomes, however, a practical problem arises during activity measurements. The kinetic profiles are severely disturbed due to a time-dependent increase in turbidity of the reaction mixture, as a result of aggregation phenomena induced by the positively charged cytochrome *c*.[49,50] (The isoionic point of horse heart cytochrome *c* is 10.04[51]). To avoid these complications, we preferred to use the zwitterionic dimyristoylphosphatidylcholine (DMPC) in the reconstitution experiments. The magnetite loading with DMPC was identical to that with DMPG (phospholipid/Fe_3O_4 ≈0.8 to 0.9).[34,52]

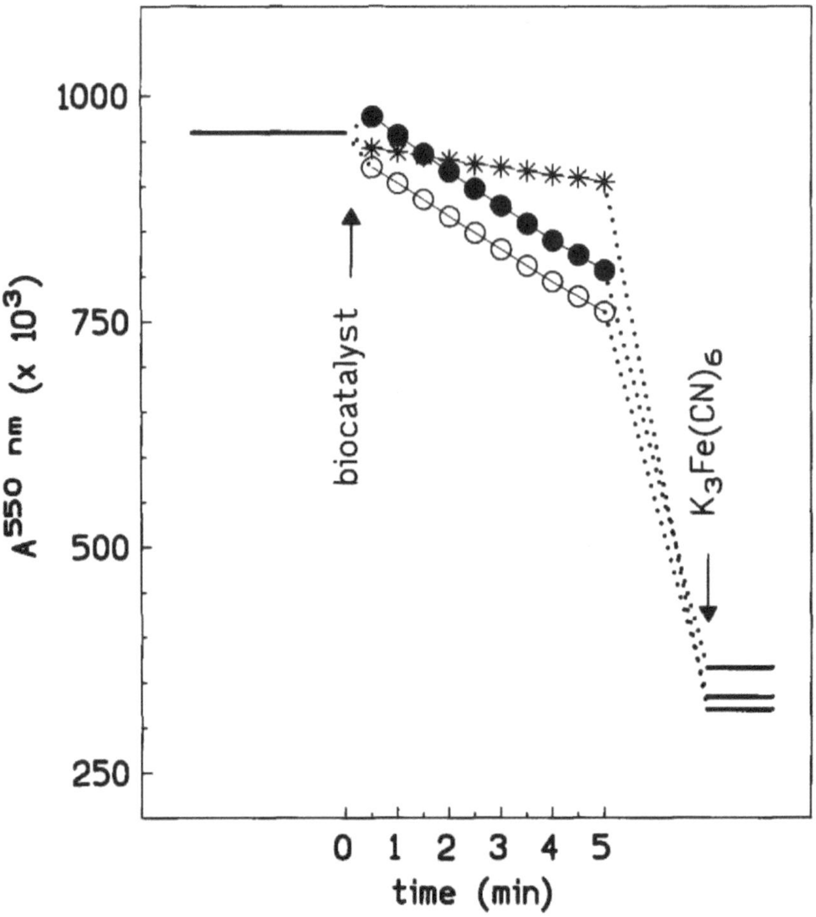

Figure 8 Kinetic profiles at 25°C for lipid-depleted cytochrome *c*-oxidase (*) and after its incorporation into DMPC vesicles (○) and DMPC magnetoliposomes (●). The enzyme preparations were sonicated for 2 min (see text). At t = 0, 60 µl of the biocatalyst solution is added to 3 ml of the cytochrome *c* substrate solution. At t = 5 min, fully oxidized cytochrome *c* was produced by the addition of a grain of $K_3Fe(CN)_6$. As compared with the adsorption values found with the proteovesicles, those obtained with the magnetoproteoliposomes are somewhat higher as a result of the dark color of the iron oxide colloids.

After a brief sonication for 2 min (see below), the k_1 value calculated for cytochrome *c*-oxidase in its lipid-depleted form equaled 0.013_4 min^{-1}. If the sonication step was performed in the presence of DMPC magnetoliposomes the k_1 value increased 5 to 6 times ($k_1 = 0.072_7$ min^{-1}). Interestingly, the same degree of reactivation was found when SUVs, also made with DMPC, were used ($k_1 = 0.071_3$ min^{-1}) (Figure 8). This observation tentatively points to a similarity in the microenvironment around the enzyme embedded in the two structures.

The underlying mechanism(s) which promote(s) the insertion of membrane proteins into preformed lipid bilayers still remain(s) obscure, although the creation of elastic deformations within the lipid bilayer seems to be crucial.[53] In the case of cytochrome *c*-oxidase reconstitution, a three-fold reactivation of the enzyme was detected immediately after confronting the enzyme with DMPC magnetoliposomes. An additional increase was achieved if the incubation mixture was sonicated for 2 to 4 min (executed in 30-s time periods interrupted by 30-s cooling intervals).[54,55] Further sonication resulted in a significant inactivation of the enzyme. In terms of "packing defects" in the lipid bilayer, the initial, spontaneous insertion can be rationalized by assuming a partitioning of cholate into the lipid bilayer after the transfer of a small volume of the detergent–oxidase stock solution into the magnetoliposome suspension.[56,57] Additional distortions of the acyl chains, which underlie a further increase in activity, are presumably induced mechanically by sonic energy.

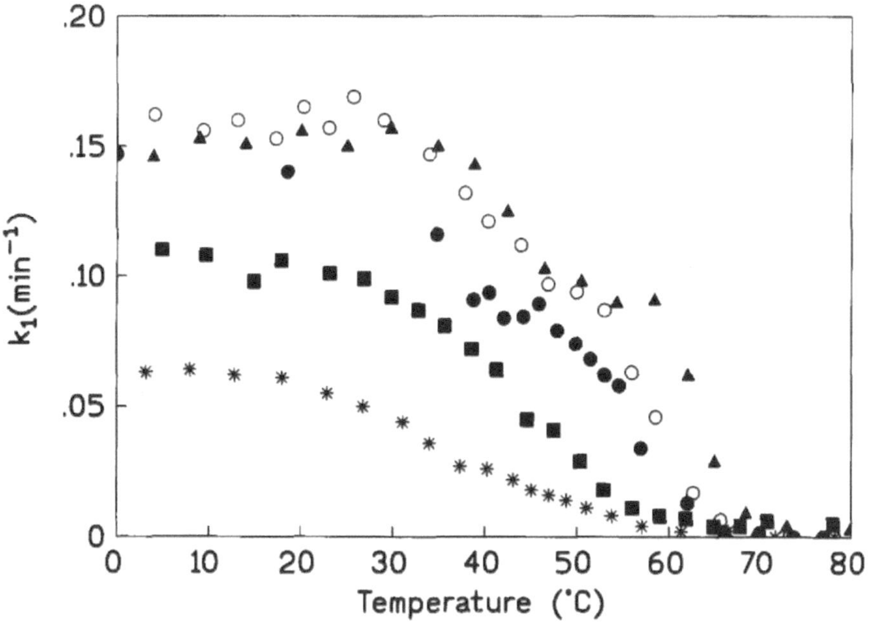

Figure 9 Inactivation of cytochrome *c*-oxidase by heat treatment. The enzyme was used either in its lipid-poor state (*) or reconstituted in monolayer-coated DMPC magnetite particles (■), DMPC vesicles (○), bilayered DMPC (●), or bilayered DOPC magnetoliposomes (▲). The experimental details are given in the text. (Redrawn from De Cuyper, M. and Joniau, M., *Biotechnol. Appl. Biochem.*, 16, 201, 1992. With permission.)

C. THERMOSTABILITY OF THE IMMOBILIZED ENZYME

Aside from the importance of the degree of (re)activation, the thermal stability of the immobilized oxidase is also an important criterion in view of its potential application in a bioreactor. In this respect, it is well known that subtle variations in the lipid matrix can drastically modify the functional stability of integral proteins.[58]

To check this in our system, lipid-depleted cytochrome *c*-oxidase was reconstituted in DMPC SUVs, in DMPC monolayer-coated Fe_3O_4 colloids, in DMPC bilayered magnetoliposomes, and in magnetoliposomes constructed with the longer chain length dioleoylphosphatidylcholine (DOPC). The consecutive practical steps for constructing the thermal inactivation curves, between 0 and 80°C, were as follows: (1) first, the enzyme preparation (60 µl) was incubated in sealed vials at the designated temperatures for exactly 5 min; (2) then the solution was cooled on ice for 1 min; (3) after which 100 µl of a 0.8% Tween 80* solution was added to dissolve any aggregates[59,60] which may have possibly been formed during the heating process; and (4) finally, the kinetics were followed for 10 min at 25°C.

The data in Figure 9 show that the thermostability was dramatically improved by the presence of phospholipid molecules. A closer look shows a dependency on the acyl chain length of lipid molecules used. The inflection point in the main transition zone was shifted from 56°C in DMPC magnetoliposomes (●) to 62°C in those containing DOPC (▲). Interestingly, a similar transition temperature has been reported for the enzyme embedded in the native mitochondrial membrane.[61,62]

Another most remarkable observation is that the thermostability critically depends on the configuration of the lipid matrix. For instance, a comparison of the ● and ■ profiles in Figure 9 shows that the stability of bilayer-coated structures was significantly greater than of those coated with a monolayer ($T_{inflection}$ at 45°C). In addition, bilayers whether deposited on Fe_3O_4 cores (●) or "free" in SUVs (○) result in an almost identical thermal stability. As already deduced from the similar degree of reactivation (see Section V.B), this observation again suggests that many common structural features exist in both types of membranes. Membrane enzymes indeed are believed to be exquisitely sensitive reporters of physical and/or chemical differences in membrane matrices.

D. CYTOCHROME *c*-OXIDASE — MAGNETOLIPOSOME ASSOCIATION

Although it is obvious that both a high degree of reactivation (as documented in Section V.B) and a considerable increase in thermostability (Section V.C) of the immobilized cytochrome *c*-oxidase are

* Registered Trademark of Desitin-Werke, Carl Klinke GmbH, Hamburg, Germany.

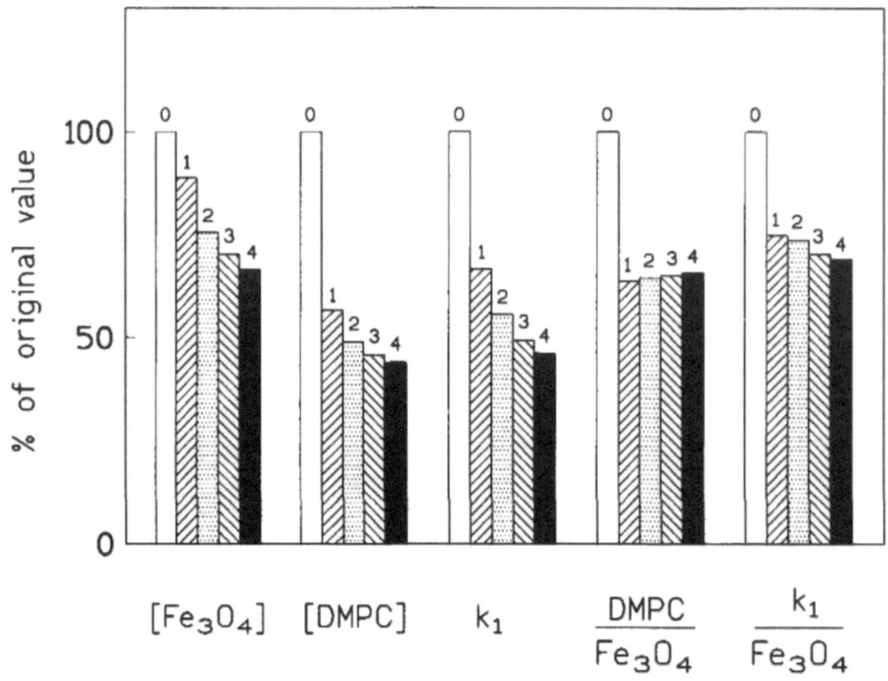

Figure 10 Retention of DMPC, magnetite, and enzymatic activity after different magnetic separation cycles of a solution of bilayer-coated DMPC magnetoproteoliposomes. Changes in the lipid content of the particles and in their activity are also given. The numbers on top of the bars refer to the original magnetoproteoliposome preparation (0) and the number of consecutive passes over the magnetic filter (1–4). The values measured are normalized with respect to the values found in the starting magnetoproteoliposome preparation. (Adapted from De Cuyper, M. and Joniau, M., *Biotechnol. Appl. Biochem.*, 16, 201, 1992. With permission.)

extremely desirable, these features alone give no information on the strength of bonding of the enzyme to the magnetoliposomal coat. A qualitative assessment can be made from a so-called recycling experiment, in which a sample of sonicated enzyme–magnetoliposome complexes was subjected to four consecutive magnetic separation cycles, each comprising attraction of the particles to the filter, a rigorous postimmobilization washing, and the subsequent release from the filter in a high velocity buffer stream (see Section III). The magnetite and phospholipid concentrations, the enzymatic activity (expressed as k_1 values), and the lipid content of the particles (expressed as DMPC/Fe_3O_4) and their catalytic quality (expressed as k_1/Fe_3O_4) were measured in both the original magnetoproteoliposomes and the different retentates. The results, shown in Figure 10, are normalized with respect to the values found in the starting magnetoproteoliposomes.

As shown, after the first separation cycle, there was clearly a discrepancy in the degree of recovery of magnetite and that of enzymatic activity: only 10% of the Fe_3O_4 particles were lost, whereas the total enzymatic activity had dropped by approximately one third. At first sight, the loss of cytochrome *c*-oxidase activity might result from loss of protein during the washing stage, due to inadequate binding to the surface. However, in view of the accompanying significant decrease in DMPC content (43%), a more plausible explanation is that, upon sonication to facilitate the incorporation of cytochrome *c*-oxidase into the magnetoliposome membrane (Section V.B), some lipid molecules were released from the lipid coat and that, concomittantly, a proportional part of the enzyme was inserted in the resulting Fe_3O_4-free lipid structures, for instance in liposomes. Actually, this effect should not seriously influence the overall activity in the starting magnetoproteoliposome suspension; as illustrated in Figure 8, the activity in vesicles and in magnetoliposomes was indeed almost identical.

In subsequent runs, the fractional loss of Fe_3O_4 and of DMPC was significantly lower and the activity of the immobilized enzyme leveled off at about 50% of the original quantity. A most interesting observation, however, was that the lipid content and the catalytic quality per particle stayed the same. In these experimental conditions in which the external medium was repeatedly completely removed, this demonstrated not only that the magnetoliposome coat remained unchanged but also that the residual enzyme molecules were not released. This firm insertion of the enzyme into the lipid coat opens promising perspectives for the use of our catalytic biocolloids in batch type bioreactors, in which recovery of the biocatalyst is necessary when the effectiveness factor reaches an intolerably low value.

E. USE OF MAGNETOPROTEOLIPOSOMES IN CONTINUOUS-FLOW BIOREACTORS

In many industrial processes, continuous-flow, fixed-bed reactor configurations are often preferred to batch type reactors because the operation is easier and higher yields of product are usually achieved. For the successful use of magnetoproteoliposomes in such a continuous-flow system, an additional requirement has to be fulfilled, namely that the whole enzyme–magnetoliposome complex must remain firmly bound to the magnetic filter fibers over prolonged periods of time and even at high flow rates of the substrate solution.

To check this point experimentally, the performance of two lab-scale bioreactor setups was compared. Each contained 2 ml magnetoproteoliposomes (enzyme concentration: 40 μg/ml), accumulated on the magnetic filter and overflowed continuously for 2 weeks with a solution of 75% reduced cytochrome *c* substrate, according to a single-pass operation regime. For the first setup, the feed flow applied was 15 ml/h (reactor I); in the second, it was 78 ml/h (reactor II). The reaction systems were cooled at 4°C in a narrow (5 mm width), homemade water-jacket (constructed in copper), which was placed in the gap between the poles of the magnet. Since the stock substrate solution was autoxidized very slowly,[63] its degree of reduction was controlled at regular times and, if need be, corrected by the addition of sodium dithionite. The effluent from both tubings was pumped through an electrically-actuated, flow-through flowpath multiposition valve (Type E6CSF1OP, Valco Instruments, Houston, TX), connected to a computer-linked, serial valve interface (SVI, Valco). This configuration allows the automatic switching, at designated times, to one of the two flow streams for spectrophotometric analysis (3 mm path way flow cell, Spectra Physics, San Jose, CA), without blocking the other one in the meantime.

As anticipated, the extent of product formation was markedly influenced by the flow rate (Figure 11). In addition, the conversion efficiencies in both systems decay as a function of time. This decrease in efficiency might be attributed to a leakage of the enzyme from the magnetoliposomes or to a slow release of entire particles from the filter fibers. According to these hypotheses, both phenomena, which are supposed to be critically dependent on prevailing shear forces, should occur to a different extent in both reactors. However, since the bioreactor performances were almost identical after reducing the flow-rate of reactor II to that applied to reactor I (Figure 11), it can be inferred that the number of enzyme molecules, and therefore the number of particles, remained identical in both reactors. Consequently, a more likely explanation for the drop in conversion efficiency is related to inactivation of the biocatalyst. Apparently this phenomenon remains the crucial "bottleneck" for the use of magnetoproteoliposomes on a time-scale of weeks. The factors which stabilize and/or inactivate immobilized enzymes, unfortunately, are not yet systematically understood.[4]

VI. CONCLUSIONS — PERSPECTIVES

Immobilization of enzymes in membranes is nature's way of obtaining stability and efficiency for the reactions they catalyze. Over the last 30 years, some of these membrane proteins have been (partly) purified and reconstituted into different types of liposomes or vesicles in order to study the biophysical properties and the physiology of biological cell membranes.[64,65] In such studies, the use of supported bilayers is becoming increasingly popular, in particular since they offer promising biotechnological applications, e.g., as components of various analytical devices, electrodes, and monitoring systems.[66]

The development of magnetoliposomes fits in well with the latter strategy. Some decisive advantageous features of magnetizable structures are obvious. For instance, because of the small dimensions of the particles, they are characterized by a favorable surface-to-volume ratio.[67] (As compared to most commercially available magnetic particles, our colloids are indeed extremely small.) In spite of the small resultant magnetic dipole, an efficient attraction from solution by high-gradient magnetic forces is still possible. On a lab scale, these forces can be easily generated, even with an inexpensive, permanent magnet. Magnetoliposomes are also very easily prepared, without the need for any special equipment. Furthermore, using a wide variety of both synthetic and natural phospholipids[68,69] or mixtures of these, magnetoliposomes can be designed to cover a very wide range of surface properties. Even after prepa-

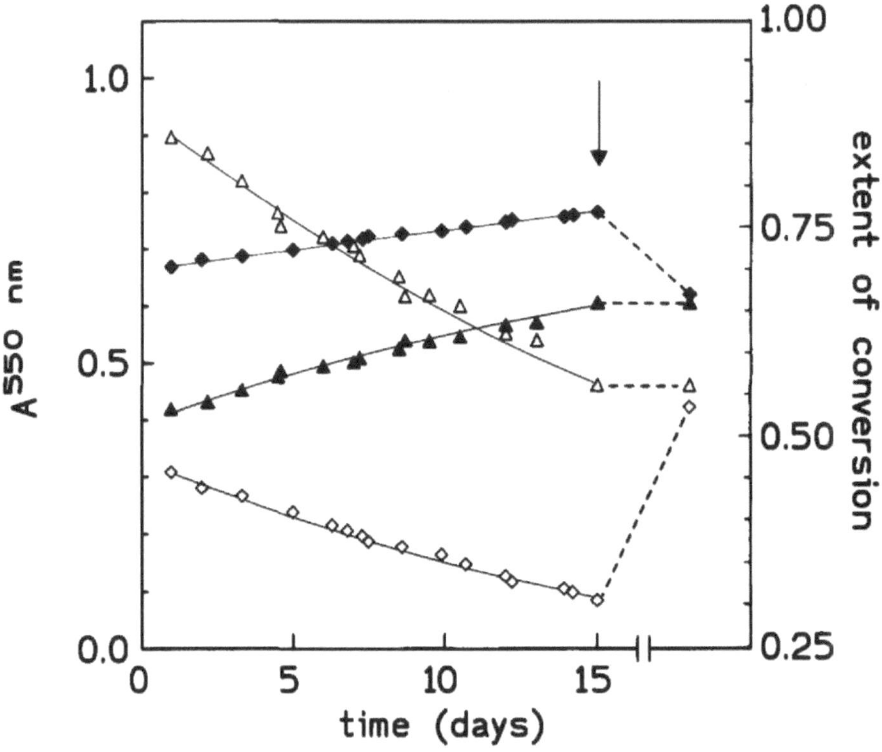

Figure 11 Time course of changes in the absorption at 550 nm and of the conversion efficiency of reduced cytochrome *c* by cytochrome *c*-oxidase, embedded in the lipidic shell of DMPC magnetoliposomes. The biocatalytic structures, which were trapped by magnetic forces on the magnetic filter device, were overflowed continuously at 4°C with a 75% reduced cytochrome *c* solution. The $A^{550\ nm}$ values are converted to a 1 cm path length of the flow cell (left side y-axis; closed symbols). The extent of substrate conversion (right side ordinate; open symbols) at time t is expressed by the following equation:

$$\frac{A_{75\%\ reduced} - A_t}{A_{75\%\ reduced} - A_{fully\ oxidized}}$$

in which *A* represents the absorption at 550 nm of the starting substrate solution ($A_{75\%\ reduced}$) and at any time t during the operation (A_t). $A_{fully\ oxidized}$ denotes the absorption of the substrate solution after the addition of a grain of $K_3Fe(CN)_6$ (see Figure 8). The feed flow was 15 (♦) and 78 ml/h (◇). After 3 weeks the flow rate in the latter setup was reduced to 15 ml/h, as indicated by the arrow.

ration, the lipidic coat can be altered by incubating them with vesicles made up with other types of phospholipids.[34,70,71] Actually, the use of readily available crude phospholipid preparations can considerably reduce production cost. In addition — although we have not yet studied this issue in detail — the presence of the phospholipid bilayer coat should make magnetoliposomes relatively nonimmunogenic and nontoxic,[72,73] so that they could be safely used as carriers for the local delivery of membrane-associated (pro)drugs[74] or as magnetic resonance imaging contrast agents.[75] With somewhat bigger Fe_3O_4 particles — procedures for their production are described in the literature[76] — it should even be possible to trap and to hold them with high efficiency at a specific site within an artery using external magnets.[77,78]

In the present overview, magnetoliposomes were used as hosts for cytochrome *c*-oxidase, which was used as a model of an intrinsic, membrane-bound enzyme. Reconstitution in membranes, however, is not peculiar to this enzyme; the insertion of other widely diverse membrane proteins or enzymes, including those enzymes which may be industrially relevant,[79] can be reasonably expected to occur. Until now, a main obstacle to the use of purified membrane enzymes in large-scale setups has been the poor yields during the different isolation steps. Rothe and Aurich's report,[7] however, mentions that some membrane enzymes have been expressed in high quantities in genetically-modified bacteria, making it probable that, in the near future, their availability will no longer be a restricting factor.

The immobilization of cytochrome *c*-oxidase in magnetoliposomes occurs under relatively mild conditions, since no chemical manipulations are employed; its binding within the phospholipid bilayer is believed to be governed by the formation of multiple, relatively weak hydrophobic, electrostatic, and hydrogen linkages.[80-82] This clearly contrasts with recipes describing the binding of water-soluble enzymes to solid supports in which covalent linkages are often formed, which may ultimately lead to a (partial) inactivation.[83]

The unique potentialities of magnetoproteoliposomes are illustrated in this chapter by the catalytic activity expressed in bioreactors. In *batch-type* reactors, for instance, the reaction mixture will be kept homogeneous by Brownian motion; consequently, stirring devices are unnecessary and rheological problems will be largely avoided. On the other hand, "classical" *packed-bed* bioreactors frequently suffer from fouling and clogging phenomena, as well as from excessive pressure drops, particularly if deformable beads of immobilized enzymes are used.[84] Such problems can be avoided in high-gradient magnetophoretic setups in which the magnetic filtration matrix consists solely of loosely-packed steel wires. The production capacity too should not be a big problem; although high-gradient magnetophoresis is a relatively new concept in industrial fractionation technology, highly sophisticated equipment for use in large-scale applications is already available.[85]

Though the results presented in this chapter show that magnetoliposomes are particularly well suited to the immobilization of hydrophobic enzymes, obviously much work still remains to be done. For instance, further studies, now in progress in our lab, will examine how the type of phospholipid can tailor the catalytic properties of the biocolloids.[86,87] The present results should also pose new challenges for the elaboration of new applications in other (bio)technological domains,[88-89] for instance in the field of immunological detection tests, or (if need be, after functionalizing the phospholipid polar headgroup) as versatile affinity-chromatography supports.[90-93] Possibly magnetoliposomes could also be successfully introduced in existing processes, which already make use of the magnetic fluid technology.[94-97] In this context, the present contribution should provide an impetus for the employment of magneto(proteo)liposomes in these areas and for an embarkment on new applications no one has yet thought of.[98]

ACKNOWLEDGMENTS

I owe a debt of gratitude to Mrs. L. Desender and Mrs. P. Rigole for their active collaboration in the experimental work. Support for this research by the *Belgian Nationaal Fonds voor Wetenschappelijk Onderzoek (Krediet aan Navorsers)* is gratefully acknowledged.

REFERENCES

1. Chemical Abstracts SEARCH, File 308-311, Field operator search, Digital Information Retrieval Services, American Chemical Society, Washington, D.C., 1994.
2. **Tewari, Y. B.,** Thermodynamics of industrially-important, enzyme-catalyzed reactors, *Appl. Biochem. Biotechnol.*, 23, 187, 1990.
3. **Cheetham, P. S. J.,** Principles of industrial enzymology: basis of utilization of soluble and immobilized enzymes in industrial processes, in *Handbook of Enzyme Biotechnology*, Wiseman, A., Ed., Ellis Horwood Ltd., Chichester, England, 1985, chap. 3.
4. **Cheetham, P. S. J.,** The application of immobilized enzymes and cells and biochemical reactors in biotechnology — Principles of enzyme engineering, in *Principles of Biotechnology*, Wiseman, A., Ed., Surrey University Press, Glasgow, 1983, chap. 7.
5. **Kheirolomoon, A., Katoh, S., Sada, E., and Yoshida, K.-I.,** Reaction characteristics and stability of a membrane-bound enzyme reconstituted in bilayers of liposomes, *Biotechnol. Bioeng.*, 37, 809, 1991.
6. **Kheirolomoom, A., Miyazato, K., Katoh, S., and Sada, E.,** Incorporation and activation of a membrane-bound enzyme in bilayers of liposomes, *Appl. Microbiol. Biotechnol.*, 35, 521, 1991.
7. **Rothe, U. and Aurich, H.,** Lipid-coated particles — A new approach to fix membrane-bound enzymes onto carrier surfaces, *Biotechnol. Appl. Biochem.*, 11, 18, 1989.
8. **Sada, E., Katoh, S., Terashima, M., Kheirolomoom, A., and Sawai, H.,** The effect of liposomization on reaction characteristics of prostaglandin synthetase, *J. Chem. Eng. Jpn.*, 20, 531, 1987.
9. **Sada, E., Katoh, S., Terashima, M., and Kikuchi, Y.,** Reaction properties of hydrophobic enzymes and their immobilization on phospholipid composite membranes, *Biotechnol. Bioeng.*, 30, 117, 1987.
10. **Adami, P., Nasser, B., and Latruffe, N.,** Interactions of the mitochondrial membrane rat liver D-3-hydroxybutyrate dehydrogenase with glass beads during adsorption chromatography — Relationships with the activation of the enzyme by phospholipids, *J. Chromatogr.*, 539, 279, 1991.

11. **De Cuyper, M. and Joniau, M.,** Behaviour of beef-heart cytochrome *c*-oxidase in reconstituted proteovesicles — a systematic evaluation of the influence of phospholipid polar headgroup and fatty-acyl side chains, *Eur. J. Biochem.*, 104, 397, 1980.
12. **Jones, O. T., Earnest, J. P., and McNamee, M. G.,** Solubilization and reconstitution of membrane proteins, in *Biological Membranes — A Practical Approach*, Findlay, J. B. C. and Evans, W. H., Eds., IRL Press at Oxford University Press, Oxford, 1987, 139.
13. **Yang, Q.,** Immobilization of liposomes in gel beads for chromatography, in *Acta Universitatis Upsaliensis*, Comprehensive Summaries of Uppsala Dissertation from the Faculty of Science, no. 457, Uppsala University, Uppsala, Sweden, 1993.
14. **Lundahl, P. and Yang, Q.,** Liposome chromatography: liposomes immobilized in gel beads as a stationary phase for aqueous column chromatography, *J. Chromatogr.*, 544, 283, 1991.
15. **Yang, Q., Wallstén, M., and Lundahl, P.,** Immobilization of phospholipid vesicles and protein-lipid vesicles containing red cell membrane proteins on octyl derivatives of large-pore gels, *Biochim. Biophys. Acta*, 938, 243, 1988.
16. **Malmström, B. G.,** Cytochrome oxidase: some unsolved problems and controversial issues, *Arch. Biochem. Biophys.*, 280, 233, 1990.
17. **Sanders, C. R., II and Prestegard, J. H.,** Magnetical orientable phospholipid bilayers containing small amounts of a bile salt analogue, CHAPSO, *Biophys. J.*, 58, 447, 1990.
18. **Maret, G. and Dransfeld, K.,** Biomolecules and polymers in high steady magnetic fields, in *Topics in Applied Physics, Vol. 57, Strong and Ultrastrong Magnetic Fields and Their Applications*, Herlach, F., Ed., Springer-Verlag, Berlin, 1985, chap. 4.
19. **Qui, X., Mirau, P. A., and Pidgeon, C.,** Magnetically induced orientation of phosphatidylcholine membranes, *Biochim. Biophys. Acta*, 1147, 59, 1993.
20. **Bean, C. P. and Livingston, J. D.,** Superparamagnetism, *J. Appl. Physics*, 30(Suppl.), 120S, 1959.
21. **Whitesides, G. M., Kazlauskas, R. J., and Josephson, L.,** Magnetic separations in biotechnology, *Trends Biotechnol.*, 1, 144, 1983.
22. **Fletcher, D.,** Fine particle high gradient magnetic entrapment, *IEEE Trans. Magn.*, 27, 3655, 1991.
23. **Rosensweig, R. E.,** Magnetic fluids, *Int. Sci. Technol.*, 48, 1966.
24. **Bacri, J.-C., Perzynski, R., and Salin, D.,** Les liquides magnétiques, *La Recherche*, 18, 1150, 1987.
25. **Reimers, G. W. and Khalafalla, S. E.,** Production of magnetic fluids by peptization techniques, U.S. Patent 3,843,540, 1974.
26. **Lee, K. M., Sorensen, C. M., Klabunde, K. J., and Hadjipanayis, G. C.,** Synthesis and characterization of stable colloidal Fe_3O_4 particles in water-in-oil microemulsions, *IEEE Trans. Magn.*, 28, 3180, 1992.
27. **New, R. R. C.,** *Liposomes — A Practical Approach*, IRL Press at Oxford University Press, Oxford, 1990, 1.
28. **Toko, K. and Yamafuji, K.,** Influence of monovalent and divalent cations on the surface area of phosphatidylglycerol monolayers, *Chem. Phys. Lipids*, 26, 79, 1980.
29. **Lewis, B. A. and Engelman, D. M.,** Lipid bilayer thickness varies linearly with acyl chain length in fluid phosphatidylcholine vesicles, *J. Mol. Biol.*, 166, 211, 1983.
30. **Marra, J. and Israelachvili, J.,** Direct measurements of forces between phosphatidylcholine and phosphatidylethanolamine bilayers in aqueous electrolyte solutions, *Biochemistry*, 24, 4608, 1985.
31. **Schwertmann, U. and Cornell, R. W.,** The iron oxides and hydroxides, in *Iron Oxides in the Laboratory — Preparation and Characterization*, VCH Verlagsgesellschaft mbH, Weinheim, Germany, 1991, 5.
32. **Dousset, N., Lapinche, L., Dousset, J.-C., and Douste-Blazy, L.,** Influence of dicarboxylic phosphatidylcholines on phosphatidylcholine liposomes as revealed by gel chromatography and electron microscopy, *Biochim. Biophys. Acta*, 692, 223, 1982.
33. **de Foresta, B., le Maire, M., Orlowski, S., Champeil, P., Lund, S., Møller, J. V., Michelangeli, F., and Lee, A. G.,** Membrane solubilization by detergent: use of brominated phospholipids to evaluate the detergent-induced changes in Ca^{2+}-ATPase/lipid interactions, *Biochemistry*, 28, 2558, 1989.
34. **De Cuyper, M. and Joniau, M.,** Magnetoliposomes — Formation and structural characterization, *Eur. Biophys. J.*, 15, 311, 1988.
35. **De Cuyper, M. and Joniau, M.,** Mechanistic aspects of the adsorption of phospholipids onto lauric-acid stabilized Fe_3O_4 nanocolloids, *Langmuir*, 7, 647, 1991.
36. **Schultz, J. S.,** Biosensors, *Sci. Am.*, 64, August 1991.
37. **Yu, C. A., Yu, L., and King, T. E.,** Studies on cytochrome *c*-oxidase — interactions of the cytochrome *c*-oxidase protein with phospholipids and cytochrome *c*, *J. Biol. Chem.*, 250, 1383, 1975.
38. **Jost, P. C., Griffith, H. O., Capaldi, R. A., and Vanderkooi, G.,** Evidence for boundary lipid in membranes, *Proc. Natl. Acad. Sci. U.S.A.*, 70, 480, 1973.
39. **Knowles, P. F., Watts, A., and Marsh, D.,** Spin-label studies of lipid immobilization in dimyristoylphosphatidylcholine-substituted cytochrome *c*-oxidase, *Biochemistry*, 18, 4480, 1979.
40. **Casey, R. P.,** Membrane reconstitution of the energy-conserving enzymes of oxidative phosphorylations, *Biochim. Biophys. Acta*, 768, 319, 1984.

41. **Robinson, N. C.,** The specificity and affinity of phospholipids for cytochrome *c*-oxidase, *Biophys. J.*, 37, 65, 1982.
42. **Huang, C.-h.,** Studies on phosphatidylcholine vesicles — formation and physical characterization, *Biochemistry*, 8, 344, 1969.
43. **Love, B., Chan, S. P. H., and Stotz, E.,** Molecular weight of two states of cytochrome *c*-oxidase, *J. Biol. Chem.*, 245, 6664, 1970.
44. **Robinson, N. C. and Talbert, L.,** Triton X-1OO induced dissociation of beef heart cytochrome *c*-oxidase into monomers, *Biochemistry*, 25, 2328, 1986.
45. **Watts, A.,** Magnetic resonance studies of phospholipid–protein interactions in bilayers, in *Phospholipids Handbook*, Cevc, G., Ed., Marcel Dekker, New York, 1993, 687.
46. **Cornea, R. L. and Thomas, D. D.,** Effects of membrane thickness on the molecular dynamics and enzymatic activity of reconstituted Ca-ATPase, *Biochemistry*, 33, 2912, 1994.
47. **Smith, L. and Conrad, H.,** A study of the kinetics of the oxidation of cytochrome *c* by cytochrome *c*-oxidase, *Arch. Biochem. Biophys.*, 63, 403, 1956.
48. **Errede, B., Haight, G. P., Jr., and Kamen, M. D.,** Oxidation of ferrocytochrome *c* by mitochondrial cytochrome *c*-oxidase, *Proc. Nat. Acad. Sci. U.S.A.*, 73, 113, 1976.
49. **Rietveld, A., Sijens, P., Verkleij, A. J., and de Kruijff, B.,** Interaction of cytochrome *c* and its precursor apocytochrome *c* with various phospholipids, *EMBO J.*, 2, 907, 1983.
50. **Nicholls, P. and Malviya, A. N.,** External and internal binding of cytochrome *c* by liposomes, *Biochem. Soc. Trans.*, 1, 372, 1973.
51. **Barlow, G. H. and Margoliash, E.,** Electrophoretic behavior of mammalian-type cytochromes *c*, *J. Biol. Chem.*, 241, 1473, 1966.
52. **De Cuyper, M. and Joniau, M.,** Immobilization of membrane enzymes into magnetizable, phospholipid bilayer-coated, inorganic colloids, *Prog. Coll. Polym. Sci.*, 82, 353, 1990.
53. **Kurrle, A., Rieber, P., and Sackmann, E.,** Reconstitution of transferrin receptor in mixed lipid vesicles. An example of the role of elastic and electrostatic forces for protein/lipid assembly, *Biochemistry*, 29, 8274, 1990.
54. **De Cuyper, M. and Joniau, M.,** Binding characteristics and thermal behaviour of cytochrome *c*-oxidase, inserted into phospholipid-coated, magnetic nanoparticles, *Biotechnol. Appl. Biochem.*, 16, 201, 1992.
55. **Scotto A. W. and Zakim, D.,** Reconstitution of membrane proteins — Spontaneous incorporation of integral membrane proteins into preformed bilayers of pure phospholipids, *J. Biol. Chem.*, 263, 18500, 1988.
56. **Warren, G. B., Toon, P. A., Birdsall, N. J. M., Lee, A. G., and Metcalfe, J. C.,** Reversible lipid titrations of the activity of pure adenosinetriphosphatase–lipid complexes, *Biochemistry*, 13, 5501, 1974.
57. **Abramovitch, D. A., Marsh, D., and Pownall, G. L.,** Activation of beef-heart cytochrome *c*-oxidase by cardiolipin and analogues of cardiolipin, *Biochim. Bipohys. Acta*, 1020, 34, 1990.
58. **Maneri, L. R. and Low, P. S.,** Structural stability of the erythrocyte anion transporter, band 3, in different lipid environments — a differential scanning calorimetric study, *J. Biol. Chem.*, 263, 16170, 1988.
59. **Rigell, C. W. and Freire, E.,** Differential detergent solubility investigation of thermally induced transitions in cytochrome *c*-oxidase, *Biochemistry*, 26, 4366, 1987.
60. **Sinjorgo, K. M. C., Durak, I., Dekker, H. L., Edel, C. M., Bieleman, A. H. L., Back, N. B. T., Hakvoort, T. B. M., and Muijsers, A. O.,** The effect of detergents on bovine cytochrome *c*-oxidase: a kinetic approach, *Biochim. Biophys. Acta*, 893, 241, 1987.
61. **Knox, B. E. and Tsong, T. Y.,** Thermal inactivation of mitochondrial electron transport chain, *Biophys. J.*, 45, 296A, 1984.
62. **Tomita, M., Knox, B. E., and Tsong, T. Y.,** Thermal inactivation of electron-transport functions and F_0F_1-ATPase activities, *Biochim. Biophys. Acta*, 894, 16, 1987.
63. **Smith, L.,** Cytochromes a, a_1, a_2 and a_3, *Methods Enzymol.*, 2, 732, 1955.
64. **Bangham, A. D.,** Liposomes: the Babraham Connection, *Chem. Phys. Lipids*, 64, 275, 1993.
65. **Cserhâti, T. and Szögyi, M.,** Interaction of phospholipids with proteins and peptides — New Advances IV, *Int. J. Biochem.*, 26, 1, 1994.
66. **Sui, S. F., Urumow, T., and Sackmann, E.,** Interaction of insulin receptors with lipid bilayers and specific and non-specific binding of insulin supported membranes, *Biochemistry*, 27, 7463, 1988.
67. **Kobayashi, H. and Matsunaga, T.,** Amino-silane modified superparamagnetic particles with surface-immobilized enzymes, *J. Coll. Interf. Sci.*, 141, 505, 1991.
68. **De Cuyper, M. and Joniau, M.,** Magnetoproteoliposomes, *J. Magn. Magn. Mat.*, 122, 340, 1993.
69. **De Cuyper, M. and Joniau, M.,** Unique possibilities for magnetoliposomes in bioreactors, in *Phosolipids: Characterization, Metabolism, and Novel Applications*, Cevc, G. and Paltauf, F., Eds., AOCS Press, Champaign, IL, 1995, chap. 10.
70. **De Cuyper, M. and Joniau, M.,** Potentialities of magnetoliposomes in studying symmetric and asymmetric phospholipid transfer processes, *Biochim. Biophys. Acta*, 1027, 172, 1990.
71. **De Cuyper, M. and Joniau, M.,** Effect of dimethylsulfoxide on the kinetics and thermodynamics of asymmetric phospholipid fluxes between magnetic and non-magnetic vesicles, *Prog. Coll. Polym. Sci.*, 84, 456, 1991.

72. **Durrani, A. A., Hayward, J. A., and Chapman, D.**, Biomembranes as models for polymer surfaces. II. The synthesis of reactive species for covalent coupling of phosphorylcholine to polymer surfaces, *Biomaterials*, 7, 121, 1986.
73. **Lasic, D. D.**, Applications in biophysics, in *Liposomes — From Physics to Applications*, Lasic, D. D., Ed., Elsevier Science Publishers, Amsterdam, 1993, 185.
74. **Van Borssum-Waalkens, M., van Galen, M., Morselt, H., Sternberg, B., and Scherphof, G. L.**, In-vitro stability and cytostatic activity of liposomal formulations of 5-fluoro-2′-deoxyuridine and its diacylated derivatives, *Biochim. Biophys. Acta*, 1148, 161, 1993.
75. **Muller, R. N., Gillis, P., Moiny, F., and Roch, A.**, Transverse relaxivity of particulate MRI contrast media: from theories to experiments, *Magn. Reson. Med.*, 22, 178, 1991.
76. **Massart, R. and Cabuil, V.**, Synthèse en milieu alcalin de magnétite colloïdale: contrôle du rendement et de la taille des particules, *J. Chimie Physique*, 84, 6159, 1987.
77. **Ishii, F., Takamura, A., and Ishigami, Y.**, Preparation and characterization of lipid vesicles containing magnetite and an anticancer drug, *J. Disper. Sci. Technol.*, 11, 581, 1990.
78. **Babincová, M.**, Microwave induced leakage of magnetoliposomes — possible clinical applications, *Bioelectrochem. Bioenerg.*, 32, 187, 1993.
79. **Sada, E., Katoh, S., and Kheirolomoom, A.**, Effects of arachidonic acid concentration on prostaglandin biosynthesis and feasibility of semibatch processes, *Appl. Biochem. Biotechnol.*, 22, 247, 1989.
80. **Wallach, D.**, Membrane lipids and the conformations of membrane proteins, *J. Gen. Physiol.*, 54, 3s, 1969.
81. **Martinek, K., Klibanov, A. M., Goldmacher, V. S., Tchernysheva, A. V., Mozhaev, V. V., Berezin, I. V., and Glotov, B. O.**, The principles of enzyme stabilization. II. Increase in the thermostability of enzymes as a result of multipoint noncovalent interaction with a polymeric support, *Biochim. Biophys. Acta*, 485, 13, 1977.
82. **Watts, A.**, Magnetic resonance studies of phospholipid-protein interactions in bilayers, in *Phospholipids Handbook*, Cevc, G., Ed., Marcel Dekker, New York, 1993, 687.
83. **Ollis, D. and Datta, R.**, Activity correlations between similarly modified soluble and immobilized enzymes, in *Methods in Enzymology*, Colowick, S. P. and Kaplan, N. O., Eds.,Vol. 44, *Immobilized Enzymes*, Mosbach, K., Ed., Academic Press, Orlando, FL, 1976, 444.
84. **Chaplin, M. F. and Bucke, C.**, *Enzyme Technology*, Cambridge University Press, Cambridge, 1990, 167.
85. **Kolm, H., Oberteuffer, J., and Kelland, D.**, High-gradient magnetic separation, *Sci. Am.*, 46, September 1975.
86. **De Cuyper, M., De Meulenaer, B., and Van der Meeren, P.**, Enzymatic activity of cytochrome c-oxidase inserted into magnetoliposomes differing in surface charge density, *Biocatal. Biotransf.*, Vol. 13, 1995, in press.
87. **De Cuyper, M.**, Effect of magnetoliposome surface charge on the enzymatic oxidation of cytochrome c, *Progr. Coll. Polym. Sci.*, 1996, in press.
88. **De Cuyper, M., De Meulenaer, B., Van der Meeren, P., and Vandereelen, J.**, Catalytic durability of magnetoliposomes captured by high-gradient magnetic forces in a miniature fixed-bed reactor, *Biotechnol. Bioeng.*, 1996, in press.
89. **Lasic, D. D.**, Applications in chemistry, in *Liposomes — From Physics to Applications*, Lasic, D. D., Ed., Elsevier Science Publishers, Amsterdam, 1993, 219.
90. **Mosbach, K. and Andersen, L.**, Magnetic ferrofluids for preparation of magnetic polymers and their application in affinity chromatography, *Nature*, 270, 259, 1977.
91. **Meutz, E. T., Havelick, J., Groman, E. V., and Josephson, L.**, Magnetic affinity chromatography: an emerging technique, *Am. Biotech. Lab.*, 4, 46, 1986.
92. **Lundahl, P. and Yang, Q.**, Liposome chromatography: liposomes immobilized in gel beads as a stationary phase for aqueous column chromatography, *J. Chromat.*, 544, 283, 1991.
93. **Haukanes, B.-I. and Kvam, C.**, Application of magnetic beads in bioessays, *Bio/Technology*, 11, 60, 1993.
94. **Bacri, J.-C., Perzynski, R., and Salin, D.**, Magnetic liquids, *Endeavour (New Series)*, 12, 76, 1988.
95. **Hirschbein, B. L., Brown, D. W., and Whitesides, G. M.**, Magnetic separations in chemistry and biochemistry, *Chemtech.*, 172, March 1982.
96. **Rosensweig, R. E.**, Fluid dynamics and science of magnetic fluids, *Adv. Elect. Electron Phys.*, 48, 103, 1979.
97. **Odenbach, S.**, Magnetic fluids, *Adv. Colloid Interface Sci.*, 46, 263, 1993.
98. **Müller-Schulte, D., Füssl, F., Leuken, H., and De Cuyper, M.**, Neuer Ansatz für die AIDS-Therapie unter Verwendung Superparamagnetischer Nanopartikel, in *Alma Mater Aquensis*, Vol. 32, Rhenish-Westphalian-Technical University Aachen, Germany, 1996, in press.

INDEX

A

B

C

D

E

F

G

H

I

K

L

M

N

O

P

Q

R

S

T

U

V

W

Y

Z

Table 4.8 Minimizing an FSR, M.

(1) reverse *M*
(2) make the result of (1) deterministic (using the subset method)
(3) reverse the result of (2)
(4) make the result of (3) deterministic (using the subset method)
The result of (4), called M_{min}, is the minimal finite state recognizer for the language recognized by *M*.

Table 4.8 specifies an algorithm that uses our reverse (Table 4.7) and subset (Table 4.6) procedures to produce a minimal machine, M_{min}, from a deterministic FSR, *M*.

We can demonstrate the operation of the above method on our example machine $M_6{}^d$ from above (we use the one with the renamed states):

Step 1: Reverse $M_6{}^d$ (Figure 4.20(b)), giving the machine shown in Figure 4.21.

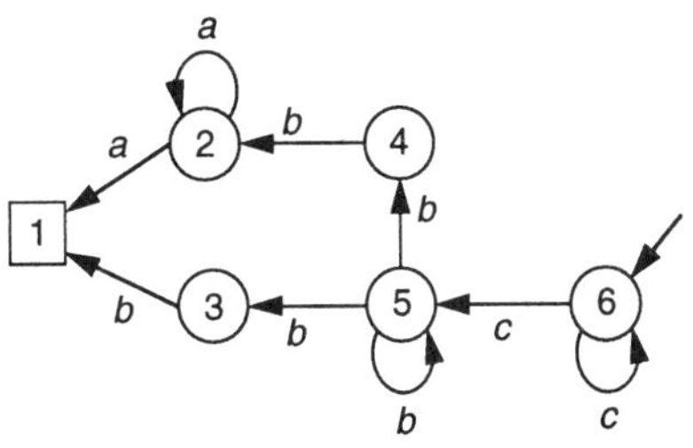

Figure 4.21 $M_6{}^d$ of Figure 4.20 reversed

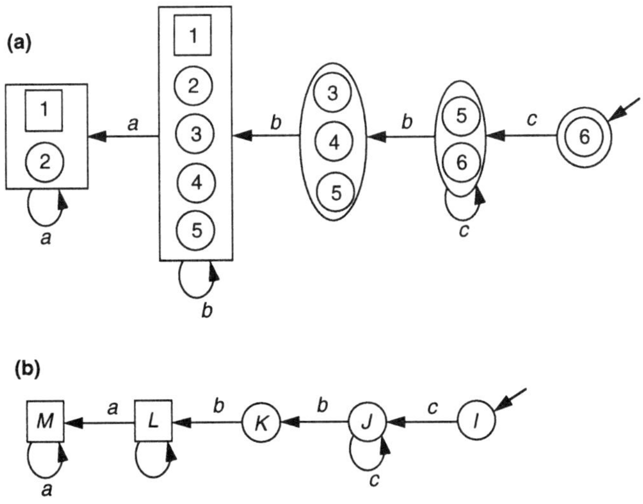

Figure 4.22 (a) The deterministic version of Figure 4.21; (b) The machine (a) with renamed states

Step 2: Make the result of step 1 (Figure 4.21) deterministic. The result of step 2, after renaming the states, can be seen in Figure 4.22(b).

Step 3: Reverse the result of step 2. The reversed version of Figure 4.22(b) can be seen in Figure 4.23.

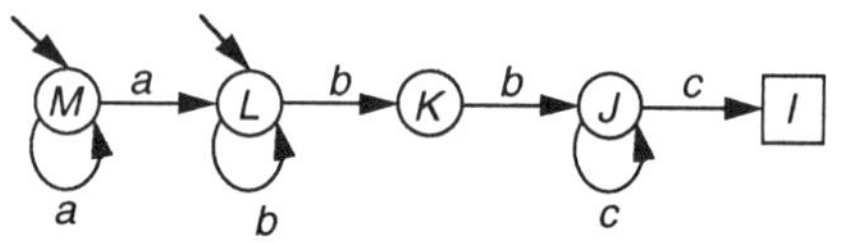

Figure 4.23 The reverse of Figure 4.22(b)

Note that the machine in Figure 4.23 has two **start states**; this is permitted in our rules for FSR behaviour, and in rule 1 of the subset algorithm. However, this represents a *non deterministic* situation, since *such a machine effectively has a choice as to which state it starts in.*

Step 4: Make the result of step 3 (Figure 4.23) deterministic. We obtain the machine depicted in Figure 4.24(a).

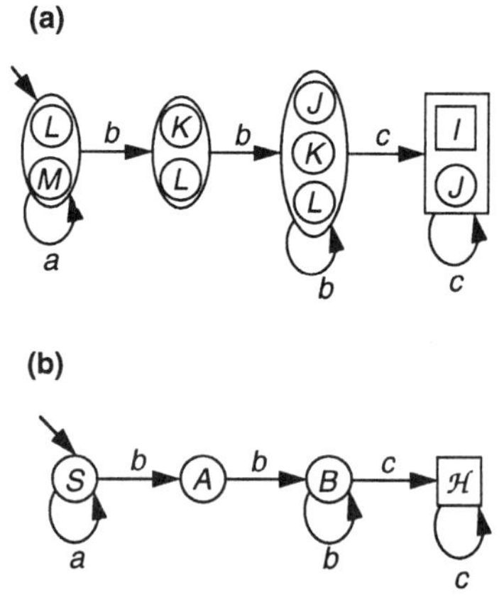

Figure 4.24 (a) Deterministic version of Figure 4.23; (b) with renamed states giving the minimal FSR for M_6, $M_{6\text{min}}$

The machine in Figure 4.24(b), resulting from renaming the states of the machine in Figure 4.24(a), has been called $M_{6\text{min}}$. It is *the minimal deterministic FSR for the language recognized by* M_6^d.

We now see clearly that the language recognized by $M_{6\text{min}}$ (and, of course, M_6 and M_6^d), is:

$\{a^i b^j c^k \mid i \geq 0, j \geq 2, k \geq 1\}$,

i.e. all sentences are of the form:

zero or more *a*s, followed by two or more *b*s, followed by one or more *c*s.

With a little thought, we can easily justify that the machine is minimal. We must have one state for the 'zero or more as' part, a further two states for the 'two or more bs' part, and an extra state to ensure the 'one or more *c*s' part.

4.6.2 How minimization works

The minimization algorithm of Table 4.8 simply uses the *subset algorithm* twice, first applying it to the *reversed form* of the original machine, and then to the *reversed form* of the *deterministic reversed form*. Let us consider why this approach produces the correct result. To do this, we focus on the example FSR in Figure 4.25.

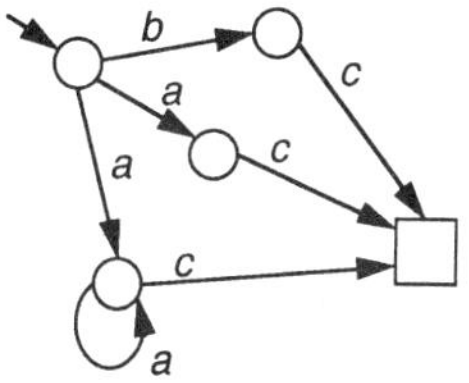

Figure 4.25 An FSR that is also non deterministic when in reverse

Making the machine in Figure 4.25 deterministic, by using our subset algorithm, we merge the two paths labelled *a* into one. We then obtain the machine in Figure 4.26.

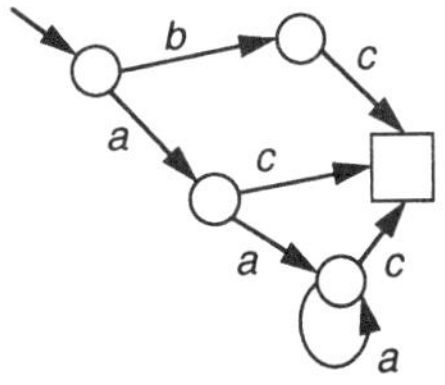

Figure 4.26 Subset method applied to Figure 4.25

Now, while the machine in Figure 4.26 is deterministic, it is *not* minimal. It seems that we ought to be able to do something about the fact that regardless of whether the first symbol is an *a* or a *b*, the machine could get to its halt state given one *c*. However, as this is not a non deterministic situation, the subset algorithm has ignored it.

Now, reversing the machine in Figure 4.26 gives us the machine in Figure 4.27.

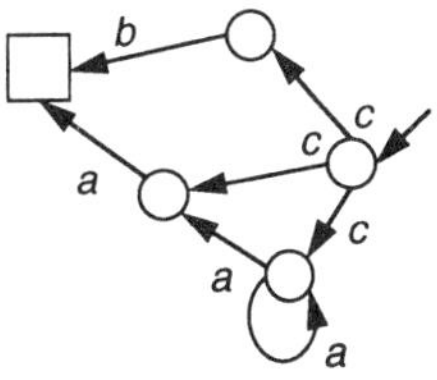

Figure 4.27 Reverse of Figure 4.26

The 'three *c*s situation' has become non deterministic in Figure 4.27. The subset algorithm will now deal with this, resulting in the machine of Figure 4.28.

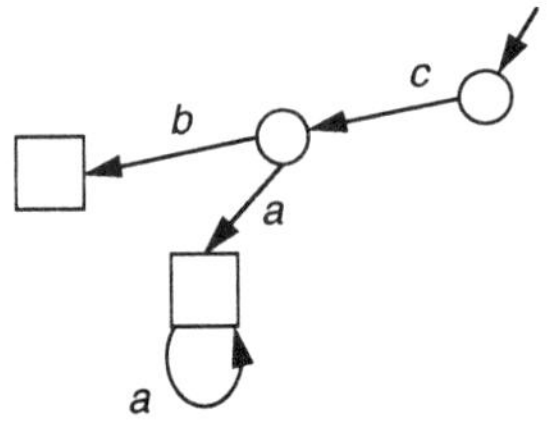

Figure 4.28 Deterministic version of Figure 4.27

While the machine in Figure 4.28 now consists of only *four* states, and is deterministic, it unfortunately recognizes our original language *in reverse*, so we must reverse it again, giving Figure 4.29.

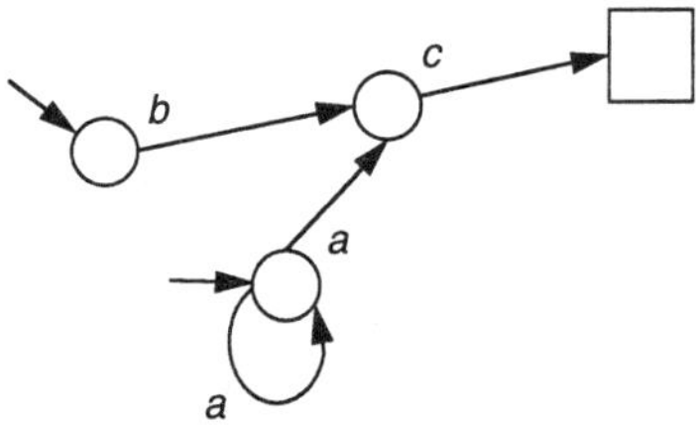

Figure 4.29 The reverse of Figure 4.28

The Figure 4.29 machine now recognizes our original language, but is *non deterministic* (it has *two* start states and *a state with two outgoing arcs labelled a*). So, we make it deterministic again, giving the machine of Figure 4.30.

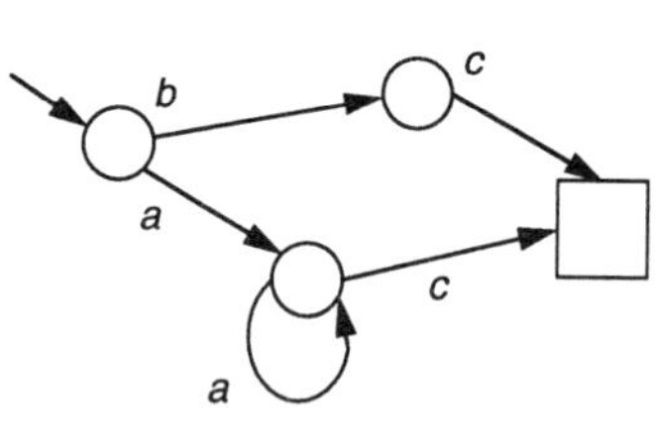

Figure 4.30 Deterministic and minimal version of Figure 4.29

The machine in Figure 4.30 is the *minimal (deterministic) machine*.

The algorithm thus removes non determinism 'in both directions' through the machine. Reversing a machine creates a machine that recognizes the original language with each and every string reversed, and the subset algorithm always results in a machine that accepts exactly the same strings as the machine to which it is applied. Thus, making the reversed machine deterministic and then reversing the result will not change the language recognized by the machine.

There are other ways of creating a minimal FSR. One of the most popular is based on *equivalence sets*. The beauty of the method we use here is that it is very simple to specify, and makes use of a method we defined earlier, i.e. the subset algorithm. Thus, the minimization method we have used here has a pleasing modularity. It uses the *subset* and *reverse* methods as *subroutines*, so to speak.

4.7 THE GENERAL EQUIVALENCE OF REGULAR LANGUAGES AND FSRS

We saw above that any regular grammar can be converted into an equivalent FSR. In fact, the converse is also true, i.e.,

every FSR can be converted into a regular grammar.

We will consider an example showing the conversion of an FSR into regular productions. The interesting thing about our example is that the particular machine in question has transitions labelled ε. These are called **empty moves** or **ε-moves**, and enable an FSR to make a transition without **reading** its input. As we will see, they produce slightly different **productions** from the ones we have seen up to now. We consider this observation in more detail after working through the following example, which is based on the machine M shown in Figure 4.31.

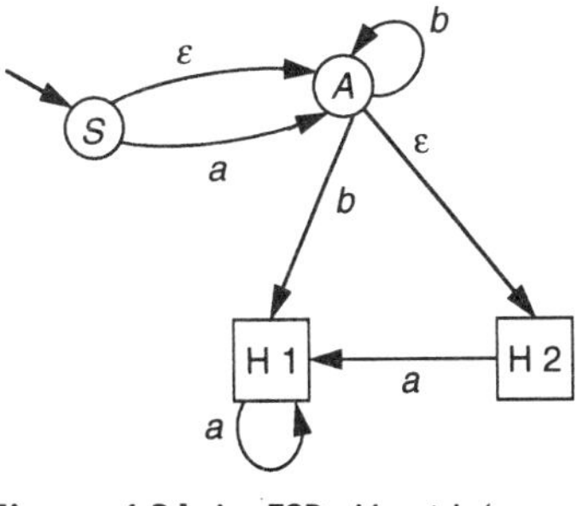

Figure 4.31 An FSR, *M*, with 'empty moves'

The grammar that we obtain from M, in Figure 4.31, which we will call G_m, is as follows:

$S \rightarrow aA \mid A$
$A \rightarrow bA \mid bX \mid Y$
$X \rightarrow aX \mid \varepsilon$
$Y \rightarrow aX \mid \varepsilon$

Some issues are raised by this construction:

- *M* can get to its halt state *H*1 without reading any input. Thus, ε is in the language accepted by *M*. As we would therefore expect, G_m generates ε ($S \Rightarrow A \Rightarrow Y \Rightarrow \varepsilon$). The empty move from *S* to *A* becomes a production $S \rightarrow A$. Productions of the form $x \rightarrow y$ where *x* and *y* are both non terminals are called **unit productions**, and are actually prohibited by the definition of regular grammar productions with which we began this chapter. However, we see in Chapter 5 that there is a method for removing these productions, leaving a grammar that *does* conform to our specification of regular grammar productions, and generates the same language as the original.
- *M* has arcs *leaving* halt states, and *has more than one halt state*. We saw above that the subset algorithm produces machines with these characteristics. In any case, there is nothing in the definition of the FSR that precludes multiple halt *or* start states. Where there is an arc leaving a halt state, such as the arc labelled *a* leaving *H*2, the machine could either halt or read an *a*. Halting in such a situation is exactly the same as an empty move, so in the grammar the *H*2 situation becomes $Y \rightarrow aX \mid \varepsilon$, where *Y* represents *H*2, and *X* represents *H*1.

I leave it to you to convince yourself that $L(G_m)$ is identical to the language accepted by *M*, and that the way the productions of G_m were obtained from *M* could be applied to any FSR, no matter how complex, and no matter how many empty moves it has. In the above machine, the start state was conveniently labelled *S*. We can always label the states of an FSR how we please; a point made earlier in this chapter. The only other potential problem that arises is for FSRs with more than one start state: in this case we can simply introduce a new *single* start state, with outgoing arcs labelled ε leaving it and directly entering each state that was a start state of the original machine (which we then no longer class as start states). Then the machine can be converted into productions, as was *M*, above.

The above should convince you that the following statement is true.

> for any FSR *M*, there exists a regular grammar, *G*, which is such that *L*(*G*), the language generated by *G*, is equal to the set of strings accepted by *M*.

Our earlier result that for every regular grammar there is an **equivalent** FSR means that if we come across a language that *cannot* be accepted by an FSR, we know that we are wasting our time trying to create a regular grammar to generate that language. As we will see in Chapter 6, there are properties of the FSR that we can exploit to establish that certain languages could not be accepted by the FSR. We now know that if we do this, we have also established that we could not define a regular grammar to generate that language. Conversely, if we can establish that there is no regular grammar to generate a given language, we know it is pointless attempting to create an FSR, or a parser of 'FSR-level-power', like the parsing program earlier in the chapter, to accept that language.

4.8 OBSERVATIONS ON REGULAR GRAMMARS AND LANGUAGES

The properties of regular languages described in this chapter make them very useful in a computational sense. We have seen that simple parsers can be written for a language generated by a regular grammar. We have also seen how we can make our parsers extremely efficient, by ensuring that they are deterministic and reduced to the smallest possible form, while reflecting a modular structure. We have also seen that there is a general equivalence between the FSRs and the regular languages.

Unfortunately, the simple structure of regular grammars means that they can only be used to represent simple constructs in programming languages, for example, basic objects such as *identifiers* (variable names) in Pascal programs. In fact, the definition of Pascal 'identifiers' given in **context free** form in Chapter 2 could be represented as a regular grammar. An alternative representation of the regular languages are symbolic statements known as the *regular expressions*. These are not only a purely formal device, but are also implemented in text editors such as the Unix 'vi' editor, and the 'emacs' text editor.

We will defer more detailed discussion of the limitations of regular grammars until Chapter 6. In that chapter we show that even a language as simple as $\{a^ib^i : i \geq 1\}$, generated by the two-production context free grammar $S \to aSb \mid ab$ (grammar G3 of Chapters 2 and 3) is not regular. If such a language cannot be generated by a regular grammar, then a regular grammar is not likely to be capable of specifying syntactically valid Pascal programs, where, for example, a parser has to ensure that the number of **begins** is the same as the number of **ends**. In fact, when it comes to programming languages, the context free grammars as a whole are much more important[2]. Context free grammars and context free languages are the subject of the next chapter.

EXERCISES

For exercises marked '†', solutions, partial solutions, or hints to get you started appear in *Solutions to selected exercises* at the rear of the book.

1 Construct **FSRs** equivalent to the following grammars. For each FSR briefly justify whether or not the machine is **deterministic**. If the machine is **non deterministic**, convert it into an equivalent deterministic machine.

(a)† $S \to aA$

$A \to aS \mid aB$

$B \to bC$

$C \to bD$

[2] Bear in mind that the class of the context free grammars **includes** the class of the regular grammars.

$D \rightarrow b \mid bB$

Note: This is the grammar from Chapter 2, Exercise 1 (a)

(b) $S \rightarrow aA \mid aB \mid aC$
$A \rightarrow aA \mid aC \mid bC$
$B \rightarrow aB \mid cC \mid c$
$C \rightarrow cC \mid c$

(c) Grammar G_1 from Chapter 2, i.e.
$S \rightarrow aS \mid bB$
$B \rightarrow bB \mid bC \mid cC$
$C \rightarrow cC \mid c$

2 Given the FSR in Figure 4.32

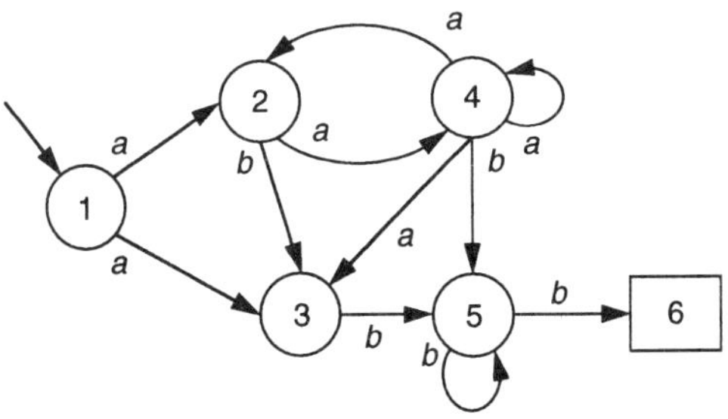

Figure 4.32 An FSR

produce a **minimal** version.

Construct an **equivalent** regular grammar.

Hint: it is often more straightforward to use the original, non deterministic machine (in this case the one above) to obtain the productions of the grammar, i.e., we write down the productions that would have given us the above machine according to our conversion rules.

3 Design suitable data structures to represent FSRs. Design a program to apply operations such as the subset method and the minimization algorithm to FSRs.

4 The **subset algorithm** is so called because each new state of M^d, the deterministic machine, produced contains a subset of the states of M. An algorithm is a procedure that always terminates for valid inputs. Consider why we can be sure that the subset procedure will terminate given any non deterministic machine as its input, and thus deserves to be called an algorithm.

5 Context free languages and pushdown recognizers

AIMS

To consider the context free grammars (CFGs) and context free languages, including learning

- how to remove the empty string (ε) from a CFG
- how to convert a CFG into Chomsky Normal form
- how to create a non deterministic PDR from a CFG
- about deterministic PDRs
- about the deterministic and non deterministic CFLs
- the implications of the fact that not all CFLs are deterministic

5.1 CONTEXT FREE GRAMMARS AND CONTEXT FREE LANGUAGES

Recall from Chapter 2 that the context free grammars (type 2 in the **Chomsky hierarchy**) are **Phrase Structure Grammars (PSGs)** in which each and every production conforms to the following pattern:

$x \rightarrow y,\ x \in N,\ y \in (N \cup T)^*$

i.e. x is a single **non terminal**, and y is an arbitrary (possibly empty) mixture of **terminals** and/or non terminals.

So, the grammar G_2 of Chapter 2, i.e.:

$S \rightarrow aB \mid bA \mid \varepsilon$
$A \rightarrow aS \mid bAA$
$B \rightarrow bS \mid aBB$

is context free. If G is a CFG, then $L(G)$, the **language generated by the grammar** G, is a **context free language (CFL)**. We will consider a further example CFG, G_7:

$S \rightarrow ABC$
$A \rightarrow aAb \mid B$
$B \rightarrow bB \mid \varepsilon$
$C \rightarrow ccC \mid B$

$L(G_7) = \{a^h b^i c^{2j} b^k : h, j, k \geq 0, i \geq h\}$

i.e. 0 or more *a*s followed by at least as many *b*s, followed by an *even* number of (or no) *c*s, followed by 0 or more *b*s as you may like to justify for yourself. You may arrive at a different, though equivalent, **set definition**; if so, check that it describes the same set as the one above.

5.2 CHANGING G WITHOUT CHANGING $L(G)$

The purpose of this section is to show how one can modify the productions of a CFG in various ways *without changing the language the CFG generates*. There are more ways of modifying CFGs than are featured here, but the two methods below are relatively simple to apply. The results of this section serve also to simplify various proofs in Chapter 6.

5.2.1 *The empty string*

Our example grammar G_7, above, features the **empty string** (ε) on the **right hand side** of one of its productions. The empty string often causes problems in **parsing**, as was noted in Chapter 3, and again in the context of **empty moves** in **finite state recognizers (FSRs)**. Ideally, then, for parsing purposes, we would not like the empty string to feature in our grammars. Indeed, you may wonder whether the empty string actually features in any 'real world' situations, or if it is included for mathematical reasons. In string concatenation for example, ε could be regarded as playing a similar role as does 0 in addition, or 1 in multiplication.

However, it is quite common to find the empty string in the definition of programming languages. Pascal, for example, allows a <statement> to be *empty*, and such 'empty' statements have their practical uses. For example, the programmer may leave empty 'actions' in a 'case' statement, or an empty 'else' part of an 'if' statement, so that additional code can be entered later. Note, however, that the empty string is not a sentence of the Pascal language. The Pascal syntax does not allow for an empty <program>, and thus, according to the definition of the language, an empty source code file should not compile.

It turns out, however, that for CFGs we can deal with the empty string in one of two ways. Firstly, for any PSG, G, we can, by applying the rules we are about to see, remove it altogether from the productions, except for one special case which we discuss later. We will need the terminology defined in Table 5.1 to discuss the empty string in CFGs.

Table 5.1 Terminology relevant to the empty string in grammars

Term	Definition	Examples from G_7
ε production	any production with ε on the right hand side	$B \rightarrow \varepsilon$
ε generating non terminal	any non terminal, X, which is such that $X \overset{*}{\Rightarrow} \varepsilon$ (i.e, any non terminal from which we can generate the empty string)	S, A, B and C since we can have: $S \Rightarrow ABC$ and $A \Rightarrow B \Rightarrow \varepsilon$ and $C \Rightarrow B \Rightarrow \varepsilon$
ε free production	any production which does not have ε on the right hand side	all productions except $B \rightarrow \varepsilon$

The method for removing ε productions from a grammar is described in Table 5.2. On the right of Table 5.2 we see the method applied to our grammar G_7.

You may like to verify for yourself that the new grammar, which in Table 5.2 has been called G_8, generates exactly the same set of strings as the grammar with which we started (G_7).

Table 5.2 How to remove ε from a CFG

Op No	Operation Description	As applied to G_7
1	Put all the ε generating non terminals into a set, call the set X	S, A, B, C – set X
2	Put all the ε free productions into another set, call the set Y	$S \to ABC$ $A \to aAb \mid B$ $B \to bB$ $C \to ccC \mid B$ – set Y
3	for each production, p, in Y, add to Y any ε free productions we can make by omitting one or more of the non terminals in X from p's right hand side	$S \to ABC \mid AB \mid AC \mid BC \mid A \mid B \mid C$ $A \to aAb \mid B$ $B \to bB \mid b$ $C \to ccC \mid cc \mid B$ – new set Y (G_8)
(4)	WE ONLY DO THIS IF S IS MEMBER OF SET X:	
	4.1 replace the non terminal S throughout the set Y by a new non terminal, A, which is not in any of the productions of Y 4.2 put the productions $S \to A \mid \varepsilon$ into Y	$S \to Z \mid \varepsilon$ $Z \to ABC \mid AB \mid AC \mid BC \mid A \mid B \mid C$ $A \to aAb \mid B$ $B \to bB \mid b$ $C \to ccC \mid cc \mid B$ – new set Y

Referring to Table 5.2, at stage 3, the productions added to the set Y were necessary to deal with cases where one or more non terminals in the original productions could have been used to generate ε. For example, we have the production $S \to ABC$. Now in G_7, the B could have been used to generate ε, so we need to cater for this by introducing the production $S \to AC$. Similarly, both the A and the C could have been used to generate ε, so we need a production $S \to B$, and so on. Of course, we still need to include $S \to ABC$ itself, because, in general, the ε **generating** non terminals may generate other things apart from ε (as they do in our example).

As for stage 4, it is important to remember that we only do this if ε is in the language generated by the grammar (in formal parlance, if $\varepsilon \in L(G)$). This occurs only when S itself is an ε generating non terminal, as it is in the case of G_7. If S is an ε generating non terminal for a grammar G, then obviously ε is in $L(G)$. Conversely, if S is not such a non terminal, then ε cannot be in the language. If we applied step 4 to a grammar for which ε is not in the corresponding language, the result would be a new grammar that did not generate the same language (it would generate the same set of strings along with ε, formally $L(G) \cup \{\varepsilon\}$).

Alternatively, if we did not apply step 4 to a grammar in which S *is* an ε generating non terminal, then we would produce a grammar which generated all the strings in $L(G)$ except for the string ε (formally $L(G) - \{\varepsilon\}$). We must always take care to preserve exactly the set of strings generated by the original grammar.

If we were to apply the above rules to a *regular* grammar, *the resulting grammar would also be regular*. That is why Chapter 4 explained that dealing with the empty string in regular grammars would be best left to this chapter.

For **parsing**, the grammars produced by the above rules are useful, since at most they contain one ε production, and if they do, it is $S \to \varepsilon$. This means that when we parse a string x:

- if x is non empty, our parser need no longer consider the empty string; it will not feature in any of the productions that could have been used in the derivation of x;
- if, on the other hand, $x = \varepsilon$, either our grammar has the one ε production $S \to \varepsilon$, in which case we accept x, or it has no ε productions, and we simply reject it.

However, while for parsing purposes our amended grammars may be useful, there may be reasons for leaving the empty string in a grammar showing us the structure of, say, a programming language. One of these reasons is readability. As we can see, G_8 consists of many more productions (16) than does the equivalent G_7 (seven), and thus it could be argued that G_7 is easier for us to read. A format for grammars useful for a parser may not necessarily support one of the other purposes of grammars, i.e. to show us the structure of languages.

To complete this section, we consider a further example of the removal of ε productions from a CFG. In this case ε is not in the language generated by the grammar. Only a sketch of the process will be given; you should be able to follow the procedure by referring to the rules in Table 5.2.

The example grammar, G_9, is as follows:

$S \to aCb$
$C \to aSb \mid cC \mid \varepsilon$

In this case, the only ε generating non terminal is C. S cannot generate ε as any string it generates *must* contain at least one a and one b. The new productions we would add are $S \to ab$ and $C \to c$, both resulting from the omission of C. Step 4 should *not* be applied in this case, as S is not ε generating. The final amended productions would be:

$S \to aCb \mid ab$
$C \to aSb \mid cC \mid c.$

A set definition of the language generated by this grammar, which, of course, is exactly the same as $L(G_9)$, is:

$$\{a^{2i+1}c^j b^{2i+1} : i, j \geq 0\}$$

i.e. a non zero *odd* number of *a*s, followed by 0 or more *c*s, followed by the same number of *b*s as there were *a*s.

5.2.2 *Chomsky normal form*

Once we have removed the empty string (ε) from a context free grammar, isolating it if necessary in one production, $S \rightarrow \varepsilon$, there are various methods for transforming the ε free part of G while leaving $L(G)$ unchanged. One of the most important of these methods results in the productions being in a form we call **Chomsky Normal Form** (CNF). This section briefly describes the method for converting a CFG into CNF by presenting a worked example. That we can assume that *any CFG can be converted into CNF* is useful in Chapter 6.

As an example grammar, we use part of G_8 (the ε free version of G_7), and assume in this case that Z is the start symbol:

$Z \rightarrow ABC \mid AB \mid AC \mid BC \mid A \mid B \mid C$
$A \rightarrow aAb \mid B \mid A$
$B \rightarrow bB \mid B \mid b$
$C \rightarrow ccC \mid cc \mid B$

The method to 'Chomskyfy' a grammar is as follows. There are three steps:

Step 1: Unit productions

Unit productions were briefly introduced in Chapter 4. A unit production is a production of the form: $x \rightarrow y$, where x and y are single non terminals. Thus, G_8 has the following unit productions:

$Z \rightarrow A \mid B \mid C$
$A \rightarrow B$
$C \rightarrow B$

To help us deal with the unit productions of G_8, we draw the structure shown in Figure 5.1, representing all derivations that begin with the application of a unit production and end with a string that does not consist of a single non terminal (each path stops the first time such a string is encountered).

The structure in Figure 5.1 is not a tree, but a **graph**. We can call the structure a **unit production graph**. It could be drawn as a tree if we repeated some nodes at lower levels (in the graph in Figure 5.1 we would repeat B beneath A and beneath C). However, such a representation would not be so useful in the procedure for creating new non unit productions described below.

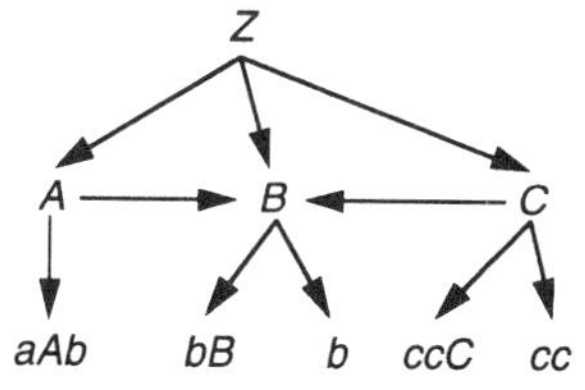

Figure 5.1 A unit production graph

Some unit production graphs we draw may lead to even more circular (known as cyclical) situations, as in the example in Table 5.3.

Table 5.3 Example productions and a corresponding unit production graph

Example productions	Resulting graph
$S \rightarrow aS \mid B$ $B \rightarrow bS \mid A \mid C$ $A \rightarrow a \mid C$ $C \rightarrow bA \mid S \mid B$	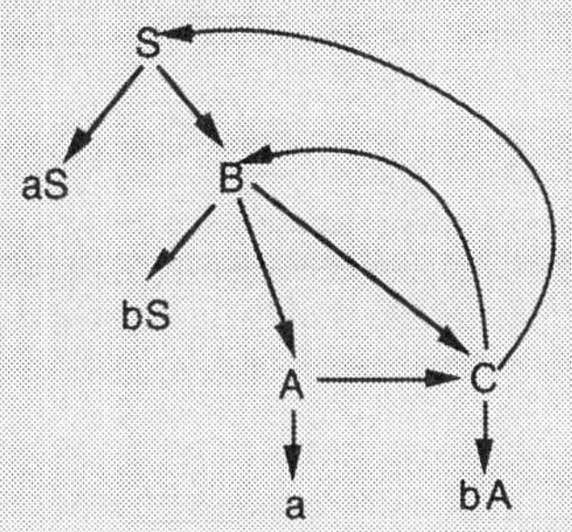

We now continue with our initial example. From the unit production graph for G_8, (Figure 5.1), we obtain:

> every possible non-unit production we can obtain by following a directed path from every non terminal to every node that does not consist of a single non terminal, where each production we make has a left hand side that is the non terminal we started at, and a right hand side that is the string where we stopped.

So, from our unit production graph for G_8 in Figure 5.1, we obtain the following productions (those crossed out are already in G_8, so we need not add them):

$Z \rightarrow bB \mid b \mid aAb \mid ccC \mid cc$	{by following all paths from Z}
$A \rightarrow$ ~~aAb~~ $\mid bB \mid b$	{by following all paths from A}
$B \rightarrow$ ~~bB~~ $\mid b$	{...from B}
$C \rightarrow$ ~~ccC~~ $\mid$ ~~cc~~ $\mid bB \mid b$	{...and from C}

We remove all the unit productions from the grammar and add the new productions to it, giving $G_{8(a)}$ (new parts are underlined):

$\underline{Z} \rightarrow ABC \mid AB \mid AC \mid BC \mid \underline{bB} \mid \underline{b} \mid \underline{aAb} \mid \underline{ccC} \mid \underline{cc}$
$\underline{A} \rightarrow aAb \mid \underline{bB} \mid \underline{b}$
$B \rightarrow bB \mid b$
$C \rightarrow ccC \mid cc \mid \underline{bB} \mid \underline{b}$

You should perhaps take a moment to convince yourself that the terminal strings we can generate from $G_{8(a)}$ (remember Z is currently assumed to be the start symbol) are exactly the same as those we can generate starting from Z in G_8. Observe, for example, that from the initial grammar, we could perform the derivation $Z \Rightarrow A \Rightarrow B \Rightarrow C \Rightarrow ccC \Rightarrow cccc$. With $G_{8(a)}$ this derivation is simply $Z \Rightarrow ccC \Rightarrow cccc$.

In Chapter 4, with reference to the conversion of an arbitrary finite state recognizer into regular grammar productions, we noted that *empty moves in an FSR correspond to unit productions*. You can now see that if we applied step 1 of 'Chomskyfication' to the ε free part of a regular grammar containing unit productions, we would obtain a grammar that was also regular.

Step 1 is now complete, as unit productions have been removed. We turn now to step 2.

Step 2: Productions with terminals and non terminals on the right hand side, or with more than one terminal on the right hand side

Our grammar $G_{8(a)}$ now contains no unit productions. We now focus on the productions in $G_{8(a)}$ described in the title to this step. These productions we call *secondary productions*:

$Z \rightarrow bB \mid aAb \mid ccC \mid cc$
$A \rightarrow aAb \mid bB$
$B \rightarrow bB$
$C \rightarrow ccC \mid cc \mid bB$

To these productions we do the following:

> replace each terminal, t, by a new non terminal, N, and introduce a new production $N \rightarrow t$, in each case

So, for example, from $A \rightarrow aAb \mid bB$, we get:

$A \rightarrow JAK \mid KB$
$J \rightarrow a$
$K \rightarrow b$

Now we can use J and K to represent a and b, respectively, throughout our step 2 productions, adding all our new productions to $G_{8(a)}$ in place of the original step 2 productions. Then we replace c by L in all the appropriate productions, and introduce a new production $L \rightarrow c$. The result of this we will label $G_{8(b)}$:

$\underline{Z} \rightarrow ABC \mid AB \mid AC \mid BC \mid b \mid \underline{KB} \mid \underline{JAK} \mid \underline{LLC} \mid \underline{LL}$
$\underline{A} \rightarrow \underline{JAK} \mid \underline{KB} \mid b$
$\underline{B} \rightarrow \underline{KB} \mid b$
$\underline{C} \rightarrow \underline{LLC} \mid \underline{LL} \mid \underline{KB} \mid b$
$\underline{J \rightarrow a}$
$\underline{K \rightarrow b}$
$\underline{L \rightarrow c}$

The new productions are again underlined. You can verify that what we have done has not changed the terminal strings that can be generated from Z.

We are now ready for the final step.

Step 3: Productions with more than two non terminals on the right hand side

After steps 1 and 2, all our productions will be of the form:

$x \rightarrow y$ or where y consists of two or more non terminals, and
$x \rightarrow z$ z is a single terminal.

As the step 3 heading suggests, we now focus on the following productions of $G_{8(b)}$:

$Z \rightarrow ABC \mid JAK \mid LLC$
$A \rightarrow JAK$
$C \rightarrow LLC.$

As an example, consider $Z \rightarrow ABC$. It will not affect things if we introduce a new non terminal, say P, and replace $Z \rightarrow ABC$ by:

$Z \rightarrow AP$
$P \rightarrow BC.$

The $G_{8(b)}$ derivation $Z \Rightarrow ABC$ would then involve doing: $Z \Rightarrow AP \Rightarrow ABC$. As P is not found anywhere else in the grammar, this will not change any of the terminal strings that can be derived. We treat the rest of our step 3 productions analogously, again replacing the old productions by our new ones, and using a new non terminal in each case. This gives us the grammar $G_{8(c)}$:

$\underline{Z} \rightarrow \underline{AP} \mid AB \mid AC \mid BC \mid KB \mid b \mid \underline{JQ} \mid LR \mid LL$
$\underline{P \rightarrow BC}$
$\underline{Q \rightarrow AK}$
$\underline{R \rightarrow LC}$
$\underline{A} \rightarrow \underline{JT} \mid KB \mid b$
$\underline{T \rightarrow AK}$
$B \rightarrow KB \mid b$
$\underline{C} \rightarrow \underline{LU} \mid LL \mid KB \mid b$
$\underline{U \rightarrow LC}$

$J \rightarrow a$
$K \rightarrow b$
$L \rightarrow c$

Changes have again been indicated by underlining.

The resulting grammar, $G_{8(c)}$, is in Chomsky normal form.

It is important to appreciate that, unlike in G_8, step 3 productions may include *right hand sides of more than three non terminals*. For example, we may have a production such as:

$E \rightarrow DEFGH.$

In such cases, we split the production up into productions as follows:

$E \rightarrow DW$	where W, X, and Y are new non terminals, not found
$W \rightarrow EX$	anywhere else in the grammar.
$X \rightarrow FY$	
$Y \rightarrow GH.$	

So the old derivation $E \Rightarrow DEFGH$ now becomes the new derivation:

$E \Rightarrow DW \Rightarrow DEX \Rightarrow DEFY \Rightarrow DEFGH.$

From the final example, you should be able to induce the general rule for creating several 'two-non-terminals-on-the-right-hand-side' productions from one 'more-than-two-non-terminals-on-the-right-hand-side' production.

If we apply steps 1, 2 and 3 to any ε free CFG, we obtain a CFG where all productions are of the form:

$x \rightarrow y$ or	where y is a single terminal,
$x \rightarrow uv$	where u and v are single non terminals.

A CFG like this is said to be in Chomsky normal form (CNF).
So, it should be clear that

any ε free CFG can be converted into CNF.

Once the ε free part of the grammar has been converted into CNF, we need to include the two productions that resulted from step 4 of the method for removing the empty string (Table 5.2), *but only if step 4 was carried out for the grammar*. In the case of G_8, step 4 *was* carried out. To arrive at a grammar equivalent to G_8 we need to add the productions $S \rightarrow \varepsilon \mid Z$ to $G_{8(c)}$, and designate S as the start symbol again.

It is very useful to know that any CFG can be converted into CNF. For example, we will use this 'CNF assumption' to facilitate an important proof in the next chapter. In a CFG in CNF, all **reductions** will be considerably simplified, as the right hand sides of productions consist of single terminals or two non terminals. A parser using such a grammar need only examine two adjacent non terminals or a single terminal symbol in the **sentential form** at a time. This simplifies things considerably, since a non-CNF CFG could have a large number of terminals and/or non terminals on the right hand side of some of its productions.

A final point is that if the CFG is in CNF, any **parse tree** will be such that every non terminal node expands into either two non terminal nodes or a single terminal node. So, for example, a parser written in Pascal could use a three field *record* to hold details of each non terminal node that expands into two non terminals (one field for the node's label, and two pointers to the two non terminals into which the non terminal expands). A single type of record structure would therefore be sufficient to represent all the non terminal nodes of the parse tree.

5.3 PUSHDOWN RECOGNIZERS

We saw in Chapter 4, that from any **regular grammar** we can build an abstract machine called a **finite state recognizer**. The machine is such that it recognizes all (and not any other) strings that are generated by the corresponding grammar. Moreover, for any finite state recognizer we invent we can produce an **equivalent** regular grammar. It was also said (though it will not be proved until Chapter 6) that there is at least one language, $\{a^i b^i : i \geq 1\}$, that *cannot* be recognized by any finite state recognizer (and therefore cannot be generated by a regular grammar), which nevertheless we have seen to be context free.

In terms of the processing of the context free languages, the problem with the finite state recognizer is that it has no memory device: it has no way of 'remembering' details about the part of the string it has processed at a given moment in time. This observation is crucial in several results considered in this book. The FSR's only memory is in terms of the state it is in, which provides limited information. For example, if a finite state recognizer wishes to 'count' n symbols, it has to have n states with which to do it. This immediately makes it unsuitable even for a simple context free language such as $\{a^i b^i : i \geq 1\}$, that requires a recognizer to be able to 'count' a number of *a*s that cannot be predicted in advance, so that it can make sure that an equal number of *b*s follows.

It turns out that with the addition of a rudimentary memory device called a **stack**, the finite state recognizer can do the job for context free languages that the basic finite state recognizer can do for regular languages. We call such an augmented finite state recognizer a **pushdown recognizer** (**PDR**). Before we examine these machines in detail, we need to be clear about what is meant by a *stack*.

5.3.1 The stack

A stack is simply a linear data structure that consists of a pile of symbols. The stack can only be accessed by putting a symbol on top of, or removing a symbol from the top of, the pile. The stack has a special marker (⊥), called the **stack bottom marker** which is always at the bottom of the pile. We use a diagrammatic representation of a stack as shown in Figure 5.2. If you prefer formal definitions, our stacks are really *strings* with the symbol ⊥ at the rightmost end.

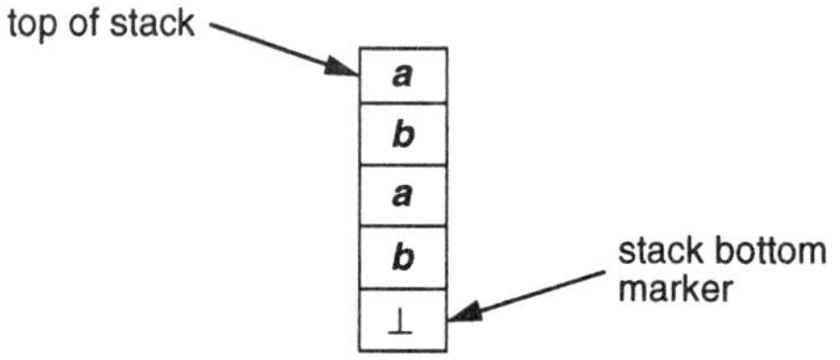

Figure 5.2 A stack

Figure 5.3 specifies the operations **push** and **pop**, which will be needed by our machines.

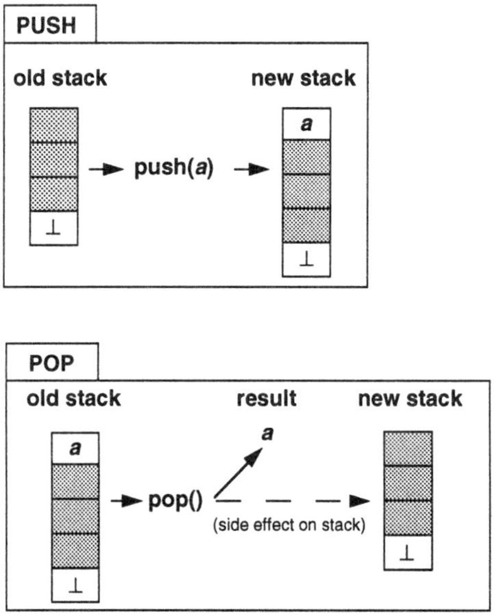

Figure 5.3 'Push' and 'Pop'

However,

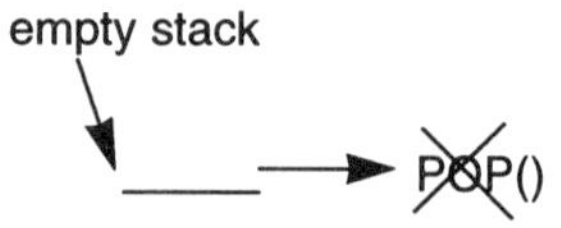

i.e. we cannot pop anything from an empty stack. Also note that we have defined a destructive pop, in that its result is the top element of the stack, but its side effect is to remove that element from the stack. Finally, observe from Figure 5.4. that pushing the empty string onto the stack leaves the stack unchanged.

Table 5.4 How to create an NPDR from a CFG

Step	Actions	Comments
1	We always begin with 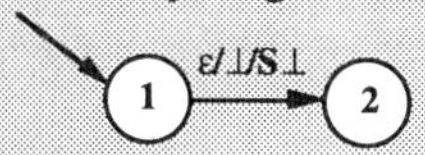 where S is the start symbol of the grammar.	Note that state 1 is the start state. The meaning of the labels on the arcs can be appreciated from the following: 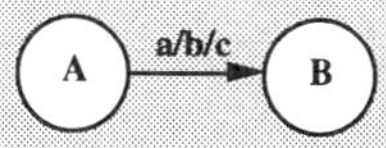 which means: if in state *A*, about to read a, and there is a b on top of the stack, a transition can be made to state B, with the symbol b being popped off the stack, and then the string c being pushed onto the stack. When strings are PUSHed, PUSH(efg) is interpreted as PUSH(g) then PUSH(f) then PUSH(e). So, the transition introduced on the left means: 'when in state 1, without reading any input, and POPping the stack bottom marker off the stack, PUSH the stack bottom marker back on the stack, then push S onto the stack, then go to state 2'
2a	For each production of *G*, $x \rightarrow y$, introduce a transition: 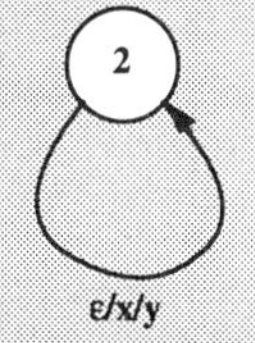	So, in state 2, we will have a situation where the left hand side (i.e. non terminal symbol) of any production of G that is at the top of the stack can be replaced by the right hand side of that production. Again, all these transitions are made without affecting the input string. As you may be able to see, the machine will operate by modelling derivations that could be carried out using the grammar.
2b	For each terminal symbol, *t*, found in the grammar *G*, introduce a transition: 	Still in state 2, we now will have a situation where if a given terminal symbol is next in the input, and also on top of the stack, we can read the symbol from the input, and remove the one from the top of the stack (i.e. the effect of POPping the symbol off and then PUSHing ε). These transitions allow us to read input symbols only if they are in the order that would have been produced by derivations made by the grammar (ensured by the transitions introduced by rule 2a).
3	Introduce a transition: 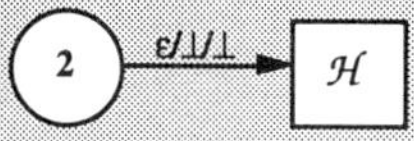	In state 2, we can move to the halt state (again represented as a box) only if the stack is empty. We make this move without reading any input, and we ensure that the stack bottom marker is replaced on the stack.

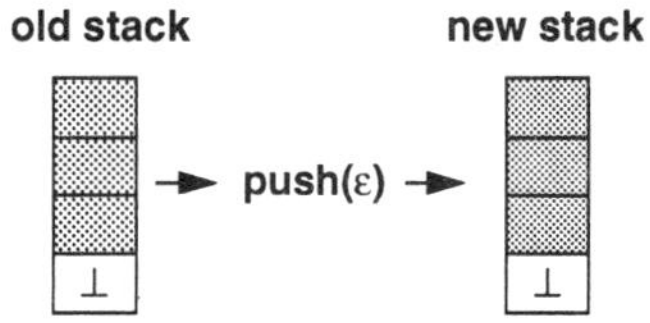

Figure 5.4 Pushing the empty string leaves the stack unchanged

The stacks in this chapter are used to store strings of terminals and non terminals, such that each symbol in the string constitutes one element of the stack. For convenience, we allow PUSH(abc), for example, as an abbreviation of PUSH(c) followed by PUSH(b) followed by PUSH(a). Given that, as stated above, our stacks are really strings, we could have defined POP and PUSH as formal string functions.

Stacks are important data structures, and have many applications. A related data structure you may encounter (though we have no cause to use it in this book) is called a queue. A **queue** is like a stack, except that, as its name suggests, items are added to one end, but taken from the other.

5.3.2 Constructing a non deterministic PDR

The rules we introduce in this section (in Table 5.4) will produce a **non deterministic PDR** (**NPDR**) from the productions of any context free grammar. As for finite state recognizers, the PDR has a pictorial representation. However, due to the additional complexity of the PDR resulting from the addition of the stack, the pictorial form does not provide such a succinct description of the language accepted by a PDR as it does in the case of the FSR. The rules in Table 5.4 convert any CFG, G, into a three-state NPDR.

The **acceptance conditions** for the NPDR, given some input string x, are specified in Table 5.5.

Table 5.5 The acceptance conditions of the NPDR

The machine begins in state 1, with the stack containing only the stack bottom marker, ⊥.
If there is some sequence of transitions by which the machine can reach the halt state $\mathcal{H}$, and on doing so, the input is exhausted and the stack contains only the stack bottom marker ⊥, then the string x is said to be acceptable.

As for FSRs, in Chapter 4 we assert that the set of all **acceptable** strings of an NPDR derived from a grammar, G, is exactly the same as $L(G)$. We are not going to prove this, however, because as you will be able to see from the examples we consider next, it is reasonably obvious that any PDR constructed by using rules in Table 5.4 will be capable of reproducing any **derivation** that the grammar could have produced, and, moreover, will reach a **halt state** only if its input string is a sentence that the grammar could derive. However, if you want to try to prove this, it makes a useful exercise.

5.3.3 *Example NPDRs, M_3 and M_{10}*

Let us now produce an NPDR from the grammar G_3. Figure 5.5 shows the **transitions** produced by each rule (the rule number applied is in the upper right hand corner of each box).

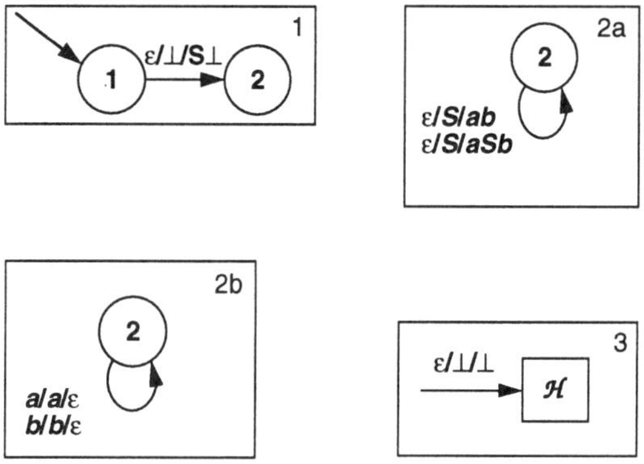

Figure 5.5 Creating a non deterministic PDR

Figure 5.6 shows the connected machine (we have simply joined the appropriate circles together[1]):

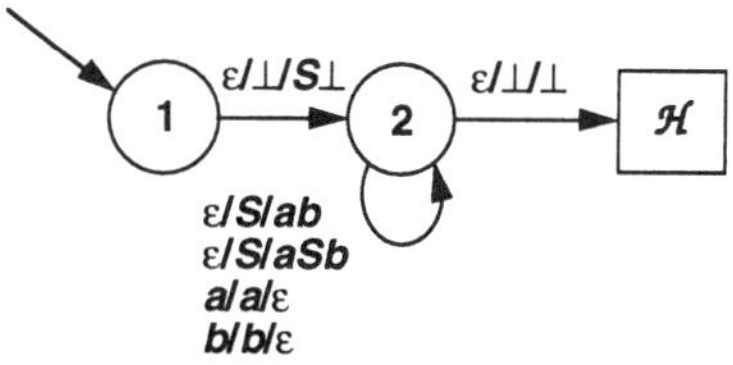

Figure 5.6 The NPDR, M_3

As in the previous chapter, we show the correspondence between the grammar and the machine by using the same subscript number in the name of both. We now investigate the operation of such a machine, by presenting M_3 with the input string a^3b^3. G_3 can generate this string, so M_3 should accept it. The operation of M_3 on the example string is shown in Table 5.6. Table 5.6(a) shows M_3's initial configuration, while Tables 5.6(b) and (c) show the machine's behaviour, eventually accepting the string in Table 5.6(c).

[1] As for the FSRs in Chapter 4, once you understand the process of constructing the machine you can draw it directly in its connected form.

Table 5.6(a) An initial configuration of the NPDR, M_3

Initial state	Initial stack	Input
	⊥	aaabbb

Table 5.6(b) Trace of M_3's behaviour (part 1)

Transition	Resulting stack	Input
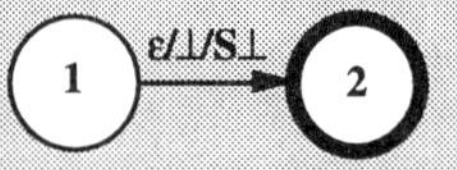	S ⊥	aaabbb
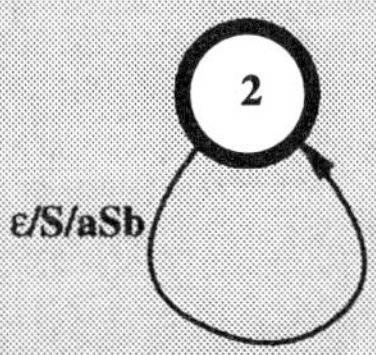	a S b ⊥	aaabbb
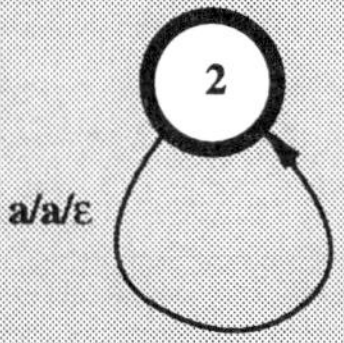	S b ⊥	aaabbb
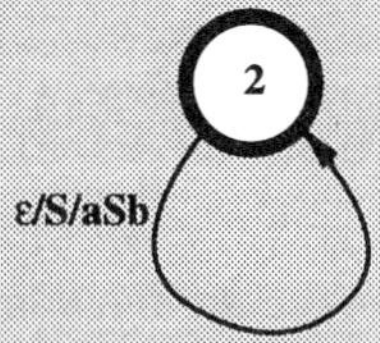	a S b b ⊥	aaabbb
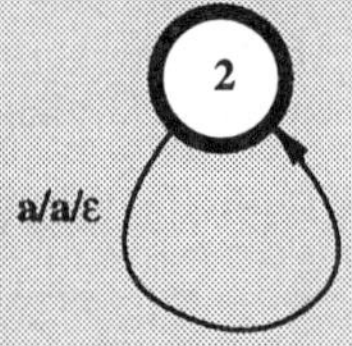	S b b ⊥	aaabbb

Table 5.6(c) Trace of M_3's behaviour (part 2)

Transition	Resulting stack	Input
ε/S/ab 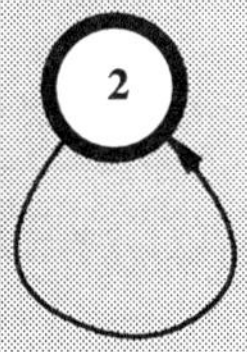	a b b b ⊥	aaabbb
a/a/ε 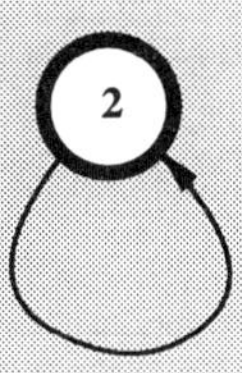	b b b ⊥	aaabbb
b/b/ε 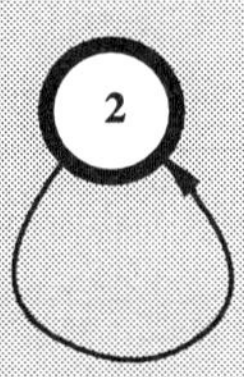	b b ⊥	aaabbb
b/b/ε 	b ⊥	aaabbb
b/b/ε 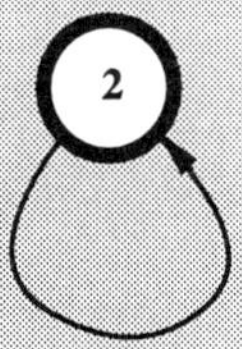	⊥	aaabbb
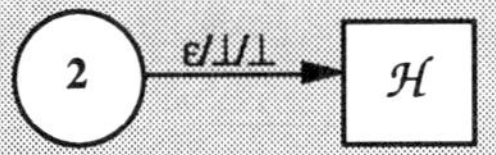 ε/⊥/⊥	⊥	aaabbb

Input exhausted, in halt state, stack empty, therefore string accepted.

Notice in Tables 5.6(b) and (c) how either of the transitions derived from the productions $S \to aSb$ and $S \to ab$ can be applied any time we have S on the top of the stack. There are in fact (infinitely) many sequences of transitions that can be followed, but most of them will lead to a non acceptance state.

We complete this section by considering a further example, this time a grammar that is **equivalent** to grammar G_2, of Chapters 2 and 3. G_2 is:

$S \to aB \mid bA \mid \varepsilon$
$A \to aS \mid bAA$
$B \to bS \mid aBB.$

However, the example, G_{10}, has been constructed by applying our rules for removing the empty string (Table 5.2) to G_2. You should attempt this construction as an exercise.

G_{10}'s productions are as follows:

$S \to \varepsilon \mid X$
$X \to aB \mid bA$
$A \to aS \mid a \mid bAA$
$B \to bS \mid b \mid aBB$

We now apply the rules for creating the equivalent NPDR, resulting in the machine M_{10}, shown in Figure 5.7.

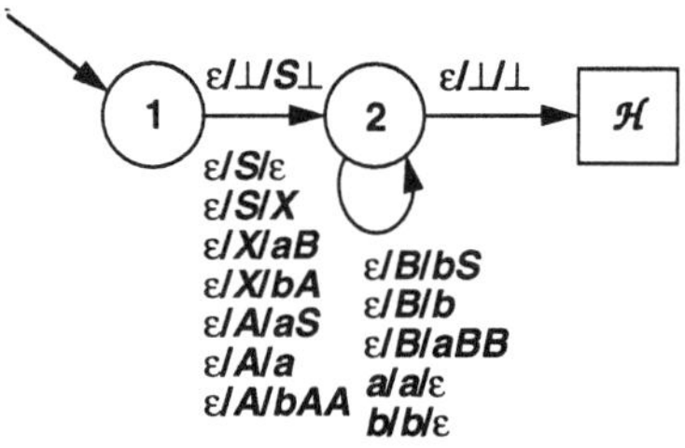

Figure 5.7 The NPDR, M_{10}

Note that, unlike M_3, M_{10} **will accept the empty string** (since it allows S at the top of the stack to be replaced by ε), as indeed it should, since $\varepsilon \in L(G_{10})$. The empty string can be derived by G_{10}. However, $\varepsilon \notin L(G_3)$, i.e. the empty string is *not* one of the strings that can be derived from the start symbol of G_3.

I now leave it to you to follow the M_{10}'s behaviour on some example strings. You are reminded that $L(G_{10})$ is the set of all strings of *a*s and *b*s in which the number of *a*s is equal to the number of *b*s.

5.4 DETERMINISTIC PUSHDOWN RECOGNIZERS

For the FSRs, non determinism was not a problem, since we discovered that for any non deterministic machine we could construct an equivalent, deterministic FSR. In Chapter 4, we also discussed the problems inherent in parsing by non deterministic methods. First, let us clarify how non determinism manifests itself in PDRs. In PDRs the next move that can be made from a given state is determined by two things:

1. The next symbol to be read from the input.
2. The symbol currently on top of the stack.

If, given a particular instance of 1 and 2, the machine has transitions which would allow it to:

- make a choice as to which state to move to
  or
- make a choice as to which string to push onto the stack
  (or both),

then the PDR is *non deterministic*.

On the other hand, a *deterministic* pushdown recognizer (DPDR), is such that

in any state, given a particular input symbol and particular stack top symbol, the machine has only *one* applicable transition.

The question thus arises: *for any NPDR, can we always construct an equivalent DPDR*? We saw in Chapter 4 that the answer to the corresponding question for FSRs was 'yes'. For the PDRs however, the answer is more like: 'in general no, but in many cases, yes'. We shall see that the 'many cases' includes *all* the regular languages and *some* of the proper CFLs (i.e. those that are *context free but not regular*). However, we shall also see that there are some languages that are certainly context free, but *cannot* be recognized by a DPDR.

5.4.1 M_3^d, *a deterministic version of* M_3

First, we consider context free languages that *can* be recognized by a DPDR. The language:

$$A = \{a^i b^i : i \geq 1\},$$

is not regular (this is proved in Chapter 6). However, A is *certainly* context free, since it is generated by the CFG G_3. Above, we constructed an NPDR (M_3) for A, from the productions of G_3 (see Figure 5.6). However, we can see that any sentence in the language A consists of a number of as followed by the same number of bs. A deterministic method for accepting strings of this form might be:

> PUSH each a in the input string onto the stack until a b is encountered, then POP an a off the stack for each b in the input string.

The above method is used by the DPDR $M_3{}^d$, shown in Figure 5.8.

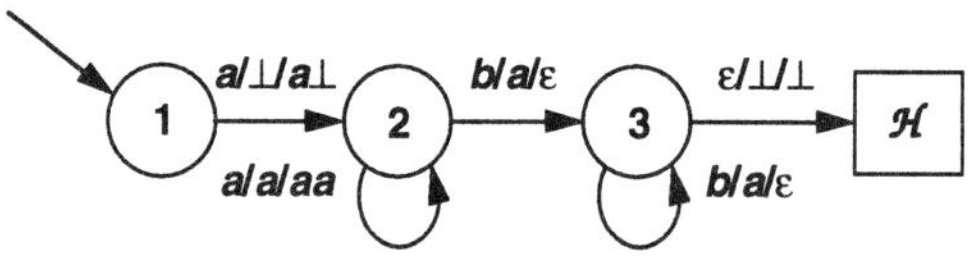

Figure 5.8 The DPDR, $M_3{}^d$

In Table 5.6 we saw a trace of the operation of the NPDR M_3 for the input a^3b^3. We now trace the operation of $M_3{}^d$ for the same input. Table 5.7(a) shows the initial configuration of $M_3{}^d$, while Table 5.7(b) shows the behaviour of the machine until it accepts the string.

Table 5.7(a) An initial configuration of the DPDR, $M_3{}^d$

Initial state	Initial stack	Input
1	⊥	aaabbb ↑

5.4.2 More deterministic PDRs

Two more example DPDRs follow (Figures 5.9 and 5.10). It is suggested that you trace their behaviour for various valid and invalid input strings.

First, Figure 5.9 shows a DPDR for the language $L(G_{10})$, the set of all strings of as and bs in which the number of as is equal to the number of bs. This is the same

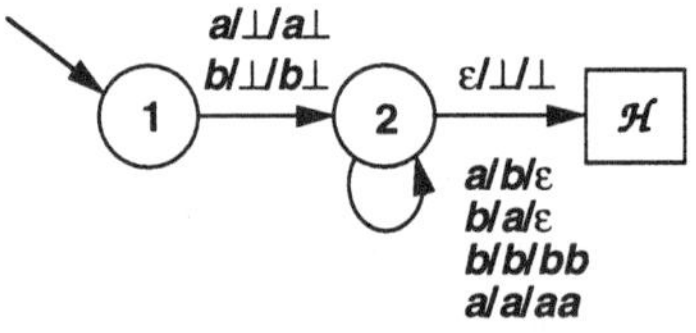

Figure 5.9 A DPDR for the language $L(G_{10}) - \{\varepsilon\}$.

Table 5.7(b) Trace of M_3^{d}'s behaviour

Transition	Resulting stack	Input
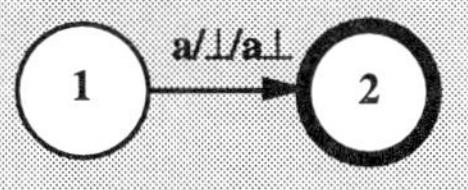	a ⊥	aaabbb
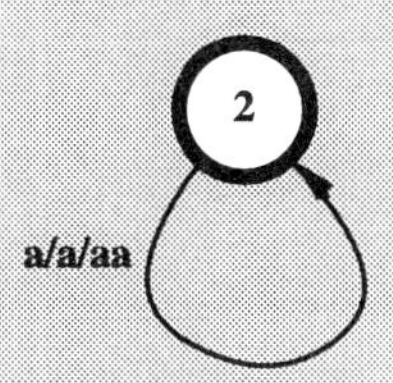	a a ⊥	aaabbb
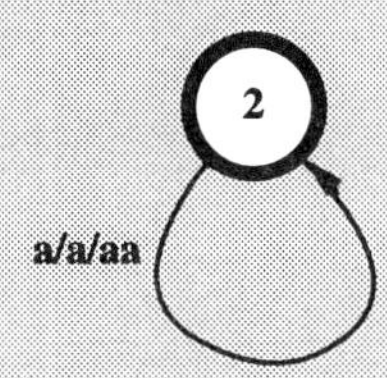	a a a ⊥	aaabbb
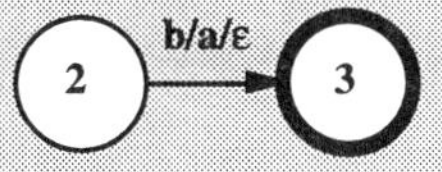	a a ⊥	aaabbb
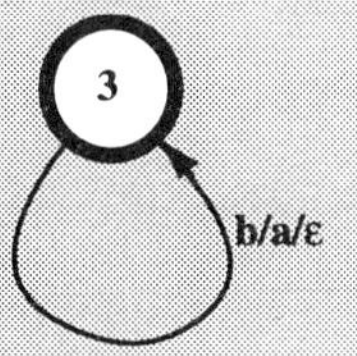	a ⊥	aaabbb
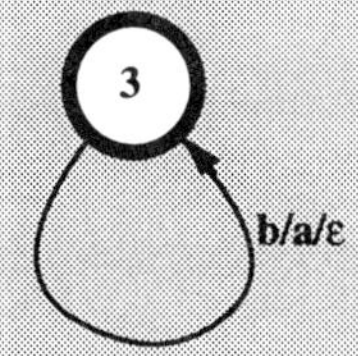	⊥	aaabbb
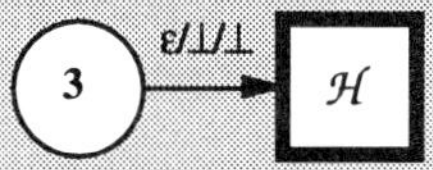	⊥	aaabbb

Input exhausted, in halt state, stack empty, therefore string accepted.

language as $L(G_2)$, as we obtained the productions for G_{10} by applying our procedure for removing ε productions to G_2. Let us consider $L(G_{10})$ without the empty string (using the **set difference** operator, $L(G_{10}) - \{\varepsilon\}$), which is obviously the set of all *non empty* strings of *a*s and *b*s in which the number of *a*s is equal to the number of *b*s. The DPDR shown in Figure 5.10 recognizes the language $L(G_9)$, which is:

$$\{a^{2i+1}c^j b^{2i+1} : i, j \geq 0\}$$

a non zero *odd* number of *a*s, followed by 0 or more *c*s, followed by the same number of *b*s as there were *a*s.

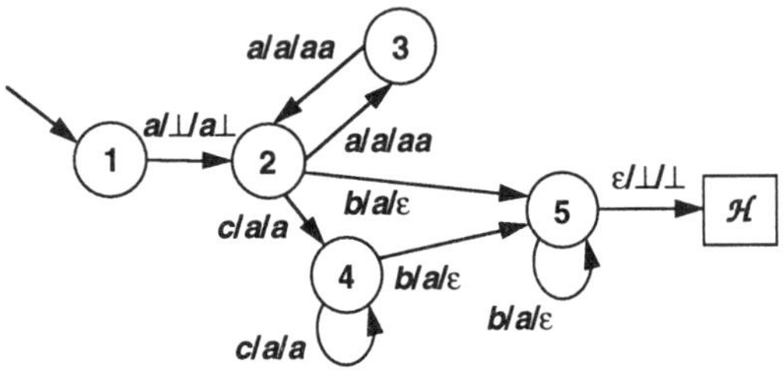

Figure 5.10 A DPDR for the language $L(G_9)$

Clearly, then, there are CFLs that can be recognized by DPDRs. Here, we have encountered but a few. However, there are, as you can probably imagine, an infinite number of them. If a CFL can be recognized by a DPDR, we call it a **deterministic context free language**.

5.5 DETERMINISTIC AND NON DETERMINISTIC CONTEXT FREE LANGUAGES

In this section, we discover that *there are context free languages that cannot be recognized by a deterministic PDR*. There is no general equivalence between the non deterministic and the deterministic PDRs, as there is for FSRs. We begin our investigation with the regular languages, which we find to be wholly contained within the deterministic CFLs.

5.5.1 *Every regular language is a deterministic CFL*

Any regular language is generated by a regular grammar without ε productions (apart from the one ε production $S \to \varepsilon$, if it is necessary). Such a grammar can be represented as a non deterministic FSR, which can, if necessary, be represented as a deterministic FSR (DFSR).

Now, any DFSR can be represented as a DPDR which effectively does not use its stack. We will use a single example that suggests a method by which any DFSR could be converted into an equivalent DPDR. The example deterministic FSR is M_5^d, from Chapter 4, shown in Figure 5.11.

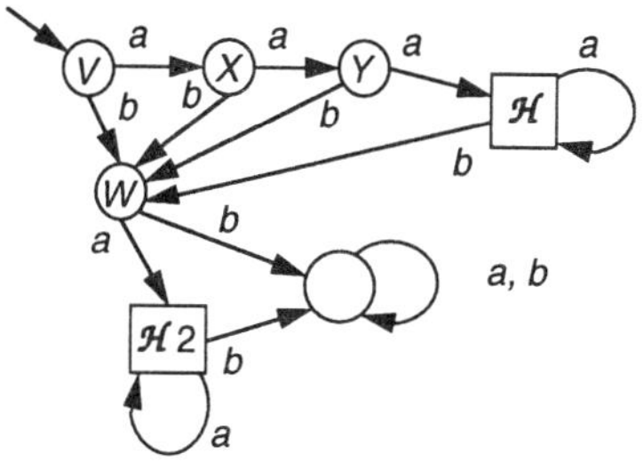

Figure 5.11 The DFSR M_5^d from Chapter 4

This produces the DPDR of Figure 5.12.

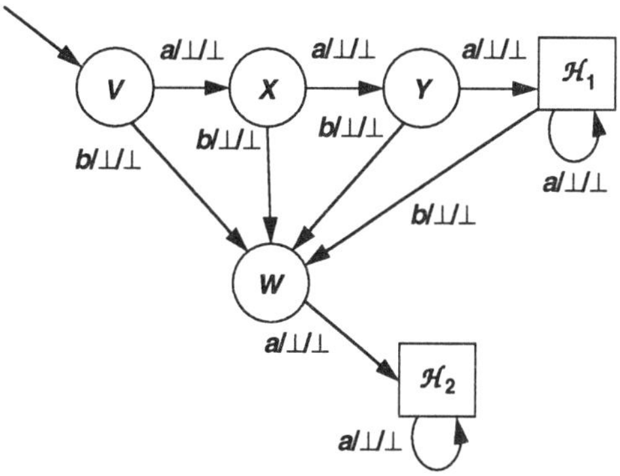

Figure 5.12 The DPDR equivalent to M_5^d (Figure 5.11)

The DPDR in Figure 5.12 has two **halt states** (we have hitherto said nothing to prohibit this). We have also ignored the **null state** of the DFSR. We could clearly amend any DFSR to produce an equivalent DPDR as we have done above.

However, for the DPDR, we do need to amend our definition of the acceptance conditions as they applied to the NPDR, which our rules ensure always has only one halt state.

The acceptance conditions for the DPDR, for an input string x, are specified in Table 5.8.

You may notice that the condition 'the stack contains only the stack bottom marker ⊥', that applied to NPDRs (Table 5.5), does not appear in Table 5.8. An exercise asks you to justify that the language $\{a^i b^j : i \geq 1, j = i \text{ or } j = 0\}$ can only be recognized by a deterministic PDR if that machine is *not* required to have an empty stack every time it accepts a string.

Table 5.8 The acceptance conditions for the DPDR

The machine begins in state 1, with the stack containing only the stack bottom marker, ⊥.
If there is a sequence of transitions by which the machine can reach one of its halt states, and on doing so, the input is exhausted, then the string x is said to be acceptable.

So far, we have established the existence of the deterministic context free languages. Moreover, we have also established that this class of languages includes the regular languages. What we have yet to do is establish that there are some CFLs that are *necessarily non deterministic*, in that *they can be accepted by an NPDR but not a DPDR*.

5.5.2 The non deterministic CFLs

Consider the following CFG, G_{11}:

$S \rightarrow a \mid b \mid c \mid aSa \mid bSb \mid cSc$

$L(G_{11})$ is the set of every possible *non empty odd length* **palindromic** string of *a*s and/or *b*s and/or *c*s[2]. We could use our procedure from Table 5.4 to produce a **NPDR** to accept $L(G_{11})$. However, could we design a **DPDR** to do the same? We cannot, as we now establish by intuitive argument. General palindromic strings are of the form specified in Figure 5.13.

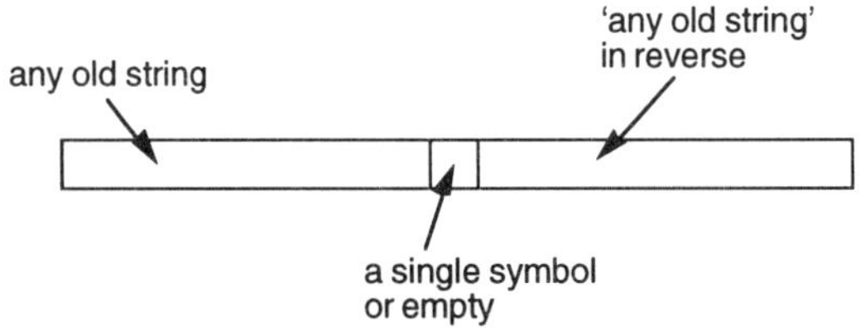

Figure 5.13 A specification of arbitrary palindromic strings

The centre box of Figure 5.13 could be empty if *even length palindromic strings* were allowed. Note that G_{11}does not generate even length palindromes[3]. An example sentence from $L(G_{11})$, *aabccccbaa* can be partitioned as demonstrated in Figure 5.14.

Let us hypothesize that a DPDR could accept $L(G_{11})$. It would have to read the input string from left to right, as that is how it works. An obvious deterministic way to do the task on that basis is as follows:

[2] A palindromic string is the same as the reverse of itself.

[3] It is very simple to amend the grammar so that it generates all (even and odd length) palindromic strings, as you may like to see for yourself.

a a b a c c	*c*	*c c a b a a*

Figure 5.14 A palindrome, partitioned according to Figure 5.13.

1. Read, and PUSH onto the stack, each input symbol encountered until the 'centre' of the string is reached.
2. Read the rest of the string, comparing each symbol to the next symbol on the stack. If the string is valid, the string on the stack following (1) will be the part of the string after the centre in reverse.

The problem with the above is in the statement 'until the 'centre' of the string is reached' in stage (1). How can the DPDR detect the centre of the string? Of course, it cannot. If there is a symbol in the centre it will be one of the symbols – *a*, *b*, or *c* – that can feature anywhere in the input string. The machine must PUSH every input symbol onto the stack, on the assumption that it has not yet reached the centre of the input string. This means that the machine must be allowed to **backtrack**, once it reaches the end of the input, to a situation where it can see if it is possible to empty its stack by comparing the stack contents with the remainder of the input. By similar argument, any such language of arbitrary palindromic strings defined with respect to an alphabet of two or more symbols is necessarily non deterministic.

We have argued that the problem vis-à-vis palindromic strings lies in the PDR's inability to detect the 'centre' of the input. If a language of palindromic strings is such that every string is of the form shown in Figure 5.15 then it can be accepted by a DPDR, and so is *deterministic*.

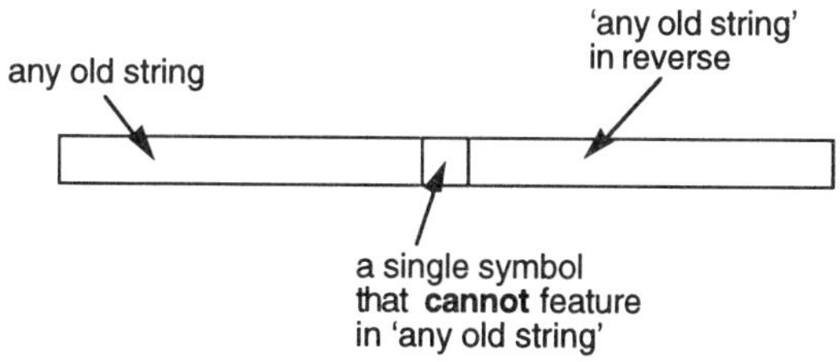

Figure 5.15 Palindromes with a distinct central symbol

As an example of a *deterministic* palindromic language, consider $L(G_{12})$, where G_{12} (a slightly amended G_{11}) is as follows:

$S \rightarrow d \mid aSa \mid bSb \mid cSc$

$L(G_{12})$ is the set of all palindromic strings of a form described in Figure 5.16.

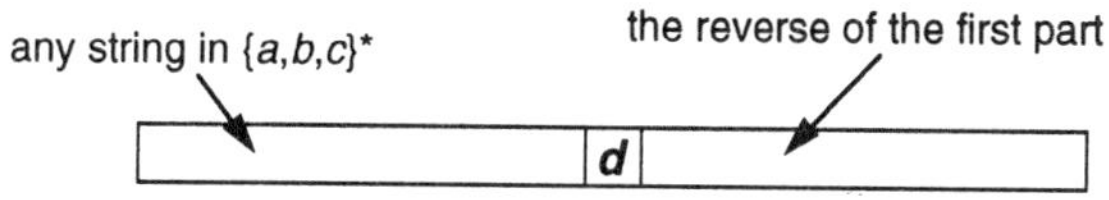

Figure 5.16 Sentences of $L(G_{10})$

$L(G_{12})$ is accepted by the DPDR, M_{12} depicted in Figure 5.17.

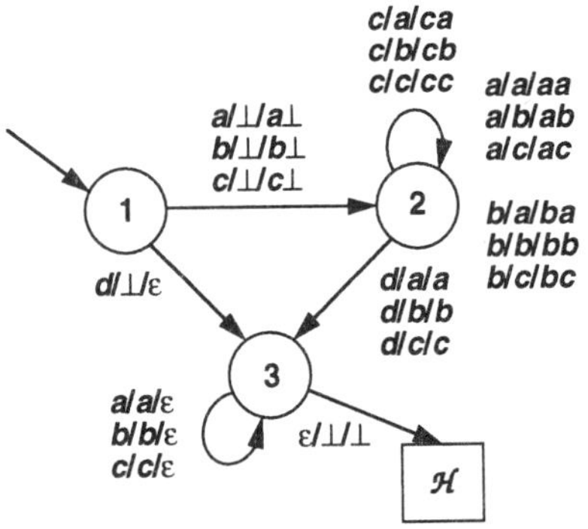

Figure 5.17 The DPDR M_{12}

5.5.3 *A refinement to the Chomsky hierarchy in the case of CFLs*

The existence of **properly** non deterministic CFLs suggests a slight refinement to the **Chomsky hierarchy** with respect to the context free and regular languages, as represented in Figure 5.18.

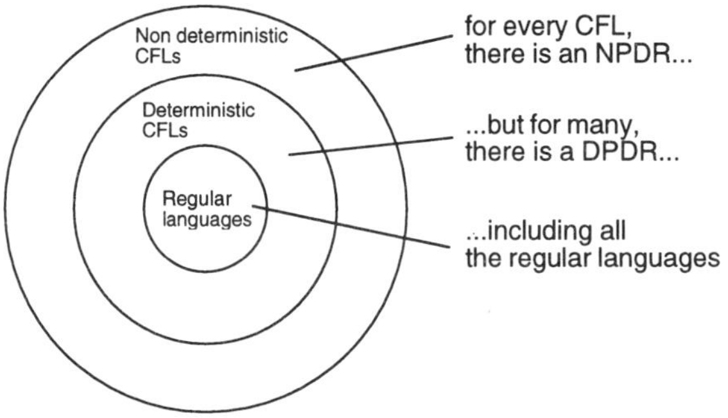

Figure 5.18 The context free languages

5.6 THE EQUIVALENCE OF THE CFLS AND THE PDRS

The situation for FSRs and regular languages, i.e. that any regular grammar can be represented by an **equivalent** FSR, and every FSR can be represented by an equivalent regular grammar, is mirrored in the relationship between PDRs and CFGs. In this chapter, we have seen that for any CFG we can construct an equivalent NPDR. It turns out that for any NPDR we can construct an equivalent CFG. This will not be proved here, as the proof is rather complex, and it was considered more important to investigate the situation pertaining to the PDR as the recognizer for the CFLs, rather than the converse result.

One way of showing that every NPDR has an equivalent CFG is to show that from any NPDR we can create CFG productions that generate exactly the set of strings that enable the NPDR to reach its halt state, leaving its stack empty. The details are not important, but the result is powerful, as it shows us that any language that cannot be generated by a CFG (and we see two such languages in the next chapter) needs a machine that is more powerful than a PDR to recognize it. Later in the book we consider abstract machines as *computers*, rather than language recognizers, and we see the close relationship between computation and language recognition.

Consider the 'multiplication language'

$$\{a^i b^j c^{i \times j} : i, j \geq 1\}$$

introduced in Chapter 2. This is shown to be not a CFL at the end of Chapter 6, a result that means that the language cannot be recognized by a PDR. This leads us to speculate that we need a more powerful machine than the PDR to model many computational tasks. We return to this theme in the second part of the book.

5.7 OBSERVATIONS ON CONTEXT FREE GRAMMARS AND LANGUAGES

The lack of general equivalence between deterministic and non deterministic context free languages is problematic for the design of programming languages. Efficient parsers can be designed from deterministic PDRs. As argued in Chapter 4, non determinism is undesirable, as it is computationally expensive. Moreover, unnecessary non determinism is simply wasteful in terms of resources. It turns out that any deterministic context free language can be parsed by a pushdown machine that never has to look ahead more than one symbol in the input to determine the choice of the next transition.

Consider, for example, sentences of the language $\{a^i b^i : i \geq 1\}$. If we look at the next symbol in the input, it will either be an a or a b, and, like our deterministic machine M_3, above, we know exactly what to do in either case. Now consider strings of arbitrary palindromes, with no distinct central 'marker' symbol. In this case, for any machine that could look ahead a *fixed* number of symbols, we could simply *present that machine with a palindromic string long enough to ensure that the machine could not tell when it moved from the first to the second 'half' of the input*. Languages that can be parsed by a PDR (and are therefore context free) that never has to look ahead more than a fixed number, say k, of symbols in the input are called **LR(k) languages** (**LR** stands for **L**ook ahead **R**ight). The **LR(k)** languages *are* the deterministic context free languages. In fact, the deterministic languages are all **LR(1) languages**.

Languages of arbitrary length palindromic strings with no distinguishable central marker are clearly not **LR(k)** languages, for any fixed k.

The preceding chapter closed with the observation that the language $\{a^i b^i : i \geq 1\}$ is not regular. It was promised that this would be proved in Chapter 6. Now we make a similar observation with respect to the context free languages, in that we argue that there are languages that are not context free, and cannot therefore be generated by a context free grammar, or recognized by an NPDR. They therefore must be either **type 1** (**context sensitive**) or **type 0** (**unrestricted**) languages (or perhaps *neither*).

In Chapter 6, we will see that both the 'multiplication language' and the language:

$$\{a^i b^i c^i : i \geq 1\}$$

i.e. the set of all strings of the form *as* followed by *bs* followed by *cs*, where the number of *as* = the number of *bs* = the number of *cs*, are not context free. Given that the language $\{a^i b^i : i \geq 1\}$ *is* context free, it may at first seem strange that $\{a^i b^i c^i : i \geq 1\}$ is not. However, this is indeed the case. Part of the next chapter is concerned with demonstrating this.

EXERCISES

1 Given the following grammar:

$S \rightarrow ABCA$
$A \rightarrow aA \mid XY$
$B \rightarrow bbB \mid C$
$X \rightarrow cX \mid \varepsilon$
$Y \rightarrow dY \mid \varepsilon$
$C \rightarrow cCd \mid \varepsilon$

produce an equivalent **Chomsky normal form** grammar.
Note: remember that you first need to apply the rules for removing the empty string.

2† $\{a^i b^j : i \geq 1, j = i \text{ or } j = 0\}$ is a **deterministic CFL**. However, as mentioned earlier in this chapter, it can only be accepted by a **deterministic PDR** that does not always clear its **stack**. Provide a DPDR to **accept** the language, and justify the preceding statement.

3 Argue that we can always ensure that an NPDR clears its stack, and that NPDRs need have only one **halt state**.

4 Amend $M_3{}^d$ from earlier in the chapter, to produce DPDRs for the languages $\{a^i b^i c^j : i, j \geq 1\}$ and $\{a^i b^j c^j : i, j \geq 1\}$.
Note: in Chapter 6 it is assumed that the above two languages are deterministic.

5† Justify by intuitive argument that $\{a^i b^j c^k : i, j \geq 1, i = j \text{ or } j = k\}$ (from exercise 1 in Chapter 3) is a non deterministic CFL.

6† Design a program to simulate the behaviour of a DPDR. Investigate what amendments would have to be made to your program so that it could model the behaviour of an NPDR.

7 Design DPDRs to accept the languages $\{a^i b^j c^{i+j} : i, j \geq 1\}$ and $\{a^i b^j c^{i-j} : i, j \geq 1\}$. Note: consider the implications of these machines in the light of the remarks about the *computational power of the PDR*, above.

6 Important features of regular and context free languages

AIMS

To learn about

- closure, which means seeing if operations applied to languages of a given class always result in a language of the same class
- two theorems relating to the Chomsky hierarchy: the repeat state theorem for finite state recognizers and the *uvwxy* theorem for context free languages
- how to use the two theorems in relation to context free, context sensitive and unrestricted languages

6.1 CLOSURE PROPERTIES OF LANGUAGES

In terms of the **Chomsky hierarchy**, the notion of **closure** applies as follows:

1. Choose a given type from the Chomsky hierarchy.
2. Choose some operation that can be applied to any language (or pairs of languages) of that type, and which yields a language as a result.
3. If it can be demonstrated that when applied to each and every language or pair of languages of the chosen type, the operation yields a resulting language which is also of the chosen type, then the languages of the chosen type are said to be **closed under** the operation.

For example, take the **set union** operator, $\cup$, and the regular languages. If for every pair L_1, L_2, of **regular languages** the set $L_1 \cup L_2$ was also a regular language, then the regular languages would be closed under union (in fact they are, as we see, shortly).

In the following two sections, we investigate various closure properties of the regular and **context free languages**, with respect to the following operations: **complement, union, intersection**, and **concatenation**. As usual, we will be content with intuitive argument.

6.2 CLOSURE PROPERTIES OF THE REGULAR LANGUAGES

6.2.1 Complement

As we discovered early on in the book, given some **finite** non empty **alphabet** A, A^* denotes *the infinite set of all possible strings that can be formed using the symbols of that alphabet*. Given this basis, a formal language was defined to be *any subset of such a set*. Take for example the language

$$R = \{a^i b^j c^k : i, j, k \geq 1\},$$

which is a **subset** of $\{a, b, c\}^*$.

Now, $\{a, b, c\}^* - R$ denotes the set of all strings that are in $\{a, b, c\}^*$ *but not in* R ('–' is called the *set difference* operator) i.e.

> all possible strings of *as* and/or *bs* and/or *cs* *except* for those of the form 'one or more *as* followed by one or more *bs* followed by one or more *cs*'.

$\{a, b, c\}^* - R$ is called the *complement* of R, and is, of course, a formal language, as it is a subset of $\{a, b, c\}^*$.

In general, then, if L is a language in A^*, $A^* - L$ is the complement of L.

We are now going to see that:

If L is a regular language in A^*, then so is $A^* - L$.

In other words, the regular languages are closed under complement. This is best demonstrated in terms of **deterministic finite state recognizers (DFSRs)**. Our argument will be illustrated here by using the DFSR for R, the language described above, as an example.

We start with a DFSR that accepts R, as shown in Figure 6.1.

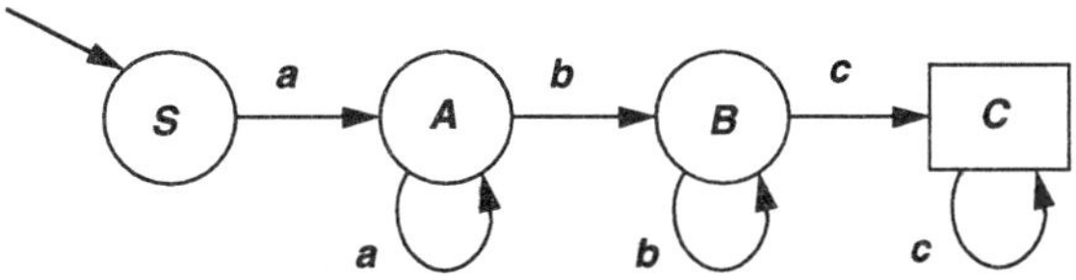

Figure 6.1 A DFSR for the language, *R*

To the machine in Figure 6.1 we add a **null state** to make the machine completely deterministic. The resulting machine can be seen in Figure 6.2.

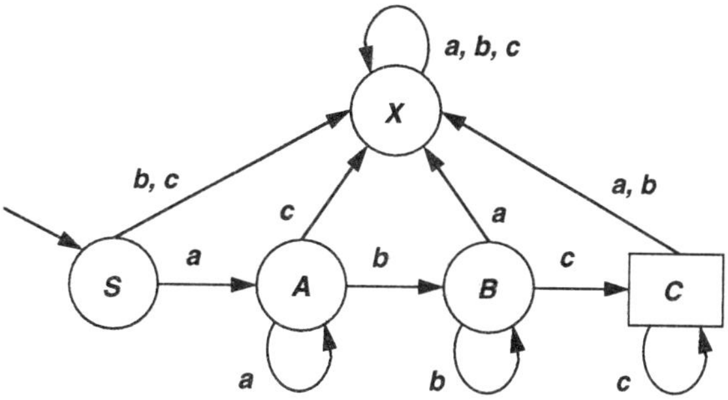

Figure 6.2 The FSR of Figure 6.1 with a null state

We take the machine in Figure 6.2, and make all its non halt states into **halt states**, and all its halt states into non halt states, achieving the machine shown in Figure 6.3.

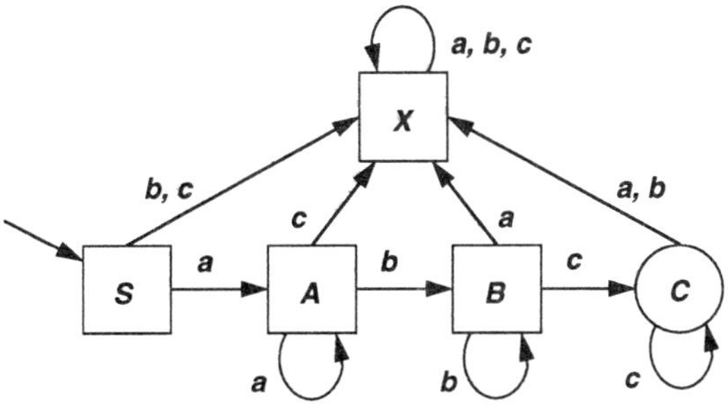

Figure 6.3 The complement of the machines in Figures 6.1 and 6.2

The FSR in Figure 6.3 will reach a halt state *only* for strings in $\{a, b, c\}^*$ for which the original machine (Figure 6.1) *would not*. It thus accepts the *complement* of the language accepted by the original machine. Note that, since *the start state is also a halt state*, the final machine also accepts as a valid input the empty string, as it should, since the original machine did not.

It should be clear that the same method could be applied to *any* DFSR. As we saw in Chapter 4, there is a general equivalence between DFSRs and regular languages. What applies to all DFSRs, then, applies to *all* regular languages.

6.2.2 Union

We now see that:

If L_1 and L_2 are regular languages, then so is $L_1 \cup L_2$.

Consider the DFSR for R, in Figure 6.1 above, and the DFSR for the language

$$\{a^i b^{2j} : i \geq 0, j \geq 1\},$$

which is shown in Figure 6.4.

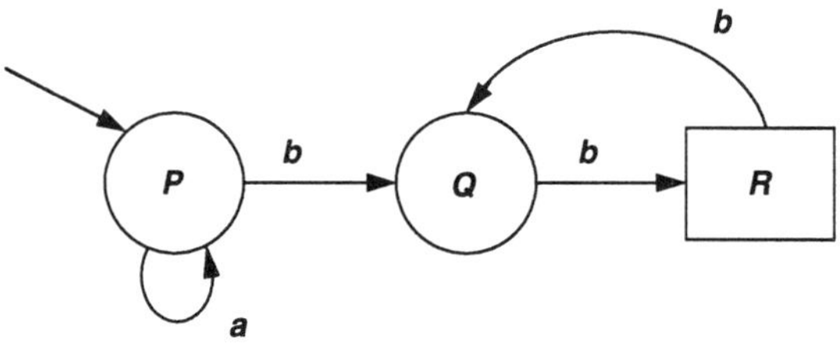

Figure 6.4 A DFSR

We produce a composite machine of the machines from Figures 6.1 and 6.4. As can be seen in Figure 6.5, we have introduced an additional **start state**, I, and for each **outgoing arc** from a start state of either original machine we draw an **arc** from I to the destination state of that arc.

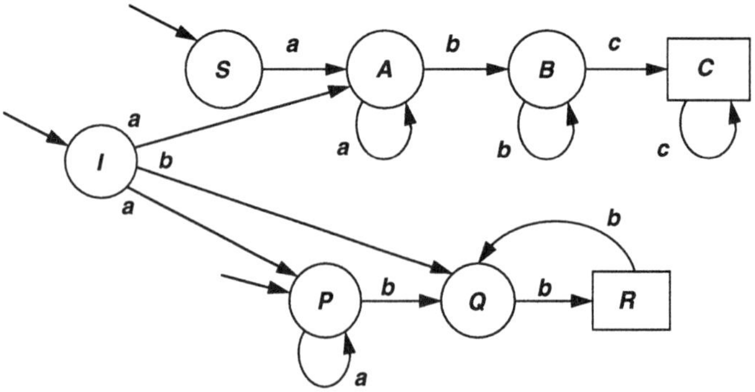

Figure 6.5 The non deterministic union machine for the FSRs in Figures 6.1 and 6.4

The machine in Figure 6.5 has three start states, but if we use our **subset algorithm** from Chapter 4 to make it deterministic, we obtain the FSR in Figure 6.6.

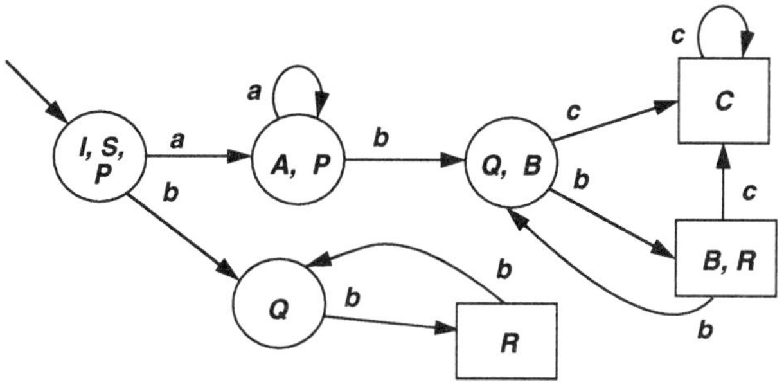

Figure 6.6 A deterministic version of the FSR in Figure 6.5

The machine in Figure 6.6 accepts the language:

$$\{a^i b^j c^k : i, j, k \geq 1\} \cup \{a^i b^{2j} : i \geq 0, j \geq 1\},$$

i.e. all strings accepted by *either*, *or both*, of the original machines.

You should again satisfy yourself that the suggested method can be applied to any pair of DFSRs.

6.2.3 Intersection

Now we establish that:

If L_1 and L_2 are regular languages, then so is $L_1 \cap L_2$.

A basic law of set theory is one of de Morgan's laws, which states:

$$L_1 \cap L_2 = \overline{\overline{L_1} \cup \overline{L_2}}$$

i.e. the intersection of two sets is the same as the complement of the union of the complements of the two sets. Now, we have to do little to establish that if L_1 and L_2 are regular languages, then so is $L_1 \cap L_2$, since:

- we saw above how to do the complement,

and

- we also saw how to do the union,

so therefore

- if we take the two DFSRs for L_1 and L_2, say M_1 and M_2:
  a. we create the 'complement machines' for each of M_1 and M_2
  b. we create the 'union machine' for the results of (a)
  c. we create the 'complement machine' from the result of (b).

We know that *the complement of a regular language is always regular, and the union of two regular languages is also always regular*. The language recognized by the FSR that results from doing (a) and then (b) and then (c) will therefore also be regular.

6.2.4 Concatenation

As two *strings* can be concatenated together, so can two *languages*. To concatenate two languages A and B together, we simply make a set of all of the strings which result from concatenating each string in A in turn, with each string in B. So, for example, if

$$A = \{a, ab, \varepsilon\}$$

and

$$B = \{bc, c\},$$

then A concatenated to B, which we write AB, or in this case $\{a, ab, \varepsilon\}\{bc, c\}$ would be

$$\{abc, ac, abbc, bc, c\}.$$

Analogously to the way we use the superscript notation for strings, so we can use the superscript notation for languages, so that A^3 represents the language A concatenated *to itself, and the result of that concatenated to A again* (i.e. AAA), or in this case (given the language A as defined above):

$$A^3 = \{a, ab, \varepsilon\}\{a, ab, \varepsilon\}\{a, ab, \varepsilon\} = \{aa, aab, a, aba, abab, ab, \varepsilon\}\{a, ab, \varepsilon\}, \text{ etc.}$$

The same applies to **infinite languages**. For example, we could write:

$$\{a^ib^j : i, j \geq 1\}\{c^i : i \geq 1\}$$

to mean *the set containing each and every string in the first set concatenated to each and every string in the second*. In this case, the language described is R from earlier in this chapter, i.e.

$$\{a^ib^jc^k : i, j, k \geq 1\}.$$

The next thing we are going to show is:

If L_1 and L_2 are regular languages, then so is L_1L_2.

This is easier to show by using *grammars* than FSRs. Table 6.1 describes the combining of two regular grammars G_{13} and G_{14}, to produce a grammar, G_{15}, such that $L(G_{15}) = L(G_{13})L(G_{14})$.

Table 6.1 Combining two regular grammars to create a regular grammar that generates the concatenated language

G_{13}	G_{14}
$S \to aS \mid a \mid bB$	$S \to aS \mid aA$
$B \to b \mid c \mid bB$	$A \to dA \mid d$
so that:	
$L(G_{13}) =$	$L(G_{14}) =$
$\{a^i b^j x : i \geq 0, j \geq 1, x = b \text{ or } x = c\}$	$\{a^i d^j : i, j \geq 1\}$

To produce a composite grammar which generates $L(G_{13})L(G_{14})$:

Step	Description	Result
1.	Rename the non terminals in the second grammar so that the two grammars have no non terminals in common	G_{14} is now: $X \to aX \mid aA$ $A \to dA \mid d$ with X as the start symbol
2.	In the first grammar, replace each production of the form $x \to y$, where y is a single terminal, by a production $x \to yZ$, where Z is the start symbol of the second grammar	G_{13} is now: $S \to aS \mid aX \mid bB$ $B \to bX \mid cX \mid bB$
3.	Put all the productions together. They now constitute one grammar, the start symbol of which is the start symbol of the first grammar. The resulting grammar generates the language which is the language generated by the first grammar concatenated to the language generated by the second.	Composite grammar, L_{15}: $S \to aS \mid aX \mid bB$ $B \to bX \mid cX \mid bB$ $X \to aX \mid aA$ $A \to dA \mid d$

The composite grammar, G_{15}, shown in Table 6.1 is such that $L(G_{15}) = L(G_{13})$ $L(G_{14})$, i.e.

$$\{a^i b^j x : i \geq 0, j \geq 1, x = b \text{ or } x = c\}\{a^i d^j : i, j \geq 1\}$$
$$= \{a^h b^i x a^j d^k : h \geq 0, i, j, k \geq 1, x = b \text{ or } x = c\}.$$

It is obvious that the method in Table 6.1 would work for any pair of regular grammars. Note that the method could also be used repeatedly, beginning with a grammar G, to create a grammar that generated $L(G)^{n,}$ for any finite n.

6.3 CLOSURE PROPERTIES OF THE CONTEXT FREE LANGUAGES

In this section, we investigate the closure properties of the **context free languages** (**CFLs**) in general. We find that the situation is not so straightforward for the CFLs as for the regular languages. We also find that there are differences between the closure properties of the CFLs in general, and the **deterministic CFLs**, the latter being described in Chapter 5 as those CFLs recognized by *deterministic* PDRs. For the CFLs to be closed under an operation means that the language resulting from the application of the operation to any CFL (or pair of CFLs, if the operation applies to two CFLs) is also a CFL. Moreover, for the deterministic CFLs to be closed under an operation means that the result must always be a deterministic CFL.

To support the arguments, we will use the following languages:

$$X = \{a^i b^i c^j : i, j \geq 1\}$$

and

$$Y = \{a^i b^i c^j : i, j \geq 1\}$$

First of all, let us observe that the two languages are context free. In fact, X and Y are both *deterministic* CFLs, as we could design DPDRs to accept each of them, as you were asked to do in the exercises at the end of Chapter 5.

In any case, they are certainly context free, as we can show by providing CFGs for each of them. The two grammars, G_x and G_y, are shown in Table 6.2.

Table 6.2 Two example grammars, Gx and Gy

G_x, which generates X	G_y, which generates Y
$A \rightarrow BC$ {note: start symbol is A}	$D \rightarrow EF$ {note: start symbol is D}
$B \rightarrow aBb \mid ab$	$E \rightarrow aE \mid a$
$C \rightarrow cC \mid c$	$F \rightarrow bFc \mid bc$

To simplify our constructions, the two grammars in Table 6.2 have no **non terminals** in common.

6.3.1 *Union*

The first result is that the CFLs are closed under **union**, i.e.

If L_a and L_b are context free languages, then so is $L_a \cup L_b$.

Suppose that for L_a and L_b there are CFGs, G_a and G_b, respectively. Then to produce a grammar that generates the *union* of L_a and L_b, we carry out the process described

Table 6.3 Constructing a CFG to generate the union of two CFLs (G_x and G_y are defined in Table 6.2)

Step	Description	Result
1	arrange it so the two grammars have no non terminals in common	G_x and G_y are already like this
2	collect all the productions of both grammars together	$A \to BC$ $B \to aBb \mid ab$ $C \to cC \mid c$ } G_x $D \to EF$ $E \to aE \mid a$ $F \to bFc \mid bc$ } G_y
3	add the two productions $S \to P \mid Q$ where S is a new non terminal and P and Q are the start symbols of the two grammars	$S \to A \mid D$ $A \to BC$ $B \to aBb \mid ab$ $C \to cC \mid c$ $D \to EF$ $E \to aE \mid a$ $F \to bFc \mid bc$

in Table 6.3 (the example on the right of Table 6.3 uses the two grammars G_x and G_y from Table 6.2.).

Referring to Table 6.3, you should be able to see that the grammar resulting from step 3 generates *all the strings that could be generated by one grammar or the other.* Step 1 is necessary so that any derivation uses only productions from one of the original grammars. If G_a and G_b are the original grammars, the resulting grammar generates $L(G_a) \cup L(G_b)$. You can also see that *if the two starting grammars are context free, the resulting grammar will also be context free.*
Therefore,

the context free languages are closed under union.

Now we focus on our particular example. The new grammar that we obtained from G_x and G_y generates $X \cup Y$, where X and Y are the languages above. This set, which will be called Z, is:

$$Z = X \cup Y = \{a^i b^j c^k : i, j, k \geq 1, i = j \text{ or } j = k\}$$

Now, while this is a CFL, and represents the union of two *deterministic* CFLs, it is not itself a deterministic CFL. To appreciate this, think about a PDR to accept it: such a machine could not be deterministic, as when it was reading the *a*s in the input it would have to assume that it was processing a string in which the number of *a*s was equal to the number of *b*s, which may not be the case (it may be processing a string in which the number of *b*s was equal to the number of *c*s). Thus there would be a need for *non* determinism. For a class of languages to be closed under an operation, the result of every application of the operation must also be a language of that class. So our example (and there are many more) shows us that

the deterministic context free languages are not closed under union.

However, as we showed above, in general, the CFLs *are* closed under union.

6.3.2 Concatenation

In this section, we see that:

If L_a and L_b are context free languages, then so is L_aL_b.

Remember that, if X and Y are languages, then XY denotes the language resulting from the **concatenation** of each and every string in X with each and every string in Y. Table 6.4 shows a simple method for taking two CFGs, G_a and G_b, and producing from them a new CFG, G, which generates $L(G_a)L(G_b)$. As before, we use our grammars G_x and G_y, from Table 6.2 to illustrate.

The new grammar created in Table 6.4 ensures that *any terminal string that could be derived by the first grammar is immediately followed by any terminal string that the second grammar could produce*. Therefore, the new grammar generates *the language*

Table 6.4 Constructing a CFG to generate the concatenation of two CFLs (G_x and G_y are defined in Table 6.2)

Step	Description	Example
1	arrange it so the two grammars have no non terminals in common	G_x and G_y are already like this
2	for each production in the first grammar with S (the start symbol of the grammar) on its left hand side, put R (the start symbol of the second grammar) at the end of the right hand side of that production.	New production created: $A \rightarrow BCD$ (only one created, since G_x has only one production with A, its start symbol, on the l.h.s.)
3	Add the rest of the first grammar's productions, and all of the second grammar's productions to the productions created in 2. The start symbol of the first grammar is the start symbol of the new grammar.	New grammar: $A \rightarrow BCD$ $B \rightarrow aBb \mid ab$ } rest of G_x $C \rightarrow cC \mid c$ $D \rightarrow EF$ $E \rightarrow aE \mid a$ } rest of G_y $F \rightarrow bFc \mid bc$

generated by the first grammar concatenated to the language generated by the second grammar. For the example in Table 6.4. the new grammar generates the set:

$XY = \{a^i b^i c^j a^s b^t c^t : i, j, s, t \geq 1\}.$

The Table 6.4 construction could be applied to any pair of context free grammars, and so **the context free languages are closed under concatenation**.

However, the deterministic CFLs are *not* closed under concatenation. This is not demonstrated by our example, because XY happens to be deterministic in this case. The two deterministic CFLs

$\{a^i b^j : i \geq 1, j = 0 \text{ or } j = \text{i}\}$

and

$\{b^i c^j : i = 0 \text{ or } i = j, j \geq 1\}$

concatenated together result in a language which is a *non* deterministic CFL. You are asked in the exercises to argue that the concatenation of the above two languages is a non deterministic CFL.

Our concatenation result for the deterministic CFLs is thus:

the deterministic context free languages are not closed under concatenation.

6.3.3 Intersection

Thus far, we have seen that the closure situation is the same for the CFLs as for the regular languages, with respect to both union and concatenation. However, in terms of **intersection**, the CFLs differ from the regular languages in that the CFLs are *not* closed under intersection. To appreciate this, consider the two sets we have been using in this section, i.e.

$X = \{a^i b^i c^j : i, j \geq 1\}$

and

$Y = \{a^i b^j c^j : i, j \geq 1\}.$

What is the intersection ($X \cap Y$) of our two sets? A little thought reveals it to be the set of strings of *a*s followed by *b*s followed by *c*s, in which the number of *a*s equals the number of *b*s and the number of *b*s equals the number of *c*s, i.e.:

$X \cap Y = \{a^i b^i c^i : i \geq 1\}.$

This language was stated to be not context free at the end of the last chapter, and this will be demonstrated later in this chapter. Moreover, both X and Y are deterministic CFLs, so this counter example shows us that:

both the deterministic and the non deterministic context free languages are *not* closed under intersection.

6.3.4 Complement

We can argue that it is also the case that the CFLs are not closed under **complement** (i.e. that the complement of a CFL is not necessarily itself a CFL). You may recall that above we used the following law to show that the *regular* languages were closed under intersection:

$$L_1 \cap L_2 = \overline{\overline{L_1} \cup \overline{L_2}}.$$

We can use the same law to show that the CFLs *cannot* be closed under complement. If the CFLs were closed under complement, the above law would tell us that they were closed under intersection (as we showed that they were closed under union earlier on). However, we have already shown that the CFLs are not closed under intersection, so to show that this is the case is absurd. We must therefore reject any assumption that the CFLs are closed under complement.

The above argument is our first example of proof by a technique called ***reductio ad absurdum***. We simply assume the logical negation of the statement we are trying to prove, and on the basis of this assumption follow a number of logically sound arguments that eventually lead to a nonsensical conclusion. Since the steps in our argument are logical, it must be our initial assumption that is false. Our initial assumption is false, but it is actually the negation of the statement we are trying to prove. Therefore, the statement we are trying to prove is indeed true.

In this case, what our reductio ad absurdum proof shows us is that

the context free languages are not closed under complement.

This result is very important, at least with respect to the pushdown recognizer. Imagine you are writing a **parser** for a programming language that you have designed. Suppose you implement your parser to be equivalent in power to a non deterministic pushdown recognizer. It turns out, unknown to you, that your language is one for which the complement is not a context free language. Now, this is not likely to cause problems for your parser if its input consists entirely of syntactically correct programs. This is highly unlikely, as I'm sure you will agree if you have ever done any programming! However, there will be one (or more) syntactically incorrect 'programs' for which your parser will not be able to make any decision at all (it will probably enter an infinite loop).

If for every string we input to the parser, the parser (which remember is no more powerful than a PDR) could output an indication of whether *or not* the program is syntactically legal, the same parser could be used to accept the complement of our original language (i.e. we simply interpret the original parser's 'yes' as a 'no', and vice versa). However, the fact that the CFLs are not closed under complement tells us that for some languages a parser equivalent in power to a PDR will be unable to make the 'no' statement. I suggest you re-examine the **acceptance conditions** for non deterministic PDRs in the previous chapter. Now you can probably see why there is no condition that dictates that the machine must always halt, only that it must *always* halt in the case of **sentences** of the language.

The issue of whether or not abstract machines eventually stop at some time during a computational process is central to computer science. However, we defer consideration of such issues until much later in the book.

All is not lost, however, for two reasons. Firstly, if your programming language is a *deterministic* CFL, its complement will also be a *deterministic* CFL. Remember, at the end of Chapter 5 we saw that a deterministic CFL was as a language for which a pushdown machine did not have to look more than one symbol ahead in the input. We saw that deterministic CFLs are called **LR(1) languages** for that reason. In such cases there are only a finite number of symbols the machine expects to encounter when it looks ahead to the next one. They form two sets at each stage: the symbols the machine expects to see next, and the symbols it does not expect to see next. If at any stage, the machine looks ahead and finds one of the symbols which at that stage it *does not* expect, the machine simply rejects the input as invalid (moves to a *halt state that denotes rejection,* as opposed to a *halt state that denotes acceptance*). We can make sure, then, that our machine always halts with 'yes' or 'no'. We can then simply exchange the acceptance/rejection status of our machine's halt states, and we have a machine that accepts the complement of the original language.

So,

the deterministic context free languages are closed under complement.

6.4 ESTABLISHING THAT CHOMSKY'S HIERARCHY IS A PROPER HIERARCHY

In this section, we sketch out two fundamental theorems of formal languages. One is for the regular languages, though it actually applies to the FSR, and is called the **repeat state theorem**. The other, for the context free languages, is in some ways similar to this, and is called the **uvwxy theorem**. The usefulness of these theorems is that they apply to every language in their class, and can therefore be used to show that a given language is not in the class. To appreciate this point, think about it like this: if every element in a set *must* possess a particular property, and we come across an object that cannot have that property, then the object cannot belong to our set. If it is true that every bird has feathers, and we come across a creature that is featherless, then that creature is certainly not a bird.

So, if we can use the *repeat state theorem* to show that a certain language cannot be regular, but we can show that the same language is context free, then we know that the regular languages are a **proper subset** of the context free languages (since we know that every regular language is context free, we have then also shown that there are context free languages that are not regular). Moreover, if we use the *uvwxy theorem* to show that a given language cannot be context free, and we then show that it is **type 0**, we know that the context free languages are a *proper subset* of the type 0 languages[1]. We have then established that **Chomsky's hierarchy** is a real hierarchy.

Some students experience problems in thinking in terms of *proper* subsets. This is because when you say to them 'all *x*'s are *y*'s', they automatically assume that you also mean 'all *y*'s are *x*'s'. To demonstrate the silliness of this inference, consider *parrots* and *birds*. If I said to you 'all parrots are birds', you would not then think 'all birds are parrots'. The set of all regular languages corresponds to the set of all parrots, and the set of all context free languages corresponds to the set of all birds. One set is *properly contained in the other.*

6.4.1 The 'repeat state theorem'

Any **FSR** that has a **loop** on some path linking its **start** and **halt states** recognizes an **infinite language**.

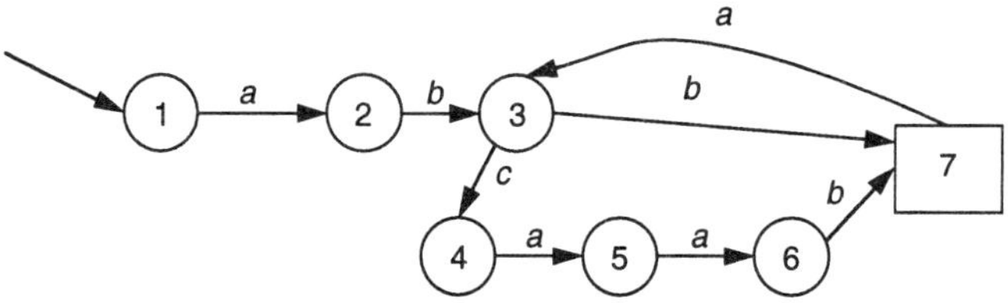

Figure 6.7 The FSR, *M*

Consider the example FSR, *M*, in Figure 6.7.

M of Figure 6.7 has seven states. Consider any string **acceptable** to *M* that is greater than *six* symbols in length. That string must have resulted in the machine passing through one of its states more than once. After all, *M* has only seven states, and it would need *eight* states to accept a string of length 7, if no state was to be visited more than once. Consider, as a suitable example the string *abbacaab*. This string is eight symbols in length, and the sequence of states visited during the acceptance of this string is as shown by the numbers placed next to *M*'s states in Figure 6.8.

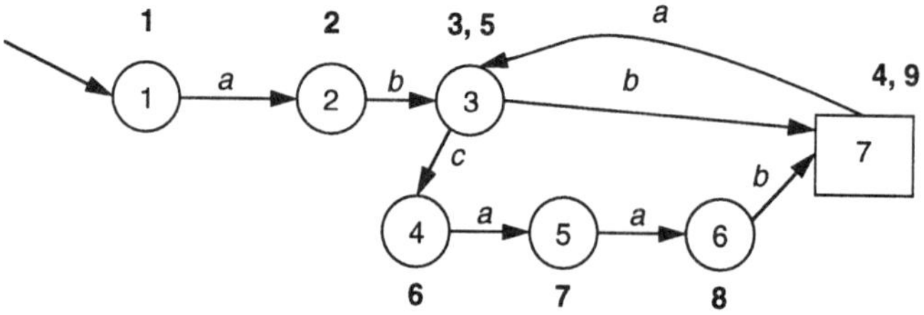

Figure 6.8 The sequence of states visited by *M* for the input *abbacaab*

[1] We omit the **context sensitive (type 1)** languages from this discussion. Only the **ε-free CFGs** are context sensitive, as context sensitive grammars are not allowed to have any ε productions.

Let us choose any state that *M* visited more than once in accepting our string, say, state 3. State 3, as the diagram shows, was the third and fifth state visited. Consider the **substring** of *abbacaab* that caused *M* to visit state 3 twice, i.e. *ba*. Where *ba* occurs in *ab**ba**caab*, we could copy it **any** number of times, and we would still obtain a string that *M* would accept. As an example, we will copy it four times: *ab**babababa**-caab*. Furthermore, we could omit *ba*, giving *abcaab*, a string that is also accepted by our machine.

The other state that was visited twice in the acceptance of *abbacaab* was state 7 (the halt state). We now choose this as our example repeated state. The substring of *abbacaab* that caused state 7 to be visited twice was *acaab*. As before, our machine *M* would accept the original string with *no* copies of *acaab* in it: *abb*. It would also accept the original string with any number of copies inserted after the first occurrence, for example three copies: *abbacaabacaabacaab*.

To appreciate the generality of the above, consider any FSR that accepts a string of length greater than or equal to the number of states it has. As we said before, one or more of its states was visited twice. We consider any of those 'visited-twice' states. The string that was accepted can be expressed as three **concatenated** strings:

- a possibly empty first part, representing the states visited before the machine got to its 'visited twice' state,
- a non empty second part, representing the part of the string that caused the machine to visit the 'visited-twice' state again. This part of the string can be copied 0 or more times in between the first and third parts, and the resulting string will also be acceptable to the machine,
- a possibly empty third part, representing the states visited after the machine has left the 'visited-twice' state loop.

In our examples for the string *abbacaab*, above, when state 3 was the chosen 'visited-twice' state:

- *first part* was *ab*,
- *second part* was *ba*,
- *third part* was *caab*.

Alternatively, when state 7 was the chosen 'visited-twice' state:

- *first part* was *abb*,
- *second part* was *acaab*,
- *third part* was empty.

For many machines and strings, *first part* may be empty, if the loop began at the start state. In other cases, both *first* and *third parts* may be empty if the loop encompasses every state in the machine.

The repeat state theorem is expressed formally in Table 6.5.

Table 6.5 The repeat state (*vwx*) theorem for regular languages

For any deterministic FSR, *M*:
if *M* accepts a string *z*, such that $| z | \geq n$, where n is the number of states in *M*,
then *z* can be represented as three concatenated substrings:
$z = vwx$ in such a way that all strings vw^ix,
for all $i \geq 0$ are also accepted by *M*.

As was established in Chapter 4, *any language accepted by a DFSR is a regular language*. Therefore, *our theorem applies to any infinite regular language*.

6.4.2 *A language that is context free but not regular*

Now, as promised at the end of Chapter 4, we turn our attention to the language

$\{a^ib^i : i \geq 1\}$.

We know that this language is context free, because it is generated by the context free grammar G_3, of Chapter 2:

$S \rightarrow aSb \mid ab.$

The question is, could it also be a regular language? We shall now use our new theorem to show that it cannot be a regular language, using the technique of ***reductio ad absurdum*** introduced earlier in this chapter.

Our assumption is that $\{a^ib^i : i \geq 1\}$ is a regular language. If it is regular it is accepted by some FSR, *M*. *M* can be assumed to have *k* states. Now, *k* must be a definite number, and clearly our language is infinite, so we can select a sentence from our language of the form a^jb^j, where $j \geq k$. The *repeat state theorem* then applies to this string.

The repeat state theorem applied to our language tells us that our sentence a^jb^j can be split into three parts, so that the middle part can be repeated any number of times within the other two parts, and that this will still yield a string of *a*s followed by an equal number of *b*s. But how can this be true? Consider the following:

the 'repeatable substring' *cannot* consist of *a*s followed by *b*s, as when we repeated it we would get *a*s followed by *b*s followed by *a*s... Our machine would accept strings that are not in the language $\{a^ib^i: i \geq 1\}$.

This means that the repeatable substring must consist of *a*s alone or *b*s alone. However:

the repeatable substring *cannot* consist of *a*s alone, or when we repeated it more than once, the *a*s in our string would outnumber the *b*s, and our machine would accept strings not in the language $\{a^ib^i: i \geq 1\}$.

A similar argument obviously applies to a repeatable substring consisting entirely of *b*s.

Something is clearly amiss. Every statement we have made follows logically from our assumption that $\{a^ib^i: i \geq 1\}$ is a regular language. We are therefore forced to go right back and reject the assumption itself. Therefore $\{a^ib^i: i \geq 1\}$ is not a regular language.

Our result tells us that an *FSR could not accept certain strings in a language like* $\{a^ib^i: i \geq 1\}$ *without accepting many more that were not supposed to be acceptable.* We simply could not constrain the FSR to accept *only* those strings in the language. That is why an FSR can accept a language such as $\{a^ib^j : i, j \geq 1\}$, as you will find out if you design an FSR to do so, but the language $\{a^ib^i: i \geq 1\}$, which is a proper subset of $\{a^ib^j : i, j \geq 1\}$ is beyond the computational abilities of the FSR.

One interesting outcome of discovering that $\{a^ib^i: i \geq 1\}$ is not regular is that we can use that fact to argue that a language such as Pascal is not regular. The set of Pascal <compound statement>s is of the form:

$\{\textbf{begin}^i x \textbf{end}^i : x$ is a sequence of Pascal <statement>s$\}$.

The repeat state theorem tells us that many strings in this part of the Pascal language cannot be accepted by an FSR unless the FSR accepts many other strings in which the number of **begins** does not equal the number of **ends**. By extension of this argument, *the syntax of Pascal cannot be represented as a regular grammar.* However, as mentioned in Chapter 3, the syntax of Pascal can be expressed as a context free grammar. The same applies to most programming languages, including, for example, Lisp, with its use of arbitrary length strings of balanced parentheses.

You are asked in the exercises to provide similar justifications that certain other context free languages cannot be regular. It is thus clear that *the regular languages are properly contained in the class of context free languages.* In fact, we have shown two things about Chomsky's hierarchy:

1. the regular languages are a proper subset of the CFLs in general,
and, more specifically,
2. the regular languages are a proper subset of the deterministic CFLs.

Point 2 is demonstrated by considering that $\{a^ib^i: i \geq 1\}$ is not regular, but is a CFL, and is also deterministic (as we saw in Chapter 5). So there is at least one language that is a deterministic CFL that is not regular. In fact, none of the deterministic CFLs we considered when establishing the closure results for CFLs in this chapter is regular. You might like to convince yourself of this by using the repeat state theorem.

6.4.3 The uvwxy theorem for context free languages

The theorem described in this section is a context free language counterpart of the repeat state theorem described above. However, we establish our new theorem by using properties of context free grammars and derivations rather than the corresponding abstract machines (PDRs).

Let us assume that we have an ε **free CFG** that is in **Chomsky normal form** (**CNF**). That we can assume that any CFG is so represented was justified in Chapter 5[2]. As an example, we will use a CNF version of G_3 ($S \rightarrow aSb \mid ab$), which we will simply call G:

$S \rightarrow AX \mid AB$
$X \rightarrow SB$
$A \rightarrow a$
$B \rightarrow b.$

G has four non terminal symbols. So if we do a **derivation tree** for a sentence in which there is a path from *S* with more than four non terminals on it, one of the non terminals will appear more than once.

Now, consider the derivation tree for the sentence a^4b^4, shown in Figure 6.9. and labelled DT 1.

In particular, we focus on the path in DT 1 proceeding from the top of the tree to the non terminal marked 'low X', in DT 1. Sure enough, this path features a repeated non terminal. In fact, both *S* – three times – and *X* – also three times – are repeated on this path. We can choose any pair of the repeated non terminals, so we will choose the two instances of the non terminal *X* that have been labelled 'low *X*' and 'high *X*'.

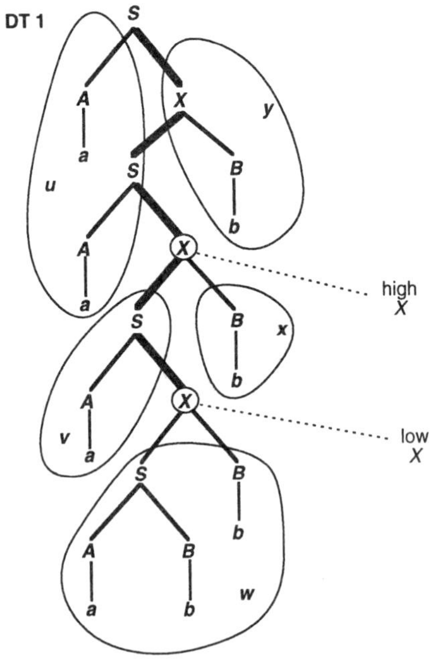

Figure 6.9 The derivation tree, DT I

[2] We leave out the single ε production, if it exists, and concentrate only on the ε free part of the grammar.

Five substrings, that concatenated together in order make up the sentence have been identified and labelled in Figure 6.9 as follows:

u this is the string derived in the part of the tree from *S* to the left of high *X*, and is the string *aa*
v this is the string derived from high *X* to the left of low *X*, and is the string *a*
w this is the string derived from low *X*, i.e. abb
x this is the string derived from high *X* to the right of low *X*, i.e. *b*
y this is the string derived from *S* to the right of high *X*, and is *b*.

Concatenating all the above strings in order, i.e. *uvwxy*, gives us our sentence a^4b^4.

Careful thought may now be required. The derivation from high *X* in DT 1 is simply the application of a sequence of productions, the first having *X* as its left hand side. *G* is a CFG, so the derivation we did from high *X* in DT 1 could be repeated from low *X*. We would still have a sentence, as the whole derivation begins with *S*, and a terminal string is derived from high *X* in DT 1.

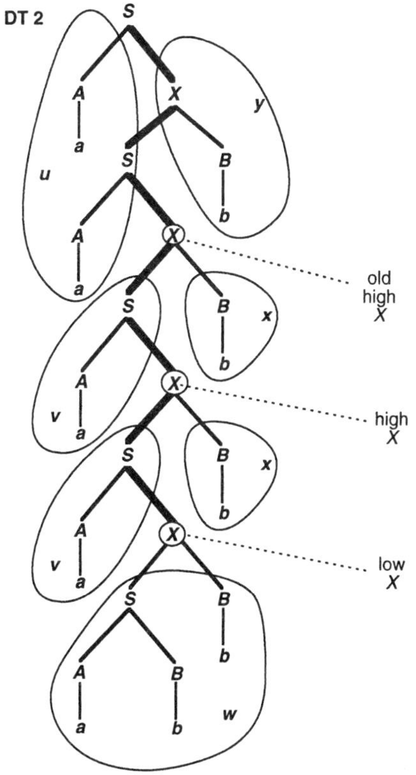

Figure 6.10 The derivation tree, DT 2

Doing the above gives us DT 2, as shown in Figure 6.10. DT 2 is the derivation tree for the sentence a^5b^5. As DT 2 shows, the way we have obtained the new derivation ensures that we now have *two* copies of the substrings v and x, where there was one copy of each, respectively, in the DT 1 sentence. Expressed in an alternative way, we have:

$$uv^2wx^2y \ (= a^5b^5).$$

What we have just done could be repeated, i.e. we could do the high X derivation we did in DT 1 from the new low X near the bottom of DT 2, in Figure 6.10. Then we would have uv^3wx^3y. Moreover, we could repeat this process indefinitely, i.e. we can obtain uv^iwx^iy, for all $i \geq 1$.

Finally, referring back to DT 1 (Figure 6.9), the derivation originally done from low X could have been done from high X instead. If we do this, we obtain DT 3, as shown in Figure 6.11. DT 3 is the derivation tree for the sentence a^3b^3. As can be seen from Figure 6.11, this time there are only *three* of our substrings, namely u, w and y. We have $uwy = a^3b^3$.

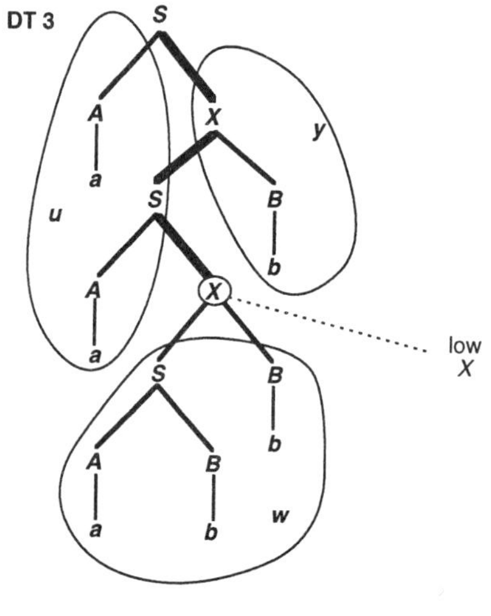

Figure 6.11 The derivation tree, DT 3

As stated in Chapter 2, for any string, s, $s^0 = \varepsilon$, the empty string, and for any string z, $z\varepsilon = z$, and $\varepsilon z = z$. uwy can therefore be represented as uv^0wx^0y.

The constructions used to obtain the three trees, DT 1–3 (Figures 6.9 – 6.11) and our accompanying discussion tell us that, for the language $L(G)$, i.e. $\{a^ib^i : i \geq 1\}$, we have found a string a^4b^4 which is such that:

> a^4b^4 can be represented as concatenated substrings u, v, w, x, and y, such that $uvwxy = a^4b^4$, and for all $i \geq 0$, uv^iwx^iy is in $L(G)$.

We initially chose the two Xs called high X and low X in DT 1. However, *any other equal pair of non terminals on the path could have been chosen, and a similar argument could have been made*. For example, we could have chosen two of the repeated S nodes. A possible arrangement in this case is shown in DT 1a (Figure 6.12).

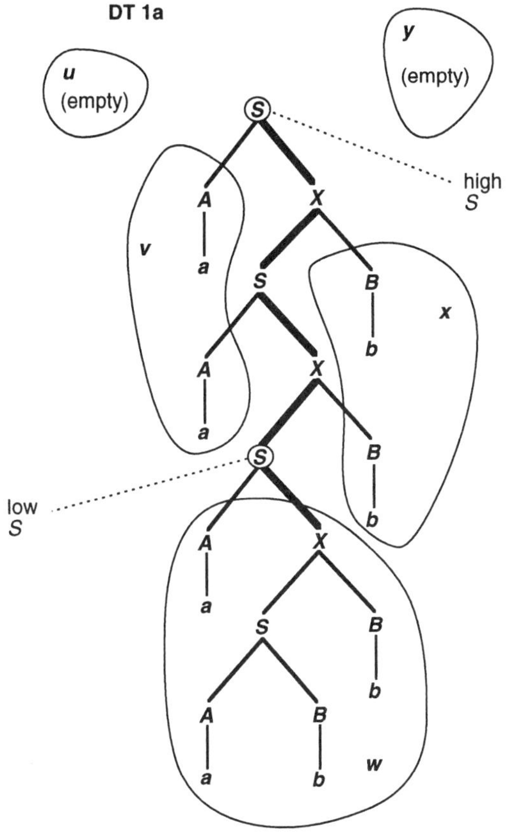

Figure 6.12 The derivation tree, DT 1a

In Figure 6.12 there is no string u to the left of the top S down to high S (as high S *is* the top S), and similarly there is no string y to its right, and both u and y are empty, i.e. $u = y = \varepsilon$. So we now have:

$u = \varepsilon$
$v = aa$
$w = aabb$
$x = bb$
$y = \varepsilon$.

This time, if we duplicated the original high S derivation from low S in Figure 6.12, we would have uv^2wv^2x, i.e., ε concatenated to aa concatenated to aa concatenated to *aabb* concatenated to *bb* concatenated to *bb* concatenated to ε, which is a^6b^6. The next time we repeated it we would obtain uv^3wx^3y, i.e. a^8b^8, and so on. If we duplicated the low S derivation from high S, we would obtain uv^0wx^0y. Since all the substrings except for w is empty, this leaves us with w, i.e. *aabb*. So again, our statement above:

> a^4b^4 can be represented as concatenated substrings u, v, w, x, and y, such that $uvwxy = a^4b^4$, and for all $i \geq 0$, uv^iwx^iy is in $L(G)$,

is true for this case also, as long as we permit u and y to be empty.

How general is this *uvwxy* type construction? It can be argued to apply to *any* context free grammar that generates a sentence long enough so that its derivation tree contains a path from S that contains two (or more) occurrences of one non terminal symbol. Now, any CFG that generates an infinite language must have a path like this. Such a CFG, as we showed in Chapter 5, can have its ε free part converted into CNF. If there are n non terminals in a CNF grammar, by the time we generate a sentence that is greater in length than 2^{n-1} (numerically speaking), we know we have a path of the type we need. To appreciate this, suppose that a CNF grammar has three non terminals, i.e. $n = 3$. The shallowest tree for a sentence of length equal to 2^{n-1} (4) is shown in Figure 6.13.

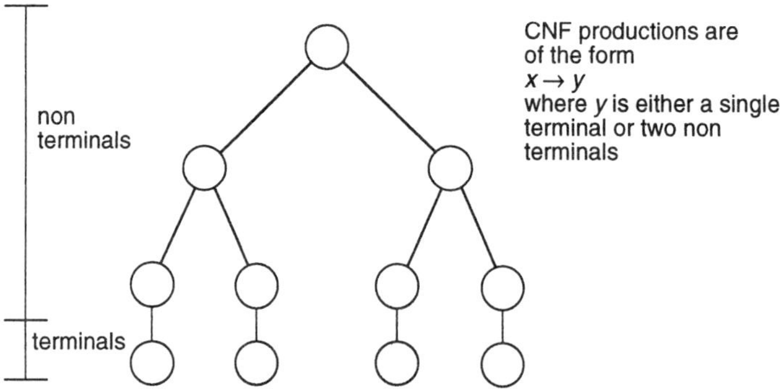

Figure 6.13 The shallowest derivation tree for CNF productions to generate a sentence of length 4

Notice that the longest paths in the tree in Figure 6.13 have length, or number of arcs, 3, and, since the last node on the path would be a terminal, such paths have only three non terminals on them. Such paths are not long enough to ensure that they feature repeated non terminals. Our hypothetical grammar has three non terminals, so to be absolutely sure that our tree contains a path of length greater than 3 we must generate a sentence of length *greater than 4*. You may need to think about this; perhaps drawing some possible extensions to the above tree will convince you: remember, the grammar is assumed to be in CNF. You should also convince yourself that the 2^{n-1} argument applies to any CNF grammar.

Of course, it is not always necessary to generate a long sentence to obtain our 'repeated non terminal path'. Our grammar G can be used to create derivation trees for sentences shorter in length than 8 (2^3, since G has four non terminals) that have paths of the type we require. However, here we are interested in defining conditions that apply in general, which requires us to be *sure* that we can obtain a path of the type required.

Let G stand for *any* ε free context free grammar. We now know that if we generate a sentence of sufficient length, the derivation tree for that sentence will have a path, starting from the S at the root of the tree, upon which the same non terminal, say X, appears twice or more. We choose two occurrences of X on that path, and call one high X and one low X. The sentence can be split into five concatenated substrings as before (use the derivation trees DT 1–3 and DT 1a in Figures 6.9 – 6.12 for illustration, if you wish):

- u the string derived from S at the root and to the left of high X. This string may be empty, either because S = high X (as in DT 1a above), or because our path proceeds directly down the left of the tree.
- v the string derived from high X to the left of low X. This string may be empty if the path from high X to low X is on the left. However, if this string is empty, x (see below) will not be. As this is a CNF grammar, high X must expand initially to two non terminals. Either (a) one of these is low X, or (b) must eventually expand into two non terminals of which one is low X. The other one expands into part of v or x, depending on the particular tree.
- w the string derived from low X. This string will not be empty, as the grammar is ε free.
- x as for v but derived from high X to the right of low X. The same argument as for v applies here. In particular, cannot be empty if v is.
- y as for u but to the right of high X. This string may also be empty, for analogous reasons to those given above, for u.

As was shown by our example derivation trees DT 1–3, above, we can always repeat the 'deriving what was derived from the high X in the original tree from low X' routine as many times as we wish, thus obtaining the characteristic uv^iwx^iy form, for all $i \geq 1$. Furthermore, we can derive what was derived from low X in the original tree from high X, giving us the uv^0wx^0y case.

Enough has been said to demonstrate the theorem formally stated in Table 6.6.

Table 6.6 The *uvwxy* theorem for context free languages

Any sufficiently long sentence, z, of an infinite context free language, L, can be represented as five concatenated substrings u, v, w, x, and y, such that $z = uvwxy$, and uv^iwx^iy is also in L, for all $i \geq 0$.

6.4.4 $\{a^ib^ic^i : i \geq 1\}$ *is not context free*

As for the **repeat state theorem** for regular languages, we can use our *uvwxy* theorem to show that a given infinite language *cannot* be context free. As promised at the end of Chapter 5, we now demonstrate that the language

$$\{a^ib^ic^i : i \geq 1\}$$

is not context free.

The proof of this is very simple. Again, we use ***reductio ad absurdum*** as our technique. We assume that $\{a^ib^ic^i : i \geq 1\}$ is a CFL. The language is clearly infinite, so the *uvwxy* theorem applies to it. Our sufficiently long sentence is going to be one greater in length than 2^{n-1}, where n is the number of non terminals in a Chomsky normal form grammar that generates our language.

We can assume that our sufficiently long sentence is $a^kb^kc^k$, for some k, where $k > 2^{n-1}$. The theorem tells us that this sentence can be represented as five concatenated substrings *uvwxy*, such that uv^iwx^iy are also in $\{a^ib^ic^i : i \geq 1\}$ for all $i \geq 1$. However, how are substrings of $a^kb^kc^k$ to be allocated to u, v, w, x, and y in this way?

> neither v nor x can be strings of *a*s and *b*s, or *b*s and *c*s, or *a*s, *b*s and *c*s, as this would result in symbols being in the wrong relative order when v and x were repeated.

So this means that the repeatable strings v and x must consist of strings which are either all *a*s, all *b*s or all *c*s. However,

> this would also be inappropriate, as even if, say, v was *aaa* and x was *bbb*, when v and x were repeated, though the number of *a*s would still equal the number of *b*s, the number of each would exceed the number of *c*s. A similar argument can be used to show that no other allocation of substrings to v and x would work.

We must again reject our original assumption that $\{a^ib^ic^i : i \geq 1\}$ is a context free language.

However, $\{a^ib^ic^i : i \geq 1\}$ is a *context sensitive* language, since it is generated by the following **context sensitive grammar**, G_{16}:

$S \to aSBC \mid aBC$
$CB \to BC$
$aB \to ab$
$bB \to bb$
$bC \to bc$
$cC \to cc$

as you might like to justify for yourself.

The arguments in the following section may require some effort to appreciate. The section can be omitted, if required, though the result, as stated in the section's title, should be appreciated.

6.4.5 The 'multiplication language' is not context free

Now we consider one final refinement to our *uvwxy* theorem. We focus on a derivation tree for a sentence of length greater than 2^{n-1}, as described above. We choose the longest path (or one of them, if there are several). Now, as our high X and low X, we pick the lowest two repeated nodes on this path, a constraint that was not enforced in our earlier examples. The string derived from high X (i.e. the one we call vwx) is derived by a tree with high X as its root in which the path that passes through low X is the longest path. We selected the two Xs because they were the lowest repeated nodes on our longest path, therefore high X and low X are the only repeated nodes on our 'sub-path'. Therefore, the path must have maximum length $n + 1$, with the last node on the path being a terminal. Furthermore, the longest string that could be generated from high X (the one we call vwx) must be such that its length is less than or equal to 2^n (formally, $|\, vwx \,| \leq 2^n$).

We have established that for any sentence, z, such that $|\, z \,| > 2^{n-1}$, where there are n non terminal symbols in a Chomsky normal form grammar that generated z, we can define the *uvwxy* substrings so that $|\, vwx \,| \leq 2^n$. To see why this can be useful, we consider a language introduced at the end of Chapter 2, i.e.

$$\{a^i b^j c^{i \times j} : i, j \geq 1\},$$

which was generated by our **type 0 grammar**, G_4.

It was stated at the close of Chapter 5 that the above language is not context free. We now use the *uvwxy* theorem, along with our newly established result obtained immediately above, to demonstrate this.

Again, we use *reductio ad absurdum*. We assume that $\{a^i b^j c^{i \times j} : i, j \geq 1\}$ is a context free language. This means that the *uvwxy* theorem applies to a sufficiently long string. Consider the following string, z, in our language: $a^{2^n} b^{2^n} c^{(2^n)^2}$, where n is the number of non terminals in a Chomsky normal form grammar for the language. Obviously, the length of z is greater than 2^{n-1}. We can then, as shown in the preceding paragraph, establish a *uvwxy* form for this sentence, such that $|\, vwx \,| \leq 2^n$ (n being the number of non terminals). Now, by a similar argument to that applied in the case of the language $\{a^i b^i c^i : i \geq 1\}$ above, we can certainly justify that v and x *cannot be mixtures of different terminals*. This means that v must consist of either all as or all bs, and x must be all cs (otherwise when we increase the number of as or bs for v^i, the cs would not increase accordingly). Therefore,

suppose v is a string of as. Then, since x must consist of cs only, w must include all of the bs. This cannot be so, as both v and x are non empty, so $|\, vwx \,|$

exceeds 2^n (since there are 2^n *b*s in w alone).

v cannot be a string of *a*s then, so:

> v must be a string of *b*s. Now, suppose $v = b^k$, for some $k \geq 1$. Then x must be $c^{2^n \times k}$. This is because each time uv^iwx^iy inserts a new copy of v (b^k) into the *b*s, it inserts a new copy of x into the *c*s. To ensure that the number of *c*s is still the same as the number of *a*s times the number of *b*s, we must insert 2^n *c*s (2^n being the number of *a*s) for each of the k *b*s. However, again this cannot be so, because if $| x | = 2^n \times k$, then $| vwx |$ is certainly $> 2^n$.

If the language had been context free, we could, for our sufficiently long sentence, have found a *uvwxy* form such that $| vwx | \leq 2^n$. By assuming the existence of this form, we are led to a contradiction. We are therefore forced, once more, to reject our original assumption, and therefore conclude that $\{a^ib^jc^{i \times j} : i, j \geq 1\}$ is not a context free language.

Note that the addition of the length constraint to the *uvwxy* theorem makes it especially powerful. It would have been difficult to establish that $\{a^ib^jc^{i \times j} : i, j \geq 1\}$ is not a context free language by arguing simply in terms of *uvwxy* patterns alone. You may wish to try this to appreciate the point.

6.5 PRELIMINARY OBSERVATIONS ON THE SCOPE OF THE CHOMSKY HIERARCHY

The fact that $\{a^ib^jc^{i \times j} : i, j \geq 1\}$ is not context free suggests that we need more powerful machines than pushdown recognizers to perform computations such as multiplication. This is indeed the case. However, the **abstract machine** that is capable of multiplication (of arbitrary length numbers), is not that far removed from the pushdown machine. The study of the machine, the **Turing machine**, will be a major part of the remainder of this book. It also turns out that the Turing machine is the recognizer for the context sensitive (type 1) and unrestricted (type 0) languages, which is how we introduce it in the next chapter.

In the preceding sections we have seen that there is indeed a hierarchy of formal languages, with each class being properly contained within the class immediately above it. The question arises: *are the type 0 languages a proper subset of the set of all formal languages?* Remember, a **formal language** is simply any set of strings. In Chapter 11, we see that there are indeed formal languages that are outside the class of type 0 languages, and thus cannot be generated by any grammar specified by the **Chomsky hierarchy**.

EXERCISES

1 Produce a **DFSR** to **accept** the language $\{a^{4i} : i \geq 1\}$, then use the **complement** construction described earlier in the chapter to produce a machine that accepts the language $\{a^{4i+j} : i \geq 0, 1 \leq j \leq 3\}$.

2 As mentioned in the text above, if $L_1 = \{a^i b^j : i \geq 1, j = 0 \text{ or } j = i\}$ and $L_2 = \{b^i c^j : i = 0 \text{ or } i = j, j \geq 1\}$, both L_1 and L_2 are **deterministic CFLs**, but the language $L_1 L_2$ is not. Justify this.

3† As expressed above, the first sentence of the section on the **repeat state theorem** (section 6.4.1) is not generally true. Rewrite the sentence so that it is.

4† Use the *repeat state theorem* to show that $\{a^i b^j c^{i+j} : i, j \geq 1\}$ and $\{a^i b^j c^{i-j} : i, j \geq 1\}$, both shown to be deterministic by exercise 7 of Chapter 5, are not **regular languages**.

5† Use the ***uvwxy* theorem** to show that $\{xx : x \in \{a, b\}^*\}$ is not **context free**.

7 Phrase structure languages and Turing machines

AIMS

To introduce the Turing machine (TM) and to look at TMs as language recognizers, learning that

- TMs can do anything that Finite State Recognizers (FSRs) and Pushdown Recognizers (PDRs) can do
- TMs are more powerful than non deterministic PDRs
- TMs are the recognizers for the phrase structure (type 0) languages in general

7.1 THE ARCHITECTURE OF THE TURING MACHINE

The **abstract machine** we consider in this chapter, and for much of the remainder of this book, is called the **Turing machine (TM)**. Turing was the name of the English mathematician who first described such a machine in 1936. A TM is essentially a finite state machine, like a **finite state recognizer**, but with some very simple additional facilities for input and output, and a slightly different way of representing the **input sequences**.

7.1.1 'Tapes' and the 'read/write head'

The *input* to a TM is regarded as being contained on a '**tape**'. The tape is divided into '**tape squares**', so that each symbol of the input sequence occupies one square of the tape. We assume that the tape is not restricted to any particular length, so that we can place an input sequence of any length we like on the tape (we assume that any unused part of the tape consists of **blank** tape squares). For example, the input sequence *aabbaa* may occupy part of a tape such as that depicted in Figure 7.1.

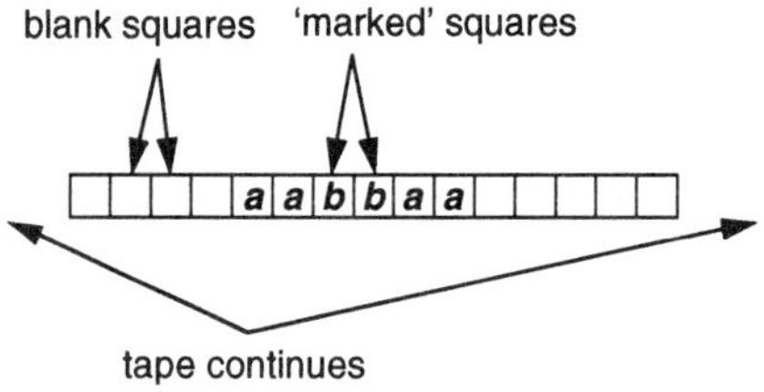

Figure 7.1 The tape of a Turing machine

Hitherto, our abstract machines have been assumed to read their input sequences from left to right, one symbol at a time. However, a TM can move along its tape in either direction, examining, or 'reading' one tape square at a time.

Let us now consider the *output* of a TM. The TM simply uses its tape for output as well as input. On reading an input symbol from a particular tape square in a particular state, the TM can replace the input symbol on that particular square with any designated symbol. As a simple example, consider a tape containing a sequence of *a*s and/or *b*s, where we require as output a sequence representing the input sequence, but with all *b*s replaced by *a*s, and vice versa.

An example tape is shown in Figure 7.2, as the hypothetical TM described immediately above has completed its processing of the first three symbols (assume that the tape is being processed from left to right).

In order to clarify the notion of a TM 'examining' one square of a tape, and placing a symbol on a tape, we say the TM possesses a **read/write head** (we sometimes simply say **r/w head** , or even just **head**). The r/w head is at any time located on one

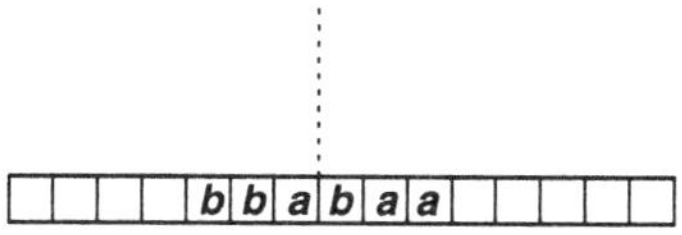

Figure 7.2 A partly processed TM tape

of the tape squares, and is assumed to be able to 'sense' which symbol of some finite alphabet occupies that square. The head also has the ability to 'write' a symbol from the same finite alphabet onto a tape square. When this happens, the symbol written replaces the symbol that previously occupied the given tape square. The read/write head can be moved along the tape, one square at a time, in ways we discuss later.

7.1.2 Blank squares

In Figure 7.2, we examined the state of the tape of an imaginary TM that replaced *b*s by *a*s (and vice versa), when the machine had dealt with only some of the input symbols on its tape. Let us now consider the example tape for Figure 7.2 when the machine has processed the last symbol (the *a* at the right hand end) of the marked portion of the tape, at which point, since the machine is replacing *a*s by *b*s, and vice versa, the tape will be as shown in Figure 7.3.

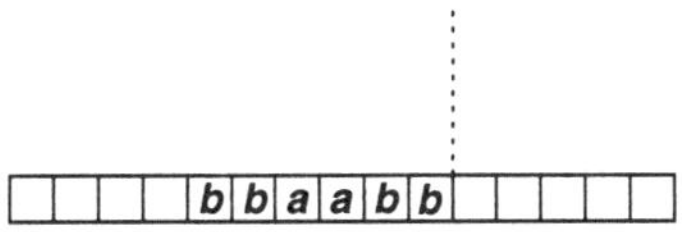

Figure 7.3 A TM reaching the end of the symbols

When our machine reaches the situation shown in Figure 7.3, it will attempt to read the symbol on the next tape square to the right, which of course is **blank**. So that our machine can terminate its processing appropriately, we allow it to make a transition on reading a blank square. In other words, we assume that a blank square can be sensed by the read/write head of a TM, and that a TM can write a symbol into a blank square and even 'blank out' an occupied tape square.

We denote the blank symbol by an empty box[1] (suggesting that it is represented as a *blank tape square*), thus: □

7.1.3 TM 'instructions'

Like FSRs (Chapter 4) and PDRs (Chapter 5), TMs can be represented as drawings, featuring states and labelled arcs. An **arc** in a TM is, like the arc between two **states** of a **PDR**, labelled with three symbols.

[1] It is sometimes represented as a 0, or a β (Greek letter *beta*).

By way of illustration, Figure 7.4 shows part of a TM. Assume that the machine is in state *A*, and that the tape and position of the read/write head are as shown on the right of Figure 7.4.

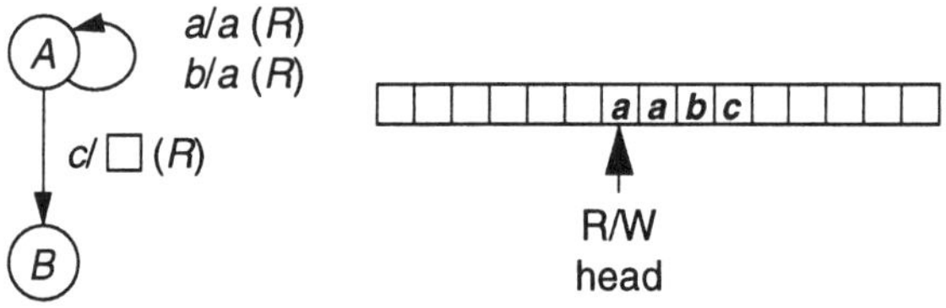

Figure 7.4 Part of a TM, and an appropriate tape set up

The TM represented in Figure 7.4 would carry out the activities described in Table 7.1.

Table 7.1 A trace of the behaviour of a simple Turing machine

Action	State of tape immediately following action
read an *a*, write an *a*, move right (R), stay in state	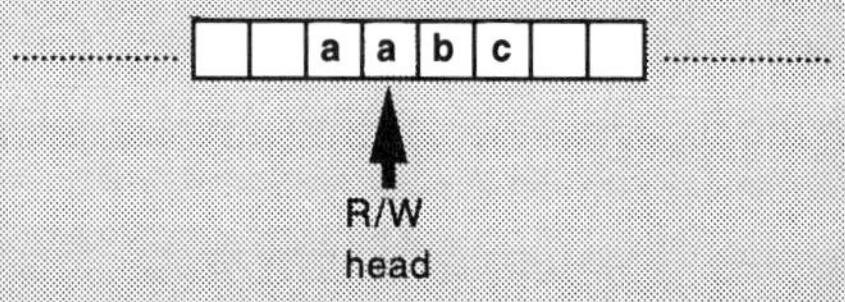
read an *a*, write an *a*, move right (R), stay in state *A*	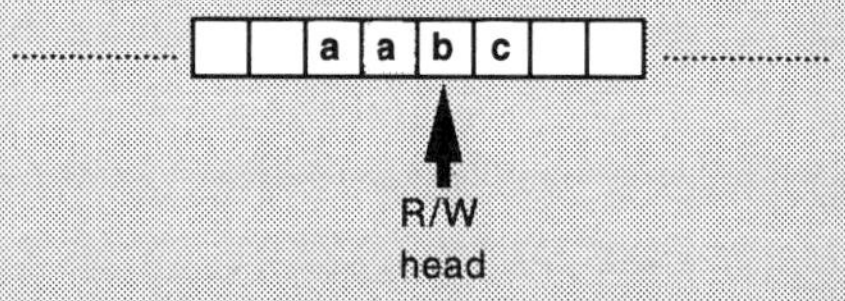
read a *b* write an *a*, move right (R), stay in state *A*	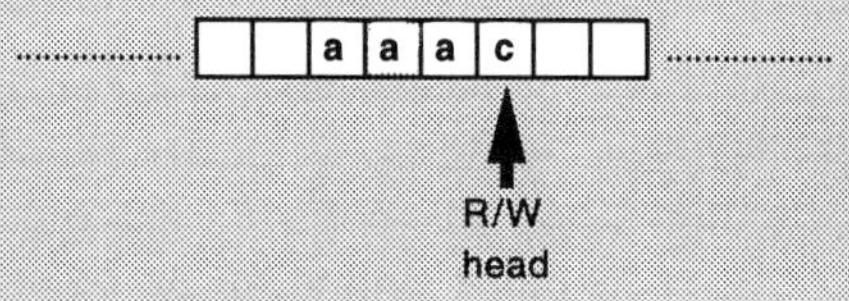
read a *c*, write a blank, move right (R), move to state *B*	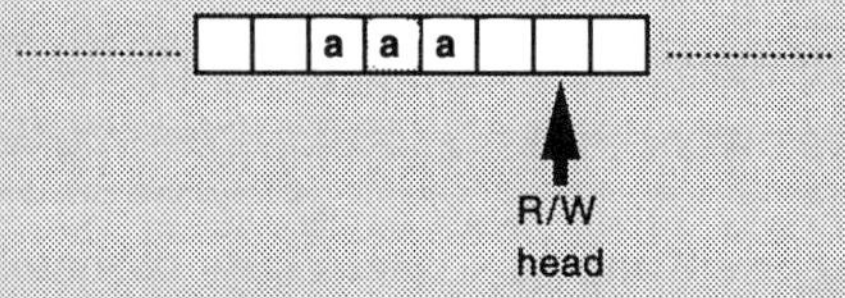

We now use the conceptual framework provided above as the basis of a more rigorous definition of TMs.

7.1.4 Turing machines defined

A TM has essentially the same pictorial formulation as the FSR, except that each arc of a TM is labelled with three instructions:

1. an *input symbol*, which is either □ or a symbol from a finite **alphabet**;
2. an *output symbol*, either □ or a symbol from the same alphabet as for (1);
3. a *direction designator*, specifying whether the read/write head is to move one square left (L), right (R), or not at all (N).

The TM can make a **transition** between states if (1) is the current tape symbol, in which case the TM replaces (1) by (2), and moves one tape square in the direction specified by (3), before moving to the state pointed at by the relevant arc.

Next, we consider a simple example TM, which is discussed in some detail. Much of the remainder of the chapter is devoted to demonstrating that TMs can perform all the tasks that can be performed by the other abstract machines we have hitherto considered, and many tasks that they could not.

7.2 THE BEHAVIOUR OF A TM

Figure 7.5 shows our first full TM, T_1.

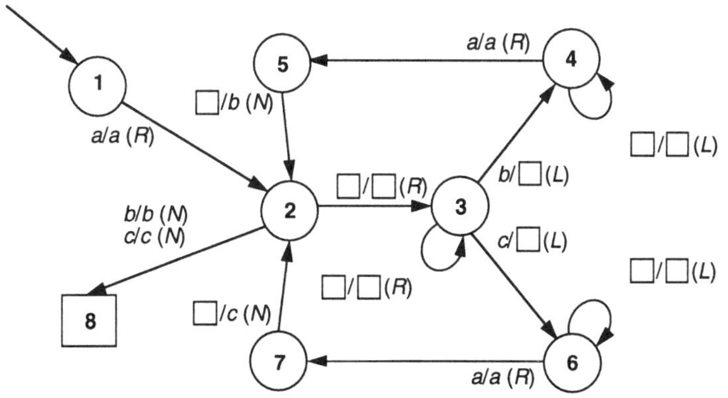

Figure 7.5 The TM, T_1

T_1 processes input tapes of the form shown in Figure 7.6.

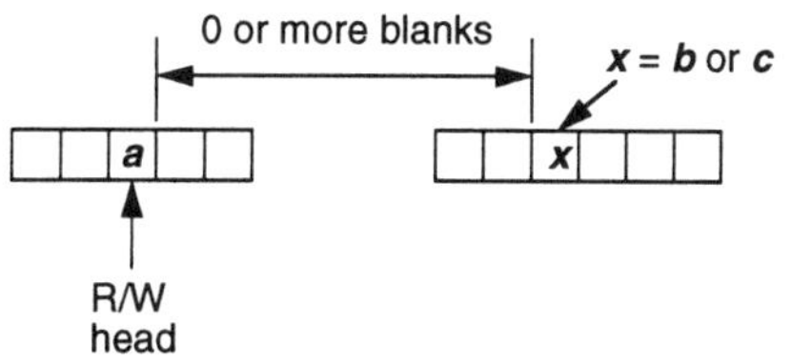

Figure 7.6 The tape set up for T_1

As shown in Figure 7.6, T_1 (Figure 7.5) expects the marked portion of its input tapes to consist of an *a* followed by zero or more blank squares, followed by either a *b* or a *c*. T_1 has its head initially positioned at the *a*, and produces as its output a tape of the form depicted in Figure 7.7.

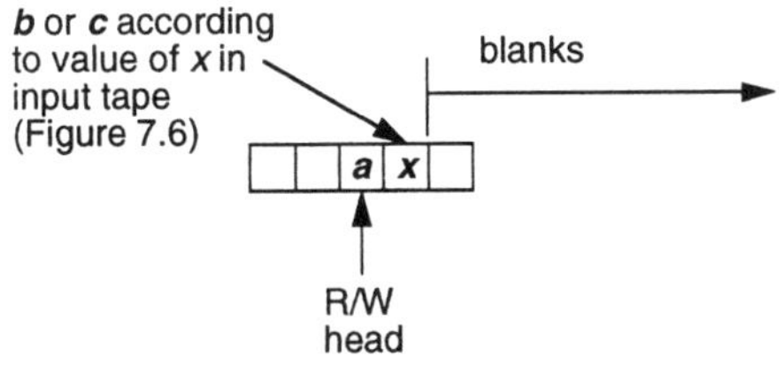

Figure 7.7 The tape when T_1 halts

As Figure 7.7 shows, the *b* or *c* of the input tape (Figure 7.6) has been 'erased' from its initial position – replaced by a blank – and placed in the tape square immediately to the right of the *a*, with T_1's head finally left positioned at that square. We will call this moving of a symbol from one part of a tape to another '**shuffling**' the symbol. We can talk of *right shuffling* and *left shuffling*. T_1 does a *left shuffle* of the symbol denoted by *x* in Figure 7.6.

We will not trace the behaviour of T_1 in detail here. Table 7.1 showed how a trace of a TM's behaviour is done; you can perform a similar analysis of T_1's behaviour if you wish. However, we will spend a short time examining the function of the states in the machine:

State 1
The start state. Assumes that the head is positioned at a square containing an *a*. If there is no *a*, or another symbol occupies the square, no transition is possible.

States 2–5 & 2–7
When state 2 is entered from state 1, a right move of the head has been made. Since there are expected to be 0 or more blanks between the *a* and the *b* or *c*, the head could at this point be pointing at a *b* or a *c* (in which case the machine has nothing else to do, so moves to its halt state leaving the head where it is). Otherwise, there are one or more blanks after *a*, and the machine moves into State 3, where it 'skips over' (to the right) any more blanks, until the head reaches the *b* or *c*. State 3 also 'erases' the *b* or *c* before moving to state 4 (for a *b*) or state 6 (for a *c*). State 4 skips back over any blanks (left) until the head reaches the *a*, at which point the head moves back to the blank square to the right of *a*, then state 5 ensures that the blank to the right of *a* is overwritten by a *b*, as the transition back to state 2 is made (note that no movement of the head is specified by this transition). States 6 and 7 perform an analogous function to states 4 and 5, but where the symbol after the blanks is *c* rather than *b*).

State 8

This state, the **halt state**, is reached only from state 2, when the head points at a *b* or a *c* (which could be either because a *b* or a *c* was initially next to the *a*, or because a *b* or a *c* had been moved there by states 3–5 or states 3–7 (see above). Note that the arcs leaving states 5 and 7 could have been drawn directly to state 8 if wished.

T_1 will reach its halt state, producing an output tape as described in Figure 7.7, when it is given an **input configuration** as specified in Figure 7.6. If the input configuration is not set up correctly, then T_1 may:

(a) *halt in a non halt state*, as would result from the input tape configuration shown in Figure 7.8.

(b) *never halt*, which would be the result of the input tape configuration shown in Figure 7.9.

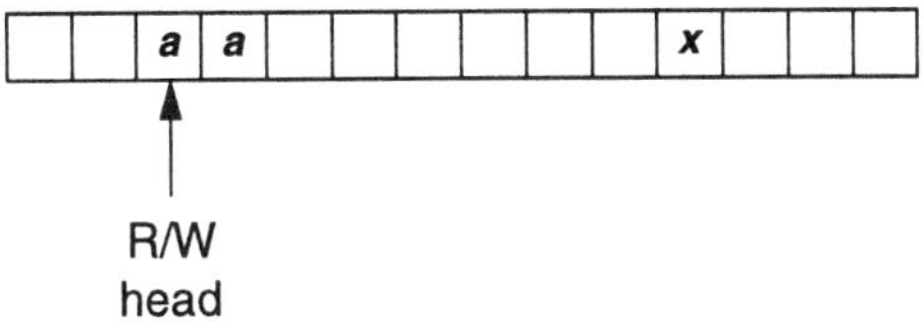

Figure 7.8 A tape that causes problems for T_1

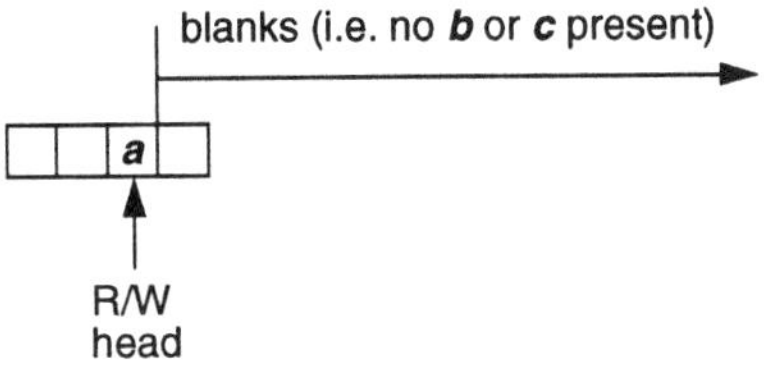

Figure 7.9 A tape that would cause even bigger problems for T_1

You may like to satisfy yourself that (a) and (b) are true for the input configurations of Figures 7.8 and 7.9, respectively. You should also observe that a third general possibility for a Turing machine's halting situation:

(c) *reaches halt state, but produces output not conforming to specification,*

is not possible in the case of T_1, if only blanks lie to the right of the *b* or *c*. The latter comment relates to the fact that *the output specification requires a totally blank tape to the right of the head when* T_1 *halts.*

The immediately preceding discussion indicates the need for care in describing a TM's expected input configuration (i.e. the marked portion of its tape, and initial location of the read/write head). If we carefully specify the input conditions, then we need not consider what happens when an input situation that does not meet our specification is presented to the machine. Of course, we should also specify how the **output configuration** (the state in which the machine halts, the contents of its tape, and the final location of its read/write head) is to represent the 'result'.

The remainder of this chapter is devoted to demonstrating that for language recognition, TMs are indeed more powerful than the other abstract machines we have studied thus far.

7.3 TURING MACHINES AS LANGUAGE RECOGNIZERS

7.3.1 Regular languages

We do not have to expend much effort to demonstrate that TMs are at least as powerful as **finite state recognizers** (**FSRs**). Recall from Chapter 4 that for each **regular language**, there exists a **deterministic** FSR (DFSR) to recognize that language. A DFSR is simply *a TM that always moves its read/write head to the right, replacing each symbol it reads by that same symbol, on each transition.*

Thus, the DFSR, M_5^d, from Chapter 4, shown here in Figure 7.10, becomes the TM, T_2, shown in Figure 7.11.

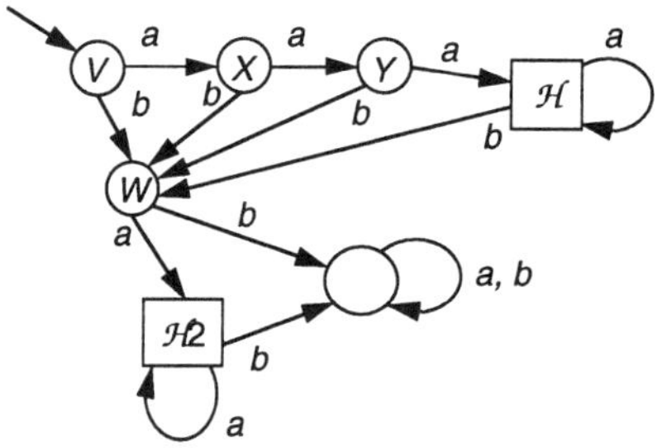

Figure 7.10 The DFSR M_5^d from Chapter 4

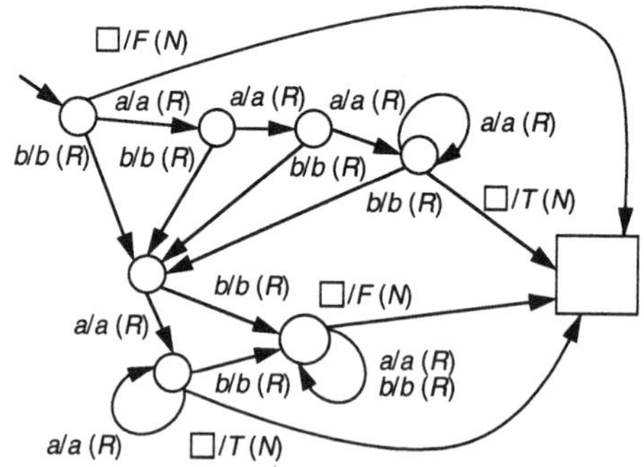

Figure 7.11 The TM, T_2

We have taken advantage of the TM's ability to write, as well as read, symbols, and its ability to detect the end of the input string when reaching a blank square, to ensure that T_2 has two useful features.

1. It prints T for any *valid* string, and F for any *invalid* string.
2. It has *only one halt state.*

We henceforth assume that all our TMs have only one halt state. An exercise asks you to convince yourself that this is a reasonable assumption.

An important point should be noted. T_2 prints F and goes directly to its halt state if the head is initially located on a blank square. In a sense, this reflects an assumption that *if the head is initially at a blank square then the tape is entirely blank.* We will call this assumption the **blank tape assumption**. We will further assume that a blank tape represents the **empty string**, ε.

For our example machine T_2, of Figure 7.11, as ε is not in the language accepted by $M_5{}^d$, T_2 prints an F if its head is initially on a blank square. Of course, T_2 can never be *sure* that the tape is empty. No TM can ever establish beyond doubt that its tape is unmarked. In view of the fact that we assume that a tape has no definite ends in either direction, i.e. the blank squares at either end of the marked portion extend indefinitely, it should be clear by now why we have not designed T_2 so that when its head points initially at a blank square, it searches its tape to see if the input string is located somewhere else on the tape.

You will appreciate that we could do the same type of DFSR-to-TM conversion for any DFSR.

7.3.2 Context free languages

Recall from Chapter 5 that the abstract machine associated with the **context free languages** is the **non deterministic pushdown recognizer (NPDR)**. Recall also that some of the context free languages are **deterministic** (can be recognized by a DPDR) while some are not (can be recognized by an NPDR, but *not* a DPDR).

It can be shown that any DPDR can be converted into a TM. We will not consider this here, but merely point out that the basic method is for the TM to use part of its tape to represent the **stack** of the PDR, thus working with two parts of the tape, one representing the input string, the other representing the stack. However, one usually finds that a TM which must do the job done by a DPDR is better designed from scratch. Here, we consider T_3, which is equivalent to $M_3{}^d$ of Chapter 5, the DPDR that accepts the language

$$\{a^ib^i : i \geq 1\},$$

which we know from Chapter 6 to be **properly** context free, i.e. context free but not regular. T_3 is shown in Figure 7.12.

T_3 expects on its tape an arbitrary string of *a*s and/or *b*s (using the notation of Chapter 2, a string in $\{a,b\}^*$), finally printing T if the string is in the language $\{a^ib^i :$ $i \geq 1\}$, and F if it is not. Note that, like T_2, T_3's read/write head begins on the leftmost symbol of the input. Moreover, T_3 makes the same *blank tape assumption* as did T_2. T_3

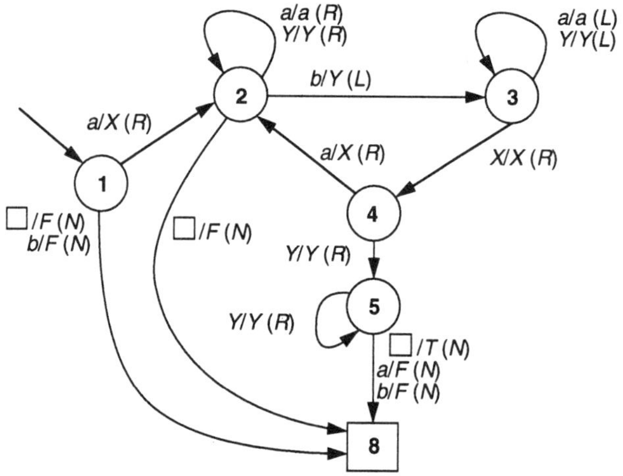

Figure 7.12 The TM, T_3

uses the symbols X and Y to 'tick off' the *a*s and corresponding *b*s, respectively. You should trace T_3's behaviour on example **sentences** and non sentences. Such a trace is shown in Table 7.1.

Now for the *non deterministic* context free languages. Recall from Chapter 5 that some context free languages can be accepted by a non deterministic PDR (NPDR), but not a deterministic PDR. Later, in Chapter 11, we briefly consider the conversion of NPDRs into TMs. However, to demonstrate that TMs are more powerful than NPDRs, we see in Figure 7.13 a TM, T_4, that recognizes the language of possibly empty **palindromic strings** of *a*s and *b*s. Recall that in Chapter 5 such languages were shown to be non deterministic CFLs, and therefore beyond the processing power of the DPDR.

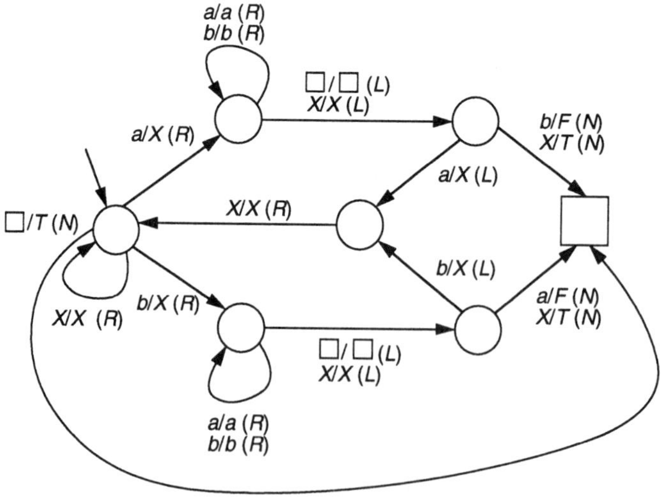

Figure 7.13 The TM, T_4

T_4 accepts only those strings in $\{a, b\}^*$ that are palindromic, and, what is more, does the job deterministically.

T_4 uses only one marker, X, and works from the outside in, 'ticking off' matching symbols at the extreme right and left ends of the string. This time, the *blank tape assumption* results in a T being printed if the read/write head of T_4 starts on a blank, as the empty string is in $\{a, b\}^*$, and is palindromic. Again, you should make sure you understand how the machine works, since more complex TMs follow.

7.3.3 Turing machines are more powerful than PDRs

The above implies that Turing machines are as powerful, in terms of language processing, as FSRs and (deterministic *and* non deterministic) PDRs. This claim is justified later in this chapter. However, by using a single example we can show that TMs are actually *more* powerful than PDRs. Consider the Turing machine, T_5, shown in Figure 7.14, which processes the language

$$\{a^i b^i c^i : i \geq 1\},$$

demonstrated in Chapter 6 to be properly **context sensitive** (i.e. context sensitive but not context free).

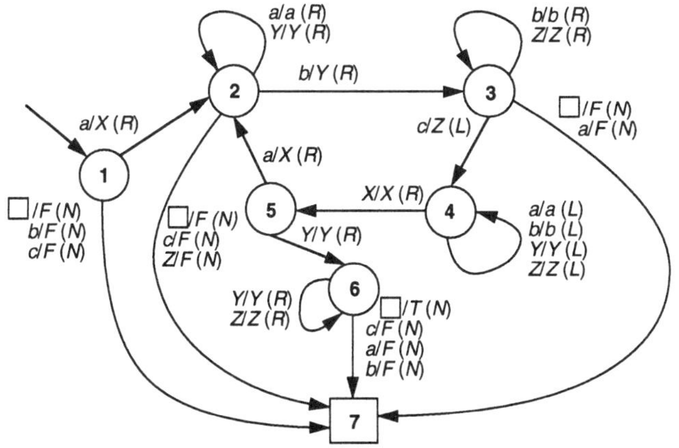

Figure 7.14 The TM, T_5

T_5 requires **input configurations** of the form shown in Figure 7.15.

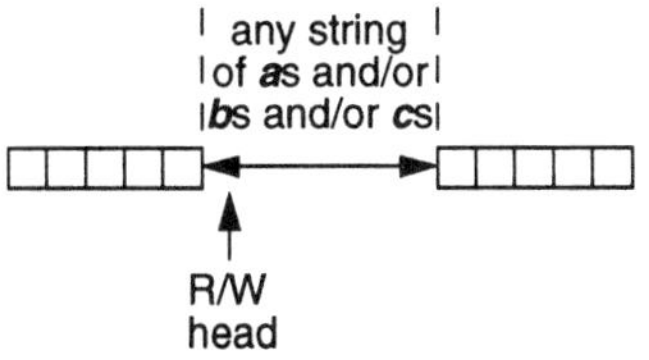

Figure 7.15 An initial tape set up for T_5

It reaches its halt state for valid or invalid strings. For valid strings (i.e. strings of one or more *as* followed by the same number of *bs* followed by the same number of *cs*), the output configuration is as specified in Figure 7.16.

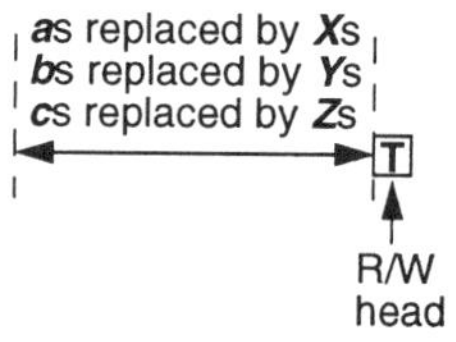

Figure 7.16 The tape after a valid string has been processed by T_5

The input sequence of Figure 7.15 is altered by T_5 to produce a tape of the form described in Figure 7.16 (the symbols X, Y, and Z are used as auxiliary *markers*). Again, T indicates a sentence.

For invalid strings (i.e. any string in $\{a, b, c\}^*$ which is not in $\{a^i b^i c^i : i \geq 1\}$), however, when the halt state is reached, the configuration will be as shown in Figure 7.17.

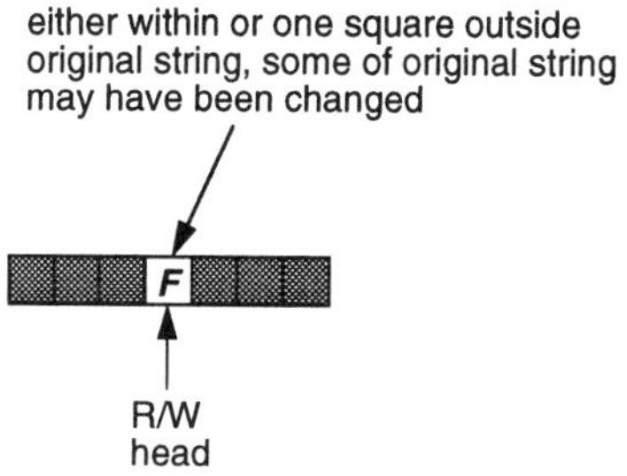

Figure 7.17 The tape after an invalid string has been processed by T_5

In the case of invalid strings, the read head will be left pointing at an F, as shown in Figure 7.17.

When students design T_5 and similar machines, their machines often fail to check for the possible occurrence of extra symbols at the end of an otherwise valid string. For the language $\{a^i b^i c^i : i \geq 1\}$, such machines will indicate that a valid string immediately followed by another string in $\{a, b\}^+$ is valid, when, of course, it is not.[2]

However, students are not the only guilty parties when it comes to failing to check for invalid symbols at the end of an otherwise valid input. I have encountered a Pascal compiler that simply ignores all the source file text that occurs after an end followed by a full stop (in Pascal, the only **end** terminated by a full stop is the last one in the program, and corresponds to a **begin** before the first statement). This caused severe difficulties to a novice who had accidentally put a full stop after an **end** halfway down their program, and happened to have an extra **begin** near the beginning. The program

[2] Recall from Chapter 2 that $\{a, b\}^+$ is the set of all non empty strings of *as* and/or *bs*.

compiled perfectly, but when run only carried out the first part of what was expected by the programmer! The Pascal compiler in question was *accepting an invalid input string*.

7.4 INTRODUCTION TO (TURING MACHINE) COMPUTABLE LANGUAGES

We have seen that TMs are more powerful language recognizers than finite state and pushdown recognizers. Regular languages are associated with finite state recognizers in the same way as context free languages are associated with pushdown recognizers. It seems reasonable to ask, then, if there is a class of languages that are associated with TMs in a similar way. In other words, *is there a class of languages, as defined by a type of grammar in the* **Chomsky hierarchy**, *so that for any language in that class, there is some TM that can recognize that language, and for any TM that recognizes a language, we can be sure that the language is in the class?* In fact, there is such a class of languages, and it turns out that this class is exactly that defined by the most general Chomsky classification, i.e. the general **phrase structure languages** themselves, these being those generated by the **type 0**, or **unrestricted grammars**. Type 0 grammars were defined in Chapter 2 as being those with productions of the form:

$$x \to y,\ x \in (N \cup T)^+,\ y \in (N \cup T)^*$$

i.e. a non empty arbitrary string of terminals and non terminals on the left hand side of each production, possibly empty arbitrary string of terminals and non terminals on the right hand sides.

Let us call the languages that can be recognized by a TM the Turing Machine **Computable languages.** For convenience, and with some justification, as we see later in the book, we will refer to them as computable languages. We eventually discover that the computable languages include, *but are not restricted to*, all the regular languages, all the context free languages, and all of the context sensitive languages.

In Chapter 4, the correspondence between regular languages and finite state recognizers was specified. We saw how we can take any regular grammar and construct an FSR to accept exactly the same set of strings as that generated by the grammar. We also showed that we could take any FSR and represent it by a regular grammar that generated the same set of strings as that accepted by the FSR (we did a similar thing with respect to NPDRs and the context free languages in Chapter 5). Analogously, to show the correspondence between TMs and unrestricted languages we would have to show how to create a type 0 grammar from any TM, and vice versa. However, this is rather more complicated than is the case for the more restricted grammars and machines, so the result will merely be sketched here.

We first consider the context sensitive languages, as their defining grammars, the **context sensitive** grammars, do not have the empty string on the right hand side of productions. We then generalise our method to deal with the type 0 grammars in general. The definition of context sensitive productions is as for type 0 productions given above, with the additional restriction that the left hand sides of context sensitive productions are not allowed to be shorter in length than the right hand sides.

7.5 THE TM AS THE RECOGNIZER FOR THE CONTEXT SENSITIVE LANGUAGES

The machine T_5, shown in Figure 7.14, recognizes the language

$\{a^i b^i c^i : i \geq 1\}$.

In Chapter 6 we established that this language is not context free. We also saw that it is, in fact, a context sensitive language, as it is generated by the following grammar, G_{16}:

$S \rightarrow aSBC \mid aBC$
$CB \rightarrow BC$
$aB \rightarrow ab$
$bB \rightarrow bb$
$bC \rightarrow bc$
$cC \rightarrow cc$

We wish to establish that for any context sensitive language we can provide an equivalent TM. T_5 itself is not enough to convince us of this, as it was designed for a specific language, and not constructed by a method for creating TMs from the productions of the corresponding grammar. We therefore require *a method for constructing TMs for context sensitive grammars that we could apply to any such grammar.* Here, we will consider the application of such a method to the grammar G_{16}. I then leave it to you to convince yourself that the method could be applied to any context sensitive grammar.

7.5.1 Constructing a non deterministic TM for reduction parsing of a context sensitive language

The TM, T_6, that has been constructed from the productions of G_{16}, is partly depicted in Figure 7.18.

You will note that T_6 is in three sections, namely sections A, B, and C, all of which have state 1 in common. Sections A and B are shown in full, while section C is only partly specified. The machine is far more complex than T_5 (Figure 7.14), a TM that

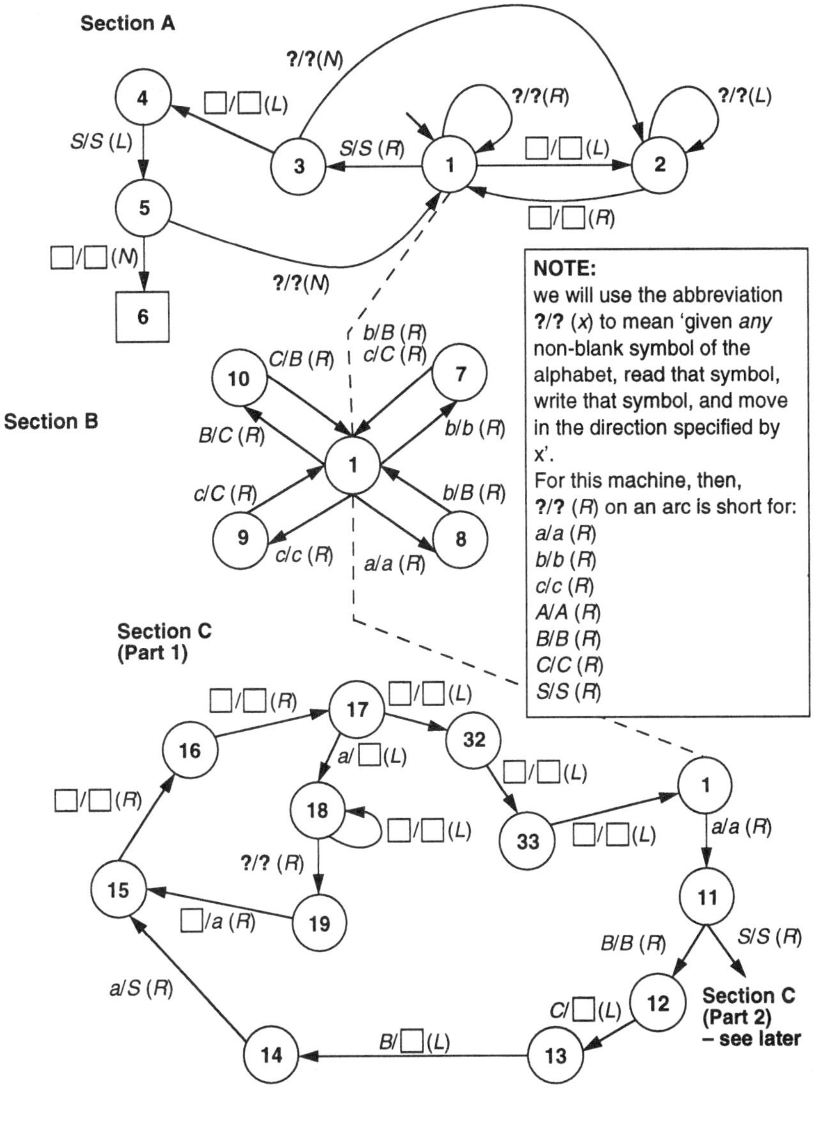

Figure 7.18 The TM, T_6. Part of sections A, B, and C

was designed to recognize the same language. T_6 was constructed from the productions of a grammar (G_{16}), so its purpose is only to demonstrate a general method for constructing such TMs.

The first thing to note is that T_6 is **non deterministic**. Its overall process is to model the **reduction** method of **parsing** introduced in Chapter 3. Beginning with a **terminal string**, we repeatedly substitute **substrings** within that string that are found on the **right hand sides** of productions with the **left hand sides** of the productions. The process is complete if we manage to reduce the string to the **start symbol** S alone, i.e., we finish up with a tape containing only S.

T_6 carries out *reduction parsing* using the grammar G_{16}, by effectively 'scanning' the input string repeatedly from left to right, then right to left, attempting to find a sequence of reductions that will leave only the symbol S on the tape. The process is non deterministic since many reductions may apply at any one time, and thus we may often have to **backtrack** to find alternative reduction sequences. As for all of the other non deterministic machines we have studied in this book, we assume that *the string is accepted if some sequence of* **transitions** *will lead to the machine's* **halt state** (in the case of T_6, this is state 6). T_6 reaches its halt state when only S, the start symbol of the grammar, remains on its tape, i.e. if it finds a sequence of reduction applications that enable the input string to be reduced to S. You may like to convince yourself that T_6's halting conditions will only be met if its input tape initially contains a string from $\{a^ib^ic^i : i \geq 1\}$.

To establish that we could produce a 'T_6-like' TM from any context sensitive grammar, we briefly consider the functions of the three sections shown in the diagram of T_6, in Figure 7.18, and also the correspondence between parts of the machine and the productions of the grammar.

Section A

This section of the machine repeatedly scans the input string (state 1 scans from left to right, while state 2 scans from right to left). States 3 – 5 test if the symbol S alone remains on the tape (in which case the machine can halt in state 6).

Section B

This section contains a state *for each production in* G_{16} *that has a left hand side of the same length as its right hand side* (e.g. $CB \to BC$). For example, state 1 and state 10 replace BC by CB, i.e. carrying out the reduction based on the production $CB \to BC$, while state 1 and state 7 perform an analogous function for the productions $bB \to bb$ and $bC \to bc$, respectively.

Section C

This section also carries out reduction substitutions, but *for productions whose right hand sides are greater in length than their left hand sides* (in G_{16}, these are $S \to aSBC$ and $S \to aBC$). In such cases, a substitution uses up less tape than the sequence of symbols being replaced, so the extra squares of tape are replaced by **blanks**, then any other symbols to the right of the new blanks are **shuffled** up one by one until the occupied part of the tape is packed. Figure 7.18 shows only Part 1 of section C. We first discuss this part, before examining Part 2 (Figure 7.21) below.

Part 1

This section of T_6 deals with the production $S \to aBC$, i.e. 'reduces' aBC to S.

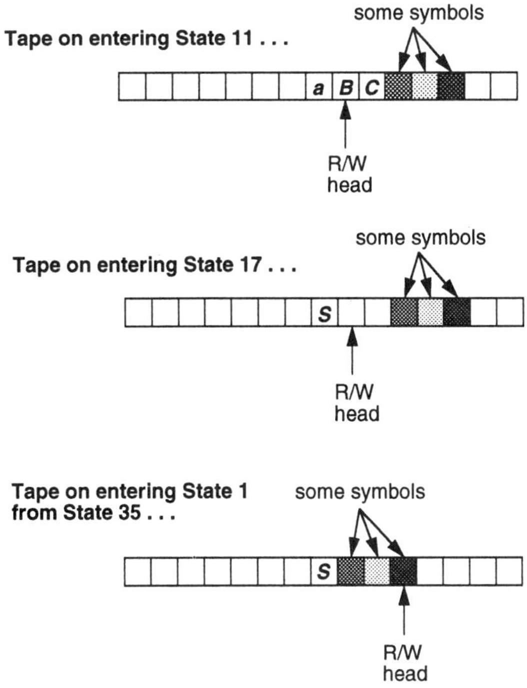

Figure 7.19 Reducing *aBC* to *S* (for states refer to Figure 7.18)

To illustrate, Figure 7.19 shows what happens when we are in state 11 with the tape configuration as shown.

Note that, for simplicity, Figure 7.18 shows only the **loop** of states (states 17, 18, 19, 15, 16) that shuffle an *a* left so that it goes immediately to the right of the next non blank symbol to its left. We have omitted the 12 states that would be needed to shuffle any other symbols of the alphabet (i.e. *b*, *c*, *A*, *B*, *C*, *S*) that may be required. To consider why two states are needed to deal with each symbol, consider the part of T_6 that does the shuffling job for *a*, depicted in Figure 7.20.

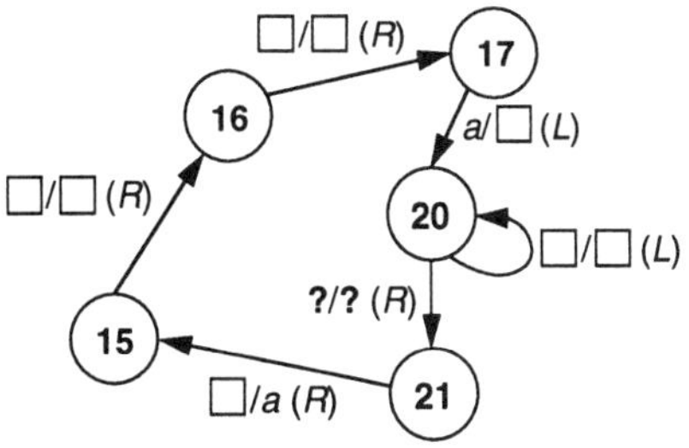

Figure 7.20 Shuffling an '*a*' along

So, states 15, 16, and 17 are common to the shuffling loop for each symbol in the alphabet. We need, say, states 22 and 23 to shuffle any *c*s, states 24 and 25 to shuffle any *A*s, states 26 and 27 to shuffle any *b*s, states 28 and 29 to shuffle any *C*s, and states 30 and 31 to shuffle any *S*s. That is why the state in Figure 7.20 to the right of state 17 is named state 32.

Part 2

This second part of Section C of T_6, which is depicted in Figure 7.21, deals with the production $S \rightarrow aSBC$, i.e. 'reduces' $aSBC$ to S. It is very similar to Part 1 (Figure 7.18), which deals with $S \rightarrow aBC$, and was discussed immediately above.

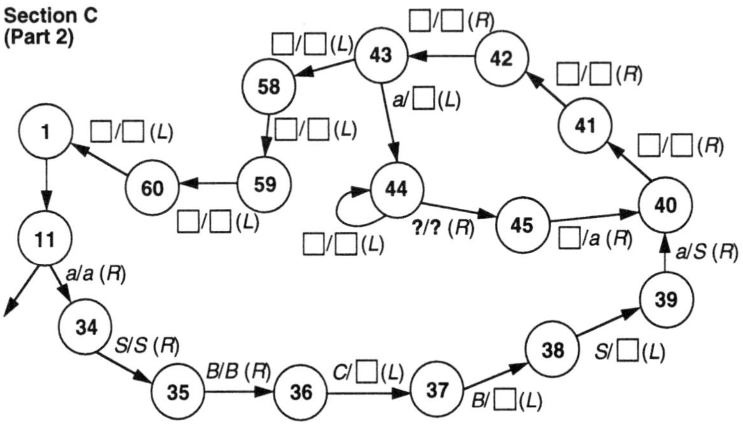

Figure 7.21 Reducing *aSBC* to *S* (for Part 1 see Figure 7.18)

As for Section C Part 1, we have only included the pair of states (state 43 and state 45) between states 43 and 40 that shuffle the *a* to the left (note that in this case, shuffling involves *three* blanks, as the right hand side of $S \rightarrow aSBC$ has *four* symbols). As for Part 1, we assume that there is a similar pair of states linking states 43 and 40 for each of the other alphabet symbols (*b*, *c*, *S*, *A*, *B*, *C*). Thus we need 12 more states (assumed to be numbered 46-57), which is why the state leaving the loop has been labelled 58. We have taken advantage of the fact that both productions dealt with by Section C have *a* as the leftmost symbol of the right hand side. This was not necessary (as our machine is non deterministic), but was done for convenience.

7.5.2 *The generality of the construction*

To see the generality of the construction that enabled us to produce T_6 from G_{16}, consider that, from the definition of context sensitive productions, *the left hand side of each production does not exceed, in length, the right hand side* (i.e. productions are of the form $x \to y$, where $|x| \leq |y|$). Therefore we can represent any context sensitive grammar by a Turing machine with:

a section A part
To scan back and forth over the input between the start state and another state, enabling the halt state to be reached only when the start symbol, S, appears alone on the tape.

a section B part
That consists of a distinct path *for each production in the grammar with equal length left and right hand sides*, each path leaving and re-entering the start state. The path for each production performs the reduction associated with that production, i.e. when the right hand side of a production appears anywhere on the input tape, it can be replaced by the left hand side of that production.

a section C part
That consists of a distinct path *for each production in the grammar with left hand side shorter in length than the right hand side*, each path leaving and re-entering the start state. As in Section B, the path for each production performs the reduction associated with that production. However, in section C cases the newly placed symbols occupy less tape space than the symbols replaced, so each path has a 'shuffling routine' to ensure that the tape remains in packed form. In G_{16}, only the productions $S \to aSBC$ and $S \to aBC$ are of Section C type, and both represent reductions where the replacement string is a single symbol. In general, of course, a context sensitive production could have several symbols on the left hand side. This represents no real problems however, as demonstrated by the sketch of part of a TM that does the reduction for the context sensitive production $XY \to PQR$, shown in Figure 7.22.

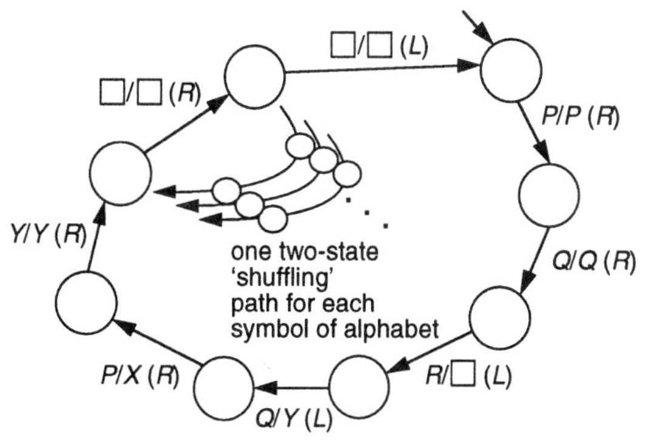

Figure 7.22 Reducing *PQR* to *XY*

As you can see from Figure 7.22, in this example we need only to **skip** over one tape square to the right of the newly written *XY*. This is because the right hand side of the production $XY \rightarrow PQR$ is one symbol longer than the left. We thus replace the PQ with XY, and the R with a blank. We then need to move over that blank to see if there is a symbol there to be shuffled up, and so on.

In Chapter 5, the non deterministic pushdown recognizers (NPDRs) that we created from the productions of context free grammars recognized the language generated by the grammar by being able to model any derivation that the grammar could carry out. Our non deterministic Turing machines operate by modelling every possible sequence of reductions for the corresponding context sensitive grammar. Such TMs reach their halt state only if the string on the input tape can be reduced to S. The construction sketched out above ensures that this would happen only for inputs that were sentences of the language generated by the original grammar.

Enough has been said to establish the following:

The recognizer for the context sensitive languages is the Turing Machine. Context sensitive languages are computable languages.

You may be wondering why the statement immediately above does not say 'non deterministic Turing machine'. In Chapter 10 we see that *any non deterministic TM can be simulated by a deterministic TM*. Hence, the non deterministic TMs of this chapter could be replaced by equivalent deterministic TMs.

7.6 THE TM AS THE RECOGNIZER FOR THE TYPE 0 LANGUAGES

Now, the sole difference between **context sensitive** productions and **unrestricted**, or **type 0**, productions is that the latter are *not* restricted to having right hand sides that are greater than or equal in length to the left hand sides. This means that, say, a production such as $WXYZ \rightarrow Pa$, which is *not* a valid context sensitive production, is a perfectly reasonable type 0 production. Moreover, type 0 grammars can contain ε **productions**. Productions like $WXYZ \rightarrow Pa$ do not present a severe problem; we simply add another section to our TMs to deal with these types of production, as discussed next (we return to ε productions shortly).

7.6.1 *Amending the reduction parsing TM to deal with type 0 productions*

We assume that we already have sections A, B and C as defined for our TMs derived from context sensitive grammars (for example, Figure 7.18). We add to such a construction a new section, as follows:

Section D

Consists of *a distinct path for each production in the grammar with left hand side longer in length than the right hand side*, each path leaving and re-entering the start state. As in sections *B* and *C*, the path for each production performs the reduction associated with that production. However, in such cases the newly placed symbols occupy more tape space than the symbols replaced, so each path has a '**shuffling** routine' to ensure that all symbols in the way are each moved to the right an appropriate number of spaces.

Figure 7.23 shows part of the section D component of a TM, in this case the part dealing with the type 0 example production $WXYZ \rightarrow Pa$.

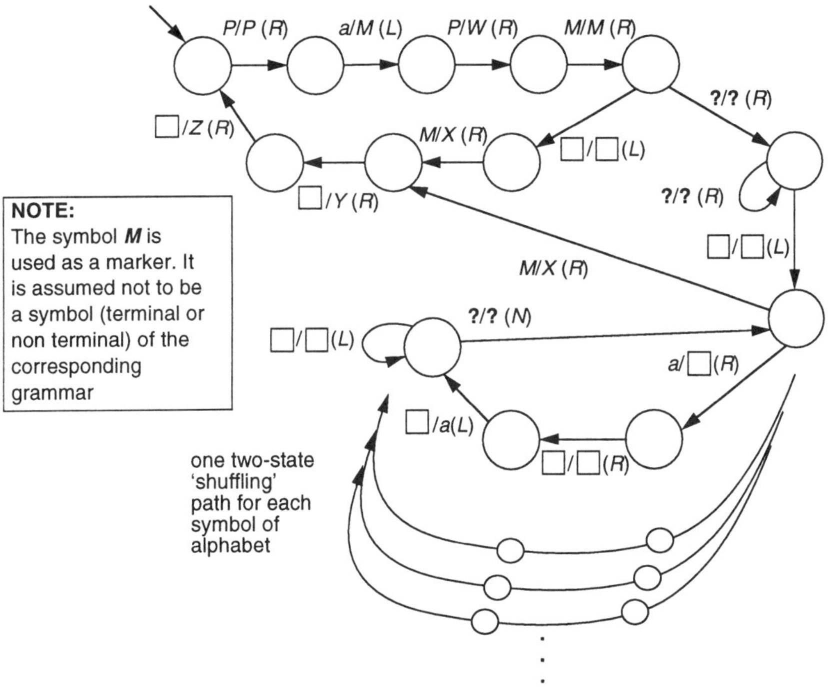

Figure 7.23 Replacing *Pa* by *WXYZ*

For our example production, there are two extra symbols to insert when performing the associated reduction, as the right hand side of our production – *Pa* – has two fewer symbols than the left hand side – *WXYZ*. A machine of the type illustrated in Figure 7.23 therefore has to shuffle all symbols (if any) that were to the right of the *Pa* in question two tape squares to the right, leaving two blanks immediately to the right of *Pa* (which is eventually changed to *WX*), so that *YZ* can be inserted. The symbol *M* is used as a marker, so that the machine can detect when to stop shuffling (when the *M* has a blank next to it, either there are no symbols to its right, or we have shuffled the symbol next to it to the right, and so no more symbols need shuffling). You should convince yourself that a Section D-type construction of the type shown in the example could be carried out for any production where the right hand side is non-empty and of smaller length than the left hand side.

7.6.2 *Dealing with the empty string*

As promised above, we now consider *ε productions.* In fact, we can treat these as a special case of the section D-type construction. We simply allow for the fact that *on either side of any non-blank symbol on the tape, we can imagine there to be as many εs as we like.* This means we can insert the left hand side (i.e. after shuffling to make sufficient room, in a section D-type way) of a type 0 ε production *anywhere we like on the marked portion of the tape.* As you may recall, in earlier chapters we noted that this was a cause of problems in **parsing**. Our machine should only be able to reach its halt state if it inserts the left hand sides of ε productions in the appropriate places, i.e. in terms of a correct reduction parse of the input string being processed. So, we complete our specification of the TM to represent any type 0 grammar by introducing the following, where necessary, to the machine sketched out in Figure 7.18:

Section E

Consists of *a distinct path for each production in the grammar with ε on the right hand side*, each path leaving and re-entering the start state. As in sections *B*, *C*, and *D*, the path for each production performs the reduction associated with that production. Figure 7.24 shows a sketch of a Section E construction for the type 0 production *XYZ* → ε.

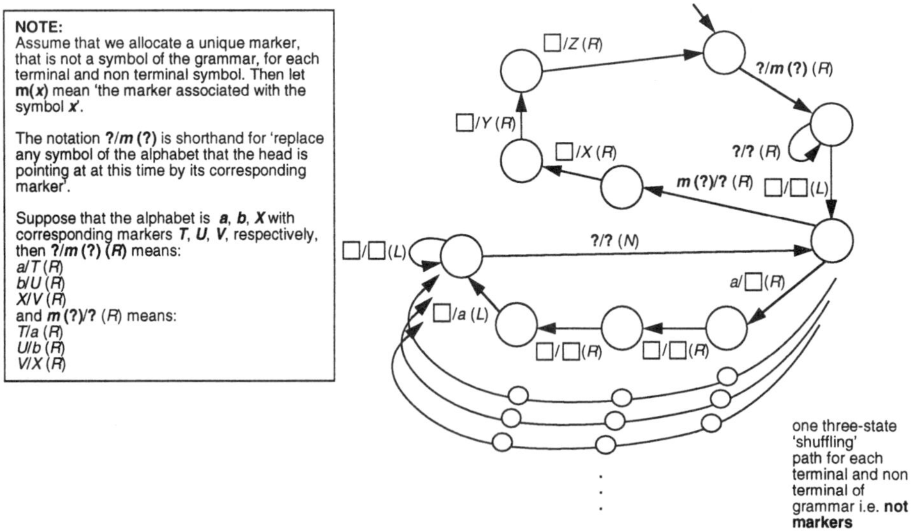

Figure 7.24 Placing *XYZ* into the tape (the reduction associated with the type 0 production *XYZ* → ε)

The section E construction, as illustrated in Figure 7.24, inserts the left hand side of an ε production after any symbol on the tape. There are two problems remaining. Firstly, we must deal with an insertion *before* the leftmost symbol. Secondly, we need to cater for an insertion on a blank tape. Considering the second case first, if ε is in the language generated by the grammar, the corresponding machine should be able to reach its halt state appropriately. Our **blank tape assumption** dictates that we represent ε as the input string by setting up our machine so that its head points initially to a blank square. We can solve both problems with very similar constructions to Section E, though there are a few differences. I leave you to sketch the construction yourself.

7.6.3 The TM as the recognizer for all types in the Chomsky hierarchy

Enough has now been said to support the following claim[3]:

> *The recognizer for the phrase structure languages in general is the Turing Machine. Phrase structure languages are computable languages.*

We argued above that any DFSR can be converted into an equivalent TM. It was then implied that TMs could recognize any of the context free languages. However, by showing that any type 0 grammar can be represented by a TM that accepts the same language, we have also shown that the same applies to the other types of language in Chomsky's hierarchy. This is because *any production of any grammar of type 0, type 1, type 2 and type 3 is catered for in one of the sections of our construction.*

However, if we were to implement the non deterministic TMs that are produced by our construction, we would have very inefficient parsers indeed! That is the whole point of defining restricted types of machine to deal with restricted types of language.

7.7 DECIDABILITY: A PRELIMINARY DISCUSSION

We close this chapter by briefly discussing one of the most important concepts related to language processing, i.e. **decidability**. In doing this, we anticipate a more thorough treatment of the topic in Chapter 11.

7.7.1 Deciding a language

Consider the TMs T_2, T_3, T_4 and T_5 in this chapter. Each accepts an input tape containing a **terminal string** from the appropriate **alphabet**. If each of these machines is

[3] Remember that in Chapter 10 we are going to see why we can omit 'non deterministic' from the statement.

UNIVERSITY OF LINCOLNSHIRE AND HUMBERSIDE

started with a valid **input configuration**, it will halt in a **halt state**, and as it reaches its halt state it will:

print T if the input sequence was a **sentence** of the language, otherwise print F.

A language-recognizing TM that conforms to the above behaviour is said to **decide** the corresponding language (because given any string of the appropriate terminals it can say 'yes' if the string is a sentence *and* 'no' if the string is *not* a sentence).

For example, consider some of the decisions made by T_5, which decides the language $\{a^ib^ic^i : i \geq 1\}$, as shown in Table 7.2.

Table 7.2 T_5's output on accepting and rejecting various input sequences

Input string	**T_5 input and output**	
aabbcc (valid string)	initial configuration:	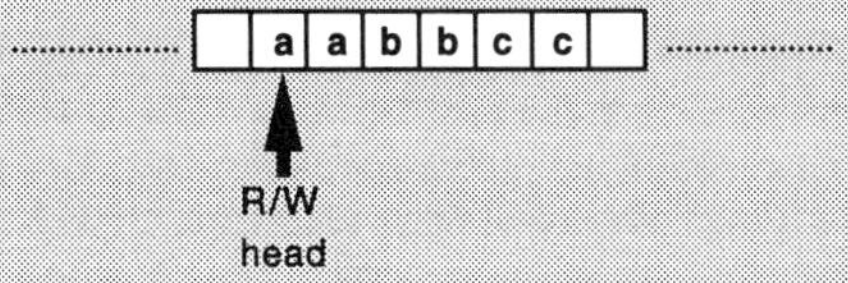
	final configuration:	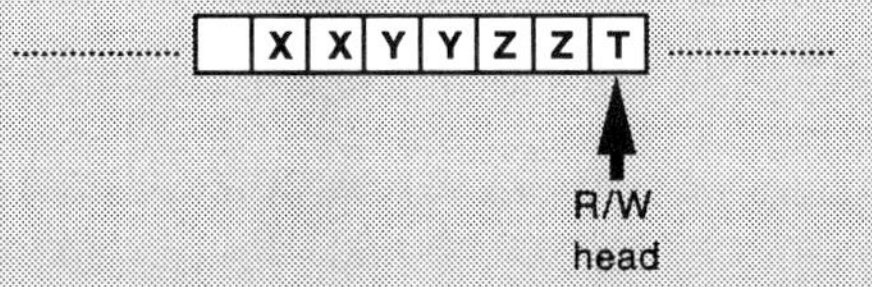
aaabbcc (invalid string – too many *as*)	initial configuration:	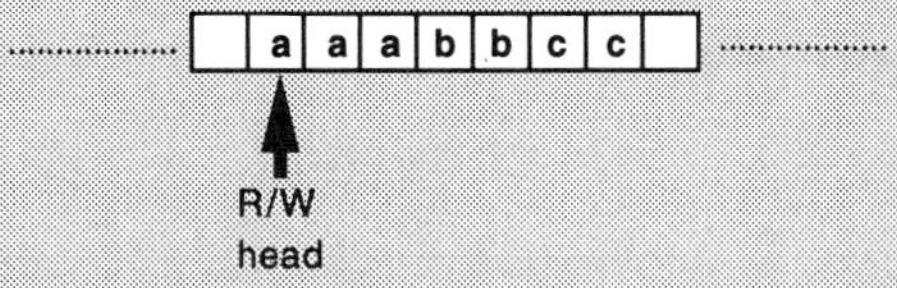
	final configuration:	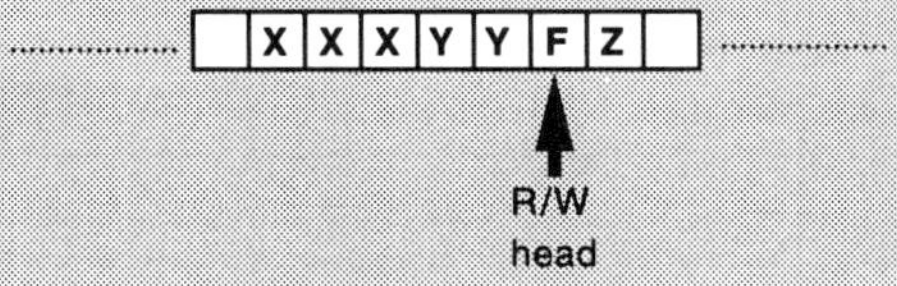
caaaa (invalid string – starts with *c*)	initial configuration:	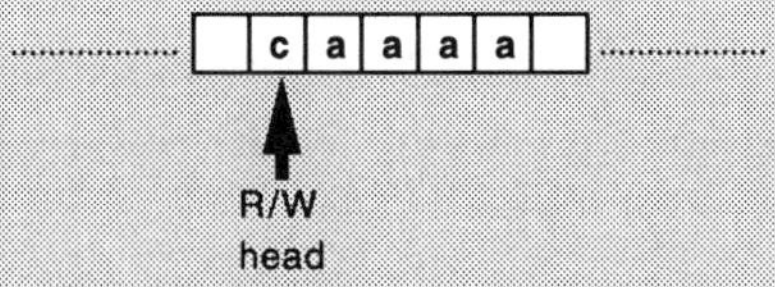
	final configuration:	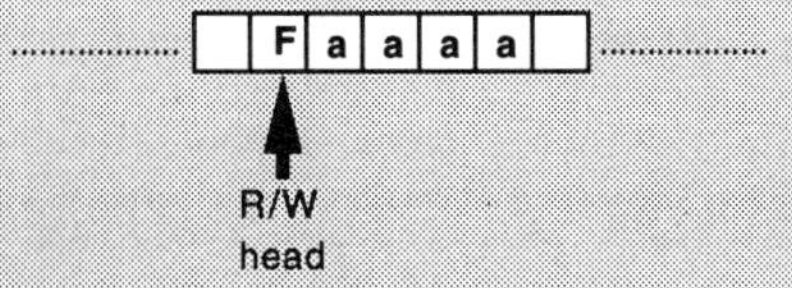

The formal definition of deciding a language is given in Table 7.3.

Table 7.3 Deciding a language

Given some language L, with alphabet A, a TM decides that language if for any string, x, such that $x \in A^*$: if $x \in L$ the machine indicates its acceptance, and if $x \notin L$ the machine indicates its rejection of that string. A language for which there is a TM that decides that language is called a decidable language.

7.7.2 Accepting a language

Suppose we amend T_5 (of Figure 7.14) to produce the machine T_7, depicted in Figure 7.25.

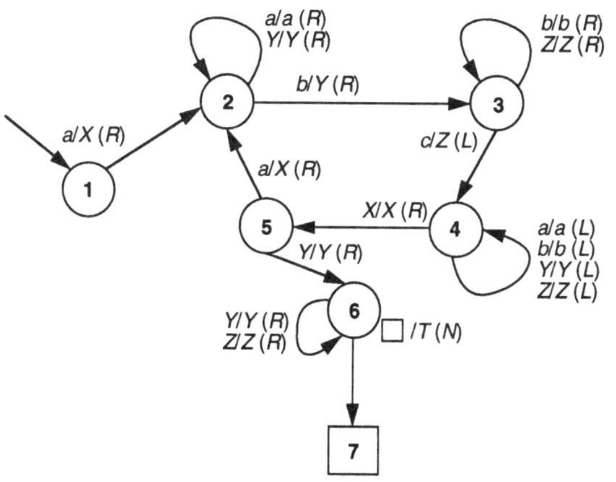

Figure 7.25 The TM, T_7

We have simply removed from T_5 all arcs entering the halt state which result in writing the symbol F, and removed the instructions that write an F from the arc between states 6 and 7. Like T_5, T_7 will now reach its halt state and write a T for valid input strings. However, unlike T_5, for invalid strings T_7 will eventually stop in one of its non halt states, and make no F indication at all. For example, given the input *aaabbcc*, (the second example used for T_5 in Table 7.2), T_7 would eventually come to a stop in state 2, being unable to make any move from that state in the way that T_5 could.

A machine that can output T for each and every valid string of a language (and never outputs T for any *in*valid strings) is said to **accept** that language. Compare this carefully with the definition of *deciding*, from Table 7.3. Clearly, *a machine that decides* *a language also accepts it,* but the converse is not necessarily true. T_5, T_6, and T_7 all *accept* the language $\{a^ib^ic^i : i \geq 1\}$, but *only* T_5 *decides* it.

The formal definition of accepting a language is given in Table 7.4.

Table 7.4 Accepting a language

Given some language L, with alphabet A, a machine accepts that language if for any string, x, such that $x \in A^*$,

if $x \in L$ the machine indicates its acceptance of that string.

A language for which there is a TM that accepts that language is called an acceptable language.

7.8 CONCLUSION

In this chapter, we have encountered *computable* languages, *decidable* languages, and now *acceptable* languages. From the way in which the construction of TMs from type 0 grammars was developed, it can be seen that such TMs would **accept**, rather than **decide**, the language generated by the grammar from which the machine was constructed. In other words, computable languages are acceptable languages. However, we have also seen that some computable languages are decidable (since TMs T_2, T_3, T_4 and T_5 all *decide* their respective languages). A question addressed in Chapter 11 asks whether all computable languages are also decidable. We find that in general they are not, though the more restricted languages of the **Chomsky hierarchy** are.

In Chapters 2 and 6, we saw the type 0 grammar, G_4, that generates what we call the multiplication language:

$$\{a^ib^jc^{i \times j} : i, j \geq 1\}.$$

In discussing this language in Chapters 2 and 5, it was remarked that we could regard **phrase structure grammars** as *computational,* as well as *linguistic,* devices. Subsequently, in Chapter 6, we saw that the multiplication language could not be generated by a context free grammar, which tells us that a pushdown recognizer is insufficiently powerful to accept the language. From the present chapter we know that we could define a TM, from the productions of G_4, that would accept this language. This seems to imply that the TM has a certain computational ability that is worthy of clearer definition and exploration. Such an enterprise has played a central part in computer science, and it is a major aim of the second part of this book to discuss this.

However, this does not mean that we have finished with languages. Far from it. It can in fact be shown that *the computation of any TM can be modelled by some type 0 grammar.* The proof of this is complex, and we will not consider it in this book, but the result is very important, since again it establishes a general **equivalence** between a class of languages and a class of abstract machines. More than this, since the TM is the acceptor for the type 0 languages, and thus for all the subordinate language types, such a result establishes a general equivalence between TMs and the entire Chomsky hierarchy.

The TM is the first machine we have seen in this book that is capable of producing a string of symbols (on its tape) as a *result*, when given a string of symbols as its *input*. This means that the TM can carry out forms of computation that are not restricted to expecting 'yes/no' answers. The second part of this book investigates the nature of the computational power of the TM. In the process of this, we discover that the TM is more powerful than any digital computer. In fact, the TM is the most powerful computational model we can define. Eventually, our investigations bring us full circle, and enable us to precisely specify the relationship between formal languages, computable languages, and abstract machines. In doing so, we are able to answer a question that was asked at the end of Chapter 6: are there formal languages that are a **proper superset** of the type 0 languages?

EXERCISES

1† Sketch out a general argument to demonstrate that any **Turing machine** needs only one **halt state**, and also that that halt state needs have no **outgoing arcs.**

2 Design Turing machines to **decide** the following languages:

(a) $\{a^{2i+1}c^jb^{2i+1} : i, j \geq 1\}$
(b) $\{xx : x \in \{a, b\}^*\}$
(c) $\{a^ib^jc^{i+j} : i, j \geq 1\}$
(d) $\{a^ib^jc^{i-j} : i, j \geq 1\}$
(e) $\{a^ib^jc^{i \times j} : i, j \geq 1\}$
(f) $\{a^ib^jc^{i \text{ div } j} : i, j \geq 1\}$ (div is integer division)

Hint: in all cases, first design the machine so that it deals only with **acceptable** strings, then add additional arcs so that it deals appropriately with non acceptable strings.

3 The following is a *recursive definition* of 'well formed parenthesis' strings (WPEs):

() is a WPE
if X is a WPE, then so is (X)
if X and Y are WPEs, then so is XY

(a) Define a TM to decide the language of all WPEs.
(b) The language of all WPEs is **context free**. Show this by defining a **context free grammar** to generate it.

Hint: the recursive definition above converts almost directly into a CFG.

PART 2

From computation to languages

8 Finite state transducers

AIMS

To learn about

- the types of computation that can be carried out by finite state transducers
- the various limited tasks involving memory of input sequences carried out by FSTs
- the addition and subtraction of arbitrary length numbers by FSTs
- the restricted multiplication and modular division performed by FSTs
- the inability of FSTs to do multiplication of arbitrary length numbers
- FSTs as poor models of real computers

8.1 FINITE STATE TRANSDUCERS

A **finite state transducer** (**FST**) is a **deterministic finite state recognizer** (DFSR) with the ability to output symbols. Alternatively, it can be viewed as a **Turing machine** (TM) with two **tapes** (one for *input*, one for *output*) that

- always moves its head to the right
- cannot recognize **blank** squares on its tape.

As the machine makes a **transition** between two **states** (according to the input symbol it is reading, of course), it outputs one symbol. In addition to performing recognition tasks, such a machine can also carry out *computations*. The purpose of this chapter is to investigate the scope of the FST's computational abilities. We shall see that we need a more powerful type of machine to carry out all the computations we usually expect machines (i.e. computers) to be able to do. As we will see, this more powerful machine is the one we encountered in Chapter 7. It is the TM.

8.2 FINITE STATE TRANSDUCERS AND LANGUAGE RECOGNITION

First of all, you should appreciate that the FSRs used for regular language recognition in Chapter 4 are simply a special case of FSTs. We could have defined FSRs so that on each transition they produce an output symbol, and we could have defined these output symbols to be significant in some way. In the following example machine, we assume that, when the input is **exhausted**, if the last symbol the machine outputs is a 1, then the input string is **accepted**, and if the last symbol output is a 0, then the string is rejected. Figure 8.1 shows an example of such a machine.

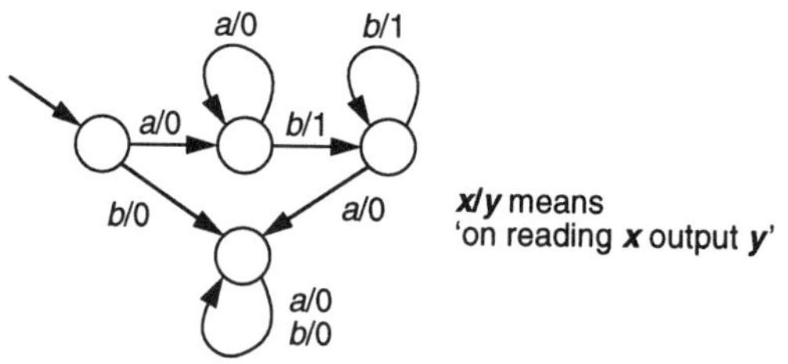

Figure 8.1 A finite state transducer for recognizing a regular language

The FST in Figure 8.1 accepts the language $\{a^i b^j : i, j \geq 1\}$. It indicates acceptance or rejection of an input string by the last symbol it outputs as the last symbol of the input is read (1 = accept, 0 = reject). Some examples of the machine's output are shown in Table 8.1. The significant symbol of the output (the last) is underlined in each case.

Table 8.1 Outputs from the finite state transducer of Figure 8.1

input string	output string	ACCEPT or REJECT
aaab	0001	ACCEPT
aa	00	REJECT
aabba	00110	REJECT
aabbb	00111	ACCEPT

As you can see, the machine in Figure 8.1 has no state designated as a **halt state** or accept state. This is because *the output is now used to tell us whether or not a string is accepted.*

8.3 FINITE STATE TRANSDUCERS AND MEMORY

An interesting application of FSTs is as 'memory' devices. For example, FST_1, in Figure 8.2 processes inputs that are strings of binary digits. As output, the machine produces strings of binary digits, representing the input string 'shifted' two digits to the right.

Figure 8.2 A binary two digit 'shift' FST, FST_1

You may think that the names of FST_1's states look strange. Of course, the state names do not affect the operation of the machine, but they have been chosen for symbolic reasons, which should become apparent as the chapter continues.

The next machine, FST_2, does a similar task, but with a right shift of three digits. FST_2 is shown in Figure 8.3.

Now consider FST_3, shown in Figure 8.4, which does a 'shift two to the right' operation for inputs consisting of strings in $\{0, 1, 2\}^+$ (remember from Chapter 2 that A^+ denotes the set of all *non empty* strings that can be taken from the **alphabet** A).

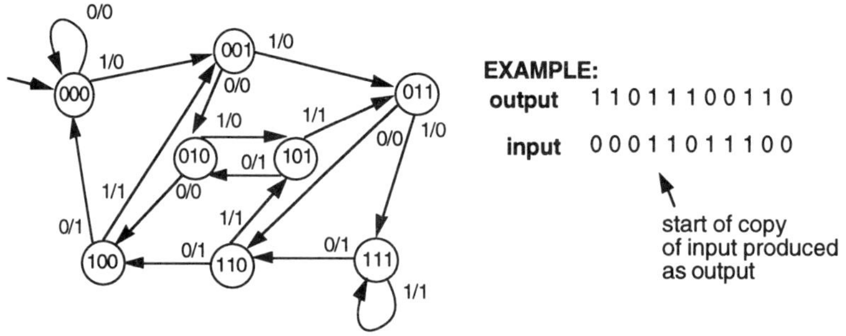

Figure 8.3 A binary three digit 'shift' FST, FST_2

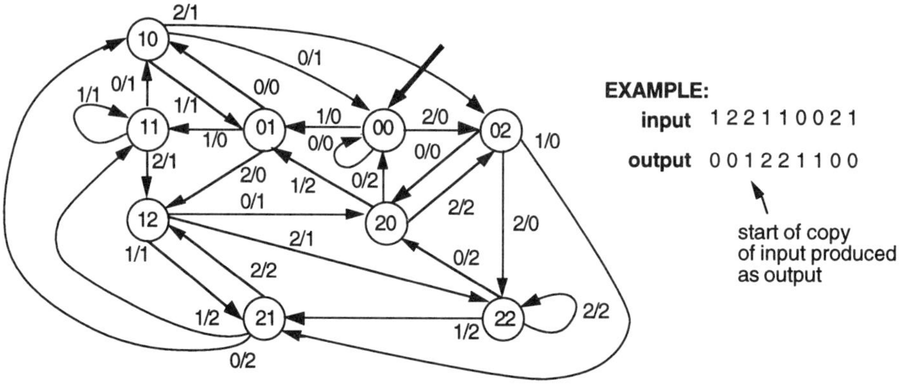

Figure 8.4 A 'trinary' two digit 'shift' FST, FST_3

Machines like FST_1, $_2$ and $_3$ are sometimes called 'memory machines' as they perform a simple memory task. Imagine that a 'clock' governs the processing of the input, a concept we return to in Chapter 10. At each 'tick' of the **clock**, the next symbol is read, and the appropriate symbol is output. Then:

FST_1 stores a symbol of a two symbol alphabet for two 'ticks' and then outputs it;

FST_2 stores a symbol of a two symbol alphabet for three 'ticks' and then outputs it;

FST_3 stores a symbol of a three symbol alphabet for two 'ticks' and then outputs it.

You should be able to see that we could design memory machines for an alphabet of any number, *s*, of symbols, and for any number, *t*, of clock ticks, as long as we know in advance the values of *s* and *t*. You should also be able to see that we need to increase the number of states as the values of *s* and *t* increase.

In fact, and this is where the names of the states in our three machines are significant, there is a simple numeric relationship between the number of symbols, *s*, in the alphabet, the number of ticks, *t*, that each input symbol has to be 'remembered' for, and the minimum number of states required to do the job. Since there are s^t (numerically speaking) distinct sequences of length t when the alphabet has s symbols, and the only way the FST can remember which of the s^t sequences it has just read is to have a state for each one, the FST needs a minimum of s^t states. As you can see, the state names in the machines in this section represent, for each machine, the s^t sequences of length t we can have for the s-symbol alphabet in question.

So,

FST_1 where $t = 2$ and $s = 2$, has $s^t = 2^2 = 4$ states;
FST_2 where $t = 3$ and $s = 2$, has $s^t = 2^3 = 8$ states;
and
FST_3 where $t = 2$ and $s = 3$, has $s^t = 3^2 = 9$ states.

8.4 FINITE STATE TRANSDUCERS AND COMPUTATION

In this section, we shall see that FSTs are capable of some reasonably sophisticated computation (subject to some creativity in defining the way in which the input and/or output sequences are interpreted). As an example of such 'creativity', we will first look in a new way at the tasks carried out by our machines FST_{1-3} (Figures 8.2–4), in the preceding section.

It should be pointed out that there is no trickery in the way we are going to interpret the input and output of our machines. We are simply defining the type of input expected by a particular machine, and the form which the output will take. This, of course, is a crucial part of problem solving, whether we perform it with formal machines or real programs.

8.4.1 Simple multiplication

Beginning with FST_1, the machine that shifts a binary sequence two digits to the right, let us consider that the sequence of 1s and 0s presented to the machine is first subjected to the following (using 1011 as an example):

1. The string 00 is appended to the rightmost end (giving 101100, for the example).
2. The resulting sequence is interpreted as a binary number *in reverse* (for the example we would then have 001101, which is 13 in *decimal* (base 10).

The output sequence is then also interpreted as a binary number in reverse.

For the example sequence 101100, FST_1 would produce as output 001011, which in reverse (110100) and interpreted as a binary number is decimal 52.

You can convince yourselves that under the conditions defined above, FST_1 performs multiplication by 4.

For FST_2, if we apply similar conditions to the input, except that for condition (1) we place three 0s at the rightmost end of the input sequence, FST_2 then performs multiplication by 8.

For FST_3, the same conditions apply as those for FST_1, except that the input and output number are interpreted as being in base 3, in which case FST_3 performs multiplication by 9. Thus, given the input 121100 (i.e. 001121, or 43 decimal), FST_3 would produce as output 001211 (i.e. 112100) which, according to our specification, represents 387 (i.e. 43×9).

So, we have seen that FSTs can carry out certain restricted multiplication tasks. Later in this chapter we see that there are limitations to the abilities of FSTs with respect to multiplication in general.

8.4.2 *Addition and subtraction*

FSTs are capable of adding or subtracting binary numbers of any size. However, before we see how this is done, we need to define the form of our input and output sequences.

The input

When we add or subtract two numbers (assuming that the numbers are too long for us to mentally perform the calculation), we usually work from right to left (least significant digits first). It seems reasonable, then, in view of the need to deal with any 'carry' situations that may arise, that our machines could do likewise. In the previous section we came across the notion of the input being in reverse. Clearly, this method can also be used here, and will then represent the machine processing the pairs of corresponding digits in the way that we usually do.

Apart from some minor problems which will be addressed shortly, the remaining major problem is that we have *two* numbers to add or subtract, whereas *our machines only accept one sequence as input.* Our solution to this problem is to present the two binary numbers (to be added or subtracted) as one sequence, in which the digits of one of the numbers are interleaved with the digits of the other. This, of course, means that the numbers must be of the same length. However, we can ensure that this is the case by placing sufficient 0s at the leftmost end of the shorter of the two numbers before we present the numbers to our machines.

Table 8.2 summarizes the preparation of input for the FSTs for addition and subtraction which we will define shortly.

We now turn to the output.

Table 8.2 Data preparation for binary adding and subtracting FSTs

To prepare two binary numbers for input to the addition or subtraction FSTs:	
1 Ensure numbers are the same length by padding the shorter number with leading 0s	first number: 1 2 3 4 5 6 7 second number: 1 2 3 4 5 6 7
1a. (For addition only) Place an extra leading 0 on each number, to accommodate possible final carry	
2 Reverse both numbers	reverse of first number: 7 6 5 4 3 2 1 reverse of second number: 7 6 5 4 3 2 1
3 Make one sequence by interleaving the corresponding digits of each reversed number (Note: for subtraction FST, begin with second number)	interleaved digits of reversed numbers: 7 7 6 6 5 5 4 4 3 3 2 2 1 1

The output

For our addition and subtraction FSTs to model our behaviour, they must process the two numbers by examining pairs of corresponding digits, in least to most significant digit order of the original numbers. However, as an FST can only read one digit at a time, it will not be able to output a 0 or 1 representing a digit of the answer until it has examined the second digit of each pair of corresponding digits. Moreover, our FSTs *must output a symbol for each input symbol encountered.* For the first digit in each corresponding pair then, our FSTs will output the symbol *. Note that there is no special significance to this symbol, it is simply being used as a separator.

Thus, given input of the form indicated by the diagram accompanying step 3 in Table 8.2, our FSTs will produce output as represented in Figure 8.5.

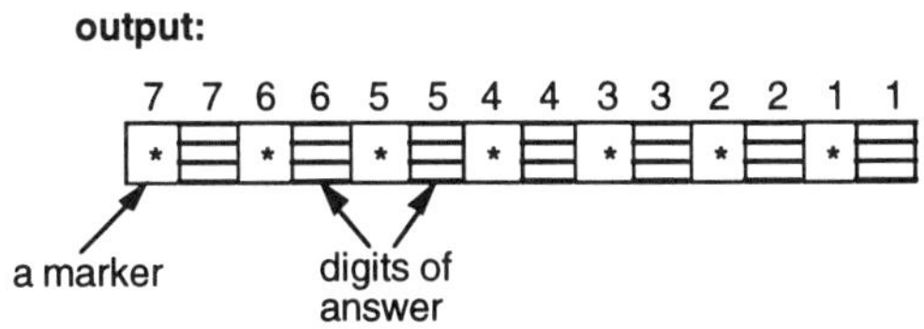

Figure 8.5 Specification of output for arithmetic FSTs

The output shown in Figure 8.5 represents the digits of the answer in reverse, with each digit preceded by *. The *s are not part of the answer, of course, and so we read the answer as specified in Figure 8.6.

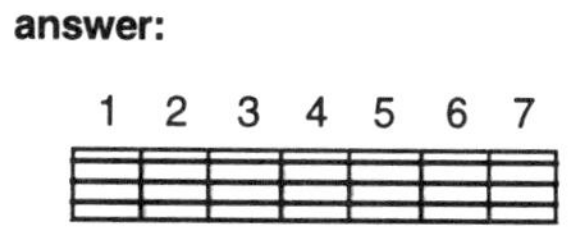

Figure 8.6 How we read the answer from the output specified in Figure 8.5

The machines

We are now ready to present the two machines, FST_4, the binary adding FST (Figure 8.7), and FST_5, the binary subtractor (Figure 8.8). Each machine is accompanied by an example showing an input and the resulting output. I leave it to you to convince yourself that the machines operate as claimed.

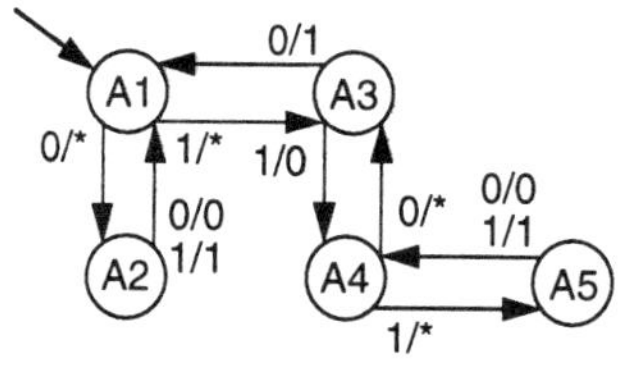

EXAMPLE:
0 1 1 0 1 + 0 0 1 1 0
(i.e. 13 + 6)
is presented to the machine as:
0 1 1 0 1 1 0 1 0 0
(i.e. numbers in reverse, one digit of each number in turn, doesn't matter which one first)
This results in the output:
*** 1 * 1 * 0 * 0 * 1**
which is also in reverse, so it represents
1 0 0 1 1 (i.e. 19)

Figure 8.7 A binary adder FST, FST_4

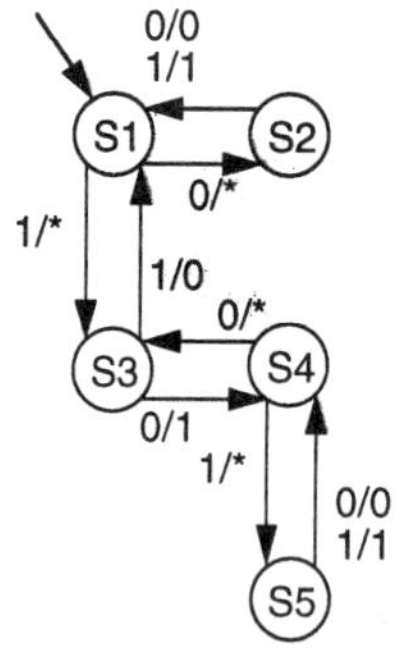

EXAMPLE:
1 1 0 1 − 0 1 1 0
(i.e. 13 − 6)
is presented to the machine as:
0 1 1 0 1 1 0 1
(i.e. numbers in reverse, one digit of each number in turn, beginning with second number)
This results in the output:
*** 1 * 1 * 1 * 0**
which is also in reverse, so it represents
0 1 1 1 (i.e. 7)

Figure 8.8 A binary subtractor FST, FST_5

It is important to note that *for both machines there is no limit to the length of the input numbers*. Pairs of binary numbers of any length could be processed. It is clear, therefore, that *the FST is capable of adding and subtracting arbitrary length binary numbers*.

8.4.3 *Simple division and modular arithmetic*

Certain restricted forms of division are also within the capabilities of the FST. For example, FST_6, in Figure 8.9, performs integer division of a binary number by 3, returning the result as a binary number. In some programming languages integer division is represented by the *infix* operator **div**, so FST_6 actually does x **div** 3, where x is the input number[1]. However, FST_6 does not require the input sequence in reverse. We usually perform long division from the most significant end of the number, and so does FST_6. The machine also outputs the result in the correct order. Figure 8.9 includes an example input sequence, and its corresponding output sequence.

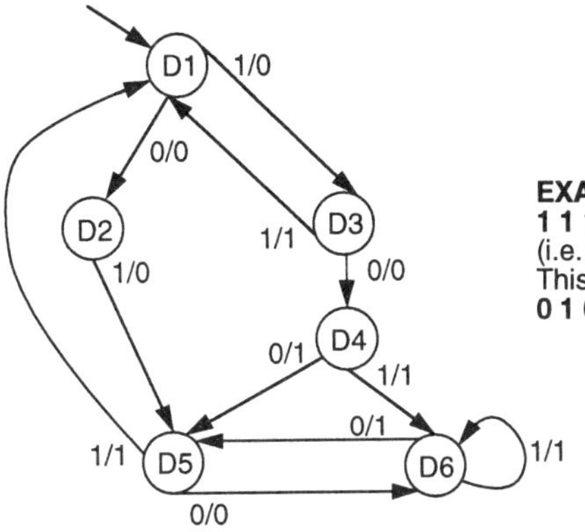

EXAMPLE:
1 1 1 0 0 0 1 0 1 0 1
(i.e. 1,813)
This results in the output:
0 1 0 0 1 0 1 1 1 0 0 (i.e. 604 = 1,813 div 3)

Figure 8.9 A binary 'div 3' FST, FST_6

We can quite easily use FST_6 as the basis of a machine that computes the quotient *and* the remainder by making the following observations about states of FST_6 :

*D*5 represents a remainder of 1 (i.e. a carry of 1);
*D*6 represents a remainder of 2 (i.e. a carry of 2);
*D*1–4 represent a remainder of 0 (i.e. no carry).

Now, the remainder resulting from a division by 3 requires a maximum of two binary digits, as it can be 0, 1, or 10. We could make sufficient space for the remainder by placing *** (three asterisks) after the input sequence. I leave it to you to decide how to amend the machine to ensure that the quotient is followed by the binary representation of the remainder. The first of the three asterisks should be left unchanged by the machine, to provide a separator between quotient and remainder.

We could also amend our machine to provide the remainder *only*. For this we would ensure that all the outputs of FST_6 were changed to '*', and then carry out the amendments discussed in the preceding paragraph. Then our machine would carry out an operation similar to that called **mod** in programming languages, except it would be restricted to **mod** 3.

[1] An *infix* operator is written in between its operands. For example, '+' is thus an infix operator.

8.5 THE LIMITATIONS OF THE FINITE STATE TRANSDUCER

We have seen that FSTs can be defined to do restricted forms of multiplication and division, and also that FSTs are capable of unrestricted addition and subtraction. You will note that the preceding sentence does not include the word 'binary'. This is because we could define adders and subtractors for *any* number base (including base 10), since there are only a finite number of 'carry' cases for addition and subtraction in any number system. Similarly, we can define FSTs to do multiplication or division by any fixed multiplicand or divisor, in any number base. However, the binary machine is much simpler than would be the corresponding FST for, say, operations on base 10 numbers. Compare the complexity of FST_1 (Figure 8.2) and FST_3 (Figure 8.4), for example. I will leave it to you to consider the design of FSTs for number bases from three upwards.

In fact, our intuition tells us that the binary number system is sufficient; after all, we deal every day with a machine that performs all its operations on representations that are essentially binary numbers, and we never use that as the basis of an argument that the machine is insufficiently powerful! We return to the issue of the universality of binary representations several times in subsequent chapters.

For now, our purpose is to establish that the FSTs are not a sufficiently powerful formalism for modelling all computational tasks that we may wish an abstract machine to model. We have seen that the FST can carry out unrestricted addition and subtraction. The question arises, *can we design an FST to do unrestricted multiplication*? We shall eventually demonstrate that the answer to this last question is 'no'. This is obviously a critical limitation, since the ability to perform multiplication is surely one of the basic requirements of useful computing systems. First, however, we show that there are some multiplication operations we *can* perform with FSTs.

8.5.1 Restricted FST multiplication

Earlier in this chapter, we saw that FSTs are capable of multiplying a base n number by some constant c, where c is a power of n (for example, FST_1 was used to multiply a binary – base 2 – number by 4, i.e. 2^2). Suppose we wished to multiply a number, say a binary number, by a number that cannot be expressed as an integral power of 2, say 3? One way to achieve such tasks is to effectively 'hardwire' the multiplicand 3 into the machine, in a similar way that the divisor 3 was represented in the **div** 3 machine, FST_6 (Figure 8.9).

FST_7, shown in Figure 8.10, multiplies an arbitrary length binary number by 3.

To see how FST_7 works, consider its operation in terms of writing down a binary number three times, and adding the three copies together. Figure 8.11 represents the manual version of the same example shown in Figure 8.10.

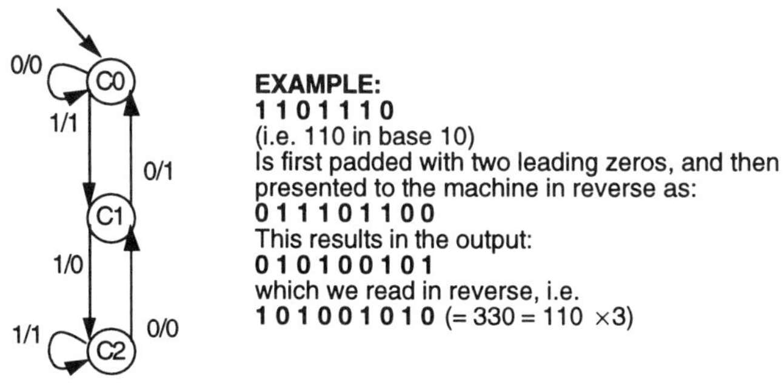

Figure 8.10 A binary '× 3' FST, FST_7

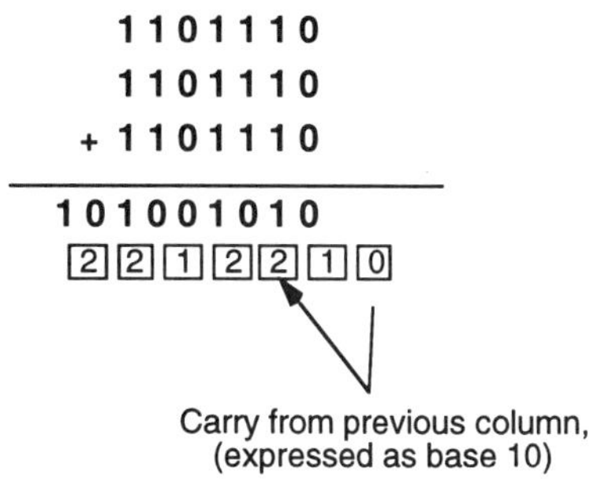

Figure 8.11 Adding in binary, with carry in decimal, to achieve multiplication by 3

In adding together three identical binary numbers, beginning from the rightmost column, we can either have three 1s (i.e. 1 + 1 + 1) = 11, i.e. 1 carry 1, or three zeros (0 carry 0), i.e. a carry of 1 or 0. If there is a carry of 0, we continue as in the previous sentence. Otherwise, we have a carry of 1 and the next column could be three 1s (i.e. 1 + 1 + 1 + 1 – adding in the carry) so we write down a 0 and carry 2 (decimal). If the next column is three 0s, we have 0 + 0 + 0 + 10 = 10, i.e. 0 carry 1 – back to carry 1, which has been discussed already. Otherwise, we have three 1s and a carry of 10, i.e. 1 + 1 + 1 + 10 = 101 = 1 carry 2 – back to carry 2, which, again, we have already considered. The whole situation is summarized in Table 8.3.

Table 8.3 Carry situations, expressed in base 10, for multiplication by 3 by adding three copies of a binary number

		Column			
		1		0	
Carry	0	1	0	0	0
	1	0	2	1	0
	2	1	2	0	1
		result	new carry	result	new carry

The states in Figure 8.10 were labelled $C0-2$ to represent the three carry situations found in Table 8.3. From Table 8.3, you should be able to see why only three states are required for FST_7, but why *we could not define a machine of fewer than three states to carry out the same task.*

A similar method can be used to create a machine, FST_8 that multiplies a binary number by 5. The machine is shown in Figure 8.12, and as you can see, requires 5 states.

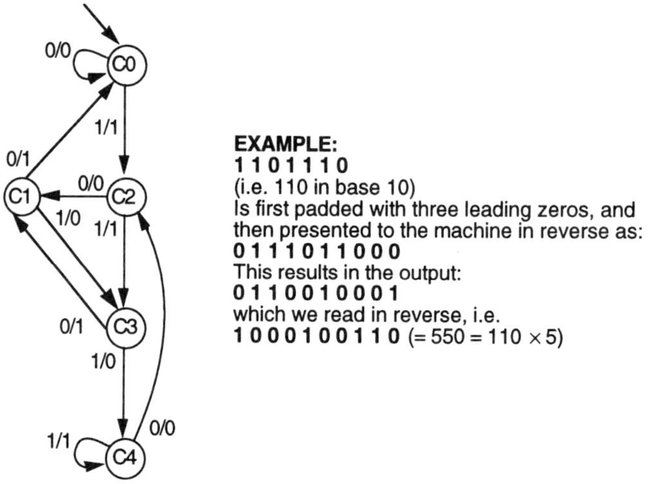

Figure 8.12 FST_8 does binary multiplication by 5

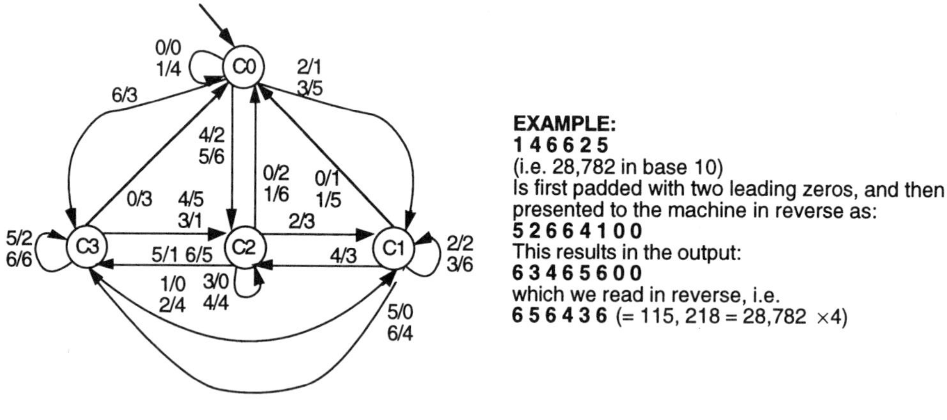

Figure 8.13 FST_9 multiplies a base 7 number by 4

Finally, we consider in Figure 8.13 the rather unusual FST_9, which multiplies a base 7 number by 4 (!).

I leave it to you to justify to yourself that the above machines, FST_{7-9}, operate as claimed.

The above multipliers, FST_{7-9}, reflect a construction suggesting that a machine to multiply a number in any base by a constant c needs no more states than c (i.e. one for each of the possible base 10 carry values). As we saw above:

FST_7, where $c = 3$ had 3 states;
FST_8, where $c = 5$, had 5 states;
FST_9, where $c = 4$, had 4 states.

However, you may like to consider what happens to our construction when the numbers to be multiplied are in bases greater than 10.

8.5.2 FSTs and unlimited multiplication

In this section we see that FSTs are incapable of multiplication of arbitrary length numbers. The argument used is essentially an informal application of the **repeat state theorem** introduced in Chapter 6. Again, the method of 'proof' used is ***reductio ad absurdum***.

I will assume, for sound reasons that are discussed in later chapters, and hinted at earlier in this chapter, that it is sufficient to focus on the task of multiplying arbitrary length binary numbers. Our assumption is that *we have an FST, M, that can multiply arbitrary length pairs of binary numbers, and print out the result.* Now, we can assume that M has a fixed number of states. Say it has k states. We ask M to multiply the number 2^k by itself:

- 2^k is a 1 followed by k 0s;
- $2^k \times 2^k$ is a 1 followed by $2k$ 0s
  (example: $2^3 = 1000$, $2^3 \times 2^3 = 2^6 = 1000000$, i.e. $8 \times 8 = 64$).

Suppose M reads the corresponding digits of its input numbers simultaneously. M has only k states, and it must use at least one of them to process the input digits. However, after it has processed the input digits it still has to print out the rest of the 0s in the answer. Assuming it only uses one of its states to do the multiplication (an extremely conservative estimate!) and print out the 1 followed by the first k 0s of the answer (one digit for each of the k pairs of input digits it encounters), M has only $k - 1$ states left with which to print out the remaining k 0s of the answer.

M must therefore loop when it prints out the rest of the 0s. It is printing out the 0s without any further input of digits, so there is no way of controlling the number of 0s that M prints. The *only* way to make an FST generate a fixed number, k, of symbols without input is to have a sequence of $k + 1$ states to do it, a notion that was discussed in the context of **finite state recognizers** in Chapter 4.

We are forced to reject our initial assumption: the machine M cannot exist. *Multiplication of arbitrary length binary numbers is beyond the power of the FST.* By extension of this argument, multiplication of arbitrary length numbers in any number base is also beyond its power.

8.6 FSTs AS UNSUITABLE MODELS FOR REAL COMPUTERS

It is unfortunate that FSTs are not powerful enough to perform all the operations that we expect computers to be able to carry out. As you can probably guess from considering the parsing routine for the FSR in Chapter 4, FSTs also lend themselves to conversion into very simple programs. However, at the very least, we expect any reasonable model of computation to be able to deal with arbitrary multiplication and division. You have seen that such basic requirements are beyond the capabilities of FSTs.

You may think you detect a flaw in an argument that assumes that real computers can perform operations on arbitrarily large data. Most of us who have ever written any non trivial programs have experienced having to deal with the fact that the computer we use is *limited,* not limit*less*, in terms of its storage and memory capacity. Since a computer is, in terms of available storage at any given point in time, finite, it could be regarded as a (huge) FST. As such, for any given computer, there is, even for basic operations such as multiplication, a limit to the size of the numbers that can be dealt with. In this sense, then, a computer is an FST.

It turns out, however, that treating a digital computer as if it were an FST is not, in general, a useful position to adopt, either theoretically or practically[2]. Let us look at the practical issues first. When a deterministic FST enters the same state (say q) twice during a computation, it is looping. Moreover, once the machine reaches state q, repeating the input sequence that took it from q back to q (perhaps via several other states) will result in the same behaviour being exhibited again. This was the basis of the *repeat state theorem* for FSRs in Chapter 6.

Suppose a program is performing some calculations, subsequently to print some output at the terminal, and receives no output from external sources once it has begun (i.e. it operates on data stored internally). The *state* of the computer running the program *is its entire internal configuration*, including the *contents of its memory*, its *registers*, its *variables*, its *program counter*, and so on. Viewed in this way, if such a state of the computer is ever repeated during the execution of the program, the program will repeat the whole cycle of events that led from the first to the second occurrence of that configuration over and over again, *ad infinitum.* If this happens we have obviously made a (fairly common) mistake in our coding, resulting in what is often called an *infinite loop*.

It would be useful if, before we marketed our programs, we could test them to make sure that such infinite loops could never occur. Let us suppose that we have some way of examining the internal state of the machine each time it changes, and some way of deciding that we have reached a state that has occurred before. In theory, this is

[2] This is not altogether true. The FST is used, for example, to model aspects of computer networks (see Tanenbaum, 1988, in the *Further Reading* section). Here, I am referring to its ability to model *arbitrary computations*.

possible because, as stated above, a computer at a given point in time is a finite state machine.

Suppose that our computer possesses 32 kilobytes of total computer storage, including data, registers, primary memory, and so on. Each individual bit of the 32K can be either 1 or 0. Thus we have a total of 262,144 bits, each of which can be either 1 or 0. This means that our machine can, at any point, be in one of $2^{262,144}$ states, $2^{262,144}$ being about $10^{79,000}$. This is an *incredibly* large number. In Chapter 10 we come across a really fast (over three million million instructions per second) hypothetical machine that would take 10^{80} *years* (a 1 followed by eighty 0s) to solve a problem involving passing through 10^{100} states. Our computer may thus have to go through a huge number of internal states before one of these states is repeated. Waiting for even a tiny computer, like our 32K model here, to repeat a state because it happens to be an FST is thus a rather silly thing to do!

In the remainder of this book, we study the deceptively simple abstract machine that was introduced in Chapter 7, i.e. the **Turing machine** (**TM**). The TM is essentially an FST with the ability to move left *and* right along a single input tape. This means that the TM can overwrite symbols on its tape with other symbols, and can thus take action on the basis of symbols placed there earlier in its activities. We find in Chapters 9 and 10 that this simple machine is actually more powerful than any digital computer. In fact, in some senses the Turing machine is the most powerful computational device of all. In Chapter 11, however, we find that there is something about the problem of whether or not a certain computation will eventually stop; discussed in terms of the FST above, that puts its general solution beyond *any* computational device.

EXERCISES

1 **FSTs** can be used in logic design[3]. Design FSTs to perform bitwise manipulation of their binary number inputs. For machines needing two input values, assume an input set up similar to that defined for FST_4 and FST_5, the addition and subtraction machines of Figures 8.7 and 8.8, respectively. The operations you may wish to consider are:

- and
- or†
- not
- nand
- nor

[3] See Hutchison and Silvester (1987), in the *Further Reading* section.

9 Turing machines as computers

AIMS

To learn

- how TMs can perform computational tasks such as multiplication and division
- how TMs can perform computer operations such as logical operations, memory accessing operations and control operations
- about Turing's thesis

9.1 TURING MACHINES AND COMPUTATION

In Chapter 7, **Turing machines** (**TMs**) for language recognition were introduced. We saw that the TM is the **recognizer** for *all* the languages in the Chomsky hierarchy. Chapter 8 demonstrated that the **finite state transducer** (**FST**), which is essentially a TM that can move only to the right on its **tape**, is capable of some fairly sophisticated computational tasks, including certain restricted types of multiplication and division.

Coupled with the fact that any given digital computer is essentially finite in terms of storage capacity at any moment in time, this seemed to suggest that seeing the computer as an FST might be a useful perspective to take, especially since the FST lends itself to conversion into very simple programs. However, we saw that the huge number of internal states possible in even a smallish computer meant that notions typically applied to FSTs, such as the **repeat state theorem**, were of limited practical use in this context.

In this chapter, we are going to see that, in terms of computation, the Turing machine is a much better counterpart of the computer than is the FST. Simple as it is, and resulting as it does from addition of a few limited facilities to the FST, *the TM actually exceeds the computational power of any single computer, of any size*. In this sense, the TM is an abstraction over all computers, and the theoretical limitations to the power of the TM are often the practical limitations to the digital computer, not only at the present time, but at any time in the future. However, we shall descend from these heady philosophical heights until the next chapter. For now, we concentrate on demonstrating that the TM is capable of the basic computational tasks we expect from any *real* computer.

9.2 TURING MACHINES AND ARBITRARY BINARY MULTIPLICATION

Chapter 8 established that FSTs are not capable of multiplication of arbitrary length numbers in any number base. In this section, we sketch out a TM to perform multiplication of two binary numbers of any length. I will leave it to you to realize that similar TMs could be defined for any number base (cf. the treatment of differing number bases in Chapter 8). However, here we choose the binary number system, for obvious reasons.

9.2.1 *Some basic TM operations*

A side effect of defining a TM for multiplication is that we will appreciate a notion actually formulated by Turing himself in conjunction with his machines, and one which occurs in some form in many programming languages and computer environments. Essentially, we shall build up our binary multiplier by creating a 'library' of TM 'subroutines'. These subroutines, like subroutines in typical programming languages ('procedures' in Pascal), represent operations that we wish to carry out many times in a machine. However, since there are no straightforward mechanisms for *parameter passing* in TMs, our subroutines are more of an analogy to *macros*, i.e. we have to imagine that the 'sub-machine' that constitutes the macro is actually inserted into the appropriate place in the overall machine. This means that we have to 'call' our sub-machine by ensuring that the 'calling' machine configures the tape and r/w head appropriately.

Some operations needed by our multiplier Turing machine are now specified.

Our multiplication routine will work by repeated addition, based on the common 'shift and add' long multiplication method (we examine this in more detail in the next section). Clearly, then, we need an ADD routine. Our ADD routine will make use of two sub-machines, WRITE-A-ZERO and WRITE-A-ONE, as specified in Figure 9.1.

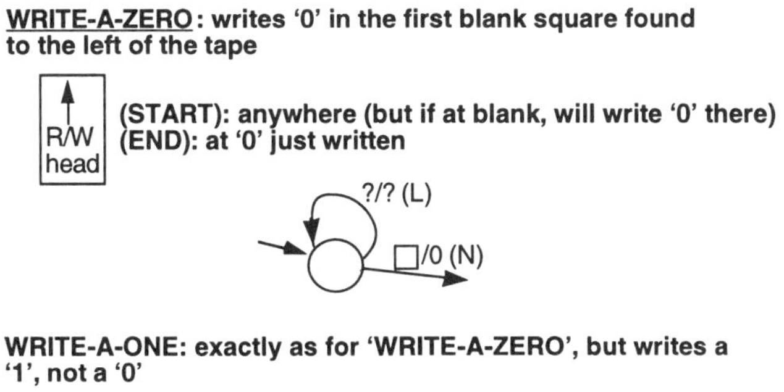

Figure 9.1 'WRITE-A-ZERO' and 'WRITE-A-ONE'

Referring to Figure 9.1, recall from Chapter 7 that we use ?/? to mean 'replace any non **blank** symbol of the alphabet for the machine by the same symbol'.

Our ADD routine adds two arbitrary length binary numbers, adding pairs of digits in least to most significant order (as we usually do). The machine assumes that the two numbers are initially represented on a tape as specified in Figure 9.2.

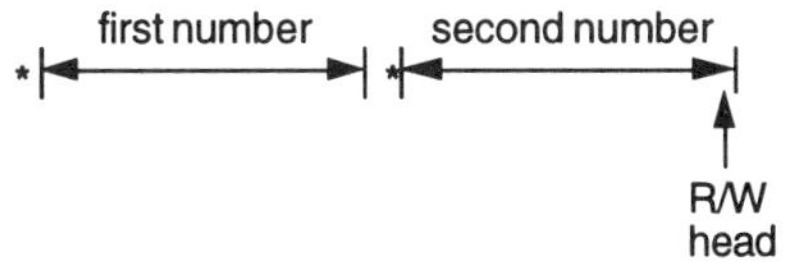

Figure 9.2 The tape set-up for the ADD TM

So, ADD expects on its tape a * followed by the first number, followed by a * followed by the second number. Its **head** is initially located on the least significant digit of the second number. ADD will not be concerned with what is to the left of the relevant part of the input tape, but will assume a blank to the right of the second number. The left asterisk will be used by ADD to indicate the current carry situation during the computation (* = no carry, C = carry). The machine will write the result of the addition to the left of the currently occupied part of the input tape (and thus assumes that the first blank found to the left of the left * is the start of the unoccupied part of the tape, i.e. ADD assumes a **packed** tape).

The description of ADD will be supported by an example input tape, and we will examine the effects of each major part of ADD on the contents of this tape. The example input tape we will use is shown in Figure 9.3.

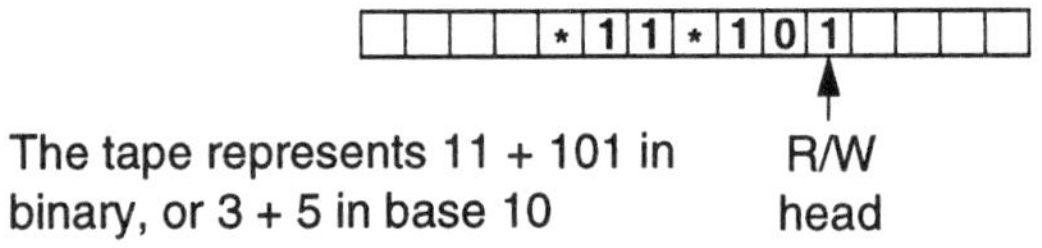

Figure 9.3 An example input configuration for the ADD TM.

9.2.2 *The 'ADD' TM*

Figure 9.4 shows Part 1 of the ADD TM.

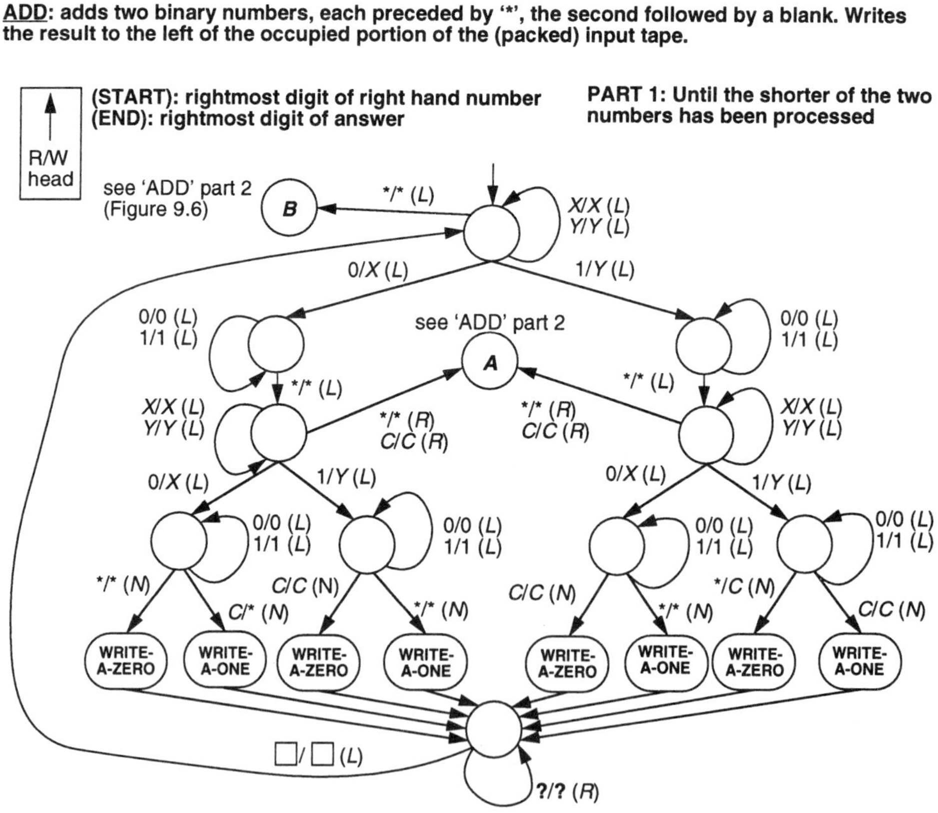

Figure 9.4 The ADD TM, Part 1

Part 1 of ADD, shown in Figure 9.4, deals with *the adding of corresponding digits of the two numbers until the digits of one or both of the numbers have been used up.* Note that *X* and *Y* are used as markers for 0 and 1, respectively, so that ADD can detect how far along each number it has reached, at any stage, and thus avoid processing the same digit more than once. When the final digit of one or both of the numbers has been reached, Part 2 of ADD (shown in Figure 9.6.) takes over. When we are ready to enter ADD Part 2, our example tape is configured as shown in Figure 9.5.

As can be seen in Figure 9.6, there are two entry points *A* and *B*, to ADD Part 2.

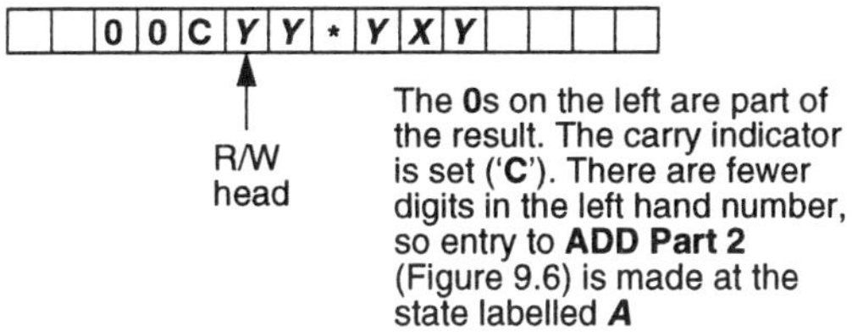

Figure 9.5 Output from ADD Part I (Figure 9.4) for the example input in Figure 9.3

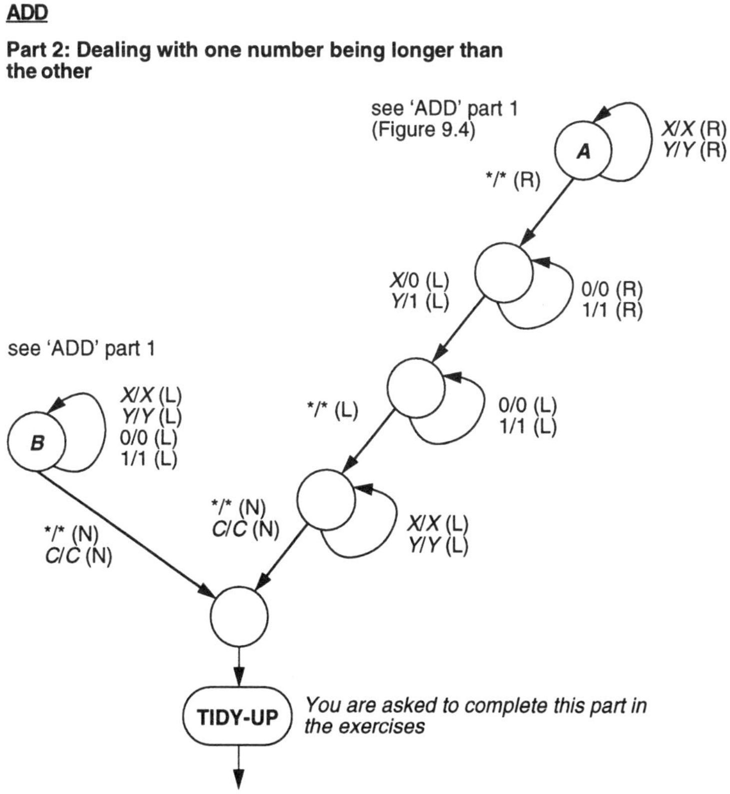

Figure 9.6 ADD, Part 2

Entry point *A*
is entered from ADD (Part 1) when *there are no digits remaining in the left hand number, but a digit was read from the right number*. Since ADD has just overwritten either a 0 (by *X*) or a 1 in the right hand number (there must have been at least one more digit in the right number than in the left number or we would not have reached entry point *A*), this digit (represented by the leftmost *X* or *Y* in the right number) is set back to its original value, so that it can be processed as an extra digit by the sub-machine TIDY-UP, the definition of which is left as an exercise.

Entry point *B*
is entered from ADD (Part 1) *when there are no digits remaining in the right number.* This case is simpler than that for entry point *A*, as no extra digits have been marked off (ADD always looks at the digit in the *right* number first, before it attempts to find the corresponding digit in the left number).

Entry points *A* and *B* both ensure that the head is located at the square of the tape representing the *carry* situation. Remember that this is the square immediately to the left of the left number, and is occupied with a * for no carry, and a *C* for carry.

Our example tape would have resulted in entering add at entry point *A*, since the right hand number ran out first. Immediately before the machine enters the component labelled TIDY-UP, which you are invited to complete as an exercise, our example tape from Figure 9.3 would be as described in Figure 9.7.

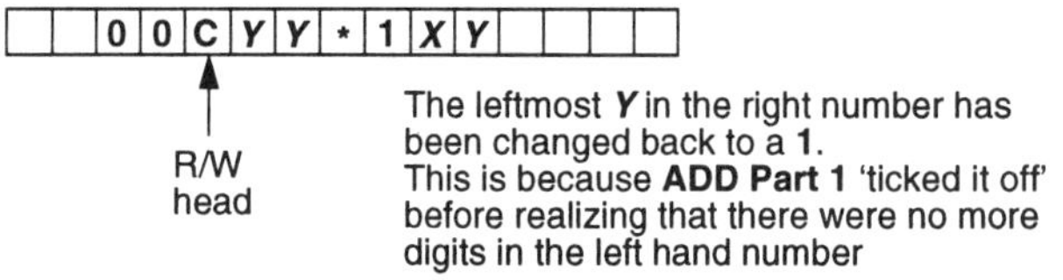

Figure 9.7 The input from Figure 9.3 on entry to TIDY-UP in ADD Part 2 (Figure 9.6)

ADD should now complete its computation, the result being the binary number at the left of the occupied portion of the tape. For our example tape, Figure 9.8 shows an appropriate outcome from TIDY-UP, and thus an appropriate representation of the result of ADD, given the input tape of Figure 9.3.

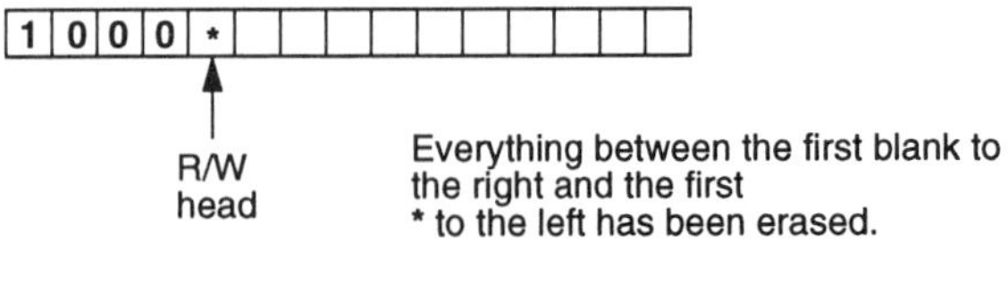

Figure 9.8 The tape of Figure 9.7 after processing by TIDY-UP of ADD Part 2 (Figure 9.6)

9.2.3 The 'MULT' TM

We are now ready to sketch out the description of a TM, MULT, to multiply two arbitrary length binary numbers. As mentioned earlier, our machine uses the 'shift and add' method. For example, suppose we wish to compute 1110 × 101. This would be carried out as shown in Figure 9.9.

```
   1110
   ×101
 ------
 111000
  +1110
-------
1000110
-------
```

Figure 9.9 The 'shift and add' multiplication method: binary version

An example of the same method being employed for the multiplication of two base 10 numbers may illustrate the method in more familiar terms. Figure 9.10 shows the method being used for the base 10 computation 345 × 23.

```
  345
  ×23
 ----
 6900
+1035
 ----
 7935
 ----
```

Figure 9.10 The 'shift and add' multiplication method: decimal version

The 'shift and add' process is simpler for binary numbers, since the same rule applies for placing zeros at the end of each stage, but we only ever have to multiply at intermediate stages by either 1 or 0. Multiplying by 1 involves simply copying the number out exactly, while multiplying by zero can, of course, be ignored. An algorithm to do 'shift and add' binary multiplication is given in Table 9.1.

Table 9.1 An algorithm to perform 'shift and add' multiplication of binary numbers

```
{n1 and n2 are the two numbers to be multiplied}
{let d_i be the ith digit from the right of n2}
if d_1 = 1
then ans := n1
else ans := 0
endif;
for each digit d_i, where i ≥ 2 in n2 do
      if d_i ≠ 0
      then ans:= ans + (n1 with i – 1 0s appended to its rightmost end)
      endif
endfor
```

Our MULT machine will be based on the 'shift and add' algorithm shown in Table 9.1. The detailed implementation has been left to the exercises. The MULT machine will expect its input to be configured exactly as that for the ADD TM, as specified in Figure 9.2.

An example input configuration for MULT is shown in Figure 9.11. Note that we have decided that the read/write head should be positioned initially on the rightmost digit of the right hand number, this number being called $n2$ in the algorithm of Table 9.1.

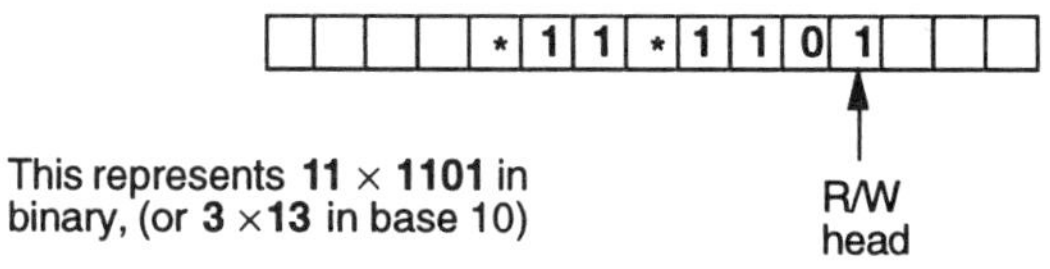

Figure 9.11 An example input tape for the MULT TM

We will now sketch out the actions to be carried out by MULT, in terms of the algorithm of Table 9.1. The first part of the algorithm is the 'if' statement.

if $d_1 = 1$
then ans := $n1$
else ans := 0

The head is currently over the first (rightmost) digit of n_2, which in the algorithm is called d_1. If d_1 is a 0, MULT initializes the answer to 0, otherwise the answer is initialized to $n1$. MULT will use the area to the left of $n1$ as the 'work area' of its tape, and leave the answer there when it terminates, separated by a single blank from the * preceding $n1$. ans := $n1$ thus means 'copy $n1$ into the answer area'. For this we need a copy routine, which we will call COPY-L. COPY-L is defined in Figure 9.12.

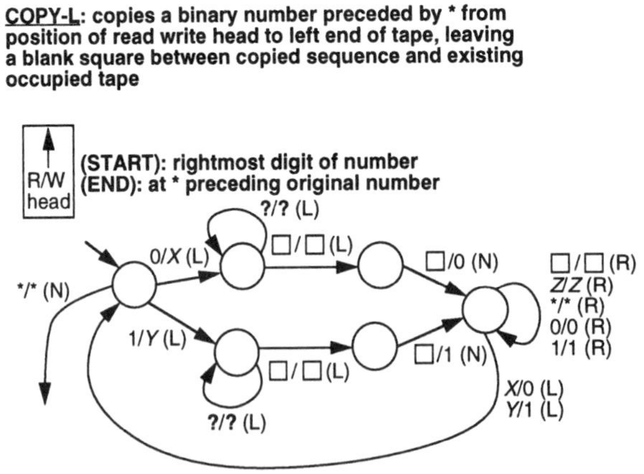

Figure 9.12 The TM, COPY-L for copying a binary number to the left hand end of the marked tape

COPY-L needs little explanation. The symbols X and Y are again used as temporary markers for 0 and 1, respectively. The machine leaves the copied sequence exactly as it finds it. Clearly, the machine is designed to expect an alphabet of (at least) {*, Z, X, Y, 0, 1}, which is due to the requirements of the multiplier[1]. The machine could easily be amended to deal with other alphabets. It would also be a simple matter to create a corresponding COPY-R ('copy right') machine, if required. If $d_1 = 0$, then MULT simply writes a 0 in the answer area. In our example here, $d_1 = 1$, so MULT must move its head to the rightmost digit of $n1$, then enter COPY-L. At this point, MULT could also place a * at the left hand end of the newly written answer sequence. The example tape would then be as shown in Figure 9.13.

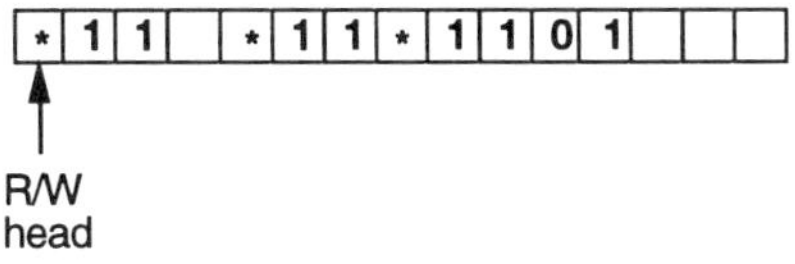

Figure 9.13 The state of the input tape from Figure 9.11 after the statement ans: = $n1$ (Table 9.1)

MULT will now deal with the next part of the algorithm, i.e. the 'for loop'.

```
for each digit d1, where i ≥ 2 in n2 do
    if di ≠ 0
    then ans := ans + (n1 with i – 1 0s appended to its rightmost end)
    endif
endfor
```

Now that d_1 has been processed, MULT turns its attention to the remaining digits of n_2. We have seen examples of the type of processing involved here earlier in this book; TMs use auxiliary symbols to 'tick off' each symbol (i.e. temporarily replace it by another symbol) so that they can return to that symbol, thus being able to locate the next in the sequence, and so on. For the algorithm above, MULT must examine all the digits, d_i, of n_2 (for $i \geq 2$). If a d_i is 0, MULT simply proceeds to the next digit (if there is one), else (d_i being a 1), MULT must write 0 at the left hand end of its tape for each digit in n_2 to the right of d_i (this corresponds to us placing zeros at the end of a number in the 'shift and add' method), and then it must do a COPY-L to copy n_1 immediately to the left of the leftmost 0 just written. If MULT then places a * to the left of the copied n_1 and moves the head right until it reaches a blank, the tape is now set up so that our ADD TM can take over. Via ADD, MULT is now carrying out the

ans:= ans + ($n1$ with $i-1$ 0s appended to its rightmost end)

part of the algorithm.

[1] Since MULT is going to use ADD, the sub-machines we define for MULT may need to take account of symbols used by ADD, such as C, the *carry indicator*, and so on.

For our example tape, $d_2 = 0$, but d_3, the leftmost digit of $n2$, is 1. MULT must then place two zeros to the left of the answer field, then copy $n1$ to the left of those zeros, meaning that the tape on input to ADD would be as shown in Figure 9.14.

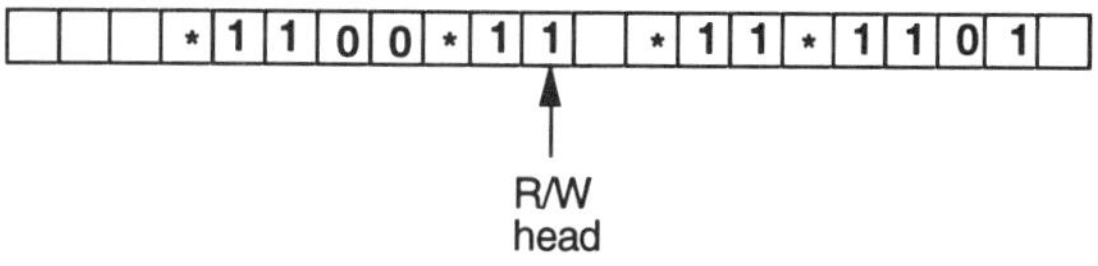

Figure 9.14 The example input tape from Figure 9.11, immediately before 'ans := ans + (*n*1 and two zeros appended to its rightmost end)' (Table 9.1)

ADD was designed so that it erases its tape from the right of the answer as far as the first blank to the right. This means that ADD would not erase MULT's input numbers, since a blank separates the input to MULT from the answer. MULT would then need to prepare the tape for the next digit of $n2$ (d_4 in our example) to be processed. This could involve simply '**shuffling**' the answer sequence up so that only one blank separates it from MULT's input sequence. Our example tape would now be as displayed in Figure 9.15 (subject to some slight repositioning of the *).

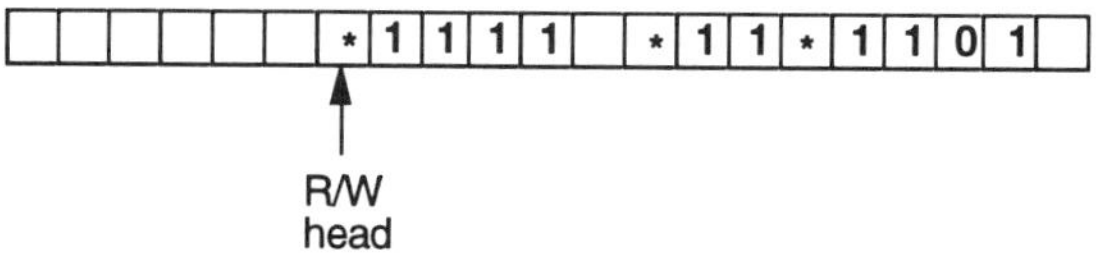

Figure 9.15 The example input tape from Figure 9.11, immediately after 'ans := ans + (*n*1 with two zeros appended to its rightmost end)' (Table 9.1)

MULT would now deal with d_4 of $n2$ in an analogous way to its treatment of d_3, described above. This time, three zeros would be needed, so the tape on input to ADD would be as shown in Figure 9.16.

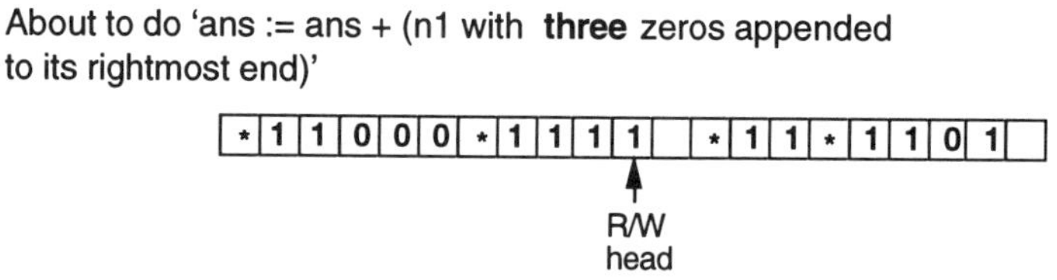

Figure 9.16 The example input tape from Figure 9.11, immediately before 'ans := ans + (*n*1 with three zeros appended to its rightmost end)' (Table 9.1)

Figure 9.17 shows a suitably tidied up form of the final result.

Sufficient detail has been presented to demonstrate that such a MULT machine could be implemented. The exercises ask you to do this. It is clear that such a MULT machine would operate on binary numbers of any length. We established in the preceding chapter that multiplication of arbitrary length numbers is beyond the capabili-

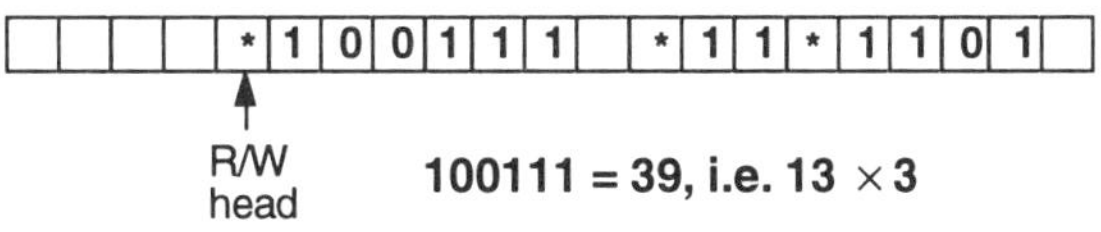

Figure 9.17 The example input tape from Figure 9.11, after the algorithm of Table 9.1 has been carried out by the MULT TM

ties of the FST. We now know, therefore, that *the TM possesses greater computational power than does the FST*.

9.3 TURING MACHINES AND ARBITRARY INTEGER DIVISION

As we can do *multiplication by repeated addition*, so we can do *division by repeated subtraction*: we keep subtracting one number from the other until we obtain zero or a negative result. We keep a count of how many subtractions have been carried out, and this count is our answer. In this section, we sketch a TM that does arbitrary integer division. The actual construction is left as an exercise.

Our machine, DIV, will perform division by repeated subtraction. We therefore require a SUBTRACT sub-machine for our DIV machine (as we needed an ADD machine for MULT).

9.3.1 The 'SUBTRACT' TM

It turns out that we can do subtraction in terms of addition, and you may be familiar with the approach considered here. For example, 8 – 5 is really the same as saying 8 + (–5). If we have a way of representing the negation of the second number, we can then simply add that negation to the first number. The usefulness of this method for binary numbers relates to the fact that the negation of a number is simple to achieve, using a representation called the *twos complement* of a binary number.

The process of obtaining the twos complement of a binary number is described in Figure 9.18.

To compute $x - y$, for binary numbers x and y, we compute the twos complement of y and add it to the first number.

Referring to the example of 8 – 5 mentioned above, we can represent this in binary as 1000 – 101. The twos complement of 101 is 011 (the inverse is 010, and 010 + 1 = 011). 1000 + 011 = 1011. Now, this result appears to be 11 (eleven) in base 10. However, after the addition we ignore the leftmost digit of the result, and thus our answer is 011 in binary (i.e. 3 in base 10), which is the correct result.

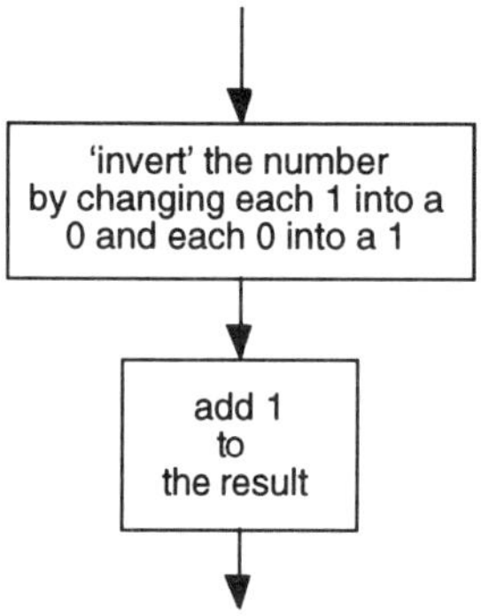

Figure 9.18 Obtaining the twos complement of a binary number

If you have not come across the method before, you should convince yourself that twos complement arithmetic works, and also discover for yourself why the leftmost digit of the result is ignored.

In terms of the development of the SUBTRACT machine, the subtraction method we have just described can be done in terms of the ADD TM described earlier in the chapter. We can even use ADD in the sub-machine that computes the twos complement, since once we have inverted a binary number, we then need to add 1 to the result. A TM, TWOS-COMP that embodies such an approach is described in Figure 9.19.

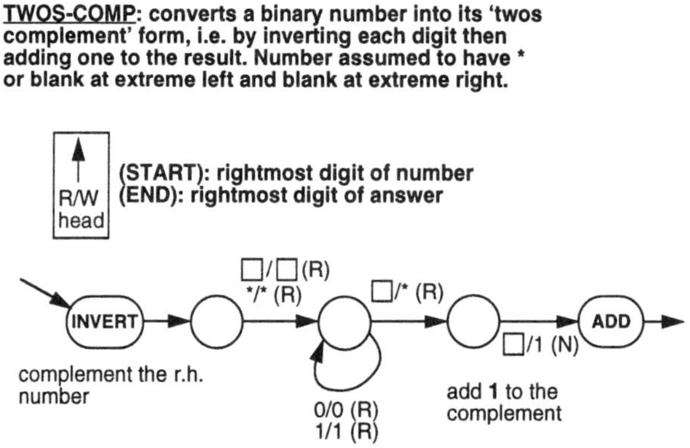

Figure 9.19 The TWOS-COMP TM, to compute the twos complement representation of a binary number

As you can see from Figure 9.19, TWOS-COMP makes use of the TM INVERT, which is defined in Figure 9.20.

You can study the behaviour of TWOS-COMP for yourself, if you wish.

We could set up our SUBTRACT TM so that it uses the same input format as that used for ADD, and shown in Figure 9.21.

We assume that the 'second number' of Figure 9.21 is to be subtracted from the 'first number'.

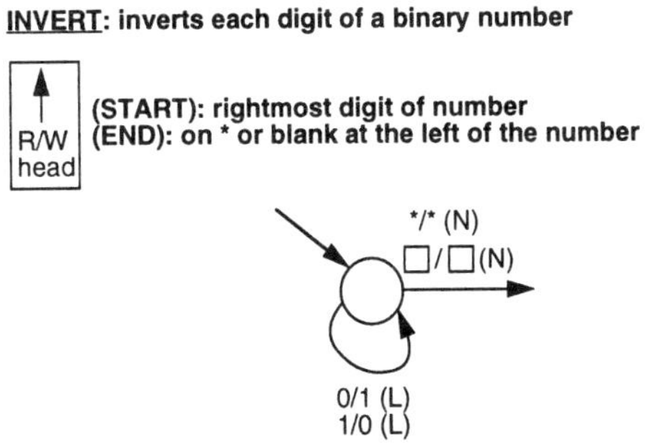

Figure 9.20 The INVERT TM, used by TWOS-COMP (Figure 9.19)

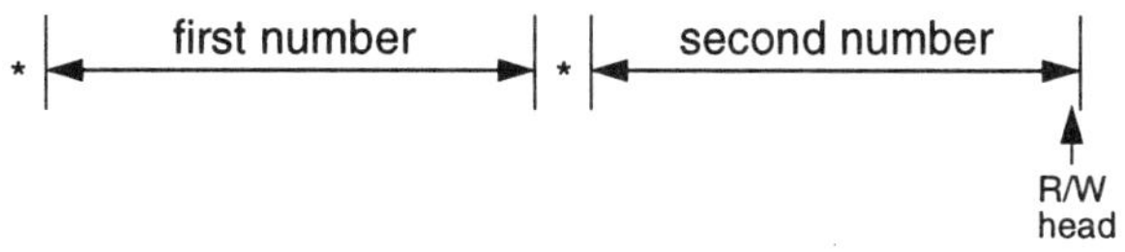

Figure 9.21 The tape set-up for the SUBTRACT TM

The exercises ask you to complete the SUBTRACT machine. This simply involves using the ADD machine to add the twos complement of one number to the other number. The main problems, as for the MULT machine, concern shuffling the various intermediate sequences around to ensure that essential information is not corrupted; but it is not too difficult to achieve this.

9.3.2 The 'DIV' TM

We now sketch a TM for performing integer division by repeated subtraction. Again, its implementation is left as an exercise. An algorithm to carry out the task is specified in Table 9.2.

Table 9.2 A 'repeated subtraction' algorithm for division, that makes use of TMs defined earlier

```
count := 0;
temp := n1;

while n2 ≤ temp do
    temp := SUBTRACT(temp,n2);     { T(x,y) is a functional representation of the
                                     statement: 'apply the TM, T, to the tape
                                     appropriately configured with the numbers x
                                     and y' }
    count := ADD(count,1)
endwhile;

count is the result
```

As for the MULT and SUBTRACT TMs, the main difficulty in implementing the DIV TM is in ensuring that the sub-machines are 'called' properly, and do not accidentally access inappropriate parts of the marked tapes. This needs careful consideration of where each sub-machine expects asterisks or blanks as delimiters, and which parts of the tape a sub-machine erases after completing its computation. One way is to use different areas of the tape (separated by one or more blanks) for different purposes. For example, you might represent the *count* variable from the algorithm in one place, the original $n1$ and $n2$ in another place. Sometimes, you may have to shuffle up one marked area to make room for the result from another marked area. However, as the main operations that increase the tape squares used are *copying* and *adding*, your machine can always be sure to create enough room between marked areas to accommodate any results (for example, if we add two binary numbers we know that the length of the result cannot exceed the length of the longer of the two numbers by more than one digit). All this *shuffling* can be quite tricky, and it is more important to convince yourself that it can be done, rather than to actually do it.

You will notice that the division algorithm in Table 9.2 includes the ≤ test. As yet we have not designed a TM to perform such logical comparisons. Therefore, in the next section we develop a generalized COMPARE TM. COMPARE may also be of use to the SUBTRACT machine (to detect if the number being subtracted is larger than the number from which it is being subtracted).

9.4 LOGICAL OPERATIONS

The generalized comparison TM, COMPARE tells us whether two arbitrary length binary numbers are *equal*, or whether one number is *greater than*, or *less than* the first number. COMPARE's **input configuration** will follow that for ADD, MULT and SUBTRACT, and is shown in Figure 9.22.

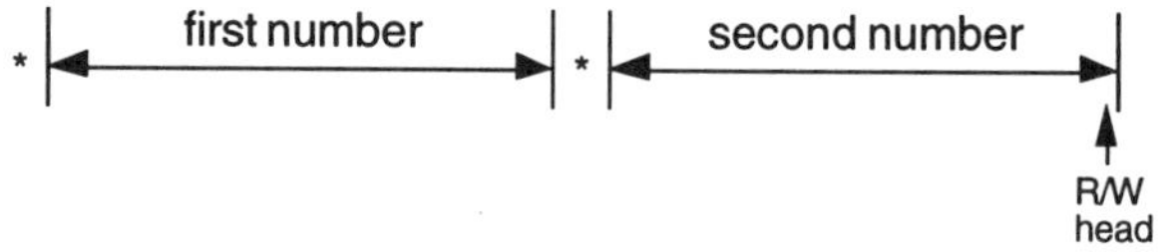

Figure 9.22 The tape set-up for the COMPARE TM

COMPARE will write the symbol =, <, or > in place of the * between the two numbers in Figure 9.22 according to whether the 'second number' is equal to, less than, or greater than the 'first number', respectively. Thus, for example, given the example input tape configuration shown in Figure 9.23, COMPARE will produce an **output configuration** as shown in Figure 9.24.

The COMPARE TM is shown in Figures 9.25 (Part 1) and 9.26 (Part 2). Part 1 deals with the comparison until all the digits of one number have been processed.

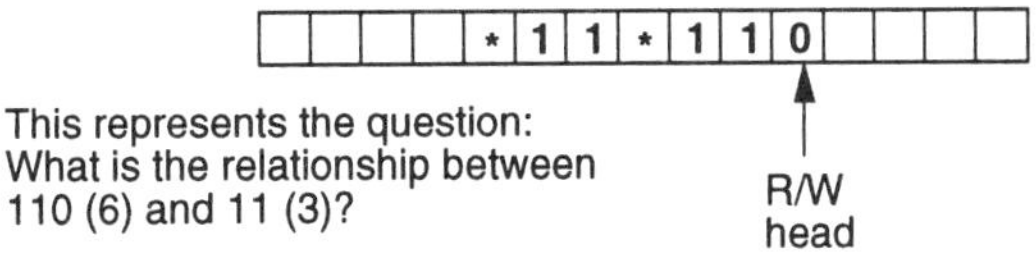

Figure 9.23 An example input tape for the COMPARE TM

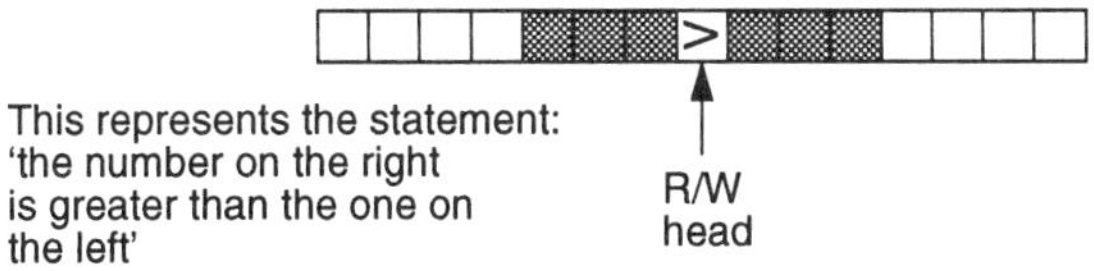

Figure 9.24 Suitable output from COMPARE, for the input tape from Figure 9.23

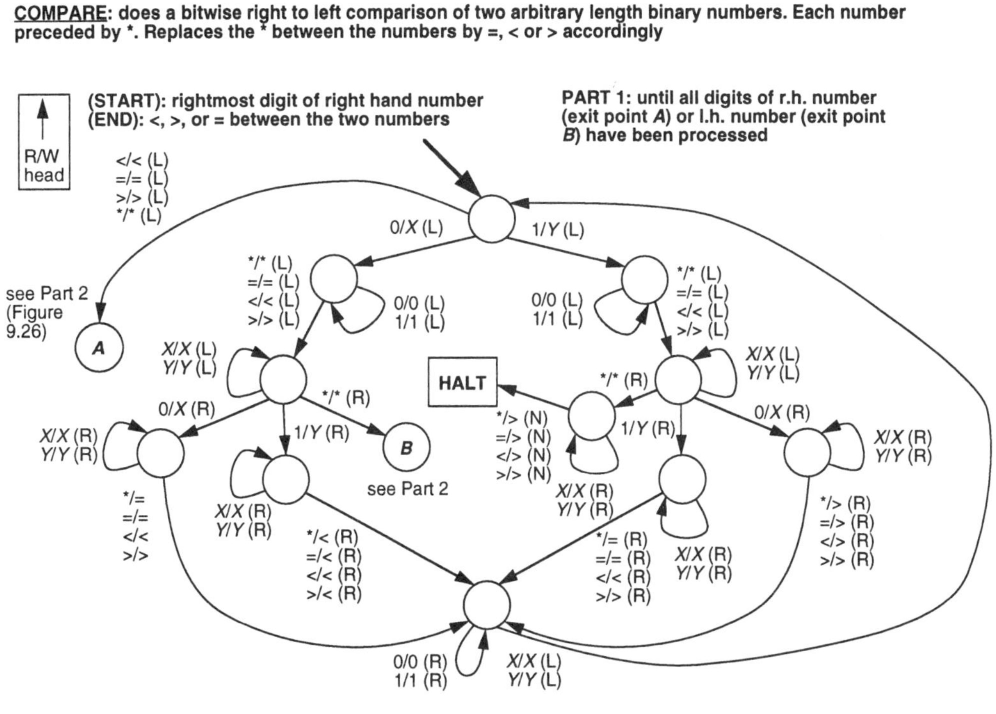

Figure 9.25 The COMPARE TM (Part 1)

When all the digits of one of the numbers have been processed, Part 2 of COMPARE (Figure 9.26) takes over.

I leave it to you to follow the behaviour of COMPARE on some examples, and to convince yourself that it operates as claimed. As part of the DIV machine (see the exercises), you could use COMPARE to carry out the test to determine when the repeated subtractions should stop.

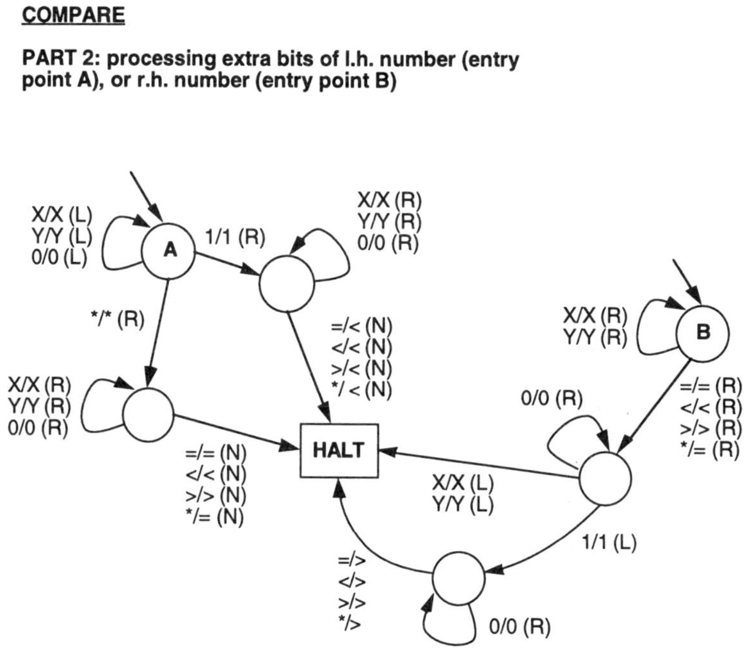

Figure 9.26 The COMPARE TM (Part 2), dealing with the rest of the longer of the two input numbers

9.5 TMS AND THE SIMULATION OF COMPUTER OPERATIONS

TMs are extremely primitive machines, with a one-dimensional data storage device. It thus may seem strange to claim, as we did at the beginning of this chapter, that TMs are actually as powerful as, and in some senses more powerful than, *any* digital computer can be. To provide evidence for this claim, we have seen TMs that do *multiplication* and *division* of *arbitrary length binary numbers*. Such operations are beyond the capabilities of any machine with a fixed amount of storage, such as any real computer at a given point in time. We have also seen how *logical operations* such as comparison of (again, arbitrarily large) numbers can be carried out by our TMs.

In this chapter, we considered only the *binary* number system. You should not see this as a limitation, however, as your experience with computers tells you that ultimately, everything that is dealt with by the computer, including the most complex program, and the most convoluted data structure, is represented as binary codes. Even the syntactic rules (i.e. grammar) of our most extensive programming languages are also ultimately represented (in the compiler) as binary codes, as is the compiler itself.

In the next section, we take our first look at **Turing's thesis**, concerning the computational power of TMs. Before this, we consider more evidence supporting the claim that TMs are as powerful as any computers by briefly mapping out how many internal functions performed by, and many representations contained within, the digital computer can be modelled by a TM. This is not an exhaustive map, but a sketch of the forms which the TM implementation of computer operations could take. I leave it to you to convince yourself of the general applicability of TMs to the modelling of any other operations found in the modern computer.

Tables 9.3–9.6 show, for each type of internal operation, a number of examples, and for each example, either a TM from this chapter that can model that operation, or a suggestion as to how that operation may be modelled by a TM. In each case, certain limitations are given that would apply to any digital computer at any given moment, but do not apply to TMs.

Table 9.3 suggests how *arithmetic operations* can be modelled by TMs. In Table 9.4, logical operations are considered. *Memory and array representation and accessing* feature in Table 9.5, while Table 9.6 briefly considers the TM *simulation of program execution* (a subject which is given much more attention in Chapters 10 and 11).

It is worth focusing briefly on memory and array operations (Table 9.5), as these did not feature earlier in this chapter. In the examples given in Table 9.5 we have used memory structures where each element was a single binary number. This is essentially the memory organization of the modern digital computer. I leave it to you to appreciate how complex data items (such as *records*, etc.) could also be incorporated into the scheme.

9.6 TURING'S THESIS

We close this chapter by briefly looking at what has been called **Turing's thesis**. What you have to appreciate about a *thesis* is that it is not a *theorem*. A *theorem* is a statement that can be proved to be true by a step by step argument that follows valid logical rules, just as the **repeat state theorem** for finite state recognizers was 'proved' in Chapter 6. A *thesis*, on the other hand, is something that somebody asserts, for some reason, as being true. In other words, anybody can state any assertion is a thesis, but it serves little use as a thesis if someone else immediately shows that it cannot be true.

However, Turing's thesis, which was initially posited in the 1930s, has yet to be disproved. In fact, not only has it not been disproved, but related work carried out by other mathematicians has led to essentially the same conclusions as those reached by Turing himself[2]. For example, an American mathematician, Alonzo Church, working at around the same time as Turing, described the theory of recursive functions. Similarly, Emile Post, also a contemporary of Turing and Church, and like Church, an

[2] References to the research in this section can be found in the *Further Reading* section.

Table 9.3 TM simulation of arithmetic operations

Example operations	TM implementation	Computer limitations
Division	DIV	Size of operands/ size of result
Multiplication	MULT	"
Addition	ADD	"
Subtraction	SUBTRACT	"
Power (x^y)	result := 1 {write a 1 in the result area of tape} while $y \neq 0$ {COMPARE(y,0) in another area of the tape} result := result $\times$ x {MULT(result,x) in another area of tape} $y := y - 1$ {SUB(y,1) in another area of tape} endwhile;	"

Table 9.4 TM simulation of logical operations

Example operations	TM implementation	Computer limitations
<, >, =	COMPARE	length of operands
≥, ≤, ≠	e.g. ≠ 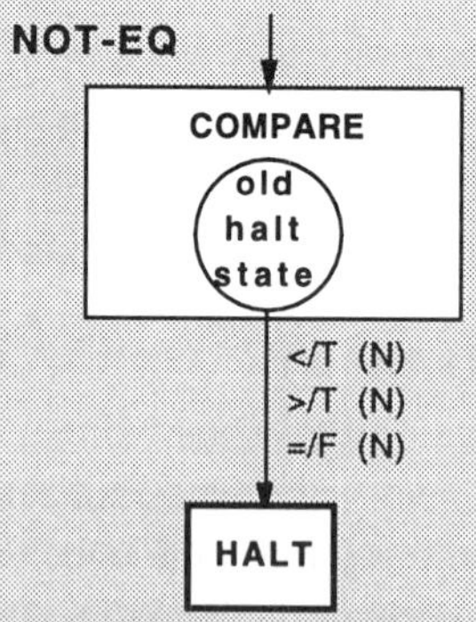 	"
bitwise not	INVERT	length of operand

Table 9.5 TM simulation of memory/array structure and access operations

Example operation	TM implementation	Computer limitations
1-D ARRAY	ACCESSING THE MEMORY LOCATION 'address' 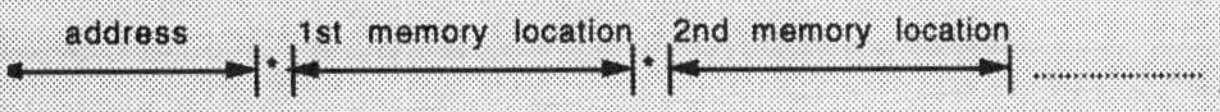 Start: Proceed right along tape, until reach a * or a blank: if a * was found, change it to X, go back and ADD 1 to a count (initially set to zero, and somewhere to the left of address); if a blank was reached address was too high, so halt. COMPARE count to address: if they are equal, go right to the first * at the right of address and copy the number that is to the left to the leftmost end of the marked tape, set all Xs back to *s, and halt; else find rightmost X, and goto Start	word length address space memory/ array size
INDIRECT ADDRESS	As above, but use found value to replace address then start again.	
MULTI-DIMENS-IONAL ARRAYS	EXAMPLE: THREE DIMENSIONAL ARRAY STRUCTURE 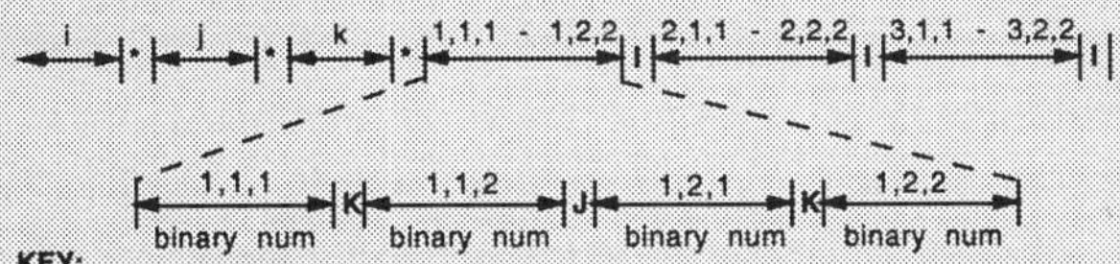 KEY: i, j, k: ADDRESS – 'access the kth element of the jth array in the array' K: marker for increment of k subscript J: marker for increment of j subscript, and to set k subscript to 1 I: marker for increment of I subscript, and to set both j and k subscripts to 1	as for 1-D array above, plus possible limit to number of dimensions in an array

Table 9.6 TM simulation of program execution

Example operations	TM implementation	Computer limitations
Running a program	Use binary numbers to represent codes of operations such as ADD, MULT, memory access, and so on. Set up data and memory appropriately on the tape. Then a portion of tape can be used to represent a binary coded version of the program, etc.	size of program amount of data number of operations

American, described in 1936 a class of abstract machines consisting of an infinite tape on which an abstract problem-solving mechanism could either place or erase a mark, and also described the result as being the marked portion of the tape on completion of the process. All these formalisms, and later ones such as *Wang's machines* (like TMs but which have only the alphabet blank and *), and the *register machines* of Sheperdson and Sturgis (closely resembling a simple assembler code-based system but with registers that can store arbitrarily large integers), have been shown to be:

(a) equivalent in computational power to Turing machines, and therefore,
(b) more powerful than any machine with limited storage resources.

Turing's thesis concerns the computational power of the primitive machines which he defined, and which we now call Turing machines. Essentially, in its broadest sense, Turing's thesis says this:

> Any well defined information processing task can be carried out by some Turing machine.

However, in a mathematical sense, the thesis says:

> Any computation that can be realized as an **effective procedure** can also be realized by some Turing machine.

This last statement of the thesis is probably not particularly helpful, as we have yet to specify what constitutes an effective procedure. Turing was working in an area of pure mathematics which had, for many years, studied the processes involved in 'doing' mathematics itself, i.e. the procedures that were used by mathematicians to actually carry out the calculations and reasoning that mathematicians do. At the time Turing formulated the notion of TMs, the 'computers' that his machines were intended to emulate were thus *humans engaged in mathematical activity*. An effective procedure, then, is a step by step following of a set of unambiguous instructions to carry out a task, *but specified to such a level of detail that they could be executed by a machine*. In this context, as far as we are concerned, Turing's thesis tells us this:

> There is no machine (either real or abstract) that is more powerful, in terms of what it can compute, than the Turing machine.

Therefore:

> *The computational limitations of the Turing machine are the least limitations of any real computer.*

Turing machines are radically different from the computers with which you and I interact in our everyday lives. These real computers have interfaces; they may display detailed graphics, or even animations and video, or play us music, or enable us to speak to, or play games with them. The computer may run an artificial intelligence program, such as an expert system that solves problems that even 20 years ago would have been impossible on anything but the largest mainframe. Yet, in a formal sense, everything that happens on the surface, so to speak, of the computer system, is merely a re-presentation, an alternative encoding, of the results of symbol manipulation procedures that can be modelled by a simple machine that merely repeatedly reads a single symbol, writes a single symbol, and then moves to examine the next symbol to the right or left. Moreover, this simple abstract machine was formulated at a time when 'real computers' were things like mechanical adders and Hollerith card readers.

The Turing machine represents a formalism that allows us to study the nature of computation in a way that is independent of any single manufacturer's architecture, machine languages, high level programming languages or storage characteristics and capacity. TMs are thus a single formalism for application to the study of topics such as the complexity of automated processes and languages.

In terms of languages, we have already seen the close relationship between TMs and the **Chomsky hierarchy**, and we refine this still further in Chapter 11. In the next chapter we study the notion of **complexity** as it applies to TMs. However, an even more exciting prospect is suggested by our statement that the limitations of the TM are the least limitations of any real computer, no matter how powerful that real computer is (and only a real computer to which we could constantly add additional storage as required could approach the power of the TM). A discussion of some of these limitations, and their ramifications in the real world of computing, form a major part of the remaining two chapters of this book.

EXERCISES

1 Define the **TM** TIDY-UP for the ADD machine, above.

2 Complete the definitions of the TMs MULT (using the ADD TM), SUBTRACT and DIV (perhaps using COMPARE).

3 Design a TM that accesses elements of a two dimensional array, following the specification given in Table 9.5.

4† What **function** is computed by the TM in Figure 9.27, assuming that its input tapes are configured as specified in Figure 9.28?

In Figure 9.28, y is an arbitrary string of the form 1^n, $n \geq 1$, representing the integer n, followed by a marker, A. The output is assumed to be the integer represented by m, where m is the number of 1s to the right of the A when the machine halts.

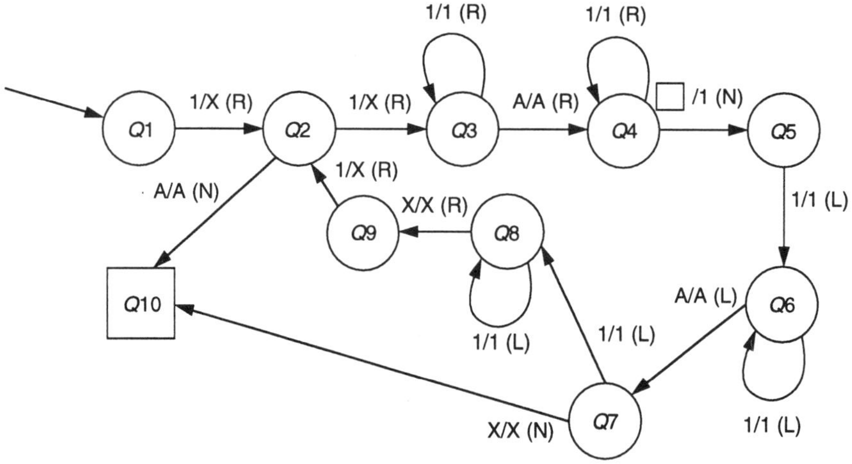

Figure 9.27 What function is computed by this TM?

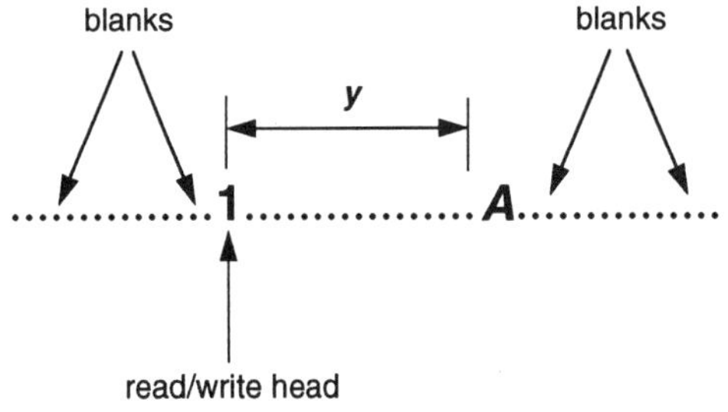

Figure 9.28 An appropriate input configuration for the TM of Figure 9.27

5 The Turing machine, T, is as depicted in Figure 9.29. The Initial tape is of the form shown in Figure 9.30.

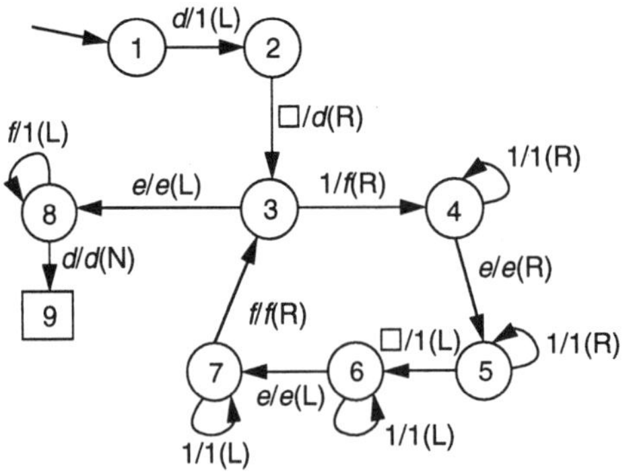

Figure 9.29 A TM for language recognition and computation

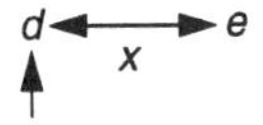

Figure 9.30 A specification of the input configuration for the TM of Figure 9.29

Referring to Figure 9.30, note that:

d, *e*, and *f* are alphabet symbols;
x is a (possibly empty) sequence of 1s;
the arrow indicates the initial position of read/write head.

(a)† Briefly describe the relationship between the tape on entering the **loop** of states 3–7, and the tape when all iterations through the loop have been completed.

(b)† If the **arc** between state 3 and state 8 labelled '*e*/*e* (*L*)' was altered to '*e*/1 (*L*)', what function, expressed in terms of *x*, would the amended machine compute, assuming that $|x|$ (i.e., the number of 1s on the initial tape) represented an integer, and that the result was the number of 1s to the right of *d* when the machine halted?

(c) Delete state 1 from *T*, and by amending arcs as necessary, draw a machine, T_1, which **accepts** the language $\{d1^ne1^n \mid n \geq 0\}$. State 2 of *T* should become the start state of T_1 and state 9 of *T* its halt state. Additional alphabet symbols may be required, but you should introduce no additional arcs.

10 Dimensions of computation

AIMS

To learn about

- function (what is computed), space (how much storage computation takes) and time (how long computation takes)
- the addition of facilities and how they do not increase the computational power of the TMs
- the non-deterministic TM as a parallel computer
- whether a non deterministic TM can be replaced by an equivalent deterministic TM that does the job as efficiently

10.1 ASPECTS OF COMPUTATION: SPACE, TIME AND COMPLEXITY

The last chapter ended by arguing that the limitations, in terms of computational power, of the **Turing Machine (TM)** are also the limitations of the modern digital computer. We argued that, simple as it is, the TM is actually more powerful than any real computers unless we allow for an arbitrary amount of storage to be added to the real computer, as required, during the course of its activity. Even then, the computer would not be *more* powerful than the TM. However, claims about the power of the TM must be appreciated in terms of the TM being an **abstract machine**. In Chapter 8, when discussing the **FST**, we presented so called memory machines, that 'remember' an input symbol for a fixed number of state transitions before that symbol is output. In a sense, by doing this we were using an implicit model of *time*, and the basic assumption of that model was that the **transition** between two states of a machine took exactly one unit of this time, which we will now call a **moment**.

Turing's thesis, as discussed in Chapter 9, reflects a functional perspective on computation: it focuses on *what* the TM can perform. However, as users and programmers of computers, we are not only interested in the functions our computer can carry out. At least two other fundamental questions concern us, these being:

- how much *memory* and *storage* will our task require (**space** requirements), and
- *how long will it take* for our task to be completed (**time** requirements)?

In other words, the dimensions of computations that occupy us are **function**, space and time. It is clear then that reference to *computational power*, a term used freely in the last chapter, should be qualified in terms of those dimensions of computational power to which reference is being made.

Now, in terms of space requirements, our binary TMs of Chapter 9 seem to require an amount of memory (i.e. occupied **tape** squares) that does not differ significantly from that required by a computer carrying out a similar task on a similar set of data. For example, consider the descriptions of the memory accessing TMs towards the end of Chapter 9. Apart from the binary numbers being processed, the only extra symbols that our machines required were special markers[1] to separate one number from the next. However, these markers were needed because we had implicitly allowed for variable memory 'word' (or array element) length. If we had introduced a restriction on word length, as is the case in most memory schemes, we could have dispensed with the need for such markers. However, if we had dispensed with the markers between memory items, we would have had to 'program in' the knowledge of where the boundaries between items occurred, which in terms of TMs means we would have had to introduce more states to deal with this.

[1] The use of the word 'marker', as used throughout this chapter, refers to the way in which the TM uses a symbol. Such a 'marker' symbol is simply a symbol of the TM's alphabet.

The discussion above seems to imply that *we can reduce the number of states in a (well-designed) TM at the expense of introducing extra alphabet symbols, and we can reduce the number of alphabet symbols at the expense of introducing extra states*. In fact, this relationship has been quantified, and it has been shown that

> any TM that has more than two states can be replaced by an **equivalent** TM that has only two states.

More importantly, from our point of view, it has been shown by the information theorist Claude Shannon that

> any TM that uses an alphabet of more than two symbols can be replaced by an equivalent TM that uses only two.

The fact that any TM can be replaced by an equivalent TM using only a binary alphabet is of fundamental importance to the practical use of computers as it indicates that any shortfalls in functional power that real computers have is certainly not a consequence of the fact that they only use a two-symbol alphabet.

Due to the reciprocal relationship outlined above between the number of states in a TM and the number of alphabet symbols it needs, the product of the number of states and the number of alphabet symbols is often used as a measure of the **complexity** of a TM. However, there are other aspects of TMs that could be said to relate to their complexity. One of these aspects is **non determinism**, which, with respect to TMs we encountered in Chapter 7. We have already seen that non determinism is computationally expensive, and therefore we would like to do without it, if possible. For the finite state recognizers, it is possible to dispense with non determinism, but for the **pushdown recognizers** it is not always possible to do so. As we will discover, non determinism is not necessary in the case of the Turing machine. However, as we see later in the chapter, there are implications for the *space* and *time* dimensions of computation if we remove it.

10.2 NON DETERMINISTIC, DETERMINISTIC AND MULTI-TAPE TMs

In Chapter 7, we saw that the non deterministic TM was the abstract machine counterpart of the **type 0**, or **unrestricted grammars**, the most general classification of grammars of the **Chomsky hierarchy**. We saw a method for constructing non deterministic TMs from the productions of type 0 grammars. Now we return to the notion of non deterministic TMs to demonstrate that non determinism in TMs provides us with no extra functional power (although we see later in the chapter that things change if we also consider the time dimension of computation). This result has implications for *parallel computation*, to which we return later in this chapter. Only a sketch will be given of the method for converting any non deterministic TM into an equivalent deterministic TM. However, some of the techniques introduced are fundamental to discussions later in the chapter.

10.2.1 Quintuples

Hitherto, this book has used a pictorial representation of TMs. However, many books featuring TMs also use a representation known as **quintuples**, where each quintuple represents one arc in a TM, and is written as a five element list as follows:

(<current state>, <read symbol>, <write symbol>, <move>, <new state>).

Thus, the pictorial TM on the left of Figure 10.1 is represented by the quintuples shown on the right of the figure.

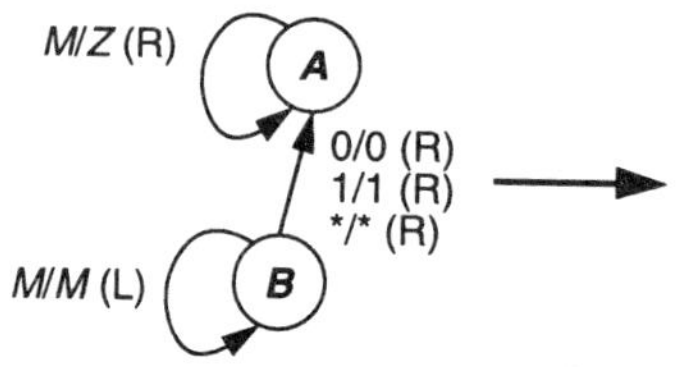

(*A*, *M*, *Z*, *R*, *A*)

(*B*, 0, 0, *R*, *A*)

(*B*, 1, 1, *R*, *A*)

(*B*, *, *, *R*, *A*)

(*B*, *M*, *M*, *L*, *B*)

Figure 10.1 Diagrammatic representation as quintuples (right)

10.2.2 The problem solving behaviour of the non deterministic TM

If a TM is non deterministic, and it correctly computes some function (such as **accepting** the strings of a language), that TM will be capable of reaching its **halt state**, leaving some kind of result on the tape. For a non deterministic machine this is alternatively expressed as 'there is some sequence of transitions that would result in the machine reaching its halt state from one of its start states'. In terms of non deterministic TMs the phrase:

'there is some sequence of *transitions*'

can now be stated as:

'there is some sequence of *quintuples*'.

Suppose, then, that we have a non deterministic TM that attempts to solve some problem, or produce some result. If there is a result, and it can try all possible applicable sequences of quintuples *systematically* (we will return to this in a moment), then it will find the result. What we are going to discover is that there is a deterministic method for trying out every possible sequence of quintuples (shorter sequences first), and stopping when (or rather *if*) a sequence of quintuples is found that leads to a halt state of the original machine. Moreover, it will be argued that the method described could be applied by a deterministic TM. The construction described represents an exciting development, because it is our first example where one TM models the behaviour of another TM by executing its quintuples as if they were a 'program'. The possibility of this was briefly alluded to towards the end of Chapter 9.

In order to simplify the description of our deterministic version of a non deterministic TM, we shall assume that we have a special TM with four tapes. Then, in the next section, we shall show that any deterministic TM with more than one tape can be replaced by an equivalent deterministic machine with only one tape!

10.2.3 *Multi-tape TMs*

A four-tape TM is a special case of the notion of a **multi-tape TM**, this being a TM with more than one tape, and a corresponding **read/write head** for each tape. The tapes of a k-tape TM are numbered 1, 2, .., k. On each of the arcs of a multi-tape TM we find an extra designator (an integer between 1 and k) that specifies the tape on which the next operation is to occur. The machine is assumed to begin its operation on tape 1, and each time processing resumes on a given tape, the read/write head for that tape is assumed to be located where it was when that tape was last used. Multi-tape TMs are represented as **sextuples** rather than quintuples, as demonstrated by the fragment of a two-tape machine shown in Table 10.1. The table shows a diagrammatic representation, the sextuples, and a description of the behaviour specified by the machine fragment. Assume that in Table 10.1 the machine is currently in state A, and using tape 1.

Table 10.1 A k-tape TM fragment: diagram, sextuples, behaviour

Diagram	Sextuples	Behaviour
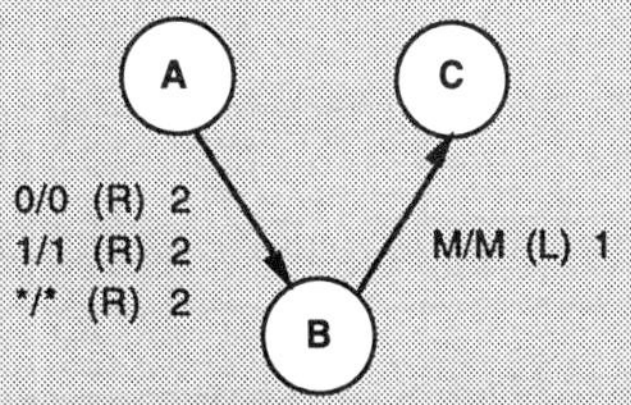	$(A, 0, 0, R, B, 2)$ $(A, 1, 1, R, B, 2)$ $(A, *, *, R, B, 2)$	From state A: if the symbol at the read/write head on tape 1 is a 0, a 1 or a *, replace it by the same symbol, move right on tape 1, then switch to tape 2, and enter state B.
	$(B, M, M, L, C, 1)$	From state B: if the symbol at the read/write head on tape 2 is M, replace it by M, move left on tape 2, then switch to tape 1, and enter state C.

10.3 CONVERTING A NON DETERMINISTIC TM INTO A FOUR-TAPE DETERMINISTIC TM

We now describe the conversion of a non deterministic TM into a four-tape deterministic TM. Let us call the **N**on deterministic TM *N*, and the corresponding **D**eterministic 4-tape machine, *D*.

10.3.1 The four tapes of the deterministic machine, D

We begin by describing the use of *D*'s four tapes, beginning with tape 1, which is shown in Figure 10.2.

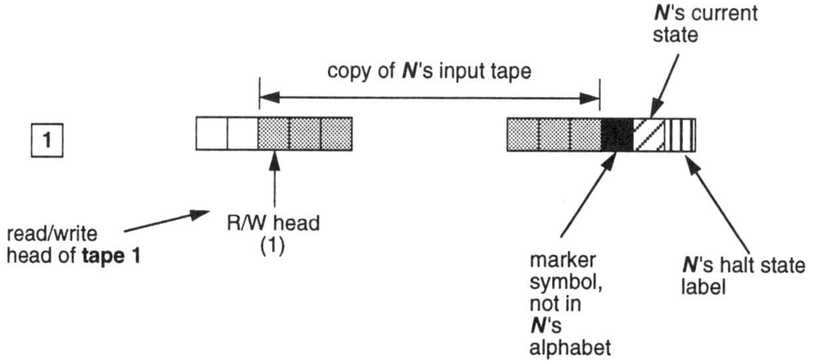

Figure 10.2 Tape 1 of the four-tape deterministic TM, *D*

As shown by Figure 10.2, tape 1 of *D* stores:

- a copy of the *marked portion of the tape that N would have started with* (we can assume for simplicity that *N*'s computation would have begun with its read/write head on the leftmost symbol of its input);
- *the label of N's current state*, initially marked with *N*'s start state label (let us assume that the start state is always labelled *S*);
- *the label of N's halt state* (we can assume that *N* has only one halt state: cf. exercises, Chapter 7), which in Figure 10.2 is *H*.

Note that we are making the assumption that labels such as state names and alphabet symbols each occupy one square of a tape, i.e. are *single symbols*. In the formal world, there is no concept of a 'character set' in which we 'run out' of single symbol labels: there are a potentially infinite number of available single symbol labels. Thus, *a TM of any size can be assumed to require state and alphabet labels each consisting of one symbol.*

Now we can describe tape 2, schematically represented in Figure 10.3.

Figure 10.3 tells us that tape 2 of *D*, the deterministic four-tape TM, contains the quintuples of *N*, each occupying five consecutive tape squares, each preceded by a unique single symbol that is not a symbol of *N*'s alphabet. To illustrate, Figure 10.4 shows a quintuple from the TM fragment in Figure 10.1, along with its corresponding representation according to the scheme specified in Figure 10.3.

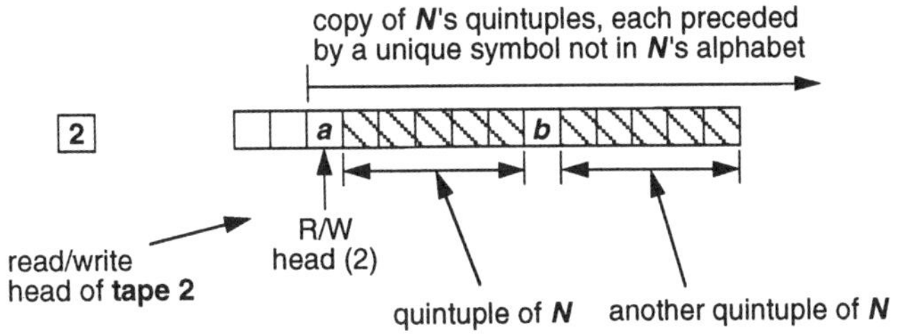

Figure 10.3 Tape 2 of the four-tape deterministic TM, *D*

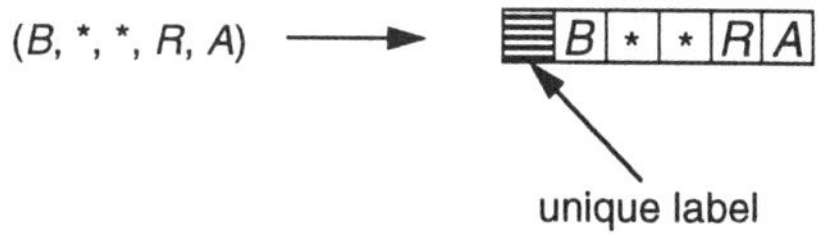

Figure 10.4 The representation of a single quintuple of the non-deterministic TM, *N*, on Tape 2 of *D* (Figure 10.3)

Next, tape 3 of *D*. Tape 3 is where *D* will repeatedly perform the systematic application of *N*'s quintuples, starting from a copy of *N*'s input tape (from tape 1, specified in Figure 10.2). *D*'s first activity, then, is to set up tape 3 as shown in Figure 10.5.

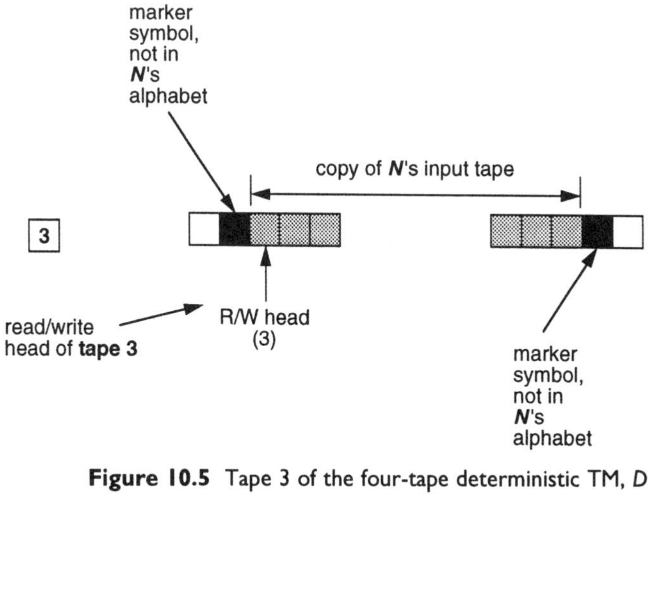

Figure 10.5 Tape 3 of the four-tape deterministic TM, *D*

Note that D places special 'markers' at either end of the copied input. These are necessary, as after the application of a sequence of N's quintuples that leads to a dead end, D will have to erase the marked portion of tape 3, and copy the original input from tape 1 again. The potential problem is that N's computation may result in blank squares in between marked symbols, resulting in a tape that is no longer **packed**, so D needs a way of detecting the extreme ends of the marked portion of the tape.

The existence of the markers causes a slight complication in the operation of D. One of N's quintuples may result in D's tape 3 head moving onto the marker. When this happens D must simply move the marker one square to the right or left (depending on whether the marker was encountered on a right or left move, respectively), i.e. replace the marker by a blank, and then overwrite the blank on the right or left, respectively, with the marker. D must also then move the head back one square to where the marker originally was (this corresponds to the position in the tape expected by the next applicable quintuple of N). Every time D makes a left or right move on tape 3, then, it must check if the next symbol is the marker; if so, it must take the appropriate action before attempting to apply the next quintuple of N.

10.3.2 *The systematic generation of the strings of quintuple labels*

The whole process hinges on the statement, made more than once in the above, that D *will apply the quintuples of N in a systematic way*. As N can consist of only a finite number of quintuples, the set of symbols used to label the quintuples of N (on tape 2) is an *alphabet*, and a sequence of symbols representing labels for some sequence of quintuples of N is a string **taken from** that alphabet. So, the set of all possible sequences of quintuples that we may have to try starting from the initial tape given to N is denoted by the set of every possible non-empty string formed using symbols of the alphabet of labels.

Let us call the alphabet of quintuple labels Q. Recalling concepts from Chapter 2, the set of every possible non-empty string formed using symbols of the alphabet Q is denoted by Q^+. Now, Q^+, for any alphabet Q, is not only infinite but also enumerable. Q^+ is a **countably infinite** set. Now, by Turing's thesis, if a program can enumerate Q^+, then a TM can do it[2]. That is exactly what we need as part of D:

> a sub-machine that will generate sequences of quintuple labels, one by one, so that the sequence of N's quintuples denoted by the sequence of symbols can be executed by D, and we are sure that every possible sequence is going to be tried, if it proves necessary to do so.

To be prepared to try every possible sequence of quintuples is a rather brute force approach, of course. Many (in fact probably most) of the strings of labels generated

[2] A pseudo-code program was given in Chapter 2 to perform the same task.

will not denote 'legal' sequences of N's quintuples for the given tape, or for any tape at all. Moreover, the method will try any sequence that has a prefix that has already led nowhere (there are likely to be an infinite number of these, also). However, at least we know that if there *is* a solution that N could have found, our machine D will eventually come across the sequence of quintuples that will model such a solution. We should also point out that if there is not a solution, then D will go on forever trying different sequences of quintuples (but then, if a solution does not exist, N would also be unable to find it!).

We finally come to tape 4 of D. Tape 4 is where D is going to repeatedly generate the next sequence of quintuple labels (strings in the set Q^+) that are going to be tried. Only an overview of one method will be given here. You are invited to construct the actual machine as an exercise.

To demonstrate the method, we will assume an alphabet $\{a, b, c\}$. The approach will be:

generate all strings of length 1:
a, *b*, *c*
generate all strings of length 2:
aa, *ab*, *ac*, *ba*, *bb*, *bc*, *ca*, *cb*, *cc*
generate all strings of length 3:
aaa, *aab*, *aac*, *aba*, *abb*, *abc*, *aca*, *acb*, *acc*, *baa*, *bab*, *bac*, *bba*, *bbb*, *bbc*, *bca*, *bcb*, *bcc*, *caa*, *cab*, *cac*, *cba*, *cbb*, *cbc*, *cca*, *ccb*, *ccc*

.
.
.

You may notice from the order in which the strings are written that we obtain all of the strings of length n by copying each of the strings of length $n-1$ three times (as there are three alphabet symbols) and appending a different symbol of the alphabet to each one. For example, from *aa* we generate *aaa*, *aab*, and *aac*.

Let us assume that the machine initially has the three alphabet symbols somewhere on its tape. To generate the three strings of length 1 it simply copies each symbol to the right, each being followed by a blank, with the final blank followed by some marker (say *) as shown in Figure 10.6.

The blanks following each of the strings in Figure 10.6 can be used both to separate the strings and to enable an auxiliary marker to be placed there, thus allowing the machine to detect the next string to be copied.

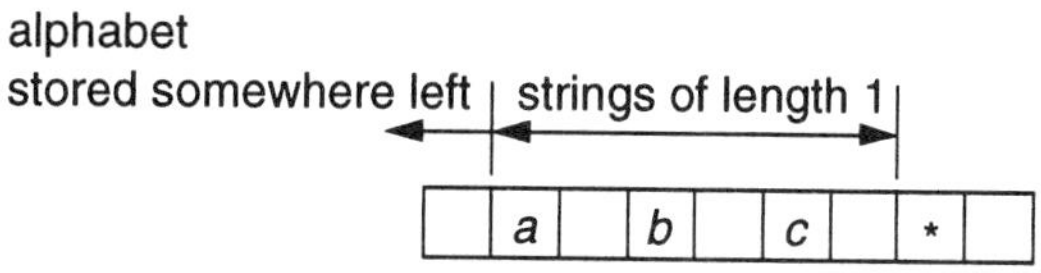

Figure 10.6 Generating all strings in $\{a, b, c\}^+$ of length 1

The next stage, shown in Figure 10.7, involves copying the first string of length 1 (a) three times to the right of the *. The number of times the string is copied can be determined by 'ticking off', one by one, the symbols of the alphabet, which, as shown in Figure 10.6, is stored somewhere to the left of the portion of tape being used to generate the strings.

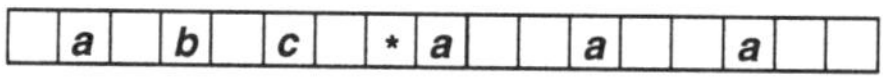

Figure 10.7 The first stage of generating all strings in $\{a, b, c\}^+$ of length 2

From the situation in Figure 10.7, the next stage is to append each symbol of the alphabet in turn to the three copied as, giving the situation in Figure 10.8.

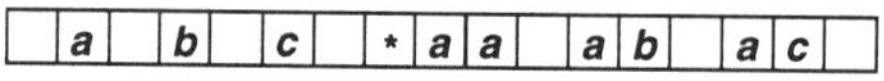

Figure 10.8 The second stage of generating all strings in $\{a, b, c\}^+$ of length 2

The next string of length 1, (b) is then copied three times to the right, as illustrated in Figure 10.9.

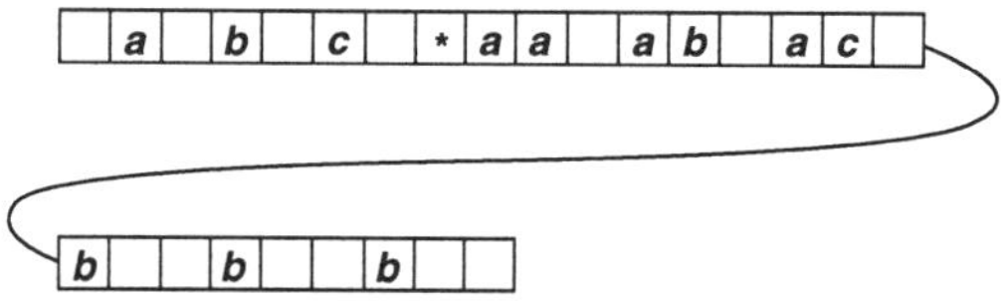

Figure 10.9 The third stage of generating all strings in $\{a, b, c\}^+$ of length 2

From Figure 10.9, each symbol of the alphabet is appended in turn to the strings just copied, yielding the situation in Figure 10.10.

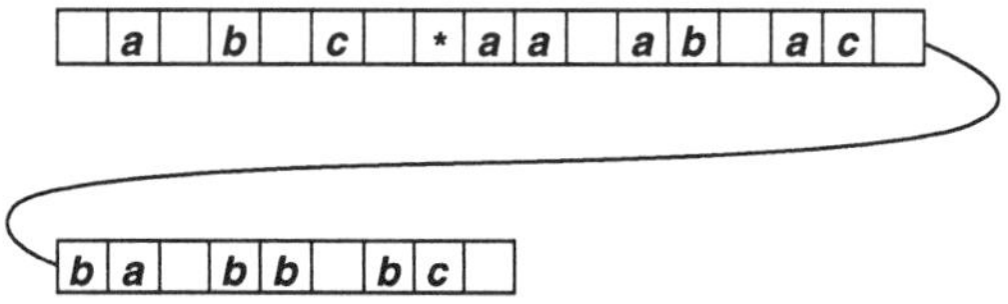

Figure 10.10 The fourth stage of generating all strings in $\{a, b, c\}^+$ of length 2

The next stage from Figure 10.10, is to copy the next string of length 1 (c) three times to the right, leading to the set up shown in Figure 10.11.

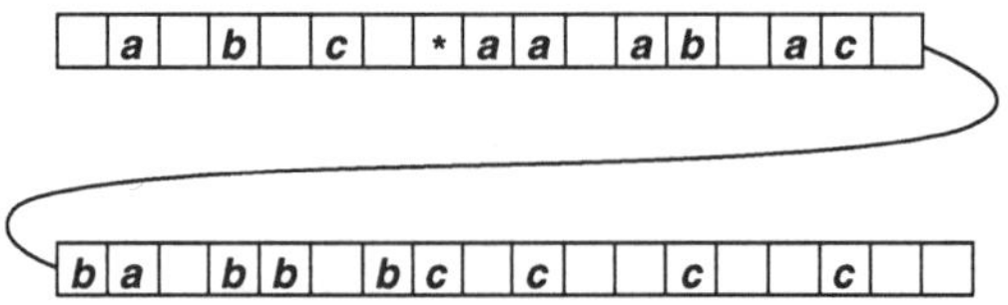

Figure 10.11 The fifth stage of generating all strings in $\{a, b, c\}^+$ of length 2

From Figure 10.11, each symbol of the alphabet is appended to each string that was copied in the previous stage, giving a situation shown in Figure 10.12.

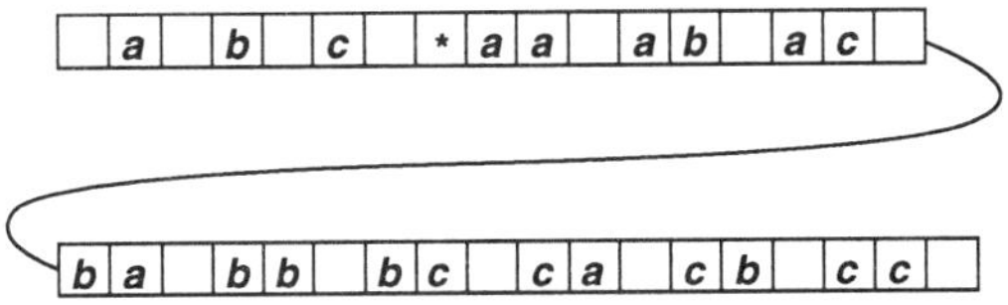

Figure 10.12 The sixth stage of generating all strings in $\{a, b, c\}^+$ of length 2

Now, when the machine returns along the tape, then attempts to copy the next string of length 1, it encounters the *. The machine thus detects that it has finished generating all strings of length 2. The machine can place a * at the end of the strings it created (Figure 10.12), as depicted in Figure 10.13.

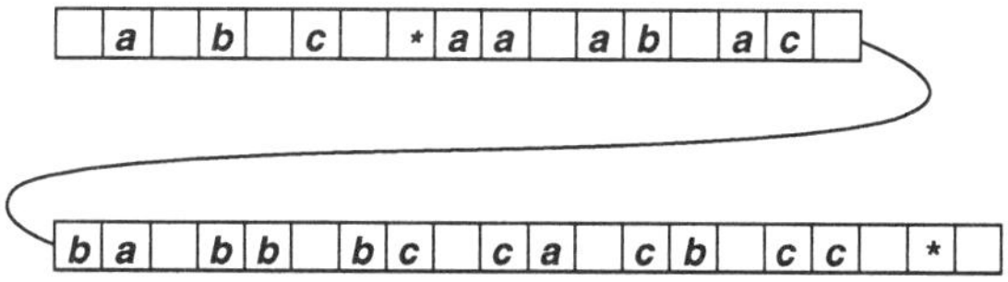

Figure 10.13 The seventh and final stage of generating all strings in $\{a, b, c\}^+$ of length 2

From Figure 10.13, the machine is ready to begin the whole process again, i.e. copying each string between the two *s three times to the right, each time appending each symbol of the alphabet in turn to the newly copied strings, and so on.

Of course, it will be necessary for *D* to try the quintuple sequences each time a certain number of them have been generated (since the generation process will go on forever, otherwise). It could, for example, try each new sequence as it was created, i.e. each time a symbol of the alphabet is appended to a recently copied sequence. Alternatively, it could generate all of the sequences of a given length, and then try each of those sequences in turn.

10.3.3 The operation of D

Here then, is a sketch of the method used by the four-tape TM, *D*, to deterministically model the behaviour of the non deterministic TM, *N*:

D is assumed to be initially configured so:
Tape 1 contains a copy of *N*'s input
Tape 2 contains the labelled quintuples of *N*
Tape 4 contains the quintuple label symbols
} as described above

D operates as shown in Figure 10.14.

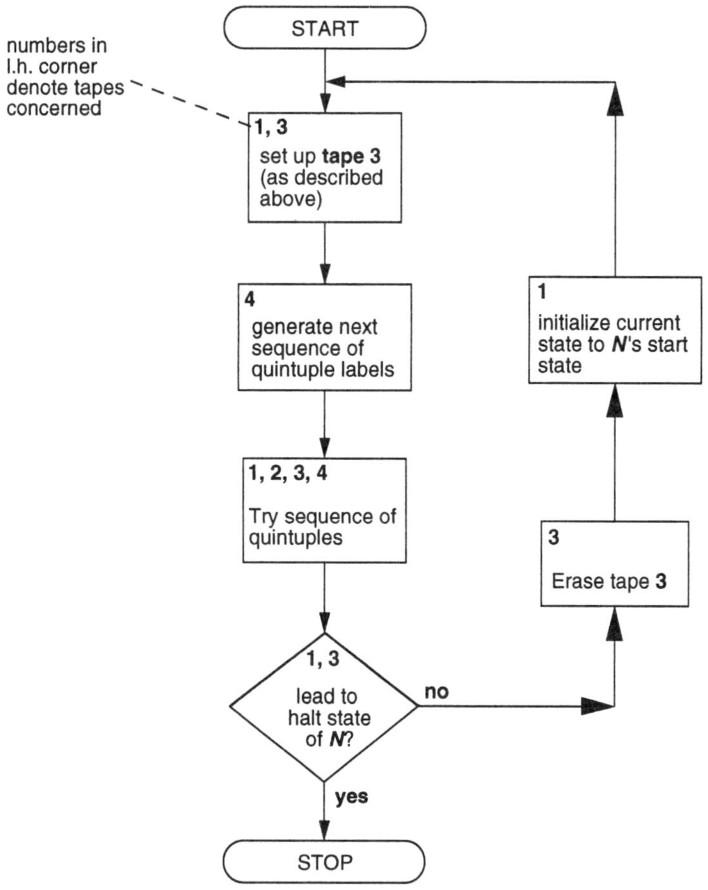

Figure 10.14 The operation of the four-tape deterministic TM, *D*

The most complicated process in the method specified in Figure 10.14 is 'Try sequence of quintuples'. This is expanded in Figure 10.15.

It may not be obvious how the machine *D* carries out the instructions specified by a quintuple of *N*. By way of illustration, suppose the label for the quintuple of *N* (the non deterministic machine) that is next to be tried in the sequence of quintuples is the label for the quintuple.

$(A, 1, 0, R, B)$,

Imagine that *D* has located the quintuple label on tape 2 (the tape on which the quintuples of *N* are represented – see Figure 10.3) as shown in Figure 10.16.

D now has to examine the square on tape 1 that holds the symbol representing the 'current state' of *N* (see Figure 10.2 for the organization of tape 1). The quintuple $(A, 1, 0, R, B)$ will be applicable if the current state symbol on tape 1 is *A*.

Suppose that tape 1 is configured as in Figure 10.17.

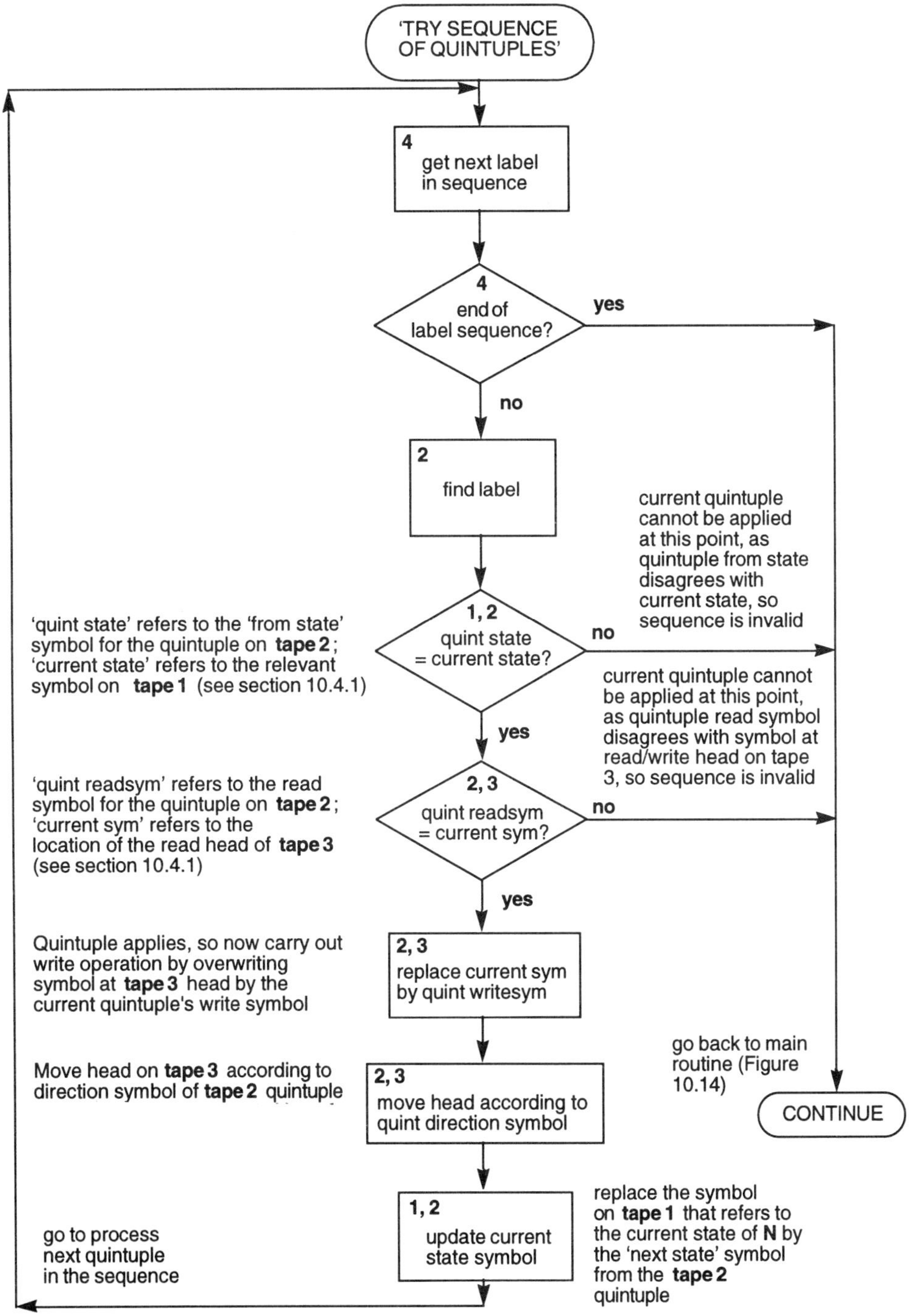

Figure 10.15 How *D* tries the sequence of *N*'s quintuples

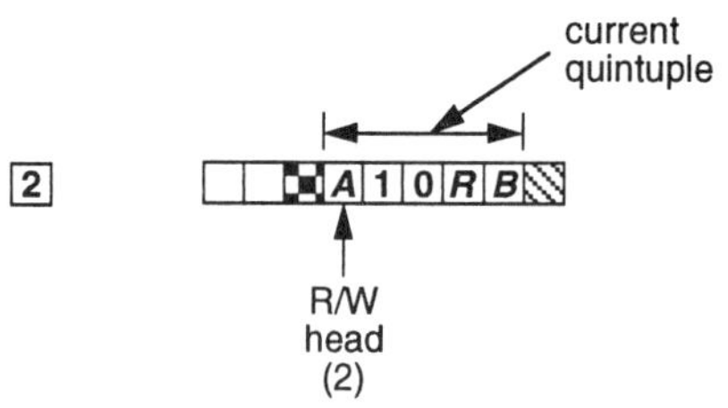

Figure 10.16 Tape 2 of D indicating that the current quintuple of N is '(A, 1, 0, R, B)'

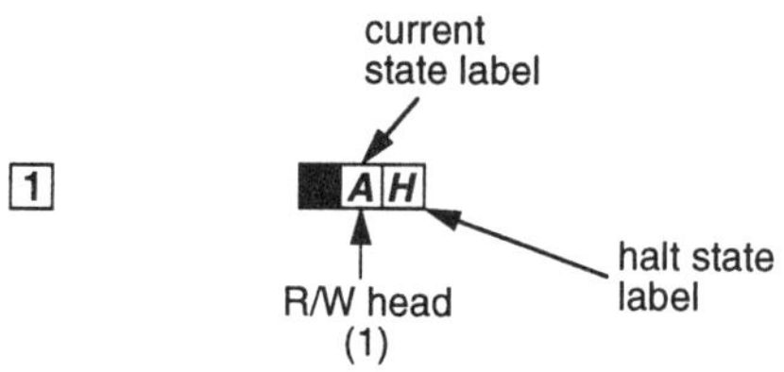

Figure 10.17 Tape 1 of D indicating that the current state of N is A

Figure 10.18 Tape 3 of D indicating that the current read symbol of N is 1

Now, D must examine the symbol at the r/w head of tape 3, since this tape represents the current state of N's tape at the stage in the current sequence of quintuples that D is applying (see Figure 10.5).

Let us suppose that tape 3 of D is as shown in Figure 10.18.

As the current configuration of tape 3 in Figure 10.18 indicates, the quintuple (A, 1, 0, R, B) is indeed applicable. D must therefore:

1. replace the 1 on tape 3 by 0 (Figure 10.18)
2. move the r/w head of tape 3 right (Figure 10.16)
3. overwrite the A on tape 1 with a B, indicating that the current state of N in the simulated computation would be B (Figure 10.17)
4. get the next quintuple label symbol, if one exists, find the quintuple associated with that label on tape 2 (Figure 10.16), and repeat the process above for that quintuple, and so on.

The exercises ask you to sketch out a sub-machine that would use the four-tape set up to simulate the application of this quintuple. From the exercise, you should be able to appreciate that the construction could be applied to any similar situation.

10.3.4 The equivalence of non deterministic, and four-tape deterministic, TMs

This section has hopefully convinced you that:

> any non deterministic TM, *N*, can be represented by a four-tape deterministic TM, *D*, in such a way that *D* and *N* are equivalent, in the sense that, given any input tape to *N*, and assuming that *D*'s tapes are initially configured as described above, if *N* could have found a solution then so will *D*, and if *N* would never find a solution then neither will *D*.

However, multi-tape TMs are not the 'standard' TMs that featured in the previous chapter. It would be problematic if, to solve certain problems, or to compute certain functions, we would have to introduce extra tapes. For this would mean we would have to discover which type of problem required, say a three-tape machine, and which other type of problem required, say, only a two-tape machine, and so on. This would not be as useful as having a single, simple formalism which we knew was capable of carrying out any of our computational tasks. However, we see in the next section that the conventional single tape, deterministic TM is that single, simple formalism.

10.4 CONVERTING A MULTI-TAPE TM INTO A SINGLE-TAPE TM

A multi-tape TM, as described above, is a TM with two or more tapes, each of which has its own read/write head. The instructions on the arcs of such TMs contain an extra designator, this being the number of the tape on which the next operation is to take place. Any such machine can be modelled by a single-tape TM, as will now be demonstrated.

Let us call the multi-tape machine *M*, and resulting single-tape machine *S*. The conversion method we consider here consists of representing the symbols from *M*'s many input tapes on *S*'s single tape in such a way that a **sextuple** of *M* can be represented as several quintuples of *S*. We will use an example involving the conversion of a three-tape machine into a single tape machine to demonstrate the method.

10.4.1 Example: representing three tapes as one

Let us suppose we have a three-tape machine, *M*, with its tapes initially set up as in Figure 10.19.

We use a scheme that places the symbols from the three tapes of *M* on the single tape of *S*. From the three tapes above, we present the tape to *S* that is shown in Figure 10.20.

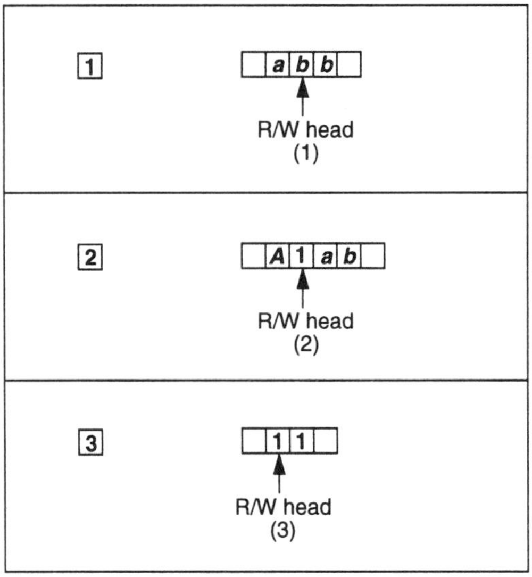

Figure 10.19 The three tapes of a three-tape machine, *M*

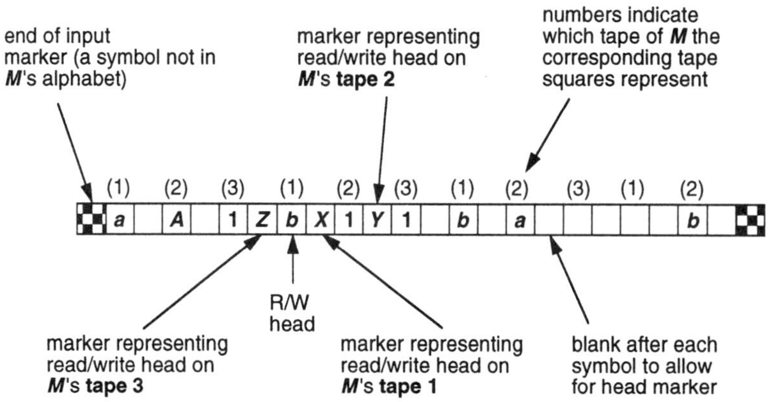

Figure 10.20 How the three tapes of Figure 10.19 could be represented on a single tape

Thus, from left to right, and taking each tape of *M* in turn, we place the leftmost symbol from tape 1, followed by a blank (or a special marker if this symbol was pointed at by the read/write head of tape 1), followed by the leftmost symbol from tape 2, followed by a blank (or a special marker, if necessary, representing the read/write head of tape 2), and so on. It will be noted that tape 3 had the least marked squares, so pairs of blanks are left where tape 3 symbols would have been found on *S*'s tape. Likewise for tape 1, where two blanks were left so that the final symbol of tape 2, the tape with the most marked squares, could be placed on the single tape.

The end of input tape markers, right and left, and the three read/write head markers (*X*, *Y*, and *Z*, for tapes 1 – 3, respectively) are required to be symbols that are *not* in the alphabet of *M*.

The encoding of the three tapes of *M* as a single tape of *S* represented in Figure 10.20 captures all of the information from the original three tapes. For example, consider the second tape square marked (1) from the left. This contains a *b*, which tells us the second tape square from the left of tape 1 of *M* should also contain a *b* (as it does). Moreover, the square to the right of the *b* in question, on *S*'s tape, contains *X*, the read/write head marker representing the head position on tape 1 of *M*. This tells us that the read/write head of *M*'s tape 1 should be pointing at the second symbol from the left, i.e. the the first *b* from the left (as it is). All other information from the initial three tapes can be similarly recovered from our single tape representation.

10.4.2 The operation of the single-tape machine, S

In the machine *S*, single instructions (sextuples) of the multi-tape machine, *M*, become several instructions (quintuples) of the single tape machine, *S*, that operate on a single tape set-up representing the multiple tapes of *M*, as specified in Figure 10.20.

Consider as an example the sextuple (1, *a*, *A*, *R*, 2, 2). The meaning of this sextuple is:

> if the symbol at the read/write head of the current tape is *a*, replace it by *A*, move right one square on the current tape, switch processing to tape 2 and move into state 2.

Let us assume that the alphabet of tape symbols of *S* is {*a*, *b*, *A*, 1}, and that (as in Figure 10.20) the markers used on the single tape to represent the read/write heads on the three tapes, tapes 1, 2, and 3, are *X*, *Y*, and *Z*, respectively. In terms of the corresponding single-tape machine S, and assuming that the read/write head of *S* is currently located at the tape square representing the appropriate marked square of *M*'s tape 1, the sextuple (1, *a*, *A*, *R*, 2, 2) is interpreted by the single tape machine as explained in Table 10.2.

Every single instruction (i.e. *sextuple*) of the multi-tape machine has to be represented in a similar way to that described above. For a left move instruction a very similar construction could be used, except that the single tape machine would have to move left six squares from the symbol to the left of the current tape's read/write head marker. For a *no* move instruction, no movement of the read/write head marker would be necessary. For a sextuple that specified that the next operation was to take place on the same tape, there would be no need to search for the marker representing the read/write head of another tape.

Like our four-tape deterministic machine *D* earlier in the chapter, *S* uses 'marker' symbols to indicate the extremes of the marked portion of its tape. Thus, like *D*, *S* has to check for this 'end of tape' marker each time it moves its head to the right or left in the portion of tape representing the tapes of the multi-tape machine (Figure 10.20).

Table 10.2 Overview of the simulation of a sextuple for a three-tape TM, *M*, by a 1-tape TM, *S*

M instruction	*S* interpretation
if the symbol at the read/write head of tape 1 is *a*, replace it by *A*…	Since we are assuming that *S* is already pointing at the tape square representing the current symbol of *M*'s tape 1 we just test if the current symbol is an *a*, replacing it by *A* if it is.
…move right one square on tape 1…	We have three tapes encoded as one tape for *S*, with each tape square of *M* being represented by two tape squares of *S*'s tape (see tape coding scheme above), so moving 'one square' on a tape of *M*, involves moving six squares on *S*'s tape. *S* also has to move the marker (*X*) representing *M*'s tape 1 read/write head.
…then switch processing to tape 2 and move into state 2	This involves searching for the marker that represents the read/write head of *M*'s tape 2 (in our example, *Y*)

Such situations correspond to where the multi-tape machine would move its head onto a blank square outside of the marked portion of one of its tapes. In such circumstances, *S* must **shuffle** the 'end of tape marker' to the left or right by $2k$ squares, where k is the number of tapes of the simulated machine, *M*. For example, a 'move right' on a single tape representing a 10-tape TM would involve moving 20 squares to the right (shuffling the 'end of tape marker' right appropriately if necessary), and we would of course need 10 distinct read/write head markers, and so on.

You should be able to see how the construction of both the tape configuration (Figure 10.20) and the quintuples of the single-tape machine (Table 10.2) could be adapted for multi-tape machines with any number of tapes. You are asked to add more detail to the construction in an exercise at the end of this chapter.

10.4.3 The equivalence of deterministic multi-tape, and deterministic single-tape TMs

The preceding discussion justifies the following statement:

> Any deterministic multi-tape TM, *M*, can be replaced by a deterministic single-tape TM, *S*, that is equivalent to *M*.

It is worth pointing out some reasons why multi-tape machines are useful. Recall the previous chapter, when we described TMs such as MULT and DIV. These machines featured as 'sub-machines' other TMs we had defined, such as ADD, and SUBTRACT. We were forced to use different areas of the tape to deal with the computations of different sub-machines, which led to some tricky *copying* and *shuffling*.

A multi-tape approach would have enabled us to use certain tapes in a similar way to the 'registers' or 'accumulators' of the CPU, to store temporarily the results of sub-machines that would be subsequently needed.

A second related feature of multi-tape machines is that they permit a modularity of 'memory organization', a memory 'architecture', so to speak, which has more in common with the architecture of the real computer system than the single-tape TM. This can be appreciated by considering TMs such as the deterministic multi-tape machines we specified in the previous section, where the *program* (i.e. the quintuples of another machine) was stored on one tape, and the *data* was stored on another. Moreover, the machine used a further tape as a 'working memory' area, and another tape as a kind of *program counter* (the tape where the sequences of quintuple labels were generated).

The modularity facilitated by a multi-tape TM architecture can be useful for specifying complex processes, enabling what is sometimes called a 'separation of concerns'. For example, there should be no need for MULT to be concerned with the internal workings of ADD. It needs only to gain access to the result of each addition. It would therefore be appropriate if MULT simply copied the two numbers to be added onto a particular tape, then passed control to ADD, which would then leave the result in an agreed format (perhaps on another tape), where it would be collected by MULT ready to set up the next call to ADD, and so on.

That a multi-tape TM can be replaced by an equivalent single-tape construction means that we can design a TM that possesses any number of tapes, in the knowledge that our machine could be converted into a single-tape machine. We take advantage of this when we describe what is probably the most significant TM of all, in the next chapter.

10.5 THE IMPLICATIONS OF THE EQUIVALENCE OF NON DETERMINISTIC AND DETERMINISTIC TMs

We began by showing that *from any non deterministic TM we could construct an equivalent four-tape deterministic TM*. We then showed that *from any multi-tape deterministic machine we could construct an equivalent deterministic TM*. We have thus shown the following.

For any non deterministic TM there is an equivalent deterministic TM.

Moreover, in Chapter 7 we showed that the non deterministic TM is the corresponding abstract machine for the **type 0 languages** (in terms of the ability to accept, but not necessarily to decide those languages, as we see in Chapter 11).

We can now say:

> The deterministic TM is the **recognizer** for the **phrase structure languages**, in general.

This is so because:

> any phrase structure grammar can be converted into a non deterministic TM, which can then be converted into a deterministic TM.

(via the conversion of the non deterministic TM into an equivalent four-tape deterministic TM, and *its* conversion into an equivalent deterministic single-tape TM).

10.5.1 Non deterministic TMs viewed as parallel processors

We saw earlier in this chapter that computation is concerned not only with **function**, but also with **space** and **time**. When we make statements about the **equivalence** of Turing machines, as we have in this chapter, we are usually talking about equivalence of *function*. For example, we say that the result of our 'non deterministic to deterministic TM conversion' is 'equivalent' to the original machine. However, there are different ways of viewing the operation of a non deterministic machine. One is to imagine that the machine attempts to follow a sequence of transitions until it finds a solution or reaches a 'dead end', as none of its available instructions can be applied. In the latter case, we have assumed that it then tries an alternative sequence. When combined with our notion of time as a measure of the number of transitions made by the machine, this perspective adopts a *serial* model of computation. Our earlier discussions regarding **non deterministic finite state recognizers** (Chapter 4) and **pushdown recognizers** (Chapter 5) also reflect this view, when considering concepts such as **backtracking**, for example.

As an alternative to a serial model of execution, we can assume that all possible sequences of transitions to be considered by the non deterministic TM are investigated *in parallel*. When the machine is ready to start, imagine that a signal arises, analogously to the way in which the internal 'clock' governs the state changes of the CPU. On this signal, *each and every applicable transition is carried out simultaneously* (each transition being allowed to take with it its own copy of the input tape, since each sequence of transitions may result in different configurations of the tape). Figure 10.21 shows an example starting situation with three applicable transitions.

Figure 10.21 is intended to represent the situation that arises on the first 'clock tick', which we assume results in the execution of the three applicable transitions in parallel (the result of each being shown by the dashed arrows). Note that the new tape configuration for the three cases includes a small state circle showing which state the machine has reached after making that transition. A description of the marked portion

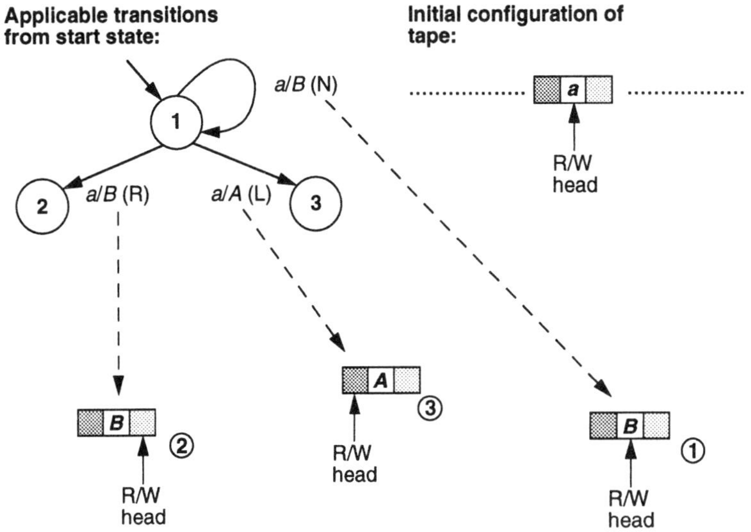

Figure 10.21 The operations of a non deterministic TM viewed as parallel processes

of a tape, the location of the read/write head and the current state of a machine at any stage in a TM's computation is sometimes called an **instantaneous description**.

The next clock tick would result in the simultaneous execution of:

- each applicable transition from state 2 given the tape configuration on the left of Figure 10.21,
- each applicable transition from state 3 given the tape configuration in the centre of Figure 10.21, and,
- each applicable transition from state 1 given the tape configuration on the right of Figure 10.21.

Then the next clock tick would result in the simultaneous execution of each applicable transition from each of the situations created at the previous clock tick.

Clearly, if a solution exists this method will find it. In fact, the first sequence of transitions we find that leads us to a solution will be the shortest such sequence that can be found[3]. Now, note that the equivalent deterministic four-tape machine that we could construct using the method described earlier in this chapter is also guaranteed to find the shortest sequence of transitions the parallel non deterministic machine would have found. This is because the deterministic version first tries all single quintuples, then all length 2 sequences of quintuples, then all sequences of three quintuples, and so on, and will stop as soon as one of the sequences of quintuples results in reaching the non deterministic machine's halt state.

[3] The shortest solution sequence may not be unique. In this case, our parallel machine will find all the shortest solutions simultaneously. It will simplify subsequent discussions if we talk as if there is only one shortest solution.

10.5.2 Parallel computations and time

Now our model of time enters the analysis. If there is a solution that requires the execution of n instructions (i.e. n quintuples), the parallel machine will find it in n moments. In our example above, if any applicable sequence of transitions leads directly to the halt state from any of the resulting configurations shown, our parallel machine would have taken two moments to find the solution. The first transitions would be simultaneously executed in one moment, then all the applicable transitions from the three resulting configurations would be similarly executed in the next moment.

However, consider how long it takes our deterministic version to find a solution requiring the execution of n instructions. The deterministic machine will have tried:

- all single instructions
- all sequences of two instructions

  .

  .

- some or all of the sequences of n instructions.

Now, think back to our earlier discussion of the order in which our deterministic version would generate all possible non empty strings taken from the alphabet of quintuple labels $\{a, b, c\}$. There would be three strings of length 1, nine strings of length 2, 27 strings of length 3, which are:

aaa, *aab*, *aac*, *aba*, *abb*, *abc*, *aca*, *acb*, *acc*, *baa*, *bab*, *bac*, *bba*, *bbb*, *bbc*, *bca*, *bcb*, *bcc*, *caa*, *cab*, *cac*, *cba*, *cbb*, *cbc*, *cca*, *ccb*, *ccc*

At each stage, the strings for the next stage were derived by appending each of the three symbols of the alphabet to each of the strings obtained at the previous stage. Thus, for our example here, there would be 81 (i.e. 27×3) strings of length 4, then 243 (81×3) strings of length 5, and so on.

In general, the number of strings of length n from an alphabet of k symbols is k^n (in the example above, where $n = 4$ and $k = 3$ we had 3^4 or 81 strings).

To return to our main discussion, consider a *three-quintuple non deterministic parallel TM*, *N*. Suppose the shortest solution that this machine could find, given a certain input configuration, involves *four instructions*. *N* will take four moments to find it. Now consider the deterministic version, *D*, constructed using our method described earlier. We will simplify things considerably by considering the execution time of *D* in terms of the number of quintuples of *N* it tries. We will assume that it takes one moment to try each quintuple. We will also ignore the fact that a large number of quintuple sequences will be illegal, and thus aborted by *D* (as specified in Figure 10.15) before the entire sequence has been applied. It is reasonable to do this, as *D* has to generate each sequence of quintuple labels, even if the corresponding sequence of quintuples subsequently turns out to be inapplicable. This machine first tries:

- three sequences of instructions of length 1 (three moments);
- nine sequences of instructions of length 2 (18 moments);
- 27 of length 3 (81 moments);
- between one and 81 sequences of length four (between four and 324 moments).

Adding all these up, we get 3 + 18 + 81 + 4, or 106 moments in the best case (i.e. when the first four-quintuple sequence tried results in a solution), or 3 + 18 + 81 + 324, or 426 moments in the worst case (when all the four-quintuple sequences have to be tried, the last one resulting in the solution). In general, given k quintuples in the non deterministic machine, and a shortest solution requiring n instructions, the parallel TM, N, will take n moments to find the solution, while the corresponding deterministic machine, D, will take somewhere between:

$$\begin{aligned}&(k \times 1) + (k^2 \times 2) +..+ (k^{n-1} \times n-1) + (k^n \times n)\\ =\ &k + 2k^2 +..+ n-1k^{n-1} + nk^n \text{ moments in the worst case,}\end{aligned}$$

and

$$\begin{aligned}&(k \times 1) + (k^2 \times 2) +..+ (k^{n-1} \times n-1) + n\\ =\ &k + 2k^2 +..+ n-1k^{n-1} + n \text{ moments in the best case.}\end{aligned}$$

Now, as we have already seen, even simple Turing machines such as those we have seen in this book, consist of many more that three quintuples, and the solutions of even the simplest problems require the application of sequences of considerably more than four quintuples. For example, if a 10 quintuple non deterministic parallel TM (a very simple machine) found the solution in 11 moments (a very short solution), the best the corresponding deterministic TM constructed by our method could hope to do (as predicted by the above formula) is find a solution in 109,876,543,221 moments. To consider the extent of the difference in magnitude here, let us suppose a moment is the same as 1 millionth of a second (a reasonably slow speed for a modern machine to carry out an operation as simple as a Turing machine transition). Then:

- the non deterministic machine executes the 11 transitions in slightly over $\frac{1}{100{,}000}$ th of a second, while
- its deterministic counterpart takes around 30.5 hours to achieve the same solution!

This example may seem rather contrived, but consider that a parallel algorithm is, say, updating a complex graphical image, and millions of individual operations are being carried out simultaneously. Such computations would be impossible to carry out in a reasonable time by a serial processor. We would not wish to wait 30.5 hours for an image in an animation to change, for example.

10.6 A BRIEF LOOK AT AN UNSOLVED PROBLEM OF COMPLEXITY

The preceding discussion leads us to an as yet unsolved problem of computer science. You may, on reading the discussion, think that the disparity between the non deterministic TMs and their four-tape deterministic counterparts is entirely due to the rather simple-minded conversion scheme we have applied. You may think that there is a more 'efficient' way to model the behaviour of the non deterministic machine than by attempting to execute every possible sequence of its quintuples. We have shown that for any non deterministic TM we can construct a deterministic TM that does the same job. The question that now arises is: *can we construct one that does the same job as efficiently*? In order to discuss whether this question can be answered, we take a slightly more detailed look at complexity.

10.6.1 A beginner's guide to the 'big O'

As far as we are concerned here, the theory of **complexity** examines the relationship between the *length* (number of symbols) *of the input* to a TM and *the number of instructions it executes to complete its computation.* The same applies to programs, where the relationship concerned is between the *number of input elements* and the *number of program instructions executed.* In considering this relationship, we usually take account only of the components of the program that dominate the computation in terms of time expenditure.

Suppose we are to sum all the elements in a two dimensional array of numbers, then print the result, and we can only access one element of the array at a time. We must visit each and every element of the array to obtain each value to be added in to the running total. The time taken by the program is dominated by this process. If the array is 'square', say, $n \times n$, then we are going to need n^2 accessing operations. The times taken by the other operations (e.g. adding the element to the running total in the loop, and printing out the result) we can ignore, as they do not change significantly whatever the size of n.

In overall terms, our array summing program can be said to take an amount of time *approximately proportional to* $n^2 + c$. Even if we had to compute the array subscripts by adding in an offset value each time, this would not change this situation (the computation of the subscripts would be subsumed into c). Obviously, since c is a constant measure of time, the amount of time our algorithm takes is dominated by n^2 (for large values of n). n^2 is called the *dominant expression* in the equation $n^2 + \text{c}$. In this case, then, we say our array processing program takes time *in the order of* n^2. This is usually written, using what is called big O – 'Oh', not nought or zero – notation, as $O(n^2)$. Similarly, summing the elements of a one-dimensional array of length n would take time $O(n)$. There would be other operations, but they would be assumed to take constant time, and so n would be the dominant expression. In a similar way, summing up the elements of a *three* dimensional $n \times n \times n$ array could be assumed to take time $O(n^3)$.

Programs that take times such as $O(n^2)$ and $O(n^3)$, where n is some measure of the length of the input, are said to take **polynomial time** in terms of the input data. A special case of this, as in our one-dimensional array summing program, is when the time is $O(n)$, and in this case we say the program takes **linear** time. If a program must process all its input, then the closer it comes to taking linear time, the more efficient it must be. In general, the goal is to design programs so that the index x in the dominant expression n^x is as small as possible to solve the given problem.

Now, if the dominant expression in an equation denoting the running time of a program is $O(x^n)$, where, again, n is some measure of the length of the input, the amount of execution time rises dramatically as the input length increases. Programs like this are said to take **exponential time** in their execution. Consider the example of the 4-tape deterministic version of the non deterministic TM in the previous section. Simplifying somewhat, we can regard this as taking time $O(nk^n)$, where k is the number of quintuples in the non deterministic machine, in order to find a solution of length n. We saw the striking difference between the time taken by the non deterministic parallel machine, which was $O(n)$, i.e. of linear relationship to the length of the solution, and that taken by the deterministic version, for even small values of n (3 and 11), and small values of k (3 and 10).

We have considered the difference in the time taken by two machines (a non deterministic TM and its four-tape deterministic counterpart) in terms of the length of a given solution. However, the disparity between the time taken by the two machines becomes even greater if we also consider the relationship between the length of the *input* and the length of the solution. For example, suppose the non deterministic machine solves a problem that has minimum length solution n^3, where n is the length of the input. Then the non deterministic machine will find that solution in time $O(n^3)$. Suppose $n = 11$, then, simplifying things grossly, our deterministic machine will find a solution in the best case in a time $O(k^{n^3-1})$, which represents a time of 10^{1330} moments. The point being made is that when the non deterministic machine requires *polynomial* time to find the shortest solution, the deterministic machine constructed from it by our method requires *exponential time*.

10.6.2 The implications of exponential time processes

If you find big O a little daunting, think about it in this way: for our hypothetical problem, our non deterministic machine, N, has 11 squares marked on its input tape. Since $n = 11$, and we know that the shortest solution is n^3 instructions in length, N is thus searching for a solution of 11^3 instructions. If N is a *parallel* machine, as described above, it will find the solution in 11^3 moments.

Now consider the behaviour of D, the deterministic version of N. To get to the first solution of length 11^3, *D will have to first generate all sequences of quintuple labels of length* $11^3 - 1$. If there are 10 quintuples, there are 10^{11^3-1} distinct sequences of labels of length $11^3 - 1$, i.e. 10^{1330}.

Let us now consider a simpler example, so that we can appreciate the size of the numbers we are dealing with here. Suppose the shortest sequence of instructions that a particular 10-quintuple non deterministic machine can find is 101 instructions, for an input of 10 symbols. Before finding the 101 quintuple sequence that yields the solution, the deterministic version will generate all sequences of quintuples of length 100. There are 10^{100} of these sequences, since there are 10 quintuples. The time taken by our deterministic machine is around 10^{100} moments, by the most gross simplification imaginable (i.e. imagining that the generation of each sequence takes only a moment, and ignoring the fact that our machine will have generated all the sequences of length 99, of length 98, and so on).

At this point we need to buy a faster deterministic machine. We go to the TM store, and purchase a machine that does 10^{20} operations in a year. This machine does in excess of *three million million operations each second.* If we used it to find the solution to the problem above that took our original non deterministic 'one-million-instructions-a-second' model 30.5 hours to find, it would take our new machine about $\frac{3}{100}$ of a second! However, our new machine would still take in the order of 10^{80} *years* to solve our new problem, while our non deterministic machine, at a similar instruction execution rate, would take $\frac{1}{30{,}000{,}000{,}000}$th *of a second.* Given that, by some estimations, the Earth is at this time around 400 million years old, if we had started off our deterministic machine at the time the Earth was born, it would now be about:

$$\frac{1}{250{,}000}\text{th}$$

of the way through the time it needs to find the solution.

10.6.3 'IS P EQUAL TO NP?'

From a purely formal point of view, the amount of time taken to solve a problem is not important. The requirements are that a solution be found in a finite number of steps, or, put another way in a finite period of time. A time period like 10^{80} years may be to us an incredibly long period of time, but nevertheless it is finite. The purpose of the construction we used to create our deterministic machine was to show *functional* equivalence, i.e. that whenever there is a solution that the non deterministic machine *could* find, the deterministic machine *will* find it (eventually!).

In a more practical vein, in creating a deterministic machine that models one that is non deterministic, we may also wish to maintain invariance between the two machines in the other dimensions of computational power (*space* and *time*) as well as invariance in *functionality*. Given a non deterministic TM that finds solutions in **polynomial** time, we would ideally like to replace it by a deterministic machine that also takes polynomial time (and not **exponential** time, as is the case for the construction method we have used in this chapter). Clearly, this can be done in many many cases. For example, it would be possible to model an array summing program (of the type discussed earlier) deterministically by a polynomial time TM.

The above discussion leads to a question of formal computer science that is as yet unanswered. The question is:

> For each polynomial time non deterministic TM, can we construct an equivalent polynomial time deterministic TM?

This is known as the 'IS *P* EQUAL TO *NP*?' problem, where *P* is the set of all *polynomial time TMs*, and *NP* is the set of all *non deterministic polynomial time TMs*. As yet, no solution to it has been found. Given our analogy between non deterministic TMs and parallel processing, *the lack of a solution to this problem indicates the possibility that there are parallel processes that run in polynomial time, that can only be modelled by serial processes that run in exponential time*. More generally, it suggests that there may be some problems that can be solved in a reasonable amount of time only if solved non deterministically. True non determinism involves *random* execution of applicable instructions, rather than the systematic execution processes considered above. This means that a truly non deterministic process is not guaranteed to find a solution even if there is one.

There may therefore be problems that are such that:

- a solution may be found in *polynomial time* by a *non deterministic program* (but it may not actually find a solution even if one exists), or
- a solution will be found in *exponential time* by a *deterministic program*, if a solution exists, but it may take a prohibitive amount of time to discover it.

10.7 CONCLUSIONS

We have seen in this chapter that, for both theoretical and practical reasons it can be useful to introduce augmentations to the basic TM. We have also seen that, in a purely functional sense, these augmentations do not lend us any extra power. This is compelling evidence in support of Turing's thesis that the (one-tape, deterministic) TM is indeed capable of computing all **effective procedures**.

A central theme of this chapter has been the notion of one TM executing the quintuples of another as if those quintuples were a program. In the next chapter, we develop this notion in the description of a three-tape deterministic TM that takes as its input the description of any other TM, *T*, and carries out the computation that *T* would have performed on any given input tape. We know that the three-tape machine could then be replaced by an equivalent single-tape machine, and thus we will have implicitly specified a single, standard TM that can compute all effective procedures. This

universal TM is then used as the basis of demonstrating the existence of problems that, unlike the 'IS $P = NP$' problem, are not only still to be solved, but in fact can *never* be solved by algorithmic means.

EXERCISES

1 Define a **TM** to generate $\{a, b, c\}^+$, as outlined earlier. Sketch out a convincing demonstration that for any alphabet A, A^+ is similarly **enumerable** (i.e. we could define a TM to print out its elements in a systematic way).

2† Sketch out a TM to execute the four-tape simulation of **quintuple** (A, 1, 0, R, B) described above.

3† Sketch out the part of the single tape machine, S, that deals with the **sextuple** (1, a, A, R, 2, 2) of the three-tape machine, M, described earlier in this chapter.

4 (cf. Exercise 4 of Chapter 4.) Write a '**big *O***' expression representing the 'worst case' time for the **subset algorithm** described in Chapter 4. The input, n, is assumed to be the number of states in the **non deterministic FSR** that is input. Find an example FSR for which the subset algorithm takes its 'worst case' time.

11 The limits of computation and the scope of phrase structure languages

AIMS

To learn

- how to code a TM and present these codes to a universal TM (UTM)
- how UTM simulates the computation of the original machine
- about the halting problem and how we can decide whether or not a TM will halt on a particular tape
- how formal languages can be totally undecidable or semi-decidable

11.1 PREVIEW: THE UNIVERSAL TURING MACHINE (UTM)

A major part of this, the final chapter, will be concerned with sketching out the description of what we call a **universal Turing machine** (**UTM**). This machine will accept as its input two things:

1. (a coded version of) the **quintuples** of a TM, M
2. (a coded version of) the marked portion of an input **tape** to M.

UTM will then simulate the behaviour of M on the given input tape. The aim of this section is to describe a coding scheme which will facilitate a definition of UTM as a machine that uses the alphabet $\{0, 1\}$.

The existence of UTM not only provides further evidence in support of **Turing's thesis**, that the Turing machine is the most powerful computational machine, but also plays a part in finally establishing the relationship between *computable languages* and *abstract machines*. In the course of this, UTM also features in the demonstration of an extremely significant result in computer science: the general unsolvability of the **halting problem**.

11.2 CODING A TURING MACHINE AND ITS TAPE AS A BINARY NUMBER

We first consider a simple method of coding any TM, by building up a coding scheme for the **states**, the **alphabet**, the quintuples and the whole machine. We then see how the code devised for the machine's alphabet can be used to code any input tape for the coded machine. The coded machine and the coded input can then be used as input to UTM, introduced above.

11.2.1 Coding any TM

To describe the scheme for coding any TM as a string of 1s and 0s we will use an example TM. As we proceed, you should appreciate that the coding scheme could be applied to any TM, consisting of any number of states and using any size of alphabet.

The example machine, shown in Figure 11.1, accepts the **context free language** $\{a^ib^{2i} : i \geq 1\}$ (i.e. all strings of *a*s followed by *b*s in which there are twice as many *b*s as *a*s).

We begin by associating the *states* of M (Figure 11.1) with strings of 0s as shown in Table 11.1. It should be clear that we could code the states of any TM using the scheme.

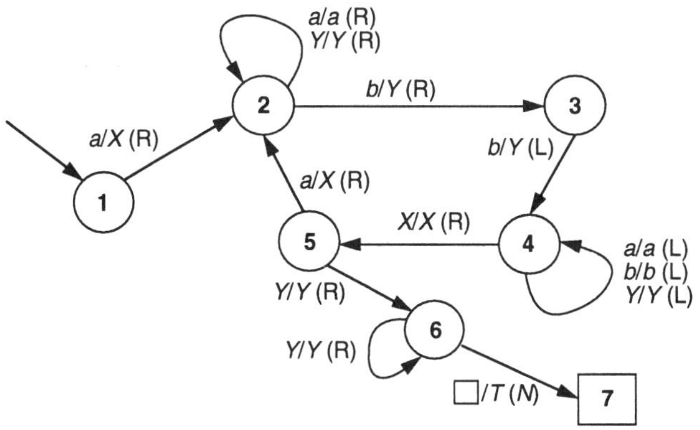

Figure 11.1 A TM, *M*, on which to practise the coding scheme

Table 11.1 Coding the states of a TM

State of *M*	Code	Comments on general case
state 1 (start state)	0	We *always* code the start state of a machine as 0
state 7 (halt state)	00	We *always* code the halt state of a machine as 00
state 2	000	We simply code the other states as sequences of 0s
state 3	0000	We code the states in any order (but we make sure that for each state we use a different length sequence of 0s), and we do not use either the sequence 0 or 00 for any of the states. Since it makes sense to use as few symbols as possible, we have used codes 000, 0000, and so on
state 4	00000	
state 5	000000	
state 6	0000000	

Next, as shown in Table 11.2, we code the *tape symbols* (alphabet) of the machine, which, for our machine *M*, is {*a*, *b*, *X*, *Y*, *T*} and the blank symbol □.

Again, we can see that this scheme could apply to a TM alphabet of any number of symbols.

Table 11.2 Coding the alphabet of a TM

State of *M*	Code	Comments on general case
a	0	
b	00	As for the states, we code the symbols in any order. We can use any number for 0s we like, providing we use a different number for each symbol. Again, we chose here a scheme that enables us to use as few 0s as possible.
X	000	
Y	0000	
T	00000	

We also code the *direction symbols L, R*, and *N*, as shown in Table 11.3.

Table 11.3 Coding the direction symbols of a TM

Symbol	Code	General Case
L	0	Always the same.
R	00	
N	000	

You will note that some of the codes representing the states (Table 11.1) are identical to those used for the alphabet (Table 11.2) and direction symbols (Table 11.3). However, this does not matter; as we will see in a moment, our coding scheme ensures that they do not become mixed up.

We have coded the *states*, the *alphabet*, and the *direction symbols*. We now code the *instructions* (quintuples) themselves. Figure 11.2(a) shows how to code any given quintuple. An example of the coding of a quintuple of the machine *M*, from Figure 11.1, is given in Figure 11.2(b).

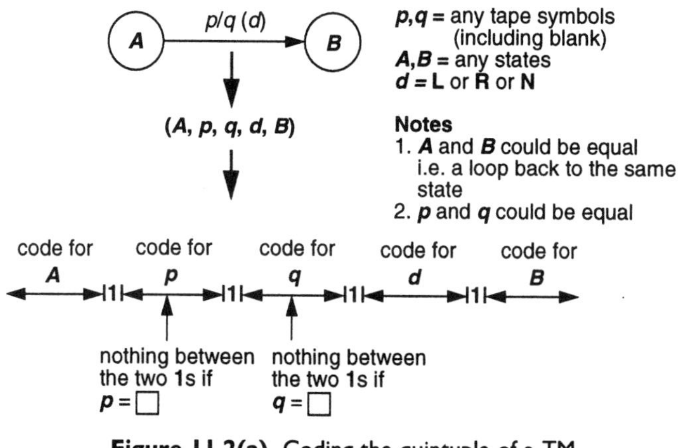

Figure 11.2(a) Coding the quintuple of a TM

Now, we simply code *the whole machine* as a sequence of quintuple codes, each followed by 1, and we place a 1 at the start of the whole code. The scheme is shown in Figure 11.3(a), and a partial encoding of the TM, *M*, from Figure 11.1, is given in Figure 11.3(b).

The above should be sufficient to convince you that we could apply such a coding scheme to *any* TM, since any TM has a finite number of states, tape symbols, and quintuples.

Such a binary code for a Turing machine could be presented as the input to *any* TM that expected arbitrary binary numbers as its input. For example, we could present the

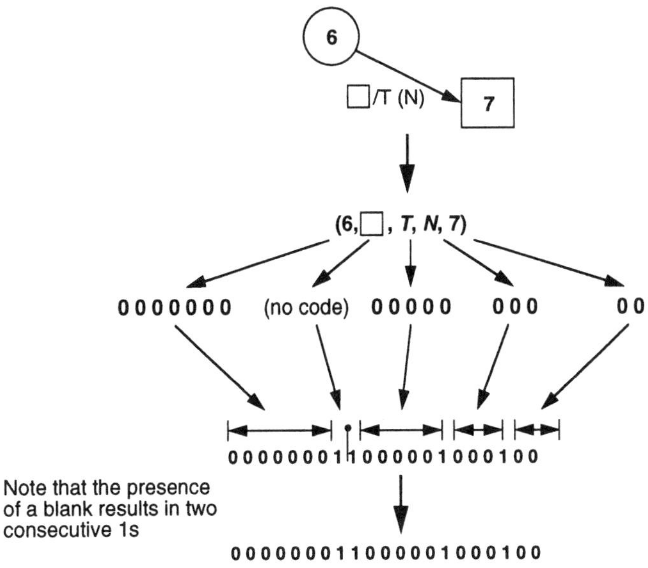

Figure 11.2(b) Coding a quintuple of *M* (Figure 11.1) according to the scheme of Figure 11.2(a)

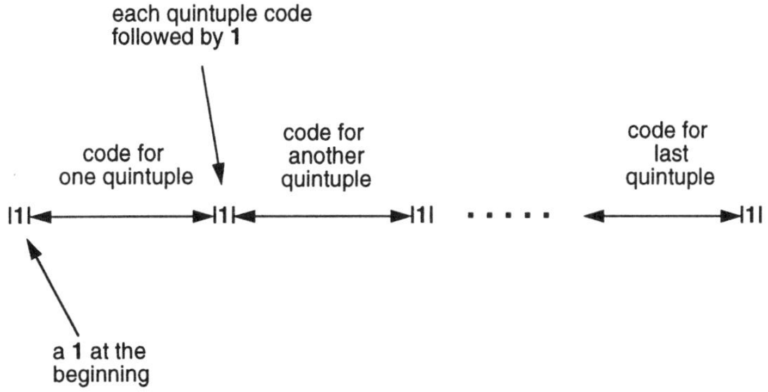

Figure 11.3(a) Coding a whole TM

code for the above machine, M, to the MULT TM (described in Chapter 9) as one of the two numbers to be multiplied. We could even present the coded version of MULT to the MULT machine as one of its input numbers, which would mean that MULT *would be multiplying its own code by another number*! All this may seem rather strange, but it does emphasize the point that codes can be viewed at many different levels. The code for M can be regarded as another representation of M itself (which is how UTM will see it), as a string in $\{0,1\}^+$ (a non empty string of 0s and 1s) or as a binary number. This is essentially no different from seeing Pascal source code as *text* (the *text editor's* view) or as a *program* (the *compiler's* view).

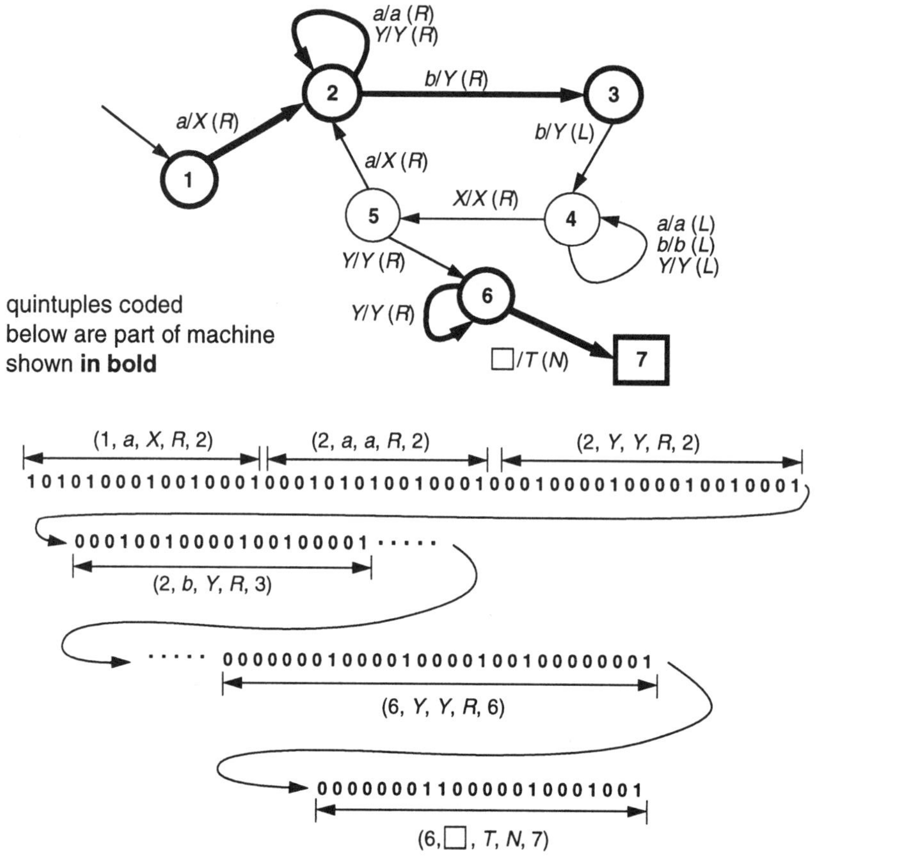

Figure 11.3(b) A partial application of the coding scheme from Figure 11.3(a) to the machine *M* of Figure 11.1

Since our UTM will require not only the coded version of a machine, but also a coded version of an input to that machine, we need to specify a scheme for coding any *marked tape* of the machine we are coding.

11.3.2 Coding the tape

The coding scheme for the input is described in Table 11.4. Note that for a given machine, we use the same codes for the symbols on the input tape as those used to code the machine's alphabet (Table 11.2).

It should be clear that any input tape to any coded machine could be similarly represented using the scheme in Table 11.4.

We have now seen how to take any TM, *M*, and code it, and any of its input tapes, with a coding scheme that requires only the alphabet {0, 1}. In the next section, we

Table 11.4 Coding a TM's input tape

General rule	Application to an example input tape
We represent each symbol using the codes (i.e. sequences of 0s) that we used for the alphabet of the TM (Table 11.2) We make sure that for each symbol of the input, we use the same code that we used when we coded the machine. As for the quintuples, we follow each code by 1, and precede the whole sequence by a 1: Note: The presence of a blank within the marked portion of *M*'s input tape would be represented by two consecutive 1s (as was the case in the quintuple codes of Figure 11.2).	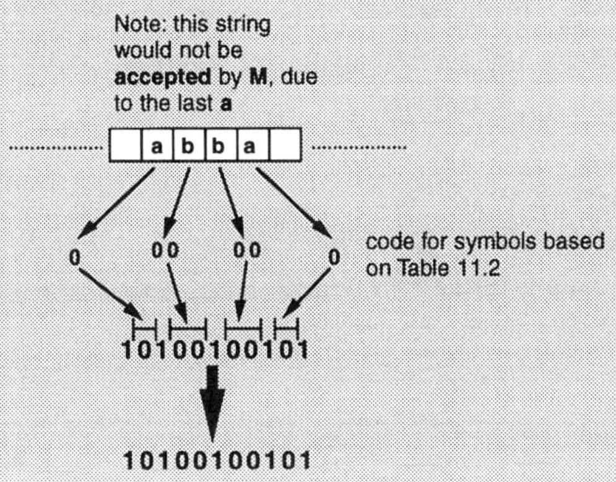

use our coding scheme as an integral component in sketching out the description of our universal Turing machine, UTM.

11.3 THE UNIVERSAL TURING MACHINE

UTM is described here as a three-tape **deterministic TM** that takes as its input a coded version of a TM, and essentially simulates the behaviour of the coded machine on a coded version of input to that machine. UTM therefore models the behaviour of the original machine on the original input, producing as its output a coded version of the output that would be produced by the original machine, which can then be decoded. We know, from the preceding chapter, that our three-tape UTM could be converted into a single-tape TM. The beauty of knowing that this can be done is that we need not actually do it.

You may at this point be concerned that there is some 'trickery' going on here. You might say that everything that UTM does is in terms of manipulating codes, not the 'real' thing. If you do think that it's trickery, then it can be pointed out that every single program you have ever written is part of the same type of 'trickery', as

- the *source file*,
- the *input*, and
- the *output*

are all similarly coded and subsequently decoded when your program is run. So if you are not prepared to believe in the representational power of coding schemes, you ought not to believe that any of your programs has ever done anything meaningful!

Next, we specify the initial set up of UTM's three tapes. Following this, we briefly consider the operation of UTM. You will appreciate that the construction of UTM described would be capable of simulating the operation of any TM, *M*, given any suitable input tape to *M*.

11.3.1 UTM's tapes

Our version of UTM is configured so that tape 1 is as shown in Figure 11.4.

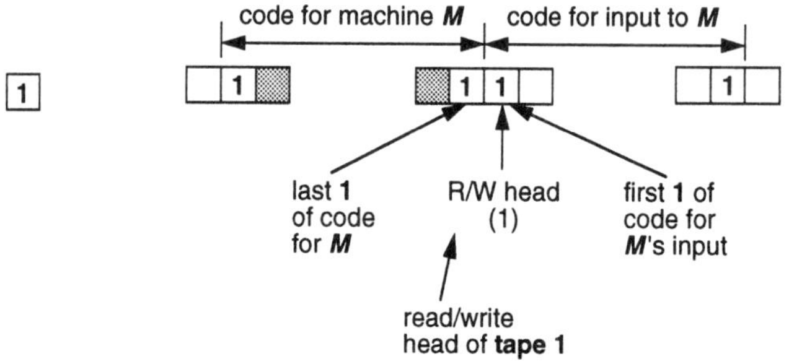

Figure 11.4 Tape 1 of UTM

As can be seen from Figure 11.4, tape 1 contains the code for the machine *M*, followed immediately by the code for *M*'s input.

UTM will use its tape 2 as a 'work tape', on which to simulate *M*'s operation on its input by carrying out the operations specified by the coded quintuples on the coded data. Thus, the first thing UTM does is to copy the coded input from tape 1 to tape 2, at which point tape 2 will be as specified in Figure 11.5.

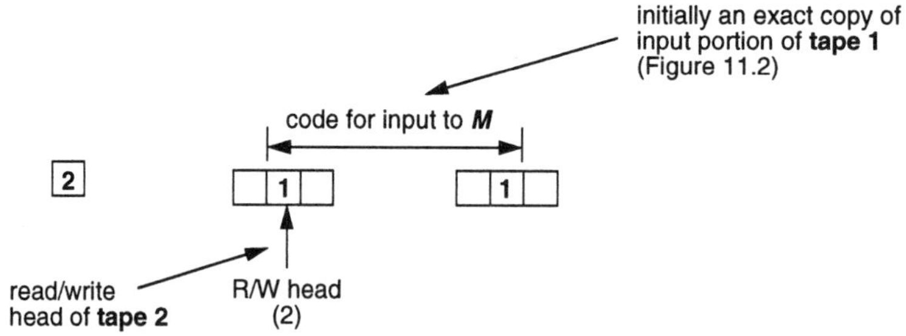

Figure 11.5 Tape 2 of UTM

Tape 3 of UTM will be used only to store a sequence of 0s denoting, at any stage, the current state of *M*. The organization of tape 3 is depicted in Figure 11.6.

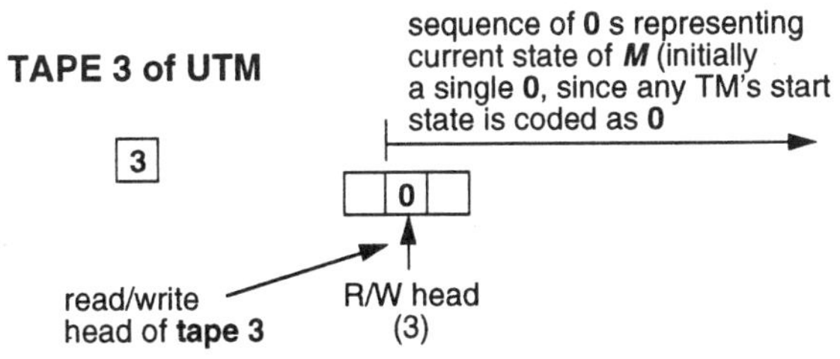

Figure 11.6 Tape 3 of UTM

To summarize:

tape 1
holds the coded *machine M*, followed by the coded input to *M*

tape 2
is used by UTM to work on a copy of *M*'s coded *input*

tape 3
is used to hold a sequence of 0s representing *M*'s *current state*

11.3.2 The operation of UTM

UTM operates as specified by the procedure in Table 11.5. I leave it to you to appreciate how the procedure could be represented as a TM. For clarification of the procedure in Table 11.5, you are advised to refer to the coding scheme and UTM's tape configurations described above.

A UTM based on the procedure in Table 11.5 is capable of carrying out the computation specified by any TM, given an initial set-up as described above. As input to such a machine, we code the TM according to our scheme, and then we can code an input tape to that TM using the same scheme. UTM will then simulate the computation performed by the original machine.

In particular:

- If *M*, the original machine, would produce a solution given a particular input tape, then UTM would *produce a coded version of that solution* (on its tape 2).
- If *M* would reach a state, *s*, in which it had no applicable quintuples for the current state and tape symbols, then UTM *would halt with the code for the state s on tape 3, and tape 2 would contain a coded version of M's tape configuration at the time M stopped*[1].
- If *M* would reach a situation where it **looped** indefinitely, repeating the same actions over and over again, then *UTM would also do the same*. We return to this last situation later in this chapter.

[1] This assumes that the 'while' loop in Table 11.5 terminates if no suitable quintuple can be found.

Table 11.5 The operation of UTM

copy code for *M*'s data from tape 1 to tape 2	tape 2 is to be used as a work tape by UTM
write a 0 on tape 3 to represent the start state of *M*	the start state of *M* is represented by 0
while tape 3 does not contain exactly two zeros do	the halt state of *M* is represented by 00
find quintuple on tape 1 for current state	this involves UTM in proceeding along tape 1, comparing the first code in each quintuple (a sequence of zeros) with the sequence of zeros on tape 3
if quintuple found	tape symbols are represented by a possibly empty sequence of zeros
then if current symbol on tape 2 = quintuple current symbol	
then overwrite the current symbol code on tape 2	this can involve shuffling, since the code being replaced may not consist of the same number of zeros as the one replacing it
move the read head on tape 2 according to the quintuple direction	the tape 2 head must be moved to the next 1 to the left or right (or not at all), as appropriate
overwrite state code on tape 3 with quintuple state code	
endif	
endif	
endwhile	

We can now appreciate that we have described a three-tape deterministic TM, UTM, that can, if we accept **Turing's thesis**, carry out any **effective procedure**. We also know that from the three-tape version we could construct a deterministic 'standard' single-tape UTM, and our subsequent discussions will assume that this has been done.

11.3.3 Some implications of UTM

The existence of UTM makes Turing's thesis that TMs are the pinnacle of functional power even more compelling. We have seen that a TM that can 'run', as a kind of 'program', any other TM needs to be no more sophisticated than a standard TM. In a sense, *UTM actually embodies the effective procedure of carrying out the effective procedure specified by another Turing machine.*

Other designs of UTM may be suggested, based on different schemes for coding the machine *M* and its input. All are equivalent in that, since each UTM is a Turing machine, a UTM itself may be coded by the scheme for one of the other UTMs, and thus presented, with a coded version of one of its input tapes (i.e. the coded version of another machine, followed by the coded version of that machine's input), as input to that other UTM. In that sense also, UTM is truly *universal*.

In a more practical vein, UTM is a closer analogy to the stored program computer than the individual TMs we have considered so far. Each TM we looked at before was essentially a purpose built machine for computing one particular function. UTM, with its ability to carry out computations described to it in some coded form, is a counterpart of real computers that carry out computations described to them in the form of codes called programs. However, to revisit our argument from previous chapters, a given computer could only be as powerful as UTM if it had unlimited storage capabilities. Even then, it could not be *more* powerful.

11.4 THE RELATIONSHIP BETWEEN FUNCTIONS, PROBLEMS, SOLVABILITY AND DECIDABILITY

We have said much about the power of the TM. We spend much of the remainder of this chapter investigating its limitations. In the later stages of this book there have been many references to functions, to solving problems, and occasionally to effective procedure, but no clear definition of these terms has been given. We are now familiar enough with TMs and Turing's thesis to be able to provide these definitions.

11.4.1 Functions and computability

A **function** associates input values with output values in such a way that for each particular arrangement of input values there is but one particular arrangement of output values. Consider the arithmetic add (+) operator. This is a *function* because for any pair of numbers, x and y, the result of $x + y$ is always associated with a number z, that represents the value obtained by adding x and y. In the light of the previous section, it should be appreciated that, since any TM can be represented in binary form, it suffices to define the values (input and output) of functions in terms of numbers. This applies equally well to the language processing machines considered in this book as much as it does to those carrying out numerically-oriented operations.

We know that we can represent any of our **abstract machines** for language recognition (**finite state recognizers**, **push down recognizers**) as TMs (Chapter 7), then represent the resulting TM in such a way that its operation is represented entirely as the manipulation of binary codes. Thus, the notion of function is wide-ranging, covering associations between *strings* and *truth values* ('true' if the string is a sentence of a given language, 'false' if it is not), *numbers* and *other numbers* (MULT, DIV, etc.), *numbers* and *truth values* (COMPARE), and so on.

Calling a function an association between input and output values represents a very high level view of a function. We are usually also interested in *how* the association between the input and output values is made, i.e. the rules by which we derive a particular arrangement of output values from a particular arrangement of input values. More specifically, we are interested in characteristics of the *program* that embodies the 'rules' and produces the output values when presented with the input values.

From time to time, we have mentioned the notion of the **effective procedure**, which we can now say describes those procedures which are:

- deterministic
- finite in time and resources (space)
- mechanical
- represented in numeric terms.

It has been argued above that such effective procedures are precisely those that can be carried out by TMs, and in fact now we can say by one TM, which we call UTM.

Having a clearer idea of the notions 'function' and 'effective procedure' enables us to define what we call the **computable functions**, in Table 11.6.

Table 11.6 'Computable functions' and 'TM-computable functions'

The computable functions are those functions for which there is an effective procedure for deriving any correct output values when presented with any appropriate input values. They can therefore be called the effectively computable functions.

A TM-computable function is a function that can be computed by some TM.

So, Turing's thesis asserts that *the effectively computable functions and the TM-computable functions are the same.*

11.4.2 Problems and solvability

Associated with any function is a **problem**. In abstract terms, the problem associated with a function is a *question*, which asks 'what is the result of the function for these particular input values?'. For the '+' function, then, an associated problem is 'what is 3 + 4?'. The previous section suggested that the sentence recognition task for a given language could be regarded as a function associating strings with truth values. The associated problem then asks, for a given string 'what is the truth value associated with this string in terms of the given language?'. This is more usually expressed: 'is this string a sentence of the language or not?'.

A problem is **solvable** if the associated function is *computable*. We know now that 'computable' is the same as saying 'TM-computable'. So, a problem is solvable if it can be coded so that a TM could compute a solution to it in a finite number of transitions, or if a program run on an unlimited storage machine could do the same.

A problem is *totally solvable* if the associated function can be computed by some TM that produces a correct result for every possible assignment of valid input values.

An example of this is the '+' operator. We could define a TM, like the ADD machine, that produces the correct result for any pair of input values, by adding the numbers and leaving the result on its tape. Another totally solvable problem is the membership problem for the **context sensitive language** $\{a^ib^ic^i : i \geq 1\}$, since in chapter 7 we defined a TM that printed T for any string in $\{a,b,c\}^*$ that was a member of the language, and F for any such string which was not. The membership problem for $\{a^ib^ic^i : i \geq 1\}$ is thus totally solvable.

A problem is *partially solvable* if the associated function can be computed by some TM that produces a correct result for each possible assignment of input values for which it halts. However, for some input values, the TM will not be able to produce any result. The critical thing is that on any input values for which the TM halts, the result it produces is correct. This concept is best considered in the context of language recognition, and you may already have appreciated the similarity between the concept of *solvability* and the concepts of **deciding** and **accepting** languages discussed in Chapter 7. This similarity is not accidental, and we return to it later in the chapter. However, we will be considering one of the most important partially solvable problems of computer science in the next section.

A problem is *totally unsolvable* if the associated function cannot be computed by any TM. It is worth mentioning at this point that in discussing the concepts of solvability we never exceed the characteristics of *effective procedures* as defined in the previous section: we are still considering *deterministic*, *finite*, *mechanical* and *numeric* processes. For an abstract, simply stated, yet totally unsolvable problem, consider this:

> Define a TM that has as its input a non-packed tape (i.e. an arbitrary number of blank squares can appear between any two marked squares) on which there are *a*s, *b*s and *c*s, in any order. The machine should halt and print T if the leftmost marked symbol on its tape is a, and should halt and print F otherwise. The read/write head is initially located on the rightmost marked tape square.

The above machine cannot be defined.

Suppose we design a TM to address the above problem that moves left on its tape until it reaches an a, then prints T and halts. Does this mean the problem is partially solvable? The answer to this is 'no', since partial solvability means that the TM must produce a correct result *whenever it halts*, and our 'cheating' TM would not (though it would 'accidentally' be correct sometimes). No TM could be designed that would even partially solve the above problem. I leave it to you to convince yourself of this.

11.4.3 Decision problems and decidability

Problems that ask a question expecting a yes/no answer are sometimes known as **decision problems**. In seeking to solve a decision problem we look for the existence of a program (TM) to resolve each of a set of *questions*, where the formulation of each question depends on the assignment of specific values to certain parameters. For example:

- given two **regular grammars**, R_1 and R_2, are they **equivalent**?
- given a number, N, is N prime?
- given a TM, M, and a tape T, will M halt eventually when started on the tape T?

A *decision problem* is *totally solvable* if we can construct a TM that will provide us with the correct 'yes/no' response for each and every possible instantiation of the parameters. Thus, for example, the second decision problem above is totally solvable only if we can construct a TM that decides, for any number, if that number is prime or not. We could do this if we wished, so the second problem is totally solvable.

Of the other two problems, only one is totally solvable, and this is the first. In Chapter 4, we saw what we now know to be called effective procedures for converting any regular grammars into **non-deterministic FSRs**, for converting non-deterministic FSRs into deterministic FSRs, and for converting deterministic FSRs into their **minimal** form. If two regular grammars are equivalent, then their associated *minimal machines* will be exactly the same, if the states of one minimal machine are renamed to coincide with those of the other. If the states of one of the minimal machines cannot be renamed to make the two minimal machines exactly the same, then the two grammars are not equivalent. We could design a (complicated!) TM to do all this, and so *the equivalence problem for regular grammars is totally solvable.*

11.5 THE HALTING PROBLEM

Let us look again at the third of the three decision problems presented in the immediately preceding section:

given a TM, M, and a tape T, will M halt eventually when started on the tape T?

This problem is known as the **halting problem**, and was briefly alluded to in the discussion of the power of the FST at the end of Chapter 8. All of us have experienced the practical manifestation of this problem. We write a program, and run it. The program seems to be taking a long time over its computation. We decide it is stuck in an 'infinite loop', and interrupt the execution of the program. We have usually made some error in the design or coding of a loop termination condition. Even commercially supplied programs can sometimes suffer from this phenomenon. Some unexpected combination of values leads to a non-terminating repetition of a sequence of instructions. It would be useful if someone would design a program that examined other programs to see if they would always terminate, i.e. for all of their permissible input values. Such a program could form a useful part of the compiler, which could then give out an error message to the effect that a given program would not terminate for certain inputs.

However, as said before, a computer can never be more powerful than a TM, so if the halting problem, as represented in TM terms, is unsolvable, then by extension it is unsolvable even by a computer with infinite memory. We see next that the halting problem is indeed unsolvable in the general sense, though it is *partially* solvable, which is where we begin the investigation.

11.5.1 UTM_H partially solves the halting problem

Consider again our **universal Turing machine (UTM)**, as described earlier in this chapter. UTM simulates the computation of any machine, *M*, on any tape *T*. From the construction of UTM, we can see (as also discussed earlier) that this simulation is so faithful to *M* as to enter an indefinite loop if that is what *M* would do on the tape *T* in question. However, one thing is certain: if *M* would have reached its halt state on tape *T*, then UTM will also reach its halt state given the coded version of *M* and the coded version of *T* as *its* input. All we would need to do then, is to modify the UTM procedure (Table 11.5) so that it outputs a '1' and halts as soon as it finds the code '00' on tape 3 (indicating that the simulated machine *M* has reached its halt state).

Let us call the new version of UTM, described immediately above, UTM_H. As we have defined it, UTM_H *partially* solves the halting problem. Given any machine *M* and tape *T*, appropriately coded, UTM_H will write a 1 (on tape 2) and halt if *M* would have reached its halt state when presented with the tape *T* as its input. If *M* would have *stopped in an auxiliary state*, because no transition was possible, then UTM_H will not reach its halt state. We could further amend UTM_H to output a 0, then move to its halt state, if no applicable quintuple of M's could be found, and therefore *M* would have halted in a non halt state. However, if *M* would *loop infinitely* on tape *T*, then as defined, UTM_H has no alternative but to do the same.

It is clear that the critical situations for the halting problem are those in which *M* loops infinitely. If, in addition to our amendments specified above, we could show how to amend UTM_H so that it halts and write a 0 when *M* would have looped infinitely on tape *T*, we would be showing that the halting problem is totally solvable.

This last amendment is difficult to imagine, since various conditions can result in infinite loops, and a 'loop' in a TM could comprise numerous states, as a loop in a program may encompass numerous statements. At this point, then, it seems pertinent to try a different approach.

11.5.2 Reductio ad absurdum applied to the halting problem

Let us assume that we have already amended UTM_H so that it deals with both the halting and non-halting situations. (You may be becoming aware at this point that this is going to be a ***reductio ad absurdum*** proof, a technique we've used several times in this book.) This new machine, which we will simply call H, is specified schematically in Figure 11.7.

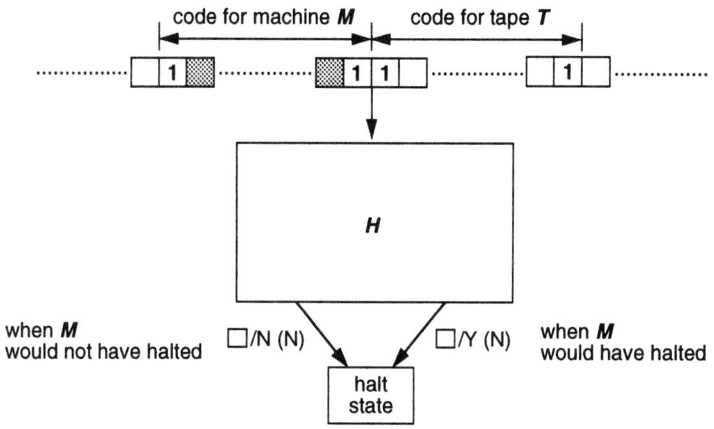

Figure 11.7 The machine H is assumed to solve the halting problem

Figure 11.7 defines H as a machine that prints Y and N rather than 1 or 0. Note that the two arcs entering H's halt state are shown as having the blank as the current symbol. We assume that at that point H has erased everything from its tape, and would thus write Y or N on a blank square before terminating, leaving the read/write head pointing at the result.

So, H is assumed to:

- print Y and halt if M would *halt* on the input tape T, or
- print N and halt if M would *not halt* on the tape T.

In other words, H is assumed to totally solve the halting problem.

Next, we make what appears to be a strange amendment to H, to create a new machine called H_1, depicted in Figure 11.8.

In order to produce H_1, we would need to make straightforward amendments to H's behaviour in its halting situations. H_1 would:

- print Y and halt if M would *not halt* on the input tape T, or
- *loop infinitely* if M would *halt* on the tape T.

You should be able to see that, if H existed, it would be straightforward to make the necessary amendments to it to produce H_1.

The next thing we do probably seems even stranger: we modify H_1 to produce H_2, depicted in Figure 11.9.

H_2 copies the code for the machine M, then presents the two copies to H_1. Our scheme codes both machine and tape as sequences beginning and ending with a 1, between these 1s being simply sequences of 0s separated by 1s, so a coded machine could clearly be interpreted as a coded tape.

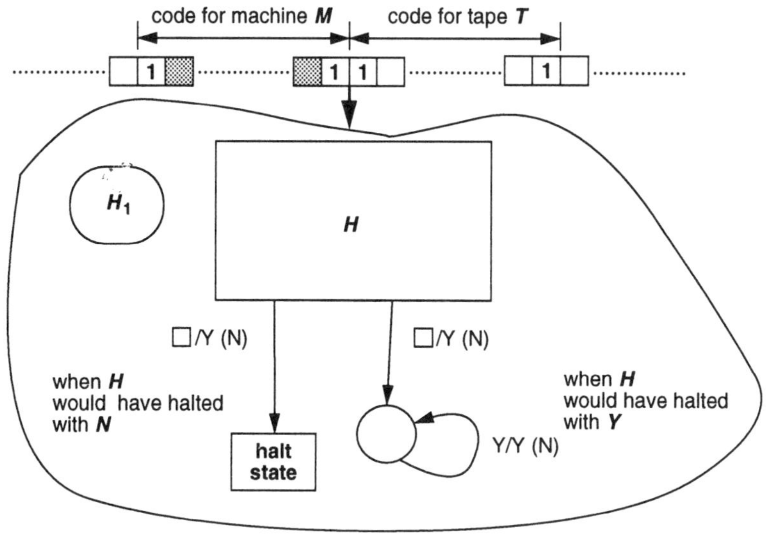

Figure 11.8 The TM H_1

Thus, instead of simulating the behaviour of M on its input tape, H_2 now simulates the behaviour of the machine M on a description of M itself (!). H_2's result in this situation will then be to:

- print Y and halt if M would *not halt* on a description of itself, or
- *loop infinitely* if M would *halt* on a description of itself.

The next step is the strangest of all. We present H_2 with the input specified in Figure 11.10.

Figure 11.10 shows us that H_2 has been presented with the binary code for itself to represent the machine, M. H_2 will first copy the code out again, then present the two copies to H_1. To H_1, the left hand copy of the code for H_2 represents a coded TM, the right hand copy the coded input to that TM. Thus, H_2's overall effect is to *simulate its own behaviour on an input tape that is a coded version of itself.* H_2 uses only the alphabet {0, 1, Y, N}, as it was derived from H_1, which also uses the same alphabet. Therefore H_2 could be applied to a description of itself in this way.

11.5.3 The halting problem shown to be unsolvable

Now we ask:

Does H_2, as configured above, halt?

Well, suppose H_2 does halt:

H_2 halts and prints Y if M would *not halt* on a description of itself

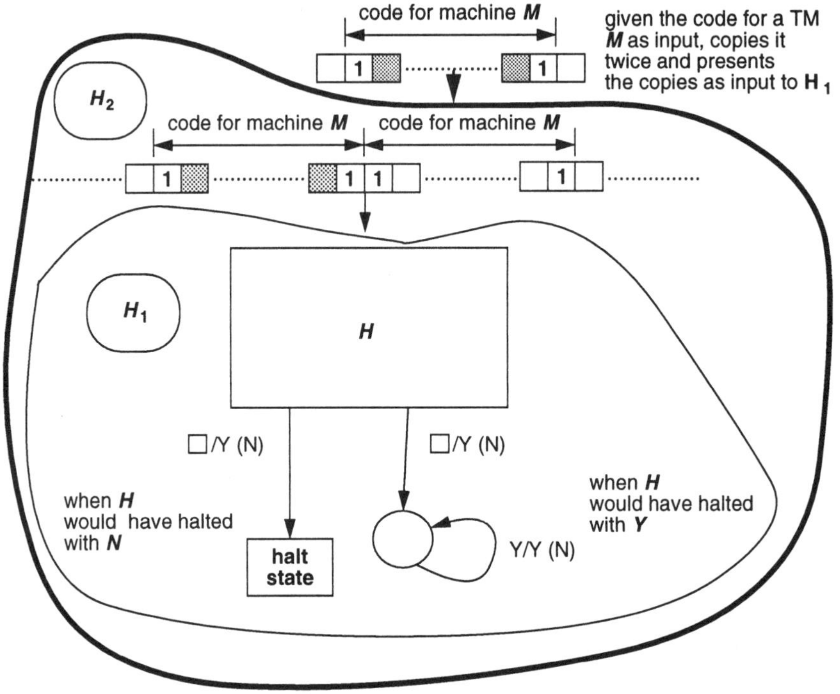

Figure 11.9 The TM H_2

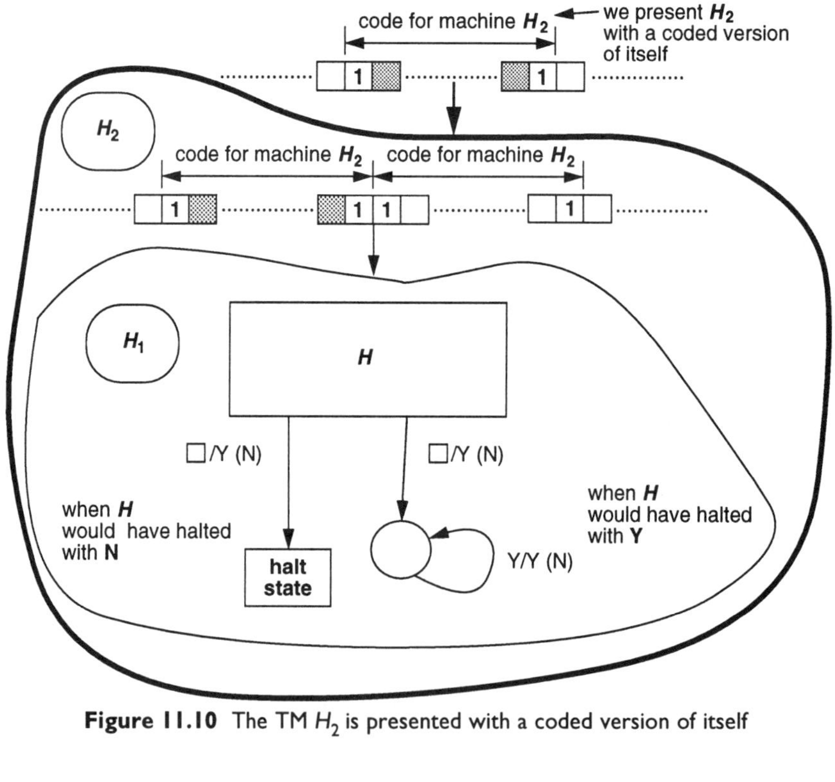

Figure 11.10 The TM H_2 is presented with a coded version of itself

However, in the situation here, M is H_2, so we have:

H_2 halts and prints Y if H_2 would *not halt* on a description of itself

To paraphrase:

H_2 halts, given a description of itself as its input, if it does not halt, given a description of itself as its input

This is clearly nonsense, so we must conclude that H_2 does *not* halt. However:

H_2 *loops infinitely* if M would *halt* on a description of itself.

Again, in our situation here, M is H_2, so we have:

H_2 *loops infinitely* if H_2 would *halt* on a description of itself.

To paraphrase once more:

H_2 does not halt, given a description of itself as its input, if it halts, given a description of itself as its input.

We have devised a machine which halts if it does not halt, and does not halt if it halts. Where was our error? Every change we have made to our machines, from H onwards, requires the addition of a few quintuples. We are forced to go right back and reject our original assumption, as it is obviously the assumption itself that is at fault. The machine H that solves the halting problem cannot exist. *The halting problem is an unsolvable problem.*

However, to reiterate, the halting problem is *partially* solvable. It is solvable (e.g. by UTM_H, as described above) in precisely those cases when M *would* halt.

Again, we are unable to reject the above argument on the basis of the indirection involved in manipulating codes representing the machines and their tapes. We *cannot*, as computer users, refuse to believe in the representational power of codes. To do so is to denounce as meaningless any activity carried out by *real* computers.

11.5.4 Some implications of the unsolvability of the halting problem

The halting problem is one of the most significant results of mathematics. The general unsolvability of the halting problem indicates the ultimate futility of defining effective criteria for determining effectiveness itself. For example, an **algorithm** is an **effective procedure** that terminates for all its input values. There can therefore be no algorithm to decide, in the general case, whether some other procedure *is* an algorithm. The unsolvability of the halting problem implies that, in general, we cannot determine by algorithmic means whether or not some well-defined process will terminate for all of its values.

The halting problem also has practical implications for the enterprise of proving the correctness of programs. A program is said to be partially correct if whenever it terminates it produces a correct result. A program is said to be totally correct if it is partially correct *and* it terminates for each and every possible input value. The goal of proving programs totally correct is thus one that cannot be attained by algorithmic means, as to be able to do so would require a solution to the halting problem.

Many unsolvable problems of computer science actually reduce to the halting problem. That there is no computational device more powerful than the standard TM implies that effective procedures for analysing the behaviour of other effective procedures are doomed ultimately to simulating those procedures, at the expense of being caught in the never-ending loops of the simulated process, if such loops arise.

As an example of a problem that reduces to the halting problem, consider the *equivalence problem* for TMs, or for programs. Two TMs, M_1 and M_2, are equivalent if for each and every input value, they both yield the same result. Now assume that E is a TM that solves the equivalence problem for any two TMs. Suppose, for argument's sake, that E simulates the behaviour of each machine by carrying out the next applicable quintuple from M_1, followed by the next applicable quintuple from M_2, on coded representations of an equivalent input tape for each machine. Suppose that the M_1 simulation reaches its halt state first, having produced a result. E must now continue with the M_2 computation until that ends in a result. But what happens if M_2 would not terminate on the tape represented? In order to output F (i.e. the two machines are not equivalent), E would need to determine that M_2 was not going to halt. *E would need to solve the halting problem.*

The halting problem also manifests itself in many **decision problems** associated with programming. Some of these are:

- will a given program terminate or not? – halting problem;
- do two programs yield the same answers for all inputs? – this is essentially the TM *equivalence problem* that is shown to be unsolvable immediately above;
- will a given program write a given value in a given memory location?
- will a given machine word be used as data or as an instruction?
- what is the shortest program equivalent to a given program?
- will a particular instruction ever be executed?

The exercises ask you whether the above decision problems of programming can be represented as TM decision problems and subsequently proved to be generally impossible to solve. In all cases, the method of proof is the same: we show that if the given problem was solvable, i.e. there was a TM to solve it, then that TM would, as part of its activity, have to solve the halting problem.

The proof of the halting problem includes what seems like a peculiar form of self-reference. A formal system of sufficient representational power is capable of being used to represent itself. If it is capable of representing itself then it can be turned in on itself, so to speak, to lead to peculiar contradictions. Turing's results are analogous to similar results in the domain of logic established by the great mathematician Kurt Gödel. Here we simplify things greatly, but essentially Gödel showed that a system of logic capable of generating all true statements (what is called *completeness*) can be used to generate a statement P, which, in essence refers to itself, saying 'P is false'[2]. Such a statement can be neither *true* nor *false*, and is therefore inconsistent (if it is true, then it is false, and vice versa). In a way, such a statement says 'I am false'. In a similar way, our machine H_2, when given its own code as input, says: 'I halt if I do not halt, and I do not halt if I halt'!

The general unsolvability of the halting problem and Gödel's findings with respect to completeness leave computer science in a curious position. A machine powerful enough to compute everything we may ask of it is powerful enough to deal with representations of itself. This leads to *inconsistencies*. If we reduce the power of the machine so that inconsistencies cannot occur, then our machines are incomplete, and will not be able to compute everything we may reasonably expect of them.

It is worth briefly discussing the wider implications of the halting problem. The inability of algorithmic processes to solve the halting problem does *not* mean that machines are necessarily less intelligent that humans. However, you may think that *we* can look at the texts of our programs and predict that this or that particular loop will not terminate, as we often do when writing programs. This does not mean we can solve the halting problem in general, but only in certain restricted situations. There must come a point at which the size and complexity of a program becomes such that we are forced to 'execute' (e.g. on paper) the program. There is no reason to disbelieve that, for the restricted cases in which we can determine that a program would not halt, we could also write a program (and thus construct a TM) to do the same. The general halting problem embodies a level of complexity beyond the capabilities of machines and human beings.

11.6 COMPUTABLE LANGUAGES

In Chapter 7 we saw how from any **phrase structure grammar** we could construct a **non-deterministic TM** to **accept** the language generated by that grammar. In Chapter 10 we saw how any non deterministic TM could be converted into an equivalent deterministic TM. Thus, we conclude that the TM is the **recognizer** for the phrase structure languages in general. The phrase structure languages include all types 0, 1, 2 and 3 of the **Chomsky hierarchy**. However, as the earlier chapters of this book have shown, the TM is a more powerful computational device than we actually need for the more restricted types (**context free** and **regular**). Nevertheless, the phrase structure languages are the *computable languages*, using the term 'computable', as we do now, to

[2] Rather like the person who says: 'Everything I say is a lie'.

mean *TM-computable*. A computable language, then, is a formal language that can be accepted by some TM.

A question that arises, then, is 'are there formal languages that are *not* computable?' In the next section, we encounter a language that is not computable, in that there can be no TM to accept the language. This, in turn, means that there can be no phrase structure grammar to generate that language (if there were, we could simply follow the rules in Chapter 7 and construct from the grammar a TM recognizer for the language). This result will enable us to refine our hierarchy of languages and abstract machines, which brings us to the end of the book.

With respect to formal languages, the analogous concept to *solvability* is called *decidability*.

A language is *decidable* if some TM can take any string of the appropriate symbols, and determine whether or not that string *is* a sentence of the language. An alternative way of putting this is to say that *the membership problem for that language is totally solvable*.

A language is *acceptable*, which we also call *computable*, if some TM can determine if any string that *is* a sentence of the language is indeed a sentence, but for strings that are not sentences cannot necessarily make any statement about that string at all. In this case, *the membership problem for the language is partially solvable*.

In the final section, we summarize the decidability properties of the various types of formal language we have encountered in this book. However, for now we return to the question asked above: *are there formal languages that are not computable*? If a language is non computable, then it is not acceptable, and by **Turing's thesis** there is no algorithmic process that can determine, for *any* appropriate string, if that string is in the language.

11.6.1 An unacceptable (non computable) language

We now establish that there is indeed a non computable formal language. As you will see, the 'proof' of this is inextricably bound up with the **halting problem**, and also takes advantage of our TM coding scheme from earlier in this chapter.

Consider the following set X:

$$X = \{1x1 : x \in \{0, 1\}^+\}$$

X is the set of all strings starting and ending in 1, with an arbitrary non-empty string of 0s and 1s in between. Now each string in this set either can, or cannot, represent the code of a TM. By considering a slight adaptation of the TM (sketched out in Chapter 10) that generates all strings in $\{a, b, c\}^+$ as part of its simulation of a non-deterministic machine, we know that the strings of this set could be generated systematically.

Now consider a **subset** Y of X defined as follows:

$$Y = \{x : x \in X \text{ and } x \text{ is the code of a TM with alphabet } \{0,1\}\}.$$

It would be straightforward to define a TM, or write a program that, given any string from X, told us whether or not that string was a valid TM code, i.e. of the format shown in Figure 11.11.

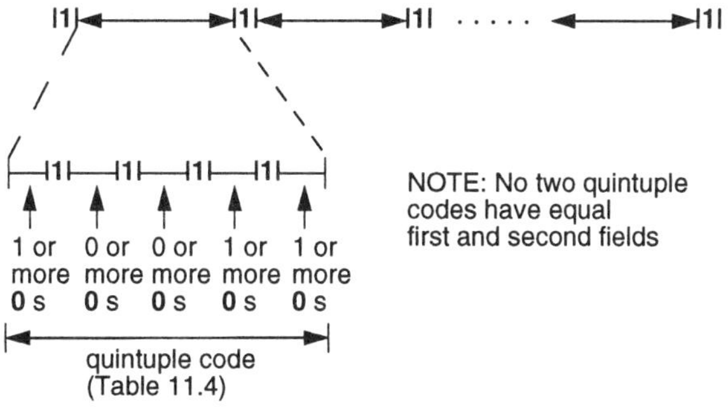

Figure 11.11 A binary code representing a TM

Note from Figure 11.11 that we are only interested in codes that represent deterministic machines, i.e. machines that have no more than one quintuple with the same 'current state' and 'current symbol'. We are also interested only in *binary TMs*, but we have already argued that all TMs can be reduced to binary form (cf UTM). This means that our 'current symbol' and 'write symbol' quintuple fields (i.e. the *second* and *third* fields of our quintuple representations) will not be allowed to contain more than two 0s.

Again, given that the set X can be generated systematically, and we can check for each string whether or not it represents a code for a TM, we can also see that there is an *effective procedure* for generating the set Y. Note that we are only interested in whether the codes are in the correct form to represent some TM, not in what the TM does. Most of the TM codes in Y would represent machines that do nothing of real interest, or indeed, nothing at all.

Given any code from the set Y we could recover the original machine. We can assume, for argument's sake, that for the alphabet, the code 0 represents 0, 00 represents 1, and all other codes are as we defined them earlier in the chapter. If y is a string in the set Y, we will call the machine whose code is y, M_y.

Now consider the following set, Z:

$Z = \{x : x \in Y$ and M_x does *not* **accept** $x\}$

Every string in Y represents a coded TM with alphabet $\{0,1\}$, and thus every string in Y can be presented as input to the machine whose code is that string. Thus, Z is the set of *all binary alphabet machines that do not accept the string represented by their own code.*

Now, we ask:

is Z an *acceptable* language?

Suppose that it is. In this case, some TM, say M_z, is the acceptor for Z. We can assume that M_z uses only a binary alphabet. Thus, M_z's code will be in the set *Y*. Let's call M_z's code *z*.

We now ask,

is the string *z* accepted by M_z?

Suppose that it is. Then *z* is *not* in Z, since Z contains only strings that are *not* accepted by the machine whose code they represent. It therefore seems that if *z* is accepted by M_z, i.e. is in the language accepted by M_z, then it isn't in the set Z. However, this is nonsense, because M_z is the acceptor for Z, and so any string it accepts *must* be in Z.

We therefore reject our assumption in favour of its negation. We suppose that the string *z* is not accepted by M_z. M_z is the acceptor for Z, and so any string M_z does not accept is not in Z. But this also cannot be correct! If M_z does not accept the string *z*, then *z* is one of the strings in Z, as it is the code for a machine (M_z) that does not accept its own code.

It seems that the string *z* is not in the set Z if it is in the set Z, and is in the set Z if it is not! The TM, M_z cannot exist, and we must conclude that *the language Z is not acceptable*. We have found an easily defined formal language taken from a two-symbol alphabet that is not computable. Such a language is sometimes called a totally undecidable language. We could generalize the above proof to apply to alphabets of any number of symbols, in effect showing that there are an infinite number of such *totally undecidable* languages.

11.6.2 An acceptable, but undecidable, language

One problem remains to be addressed, before the next section concludes the book by presenting an overall summary of the relationship between formal languages and automata. We would like to know if there are computable languages that are *necessarily acceptable, but not decidable.* For reasons that will be presented in the next section, any such languages must be properly type 0 in the Chomsky hierarchy, since the more restricted types of language in the hierarchy *are* decidable.

Here is a computable language that is acceptable, but not decidable. Given the sets *X* and *Y*, as specified above, we define the set *W* as follows:

$W = \{x : x \in Y \text{ and } UTM_H \text{ halts when it applies } x \text{ to } x\}$[3].

The set *W* is the set of codes of TMs that halt when given their own code as input. *W* is an *acceptable* language. *W* is *not* a *decidable* language because a TM that could tell us which valid TM codes were not in *W* (i.e. represent the code for machines that

[3] UTM_H was introduced earlier in the chapter. It halts and prints 1 when the simulated machine would have halted on the given tape.

do not halt when given their own code as input) would have to solve the *halting problem*.

Languages, like *W*, that are acceptable but not decidable are sometimes called *semi-decidable languages*.

11.7 COMPUTABLE LANGUAGES AND ABSTRACT MACHINES: THE BEGINNING

As implied above, associated with any language is what is called the membership problem for that language. Given a grammar *G* and a string *s*, the membership problem says: 'is *s* a sentence of the language *L*(*G*)?' From our discussion of **problems** and **solvability** earlier in this chapter, it should be clear that the membership problem for a language is solvable if the associated membership function for that language is (TM-) computable.

For all the phrase structure languages, the associated membership function is computable. *All phrase structure languages are computable (acceptable) languages.* However, we have seen that some type 0 languages are acceptable, but not decidable. However, we shall see that the other types (1, 2, 3) are *all* decidable.

Table 11.7 summarizes the decidability properties of the **regular, context free, context sensitive**, and **unrestricted** languages. In fact, since we would like to show only that types 1, 2, and 3 are decidable, we can specify the same (very inefficient!) method for all three types, assuming some initial minor amendments to the type 2 and 3 grammars.

In terms of formal languages, as presented in this book, we described each class of language in the Chomsky hierarchy, working from the regular languages up. We simultaneously worked up to the TM, by adding extra functionality to one type of abstract machine to yield a more powerful type of machine. For example, we showed how a machine that is essentially an FSR with a **stack** was more powerful than an FSR. We finally developed the Turing Machine, and showed that we need go no further. *There is no more powerful device for computation or language recognition than the TM.*

We might, as an alternative, have approached the subject by first presenting the TM, and then discussing how the TM gives us more power than we need for the more restricted languages and computations (such as *adding* and *subtracting*, which we saw being done by FSTs in Chapter 8). Then the abstract machines for the more restricted languages could have been presented as TMs that were restricted in some appropriate way. Given what we now know about the power of the TM, it seems only fitting to conclude this book by presenting, in Figure 11.12, this alternative view of *computable languages and abstract machines*.

Table 11.7 Decidability properties for the languages of the Chomsky hierarchy

Grammar type	General decidability	Justification
regular (type 3)	decidable	If necessary, remove ε from grammar using method described in Chapter 5.
context free (type 2)	decidable	Then apply the method below to the ε free part of the grammar.
context sensitive (type 1)	decidable	All productions are $x \rightarrow y$ where $\mid x \mid \leq \mid y \mid$. Thus, each step in a derivation either increases or leaves unchanged the length of the sentential form. Construct a non-deterministic TM to generate in parallel all sentential forms that do not exceed in length the length of the target string *s*. If s is one of the generated strings, print *T* and halt, if it is not, print *F* and halt. Convert the non deterministic TM into an equivalent deterministic TM.
unrestricted (type 0)	semi-decidable	Some productions may be of form $x \rightarrow y$ where $\mid x \mid \leq \mid y \mid$. The sentential form may thus lengthen and shorten during a derivation. There may therefore be no obvious point at which to give up if we do not manage to generate *s*, the target string. However, we know from Chapter 7 that all type 0 languages are acceptable.

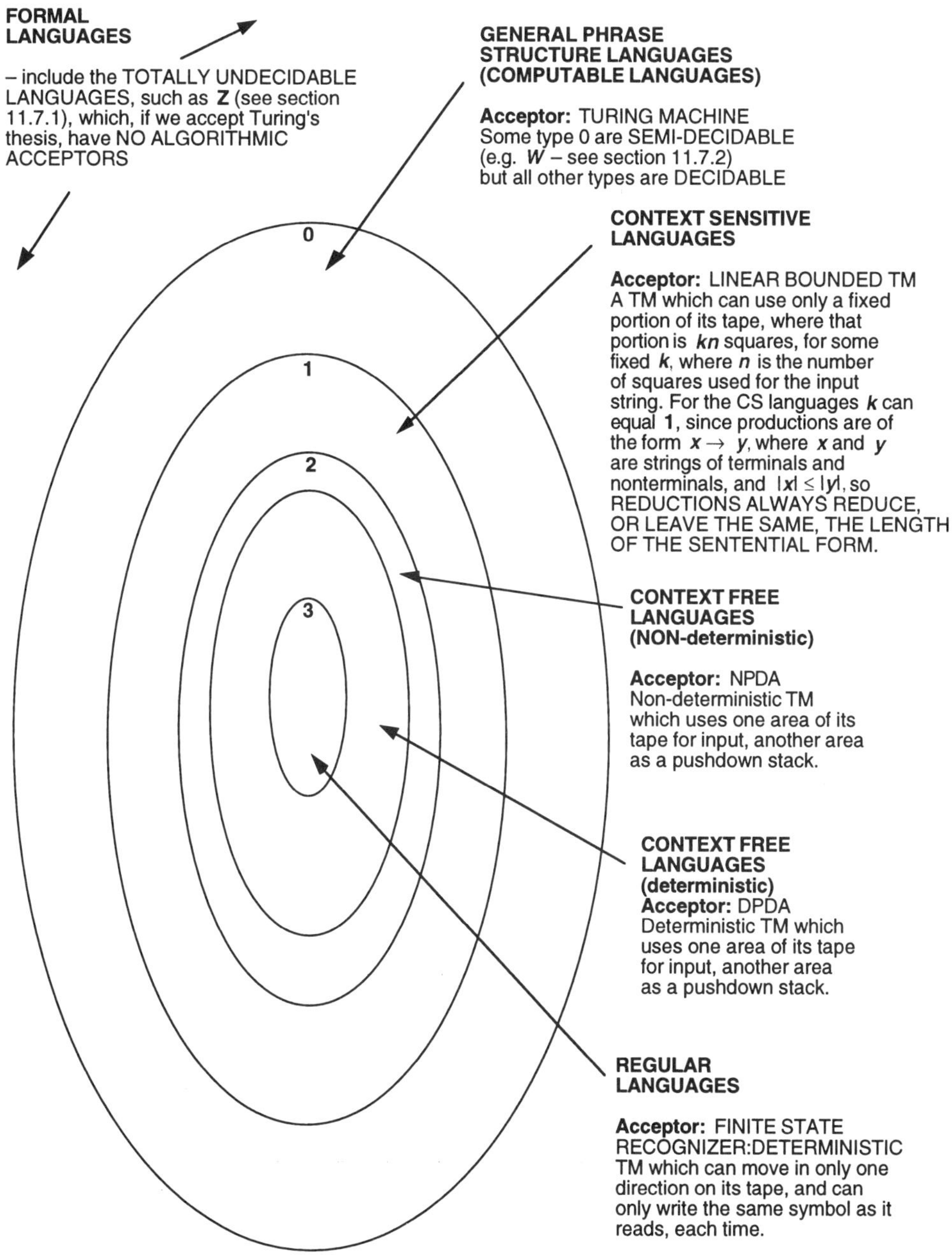

Figure 11.12 Computable languages and abstract machines

EXERCISES

1 Write a program representing a general **Turing machine** simulator. Your machine should accept as its input the **quintuples** of a Turing machine, and then simulate the behaviour of the TM represented by the quintuples on various input tapes. In particular, you should run your simulator for various TMs found in this book, and the TMs specified in Exercise 2 of Chapter 7.
Hint: the advice given for Exercise 6 of Chapter 5, in *Solutions to Selected Exercises* is also very relevant to this exercise

2† The printing problem asks if a TM, M, will print out a given symbol s, in M's alphabet, when started on a specified input tape. Prove that the printing problem is, in general, **unsolvable**.
Hint: this is a problem that reduces to the halting problem, as discussed earlier.

3 Show that the following (all taken from this Chapter) are all unsolvable problems. In each case first transform the statement of the problem so that it is couched in terms of Turing machines rather than programs. Then show that in order to solve the given problem, a TM would have to solve the **halting problem**.

(a) will a given program terminate or not? – halting problem
(b) will a given program write a given value in a given memory location?
(c) will a given machine word be used as data or as an instruction?
(d) what is the shortest program equivalent to a given program?
(e) will a particular instruction ever be executed?

4 Express the totally unsolvable problem in 11.4.2 in terms of programs, rather than TMs.
Hint: consider a non terminating program such as an operating system, for which the input can be regarded a stream of characters.

Glossary of terms

abstract machine

An abstract machine is a **function** or a collection of functions. For example:

> the **FSR** associates (symbol, **state**) pairs with sets of states, as in the tabular representation of the FSR suggested in Solutions to selected exercises, the comment on Exercise 3, Chapter 4.

The view of abstract machines as 'machines' rather than as functions enables us to consider the processing of sequences of input symbols over time, under the control of a **clock**. Thus, we can use them as the basis of a study of real computing machines.

accept

A **recognizer** essentially tells us if its input string is a member of a particular set. A string that is adjudged to belong to the set is said to be accepted by the machine.

acceptable language

A **formal language** which has an associated **TM** that, given any **sentence** of the language, terminates with an indication that the sentence is indeed a sentence of the language.

acceptance conditions

The configuration of an **abstract machine** that we expect to see when the machine halts, if that machine was given as its input an acceptable string.

- **FSR**: the acceptance conditions are that the input is **exhausted** and the machine is in a **state** designated as a **halt**, or **acceptance**, **state**
- **PDR**: as for FSR, but we may also specify the required conditions of the machine's stack
- **TM**: may include halt state and output configuration.

acceptance state

A **state** which an **FSR** or **PDR** must reach when it is signifying that it **accepts** a string.

algorithm

An **effective procedure** that terminates with a valid output for every appropriate input.

alphabet

A non-empty **finite** set of **atomic elements**.

ambiguity

A property possessed by a **grammar**, G, that is such that $L(G)$ contains a **sentence** for which G can be used to construct two or more **derivation trees**.

arc

Specifies the successor relationship between **states** of an **abstract machine**, or **nodes** in a **tree**, and that a relationship exists between two nodes in a **graph**.

atom, atomic

A component of a symbolic expression that we do not attempt to analyse structurally, i.e. the building block of symbolic expressions (e.g. **terminal symbols** in **grammars**, symbols in **alphabets**).

backtrack

If a procedure reaches a dead end and recreates an earlier situation, when a choice between alternatives was taken, in order to explore one of those alternatives, this is known as backtracking. For example, a **non deterministic PDR** may return to a particular (**state**, input symbol, **stack**) configuration in order to try an alternative sequence of **transitions**.

Backus-Naur form (BNF)

An alternative representation of **context free grammars**. Useful for defining programming languages, as it enables **non terminals** (called syntactic categories in BNF) to have arbitrary length names. For examples of BNF notation in the definition of the Pascal language, see Chapters 2 and 3, and the solution to Exercise 7 of Chapter 2 in *Solutions to selected exercises.*

big O

A notation for expressing an approximation of the **time** taken or **space** used by a program or machine when carrying out a given task. O means 'in the order of'.

blank

The status of every **tape square** of a **TM's** tape that is not occupied by a symbol. TMs can recognize blank squares, and can 'blank out' marked squares.

blank tape assumption

The assumption that if, in a TM's **input configuration**, the **head** is located on a **blank** tape square, the tape is occupied only by blanks.

bottom-up parsing

A bottom-up method attempts to build a **parse tree** from its **leaf nodes** (the **terminal string**) upwards, terminating when the tree is fully formed, and has S, the **start symbol** as its root. **Reduction parsing** is a bottom-up approach.

Chomsky hierarchy

A formal specification of the relationship between **phrase structure grammars**, in terms of four types, **type 0**, **1**, **2** and **3**. Each type, from 1 to 3, is properly contained within the type numerically preceding it.

Chomsky normal form (CNF)

A format into which the **productions** of any ε **free context free grammar** can be converted, without changing the **language generated by the grammar.** Formally, a CFG is in CNF if every production is of the form:

$x \to y$, where $x \in N$, and either $y \in T$ or $y = vw$, where $v,w \in N$

clock

Each **transition** made by an **abstract machine** is assumed to take place on receiving a signal emitted by an imaginary device. As we are interested in situations where the signals are 'transmitted' at regular intervals of time, we call the imaginary device a clock.

closed under an operation, closure

If we have a **set** of objects, and we show that a given operation applied to any **member** of the set produces an object that is also in the set, we say that the set is closed under that operation. For example, the set of natural numbers {1, 2, 3, 4,..} is 'closed under the operation +'.

complement

A **formal language**, L, is a **subset** of A^*, for some **alphabet** A. The complement of L, with respect to A, is the set $A^* - L$, i.e. the set of all **strings** in A^* that are not in L.

complexity

In this book, the term is used in two different ways:

- The complexity of an **abstract machine** is the number of **states** it consists of (for an **FSR**) or the number of **states** multiplied by the number of **alphabet** symbols (for a **TM**).
- The **time** complexity of a program is the number of basic operations the program carries out in relation to the length of the input.

computable

See **computable function**

computable function

A **function** that can be computed by a TM.

computable language

A language that can be **accepted** by a **TM**. Synonyms: **acceptable** language, **phrase structure language**, **type 0/unrestricted language**.

computation

A **function** associates input with output. Computation involves taking some input and actually producing the output specified by the function, by carrying out an **effective procedure**.

concatenation of strings

The concatenation of two **strings**, x and y, written xy, is the string formed by appending y onto the end of x. If $x = ab$ and $y = bcd$, then $xy = abbcd$.

concatenation of languages

The concatenation of two **formal languages**, L_1 and L_2, written L_1L_2, is the language resulting from concatenating every string in L_1 with every string in L_2, in turn. If $L_1 = \{a, bc, cd\}$ and $L_2 = \{a, ca\}$, then $L_1L_2 = \{aa, aca, bca, bcca, cda, cdca\}$.

context free (type 2)

A context free grammar has **productions** of the form $x \rightarrow y$, where $x \in N$ and $y \in (N \cup T)^*$.

A context free language is any language generated by a 'context free grammar'.

The **recognizer** for the context free languages is the **pushdown recognizer**.

context sensitive (type 1)

A context sensitive grammar has **productions** of the form $x \rightarrow y$, where $x \in (N \cup T)^+$ and $y \in (N \cup T)^*$, and $|x| \leq |y|$.

A context sensitive language is any language generated by a 'context sensitive grammar'.

The **recognizer** for the context sensitive languages is a TM that only needs to use the area of its **tape** occupied by the input string to process that string.

countably infinite

A set is countably infinite if it is **infinite**, but a program (TM) can print out its elements one by one in such a way that each element appears only once and each element is guaranteed to appear within a finite period of time.

decide, decidability, total d., partial d, und., semi-d.

A machine decides a language if it can indicate 'yes' for every string that is a sentence, and 'no' for every string in the **complement** of the language.

If a **TM** decides a language as above, the language is called a totally decidable language. **Types 1**, **2** and **3** of the **Chomsky hierarchy** are totally decidable languages. The type 0 languages are, in general, semi-decidable.

See also **solvability**.

decision problem

A **problem** that requires an answer for every **member** of a **set** of objects. The **halting problem** is a decision problem.

decision program

A program that can **decide** a language.

derivation, derive

The application of a sequence of **productions**.

derivation tree

A form of representing **derivations** using **context free grammars**. Since context free grammar **productions** have single **non terminals** on the **left hand sides**, each non terminal expands to a string of one or more **terminals** and/or non terminals. Derivations can thus be represented in **tree** form.

deterministic (FSR, PDR, CFL, TM)

A deterministic machine never has to make a choice about which **transition** to make, in a given situation.

A deterministic **context free language** (**CFL**) is a CFL for which a deterministic PDR can be defined.

ε generating non terminal

A **non terminal** appearing on the **left hand side** of a **production** that can be used as the first in the derivation of the **empty string**.

ε-free grammar

A grammar containing no ε **productions**.

ε-free production

A **production** that does not have the **empty string** on its **right hand side**.

ε production

A **production** with the **empty string** on its **right hand side**.

effective procedure

A procedure that:

- is deterministic
- involves finite time and resources (space)
- can be specified in terms of unambiguous rules
- is represented in numeric terms.

element (of a set)

An object that is a **member** of a **set**.

empty move, ε move

A **transition** made by a **PDR** or **FSR** without **reading** a symbol from the input string.

empty set, {}, ∅

A **set** containing no **elements**.

empty string, epsilon, ε, lambda, λ

The **string** that contains no symbols, and has a **length** of 0.

enumeration

The pairing off of the **elements** in a **set** with the natural numbers, in such a way that each element is associated with exactly one number.

equivalent

Of grammars and **recognizers**: a grammar, G, is equivalent to a recognizer, M, if $L(G)$ is exactly the same set as the set of **strings accepted** by M.

Of programs or TMs: two TMs, M_1 and M_2, are equivalent if for each and every input value, they both yield the same result.

exhausted

The input is exhausted when the **FST** or **PDR** has **read** every symbol.

exponential time

When an algorithm takes time $O(x^n)$ where n is the number of elements in the input (see **big O**).

finite

Limited in number. The **elements** of a finite **set** can be written out by a program (**TM**) that terminates, having written out all the elements.

finite language

A language consisting of a **finite** number of **strings**.

finite state recognizer (FSR), FS generator

An **abstract machine** that consists of a **finite** number of **states**, and makes a **transition** between states purely on the basis of the next symbol in the input string.

finite state transducer (FST)

A **finite state recognizer** with the additional requirement that it outputs a single symbol for each one it encounters in the input string.

formal grammar

Used in this book as synonymous with **phrase structure grammar**.

formal language

Any **subset** of the set A^*, where A is any **alphabet**.

function

An association between input values and output values, such that each configuration of input values is associated with only one configuration of output values. A function is essentially a specification of *what* is to be computed, but not necessarily *how* it is to be computed (see **computation**).

generator

An **abstract machine** that produces output without receiving any input.

graph

A **finite** set of ordered pairs of **nodes** of the form (x, y). For such a pair, we say there is an **arc** from x to y.

halt state

A **state** of an **abstract machine** in which the machine is expected to terminate its processing for valid inputs (or, in the case of **FSRs** and **PDRs**, for **acceptable strings**). For FSRs and PDRs, in this book the terms halt state and **acceptance state** are regarded as synonymous.

halting problem

The unsolvable **decision problem** that asks: can we provide a TM that will decide, for any **TM**, M, and input tape, T, whether or not M will halt if started on the **tape** T?

head

See **read/write head**.

index (of a string)

When we refer to the ith symbol in a **string** of **length** $\leq i$, the i is called the index. If x is a string we write it x_i, so if $x = abcded$, $x_4 = d$.

infinite

Limitless in number. Many infinite **sets** are **countably infinite**. Some sets, like the set of all **subsets** of the natural numbers, are not.

infinite language

A language that contains an **infinite** number of **sentences**.

ingoing arc

A directed **arc** entering a **state**.

input configuration of a Turing machine

A specification of:

- the position of the **read/write** head
- constraints applying to the **tape** symbols

of a **TM** before it begins its **computation**.

input string, input sequence

The **terminal string** presented to a **recognizer**. More generally, the initial sequence of symbols to be processed by an **abstract machine**.

instantaneous description

A description of:

- the contents of the **tape**
- the position of the **read/write** head
- the current **state**

of a **TM** at any given point in its **computation**.

intersection, $\cap$

Of two **sets**: the set of **elements** that occur in *both* of the sets.

language generated by grammar G [$L(G)$]

The set of **terminal strings** that can be derived using the **productions** of G, each **derivation** beginning with a production which has S, **the start symbol** of G, as its **left hand side**.

leaf node

In a **tree**, any **node** that has no successors (children).

left hand side

Of a **production**: the **string** to the left of the production arrow, $\rightarrow$.

length

Of a **string**: the number of symbols it contains (duplicates being counted). If x is a string, $|x|$ denotes its length, e.g. $|abcddadd| = 8$.

linear time

When an **algorithm** takes time $O(n)$, where n is the number of elements in the input (see big O).

loop

A sequence of **states** or **nodes** along a directed path, the first being the same as the last. Also used as a verb.

LR(1) language

A special case of the **LR(k)** languages. The **deterministic CFLs**.

LR(k) language

Languages that can be recognized by a **DPDR** (and are therefore **context free**) if that machine is allowed to look ahead in the input by no more than a fixed number, *k*, of symbols.

member

Any **element** of a **set** is a member of that set.

minimal FSR

A **deterministic FSR**, *M*, that is such that no deterministic FSR that **recognizes** the same language can have fewer **states** than *M*.

moment

An unspecified but finite period of time, used in discussing the time taken by **abstract machine** computations. Each signal from the **clock** is assumed to occur one moment after the previous signal.

multi-tape TM, k-tape TM

A TM that uses *k* **tapes** ($k > 1$), each of which has its own **read/write head**. Such a TM has instructions in the form of **sextuples** (rather than **quintuples**).

node

An object in a **tree** or a **graph** that is linked by **arcs** to other objects (nodes).

non deterministic (FSR, PDR, CFL, TM)

An **abstract machine** that can have a choice of applicable **transitions** in a given situation. A non deterministic CFL is a CFL for which a **deterministic PDR** cannot be found.

non proper, non properly

Of **sets**: the non proper **subsets** of a set include the set itself.

Of types of **formal language** in the **Chomsky hierarchy**: a language is not properly of a given type if it is possible to define a **grammar** of a subordinate type to generate that language.

non terminal symbol

A symbol that can appear in the **productions** of a **grammar**, but cannot appear in **sentences**.

null state

The **state** in a **deterministic FSR** which is added by the **subset method** to ensure that the machine is completely deterministic. From each state of the machine, for each **terminal symbol** for which there is not an **arc** going to a non null state, there will be an **outgoing arc** labelled with that terminal going to the null state. Then there is an arc that is both **outgoing** and **ingoing** to the null state for each terminal symbol.

outgoing arc

A directed **arc** leaving a **state**.

output configuration of a Turing machine

A specification of:

- the position of the **read/write** head
- constraints applying to the **tape** symbols

of a **TM** when it completes its **computation**.

packed tape

A **tape** on which no **blanks** appear between marked squares. Almost all our **TMs** in this book assume an initial configuration in which the input tape is packed.

palindrome

A **string** that equals the reverse of itself.

parse tree

A **derivation tree** that is built up in the process of **parsing**.

parser

A procedure that produces a structural account of a **string** according to some **grammar**.

parsing

The activity carried out by a **parser**.

phrase structure grammar (PSG), PS language (PSL)

A PSG is a **tuple** (N, T, P, S) where

- N is a **finite set** of **non terminal** symbols,
- T is a finite set of **terminal** symbols,
- $N \cap T = \{\}$,

- P is a set of **productions** defined as for **unrestricted grammar**,
- S is a **member** of N called the **start symbol**.

A PSL is a language generated by a PSG.

polynomial time

When an **algorithm** takes time $O(n^x)$, where n is the number of elements in the input (see *big O*).

pop

A **function** applied to **stacks**, which returns the symbol at the top of the stack. Actually two functions, as it also removes the top symbol. Used by the **push down recognizer**.

power

Of **strings**: a **function** to **concatenate** a string to itself a specified number of times. If x is a string, $x^0 = \varepsilon$, $x^1 = x$, and $x^i = xx^{i-1}$, all $i > 1$. So if $x = abcc$, $x_3 = abccabccabcc$.

problem

The question associated with a **function**, f, that asks 'what is the value of $f(x)$?'

production arrow, →

The symbol in between the **left hand** and **right hand** sides of **productions**.

production, production rule

A component of a **grammar** consisting of a **string** followed by a production arrow, followed by another string. A production $x \rightarrow y$ means 'the string x can be replaced by the string y'.

proper, properly

Of **sets**: A proper **subset** of a set cannot be the whole of the set.

Of types of **formal language** in the **Chomsky hierarchy**: A language is properly of a given type if it is not possible to define a **grammar** of a subordinate type to generate that language.

push

A **function** applied to **stacks**, which places a new symbol on top of the stack. Used by the **push down recognizer**.

pushdown recognizer

A **finite state recognizer** equipped with a **stack** that has **arcs** labelled with three components:

1. a symbol to be read from the input **string**
2. a symbol that may be found on the top of the stack
3. a string to be **pushed** onto the stack

The machine can make the **transition** specified by that arc only if (1) is the next symbol in the input string, and (2) is the symbol currently on the stack. The machine can then **pop** (2) from the stack, replacing it with (3), and make the transition. The push down recognizer is the recognizer for the **context free languages**.

queue

A data structure like a **stack**, but in which items are added to one end and removed from another.

quintuples

In general, a quintuple is a five element list. In this book, quintuples are used as an alternative representation of the **TM**, where each quintuple is of the form:

(current **state**, read symbol, write symbol, direction, new state).

However, **PDRs** can also be represented as quintuples (see the hints for Exercise 6, Chapter 5, in *Solutions to selected exercises*).

read/write head, r/w head

A **TM** is assumed to be equipped with a read/write head, which can sense the symbol on the current square of the **tape**, and overwrite that symbol before moving one tape square to the left or right.

reading

The **abstract machine** typically makes a **transition** for each symbol in the input **string**. We call this process reading the input.

recognizer

A machine that has the task of telling us if an input string belongs to a particular **set** is called a recognizer, or an **acceptor**. In this book, recognizing and accepting are regarded as synonyms, and are used interchangeably.

reductio ad absurdum

A proof technique. To prove a proposition, p, one assumes $\neg p$ to be true. By following a logically sound chain of inference from $\neg p$, one arrives at a contradiction.

Since a contradiction has been reached by following logically sound arguments from an original assumption, the original assumption must be false, i.e. one has shown $\neg\neg p$, which is p.

reduction parsing

A **bottom-up parsing** technique which exchanges the **left** and **right hand sides** of the **productions** of the **grammar**, resulting in what are called reductions. Parsing then begins with the input **string**, and attempts to find a sequence of reductions that will result in S, **the start symbol**, being the only remaining symbol.

reductions

See **reduction parsing**.

regular (type 3)

A regular grammar has **productions** of the form $X \rightarrow yY$, or $X \rightarrow z$ where $X, Y \in N$, $y \in T$ and $z \in T \cup \{\varepsilon\}$.

A regular language is any language generated by a regular grammar.

The **recognizer** for the regular languages is the **deterministic FSR**.

reject

A **recognizer** that has terminated without accepting an input **string** is assumed to have rejected the string.

repeat state theorem, vwx theorem

A theorem applying to **FSRs** and **FSTs** essentially saying that any FSR that processes a long enough **string** must go into a **loop**, and will continue to loop if the conditions that caused it to pass through the loop once are recreated.

right hand side

Of a **production**: the **string** to the right of the production arrow, $\rightarrow$.

semantics

Of an expression: what the expression means.

Of a program: the **computation** carried out by the program.

sentence

A **string** from a **formal language**. A **terminal string** generated by a **grammar**.

sentential form

Any **string** in $(N \cup T)^*$ that can be derived through the application of **productions** of a **grammar**, the first production applied being one with only S, **the start symbol** on its **left hand side**.

sequence

An ordered collection of **elements**, that permits duplicated elements. A **string** is a form of sequence.

set

An unordered collection of **elements** that does not allow duplicated elements.

set definition

The specification of a **set**, most commonly by specifying a **function** for generating objects along with a specification of the range of values to which the function is to be applied.

In this book set definitions are usually used to specify **formal languages**. For example, $\{a^i b^{i \text{ div } 3} : i \geq 1\}$ denotes the **infinite** set of **strings** of 1 or more *a*s followed by a number of *b*s which is the number of *a*s integer divided by 3 ({*a*, *aa*, *aaab*, *aaaab*, *aaaaab*, *aaaaaabb*, ..}).

set difference, (–), (\)

$A - B$ is the **set** of **elements** that are in A but not in B.

For example, $\{a, b, c\} - \{b, e, f\} = \{a, c\}$.

sextuples

The ***k*-tape TM** version of **quintuples**. As for quintuples, but with an extra designator, being a number between 1 and k, and denoting the **tape** on which the next operation is to take place.

shuffling

A task often performed by **TMs**, when a symbol is moved from one part of a **tape** to another, and all other symbols are moved to close up the **blank** that may result, and perhaps to make room for the symbol at its destination.

skipping

When a **TM** enters a **loop** that passes over all symbols, moving in one direction only, until it reaches a target symbol, we call this skipping over the symbols.

solvability

A **problem** is solvable if the associated **function** is **computable**.

space

The amount of storage required during a **computation**. For TMs, this is measured in terms of the number of **tape** squares required, perhaps expressed as a function of the number of squares required to represent the input.

stack

A popular data structure, used in this book by **push down recognizers**. Essentially a **string**, to which we add symbols at one end (called **pushing**), and remove values from the same end (called **popping**).

stack bottom marker, $\perp$

Since the **pop** operation is undefined for empty **stacks**, the stack bottom marker is a special symbol used by **PDRs** to represent the bottom of the stack.

start (sentence) symbol

For a **terminal string** generated by a **grammar** to be a **sentence**, it must have been generated by a sequence of **productions** the first of which had a lone S, **the start symbol** of the grammar, on its **left hand side**.

start state

Abstract machines considered in this book have one or more **states** designated as start states. The machines are assumed to make their first **transition** from one of these states.

state

Used in two senses in this book:

- an **abstract machine** is represented as a directed **graph**, the **nodes** in the graph being called states
- the state of a machine is sometimes regarded as the internal configuration of all relevant details of the machine at a given point in its **computation** (for example, **instantaneous descriptions** of **TMs**).

string

A **finite sequence** of **atomic** symbols taken from an **alphabet**.

subset

A **set** which is the **empty set**, is equal to, or is contained by, another set is called a subset of that other set.

subset construction (algorithm) for FSRs

An **algorithm** that takes any **non deterministic FSR** and produces an **equivalent** deterministic FSR. It works by deterministically modelling the behaviour of several non deterministic **states** of the original machine in single states of the new machine.

substring

Any part, including ε, of a **string**. Formally, let s be any string, **taken from** an **alphabet**, A. Then the **members** of the **set** $\{y : x, y, z \in A^*, xyz = s\}$ are all of the substrings of s.

superset

A **set** that contains, or is equal to, another set is a superset of that other set.

syntax

The structure of an expression according to a set of formal rules.

syntax diagram

A diagrammatic form for representing **syntax**, relevant to the **context free languages**, and programming languages in particular.

taken from

It is often said that a **string** is a **sequence** of symbols over an **alphabet.** In this book the phrase 'taken from' is used in preference to 'over'.

tape

The one dimensional data structure used by the **TM.** It is assumed to be divided into squares. Each square can be occupied by a tape symbol or be **blank**. The tape is assumed to extend indefinitely at both ends, though at any time only a **finite** portion of it is 'marked' (occupied by symbols).

tape squares

See **tape**.

terminal node

A **leaf node** in the **derivation tree** or **parse tree** for a **sentence**, i.e. a node representing a **terminal symbol** or the **empty string**.

terminal string

A **string** of **terminal symbols**. See **sentence**.

terminal symbol

The symbols that can appear in **sentences** of a language.

time

With respect to **abstract machines**, assumed to be apportioned into discrete equal length moments. The time taken by a **computation** is usually measured in terms of the number of **transitions** a machine makes.

top-down

A top-down method of **parsing** attempts to build a **parse tree** from its root node (the node representing *S*, **the start symbol**) downwards, terminating when the target **sentence** has been obtained.

transducer

An **abstract machine** that processes input and produces output, and thus can be regarded as computing a **function**. In this book, the transducers studied are the **FST** and the **TM**.

transition, state transition

When a machine changes from one **state** to another (possibly the same) state.

tree

A tree is a **graph** where one **node**, *r*, is called the root node. *r* may have any number (including 0) of **outgoing arcs**, but no **ingoing arcs**. Each outgoing arc from *r*, if there are any, is an ingoing arc to the root of another tree.

tuple

A tuple is an ordered list of elements each of which can be simple objects, such as single symbols, or more complex objects such as **sets**. For examples, consider the (*N*, *T*, *P*, *S*) definition of **phrase structure grammar**, **quintuples** and **sextuples**.

Turing machine computable function

See **computable function**.

Turing machine (TM)

An augmented **finite state transducer** that uses a **tape** for input and output, and can move right and left on the tape, overwriting symbols on the tape. See **quintuples**.

Turing's thesis

The statement that every **effective procedure** can be carried out by a **TM**.

type 0, type 1, type 2, type 3 grammars and languages

See **unrestricted**, **context sensitive**, **context free**, **regular**, and **Chomsky hierarchy**.

unambiguous

A **grammar**, G, that is such that $L(G)$ contains no **sentences** for which G can be used to construct two or more distinct **derivation trees** is called an unambiguous grammar.

union, $\cup$

The union of two **sets** is the set of **elements** in either or both sets.

unit production

A **production** of the form $X \rightarrow Y$ where $X, Y \in N$.

unit production graph

A **graph** representing **derivations** of **strings**, each derivation beginning with a **non terminal**, using only **unit productions**, and terminating the first time a non unit production is used.

universal Turing machine (UTM)

A **TM** that accepts as its input a description of any TM, M, and a description of an input **tape**, T, and simulates the behaviour of M on the tape T.

unrestricted (type 0)

An unrestricted **grammar** has **productions** of the form $x \rightarrow y$, where $x \in (N \cup T)^+$ and $y \in (N \cup T)^*$.
An unrestricted language is any language generated by an unrestricted grammar.
The **recognizer** for the unrestricted languages is the **TM**.

uvwxy theorem

The **repeat state theorem** for the **context free languages**. Essentially states that any sufficiently long **sentence**, s, of a CFL, L, is of the form $uvwxy$, where $uv^iwx^iy \in L$, all $i \geq 0$.

word

An alternative term for what in this book is called a **sentence**.

Further reading

The following are some suggested titles for further reading. Most of the items are accompanied by notes. In particular, some of the titles refer to articles that describe practical applications of some of the concepts in this book.
The bold numbers in parentheses refer to chapters in this book.

Beckman F.S. (1981) *Mathematical Foundations of Programming*. Addison-Wesley, Reading, MA.

*Encyclopaedic coverage of almost all aspects of computer science. The algorithm for minimization of FSRs (**4**) comes from Beckman. Also covers register machines and Wang machines (**9**). The programming problems relating to the halting problem are also adapted from Beckman (**11**).*

Brookshear J.G. (1989) *Theory of Computation: Formal Languages, Automata, and Complexity*. Benjamin/Cummings, Redwood City, CA.

*Chapter on complexity and Turing machines, including complexity of algorithms, and time complexity of language recognition problems (**10**). Does not mention context sensitive languages. The coding scheme for TMs, and the establishing of the undecidable languages in this book were based on Brookshear's approach (**11**). Much on LL(k) and LR(k) parsers (**5**). Uses quadruples, rather than quintuples, to describe TMs (**10**). Contains a list of NP problems in an appendix (**10**).*

Church A. (1936) An Unsolvable Problem of Elementary Number Theory. *American Journal of Mathematics*, **58**, 345–63.

*Church and Turing were contemporaneously addressing the same problems by different, but equivalent means. Hence, in books such as Harel, and Fenton and Hill we find references to the 'Church–Turing thesis', rather than 'Turing's thesis'. (**9, 10, 11**).*

Cohen D. I. A. (1986) *Introduction to Computer Theory*. John Wiley, New York.

*Covers some additional material such as: regular expressions, Moore and Mealy machines (in this book we use Mealy machines) (**4**). Discusses relationship between multi-stack PDRs (**5**) and TMs.*

Fenton N. and Hill G. (1993) *Systems Construction and Analysis: a Mathematical and Logical Framework*. McGraw-Hill, Maidenhead, UK.

*Contains chapters on formal language theory and complexity (**10**), the latter including an intuitive 'proof' of the halting problem (**11**) and discussion of P = NP problem (**10**).*

Floyd R.W. and Beigel R. (1994) *The Language of Machines: an Introduction to Computability and Formal Languages*. W.H. Freeman, New York.

*Includes a discussion of regular expressions (**4**), as used in the UNIX™ utility 'egrep'. Good example of formal treatment of minimization of FSRs (using equivalence classes) (**4**).*

Harel D. (1987) *Algorithmics: the Spirit of Computing*. Addison-Wesley, Reading, MA.

*Study of algorithms and their properties. Complexity, big-O (**10**), decidability (**11**). Discusses application of finite state machines to modelling simple systems (**8**). Focuses on 'counter programs': simple programs in a hypothetical programming language.*

Harrison M.A. (1978) *Introduction to Formal Language Theory*. Addison-Wesley, Reading, MA.

*Many formal proofs and theorems. Contains much on closure properties of languages (**6**).*

Hopcroft J.E. and Ullman J.D. (1979) *Introduction to Automata Theory, Languages and Computation*. Addison-Wesley, Reading, MA.

*Discusses linear bounded TMs for context sensitive languages (**11**).*

Hutchison D. and Silvester P. (1987) *Computer Logic: Principles and Technology*. Ellis-Horwood, Chichester, UK.

*Discusses use of FSTs as design tools for logical circuits (**8**).*

Jensen K. and Wirth N. (1975) *Pascal User Manual and Report*. Springer-Verlag, New York.

*Contains the BNF and Syntax chart descriptions of the Pascal syntax (**2**). Also contains notes referring to the ambiguity in the 'if' statement (**3**).*

Kain R.Y. (1972) *Automata Theory: Machines and Languages*. McGraw-Hill, New York.

*Formal treatment. Like Beckman develops Turing machines before other abstract machines. Discusses non-standard PDRs (**5**) applied to context sensitive languages.*

Minsky M.L. (1967) *Computation: Finite and Infinite Machines*. Prentice-Hall, Englewood Cliffs, NJ.

*A classic text, devoted to an investigation into effective procedures (**11**). More details on most things, of particular relevance is description of Shannon's two-state TM result (**10**), and reference to unsolvable problems (**11**). The 'proof' that FSTs cannot perform arbitrary multiplication (**8**) is based on the one here.*

Post E. (1936) Finite Combinatory Processes – Formulation 1. *Journal of Symbolic Logic*, **1**, 103–5.

*Post formulated a simple abstract string manipulation machine at the same time as did Turing (**9**). Cohen (see above) devotes a chapter to these 'Post' machines.*

Preece J. and Keller L. (eds.) (1990) *Human-Computer Interaction.* Prentice-Hall, London.

Human-computer interaction research and development uses grammars and state transition diagrams (Chapters 15 and 19).

Rayward-Smith V.J. (1983) *A First Course in Formal Language Theory.* Blackwell, Oxford, UK.

*The notation and terminology for formal languages we use in this book is based on Rayward-Smith. Very formal treatment of regular languages (plus regular expressions), FSRs (**4**), and context free languages and PDRs (**5**). Includes Greibach normal form (as does Floyd and Beigel), an alternative CFG manipulation process to Chomsky Normal Form (**5**). Much material on top-down and bottom-up parsing (**3**), LL and LR grammars (**5**), but treatment very formal.*

Rich E. (1984) *Artificial Intelligence.* McGraw-Hill, New York.

Artificial intelligence makes much use of representations such as grammars and abstract machines. In particular, machines called recursive transition networks and augmented transition networks (equivalent to TMs) are used in natural language processing.

Tanenbaum A.S. (1988) *Computer Networks* (2nd edn.). Prentice-Hall, London.

*Discusses FSTs for modelling protocol machines (sender or receiver systems in computer networks) (**8**).*

Turing A. (1936) On Computable Numbers with an Application to the Entscheidungsproblem. *Proceedings of the London Mathematical Society,* **42**, 230–265.

*The paper in which Turing introduces his abstract machine, in terms of computable numbers, not computable functions. Also includes his notion of a universal machine (**11**). A paper of remarkable contemporary applicability, considering that Turing was considering the human, rather than the machines, as computer.*

Winston P.H. (1984) *Artificial Intelligence* (2nd edn.) Addison-Wesley, Reading, MA. (see Rich)

Wood D. (1987) *Theory of Computation.* John Wiley, Chichester, UK.

*Describes several extensions to PDRs (**5**). Introduction to proof methods, including the pigeonhole principle (also mentioned by Harel) on which both the repeat state theorem (**6, 8**) and the uvwxy theorem (**6**) are based.*

Solutions to selected exercises

CHAPTER 2

1 (a) **Regular**. Every **production** has a lone **non terminal** on its **left hand** side, and the **right hand** sides consist of either a single **terminal**, or a single terminal followed by a single non terminal.

(b) **Context** free. Every production has a lone non terminal on its left hand side. The right hand sides consist of arbitrary mixtures of terminals and/or non terminals one of which (*aAbb*) does not conform to the pattern for regular right hand sides.

2 (a) $\{a^{2i}b^{3j} : i, j \geq 1\}$ i.e. *a*s followed by *b*s, any number of *a*s divisible by two, any number of *b*s divisible by 3

(b) $\{a^i a^j b^{2j} : i \geq 0, j \geq 1\}$ i.e. zero or more *a*s, followed by 1 or more *a*s followed by twice as many *b*s

(c) $\{\}$ i.e. the grammar generates *no strings at all*, as no derivations beginning with S produce a terminal string

(d) $\{\varepsilon\}$ i.e. the *only string* generated is the *empty string*.

3 xyz, where $x \in (N \cup T)^*$, $y \in N$, and $z \in (N \cup T)^*$

The above translates into: 'a possibly empty string of terminals and/or non terminals, followed by a single non terminal, followed by another possibly empty string of terminals and/or non terminals'

5 For an **alphabet** A, A^* is the set of all strings that can be taken from A including the empty string, ε. A regular grammar to generate, say $\{a, b\}^*$ is:

$$S \to \varepsilon \mid aS \mid bS.$$

ε is derived directly from S. Alternatively, we can derive a or b followed by a or b or ε (this last case terminates the derivation), the a or b from the last stage being followed by a or b or ε (last case again terminates the derivation), and so on …
Generally, for any alphabet, $\{a_1, a_2, .., a_n\}$ the grammar:

$$S \to \varepsilon \mid a_1S \mid a_2S \mid .. \mid a_nS$$

is regular and can be similarly argued to generate $\{a_1, a_2, .., a_n\}^*$.

7 (b) The following fragment of the BNF definition for Pascal, taken from Jensen and Wirth (1975), actually defines all Pascal expressions, not only Boolean expressions.

<expression> ::= <simple expression> | <simple expression>
    <relational operator> <simple expression>
<simple expression> ::= <term> | <sign> <term> |
    <simple expression> <adding operator> <term>
<adding operator> ::= + | – | or
<term> ::= <factor> | <term> <multiplying operator> <factor>
<multiplying operator> ::= * | / | div | mod | and
<factor> ::= <variable> | <unsigned constant> | (<expression>) |
    <function designator> | <set> | not <factor>
<unsigned constant> ::= <unsigned number> | <string> |
    <constant identifier> | nil
<function designator> ::= <function identifier> |
    <function identifier> (<actual parameter>
        {, <actual parameter})
<function identifier> ::= <identifier>
<variable> ::= <identifier>
<set > ::= [<element list>]
<element list> ::= <element> {, <element> } | <empty>
<empty> ::=

Note the use of the **empty string** to enable an empty <set> to be specified (Chapter 5).

10 Given a **finite** set of strings, any string x from the set can be generated by regular grammar productions as follows:

$$x = x_1x_2..x_n,\ n \geq 1$$
$$S \rightarrow x_1X_1$$
$$X_1 \rightarrow x_2X_2$$
$$\vdots$$
$$X_{n-1} \rightarrow x_n$$

For each string, x, we make sure the non terminals X_i are unique (to avoid any derivations getting 'crossed').

This applies to any finite set of strings, so any finite set of strings is a regular language.

CHAPTER 3

1 (b) $\{a^i b^j c^k : i, j, k \geq 1, i = j \text{ or } j = k\}$
i.e., strings of *a*s followed by *b*s followed by *c*s, where the number of *a*s equals the number of *b*s, or the number of *b*s equals the number of *c*s, or both

(c) Grammar *G* can be used to draw two different derivation trees for the sentence $a^3b^3c^3$, as shown in Figure SA.1. The grammar is thus ambiguous.

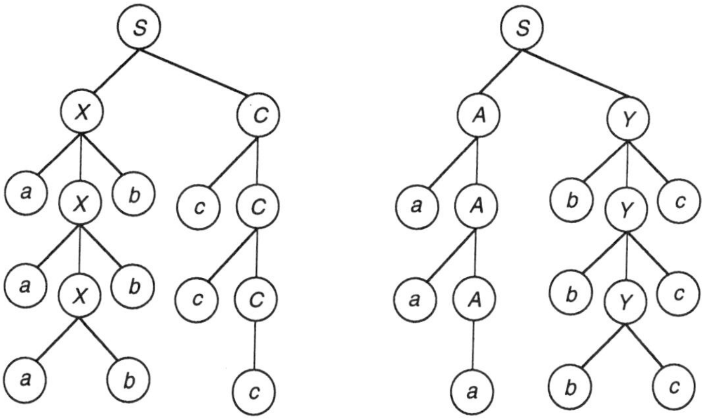

Figure SA.1

2 (b) As for the Pascal example, the *semantic implications* should be discussed in terms of demonstrating that the same statement yields different results according to which derivation tree is chosen to represent its structure.

CHAPTER 4

1 (a) The **FSR** obtained directly from the productions of the grammar is shown in Figure SA.2.

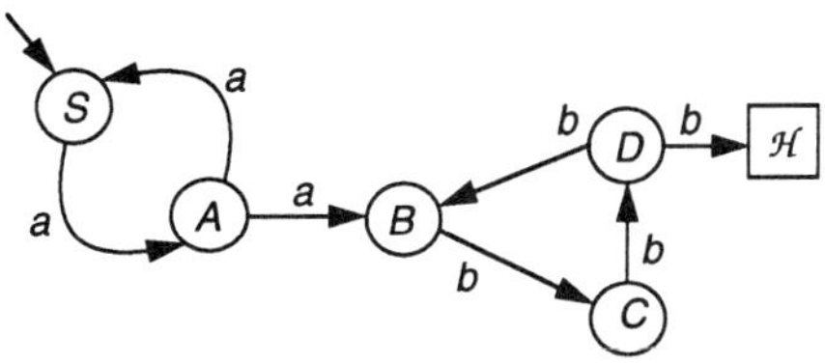

Figure SA.2

The above FSR is **non deterministic**, since it contains **states** (for example, state *A*) with more than one identically labelled outgoing arcs going to different destinations.

The **deterministic** version, derived using the subset method (null state removed) is in Figure SA.3.

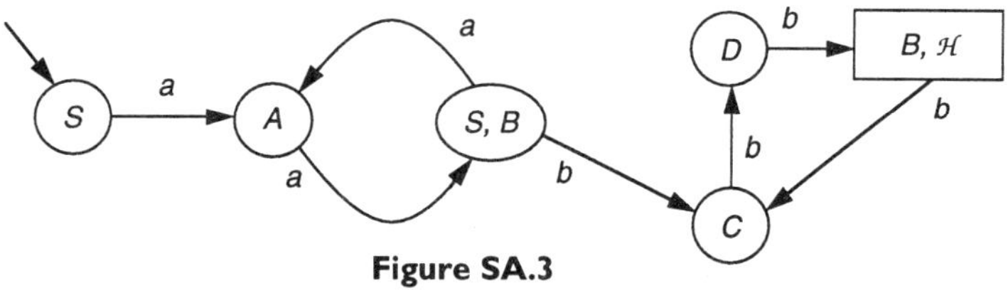

Figure SA.3

3 One possibility is to represent the FSR as a *two-dimensional* table (*array*), indexed according to (state, symbol) pairs. Table SA.1 represents the FSR of exercise 2.

Table SA.1

		TERMINALS	
		a	b
S	1	2, 3	–
T	2	4	3
A	3	–	5
T	4	2, 3, 4	5
E	5	–	5, 6
S	6	–	–

In Table SA.1, element (4, *a*), for example, represents the set of states directly reached from state 4 given the terminal *a*. Such a representation would be easy to create from the productions of a grammar that could be entered by the user, for example. The representation is also highly useful for creating the ,deterministic version. This version is also made more suitable if the language permits dynamic arrays (Pascal does not, but ADA and C, for example do). The program also needs to keep details of which states are **halt** and **start states**.

An alternative scheme represents the FSR as a *list of triples*, each triple representing one **arc** in the machine. In languages such as Pascal or ADA, this scheme can be implemented in *linked list* form. For example, consider the part of the FSR from exercise 2 shown in Figure SA.4.

This can be represented as depicted in Figure SA.5.

This representation is particularly useful when applying the *reverse* operation in the *minimization algorithm* (the program simply exchanges the first and third elements in

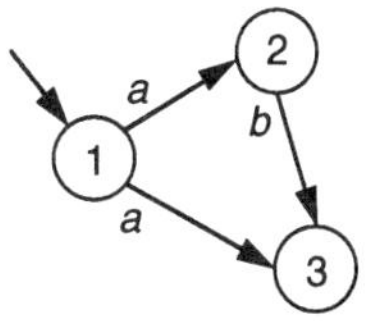

Figure SA.4

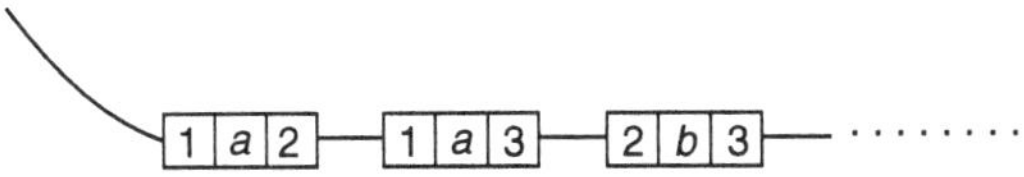

Figure SA.5

each triple). It is also a suitable representation for languages such as LISP, where the list of triples becomes a list of three element lists.

Since FSRs can be of arbitrary size, a true solution to the problem of defining an appropriate data structure would require *dynamic* data structures, even down to allowing an unlimited source of names. Although this is probably taking things to extremes, you should be aware of where you make such restrictions, and what their implications are.

CHAPTER 5

2 First of all, consider that a **DPDR** is not necessarily restricted to having *one* halt state. You design a machine, M, that enters a state that is a halt state after reading the first a, then remains in that state while reading any more as (**push**ing them on to the **stack**, to compare with the bs, if there are any). If there are no bs, M simply stops in that halt state, otherwise on reading the first b (and **pop**ping off an a) it makes a transition to another state where it can read only bs. The rest of M is an exact copy of M_3^d, of Chapter 5. If there were no bs, any as on the stack must remain there, even though M is accepting the string. Why can we not ensure that in this situation, M clears the as from its stack, but is still deterministic?

5 The reasons are similar to why arbitrary **palindromic** languages are not deterministic. When the machine reads the as and bs part of the string, it has no way of telling if the string it is reading is of the 'number of as = number of bs' , or the 'number of bs = number of cs' type. It thus has to assume that the input string is of the former type, and **backtrack** to abandon this assumption if the string is not.

6 One possibility is to represent the PDR in a similar way to the *list* representation of the FSR described above (sample answer to exercise 3, Chapter 4). In the case of the PDR, the 'current state, input symbol, next state' triples would become **quintuples** of the form:

'current state, input sym, pop sym, push string, new state'

The program could read in a description of a PDR as a list (a *file* perhaps) of such 'rules', along with details of which states were *start* and *halt* states.

The program would need an appropriate dynamic data structure (e.g. linked list) to represent a stack. It may therefore be useful to design the stack and its operations first, as a separate exercise.

Having stored the rules, the program would then be set up so that its 'current state', say *C*, was indicated to be the start state. The program would then execute the algorithm in Table SA.2.

Table SA.2

```
can-go := true
while not(end-of-input) and can-go
    if there is a quintuple, Q, such that
              Q's current state = C, and
              Q's pop sym = the current symbol 'on top' of the stack, and
              Q's read sym = the next symbol in the input
    then
                          remove the next symbol from the stack
                          set up ready to read the next symbol in the input
                          push Q's push string onto the stack
                          set C to be Q's next state
    else                  can-go := false
    endif;
endwhile;
if end-of-input and C is a halt state
then return('yes')
else return('no')
endif.
```

As the PDR is assumed to be deterministic, the program can assume that only one quintuple will be applicable at any stage, and can also halt its processing of invalid strings as soon as an applicable quintuple cannot be found. The non deterministic machine is much more complex to model: I leave it to you to consider the details.

CHAPTER 6

3 Any FSR that has a loop on some path linking its start and halt states *in which there is one or more arcs not labelled with ε*, recognizes an **infinite language**.

4 In both cases, it is clear that the *v* part of the *uvw* form can consist only of *a*s or *b*s or *c*s. Then one simply argues that when the *v* is repeated the required numeric relationship between *a*s, *b*s and *c*s is not maintained.

5 The language specified in this case is the set of all strings consisting of two copies of any string of *a*s and/or *b*s. To prove that it is not a **CFL**, it is useful to use the fact that we know we can find a ***uvwxy*** form for which $|vwx| \leq 2^n$, and *n* is the number of non terminals in a **Chomsky Normal Form** grammar to generate our language. Let $k = 2^n$. There are many sentences in our language of the form $a^ib^ia^ib^i$, where $i > k$. Consider the *vwx* form as described immediately above. I leave it to you to complete the proof.

CHAPTER 7

1 This can be done by adding an arc labelled *x*/*x* (*N*) for each symbol, *x* in the alphabet of the particular machine (including the **blank**) from each of the halt states of the machine to a new halt state. The original halt states are then designated as non halt states. The new machine reaches its single halt state leaving the **tape/head** configuration exactly as did the original machine.

CHAPTER 8

1 (The 'or' **FST**). Assuming that the two input binary strings are the same length, and are interleaved on the tape, a bitwise or FST is shown in Figure SA.6.

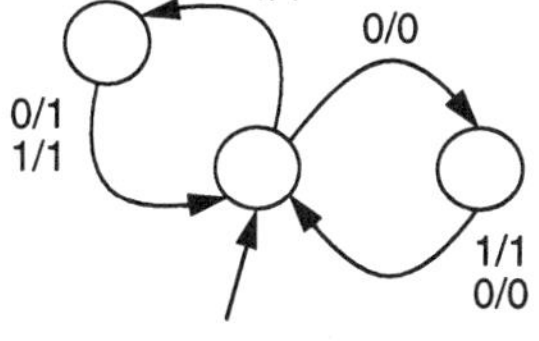

Figure SA.6

CHAPTER 9

4 **Functions** are discussed in more detail in Chapter 11. A specification of a function describes what is computed and does not go into detail about how the computation is done. In this case, then, the **TM** computes the function:

$f(y) = y \text{ div } 2, y \geq 1.$

5 (a) tape on entry:

$d1^n e, n \geq 0$ (or $d1^{x+1}e$)

tape on exit:

$df^n e1^n$

i.e. the machine copies the 1s between d and e to the right of e, replacing original 1s by fs

(b) $2x + 3$ (or $2(x + 1) + 1$)), all $n \geq 0$

CHAPTER 10

2 Figure SA.7 shows part of a TM to execute the **quintuple** (A, 1, 0, R, B). The flowchart boxes relate to those from the flowchart in Chapter 10. Remember that the numeric designator following the direction specification on the arcs represents the number of the tape on which the next operation is to be carried out.

3 The **sextuple** (1, a, A, R, 2, 2) of the three tape machine M might be represented according to the schema shown in Figure SA.8.
Notes regarding the **shuffling** of the end of input marker are given in Table SA.3.

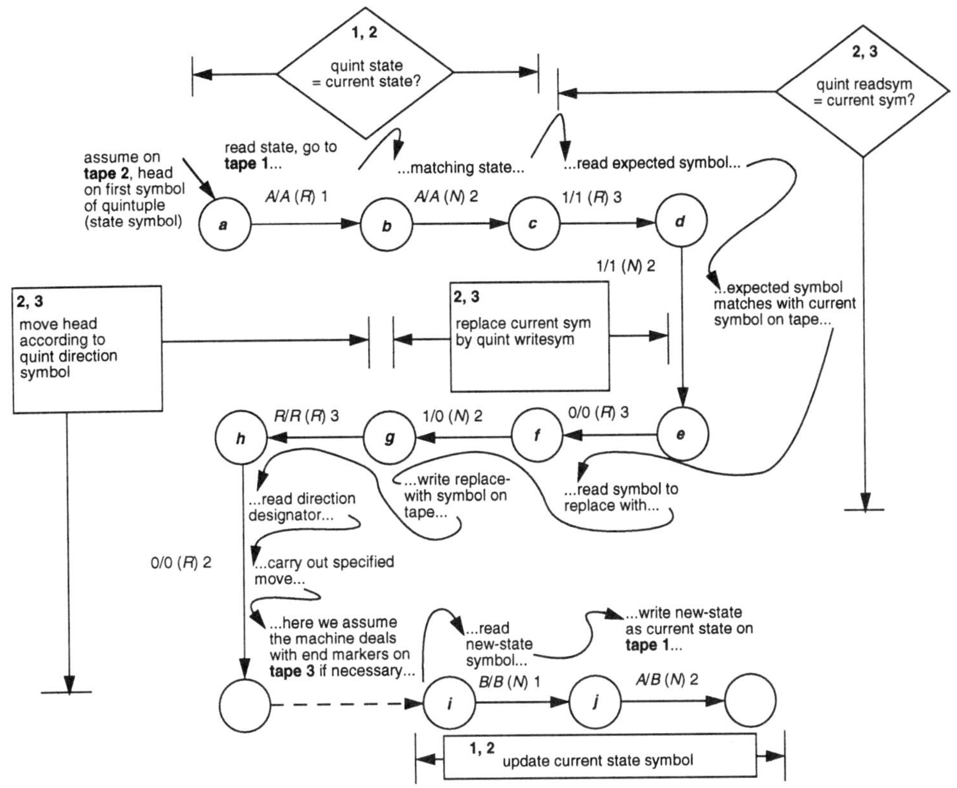

Figure SA.7

Table SA.3

States marked r, t and v, need an additional outgoing arc with current symbol ▚ leading to a sequence of states that 'shuffle' the ▚ to the right, as follows:-

state r

from here, S needs to ensure that the ▚ is moved right six squares,

state t

from here, the ▚ should be moved right four squares, and

state v

from here, the ▚ should be moved two squares to the right

Note: the number of squares the ▚ has to be shuffled along depends on the number of tapes there were on the original multi-tape machine.

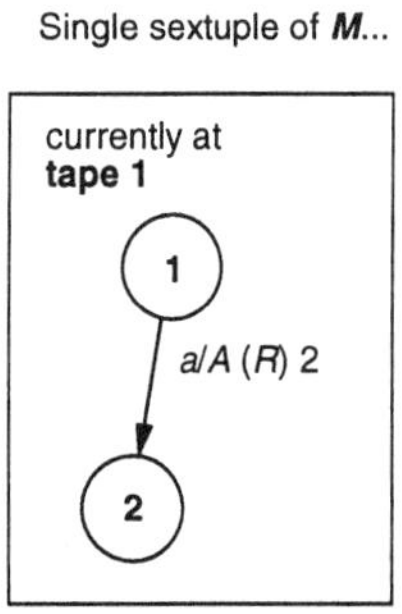

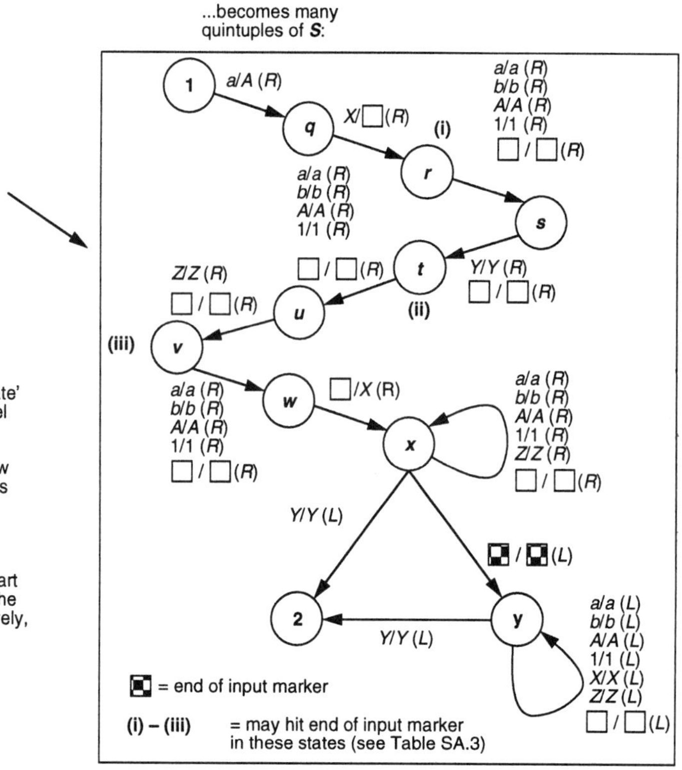

Note that we use the 'from state' label (**1**) and the 'to state' label (**2**) as the first and last states in the part of ***S*** we have constructed. We introduce new state labels for the other states of ***S*** that we create.
If we always do this, we are ensured that ***S*** will be linked exactly as ***M*** was.
In particular, what were the start and halt states in ***M*** become the start and halt states, respectively, of ***S***.

Figure SA.8

CHAPTER 11

2 Assume the TM, *P*, solves the printing problem by halting with output 1 or 0 according to whether *M* would, or would not, write the symbol *s*.

P would need to be able to solve the **halting problem**, or in cases where *M* was not going to **halt** *P* would not be able to write a 0, for 'no'.

Index